MOLECULAR SPECTROSCOPY

MOLECULAR SPECTROSCOPY

Jack D. Graybeal

Chemistry Department
Virginia Polytechnic Institute
and State University

McGraw-Hill Book Company

New York St. Louis San Francisco Auckland Bogotá Caracas Colorado Springs
Hamburg Lisbon London Madrid Mexico Milan Montreal New Delhi
Oklahoma City Panama Paris San Juan São Paulo Singapore Sydney Tokyo Toronto

MOLECULAR SPECTROSCOPY

234567890 DOCDOC 89321098

ISBN 0-07-024391-3

This book was set in Times Roman. The editor was Karen S. Misler; the cover designer was John Hite; the production supervisor was Denise L. Puryear. Project supervision was done by Harley Editorial Services.
R. R. Donnelley & Sons Company was printer and binder.

Library of Congress Cataloging-in-Publication Data

Graybeal, Jack D.
Molecular spectroscopy.

1. Molecular spectroscopy. I. Title.
QD96.M65G7 1988 543'.0858 87-2862
ISBN 0-07-024391-3

ABOUT THE AUTHOR

Jack D. Graybeal is a native of West Virginia where he began his career in chemistry with a B.S. degree from West Virginia University in 1951. He received M.S. and Ph.D. degrees in Physical Chemistry from the University of Wisconsin in 1953 and 1955. Following two years as a Member of the Technical Staff in the Radio Research Division of Bell Laboratories he returned to West Virginia University as an assistant professor in 1957. In 1968, he moved to Virginia Polytechnic Institute and State University where he has been a full professor since 1969 and associate department head for the past ten years.

Professor Graybeal has published in the areas of nuclear quadrupole resonance and rotational (microwave) spectroscopies. He was one of the first experimentalists in NQR to adopt the use of operational amplifiers for superregenerative oscillator control. His current interests are in the study of free radicals by use of microwave spectroscopy and in the use of microwave radiation for the synthesis and curing of polymers. He has had substantial teaching experience in the subjects of general chemistry, physical chemistry, chemical literature, and spectroscopy. In addition to research and teaching interests Professor Graybeal has a long-standing interest in the design and construction of chemistry facilities, having been instrumental in the building of the research building at West Virginia and currently involved in the construction of a new research laboratory at Virginia Tech.

Daniel, David, and Dale

CONTENTS

PREFACE

The primary objective of *Molecular Spectroscopy* is to present to the student a thorough introduction to the relationships among quantum mechanical formulations, experimentally determinable quantities obtained via spectroscopic methods, and physical parameters related to the structure of molecular systems. A second objective is to provide a detailed discussion of the use of matrix mechanics as it applies to spectroscopy, since this method is much less familiar to the chemist than wave mechanical methods.

In order to achieve the stated objectives the following methods will be employed: (1) The main scope of the text has been limited to the development of relationships for elementary systems so that the treatment can be in-depth and of sufficient detail that it can be followed by a student with a background of two years of college mathematics, a year of college physics, and one and one-half years of physical chemistry, including an appreciable introduction to quantum mechanics (wave mechanics) and group theory. (2) The treatment is not intended to be a survey of all areas of spectroscopy, or to discuss the forefront of research in spectroscopy but is designed to provide the student with the necessary foundation for proceeding on to advanced monographs and papers. (3) Quantum mechanics is used as a tool and is not developed fully. (4) The mathematical and quantum mechanical methods employed are either those which are normally covered in those courses listed under Part 1 or else they are summarized in appendices. (5) The relationship between theory and experimentally determined parameters is constantly emphasized by the inclusion of examples in the text. (6) The presentation is so ordered that the student can proceed through it with a minimum of referral to earlier material or to outside sources. (7) Problem sets are provided at the ends of the chapters so the student can reinforce the discussion. (8) Subjects are introduced at the elementary level, followed through to an advanced stage and frequently referred to in a later treatment at an even higher level, thus gradually building necessary relationships. (9) An effort has been made to eliminate the "it is obvious" statements that so often are not as obvious to the student. (10) A less detailed review of the spectroscopy of more complex systems is included following the introductory material to provide a more balanced view of the subject.

There are several important aspects which are characteristic of this treatment. The more predominant of these include the extensive use of matrix methods, the inclusion of discussions of nuclear quadrupole resonance, and detailed examples. Following a brief introduction to the subjects of spectroscopy, wave mechanics is reviewed by discussing the harmonic oscillator and rigid rotor. Perturbation theory is also reviewed in order to provide a background for matrix formulation of this method. In Chapter 3 the matrix method is reviewed in detail and the angular momentum and direction cosine matrix elements, which are used extensively in the book, are developed. Chapter 4 reviews point group symmetry and group theory. Chapter 5 provides a brief discussion of the electric and magnetic properties of matter. With the enhanced resolution of electronic spectra due to the use of lasers, the study of electronic spectra of diatomic molecules is becoming a more commonly used method for the determination of molecular parameters so discussion of electronic structure of diatomic molecules has been included in Chapter 6.

Following the preliminary material in the first six chapters the subject of absorption spectroscopy is introduced with the general development of the interaction of radiation and matter and the formulation of absorption coefficients.

The next three chapters provide an introduction to resonance spectroscopy methods and the formulation of relationships using matrix methods. Several concepts necessary to the discussions of spectroscopy in later chapters are introduced here. The subject of magnetic resonance is approached by dividing the subject into the areas of nuclear and electron resonance. Nuclear quadrupole resonance of solids is discussed since this subject is generally omitted from texts of this type.

The next five chapters explore the rotational, vibrational, and electronic spectra of real diatomic molecules. In these chapters dual use of wave mechanics and matrix mechanics is employed in order to ensure that the student understands their interrelationships. Following an introduction to the elementary aspects of vibrotor spectra in Chapters 11 and 12, the discussion returns in Chapters 13 and 14 to cover some of the more detailed concepts relative to the vibration and rotation of diatomic molecules. These will include perturbations due to both external and internal fields, and the Raman effect. The analysis of electronic vibronic and rovibronic spectra in Chapter 15 completes the discussion of diatomic molecules.

Once the basic treatment of rotational, vibrational, and electronic spectroscopy of diatomic molecules has been developed, Chapter 16 directs its attention to the subject of vibration of polyatomic molecules, first by considering the classical problem of normal modes followed by introduction of quantum descriptions and finally a look at the FG matrix treatment. Chapter 17 discusses the rotation of polyatomic molecules. The final chapter is devoted to a discussion of electronic spectra of polyatomic molecules, an important area relative to laser spectroscopy. It is not the intent of these latter chapters to provide either an in-depth or complete coverage of all aspects of the subjects. Instead, a few elementary topics are selected to illustrate the application of the principles developed earlier. The presentations in the latter three chapters will provide the background from which the student can proceed to the study of more complex problems involving the electronic, vibrational, and rotational spectroscopy of polyatomic systems.

The chapters contain references to two types of articles, ones of specific interest which are referenced in the chapters and ones which are sources of data and general references. At the end of the text is a set of appendices which give concise reviews of mathematical methods, classical physics, and other tangential subjects. While it is impossible to cover these subjects in great detail the presentation will give the students sufficient background to appreciate the methods, its problems, and its potential.

The material in this book is presented at a first or second year graduate level. Both the text and the problems have been used in classroom situations. At Virginia Polytechnic Institute the material from this book has been used in two sequential one-quarter courses, Chemistry 5630, Introduction to Spectroscopy, a first-year graduate course taught in the spring quarter, and Chemistry 6640, Molecular Spectroscopy, a second-year graduate course taught in the fall quarter.

The book was developed over a period of several years and my primary thanks go to the many students who, over that span of time, have provided suggestions for improvement of the presentations and have labored through the problems which are included in the chapters. I particularly wish to thank Professor C. D. Cornwell of the University of Wisconsin who was an inspiring teacher and research director and who provided many valuable suggestions during the writing of this book. I also wish to thank Professor J. C. Schug of Virginia Polytechnic Institute and State University for reviewing some of the material. Additional thanks go to the following reviewers whose comments contributed greatly to the final product: Dewitt Coffey, San Diego State University; Lawrence W. Johnson, York College of the City University of New York; Willem R. Leenstra, University of Vermont; Lee Pedersen, University of North Carolina; and Jeffrey Steinfeld, Massachusetts Institute of Technology.

Jack D. Graybeal

ACKNOWLEDGMENTS

As with most textbooks this one draws on the original literature and, in some instances, comprehensive monographs for data are used in the examples. For historical background, brief mention of concepts and experimental results, problems and referrals to a more extensive discussion of subjects, appropriate references are given at the ends of the chapters and noted in the text. Where extensive exemplary use has been made of material from a particular source the authors are acknowledged. It is with appreciation that I acknowledge the use of such material in the following chapters. (Tables and figures utilizing cited data are given in parentheses following the acknowledgment.)

Chapter 6

1 Dr. G. Herzberg for permission to use information from Tables 26, 28, 30, 31, 32 and 34 and Figure 169 from *Molecular Spectra and Molecular Structure, I. Spectra of Diatomic Molecules.* (Tables 6.4, 6.6, 6.8, 6.9, 6.10 and 6.11 and Figures 6.2, 6.12, 6.13, 6.14, 6.19, 6.23 and 6.24 and Problem 7)

Chapter 8

1 The American Institute of Physics for permission to use figures from *J. Chem. Phys.*, **26**, 1339, 1516 (1957). (Figures 8.18 and 8.19)

2 Dr. G. E. Pake and the American Institute of Physics for permission to use data and figures from *J. Chem. Phys.*, **16**, 337 (1948). (Figures 8.21 and 8.22)

3 Varian Associates for permission to use spectra No. 2 and 49 from *NMR Spectra Catalog.* (Figures 8.23 and 8.26*a*)

4 Sadtler Research Laboratories for permission to use spectra Nos. 2429 and 14670 from *Sadtler Standard NMR Spectra.* (Figures 8.15 and 8.26*b*)

Chapter 12

1 Dr. J. Hoeft for permission to use data on GeS from *Z. Naturforschg.* **20A**, 826 (1985). (Tables 12.2 and 12.3)

2 Dr. L. C. Krisher and the American Institute of Physics for permission to use data on AgCl from *J. Chem. Phys.* **44**, 391 (1966). (Tables 12.4–12.7)
3 Dr. K. N. Rao for permission to use data on HI from *J. Mol. Spect.*, **37**, 373 (1971). (Tables 12.8–12.12)

Chapter 13

1 The Canadian Journal of Physics for permission to use data from *Can. J. Phys.* **29**, 151 (1951). (Figure 13.14 and Table 13.3)
2 Dr. W. J. Jones for permission to use data from *Proc. Roy. Soc.*, ser. A, **324**, 231 (1971). (Table 13.4 and Figure 13.15)
3 Spectrochimica Acta for permission to use data from *Spectrochim. Acta*, **17**, 775 (1961). (Table 13.5)

Chapter 14

1 Dr. C. A. Burruss and the American Institute of Physics for permission to use data on CO from *J. Chem. Phys.*, **28**, 427 (1958). (Table 14.4 and Figure 14.13)
2 Mrs. Walter Gordy on behalf of the late Dr. Walter Gordy and the American Institute of Physics for permission to use data on DCl from *Phys. Rev.*, **111**, 209 (1958). (Table 14.5 and Figure 14.14)
3 Dr. C. A. Burruss and the American Institute of Physics for permission to use data on DI from *J. Chem. Phys.*, **30**, 976 (1959). (Tables 14.6 and 14.7 and Figure 14.16)
4 Dr. J. S. Gallagher and the American Institute of Physics for permission to use data on NO from *Phys. Rev.*, **93**, 729 (1954). (Tables 14.8 and Figure 14.17)

Chapter 15

1 The American Institute of Physics for permission to use the data on LiH from *Phys. Rev.*, **47**, 932 (1935). (Tables 15.3 and 15.5)
2 The American Institute of Physics for permission to use data on Li_2 from *Phys. Rev.*, **38**, 1447 (1931). (Table 15.6)
3 The Canadian Journal of Physics for permission to use data on CuH from *Can. J. Phys.*, **46**, 2291 (1918). (Table 15.8)
4 G. Herzberg for permission to employ the concepts embodied in figures such as 129 in *Molecular Structure and Molecular Spectra I, Spectra of Diatomic Molecules*

Chapter 17

1 C. H. Townes and the American Institute of Physics for permission to use data on OCS from *Phys. Rev.*, **74**, 1113 (1948). (Tables 17.1 and 17.2)
2 Mrs. Walter Gordy for the late Dr. Walter Gordy and the American Institute of Physics for permission to use data on NF_3 from *Phys. Rev.*, **79**, 513 (1950). (Table 17.4)

3 Dr. R. Varma and the American Institute of Physics for permission to use data on GeH_3 from *J. Chem. Phys.*, **46**, 1565 (1967). (Table 17.5)

Chapter 18

1 H. B. Gray for permission to use MO diagrams of AB_4 type molecules from *J. Am. Chem. Soc.*, **85**, 260 (1963).
2 Dr. K. Watanabe and the American Institute of Physics for permission to use data on H_2O from *J. Chem. Phys.*, **22**, 1564 (1954). (Figures 18.23 and 18.42)
3 Academic Press Incorporated for permission to use the spectrum of benzene from *Absorption Spectra in the Ultraviolet and Visible Region*, Ed. L. Lang, 1961, p. 367

Appendices

T Dr. E. B. Wilson, Jr., and McGraw-Hill Book Company for permission to reproduce G-Matrix elements from *Molecular Vibrations*
X Dr. Y. Tanabe for permission to reproduce Tanabe-Sugano Diagrams for O_h Symmetry from *J. Phys. Soc. Japan*, **9**, 753 (1954).

CHAPTER 1

INTRODUCTION

1.1 AN OVERVIEW

During the first half of the twentieth century the methods of thermodynamics and kinetics dominated the ways used to investigate chemical systems. Such methods provided enormous insight into the bulk behavior of matter but could not provide much detailed information regarding the properties of individual molecules. Building on the development of modern quantum mechanics, which has accelerated rapidly since the initial work of Schrödinger and Heisenberg in 1926, and employing sophisticated methods for handling electromagnetic radiation and processing data, the area of spectroscopy has achieved a status equivalent to that of thermodynamics and kinetics. The scope of spectroscopy is very extensive and is closely related in many instances to the nature of the information obtained in complementary areas such as neutron, X-ray and electron diffraction, mass spectrometry, magnetic and electric susceptibility measurements, electron microscopy, optical rotatory dispersion, and circular dichroism, to cite some major ones. These complementary methods, other than diffraction, for the most part give useful information regarding gross structures and nongeometric molecular parameters. The techniques of spectroscopy and diffraction are the primary sources of information regarding detailed molecular geometries.

The primary structural parameters which determine the geometry of a molecule are the bond lengths and interbond angles. If one can accurately measure all such parameters for a given molecule, then the geometry is known. The magnitudes of the structural parameters are frequently found to correlate with chemical properties, particularly for series of related molecules, so that reciprocal predictions of parameters and properties can be made. For many molecules, especially large ones, a complete determination of all structural parameters is often impossible or impractical. In such cases it is possible to acquire a great deal of useful information regarding the general nature of the symmetry and geometry of the system by the judicious application of selected techniques of spectroscopy. Thus, one of the primary uses for the results of spectroscopic experiments is to determine molecular geometries. In addition to geometric parameters, spectroscopic experiments yield a wide variety of information related to other features of the system being studied. Typical of such information would be parameters which are measures of electron distribution, electrical polarizability, bond strengths, intramolecular magnetic interactions, and energy level arrangements, to mention a few.

The experimental methods of spectroscopy used for the determination of molecular parameters involve the interaction of electromagnetic radiation and matter. These methods may examine radiation emitted, radiation absorbed, or radiation scattered by a system. The techniques and theories of electron, X-ray, and neutron diffraction are very valuable for the determination of molecular geometries, particularly of large molecules in the solid state, but they are sufficiently unrelated to those of spectroscopy that they will not be included in this book.

A complete discussion of the subject of spectroscopy will fill a multivolume treatise so any presentation of a reasonable size requires some judicious choices of subjects for inclusion and the adoption of a particular plan of presentation. These are three topics which contribute heavily to the subject of spectroscopy: (1) the relationships between the experimentally observed spectroscopic parameters and the individual molecular parameters, (2) the relationships between the molecular parameters and the structures of the molecules, and (3) the experimental and instrumental aspects of particular branches of spectroscopy. In this presentation the primary emphasis will be on the development, via the medium of quantum mechanics, of the analytical relationships between experimental spectroscopic data and the properties and parameters of individual molecules. The spectroscopic data will be the frequencies and intensities of radiation which is emitted, absorbed, or scattered from molecules. The development of these relationships will proceed by considering for each area of spectroscopy: (1) the specific features of quantum mechanics which are needed for the development, (2) the nature of the atomic and molecular energy-level schemes and allowed spectroscopic transitions among them, (3) the relationships between the observed transition frequencies and molecular properties and parameters, and (4) reviews of specific examples. The object will be to provide a comprehensive treatment of simple systems in order for the reader to build a fundamental background and to be able to proceed independently to more advanced works. There will be only limited development of the analytical relationships between the structure of molecules and the spectroscopically determined molecular parameters. Details regarding experimental methods will be at a minimum with any such discussions being brief and general in nature.

Having enumerated the plan for presentations of the subject but prior to examining the details of the various areas of spectroscopy a brief review of the nature of electromagnetic radiation and the general nature of spectroscopic experiments is appropriate.

1.2 ELECTROMAGNETIC RADIATION

The general nature of electromagnetic radiation was developed by Maxwell in the midnineteenth century, and many excellent discussions exist in a wide variety of textbooks on general physics, so only those aspects directly related to this presentation will be summarized. An understanding of the interaction of radiation and matter requires the recognition of the dual nature of electromagnetic radiation. In some types of interactions electromagnetic radiation is best considered as photons, each of which is equivalent to a specific amount of energy. In other interactions treatment of radiation as a wave phenomenon is required. These two types of behavior are not unrelated, however.

When considered as a wave phenomenon, the most elementary formulation is that of a plane wave traveling in one direction (polarized radiation). For this case there will be associated with the radiation oscillating electric E and magnetic H fields which are functions of the distance x, and time t.

$$E = E_0 \sin\left[2\pi(\bar{\nu}x - \nu t) + \delta\right] \tag{1.1}$$

$$H = H_0 \sin\left[2\pi(\bar{\nu}x - \nu t) + \delta\right] \tag{1.2}$$

where $\bar{\nu}$ is the reciprocal of the wavelength λ, ν is the frequency, and δ is a phase factor. The frequency and the wavelength are related to the velocity of propagation c by

$$\nu\lambda = c \tag{1.3}$$

The energy of a photon of radiation is given by

$$E = h\nu \tag{1.4}$$

The subject of spectroscopy deals with the absorption or emission of electromagnetic radiation when an atom or molecule undergoes a change from one energy state to another and as such is primarily interested in the energy content rather than the wave nature of the radiation. A discussion of the subject of diffraction, on the other hand, relies heavily on the wave nature of radiation.

The most common region of the electromagnetic spectrum, insofar as the human senses are concerned, is the visible region. Historically it was in this region that the first spectroscopic transitions were observed and measured. However the visible region is a rather small part of the total electromagnetic spectrum, and were it to be the only accessible region the study of spectroscopy would be very limited. A diagram of the electromagnetic spectrum showing the common regions and relating the values of the frequency, wavelength, and wavenumber at regular intervals is given in Fig. 1.1.

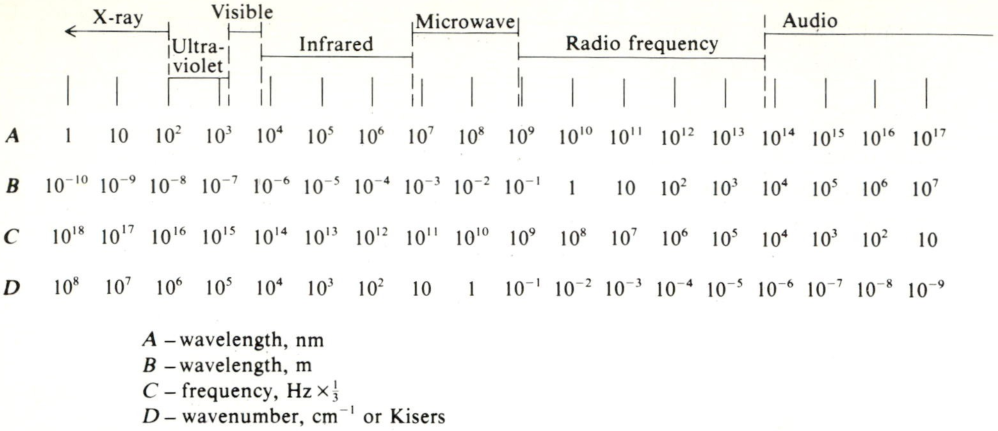

FIGURE 1.1
Electromagnetic spectrum.

1.3 SPECTROSCOPY—AN OVERVIEW OF EXPERIMENTAL METHODS

Before becoming involved in the details of various types of spectroscopy and molecular systems the general nature of spectroscopic experiments will be reviewed.

Considering a limited set of quantized molecular energy levels as shown in Fig. 1.2, it can be observed that it will require the absorption of an amount of energy, $\Delta E = E_2 - E_1$, to induce the system to go from state 1 to state 2. This energy can be provided by electromagnetic radiation of the proper frequency if there is an adequate mechanism for the system to interact with the radiation.

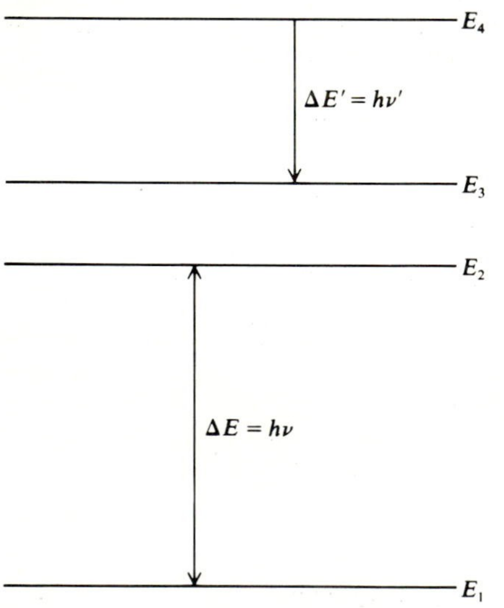

FIGURE 1.2
Quantized energy levels.

Assuming that the mechanism exists, then the frequency of the radiation absorbed is related to the energy difference by the Bohr condition

$$\Delta E = h\nu \tag{1.5}$$

If the system were to undergo a change from state 4 to state 3 then there would be an emission of energy, $\Delta E' = E_4 - E_3$. The released energy would be radiation whose frequency ν' is given by

$$\nu' = \frac{\Delta E'}{h} \tag{1.6}$$

There are two general experimental areas of spectroscopy *emission* and *absorption.* The first results from the emission of electromagnetic radiation by an excited molecule or atom as it returns from a high-energy state to a lower one. ($E_4 \rightarrow E_3$ in Fig. 1.2.) Historically emission spectra of atoms and molecules were the first to be studied. The discrete colors produced by exciting certain atoms or ions in flames were used over a century ago as a method for qualitative identification of the emitting species. The regular pattern of such emitted spectra were characterized by scientists such as Balmer, Lyman, and Ritz, in the latter part of the nineteenth century, and these studies led to the development by Bohr and Sommerfeld of the quantum theory of atomic structure. Although many of the classical analytical methods based on emission spectroscopy have been supplanted by other techniques and the amount of structural information available is limited, the area of emission spectroscopy is still an important subject and has been diversified to include such widely different areas as the monitoring of smoke stack emissions and the detection of molecules in outer space.

The general experimental method for obtaining an emission spectrum is shown in Fig. 1.3. The light from the excited source is analyzed by means of either (1) dispersive elements, such as a prism or grating, which simultaneously disperses all emitted wavelengths onto a detecting film for recording, or (2) a monochrometer which selectively directs a given wavelength of emitted radiation onto a detector for subsequent recording. In either case the resulting spectrum is a plot of emitted radiation intensity vs. wavelength.

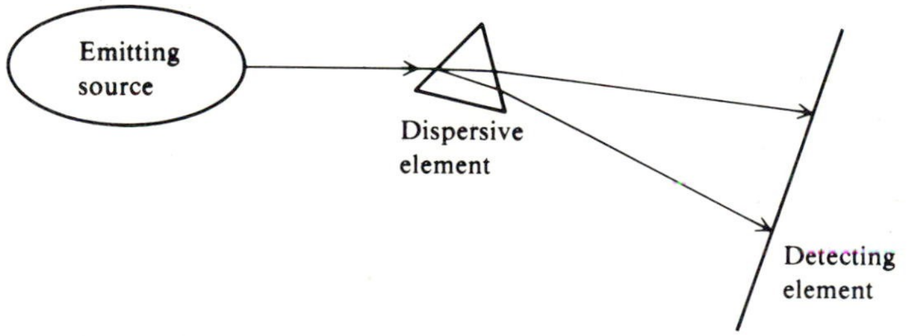

FIGURE 1.3
Principle of operation of an emission spectrograph.

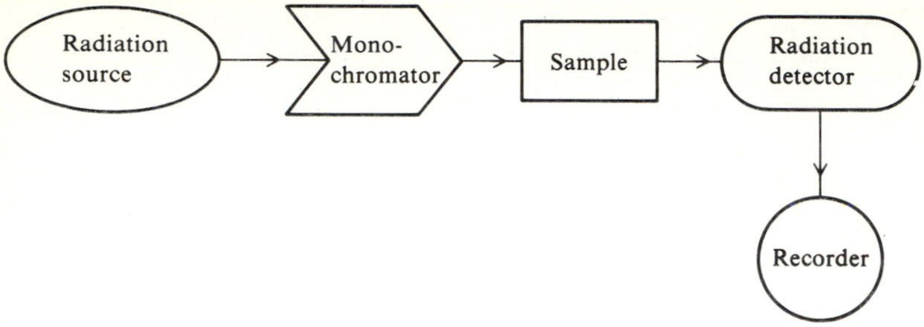

FIGURE 1.4
Principle of an absorption spectrometer.

Absorption spectroscopy, depending on the region of the electromagnetic spectrum, may involve either of two experimental methods, the first of which is illustrated in Fig. 1.4. In this experiment radiation of a given wavelength is transmitted through the sample and the intensity of the emergent beam, relative to the incident beam, is measured. When the electromagnetic radiation is at a frequency below the microwave region, the experimental arrangement is as shown in Fig. 1.5 and the absorption is detected as a current or voltage unbalance as the frequency of the applied signal corresponds to the Bohr condition for the energy levels involved. This latter method is termed a *resonance experiment.*

Another type of absorption experiment which is of paramount importance is that of *Fourier transform spectroscopy.* This is illustrated in Fig. 1.6. In this experiment the sample is irradiated with electromagnetic radiation containing a broad band of frequencies, generated either by a broad-banded source or by

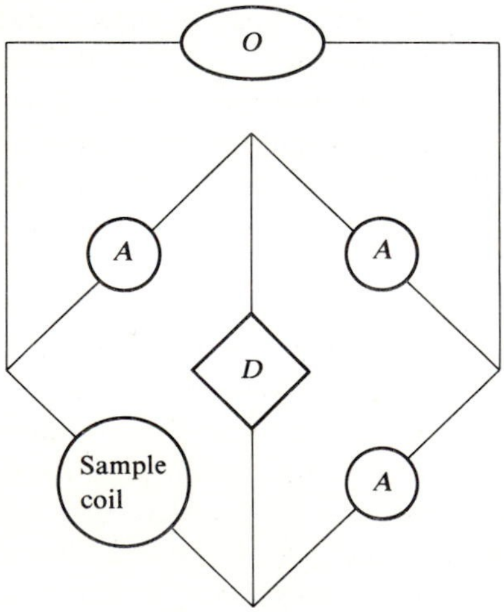

A—variable impedance elements for balancing
D—current detector
O—variable frequency oscillator

FIGURE 1.5
Principle of a resonance spectrometer.

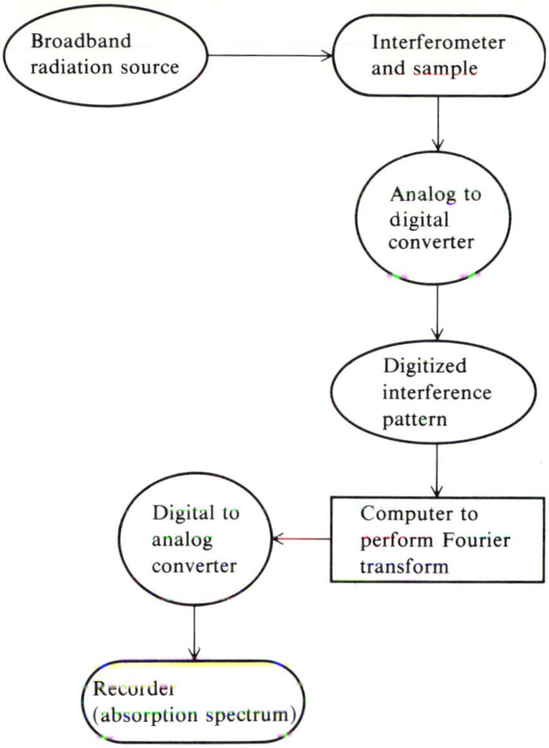

FIGURE 1.6
Principle of Fourier transform spectroscopic experiment.

pulsing a monochromatic source. Due to the absorption of specific frequencies by molecular transitions the resulting interference pattern (interferogram) will contain the same information as a normal absorption spectrum. The interferogram is the Fourier transform of the absorption spectrum so the latter is obtained from the former via a computer-generated mathematical transformation.

A brief summary of the origins of molecular energy transitions and associated electromagnetic spectral regions is given in Table 1.1. It is not possible to make a one-to-one alignment of types of molecular transitions and spectra regions, so Table 1.1 gives only major correlations.

In addition to direct absorption or emission of electromagnetic radiation there are four associated experimental phenomena which will be considered. Their relationships are shown in Fig. 1.7. The first of these phenomena is *fluorescence* which is further illustrated in Fig. 1.8. This results from an atomic or molecular system being excited into an upper state by absorption of radiation of frequency ν_0 and then decaying back to a lower state in a time less than 10^{-5} seconds. The remitted radiation may be of frequency $\nu = \nu_0$ (resonance fluorescence) or, by occurring in two or more successive transitions, may result in the emission of radiation of frequency $\nu < \nu_0$. A related phenomenon is *phosphorescence* which results when a molecule loses energy by means of a radiationless process and ends up in a high-energy, long-lived state from which

TABLE 1.1
Spectral regions and associated absorption phenomenon

Region	Absorption phenomenon	Frequency range, hz	Wavenumber range, cm^{-1}	Wavelength range
X-ray	Inner core electron transitions	3×10^{16}–3×10^{8}	10^{6}–10^{8}	10–0.1 nm
Ultraviolet	Valence electron transitions	7.5×10^{14}–3×10^{16}	2.5×10^{4}–10^{6}	400–10 nm
Visible	Valence electron transitions	4×10^{-14}–7.5×10^{14}	1.4×10^{4}–2.5×10^{4}	700–400 nm
Infrared	Vibrational transitions	6×10^{11}–4×10^{14}	20–1.4×10^{4}	0.05–7×10^{-7} m
Microwave	Rotational transitions–electron spin resonance	10^{9}–6×10^{11}	0.03–20	0.35×10^{-4}–0.05 m
Radiofrequency	Pure nuclear quadrupole resonance–nuclear magnetic resonance	5×10^{5}–10^{9}	1.7×10^{-5}–0.03	6×10^{2}–0.3 m

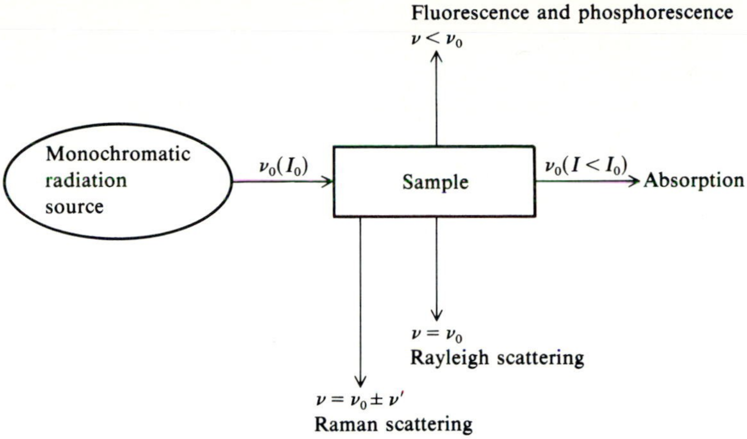

FIGURE 1.7
Interaction of radiation and matter. Principal phenomenon.

it will slowly decay back to the ground state. Often phosphorescent emission can be observed for an appreciable time following removal of the exciting source.

The other two phenomena result from the scattering of radiation as shown in Fig. 1.7. If radiation of frequency ν_0 impinges on a molecule, one can observe a scattering of radiation of the same frequency which is not the result of absorption and reemission as is resonant fluorescence but results from a different mechanism,

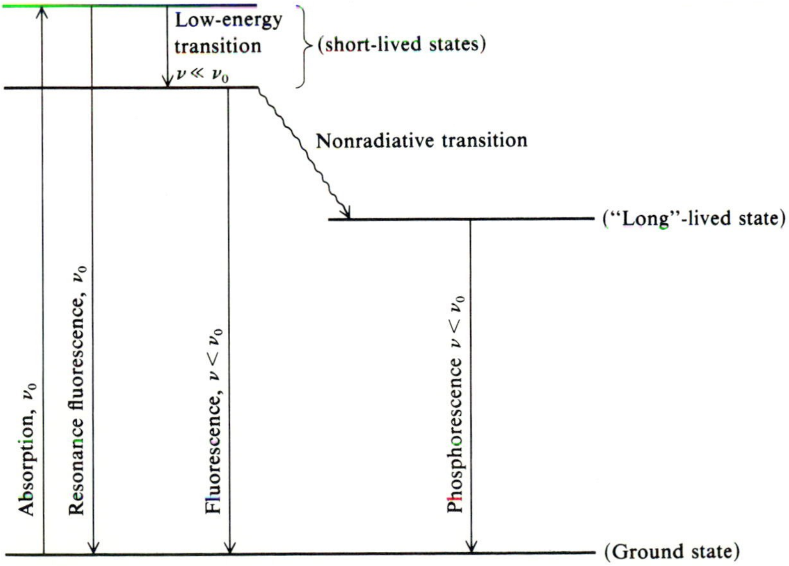

FIGURE 1.8
Mechanism of fluorescence and phosphorescence.

known as *Rayleigh scattering.* If part of the energy of the incident radiation is used by the molecular system as it undergoes a simultaneous internal energy change, $\Delta E = h\nu'$, then the scattered radiation will be of a frequency $\nu_0 \pm \nu'$. If the internal change results in a net decrease in the molecular energy the scattered radiation will be at a frequency of $\nu_0 + \nu'$, while for a molecular energy increase the reverse is true. This phenomenon is known as the *Raman effect.*

The applications of absorption spectroscopy in chemistry range from simple experiments involving the absorption of a narrow band of visible light for the determination of an ion concentration to sophisticated double resonance experiments in the microwave region and high-speed, computer-controlled Fourier transform spectrometers in the infrared and radio frequency regions. The results of absorption spectroscopy can provide very precise measurements of parameters leading to bond lengths and bond angles and can lead to the determination of molecular properties such as dipole moments, molecular quadrupole moments, nuclear quadrupole coupling parameters, and various parameters resulting from the interactions of nuclei with other nuclei and with electrons.

In the discussion of all types of spectroscopy there are two concepts which occur frequently and need to be examined. The first of these is resolution. Due to the *Heisenberg uncertainty principle* no molecular system undergoing a transition between two quantized energy states absorbs radiation at a single frequency but will absorb a band of frequencies whose intensity is a maximum at the center absorption frequency and falls off rapidly at lower and higher frequencies. Although this principle establishes a lower limit on the observed width of spectra transitions there are generally other factors which are greater contributors. These concepts will be enumerated more fully in Chapter 7. Depending on various conditions the actual line shape can be represented by either a Gaussian or a Lorentzian function characterized by a line width $\Delta\nu$. The output of a spectrometer will reproduce the absorption of the band of frequencies as shown in Fig. 1.9*a*. If a sample has two closely spaced transition frequencies as illustrated in Fig. 1.9*b*, then the resolving power $\Delta\lambda$ of the spectrometer is a measure of how close together the two transition frequencies can lie and still be recorded as individual peaks rather than a single one. The resolving power of a particular spectrometer will depend on how narrow a band of energy can be directed into the absorbing sample. For a system that uses a multifrequency source and a monochromator to select a narrow band of frequencies for transmission into the sample, the resolution will be limited by the width of an exit slit. There will always be a trade-off between slit width and power transmitted, hence between resolution and time necessary to obtain a spectrum. For a system employing a "single"-frequency source the resolution is generally determined by the stability of the source.

A second important factor is the signal-to-noise ratio (S/N). In any spectrometer there will be noise arising from random fluctuations in the electronic components which control the source and process the signals at various points. In the absence of any signal absorption due to a sample this noise will produce a background fluctuation which will result in a randomly varying output. For a

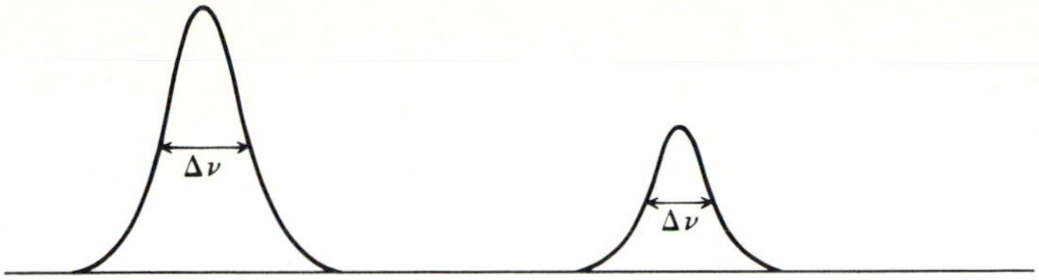

(a)

Line separation much greater than linewidth

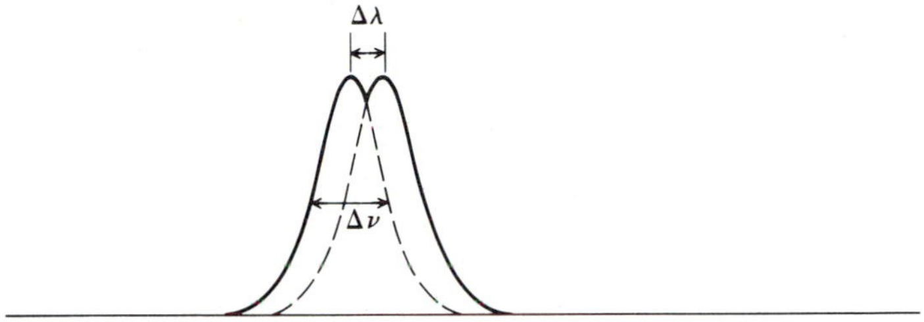

(b)

Line separation of the order of the linewidth

FIGURE 1.9
Absorption lines and resolving power.

given molecular energy transition there will be a certain absorption coefficient which will give the fraction of power absorbed per unit path length in the sample. If this coefficient is sufficiently small that the variation in output power due to absorption is less than the random fluctuations due to noise then S/N will be less than unity and the desired signal will be lost in the noise. The value of S/N can be improved by the use of: (1) high-quality stable electronics, (2) the incorporation of various modulation methods such as beam chopping in the infrared region, Stark modulation in the microwave region, and magnetic field modulation in NMR and ESR, and (3) the use of cumulative spectra techniques such as transient averaging or fast Fourier transform methods.

Following this general introduction to the subject we will proceed to develop those aspects of quantum mechanics which will be needed in the latter chapters.

CHAPTER 2

WAVE MECHANICAL METHODS

2.1 INTRODUCTION

A necessary prerequisite to the study of spectroscopy is a background in the basic aspects of quantum mechanics. This background must include a knowledge of both wave mechanics [1] and matrix mechanics [2]. Since present undergraduate physical chemistry courses devote a substantial segment of their coverage to wave mechanics and a separate senior-graduate course in this subject is frequently offered, no comprehensive treatment will be included. For this material the reader is referred to the general references given at the end of the chapter. However, there are four topics from wave mechanics which will be reviewed at this point because of their use in latter sections of this text. The solutions of the wave equations for the nonrotating harmonic diatomic oscillator and the rigid diatomic rotor provide background for the discussion of real molecules in Chap. 11. A review of the variational method precedes the discussion of the electronic structure of diatomic molecules in Chap. 6. A somewhat detailed discussion of the perturbation theory is included since this method as applied in the use of matrix mechanics will be developed more extensively in Chap. 3, and is used in the discussion of several spectroscopic applications throughout the book.

Because the subject of matrix mechanics is less extensively discussed in standard physical chemistry and quantum mechanics texts its formulation will be developed in detail in Chap. 3.

2.2 NONROTATING HARMONIC DIATOMIC OSCILLATOR

A study of the behavior of the harmonic oscillator provides the introduction to vibrational spectroscopy. The physical system considered is that of two atoms m_1 and m_2, separated by a distance r, and constrained to move in one dimension under the influence of a force which vanishes at $r = r_e$ and is proportional to the displacement from equilibrium $x = r - r_e$. The discussion of the classical oscillator in App. A shows that such a system is mechanically equivalent to that of a single mass, $\mu = m_1 m_2(m_1 + m_2)^{-1}$, constrained to move in one dimension under the influence of a force F proportional to the displacement

$$F = -kx \tag{2.1}$$

where k is a proportionality constant, denoted as the stretching force constant of the bond. The negative sign is due to the direction of the force being opposite to that of the displacement. The potential energy $V(x)$ is related to the force by

$$\frac{dV(x)}{dx} = -F(x) = kx \tag{2.2}$$

Therefore

$$V(x) = \tfrac{1}{2}kx^2 \tag{2.3}$$

The total classical energy is then

$$H = \frac{P_x^2}{2\mu} + \frac{1}{2}kx^2 \tag{2.4}$$

Substitution of the appropriate quantum mechanical operators from App. B yields the Schrödinger equation

$$-\frac{\hbar^2}{2\mu}\frac{d^2\psi(x)}{dx^2} + \frac{1}{2}kx^2\psi(x) = E\psi(x) \tag{2.5}$$

which, upon rearrangement and setting $\psi(x) = \psi$ for simplicity, becomes

$$\frac{d^2\psi}{dx^2} + \frac{2\mu}{\hbar^2}E\psi - \frac{kx^2\mu}{\hbar^2}\psi = 0 \tag{2.6}$$

In seeking the solutions to the differential equations of quantum mechanics one first tries to find a similarity between the Schrödinger equation and a standard form differential equation. The general approach is to find a change of variable which will allow the Schrödinger equation to be converted to the standard form. The choice of such a variable is not always apparent, but for the harmonic oscillator a consideration of the classical oscillator and some dimensional analysis can aid in justifying the choice.

For any quantized system the difference between two energy levels will be given by the Einstein condition

$$\Delta E = h\nu \tag{2.7}$$

Since the differences between energy levels are proportional to $h\nu$ the energies of the individual levels may be assumed to be proportional to $h\nu$ also. For a classical oscillator (App. A) the frequency is given by

$$\nu = \frac{1}{2\pi}\sqrt{\frac{k}{\mu}} \tag{2.8}$$

where ν is in s^{-1}, k is in newton m^{-1}, and μ is in kilograms. Making the assumption that the energy of a quantum mechanical oscillator will depend on the mass and force constant in a manner analogous to the classical case and recognizing that the units of energy will be the same for both cases, the energy of the quantum mechanical oscillator can be expressed by

$$E = a'h\nu = \frac{a\hbar}{2}\sqrt{\frac{k}{\mu}} \tag{2.9}$$

where a' is a proportionality constant.

The conversion of the Schrödinger equation to a standard from differential equations is made by using the variable $\eta = \beta^{1/2}x$, where the constant β is given by

$$\beta = \frac{\sqrt{k\mu}}{\hbar} \tag{2.10}$$

The dimension of β is kg^{-2}, hence η is a mass-weighted coordinate. Since β is a constant

$$\frac{d}{dx} = \frac{d}{d\eta}\frac{d\eta}{dx} = \beta^{1/2}\frac{d}{d\eta} \tag{2.11}$$

and

$$\frac{d^2}{dx^2} = \frac{d}{dx}\beta^{1/2}\frac{d}{d\eta} = \beta^{1/2}\frac{d}{d\eta}\frac{d\eta}{dx}\frac{d}{d\eta} = \beta\frac{d^2}{d\eta^2} \tag{2.12}$$

Substitution of Eqs. (2.12), (2.9), and $\eta = \beta^{1/2}x$ into Eq. (2.5) gives

$$-\frac{\hbar^2}{2}\beta\frac{d^2\psi}{d\eta^2} + \frac{1}{2}\frac{k\eta^2\psi}{\beta} = \frac{a\hbar}{2}\sqrt{\frac{k}{\mu}}\psi \tag{2.13}$$

where $a = 2a'$ and $\psi = \psi(\eta)$. Incorporating the definition of β, rearrangement gives

$$-\frac{\hbar}{2}\sqrt{\frac{k}{\mu}}\frac{d^2\psi}{d\eta^2} + \frac{\hbar}{2}\sqrt{\frac{k}{\mu}}\eta^2\psi = \frac{a\hbar}{2}\sqrt{\frac{k}{\mu}}\psi \tag{2.14}$$

which reduces to

$$-\frac{d^2\psi}{d\eta^2} + \eta^2\psi = a\psi \tag{2.15}$$

This is the eigenvalue equation discussed in App. C and has the general solutions

$$\psi_v(\eta) = N_v H_v(\eta) e^{-(1/2)\eta^2} \tag{2.16}$$

where $H_v(\eta)$ are the Hermite polynomials and

$$N_v = \left[\left(\frac{\beta}{\pi}\right)^{1/2} \frac{1}{v!2^v}\right]^{1/2} \tag{2.17}$$

A necessary condition for the solution of Eq. (2.15) to be a well-behaved function is that v be an integer. Since $a - 1 = 2v$, the energy of the quantum mechanical oscillator becomes

$$E = \frac{(2v+1)}{4\pi}\sqrt{\frac{k}{\mu}} = \left(v + \frac{1}{2}\right)\frac{h}{2\pi}\sqrt{\frac{k}{\mu}} = \left(v + \frac{1}{2}\right)h\nu_0 \tag{2.18}$$

The concept of quantization of vibrational energy has appeared as a natural consequence of the mathematics and boundary conditions and was not introduced as an original assumption. The fact that the functions $\psi_v(\eta)$ are finite, single-valued, and continuous, is shown in App. C.

The energy obtained not only differs from that obtained from the classical treatment by exhibiting quantization but also by the inclusion of the $\frac{1}{2}$. This indicates that the lowest level of the oscillator has a energy of $\frac{1}{2}h\nu_0$, rather than zero. This is in conformity with the uncertainty principle since, were the oscillator energy to be zero, then its motion would have to be zero and the simultaneous specification of its exact momentum and position would be required.

2.3 RIGID DIATOMIC ROTOR

Another system which serves to review the application of wave mechanics and at the same time introduce the use of vectors to describe motion is the diatomic rotating molecule. Not only does this provide an expression for the energy and the wavefunction of the rigid rotor but also affords to a look at the mechanics of rotation of a single body and the separation of different types of motion in a system. The definition of classical angular momentum and the relationships among Cartesian axes and spherical polar coordinates are reviewed in App. D, and a summary of vector algebra is given in App. E.

The three-dimensional rigid rotor consists of two atoms having masses m_1 and m_2 separated by a constant interatomic distance $\boldsymbol{\rho}$, free to undergo both translational and rotational motion and subject to no external forces. The system is represented by position vectors $\mathbf{r}_1$ and $\mathbf{r}_2$ as shown in Fig. 2.1. The total energy of the system is the sum of the kinetic and the potential energies, the latter being zero if the rotor is located in a field-free space. Thus

$$E = \tfrac{1}{2}m_1 v_1^2 + \tfrac{1}{2}m_2 v_2^2 = \tfrac{1}{2}m_1 \dot{\mathbf{r}}_1^2 + \tfrac{1}{2}m_2 \dot{\mathbf{r}}_2^2 \tag{2.19}$$

where $\mathbf{r}_i$ is the time derivative of $\mathbf{r}_i$, that is, the velocity.

The center of mass is located by the vector $\mathbf{R}$, and the mass points are located relative to the center of mass by the vectors $\boldsymbol{\rho}_1$ and $\boldsymbol{\rho}_2$, hence the interatomic distance $\boldsymbol{\rho}$ is the sum $\boldsymbol{\rho}_1 + \boldsymbol{\rho}_2$. Vector addition gives

$$\mathbf{r}_1 = \mathbf{R} + \boldsymbol{\rho}_1 \qquad \mathbf{r}_2 = \mathbf{R} - \boldsymbol{\rho}_2 \tag{2.20}$$

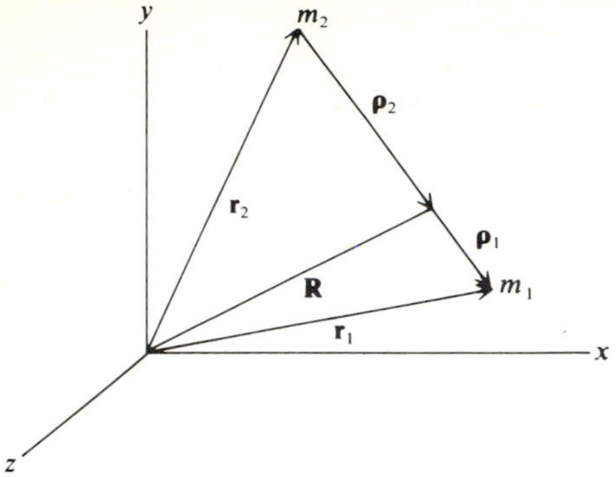

FIGURE 2.1
Vector coordinates of a diatomic rotor.

which upon differentiation yields

$$\dot{\mathbf{r}}_1 = \dot{\mathbf{R}} + \dot{\boldsymbol{\rho}}_1 \qquad \dot{\mathbf{r}}_2 = \dot{\mathbf{R}} - \dot{\boldsymbol{\rho}}_2 \tag{2.21}$$

The energy is then expressed as

$$E = \tfrac{1}{2}m_1(\dot{\mathbf{R}} + \dot{\boldsymbol{\rho}}_1)^2 + \tfrac{1}{2}m_2(\dot{\mathbf{R}} - \dot{\boldsymbol{\rho}}_2)^2 \tag{2.22}$$

Since the scalar product of a vector is $\mathbf{r} \cdot \mathbf{r} = \mathbf{r}^2$, Eq. (2.22), on expansion and rearrangement, gives

$$E = \tfrac{1}{2}m_1(\dot{R}^2 + \dot{\rho}_1^2) + \tfrac{1}{2}m_2(\dot{R}^2 + \dot{\rho}_2^2) + \tfrac{1}{2}\dot{\boldsymbol{R}} \cdot (m_1\dot{\boldsymbol{\rho}}_1 - m_2\dot{\boldsymbol{\rho}}_2) \tag{2.23}$$

Using the definition of the center of mass, $\sum_i m_i r_i = 0$, which for a two-mass system means $m_1\boldsymbol{\rho}_1 = m_2\boldsymbol{\rho}_2 = 0$, the last term becomes zero, leaving

$$E = \tfrac{1}{2}(m_1 + m_2)\dot{R}^2 + \tfrac{1}{2}m_1\dot{\rho}_1^2 + \tfrac{1}{2}m_2\dot{\rho}_2^2 \tag{2.24}$$

The first term is the translational motion of a particle having a mass of $m_1 + m_2$, located at the center of mass of the system, i.e., the three-dimensional particle in a box. The latter two terms denote the kinetic energy of rotation. Thus the motion and the energy have been separated into translational and rotational components. The ability to separate classical motion in this manner is the basis for the assumption of separability in quantum mechanical systems.

Since the interatomic distance is constant, the collinear position vectors $\boldsymbol{\rho}_1$ and $\boldsymbol{\rho}_2$ can be written as components of $\boldsymbol{\rho}$

$$\boldsymbol{\rho}_1 = a\hat{\boldsymbol{\rho}} \qquad \boldsymbol{\rho}_2 = b\hat{\boldsymbol{\rho}} \tag{2.25}$$

where $\hat{\boldsymbol{\rho}}$ is a unit vector in the direction of $\boldsymbol{\rho}$. Also

$$\dot{\boldsymbol{\rho}}_1 = a\dot{\boldsymbol{\rho}} \qquad \dot{\boldsymbol{\rho}}_2 = b\dot{\boldsymbol{\rho}} \tag{2.26}$$

and

$$\boldsymbol{\rho}_1 + \boldsymbol{\rho}_2 = (a + b)\hat{\boldsymbol{\rho}} = \boldsymbol{\rho} \tag{2.27}$$

The rotational energy can then be expressed as

$$E_r = \tfrac{1}{2}(m_1a^2 + m_2b^2)\dot{\rho}^2 \tag{2.28}$$

The term $m_1a^2 + m_2b^2$ is defined as the moment of inertia of the system and $\dot{\boldsymbol{\rho}}$ is the rate of change with time of the vector $\boldsymbol{\rho}$. Since $\boldsymbol{\rho}$ is of constant length, this change is in direction only. Hence, $\boldsymbol{\rho}$ describes the rotational motion and is the angular velocity. Therefore

$$E_r = \tfrac{1}{2}I\dot{\rho}^2 \tag{2.29}$$

The moment of inertia can also be written as

$$I = \frac{m_1 m_2}{m_1 + m_2}(a+b)^2 = \mu\rho^2 \tag{2.30}$$

The physical problem of mutual rotation of two masses about a common center is the same as the rotation of a single mass of magnitude μ at a constant distance p from a fixed point. In terms of the angular momentum P, the energy of the system is

$$E_r = \frac{I^2\dot{\rho}^2}{2I} = \frac{P^2}{2I} \tag{2.31}$$

Specifying the orientation of $\dot{\boldsymbol{\rho}}$ or $\boldsymbol{\rho}$ by the polar angles θ and ϕ and using the angular momentum operators from App. B the Schrödinger equation for the three-dimensional diatomic rotor is written as

$$\frac{-\hbar^2}{2I}\frac{1}{\sin\theta}\frac{\partial}{\partial\theta}\sin\theta\frac{\partial\psi(\theta,\phi)}{\partial\theta} + \frac{1}{\sin^2\theta}\frac{\partial^2\psi(\theta,\phi)}{\partial\phi^2} = E\psi(\theta,\phi) \tag{2.32}$$

Drawing on the experience whereby classical motion is resolved into components relative to the directions of a coordinate system, the θ and ϕ motions are assumed to be separable and the wavefunction $\psi(\theta, \phi)$ written as

$$\psi(\theta,\phi) = \Theta(\theta)\Phi(\phi) = \Theta\Phi \tag{2.33}$$

Inserting this function into Eq. (2.32), multiplying by $\sin^2\theta/\Theta\Phi$ and allowing for the independence of Θ and Φ from ϕ and θ respectively, a separable equation of the type $f(x) + f(y) = 0$ is obtained

$$\frac{\sin\theta}{\Theta}\frac{\partial}{\partial\theta}\sin\theta\frac{\partial\Theta}{\partial\theta} + \frac{2IE\sin^2\theta}{\hbar^2} + \frac{1}{\Phi}\frac{\partial^2\Phi}{\partial\phi^2} = 0 \tag{2.34}$$

Letting the separation constant be M^2, one has

$$\frac{d^2\Phi}{d\phi^2} = -M^2\Phi \tag{2.35}$$

$$\frac{\sin\theta}{\Theta}\frac{d}{d\theta}\sin\theta\frac{d\Theta}{d\theta} + \frac{2IE\sin^2\theta}{\hbar^2} = M^2 \tag{2.36}$$

The general solution of the ϕ equation is

$$\Phi(\phi) = Ne^{iM\phi} \tag{2.37}$$

The boundary condition for $\Phi(\phi)$ to be single-valued and continuous is

$$\Phi(\phi + 2\pi) = \Phi(\phi) \tag{2.38}$$

Therefore

$$Ne^{iM\phi} = Ne^{iM(\phi+2\pi)} \tag{2.39}$$

or

$$\cos M\phi + i \sin M\phi = \cos M(\phi + 2\pi) + i \sin M(\phi + 2\pi) \tag{2.40}$$

Expanding the terms on the right-hand side using conventional $\sin(x + y)$ and $\cos(x + y)$ relations and equating real and imaginary coefficients, one can express the boundary condition in the form of two equations

$$\cos M\phi = \cos M\phi \cos 2M\pi - \sin M\phi \sin 2M\pi \tag{2.41}$$

$$\sin M\phi = \sin M\phi \cos 2M\pi + \cos M\phi \sin 2M\pi \tag{2.42}$$

The necessary condition for these equations to be valid is for M to be an integer, either positive, negative, or zero. Normalization of the wavefunctions, Eq. (2.37), gives

$$\int_0^{2\pi} \Phi^*\Phi \, d\phi = N^2 \int_0^{2\pi} e^{-iM\phi} e^{iM\phi} \, d_\phi = 1 \tag{2.43}$$

or

$$N = \frac{1}{\sqrt{2\pi}} \tag{2.44}$$

The Θ equation is related to the Legendre differential equation whose solutions are well known. A brief summary of the Legendre equation and its solutions, the Legendre polynomials, are given in App. F. It is interesting to note that this is one of several differential equations, encountered in wave mechanics, which were well known to mathematicians before the development of quantum mechanics. The solution to the Θ equation is

$$\Theta_{JM} = N_{JM} P_J^{|M|} (\cos \theta) \tag{2.45}$$

where

$$N_{JM} = \left[\frac{(2J + 1)(J - |M|)!}{2(J + |M|)} \right]^{1/2} \tag{2.46}$$

$$P_J^{|M|} (\cos \theta) = \frac{(-1)^J}{2^J J!} \sin^{|M|} \theta \frac{d^{|M|}}{d (\cos \theta)^{|M|}} \frac{d^J (\cos^2 \theta - 1)^J}{d (\cos \theta)^J} \tag{2.47}$$

A necessary condition for the solutions to be finite, single-valued, and continuous is that

$$E = \frac{\hbar^2}{2I} J(J + 1) \tag{2.48}$$

where J is zero or a positive integer, and for a given value of J the value of $|M|$ cannot exceed J.

In addition to J being a quantum number which denotes the total energy it also represents the total angular momentum. Combining Eqs. (2.31), (2.33), and (2.48) one obtains

$$\frac{\mathbf{P}^2}{2I}\psi = E\psi = \frac{\hbar^2}{2I}J(J+1)\psi \tag{2.49}$$

where $\mathbf{P}$ is the angular momentum operator. Hence

$$\mathbf{P}^2\psi = J(J+1)\hbar^2\psi \tag{2.50}$$

or $h^2J(J+1)$ are the eigenvalues of the square of the angular momentum. This can be shown in a more rigorous manner by operation on the general wavefunction with the operator for $\mathbf{P}^2$. A simple demonstration of this latter approach is the application of the $\mathbf{P}_Z$ operator. Since the $\mathbf{P}_Z$ operator is independent of θ we can write

$$\mathbf{P}_Z\Theta\Phi = \Theta\mathbf{P}_Z\Phi = \frac{-i\hbar\Theta}{2\pi}\frac{de^{iM\phi}}{d\phi} = M\hbar\Theta\Phi \tag{2.51}$$

Therefore

$$\mathbf{P}_Z\psi = M\hbar\psi \tag{2.52}$$

and the eigenvalues of $\mathbf{P}_Z$ are $M\hbar$. The quantum number M gives the allowed projections of the angular momentum on a space-fixed Z axis.

2.4 PERTURBATION METHODS

Exact solutions of quantum mechanical equations are very limited. If quantum mechanics is to be a useful tool for the investigation of molecular structure and spectroscopy we must resort to the use of approximation methods to obtain solutions. Only for the very simplest of systems, such as a harmonic diatomic oscillator or a rigid diatomic rotor, can an exact solution of the Schrödinger equation be found. The general procedure followed in using approximation methods is to write the wavefunction of a system as a linear combination of the members of a known basis set of functions and vary the coefficients of the members to obtain a wavefunction, which will allow one to determine the energy of the system consistent with some prescribed criteria such as minimization. For any such method we are always seeking the best agreement with experiment when such comparisons are possible. Quite often a particular system can be considered to be a slight modification of a known one, in which case the basis set will be the eigenfunctions of the known system. Examples of this would be the anharmonic oscillator which can be considered as a harmonic oscillator to which an additional potential term has been added, or the rotational Stark effect which is the application of an external electric potential to the rigid rotor. In both cases the magnitude of the additional potential term is considerably less than the total energy of the unperturbed system.

The perturbation method for a nondegenerate system employs a Hamiltonian which is only slightly different from that of a similar but unperturbed system, and uses as its basis a set of eigenfunctions of the known unperturbed Hamiltonian, $\mathscr{H}^0$. The true wave equation for a system is written

$$\mathscr{H}\psi_k = E_k\psi_k \tag{2.53}$$

where k denotes a set of identifying quantum numbers. If we assume that the perturbation to the Hamiltonian is small compared to the Hamiltonian of the unperturbed system then the true Hamiltonian can be written

$$\mathscr{H} = \mathscr{H}^\circ + \alpha\mathscr{H}' + \alpha^2\mathscr{H}'' + \cdots \tag{2.54}$$

where α is chosen so that when $\alpha \to 0$ Eq. (2.53) reduces to

$$\mathscr{H}^\circ\psi_k^\circ = E_k^\circ\psi_k^\circ \tag{2.55}$$

where ψ_k° and E_k° are the wavefunctions and energies of the unperturbed system. The solutions of Eq. (2.55) are a set of unperturbed wavefunction, ψ_0°, ψ_1°, $\Psi_2^\circ, \ldots, \psi_n^\circ$ corresponding to energy states E_0°, E_1°, $E_2^\circ, \ldots, E_n^\circ$. The functions ψ_k° form a complete orthonormal set and can be used as the basis for expansion of an arbitrary function.

Since the perturbation is small it will cause only a small change in the total energy of the system and slight modifications of the wavefunctions. These parameters are thus written as

$$E_k = E_k^\circ + \alpha E_k' + \alpha^2 E_k'' + \cdots \tag{2.56}$$

and

$$\psi_k = \psi_k^\circ = \alpha\psi_k' + \alpha^2\psi_k'' + \cdots \tag{2.57}$$

Substituting these expansions for ψ_k, E_k and $\mathscr{H}$ into the true wave equation (2.53) and collecting coefficients of like powers of α gives

$$(\mathscr{H}^\circ\psi_k^\circ - E_k^\circ\psi_k^\circ) + (\mathscr{H}^\circ\psi_k' + \mathscr{H}'\psi_k^\circ - E_k^\circ\psi_k' - E_k'\psi_k^\circ)\alpha$$
$$+ (\mathscr{H}^\circ\psi_k'' + \mathscr{H}'\psi_k' + \mathscr{H}''\psi_k^\circ - E_k^\circ\psi_k'' - E_k'\psi_k' - E_k''\psi_k^\circ)\alpha^2 + \cdots = 0 \tag{2.58}$$

Since this series is properly convergent, that is, each successive coefficient of α^n becomes smaller, the condition for the series to be equal to zero is that each separate coefficient of α^n must vanish. Hence,

$$\mathscr{H}^\circ\psi_k^\circ = \mathscr{E}_k^\circ\psi_k^\circ \tag{2.59}$$

$$\mathscr{H}^\circ\psi_k' + \mathscr{H}'\psi_k^\circ = E_k^\circ\psi_k' + E_k'\psi_k^\circ \tag{2.60}$$

$$\mathscr{H}^\circ\psi_k'' + \mathscr{H}'\psi_k' + \mathscr{H}''\psi_k^\circ = E_k^\circ\psi_k'' + E_k'\psi_k' + E_k''\psi_k^\circ \tag{2.61}$$

Equation (2.59) is the unperturbed wave equation whose solutions were initially known. Equation (2.60) can be solved to give a first-order perturbation contribution, and Eq. (2.61) can be solved to give a second-order perturbation contribution. In principle the procedure can be continued to get higher-order contributions.

Since, to first-order, the wavefunctions $\psi_k^\circ + \alpha\psi_k'$ must be solutions to the wave equation, therefore obeying the same boundary conditions as the unperturbed functions, the unknown function ψ_k' can be written as a linear combination of the basis set, $\psi_0^\circ, \psi_1^\circ, \ldots, \psi_n^\circ$

$$\psi_k' = \sum_l a_{lk}\psi_l^\circ = \psi_k^\circ + \sum_{l \neq k} \alpha_{lk}\psi_l^\circ \tag{2.62}$$

Since

$$\mathscr{H}^\circ \psi_l^\circ = E_l^\circ \psi_l^\circ \tag{2.63}$$

it follows that

$$\mathscr{H}^\circ \psi_k' = \sum_{l=0}^{\infty} a_{lk}\mathscr{H}^\circ \psi_l^\circ = \sum_{l=0}^{\infty} a_{lk} E_l^\circ \psi_l^\circ \tag{2.64}$$

Equation (2.60) thus becomes

$$\sum_{l=0}^{\infty} a_{lk}\mathscr{H}^\circ \psi_l^\circ - E_k \sum_{l=0}^{\infty} a_{lk}\psi_l^\circ = (E_k' - \mathscr{H}')\psi_k^\circ \tag{2.65}$$

Employing Eq. (2.57) this reduces to

$$\sum_{l=0}^{\infty} a_{lk}(E_l^\circ - E_k^\circ)\psi_l^\circ = (E_k' - \mathscr{H}')\psi_k^\circ \tag{2.66}$$

Let multiplying by $\psi_k^{\circ *}$ and integrating both sides over all space gives

$$\int \sum_{l=0}^{\infty} a_{lk}(E_l^\circ - E_k^\circ)\psi_k^{\circ *}\psi_l^\circ \, dv = \int \psi_k^{\circ *}(E_k' - \mathscr{H}')\psi_k^\circ \, dv \tag{2.67}$$

However[1]

$$\int \psi_k^{\circ *}\psi_l^\circ \, dv = \delta_{kl} \tag{2.68}$$

and

$$E_l^\circ - E_k^\circ = 0 \qquad \text{for} \qquad k = l \tag{2.69}$$

so the left side vanishes. Since E_k' is a constant this reduces to

$$H_{kk}' \equiv E_k' = \int \psi_k^{\circ *}\mathscr{H}'\psi_k^\circ \, dv \tag{2.70}$$

which is the first-order perturbation correction to the energy. Note that the first-order perturbation energy terms for a nondegenerate state are obtained by averaging the first-order term of the perturbed Hamiltonian over the unperturbed states of the system.

---

[1] Kronecker delta ($\delta_{kl} = 0$ for $k \neq l$, $= 1$ for $k = l$).

It is possible to obtain the coefficients for the expansion of the wavefunction ψ'_k in terms of the basis set by multiplying Eq. (2.66) by $\psi_j^{\circ *}$ and integrating over all space. Considering the orthogonality and normalization conditions, Eq. (2.68), the coefficients in Eq. (2.63) are

$$a_{jk} = \frac{-\int \psi_j^{\circ *} \mathscr{H}' \psi_k^{\circ}\, dv}{E_j^{\circ} - E_k^{\circ}} = \frac{-H'_{jk}}{E_j^{\circ} - E_k^{\circ}} \qquad \text{for} \qquad j \neq k \tag{2.71}$$

The perturbed wavefunction is then

$$\psi_k = \psi_k^{\circ} - \alpha \sum_{j=0}^{\infty}{}' \frac{H'_{jk}}{E_j^{\circ} - E_k^{\circ}} \psi_j^{\circ} \tag{2.72}$$

where the prime on the summation indicates the omission of the term where $j = k$. It is customary to include the multiplicative constant α in the expansion for the first-order perturbation, hence the conventional expressions for perturbation to the first order are written

$$E_k = E_k^{\circ} + H'_{kk} \tag{2.73}$$

$$\psi_k = \psi_k^{\circ} - \sum_{j=0}^{\infty}{}' \frac{H'_{jk}}{E_j^{\circ} - E_k^{\circ}} \psi_j^{\circ} \tag{2.74}$$

Often first-order perturbation treatment is insufficient either due to not providing a sufficient correction to the energy or due to the absence of a first-order effect in the system. The relationships for second-order perturbation are developed from the equation obtained by equating the coefficients of α^2, Eq. (2.61). As with the case of ψ'_k, ψ''_k can be expanded in terms of the basic set of functions ψ_k°, giving

$$\psi''_k = \sum_{l=0}^{\infty} b_{lk} \psi_l^{\circ} \tag{2.75}$$

By use of this expansion along with the results of the first-order treatment one obtains E''_k. This is best done in individual steps in order to follow the logic of the process. Combining Eqs. (2.70) and (2.74) one obtains

$$E'_k \psi'_k = H'_{kk} \left[\psi_k^{\circ} - \sum_{l=0}^{\infty}{}' \left(\frac{H'_{lk}}{E_l^{\circ} - E_k^{\circ}} \right) \psi_l^{\circ} \right] \tag{2.76}$$

or

$$\mathscr{H}' \psi'_k = \mathscr{H}' \psi_k^{\circ} - \sum_{l=0}^{\infty}{}' \left(\frac{H'_{lk}}{E_l^{\circ} - E_k^{\circ}} \right) \mathscr{H}' \psi_l^{\circ} \tag{2.77}$$

Since $\mathscr{H}' \psi_l^{\circ}$ is a function which is a solution to the wave equation it can be expanded as a series of basis functions

$$\mathscr{H}' \psi_l^{\circ} = \sum_{j=0}^{\infty} H'_{jl} \psi_j^{\circ} \tag{2.78}$$

where the coefficients are the first-order terms

$$H'_{jl} = \int \psi_j^{\circ *} \mathscr{H}' \psi_l^{\circ}\, dv \tag{2.79}$$

Hence

$$\mathscr{H}'\psi_k' = \mathscr{H}'\psi_k^\circ - \sum_{i=0}^{\infty}{}' \sum_{j=0}^{\infty} \left(\frac{H_{lk}'H_{jl}'}{E_l^\circ - E_k^\circ}\right)\psi_j^\circ \tag{2.80}$$

Substituting Eqs. (2.76) and (2.80) into Eq. (2.61) gives

$$(\mathscr{H}^\circ - E_k^\circ)\left\{\sum_{l=0}^{\infty} b_{lk}\psi_l^\circ\right\} + \mathscr{H}'\psi_k^\circ - \left\{\sum_{l=0}^{\infty}{}' \sum_{j=0}^{\infty}\left(\frac{H_{lk}'H_{jl}'}{E_l^\circ - E_k^\circ}\right)\psi_j^\circ\right\} + \mathscr{H}''\psi_k^\circ$$

$$= H_{kk}'\psi_k^\circ - \sum_{l=0}^{\infty}{}'\left(\frac{H_{kk}'H_{lk}'}{E_l^\circ - E_k^\circ}\right)\psi_l^\circ + E_k''\psi_k^\circ \tag{2.81}$$

Left-multiplying by $\psi_k^{\circ *}$ and integrating over all space yields

$$\sum_{l=0}^{\infty} b_{lk}\int \psi_k^{\circ *}\mathscr{H}^\circ\psi_l^\circ\,dv - \sum_{l=0}^{\infty} b_{lk}E_k^\circ \int \psi_k^{\circ *}\psi_l^\circ\,dv + \int \psi_k^{\circ *}\mathscr{H}'\psi_k\,dv$$

$$- \sum_{l=0}^{\infty}{}' \sum_{j=0}^{\infty}\left(\frac{H_{lk}'H_{jl}'}{E_l^\circ - E_k^\circ}\right)\int \psi_k^{\circ *}\psi_j^\circ\,dv + \int \psi_k^{\circ *}\mathscr{H}''\psi_k^\circ\,dv$$

$$= H_{kk}'\int \psi_k^{\circ *}\psi_k^\circ\,dv - \sum_{l=0}^{\infty}{}'\left(\frac{H_{lk}'}{E_l^\circ - E_k^\circ}\right)\int \psi_k^{\circ *}\psi_l^\circ\,dv + E_k''\int \psi_k^{\circ *}\psi_k^\circ\,dv \tag{2.82}$$

Employing the orthogonality and normalization conditions, Eq. (2.68), considering that $\mathscr{H}^\circ\psi_l^\circ = E_l^\circ\psi_l^\circ$, and observing that the term $l = k$ is missing from the primed summations, this reduces to

$$E_k'' = \int \psi_k^\circ\mathscr{H}''\psi_k^\circ\,dv - \sum_{l=0}^{\infty}{}'\left(\frac{H_{lk}'H_{kl}'}{E_l^\circ - E_k^\circ}\right) \tag{2.83}$$

which is the second-order perturbation energy. Due to the Hermitian character of the Hamiltonian

$$\int \psi_l^*\mathscr{H}\psi_k\,dv = \int \psi_k\mathscr{H}^*\psi_l^*\,dv \tag{2.84}$$

it follows that $H_{lk}'H_{kl}'$ will always be real.

There are two contributions to the second-order energy. The integral term is the average of the second-order Hamiltonian term over the unperturbed state of the system and will vanish if the Hamiltonian contains only a single perturbation term. The sum term gives the second-order contribution of the first-order Hamiltonian term. The second-order contribution will be largest when one has near degeneracy for a pair of states ψ_l° and ψ_k°, since the denominator will become small in such a case. While the algebra is more involved, the coefficients b_{lk} can be determined in a manner analogous to that used for the evaluation of the first-order coefficients a_{lk}.

If one had a degenerate system then the $E_l^\circ - E_k^\circ$ term would go to zero and it is seen that the perturbation theory as developed for a nondegenerate system is no longer applicable. The derivation of perturbation theory for a degenerate system is beyond the intention of this treatment; however, the results will be stated and the reader referred to a standard quantum mechanics text for a detailed discussion. If one has a system containing n degenerate unperturbed states represented by $\psi_1^\circ, \psi_2^\circ, \psi_3^\circ, \ldots, \psi_n^\circ$, then the first-order perturbation energies of the perturbed states will be the roots of a secular determinant

$$\begin{vmatrix} H_{11}-E & H_{12} & H_{13} & \cdots & H_{1n} \\ H_{21} & H_{22}-E & H_{23} & \cdots & H_{2n} \\ \cdots & \cdots & \cdots & \cdots & \cdots \\ H_{n1} & H_{n2} & H_{n3} & \cdots & H_{nn}-E \end{vmatrix} = 0 \tag{2.85}$$

where

$$H_{ij} = H_{ij}^\circ + H'_{ij} \tag{2.86}$$

$$H_{ij}^\circ = E_{ij}^\circ \delta_{ij} \tag{2.87}$$

$$H'_{ij} = \int \psi_i^{\circ *} \mathscr{H}' \psi_j^\circ \, dv \tag{2.88}$$

2.5 VARIATIONAL METHOD

There are many times when the form of the Hamiltonian of a system is such that it is difficult or impossible to solve the wave equation by use of perturbation methods. In such cases the problem is often amenable to the variational method. The variational method is based on a particular property of the integral

$$\mathscr{E} = \int \phi^* \mathscr{H} \phi \, dv \tag{2.89}$$

where ϕ is an arbitrary normalized wavefunction which is finite, single-valued, and continuous throughout the space accessible to the system and $\mathscr{H}$ is the complete Hamiltonian of the system. The property of interest can be demonstrated by setting $\phi = \psi^\circ$, the true ground state wavefunction of the system, in which case

$$\int \psi^{\circ *} \mathscr{H} \psi^\circ \, dv = E^\circ \tag{2.90}$$

If ϕ is some arbitrary function it can be expanded in terms of the complete set of true wavefunctions of the system

$$\phi = \sum_{n=0}^{\infty} a_n \psi_n \tag{2.91}$$

Since the true wavefunctions are an orthonormal set we have the additional condition

$$\sum_{n=0}^{\infty} a_n^* a_n = 1 \tag{2.92}$$

Substituting this linear combination for ϕ into Eq. (2.89) gives

$$\mathscr{E} = \int \left\{ \sum_{n=0}^{\infty} a_n^* \psi_n^* \right\} \mathscr{H} \left\{ \sum_{n=0}^{\infty} a_n \psi_n \right\} dv = \sum_{m=0}^{\infty} a_n^* a_m \int \psi_n^* \mathscr{H} \psi_m \, dv \tag{2.93}$$

Since the ψ_n are the true wavefunctions of the system

$$\mathscr{H}\psi_n = E_n \psi_n \tag{2.94}$$

and

$$E = \sum_n a_n^* a_n E_n \tag{2.95}$$

Subtracting $\sum_n a_n^* a_n E_n^\circ$ from both sides, and using

$$\sum_n a_n^* a_n E^\circ = E^\circ \sum_n a_n^* a_n = E^\circ \tag{2.96}$$

one has

$$\mathscr{E} - E^\circ = \sum_{n=0}^{\infty} a_n^* a_n (E_n - E_0) \tag{2.97}$$

The product $a_n^* a_n$ is either a real positive number or zero, and since ψ° is the lowest state E_n will either equal or be greater than E°. Therefore, the right-hand side of this expression is either zero or positive and

$$\mathscr{E} - E^\circ \geqq 0 \tag{2.98}$$

or

$$\mathscr{E} \geqq E^\circ \tag{2.99}$$

We have therefore shown that the value of the variational integral is greater than or equal to the ground state energy of the system. It must be noted that Eq. (2.99) is an algebraic inequality and includes the sign of the energy as well as the magnitude. Therefore, if the energy of a system is measured downward from zero, as is the case with electrons in atoms, the value of the variational integral will have an absolute magnitude less than that of the ground state integral.

The usefulness of the variational method lies in the ability to approximate the energy on the basis of an arbitrary wavefunction. Its usefulness is amplified when one considers that if the wavefunction ϕ can be written with arbitrary coefficients, then $\mathscr{E}$ can be minimized with respect to the coefficients and the best possible value of both the energy and expansion coefficients may be obtained. This method allows the use of an arbitrary wavefunction written in terms of a known set of wavefunctions of a simpler but similar system.

A simple example of this method is illustrated by consideration of the harmonic oscillator wavefunctions in Fig. 2.2. Anharmonic motion of an oscillator destroys the symmetry of the motion, hence that of the wavefunctions, also. If a small amount of ψ_1 is added to ψ_0 for the harmonic oscillator the resulting wavefunction is no longer symmetric. The determination of the energy of the

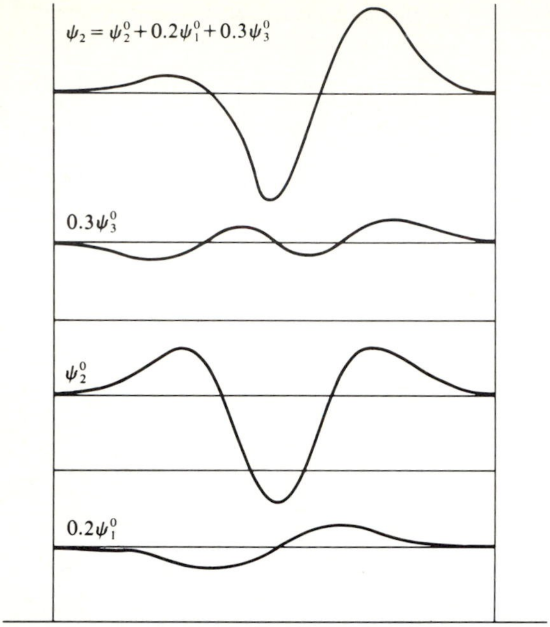

FIGURE 2.2
Addition of harmonic oscillator wavefunctions.

anharmonic oscillator could thus be approximated with the variational integral using

$$\phi = c_1\psi_0 + c_2\psi_1 \tag{2.100}$$

The problem of minimization of the variational integral will be considered using as an illustration a wavefunction with two arbitrary coefficients and then generalizing to larger systems. Since the arbitrary or trial wavefunction ϕ initially contains undetermined coefficients it is not always possible or convenient to have a normalized function. The variational energy is given in such cases by

$$\mathscr{E} = \frac{\int \phi^* \mathscr{H} \phi \, dv}{\int \phi^* \phi \, dv} \tag{2.101}$$

Using an arbitrary function of the form $\phi = c_1\psi_1 + c_2\psi_2$, where the c_n are real numbers this becomes

$$\mathscr{E} = \frac{\int (c_1\psi_1^* + c_2\psi_2^*)\mathscr{H}(c_1\psi_1 + c_2\psi_2) \, dv}{\int (c_1\psi_1^* + c_2\psi_2^*)(c_1\psi_1 + c_2\psi_2)} \tag{2.102}$$

Using the notations

$$H_{ij} = \int \psi_i^* \mathscr{H} \psi_j \, dv \qquad S_{ij} = \int \psi_i^* \psi_j \, dv \tag{2.103}$$

one has

$$\mathscr{E} = \frac{c_1^2 H_{11} + 2c_1c_2H_{12} + c_2^2 H_{22}}{c_1^2 S_{11} + 2c_1c_2S_{12} + c_2^2 S_{22}} \tag{2.104}$$

This expression has been somewhat simplified due to the Hermitian character of the Hamiltonian which gives $H_{ij} = H_{ji}^*$ and by the fact that $S_{ij} = S_{ji}^*$, since no operators are involved in the latter integrals.

To minimize the value of $\mathscr{E}$ the c_i are treated as variables and

$$\left(\frac{\partial \mathscr{E}}{\partial c_1}\right)_{c_2} = 0 \qquad \left(\frac{\partial \mathscr{E}}{\partial c_2}\right)_{c_1} = 0 \tag{2.105}$$

Taking the derivatives of Eq. (2.104) with respect to c_1 gives

$$2c_1S_{11} + c_1^2S_{11}\left(\frac{\partial \mathscr{E}}{\partial c_1}\right)_{c_2} + 2c_1c_2S_{12}\left(\frac{\partial \mathscr{E}}{\partial c_1}\right)_{c_2} + 2c_2S_{12}\mathscr{E} + c_2^2S_{22}\left(\frac{\partial \mathscr{E}}{\partial c_1}\right)_{c_2} = 2c_1H_{11} + 2c_2H_{12} \tag{2.106}$$

which rearranges to

$$\left(\frac{\partial \mathscr{E}}{\partial c_1}\right)_{c_2} = \frac{2c_1(H_{11} - S_{11}) + 2c_2(H_{12} - S_{12})}{c_1^2S_{11} + 2c_1c_2S_{12} + c_2^2S_{22}} = 0 \tag{2.107}$$

Therefore

$$c_1(H_{11} - S_{11}\mathscr{E}) + c_2(H_{12} - S_{12}\mathscr{E}) = 0 \tag{2.108}$$

Taking the derivative with respect to c_2 yields an analogous expression

$$c_1(H_{12} - S_{12}\mathscr{E}) + c_2(H_{22} - S_{22}\mathscr{E}) = 0 \tag{2.109}$$

From this pair of simultaneous equations the ratio $c_1 : c_2$ can be obtained. The independent values of c_1 and c_2 are found from the relation $c_1^2 + c_2^2 = 1$, which is an expression of the normalization condition.

A set of equations of this form is known as a set of secular equations and the necessary and sufficient condition that they have a nontrivial solution is that the determinant vanish. Therefore

$$\begin{vmatrix} H_{11} - S_{11}\mathscr{E} & H_{12} - S_{12}\mathscr{E} \\ H_{12} - S_{12}\mathscr{E} & H_{22} - S_{22}\mathscr{E} \end{vmatrix} = 0 \tag{2.110}$$

or

$$\mathscr{E}^2(S_{11}S_{22} - S_{12}^2) - \mathscr{E}(H_{11}S_{22} + H_{22}S_{11} - 2H_{12}S_{12}) + (H_{11}H_{22} - H_{12}^2) = 0 \tag{2.111}$$

This equation can be solved using the standard quadratic formula to give two roots, $\mathscr{E}_\pm$. The root corresponding to the lowest energy is the one of interest. The coefficients in the wavefunction are found from the ratio

$$\frac{c_1}{c_2} = -\frac{(H_{12} - \mathscr{E}_\pm S_{12})}{(H_{11} - \mathscr{E}_\pm S_{11})} \tag{2.112}$$

and $c_1^2 + c_2^2 = 1$.

A generalization on this treatment indicates that if the wavefunction is written as an arbitrary function of the form

$$\phi = \sum_{i=1}^{n} c_i\psi_i \tag{2.113}$$

then the variational energy would be

$$\mathscr{E} = \frac{\sum_{i=1}^{n}\sum_{j=1}^{n} c_i c_j H_{ij}}{\sum_{i=1}^{n}\sum_{j=1}^{n} c_i c_j S_{ij}} \tag{2.114}$$

and upon differentiation with respect to each c_i would give n equations of the form

$$\sum_{i=1}^{n} c_i(H_{ij} - \mathscr{E}S_{ij}) = 0 \qquad (j = 1, 2, 3, \ldots, n) \tag{2.115}$$

The solution of these equations would require that the secular determinant vanish

$$\begin{vmatrix} H_{11} - \mathscr{E}S_{11} & H_{12} - \mathscr{E}S_{12} & \cdots & H_{1n} - \mathscr{E}S_{1n} \\ H_{21} - \mathscr{E}S_{21} & H_{22} - \mathscr{E}S_{22} & \cdots & H_{2n} - \mathscr{E}S_{2n} \\ \cdots & \cdots & \cdots & \cdots \\ H_{n1} - \mathscr{E}S_{n1} & H_{n2} - \mathscr{E}S_{n2} & \cdots & H_{nn} - \mathscr{E}S_{nn} \end{vmatrix} = 0 \tag{2.116}$$

Solving for the roots of such a determinant can be a rather formidable task if n exceeds 3 and a careful choice of the ψ_n functions used to construct ϕ is generally necessary. For large values of n, the ratios of the coefficients are quite difficult to determine without high-speed computing facilities.

There are two considerations that can lead to a simplification of the secular determinant. If the set of wavefunctions ψ_n is an orthonormal set then the integrals $S_{ij} = \delta_{ij}$ where δ_{ij} is the Kronecker delta. When this is true the simple two-coefficient case given by Eq. (2.110) reduces to

$$\begin{vmatrix} H_{11} - \mathscr{E} & H_{12} \\ H_{12} & H_{22} - \mathscr{E} \end{vmatrix} = 0 \tag{2.117}$$

and

$$\mathscr{E}_{\pm} = \frac{H_{11} + H_{22}}{2} \pm \sqrt{\left(\frac{H_{11} - H_{22}}{2}\right)^2 - H_{12}^2} \tag{2.118}$$

Also

$$\frac{c_1}{c_2} = -\frac{H_{12}}{H_{11} \pm \mathscr{E}_{\pm}} \tag{2.119}$$

For the general case where the ψ_n compose an orthonormal set the secular determinant becomes

$$\begin{vmatrix} H_{11} - \mathscr{E} & H_{12} & \cdots & H_{1n} \\ H_{12} & H_{22} - \mathscr{E} & \cdots & H_{2n} \\ \cdots & \cdots & \cdots & \cdots \\ H_{1n} & H_{2n} & \cdots & H_{nn} - \mathscr{E} \end{vmatrix} = 0 \tag{2.120}$$

While this has greatly reduced the amount of mathematics necessary to set up the terms in the determinant there remains an nth order polynomial for which one must determine the roots. Once the E's and H_{ij} are known, the coefficients can be found by use of the method of determinants with Eq. (2.115).

Another simplification of the secular determinant can be obtained if the expansion set of wavefunctions ψ_n are chosen to be the eigenfunctions of some operator which commutes with the Hamiltonian. The manner in which the simplification occurs is easily demonstrated. Suppose that two of the ψ_n wavefunctions, ψ_l and ψ_m, are eigenfunctions of an operator $\boldsymbol{\alpha}$ having eigenvalues a_l and a_m. Consider the integral

$$I = \int \psi_l^* \mathscr{H} \boldsymbol{\alpha} \psi_m \, dv \tag{2.121}$$

Since

$$\boldsymbol{\alpha} \psi_m = \alpha_m \psi_m \tag{2.122}$$

the integral is equal to

$$I = a_m \int \psi_l^* \mathscr{H} \psi_m \, dv \tag{2.123}$$

However, since $\boldsymbol{\alpha}$ was chosen to commute with $\mathscr{H}$

$$\boldsymbol{\alpha} \mathscr{H} = \mathscr{H} \boldsymbol{\alpha} \tag{2.124}$$

so

$$I = \int \psi_l^* \mathscr{H} \boldsymbol{\alpha} \psi_m \, dv = \int \psi_l^* \boldsymbol{\alpha} \mathscr{H} \psi_m \, dv \tag{2.125}$$

Due to the Hermitian character of the operators

$$I = \int (\mathscr{H} \psi_m)^* \boldsymbol{\alpha}^* \psi_l \, dv = a_l \int (\mathscr{H} \psi_m)^* \psi_l \, dv = a_l \int \psi_l^* \mathscr{H} \psi_m \, dv \tag{2.126}$$

it follows that

$$a_m \int \psi_l^* \mathscr{H} \psi_m \, dv = a_l \int \psi_l^* \mathscr{H} \psi_m \, dv \tag{2.127}$$

Since $a_m \neq a_l$, they being different eigenvalues of $\boldsymbol{\alpha}$, the necessary conditions for this equality to be valid is that

$$\int \psi_l^* \mathscr{H} \psi_m \, dv = 0 \tag{2.128}$$

In this treatment we have used the subscripts on the term ψ_n to denote a set of quantum numbers. This result tells us that the H_{ij} terms in the secular equation will go to zero when any of their quantum numbers do not correspond. The net result will be a factoring of the secular determinant into blocks along the diagonal with no off-diagonal terms linking the blocks. An example of this is the consideration of the problem of the asymmetric rotor as a perturbed symmetric rotor using as the basic set for expansion the wavefunctions of the symmetric rotor. The problem resolves itself into a secular determinant which factors into blocks, each member of which has the same rotational quantum number but differing values for other quantum numbers. Each block can then be equated to zero and solved individually.

REFERENCES

A. Specific

1. Schrödinger, E., *Ann. der Physik*, **79**, 361, 489, 734 (1926); ibid., **80**, 437 (1926); ibid., **81**, 109 (1926).
2. Heisenberg, W., *Zeit. for Physik*, **33**, 879 (1925).

B. General

Anderson, J. M., *Introduction to Quantum Chemistry*, W. A. Benjamin, New York, NY, 1969.
Atkins, P. W., *Molecular Quantum Mechanics*, 2d ed., Oxford University Press, Oxford, G.B., 1983.
Davydov, A. S., *Quantum Mechanics*, 2d ed., Pergamon Press, Oxford, G.B., 1976.
Eyring, H., J. Walter, and G. E. Kimball: *Quantum Chemistry*, John Wiley and Sons, Inc., New York, NY, 1944.
Gatz, C. R., *Introduction to Quantum Chemistry*, C. E. Merrill Publishing Co., Columbus, OH, 1971.
Kauzmann, W., *Quantum Chemistry*, Academic Press Inc., New York, NY, 1957.
Levine, I. N., *Quantum Chemistry*, 2d ed., Allyn and Bacon, Boston, MA, 1974.
Pauling, L., and E. B. Wilson, Jr., *Introduction to Quantum Mechanics*, McGraw-Hill Book Co., Inc., New York, NY, 1935.
Pilar, F. L., *Elementary Quantum Chemistry*, McGraw-Hill Book Co., Inc., New York, NY, 1968.
Weissbluth, M., *Atoms and Molecules*, Academic Press Inc., New York, NY, 1978.

PROBLEMS

2.1. Show that for the harmonic oscillator $\psi_1(\eta)$ and $\psi_2(\eta)$ are orthogonal.

2.2. For the $^1H_2$ and $^{127}I_2$ molecules having force constants respectively of 570 and 17 nm^{-1} calculate both the classical energies and the quantum mechanical energies for the ground states. Assume a 1 percent displacement from equilibrium for the classical calculation.

2.3. Calculate the classical energy of rotation for the $^1H^{19}F$ and $^{127}I^{35}Cl$ molecules if their frequency of angular rotation is equivalent to the frequency corresponding to the $J = 1 \leftarrow 0$ rotational transition.

2.4. For a diatomic rotor, constrained to move in a plane, calculate the energy levels and wavefunctions when the system is perturbed by the application of a uniform electric field in one direction in the plane, i.e., the x direction. Consider the perturbing interaction between the electric dipole moment $\boldsymbol{\mu}$ and the electric field $\mathscr{E}_x$, $\boldsymbol{\mu} \cdot \boldsymbol{\mathscr{E}}_x$, to be much less than the rotational energy. Consider both first- and second-order perturbations.

2.5. Show that $P^2\psi_{31} = 12\hbar^2\psi_{31}$ where $\psi_3 = \Theta_{31}(\theta)\Phi(\phi)$ and P^2 is the quantum mechanical operator for the square of the angular momentum.

2.6. The frequencies of molecular vibrations are of the order of $5 \times 10^{13}\ \mathrm{s^{-1}}$. Assuming that three different homonuclear molecules having molecular weights of 2, 28, and 64 all had this fundamental vibrational frequency, calculate their force constants. How will the total vibrational energy of a mole of the mass 28 molecules compare to that of 0.028 kg ball oscillating on a spring at a rate of $2\ \mathrm{s^{-1}}$?

2.7. Calculate the separation between the $J = 1$ and $J = 2$ rotational levels of a planar rotor if

(*a*) $m_1 = m_2 = 10^{-26}$ kg $\qquad r = 0.2$ nm

(*b*) $m_1 = m_2 = 0.001$ kg $\qquad r = 0.02$ m

2.8. For the three-dimensional diatomic rotors $^{1}H^{19}F$ and $^{81}Br^{129}I$, calculate

(*a*) The moments of inertia in g cm^2, kg m^2 and amu A^2 for each molecule.

(*b*) The energy of the $J = 1$ rotational level for each molecule.

(*c*) The angular momentum and angular velocity in the $J = 1$ level for each molecule.

2.9. For the anharmonic diatomic oscillator in one dimension the Hamiltonian may be written

$$H = \frac{P_x^2}{2\mu} + kx^2 - k'x^3$$

Calculate the first-order perturbation energy term for the $n = 0$, 1, and 3 levels. What fraction of the total energy does this correspond to in each case? Determine the first-order perturbed wavefunction for the $n = 0$ state, ψ_0'. Obtain the necessary ground state wavefunctions by using the appropriate hermite polynomials.

2.10. Calculate the energy for the $n = 1$ state of the oscillator in Prob. 2.9 by using the variational method and representing the wavefunctions as a linear combination of the lowest three harmonic oscillator wavefunctions.

CHAPTER 3

MATRIX METHODS

3.1 INTRODUCTION

In principle, the methods of wave mechanics can be applied to the solution of spectroscopic problems. However, when dealing with systems more complicated than the harmonic diatomic oscillator and the rigid diatomic rotor their application can become mathematically very involved. In particular the wave equations for more complex systems do not have exact solutions and the introduction of approximation methods is necessary. For the spectroscopist combining elements of another approach to formulating the quantum mechanics of molecules with those of the Schrödinger method has proved to be useful. This second approach is the matrix method which was originally developed by Heisenberg [1], concurrently with the work of Schrödinger [2].

Both of these methods have as their basis the same set of three assumptions, which are justified because of their effectiveness in the formulation of theoretical results that are in agreement with experimental observations. No attempt will be made in this discussion to detail the events which lead to these assumptions, but rather the reader is asked to accept them as a starting point. The three are:

1. The properties of any dynamical variable associated with a given system can be theoretically inferred from the properties of a mathematical operator associated with the variable. More specifically, the possible numerical values obtained by an experimental measurement of a particular variable are the eigenvalues of the operator associated with that variable.

2. The quantum mechanical operators, **A** and **B**, to be associated with two variables, a and b, satisfy the quantum condition

$$\mathbf{AB} - \mathbf{BA} = i\hbar[\mathbf{A}, \mathbf{B}] \tag{3.1}$$

where $[\mathbf{A}, \mathbf{B}]$ is the operator associated with the classical Poisson bracket [3]

$$[\mathbf{A}, \mathbf{B}] = \frac{\partial \mathbf{A}}{\partial a}\frac{\partial \mathbf{B}}{\partial b} - \frac{\partial \mathbf{A}}{\partial b}\frac{\partial \mathbf{B}}{\partial a} \tag{3.2}$$

3. The states of a system are to be associated with the operands of the operators.

The wave mechanical approach is based on the mathematics of differential operators and the solutions of the resulting equations for simple systems are amenable to pictorial representations, such as in the case of atomic orbitals. The The matrix approach employs matrices as operators, is more abstract, and does not lend itself to pictorial representations. Compensating for this abstract character is the fact that the mathematics of matrices is less complex than that of differential equations. A primary asset of the matrix method lies in the ability to formulate matrix elements for angular momentum in general and then use them to find the solutions to a variety of spectroscopic problems. Another advantage of the matrix method is its easy adaptation for computer solution. In the discussion of spectroscopic problems primarily the end results of matrix mechanics will be used. However, because of its importance in so many areas of spectroscopy, the development of an appreciation of the method and the application of the results, if not a complete understanding of the origin and derivation of all quantities, will be valuable.

Soon after the publication of the original works of Schrödinger and Heisenberg it was shown [4, 5] that the two methods, although mathematically different, yielded equivalent solutions to the same problems and in fact were related in a rather elementary manner. Although it is not necessary to have a knowledge of the Schrödinger method in order to use the Heisenberg method, it is instructive for the latter to be first introduced via the relationship to wave mechanics. The matrix method will then be examined in light of its current use in solving spectroscopic problems. Following this introduction the nature of angular momentum, commutation relations, and direction cosines will be presented, and the form and application of angular momentum and direction cosine matrix elements will be discussed.

3.2 RELATIONSHIP BETWEEN THE SCHRÖDINGER AND HEISENBERG METHODS

The matrix formulation of quantum mechanics can be developed independently of the Schrödinger method. However, due to the familiarity of most students with the latter method, it is useful to begin the discussion by relating the two via a set of stated relationships. In Secs. 3.3 and 3.4 the elements of the matrix formulation will be reviewed from a more independent point of view.

The relationship will be developed by considering the one-dimensional harmonic oscillator, a case of bound one-dimensional motion in a conservative field. For this system the eigenfunctions of the time-dependent Schrödinger equation $\Psi = i\hbar(\partial\Psi/\partial t)$ are normalizable and have discrete energy levels. The development follows four steps:

1. The Heisenberg operators associated with dynamical variables are considered to be matrices which can be derived from the Schrödinger wave functions.
2. The Hamiltonian operator associated with the classical energy of the system is identified and the normalized time-dependent eigenfunctions are found. These are $\Psi_n = \psi_n e^{-iEt/\hbar}$ where the ψ_n's are solutions of the time-independent equation $\mathscr{H}\psi_n = E_n\psi_n$.
3. The Schrödinger operator associated with the dynamical variable of interest is formulated.
4. The complete set of wavefunctions, which are the solutions to the Schrödinger equation, are considered to constitute a set of basis functions in a function space. The elements for the matrix representing the operator are found from

$$\alpha_{nm} = \int \Psi_n^* \boldsymbol{\alpha} \Psi_m \, dv \tag{3.3}$$

The relationship can be formally stated as follows. The Heisenberg matrix $\underline{\boldsymbol{\alpha}}$, associated with a certain dynamical variable α, is the matrix representing the Schrödinger operator $\boldsymbol{\alpha}$ in a function space whose basis functions are the normalized time-dependent Schrödinger eigenfunctions of the Hamiltonian operator. If a set of matrices representing several Schrödinger operators or variables are all computed with respect to the same basis set of functions then they will obey the same algebraic relationships as the operators or variables. Thus, the manipulation of matrices becomes the method for the calculation of molecular properties.

To illustrate the relationship consider the one-dimensional harmonic oscillator having time-independent wavefunctions

$$\Psi_v(\eta) = N_v H_v(\eta) e^{-1/2\eta^2} \tag{3.4}$$

where $\eta = x/a$ and $a = (\hbar^2/km)^{1/4}$, and corresponding energies $E_v = (v + \frac{1}{2})h\nu_0$, where $\nu_0 = [1/(2\pi)]\sqrt{k/m}$. The time-dependent functions are

$$\Psi_v(\eta, t) = N_v H_v(\eta) e^{-(1/2)\eta^2} e^{-iE_v t/\hbar} \tag{3.5}$$

While this wavefunction is satisfactory for use in the Schrödinger formalism it must be remembered the Ψ_v is the solution of a second-order differential equation and will contain two constants of integration. One of these is the real normalization constant and the other can be expressed as a phase factor exp $(i\gamma_v)$, where γ_v is an arbitrary constant. Including this factor the wavefunction becomes

$$\Psi_v(\eta, t) = N_v H_v(\eta) e^{-(1/2)\eta^2} e^{-iE_v t/\hbar} e^{i\gamma_v} \tag{3.6}$$

These functions are normalized, well-behaved, single-valued functions. In other words, they possess all of the properties necessary to qualify them to be used as a set of basis functions for an N-dimensional function space. The matrix representing a given dynamical variable is then found using Eq. (3.3).

Taking the x coordinate as the dynamical variable of interest the elements of the x matrix are

$$x_{nm} = \int \Psi_n^* x \Psi_m \, dx \tag{3.7}$$

which, allowing for the change of variable $x = a\eta$ becomes

$$x_{nm} = a^2 \varepsilon_{nm} \sigma_{nm} N_n N_m H_n(\eta) \eta H_m(\eta) e^{-\eta^2} \, d\eta \tag{3.8}$$

where

$$\varepsilon_{nm} = e^{i(E_n - E_m)t/\hbar} \tag{3.9}$$

$$\sigma_{nm} = e^{-i(\gamma_n - \gamma_m)} \tag{3.10}$$

The recursion relations given in App. C show that

$$\int_{-\infty}^{\infty} H_n(\eta) \eta H_m e^{-\eta^2} \, d\eta = 0 \qquad m \neq n \pm 1 \tag{3.11}$$

so the only nonzero integrals are

$$\int_{-\infty}^{\infty} H_n(\eta) \eta H_{n+1}(\eta) e^{-\eta^2} \, d\eta = [(n+1) 2^{n-1} 2^{n+1} n!(n-1)! \pi]^{1/2} \tag{3.12}$$

$$\int_{-\infty}^{\infty} H_n(\eta) \eta H_{n-1}(\eta) e^{-\eta^2} \, d\eta = [n 2^{n-1} 2^{n+1} n!(n-1)! \, \pi]^{1/2} \tag{3.13}$$

Incorporating the value for N_v the only nonzero matrix elements are:

$$x_{nm} = a \left[\frac{n+1}{2}\right]^{1/2} \sigma_{nm} \varepsilon_{nm} \qquad m = n+1 \tag{3.14}$$

$$x_{nm} = a \left[\frac{n}{2}\right]^{1/2} \sigma_{nm} \varepsilon_{nm} \qquad m = n-1 \tag{3.15}$$

$$x_{nm} = 0 \qquad m \neq n \pm 1 \tag{3.16}$$

The matrix representing the x coordinate is then

$$\underline{\mathbf{x}} = \frac{a}{2} \left| \begin{matrix} 0 & 1\mathscr{L}_{01} & 0 & 0 & 0 \\ 1\mathscr{L}_{10} & 0 & 2\mathscr{L}_{12} & 0 & 0 \\ 0 & 2\mathscr{L}_{21} & 0 & 3\mathscr{L}_{23} & 0 \\ 0 & 0 & 3\mathscr{L}_{32} & 0 & 4\mathscr{L}_{34} \end{matrix} \right| \begin{matrix} \cdots \\ \cdots \\ \cdots \\ \cdots \end{matrix} \tag{3.17}$$

where $\mathscr{L}_{nm} = \sigma_{nm}\varepsilon_{nm}$. The index n denotes the row and the index m denotes the column, beginning with x_{00} in the upper left corner.

Using a similar method the matrix elements for the linear momentum matrix are found from

$$p_{nm} = -\int_{-\infty}^{\infty} \Psi_n^* i\hbar \frac{\partial}{\partial x} \Psi_m \, dx \tag{3.18}$$

which, with the variable change $x = a\eta$, gives

$$p_{nm} = -i\hbar\sigma_{nm}\varepsilon_{nm}N_nN_m \int_{-\infty}^{\infty} H_n(\eta)e^{-(1/2)\eta^2} \frac{\partial}{\partial \eta} H_m(\eta)e^{-(1/2)\eta^2} \, d\eta \tag{3.19}$$

Using the relations in Appendix C to simplify this integral leads to

$$p_{nm} = \frac{i\hbar}{a}\left[\frac{n+1}{2}\right]^{1/2} \varepsilon_{nm}\sigma_{nm} \qquad m = n+1 \tag{3.20}$$

$$p_{nm} = \frac{i\hbar}{a}\left[\frac{n}{2}\right]^{1/2} \varepsilon_{nm}\sigma_{nm} \qquad m = n-1 \tag{3.21}$$

$$p_{nm} = 0 \qquad m \neq n \pm 1 \tag{3.22}$$

showing that the p matrix is analogous to the x matrix with the multiplicative constant $a/\sqrt{2}$ being replaced with $-i\hbar/a\sqrt{2}$.

Evaluation of the matrix representing the total energy of the harmonic oscillator leads to an important point regarding the nature of its Heisenberg matrices. The elements of the energy matrix will be

$$H_{nm} = \int_{-\infty}^{\infty} \Psi_\eta^* \mathscr{H} \Psi_m \, dx \tag{3.23}$$

However, since the Ψ_n are orthogonal eigenfunctions of the Hamiltonian operator $\mathscr{H}$,

$$H_{nm} = \int_{-\infty}^{\infty} \Psi_n^* E_m \Psi_m \, dx = E_m \int_{-\infty}^{\infty} \Psi_n^* \Psi_m \, dx = E_m \delta_{nm} \tag{3.24}$$

The energy matrix is then a diagonal matrix having no time or phase factors. If the set of wavefunctions from which the matrix elements are evaluated is chosen to be the eigenfunctions of the Hamiltonian of the system then the energy matrix will be diagonal. If any other set of functions is used as a basis for evaluation of the energy matrix it will generally be found to be nondiagonal. In such cases the energy of the system may be found by diagonalization of the matrix. An example of this procedure, which will be explored in detail in Chap. 13, is the anharmonic oscillator. In this case the harmonic oscillator wavefunctions are used as a basis for evaluation of the energy matrix. The matrix obtained will be nondiagonal since these functions are not the true eigenfunctions of the anharmonic oscillator.

A valuable asset of the matrix method is the ability to generate additional matrices using matrix algebra (see App. G). For example $\underline{\mathbf{x}}^2$ is generated by matrix multiplication of $\underline{\mathbf{x}}$ with itself. The individual elements of $\underline{\mathbf{x}}^2$ are

$$(x^2)_{nm} = \sum_i x_{ni} x_{im} \tag{3.25}$$

However, $x_{ni} = 0$ for $i \neq n \pm 1$, therefore

$$(x^2)_{nm} = x_{n,n-1}x_{n-1,m} + x_{n,n+1\ n+1,m} \tag{3.26}$$

Since $x_{n-1,m} = 0$ for $m \neq n - 1 \pm 1$ and $x_{n+1,m} = 0$ for $m \neq n + 1 \pm 1$ it follows that the only nonzero elements of $\underline{\mathbf{x}}^2$ are:

$$(x^2)_{n,n-2} = x_{n,n-1}x_{n-1,n-2} \tag{3.27}$$

$$(x^2)_{n,n} = x_{n,n+1}x_{n+1,n} + x_{n,n-1}x_{n-1,n} \tag{3.28}$$

$$(x^2)_{n,n+2} = x_{n,n+1}x_{n+1,n+2} \tag{3.29}$$

Keeping in mind that the time and phase factors are zero for $n = m$, and using the $\underline{\mathbf{x}}$ elements given by Eqs. (3.14) to (3.16), the $\underline{\mathbf{x}}^2$ elements are

$$(x^2)_{nm} = \left[\frac{a^2}{2}\right][n(n-1)]^{1/2}\sigma_{nm}\varepsilon_{nm} \qquad m = n - 2 \tag{3.30}$$

$$(x^2)_{nm} = a^2(n + \tfrac{1}{2}) \qquad m = n \tag{3.31}$$

$$(x^2)_{nm} = \frac{a^2}{2}[(n+1)(n+2)]^{1/2}\sigma_{nm}\varepsilon_{nm} \qquad m = n + 2 \tag{3.32}$$

In a similar manner the elements of $\underline{\mathbf{p}}^2$ are found to be

$$(p^2)_{nm} = -\frac{\hbar^2}{a^2}\left[\frac{n(n-1)}{4}\right]^{1/2}\varepsilon_{nm}\sigma_{nm} \qquad m = n - 2 \tag{3.33}$$

$$(p^2)_{nm} = -\frac{\hbar^2}{a^2}(n + \tfrac{1}{2}) \qquad m = n \tag{3.34}$$

$$(p^2)_{nm} = -\frac{\hbar^2}{a^2}\left[\frac{(n+1)(n+2)}{4}\right]^{1/2}\varepsilon_{nm}\sigma_{nm} \qquad m = n + 2 \tag{3.35}$$

Using the classical energy equation

$$E = \frac{p^2}{2\mu} + \frac{kx^2}{2} \tag{3.36}$$

or the basis for the matrix relationship

$$\underline{\mathbf{E}} = \frac{\underline{\mathbf{p}}^2}{2\mu} + \frac{k\underline{\mathbf{x}}^2}{2} \tag{3.37}$$

and incorporating Eqs. (3.30) to (3.35) will then give us the nonzero elements of the energy matrix

$$E_{nn} = \hbar(n + \tfrac{1}{2})\left(\frac{k}{\mu}\right)^{1/2} = h\nu_0(n + \tfrac{1}{2}) \tag{3.38}$$

It is convenient to introduce a shorthand notation for discussion of matrix methods, this being the use of $\langle n|\boldsymbol{\alpha}|m\rangle$ to represent the integral or matrix element

$$\alpha_{nm} = \int \Psi_n^* \boldsymbol{\alpha} \Psi_m \, dv = \langle \psi_m | \boldsymbol{\alpha} | \psi_m \rangle = \langle n | \boldsymbol{\alpha} | m \rangle \tag{3.39}$$

This system can be expanded to accommodate multiple quantum numbers. For example, an element of $\underline{\boldsymbol{\alpha}}$ for a three-dimensional particle in a box would be $\langle n_x n_y n_z | \boldsymbol{\alpha} | n'_x n'_y n'_z \rangle$. In the same notation, wavefunctions are written as

$$\Psi_n(x) = \langle x | n \rangle \tag{3.40}$$

$$\Psi_n^*(x) = \langle n | x \rangle \tag{3.41}$$

Hence,

$$\alpha_{nm} = \int \langle n | x \rangle \boldsymbol{\alpha} \langle x | m \rangle \, dv = \langle n | \boldsymbol{\alpha} | m \rangle \tag{3.42}$$

While we have seen that the matrix representing a particular dynamical variable can be related to the Schrödinger wavefunction the carry-over of the pictorial representation of the latter is not possible. In the Heisenberg method we simply have a collection of numbers, a matrix, which represents the properties of a set of states of a system.

3.3 PRELIMINARY CONCEPTS RELATIVE TO THE USE OF THE MATRIX METHOD

In the formulation of classical mechanics the state of a system is characterized by a particular analytic function, whose arguments are the coordinates of the particles composing the system, and time. In more general terms, the state of a system is characterized by the minimal set of parameters which will provide the maximum information about the systems. For a classical system composed of N particles this set can consist of $3N$ independent Cartesian coordinates and their $3N$ time-derivatives which can be expressed either as velocities or as momenta. This set of parameters is said to represent the system and is referred to as a "good" set of parameters. The $3N$ Cartesian coordinates and momenta do not constitute the only representation of this system. For example, it could be equally well described using spherical coordinates and associated momenta. There will be the restrictions, however, that all representations of a given system possess the same number of variables.

If we attempt to apply these concepts to quantum mechanical systems one encounters a phenomenon not experienced in classical mechanics, the Heisenberg uncertainty principle. For a quantum mechanical particle, constrained to move in one dimension without bounds, it is not possible simultaneously to measure both the position and the linear momentum. Hence x and P_x do not constitute a good representation of this system. However, the energy of the particle E can be measured simultaneously with the momentum. So the pair of variables E and

P_x, constitute a good representation. Stated another way, we say that the quantum numbers denoting the values of P_x and E are good quantum numbers. Thus the good quantum numbers for a system will be the minimal set which will provide maximum characterization of information about the system. One of the objectives of the matrix formulation will be to arrive at the representation (i.e., set of quantum numbers) which provides the variables that will precisely specify the system.

An understanding of the matrix formulations of quantum mechanics will be further enhanced by introducing the concept and properties of the specific function space known as Hilbert space. The terminology employed is best established by reference to a vector in a conventional three-dimensional Cartesian space. Any arbitrary vector **G** can be expressed in terms of its Cartesian projections, G_x, G_y, G_z by

$$\mathbf{G} = G_x\hat{\mathbf{i}} + G_y\hat{\mathbf{j}} + G_z\hat{\mathbf{k}} \tag{3.43}$$

where the unit vectors, $\hat{\mathbf{i}}, \hat{\mathbf{j}}, \hat{\mathbf{k}}$, are referred to as the *basis* for **G** and are said to *span* the three-dimensional vector space. For a Hilbert space, the vector **G** is replaced with a function F, and the three unit vectors are replaced with a set of orthogonal functions, $f_1, f_2, \ldots f_N, \ldots$, which may be infinite. Such a space is quite analogous to a normal vector space and is characterized by the following properties:

1. For any two functions $f(x)$ and $g(x)$ there will exist an inner product such that

$$\langle f|g\rangle = \int_a^b f^* g\, dx \tag{3.44}$$

   if the function is defined in the interval $a \leqq x \leqq b$.
2. The square of the "length" or "norm" of any function f will be given by

$$\|f\|^2 = \langle f|f\rangle \tag{3.45}$$

3. If k is a constant and g is a function in the space then the product, kg, will also be a function.
4. For any two functions in the space, f and g, the sum, $f + g$, will also be a function.

In order to ensure that the length of any function is real, the complex conjugate of the first function in the integral of Eq. (3.44) is employed.

A set of functions which can form the basis for a Hilbert space and a set with which we already have some familiarity are the eigenfunctions of the Hamiltonian of the harmonic oscillator. In Chap. 2 it was noted that for a quantum mechanical system any arbitrary function ϕ could be expressed as a linear combination of the eigenfunctions of a system ψ_n

$$\phi = \sum_{i=1}^{\infty} c_i\psi_i \tag{3.46}$$

In this expansion the coefficient c_i exhibits a relationship to ϕ and ψ_i analogous to that among G_x, **G**, and **i** in Eq. (3.43). Analogous to the vector equation

$$G_x = \mathbf{G} \cdot \hat{\mathbf{i}} \tag{3.47}$$

we have for c_i

$$c_i = \langle \psi_i | \phi \rangle \tag{3.48}$$

Thus, the ψ_i's are the Hilbert space analogies of the three unit vectors of three-dimensional space. They can be infinite in number, and a true representation of their relationship cannot be drawn in three-dimensional space. The basis sets of functions which will be of interest in spectroscopy will generally be the eigenfunctions of the angular momentum or of the Hamiltonian, and constitute an orthonormal set such that

$$\langle \psi_i | \psi_j \rangle = \delta_{ij} \tag{3.49}$$

where δ_{ij} is the Kronecker delta. Because of the analogies between Hilbert space and three-dimensional vector space the terms eigenvector and eigenfunction are used interchangeably where a basis set is composed of the eigenfunctions of some quantum mechanical operator.

Having examined the nature of the functions which characterize Hilbert space and found that they may be a set of orthonormal eigenfunctions such as those used in Eqs. (3.4) through (3.33), we next enumerate some properties of the operators associated with quantum mechanics. Operators such as those employed in Eqs. (3.7), (3.18), and (3.23) and the operators of matrix mechanics must satisfy the mathematical property of being Hermitian, that is, for the operator **B**,

$$\int \psi_n^* \mathbf{B} \psi_m \, dV = \int \mathbf{B}^* \psi_n^* \psi_m \, dV \tag{3.50}$$

or

$$\langle \psi_n | B | \psi_m \rangle = \langle B\psi_n | \psi_m \rangle \tag{3.51}$$

There are several important properties of Hermitian operators:

1. The square of a Hermitian operator is Hermitian
2. The eigenvalues of a Hermitian operator are real
3. The eigenfunctions of a Hermitian operator, belonging to different eigenvalues, are orthogonal

3.4 ELEMENTS OF MATRIX MECHANICS

It has been noted previously that a quantum mechanical system can be characterized by a representation in which the individual states are referred to a basis set comprised of the eigenfunctions of some particular differential operator. In the matrix approach the operators and eigenfunctions are expressed as matrices and vectors. The differential equations of the Schrödinger method then became matrix equations and the mathematics of differential equations is replaced with those of linear algebra and matrices. For example, in the equation

$$\phi = \mathbf{B}\phi' \tag{3.52}$$

where ϕ, ϕ' are wavefunctions and $\mathbf{B}$ is a differential operator, the ϕ and ϕ' become column vectors and $\mathbf{B}$ is replaced with a square matrix.

The use of matrix mechanics will be dependent on the selection of the proper basis and representation for a particular system. In the study of wave mechanics it was observed that it was necessary that the wavefunctions related to a particular system satisfy the particular properties of being finite, single-valued, and continuous over the space of the system. For each quantum mechanical system one may consider there will be a related space of functions. Let us consider a set of functions, $\psi_1, \psi_2, \ldots$, which constitutes a basis for a particular Hilbert space. For example, these could be the set of eigenfunctions of the diatomic harmonic oscillator $N_v H_v(\eta)e^{-n^2/2}$, or those of the hydrogen atom $R_{nl}(r)Y_l^m(\theta, \phi)$. Since this set of functions is a basis set, any arbitrary state function f may be expanded as a series

$$f = \sum_{i=1}^{\infty} \psi_i c_i \tag{3.53}$$

or

$$|f\rangle = \sum_{i=1}^{\infty} |\psi_i\rangle\langle\psi_i|f\rangle \tag{3.54}$$

The coefficient c_i is said to represent f in the representation where the basis is $\psi_1, \psi_2, \ldots$. They are projections of f onto the basis functions $\psi_1, \psi_2, \ldots$. The coefficients, $c_1, c_2, \ldots$, have the same relationship to the state functions f as the Cartesian components G_x, G_y, G_z did to the vector $\mathbf{G}$. Just as stating the set of three G_g components defines the vector $\mathbf{G}$, stating the set of c_n coefficients defines the state function f. The matrix approach provides a method for expressing equations which involve the wavefunctions f as equations involving only the c_n coefficients.

These relationships become more evident if we consider a representative quantum mechanical equation such as

$$|\phi\rangle = \mathbf{B}|\phi'\rangle \tag{3.55}$$

where ϕ, ϕ' are state functions and $\mathbf{B}$ is some Hermitian quantum mechanical operator. Using Eq. (3.54), the state function ϕ' may be written as a linear combination of basis functions giving

$$|\phi\rangle = \sum_{i=1}^{\infty} \mathbf{B}|\psi_i\rangle\langle\psi_i|\phi'\rangle \tag{3.56}$$

Left multiplication by ψ_j, another member of the basis set, gives

$$\langle\psi_j|\phi\rangle = \sum_{i=1}^{\infty} \langle\psi_j|\mathbf{B}|\psi_i\rangle\langle\psi_i|\phi'\rangle \tag{3.57}$$

Incorporation of Eq. (3.48) and use of the shorthand notation, $B_{ji} = \langle \psi_j | \mathbf{B} | \psi_i \rangle$, gives

$$c_j = \sum_{i=1}^{\infty} B_{ji} c'_i \tag{3.58}$$

B_{ji} is the matrix representation of the operator **B** in the $\psi_1, \psi_2, \ldots$ basis set. It is alternately referred to as the matrix element which connects the two basis functions ψ_j and ψ_i. Equation (3.58), which involves only the expansion coefficients c_i and the matrix element B_{ji}, are thus equivalent to Eq. (3.56). Expanded out, Eq. (3.58) is of the form

$$\begin{pmatrix} c_1 \\ c_2 \\ c_3 \\ \vdots \end{pmatrix} = \begin{pmatrix} B_{11} & B_{12} & B_{13} \\ B_{21} & B_{22} & B_{23} \\ B_{31} & B_{32} & B_{33} \\ \vdots & \vdots & \vdots \end{pmatrix} \begin{matrix} \cdots \\ \cdots \\ \cdots \\ \end{matrix} \begin{pmatrix} c'_1 \\ c'_2 \\ c'_3 \\ \vdots \end{pmatrix} \tag{3.59}$$

Having established the fact that the quantized behavior of a system can be represented by a matrix equation, we next proceed to review some of the important properties and relations for the matrices which will be used. The fundamental mathematics of matrices is reviewed in App. G. In the remainder of this section important aspects of matrices, as they particularly relate to quantum mechanics, will be given. These relationships will be given without detailed proof since the purpose is to establish a set of working tools rather than to explore the underlying quantum mechanical theory.

When a basis set for a Hilbert space consists of the orthogonal eigenfunctions $\psi_1, \psi_2, \psi_3, \ldots$ of a Hermitian operator **B**, the matrix $\underline{\mathbf{B}}$ representing that operator is diagonal. In this situation

$$\mathbf{B}\psi_n = b_n \psi_n \tag{3.60}$$

and the matrix elements of $\underline{\mathbf{B}}$ are[1]

$$\langle n | \mathbf{B} | m \rangle = \langle n | b_m | m \rangle = b_m \langle n | m \rangle = b_m \delta_{nm} \tag{3.61}$$

with $\underline{\mathbf{B}}$ having the form

$$\underline{\mathbf{B}} = \begin{pmatrix} b_1 & 0 & 0 & \cdots \\ 0 & b_2 & 0 & \cdots \\ 0 & 0 & b_3 & \cdots \\ \vdots & \vdots & \vdots & \end{pmatrix} \tag{3.62}$$

Since the basis set consists of eigenfunctions (eigenvectors), any given eigenfunction (eigenvector) ψ_k can be expressed as a linear combination of the set,

$$| k \rangle = \sum_{i=1}^{\infty} c_i^k | i \rangle \tag{3.63}$$

---

[1] Up to this point the notation $\langle \psi_n | \mathbf{B} | \psi_m \rangle$ has primarily been used to denote the matrix elements of $\underline{\mathbf{B}}$ but in the material which follows the shorter form, $\langle n | \mathbf{B} | m \rangle$ will be used almost exclusively.

Left-multiplication by another member of the set ψ_j gives, considering that the c_i^k are constants,

$$\langle j | k \rangle = \sum_{i=1}^{\infty} c_i^k \langle j | i \rangle \tag{3.64}$$

Due to the orthogonality of the basis set this becomes

$$\delta_{jk} = \sum_{i=1}^{\infty} c_i^k \delta_{ji} = c_j^k \tag{3.65}$$

Therefore, the matrix representation of an eigenvector will be a column vector with only a single nonzero term. The basis set $\psi_1, \psi_2, \psi_3, \ldots$ will be represented by

$$\begin{pmatrix} 1 \\ 0 \\ 0 \\ 0 \\ 0 \\ \vdots \end{pmatrix} \qquad \begin{pmatrix} 0 \\ 1 \\ 0 \\ 0 \\ 0 \\ \vdots \end{pmatrix} \qquad \begin{pmatrix} 0 \\ 0 \\ 1 \\ 0 \\ 0 \\ \vdots \end{pmatrix} \qquad \cdots$$

The norms of the eigenvectors comprising the basis set will be given, from Eq. (3.45), as

$$\langle \psi_n | \psi_n \rangle = \sum_{i=1}^{\infty} |c_{in}|^2 = 1 \tag{3.66}$$

For example, in matrix notation,

$$\langle \psi_3 | \psi_3 \rangle = (00100\ldots) \begin{pmatrix} 0 \\ 0 \\ 1 \\ 0 \\ 0 \end{pmatrix} = 1 \tag{3.67}$$

We have noted that the matrix $\underline{\mathbf{B}}$, representing a particular quantum mechanical operator **B**, is diagonal when the basis set is composed of the eigenvectors of **B**. Thus, the problem of determining the eigenvalues of a particular operator reduces to that of finding a basis set which will diagonalize the operator. In general a given Hilbert space can be represented by many basis sets. The transformation from one basis set to another is analogous to the transformation of a three-dimensional vector from a representation based on one set of Cartesian unit vectors to another set which is displaced by rotation. The transformation is a unitary transformation and encompasses the following points:

1. The functions of a new basis set, $\mathscr{L}_1, \mathscr{L}_2, \mathscr{L}_3, \ldots$, are related to those of the old basis set, $\psi_1, \psi_2, \psi_3, \ldots$, by a unitary transformation $\underline{\mathbf{U}}$ such that

$$|\mathscr{L}_k\rangle = \sum_{i=1}^{\infty} |\psi_i\rangle\langle\psi_i|\mathscr{L}_k\rangle = \sum_{i=1}^{\infty} U_{ki}^* |\psi_i\rangle \tag{3.68}$$

where

$$U_{ki}^{*} = \langle \psi_i | \mathscr{L}_k \rangle \tag{3.69}$$

and

$$U_{ki} = \langle \mathscr{L}_k | \psi_i \rangle \tag{3.70}$$

2. The elements U_{ki}^{*} are the projections of the old basis set onto the new one.
3. A unitary transformation leaves the "lengths" and the "angles" between state vectors unchanged.
4. For a pair of arbitrary state vectors, ϕ and χ, which are transformed to ϕ' and χ' by a unitary transformation, the inner product is invariant, that is,

$$\langle \chi' | \phi' \rangle = \langle \chi | \phi \rangle \tag{3.71}$$

Another point to be considered in this section is the manner in which an operator will transform when the basis set is changed. The transformation equations referred to in the preceding part are

$$|\phi'\rangle = \underline{\mathbf{U}}|\phi\rangle \tag{3.72}$$

and

$$|\chi'\rangle = \underline{\mathbf{U}}|\chi\rangle \tag{3.73}$$

which, when left multiplied by $\underline{\mathbf{U}}^{-1}$, yields,

$$|\phi\rangle = \underline{\mathbf{U}}^{-1}|\phi'\rangle \tag{3.74}$$

and

$$|\chi\rangle = \underline{\mathbf{U}}^{-1}|\chi'\rangle \tag{3.75}$$

Substitution of these two equations into the general quantum mechanical equation

$$|\chi\rangle = \underline{\mathbf{B}}|\phi\rangle \tag{3.76}$$

converts it to

$$\underline{\mathbf{U}}^{-1}|\chi'\rangle = \underline{\mathbf{B}}\,\underline{\mathbf{U}}^{-1}|\phi'\rangle \tag{3.77}$$

which, upon left-multiplication with $\underline{\mathbf{U}}$, gives

$$|\chi'\rangle = \underline{\mathbf{U}}\,\underline{\mathbf{B}}\,\underline{\mathbf{U}}^{-1}|\phi'\rangle \tag{3.78}$$

Therefore, the transformed matrix $\underline{\mathbf{B}}'$, which will convert $|\phi'\rangle$ to $|\chi'\rangle$ according to

$$|\chi'\rangle = \underline{\mathbf{B}}'|\phi'\rangle \tag{3.79}$$

is

$$\underline{\mathbf{B}}' = \underline{\mathbf{U}}\,\underline{\mathbf{B}}\,\underline{\mathbf{U}}^{-1} \tag{3.80}$$

We observe that this transformation has retained the form of Eq. (3.76). For the special case where $|\phi'\rangle = |\chi'\rangle$ Eq. (3.79) is just an eigenvalue equation, and it is observed that it will be invariant to the transformation. This transformation, as expressed by Eq. (3.80), is called a *unitary-similarity transformation.*

For quantum mechanical operators, which represent observable properties, the eigenvalues are real measurable quantities and as such can not depend on the basis set chosen to represent the system. Therefore, the eigenvalues of Hermitian operators are invariant to any unitary-similarity transformation and the Hermitian character of an operator will be retained under a unitary-similarity transformation. The nature of such transformations will be reviewed more extensively in Sec. 3.5.

Before starting the discussion of the derivation of the matrix elements for the angular momentum operators a few additional comments on representations may be useful. It was noted in Sec. 3.2 that, for the harmonic oscillator, the energy matrix was diagonal. In this case the elements were determined using the harmonic oscillator wavefunctions as a basis set. This observation can be extended to any system for which the true eigenfunctions of the Hamiltonian serve as the basis set. This is referred to as the energy representation, and in this representation the Hamiltonian matrix is diagonal. For example, consider the planar rotor discussed in Chap. 2. The basis set will be the eigenfunctions $N_0, N_1 e^{-i\phi}, N_2 e^{-i2\phi}, N_3 e^{-i3\phi}, \ldots$. Using the Hamiltonian operator

$$\mathscr{H} = \frac{\hbar^2}{2I}\frac{d^2}{d\phi^2} \tag{3.81}$$

the energy matrix becomes

$$\underline{\mathscr{H}} = \frac{\hbar^2}{8\pi^2 I}\begin{pmatrix} 0 & 0 & 0 & 0 & \cdots \\ 0 & 1 & 0 & 0 & \cdots \\ 0 & 0 & 2 & 0 & \cdots \\ 0 & 0 & 0 & 3 & \cdots \\ \vdots & \vdots & \vdots & \vdots & \end{pmatrix} \tag{3.82}$$

Another example is the one-dimensional particle-in-a box for which the eigenfunctions form the basis set

$$\left(\frac{2}{a}\right)^{1/2}\sin\left[\frac{\pi x}{a}\right], \quad \left(\frac{2}{a}\right)^{1/2}\sin\left[\frac{2\pi x}{a}\right], \quad \left(\frac{2}{a}\right)^{1/2}\sin\left[\frac{3\pi x}{a}\right] \quad \cdots$$

Using the Hamiltonian operator $-(\hbar^2/2m)(\partial^2/\partial x^2)$ gives, for the energy matrix,

$$\underline{\mathscr{H}} = \frac{\hbar^2}{8ma^2}\begin{pmatrix} 1 & 0 & 0 & 0 & \cdots \\ 0 & 4 & 0 & 0 & \cdots \\ 0 & 0 & 9 & 0 & \cdots \\ 0 & 0 & 0 & 16 & \cdots \\ \vdots & \vdots & \vdots & \vdots & \end{pmatrix} \tag{3.83}$$

If one sets up matrices for other operators, using the energy representation, they will generally be nondiagonal, as was the case for the $\underline{\mathbf{x}}$ and $\underline{\mathbf{p}}$ matrices for the harmonic oscillator.

3.5 DIAGONALIZATION OF MATRICES

In the last section it was found, by choosing the basis set for the representation of a system to be composed of the eigenvectors of the Hamiltonian, that the resulting energy matrix was diagonal. This will be true for any system where the true eigenvectors of the Hamiltonian are used. However, only a relatively small number of simple systems are exactly solvable to give a complete set of energy eigenvectors. For example, although the eigenvectors of the harmonic oscillator may be readily found and serve as a basis set for a diagonal energy matrix, they are not the true eigenvectors of an anharmonic oscillator and, if used as a basis set for the determination of the energy matrix for the latter, will give a nondiagonal matrix. The approach, which is followed for more complex systems, will be to use as a basis set the true eigenvectors of a simple but similar system, establish the energy matrix using this basis set, and diagonalize the matrix to obtain the energies. The choice of a basis set to represent a particular system is not an arbitrary one. It is frequently found that the Hamiltonian of a system will consist of a sum of terms, the predominant one being the same as that for an exactly solvable system, and the remaining terms being of a magnitude so that they may be considered as perturbations. In such cases the basis set will be chosen to be the eigenvectors of the unperturbed systems. It was pointed out in Sec. 3.4 that a unitary-similarity transformation of a matrix corresponded to a change of basis sets. Hence, the diagonalization of the energy matrix for a perturbed system will correspond to a transformation from the basis set composed of the eigenvectors of a known unperturbed system to a basis set composed of the eigenvectors of the perturbed system.

Methods for the exact diagonalization of matrices are discussed in a variety of texts on applied mathematics and are readily adaptable for use with modern high-speed computers. In view of the availability of computers and software for mathematical subroutines it would appear that there is no need to pursue the subject of diagonalization further. However, there are several reasons for exploring the subject further and not just arbitrarily going to a computer:

1. Very frequently the perturbation term which distinguishes a real system from an unperturbed or ideal system contributes only a small amount to the total energy. In such cases approximation methods will often give sufficiently accurate results much quicker.
2. When there is more than one contribution to the perturbation of the energy the use of approximation methods is useful to allow one to examine these contributions separately.
3. Perturbation methods will often permit one to make quick order-of-magnitude calculations of specific energy contributions.

In the discussion which immediately follows, the development will be in terms of general matrices, although our interest in future chapters will be concerned primarily with energy and angular momentum matrices. We will review the relevant features for the approximate diagonalization of matrices, which is the matrix counterpart of the wave-mechanical perturbation method discussed in Chap. 2. Initially the method used for diagonalization will be introduced by way of a 2×2 matrix, as an example, and then the nature of first- and second-order approximations will be developed.

Let us consider a general matrix $\underline{\mathbf{A}}$ having the form

$$\underline{\mathbf{A}} \equiv \underline{\mathbf{A}}^0 + \varepsilon \underline{\mathbf{A}}^{\mathrm{I}} + \varepsilon^2 \underline{\mathbf{A}}^{\mathrm{II}} \tag{3.84}$$

where $\underline{\mathbf{A}}^0$ is a diagonal matrix, $\underline{\mathbf{A}}^{\mathrm{I}}$ and $\underline{\mathbf{A}}^{\mathrm{II}}$ are nondiagonal matrices, which are independent of ε, and ε is a small number. Such a matrix can represent a perturbed system where $\underline{\mathbf{A}}^{\mathrm{I}}$ and $\underline{\mathbf{A}}^{\mathrm{II}}$ represent the perturbations and $\underline{\mathbf{A}}^0$ represents an unperturbed system. For example

$$\underline{\mathbf{A}} = \begin{pmatrix} 2+5\varepsilon & 3\varepsilon \\ 3\varepsilon & 6+\varepsilon^2 \end{pmatrix} = \begin{pmatrix} 2 & 0 \\ 0 & 6 \end{pmatrix} + \varepsilon \begin{pmatrix} 5 & 3 \\ 3 & 0 \end{pmatrix} + \varepsilon^2 \begin{pmatrix} 0 & 0 \\ 0 & 1 \end{pmatrix} \tag{3.85}$$

Examination of this matrix shows that the eigenvalues will be 2 and 6 if $\varepsilon = 0$. For the case of ε being very small, so that the ε^2 term can be ignored, the matrix reduces to

$$\begin{pmatrix} 2+5\varepsilon & 3\varepsilon \\ 3\varepsilon & 6 \end{pmatrix} \tag{3.86}$$

The diagonalization of a matrix involves the use of a transformation matrix $\underline{\mathbf{T}}$ such that

$$\underline{\mathbf{B}} = \underline{\mathbf{T}}\underline{\mathbf{A}}\underline{\mathbf{T}}^{-1} \tag{3.87}$$

where $\underline{\mathbf{B}}$ is diagonal. A transformation matrix of the form $\underline{\mathbf{T}} = \underline{\mathbf{I}} + \varepsilon \underline{\mathbf{T}}^{\mathrm{I}}$, where

$$\underline{\mathbf{T}}^{\mathrm{I}} = \begin{pmatrix} 0 & -\frac{3}{4} \\ \frac{3}{4} & 0 \end{pmatrix} \tag{3.88}$$

and $\underline{\mathbf{I}}$ is the unit matrix is used. The reason for this particular choice will be evident following the presentation of the general case later. The reciprocal matrix $\underline{\mathbf{T}}^{-1}$ is then $(\underline{\mathbf{I}} + \varepsilon \underline{\mathbf{T}}^{\mathrm{I}})^{-1}$, which, since ε is small, can be expanded as a series to give

$$(\underline{\mathbf{I}} + \varepsilon \underline{\mathbf{T}}^{\mathrm{I}})^{-1} = \underline{\mathbf{I}} - \varepsilon \underline{\mathbf{T}}^{\mathrm{I}} + \varepsilon^2 (\underline{\mathbf{T}}^{\mathrm{I}})^2 - \cdots \tag{3.89}$$

Therefore

$$\underline{\mathbf{T}}^{-1} = \begin{pmatrix} 1 & 0 \\ 0 & 1 \end{pmatrix} - \varepsilon \begin{pmatrix} 0 & -\frac{3}{4} \\ \frac{3}{4} & 0 \end{pmatrix} + \varepsilon^2 \begin{pmatrix} -\frac{9}{16} & 0 \\ 0 & -\frac{9}{16} \end{pmatrix} \tag{3.90}$$

The transformation is then

$$\underline{\mathbf{B}} = \underline{\mathbf{T}}\underline{\mathbf{A}}\underline{\mathbf{T}}^{-1} = \begin{pmatrix} 1 & -\dfrac{3\varepsilon}{4} \\ \dfrac{3\varepsilon}{4} & 1 \end{pmatrix} \begin{pmatrix} 2+5\varepsilon & 3\varepsilon \\ 3\varepsilon & 6 \end{pmatrix} \begin{pmatrix} 1 - \dfrac{9\varepsilon^2}{16} & \dfrac{3\varepsilon}{4} \\ -\dfrac{3\varepsilon}{4} & 1 - \dfrac{9\varepsilon^2}{16} \end{pmatrix} \tag{3.91}$$

which, when multiplied out and the terms in ε^3 or higher are neglected, gives

$$\underline{\mathbf{B}} = \begin{pmatrix} 2 + 5\varepsilon - \dfrac{9\varepsilon^2}{4} & \dfrac{15\varepsilon^2}{4} \\ \dfrac{15\varepsilon^2}{4} & 6 + \dfrac{9\varepsilon^2}{4} \end{pmatrix} \tag{3.92}$$

If the terms in ε^2 are neglected

$$\underline{\mathbf{B}} = \begin{pmatrix} 2 + 5\varepsilon & 0 \\ 0 & 6 \end{pmatrix} \tag{3.93}$$

and to first-order in ε the eigenvalues are $2 + 5\varepsilon$ and 6. Since ε^2 appear in the off-diagonal terms it is not possible to use this transformation to get more exact eigenvalues.

It is possible to find a higher-order transformation matrix just as it is possible to go to higher-order wave-mechanical perturbation theory. To do this let $\underline{\mathbf{T}}$ be an undetermined unitary matrix (see App. G) of the form

$$\underline{\mathbf{T}} = \underline{\mathbf{T}}^0 + \varepsilon\underline{\mathbf{T}}^{\mathrm{I}} + \varepsilon^2\underline{\mathbf{T}}^{\mathrm{II}} + \cdots + \varepsilon^n\underline{\mathbf{T}}^N \tag{3.94}$$

and $\underline{\mathbf{A}}$ be a Hermitian matrix representing a dynamical variable of a nondegenerated perturbed system and having the form

$$\underline{\mathbf{A}} = \underline{\mathbf{A}}^0 + \varepsilon\underline{\mathbf{A}}^{\mathrm{I}} + \varepsilon^2\underline{\mathbf{A}}^{\mathrm{II}} + \cdots + \varepsilon^n\underline{\mathbf{A}}^N \tag{3.95}$$

We now wish to find the form of $\underline{\mathbf{T}}$ such that

$$\underline{\mathbf{B}} = \underline{\mathbf{T}}\underline{\mathbf{A}}\underline{\mathbf{T}}^{-1} \tag{3.96}$$

where $\underline{\mathbf{B}}$ is diagonal to ε^n and n is the desired order. This transformation represents a change in the basis set so the eigenvalues of $\underline{\mathbf{A}}$ and $\underline{\mathbf{B}}$ are identical and, since $\underline{\mathbf{B}}$ is diagonal, the terms b_{nn} will be the eigenvalues of $\underline{\mathbf{A}}$. Since we are diagonalizing a matrix containing terms in ε up to ε^n the diagonal terms of $\underline{\mathbf{B}}$ must contain terms involving ε also. Hence

$$\underline{\mathbf{B}} = \underline{\mathbf{B}}^0 + \varepsilon\underline{\mathbf{B}}^{\mathrm{I}} + \varepsilon^2\underline{\mathbf{B}}^{\mathrm{II}} + \cdots + \varepsilon^n\underline{\mathbf{B}}^N \tag{3.97}$$

where all of the $\underline{\mathbf{B}}^N$ terms are diagonal matrices.

The form of $\underline{\mathbf{T}}$ is determined by starting with the basic transformation

$$\underline{\mathbf{B}} = \underline{\mathbf{T}}\underline{\mathbf{A}}\underline{\mathbf{T}}^{-1} \tag{3.98}$$

Right-multiplication with $\underline{\mathbf{T}}$ gives

$$\underline{\mathbf{B}}\underline{\mathbf{T}} = \underline{\mathbf{T}}\underline{\mathbf{A}} \tag{3.99}$$

since $\underline{\mathbf{T}}^{-1}\underline{\mathbf{T}} = \underline{\mathbf{I}}$. Substitution of the expressions for $\underline{\mathbf{B}}$, $\underline{\mathbf{T}}$, and $\underline{\mathbf{A}}$ to terms in ε^2 and expanding gives

$$\begin{aligned} &\underline{\mathbf{B}}^0\underline{\mathbf{T}}^0 + \varepsilon(\underline{\mathbf{B}}^0\underline{\mathbf{T}}^{\mathrm{I}} + \underline{\mathbf{B}}^{\mathrm{I}}\underline{\mathbf{T}}^0) + \varepsilon^2(\underline{\mathbf{B}}^0\underline{\mathbf{T}}^{\mathrm{II}} + \underline{\mathbf{B}}^{\mathrm{I}}\underline{\mathbf{T}}^{\mathrm{I}} + \underline{\mathbf{B}}^{\mathrm{II}}\underline{\mathbf{T}}^0) \\ &\qquad = \underline{\mathbf{T}}^0\underline{\mathbf{A}}^0 + \varepsilon(\underline{\mathbf{T}}^0\underline{\mathbf{A}}^{\mathrm{I}} + \underline{\mathbf{T}}^{\mathrm{I}}\underline{\mathbf{A}}^0) + \varepsilon^2(\underline{\mathbf{T}}^0\underline{\mathbf{A}}^{\mathrm{II}} + \underline{\mathbf{T}}^{\mathrm{I}}\underline{\mathbf{A}}^{\mathrm{I}} + \underline{\mathbf{T}}^{\mathrm{II}}\underline{\mathbf{A}}^0) \end{aligned} \tag{3.100}$$

Equating like powers of ε gives three relationships

$$\underline{\mathbf{B}}^0\underline{\mathbf{T}}^0 = \underline{\mathbf{T}}^0\underline{\mathbf{A}}^0 \tag{3.101}$$

$$\underline{\mathbf{B}}^0\underline{\mathbf{T}}^{\mathrm{I}} + \underline{\mathbf{B}}^{\mathrm{I}}\underline{\mathbf{T}}^0 = \underline{\mathbf{T}}^0\underline{\mathbf{A}}^{\mathrm{I}} + \underline{\mathbf{T}}^{\mathrm{I}}\underline{\mathbf{A}}^0 \tag{3.102}$$

$$\underline{\mathbf{B}}^0\underline{\mathbf{T}}^{\mathrm{II}} + \underline{\mathbf{B}}^{\mathrm{I}}\underline{\mathbf{T}}^{\mathrm{I}} + \underline{\mathbf{B}}^{\mathrm{II}}\underline{\mathbf{T}}^0 = \underline{\mathbf{T}}^0\underline{\mathbf{A}}^{\mathrm{II}} + \underline{\mathbf{T}}^{\mathrm{I}}\underline{\mathbf{A}}^{\mathrm{I}} + \underline{\mathbf{T}}^{\mathrm{II}}\underline{\mathbf{A}}^0 \tag{3.103}$$

By inclusion of higher-order terms still further relationships could be developed.

The only known terms at this point are the elements of $\underline{\mathbf{A}}$, but these will suffice to determine the form of $\underline{\mathbf{T}}$. Examining Eqs. (3.94) to (3.97) shows that as $\varepsilon \to 0$

$$\underline{\mathbf{T}} \to \underline{\mathbf{T}}^0$$

$$\underline{\mathbf{A}} \to \underline{\mathbf{A}}^0$$

$$\underline{\mathbf{B}} \to \underline{\mathbf{B}}^0$$

and the eigenvalues of $\underline{\mathbf{B}}^0$ are the same as those of $\underline{\mathbf{A}}^0$ since the eigenvalues of $\underline{\mathbf{B}}$ and $\underline{\mathbf{A}}$ are the same. Since both $\underline{\mathbf{B}}^0$ and $\underline{\mathbf{A}}^0$ are diagonal and have the same eigenvalues they can differ at most by the ordering of the diagonal elements. We can require the ordering of $\underline{\mathbf{B}}^0$ and $\underline{\mathbf{A}}^0$ to be the same without any loss of generality therefore

$$\underline{\mathbf{B}}^0 = \underline{\mathbf{A}}^0 \tag{3.104}$$

or stated for the individual elements

$$b^0_{nn} = a^0_{nn} \tag{3.105}$$

and Eq. 3.101 becomes

$$\underline{\mathbf{A}}^0\underline{\mathbf{T}}^0 = \underline{\mathbf{T}}^0\underline{\mathbf{A}}^0 \tag{3.106}$$

However, if $\underline{\mathbf{C}}\underline{\mathbf{D}} - \underline{\mathbf{D}}\underline{\mathbf{C}} = \underline{\mathbf{0}}$ and $\underline{\mathbf{C}}$ is a nondegenerate diagonal matrix, such as $\underline{\mathbf{A}}^0$ has been established to be, then $\underline{\mathbf{D}}$ must be a unit matrix. Therefore

$$\underline{\mathbf{T}}^0 = \underline{\mathbf{I}} \tag{3.107}$$

and we have now found $\underline{\mathbf{B}}^0$ and $\underline{\mathbf{T}}^0$ in terms of $\underline{\mathbf{A}}$.

Equation (3.102) can be used to develop the matrix formulation of first-order perturbation theory. Substituting Eqs. (3.104) and (3.106) into Eq. (3.102) one obtains

$$\underline{\mathbf{A}}^0\underline{\mathbf{T}}^{\mathrm{I}} + \underline{\mathbf{B}}^{\mathrm{I}} = \underline{\mathbf{A}}^{\mathrm{I}} + \underline{\mathbf{T}}^{\mathrm{I}}\underline{\mathbf{A}}^0 \tag{3.108}$$

which rearranges to

$$\underline{\mathbf{A}}^{\mathrm{I}} - \underline{\mathbf{B}}^{\mathrm{I}} = \underline{\mathbf{A}}^0\underline{\mathbf{T}}^{\mathrm{I}} - \underline{\mathbf{T}}^{\mathrm{I}}\underline{\mathbf{A}}^0 \tag{3.109}$$

Equating the *nn*th elements of each side yields

$$a^{\mathrm{I}}_{nn} - b^{\mathrm{I}}_{nn} = \sum_i a^0_{ni}t^{\mathrm{I}}_{in} - \sum_i t^{\mathrm{I}}_{ni}a^0_{in} \tag{3.110}$$

Since $\underline{\mathbf{A}}^0$ is diagonal the only nonzero terms in the summation are those with

$i = n$ leaving

$$a^{\mathrm{I}}_{nn} - b^{\mathrm{I}}_{nn} = a^{0}_{nn}t^{\mathrm{I}}_{nn} - t^{\mathrm{I}}_{nn}a^{0}_{nn} = 0 \tag{3.111}$$

Therefore

$$a^{\mathrm{I}}_{nn} = b^{\mathrm{I}}_{nn} \tag{3.112}$$

Neglecting all terms higher than ε the eigenvalues λ_n of $\underline{\mathbf{A}}$ are

$$\lambda_n = a^{0}_{nn} + a^{\mathrm{I}}_{nn} \tag{3.113}$$

The elements of $\underline{T}^{\mathrm{I}}$ may be found by equating the nmth elements of Eq. (3.102)

$$a^{\mathrm{I}}_{nm} - b^{\mathrm{I}}_{nm} = \sum_i a^{0}_{ni}t^{\mathrm{I}}_{im} - \sum_i t^{\mathrm{I}}_{ni}a^{0}_{im} \tag{3.114}$$

Since $a^{0}_{nm} = 0$ for $n \neq m$ this becomes

$$a^{\mathrm{I}}_{nm} - b^{\mathrm{I}}_{nm} = a^{0}_{nn}t^{\mathrm{I}}_{nm} - t^{\mathrm{I}}_{nm}a^{0}_{mm} = t^{\mathrm{I}}_{nm}(a^{0}_{nn} - a^{0}_{mm}) \tag{3.115}$$

and

$$t^{\mathrm{I}}_{nm} = \frac{a^{\mathrm{I}}_{nm} - b^{\mathrm{I}}_{nm}}{a^{0}_{nn} - a^{0}_{mm}} \tag{3.116}$$

Since $a^{\mathrm{I}}_{nn} = b^{\mathrm{I}}_{nn}$ it follows that $t^{\mathrm{I}}_{nn} = 0$. Since $\underline{\mathbf{B}}$ is diagonal $b^{\mathrm{I}}_{nm} = 0$ for $n \neq m$ and

$$t^{I}_{nm} = \frac{a^{I}_{nm}}{a^{0}_{nn} - a^{0}_{mm}} \qquad (n \neq m) \tag{3.117}$$

It now becomes obvious how the transformation $\underline{\mathbf{T}}$ used in the earlier example was obtained.

Second-order perturbation expressions are found by equating the nnth elements of Eq. (3.103)

$$\sum_i b^{0}_{ni}t^{\mathrm{II}}_{in} + \sum_i b^{\mathrm{I}}_{ni}t^{\mathrm{I}}_{in} + \sum_i b^{\mathrm{II}}_{ni}t^{0}_{in} = \sum_i t^{0}_{ni}a^{\mathrm{II}}_{in} + \sum_i t^{\mathrm{I}}_{ni}a^{\mathrm{I}}_{in} + \sum_i t^{\mathrm{II}}_{ni}a^{0}_{in} \tag{3.118}$$

Using the previously evaluated or known terms

$$\begin{aligned}
t^{0}_{nm} &= 0 \qquad && n \neq m \\
b^{0}_{nm} &= 0 && n \neq m \\
b^{\mathrm{I}}_{nm} &= 0 && n \neq m \\
t^{\mathrm{I}}_{nm} &= 0 && n \neq m \\
a^{0}_{nm} &= 0 && n \neq m \\
b^{0}_{nn} &= a^{0}_{nn} \\
t^{0}_{nn} &= 1
\end{aligned}$$

this reduces to

$$b^0_{nn}t^{\mathrm{II}}_{nn} + b^{\mathrm{II}}_{nn} = a^{\mathrm{II}}_{nn} + \sum_i{}' t^{\mathrm{I}}_{ni}a^{\mathrm{I}}_{in} + t^{\mathrm{II}}_{nn}a^0_{nn} \tag{3.119}$$

The $'$ on the summation denotes omission of the term with $i = n$.

Since $b^0_{nn} = a^0_{nn}$, the first left-hand term and last right-hand term cancel, leaving

$$b^{\mathrm{II}}_{nn} = a^{\mathrm{II}}_{nn} + \sum_i{}' t^{\mathrm{I}}_{ni}a^{\mathrm{I}}_{in} \tag{3.120}$$

which, upon introduction of Eq. (3.117), gives the second-order perturbation term

$$b^{\mathrm{II}}_{nn} = a^{\mathrm{II}}_{nn} + \sum_i{}' \frac{a^{\mathrm{I}}_{ni}a^{\mathrm{I}}_{in}}{a^0_{nn} - a^0_{ii}} \tag{3.121}$$

Combination of this expression with Eqs. (3.112) and (3.113) gives the eigenvalue expression

$$\lambda_n = b^0_{nn} + \varepsilon b^{\mathrm{I}}_{nn} + \varepsilon^2 b^{\mathrm{II}}_{nn} \tag{3.122}$$

$$\lambda_n = a^0_{nn} + \varepsilon a^{\mathrm{I}}_{nn} + \varepsilon^2\left[a^{\mathrm{II}}_{nn} + \sum_i{}' \frac{a^{\mathrm{I}}_{ni}a^{\mathrm{I}}_{in}}{a^0_{nn} - a^0_{ii}}\right] \tag{3.123}$$

This is the second-order approximation of the energy. It is to be noted that the mathematics involved is simpler than that of the analogous treatment for wave-mechanical perturbation theory. Having arrived at a way of finding the energy without making the actual transformation one observes that there is little to be gained by evaluating the elements of $\underline{\mathbf{T}}^{\mathrm{II}}$. This can be done, however, by equating the *nm*th elements of Eq. (3.103) and simplifying the resulting expression. When applying this technique to spectroscopic problems the elements of $\underline{\mathbf{A}}^{\mathrm{I}}$ and $\underline{\mathbf{A}}^{\mathrm{II}}$ are generally small and the ε terms are omitted. An important point to note is that for any level of perturbation theory the equivalent eigenvalues of the $\underline{\mathbf{A}}$ and $\underline{\mathbf{B}}$ matrices depend only on the elements of the $\underline{\mathbf{A}}$ matrix, A^0_{ij}, A^{I}_{ij}, $A^{\mathrm{II}}_{ij}, \ldots$. It is unnecessary to actually evaluate the elements of the transformation matrix $\underline{\mathbf{T}}$. If there is a need to have the $\underline{\mathbf{T}}$ matrix then its elements are also obtainable by using only the known elements of the $\underline{\mathbf{A}}$ matrix. Thus the matrix method provides a method to determine the energy eigenvalues of a system by using only the classical Hamiltonian and a representation consisting of a known basis set.

3.6 ANGULAR MOMENTUM

The energy of a rotating body, be it a molecule tumbling in space or an intrinsic spin of a particle interacting with a magnetic field, can be expressed in terms of three orthogonal components of angular momentum. The understanding of the quantum mechanics of angular momentum forms a foundation for the discussion of a large segment of spectroscopy. The relationships presented in this section will be applicable to a number of later discussions with only minor changes in nomenclature. They are between components of angular momentum and will serve as a basis for the formulation of the angular momentum matrix elements.

Classical angular momentum (see App. D) is defined, in vector terminology, as

$$\mathbf{P} = \mathbf{r} \times \mathbf{p} \tag{3.124}$$

where **r** is the radius vector and **p** is the linear momentum, $m\mathbf{v}$. Throughout the literature various discussions employ different symbols to denote angular momentum, **L**, **M**, **P**, **I**, and **S**, being commonly used. The latter three have in general been used for molecular rotation, nuclear spin, and electron spin respectively. The three quantities in Eq. (3.124) are vectors and in a space-fixed axis system, X, Y, Z, with unit vectors, $\hat{\mathbf{I}}$, $\hat{\mathbf{J}}$, $\hat{\mathbf{K}}$, can be written as

$$\mathbf{P} = P_X\hat{\mathbf{I}} + P_Y\hat{\mathbf{J}} + P_Z\hat{\mathbf{K}} \tag{3.125}$$

$$\mathbf{r} = X\hat{\mathbf{I}} + Y\hat{\mathbf{J}} + Z\hat{\mathbf{K}} \tag{3.126}$$

$$\mathbf{p} = p_X\hat{\mathbf{I}} + p_Y\hat{\mathbf{J}} + p_Z\hat{\mathbf{K}} \tag{3.127}$$

The symbols $\mathbf{P}$, $\mathbf{P}_X$, $\mathbf{P}_Y$, $\mathbf{P}_Z$ denote operators while P, P_X, P_Y, and P_Z denote the magnitudes of the angular momentum and its components. These vector relationships can be used to establish the form of the angular momentum operators and to set up commutation relationships between the angular momentum components.

Expanding the vector product in Eq. (3.124) one obtains

$$\mathbf{P} = P_X\hat{\mathbf{I}} + P_Y\hat{\mathbf{J}} + P_Z\hat{\mathbf{K}} = \mathbf{r} \times \mathbf{p} = \begin{vmatrix} \hat{\mathbf{I}} & \hat{\mathbf{J}} & \hat{\mathbf{K}} \\ X & Y & Z \\ p_X & p_Y & p_Z \end{vmatrix} \tag{3.128}$$

or

$$\mathbf{P} = (Yp_Z - Zp_Y)\hat{\mathbf{I}} + (Zp_X - Xp_Z)\hat{\mathbf{J}} + (Xp_Y - Yp_X)\hat{\mathbf{K}} \tag{3.129}$$

Therefore

$$P_X = Yp_Z - Zp_Y \tag{3.130}$$

$$P_Y = Zp_X - Xp_Z \tag{3.131}$$

$$P_Z = Xp_Y - Yp_X \tag{3.132}$$

and

$$P^2 = P_X^2 + P_Y^2 + P_Z^2 \tag{3.133}$$

One should note the cyclic order of the first occurrences of each coordinate in Eqs. (3.130) to (3.132) as a nemonic device. The Heisenberg matrices representing angular momentum obey these classical relationships. The Schrödinger operators for the angular momentum components are found by substitution of $-i\hbar(\partial/\partial F)$ for p_F, where F denotes a space-fixed axis

$$\mathbf{P}_X = -i\hbar\left(Y\frac{\partial}{\partial Z} - Z\frac{\partial}{\partial Y}\right) \tag{3.134}$$

$$\mathbf{P}_Y = -i\hbar\left(Z\frac{\partial}{\partial X} - X\frac{\partial}{\partial Z}\right) \tag{3.135}$$

$$\mathbf{P}_Z = -i\hbar\left(X\frac{\partial}{\partial Y} - Y\frac{\partial}{\partial Z}\right) \tag{3.136}$$

or transformed to spherical polar coordinates (see App. D)

$$\mathbf{P}_X = -i\hbar\left(-\sin\phi\frac{\partial}{\partial\theta} - \cot\theta\cos\phi\frac{\partial}{\partial\phi}\right) \tag{3.137}$$

$$\mathbf{P}_Y = -i\hbar\left(\cos\phi\frac{\partial}{\partial\theta} - \cot\theta\sin\phi\frac{\partial}{\partial\phi}\right) \tag{3.138}$$

$$\mathbf{P}_Z = -i\hbar\frac{\partial}{\partial\phi} \tag{3.139}$$

$$\mathbf{P}^2 = -\hbar^2\left[\frac{1}{\sin\theta}\frac{\partial}{\partial\theta}\left(\sin\theta\frac{\partial}{\partial\theta}\right) + \frac{1}{\sin^2\theta}\frac{\partial^2}{\partial\phi^2}\right] \tag{3.140}$$

Referring to the solution of the rigid rotor problem in Chap. 2 it is observed that the operator for P^2 is identical to the Hamiltonian operator except for the factor of $1/2I$. Hence, the eigenvalues of $\mathbf{P}_Z$ and $\mathbf{P}^2$ are $M\hbar$ and $J(J+1)\hbar$ where m and J are restricted to integer values. Furthermore there exists a state ψ_{JM} which is a simultaneous eigenstate of both $\mathbf{P}_Z$ and $\mathbf{P}^2$. Introduction of Eqs. (3.139) and (3.140) into $\mathbf{P}_Z\mathbf{P}^2 - \mathbf{P}^2\mathbf{P}_Z$ shows that $\mathbf{P}_Z$ and $\mathbf{P}^2$ commute, that is

$$\mathbf{P}_Z\mathbf{P}^2 - \mathbf{P}^2\mathbf{P}_Z = 0 \tag{3.141}$$

Using Eqs. (3.134) and (3.135) one can develop commutation relationships for the components of angular momentum. For example,

$$\mathbf{P}_X\mathbf{P}_Y = -\hbar^2\left[Y\frac{\partial}{\partial X} + YZ\frac{\partial^2}{\partial X\partial Z} - XY\frac{\partial^2}{\partial Z^2} - Z^2\frac{\partial^2}{\partial X\partial Y} + XY\frac{\partial^2}{\partial Y\partial Z}\right] \tag{3.142}$$

$$\mathbf{P}_Y\mathbf{P}_X = -\hbar^2\left[YZ\frac{\partial^2}{\partial X\partial Z} - Z^2\frac{\partial^2}{\partial X\partial Y} - XY\frac{\partial^2}{\partial Z^2} + X\frac{\partial}{\partial Y} + XY\frac{\partial^2}{\partial Y\partial Z}\right] \tag{3.143}$$

Therefore

$$[\mathbf{P}_X, \mathbf{P}_Y] = \mathbf{P}_X\mathbf{P}_Y - \mathbf{P}_Y\mathbf{P}_X = -\hbar^2\left[Y\frac{\partial}{\partial X} - X\frac{\partial}{\partial Y}\right] = i\hbar\mathbf{P}_Z \tag{3.144}$$

Repetition of this procedure leads to three cyclic commutation relationships

$$\mathbf{P}_X\mathbf{P}_Y - \mathbf{P}_Y\mathbf{P}_X = i\hbar\mathbf{P}_Z \tag{3.145}$$

$$\mathbf{P}_Y\mathbf{P}_Z - \mathbf{P}_Z\mathbf{P}_Y = i\hbar\mathbf{P}_X \tag{3.146}$$

$$\mathbf{P}_Z\mathbf{P}_X - \mathbf{P}_X\mathbf{P}_Z = i\hbar\mathbf{P}_Y \tag{3.147}$$

These relations have a useful physical interpretation in that they tell us that it is not possible to simultaneously determine two components of angular momentum even though the total momentum and any one component can be specified. These relationships can be easily remembered because they form the coefficients of the unit vectors in the equation

$$\mathbf{P} \times \mathbf{P} = i\hbar \mathbf{P} \tag{3.148}$$

The commutator for a pair of angular momentum components may also be established directly from the Dirac general commutation relationship

$$[\mathbf{A}, \mathbf{B}] = i\hbar \sum_i \left(\frac{\partial \mathbf{A}}{\partial q_i} \frac{\partial \mathbf{B}}{\partial p_i} - \frac{\partial \mathbf{B}}{\partial q_i} \frac{\partial \mathbf{A}}{\partial p_i} \right) = \mathbf{AB} - \mathbf{BA} \tag{3.149}$$

For example,

$$[\mathbf{P}_X, \mathbf{P}_Y] = i\hbar \left[\frac{\partial \mathbf{P}_X}{\partial X} \frac{\partial \mathbf{P}_Y}{\partial p_X} - \frac{\partial \mathbf{P}_X}{\partial p_X} \frac{\partial \mathbf{P}_Y}{\partial X} + \frac{\partial \mathbf{P}_X}{\partial Y} \frac{\partial \mathbf{P}_Y}{\partial p_Y} - \frac{\partial \mathbf{P}_X}{\partial p_Y} \frac{\partial \mathbf{P}_Y}{\partial Y} + \frac{\partial \mathbf{P}_X}{\partial Z} \frac{\partial \mathbf{P}_Y}{\partial p_Z} - \frac{\partial \mathbf{P}_X}{\partial p_Z} \frac{\partial \mathbf{P}_Y}{\partial Z} \right] \tag{3.150}$$

Using the relationships, $P_X = Yp_Z - Zp_Y$ and $P_Y = Zp_X - Xp_Z$, the partial derivatives are evaluated to give

$$\mathbf{P}_X \mathbf{P}_Y - \mathbf{P}_Y \mathbf{P}_X = i\hbar (Xp_Y - Yp_X) = i\hbar \mathbf{P}_Z \tag{3.151}$$

The defining equations and the commutation relations just developed form the classical relationships for the development of the angular momentum matrix elements. The method used for the derivation of the angular momentum matrix elements will be considered following a discussion of direction cosines.

In the preceding discussion the relationships for the angular momentum components were developed relative to a space-fixed axis system, X, Y, Z. Angular momentum can also be related to a body-fixed axis system, x, y, z. The commutation relations for the body-fixed components will be developed following an introduction to the use of the Euler angle coordinates, but are presented here for completeness:

$$\mathbf{P}_x \mathbf{P}_y - \mathbf{P}_y \mathbf{P}_x = -i\hbar \mathbf{P}_z \tag{3.152}$$

$$\mathbf{P}_y \mathbf{P}_z - \mathbf{P}_z \mathbf{P}_y = -i\hbar \mathbf{P}_x \tag{3.153}$$

$$\mathbf{P}_z \mathbf{P}_x - \mathbf{P}_x \mathbf{P}_z = -i\hbar \mathbf{P}_y \tag{3.154}$$

3.7 DIRECTION COSINES AND EULER ANGLES

Since the angular momentum of a quantized system can be expressed either in terms of a set of space-fixed components or a set of body-fixed components, we need to examine the relationships between these two. In the discussion of spectroscopic phenomenon use will be made of both sets. The components of angular momentum and their associated matrix elements in the two axis systems are related by the direction cosines. A knowledge of the behavior of direction cosine matrix elements is needed to discuss the perturbation of molecular rotation

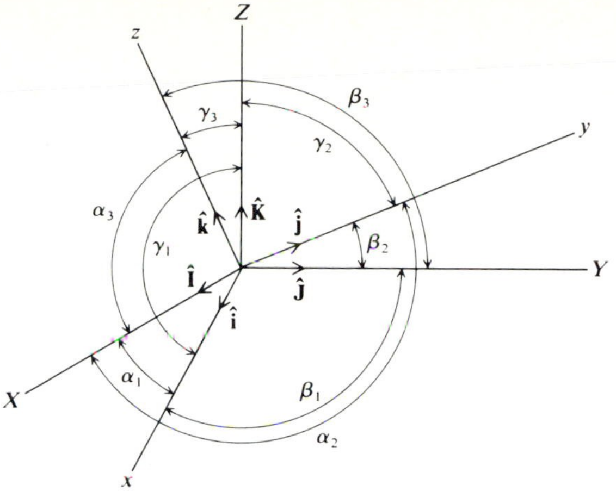

FIGURE 3.1
Direction cosine relationships.

by external fields, to establish selection rules for rotational transitions, and to examine the relationships between the components of angular momentum in the two different-axis systems.

The significance of the direction cosines is understood by the application of a few vector relationships to the two sets of unit vectors which represent the body- and space-fixed axis systems. Referring to Fig. 3.1, where X, Y, Z and x, y, z denotes the space-fixed axes and body-fixed axes, it is observed that the cosine of the angle $\alpha_{Xx} \equiv \alpha_1$ is given by

$$\cos \alpha_1 = \hat{\mathbf{I}} \cdot \hat{\mathbf{i}} \tag{3.155}$$

where $\hat{\mathbf{I}}$ and $\hat{\mathbf{i}}$ are unit vectors in the direction of the X and x axes respectively. Cos α_1 is the direction cosine relating the X and x axes. Consideration of all possible combinations between any two axes of the two systems gives a set of nine direction cosines:

$$\Phi_{Xx} = \cos \alpha_1 = \hat{\mathbf{I}} \cdot \hat{\mathbf{i}} \qquad \Phi_{Yx} = \cos \beta_1 = \hat{\mathbf{J}} \cdot \hat{\mathbf{i}} \qquad \Phi_{Zx} = \cos \gamma_1 = \hat{\mathbf{K}} \cdot \hat{\mathbf{i}} \tag{3.156–3.158}$$

$$\Phi_{Xy} = \cos \alpha_2 = \hat{\mathbf{I}} \cdot \hat{\mathbf{j}} \qquad \Phi_{Yy} = \cos \beta_2 = \hat{\mathbf{J}} \cdot \hat{\mathbf{j}} \qquad \Phi_{Zy} = \cos \gamma_2 = \hat{\mathbf{K}} \cdot \hat{\mathbf{j}} \tag{3.159–3.161}$$

$$\Phi_{Xz} = \cos \alpha_3 = \hat{\mathbf{I}} \cdot \hat{\mathbf{k}} \qquad \Phi_{Yz} = \cos \beta_3 = \hat{\mathbf{J}} \cdot \hat{\mathbf{k}} \qquad \Phi_{Zz} = \cos \gamma_3 = \hat{\mathbf{K}} \cdot \hat{\mathbf{k}} \tag{3.162–3.164}$$

The symbol Φ_{Fg} will be used to denote a direction cosine and the symbol $\mathbf{\Phi}_{Fg}$ will be used to denote a direction cosine operator. The relationships between the two sets of unit vectors are apparent from an examination of Fig. 3.1. Figure 3.2 shows in more detail the relationships between the body-fixed axis and the space-fixed systems. The projections of $\hat{\mathbf{j}}$ on the fixed-axis system will be $|\hat{\mathbf{j}}| \cos \alpha_2$, $|\hat{\mathbf{j}}| \cos \beta_2$, and $|\hat{\mathbf{j}}| \cos \gamma_2$ for the X, Y, and Z axes respectively. Since $|\hat{\mathbf{j}}| = 1$ the

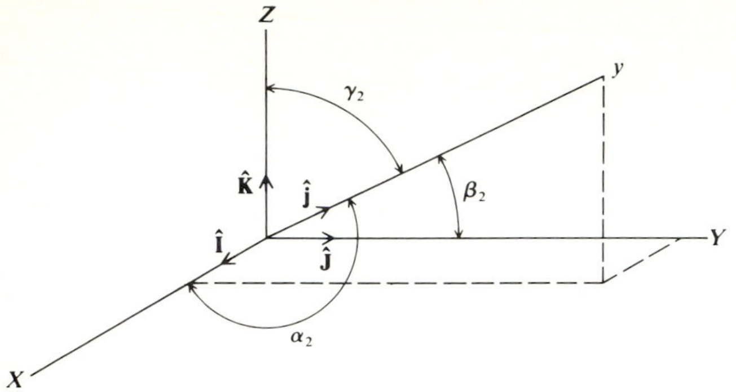

FIGURE 3.2
Unit vector relationships.

vector $\hat{\mathbf{j}}$ can be written in terms of the unit vector, $\hat{\mathbf{I}}, \hat{\mathbf{J}}, \hat{\mathbf{K}}$, of the space-fixed axis system as

$$\hat{\mathbf{j}} = \hat{\mathbf{I}} \cos \alpha_2 + \hat{\mathbf{J}} \cos \beta_2 + \hat{\mathbf{K}} \cos \gamma_2 = \hat{\mathbf{I}}\Phi_{Xy} + \hat{\mathbf{J}}\Phi_{Yy} + \hat{\mathbf{K}}\Phi_{Zy} \quad (3.165)$$

Consideration of the other two body-fixed unit vectors in this manner results in a set of three equations giving the body-fixed unit vectors in terms of the space-fixed unit vectors

$$\hat{\mathbf{i}} = \hat{\mathbf{I}}\Phi_{Xx} + \hat{\mathbf{J}}\Phi_{Yx} + \hat{\mathbf{K}}\Phi_{Zx} \quad (3.166)$$

$$\hat{\mathbf{j}} = \hat{\mathbf{I}}\Phi_{Yx} + \hat{\mathbf{J}}\Phi_{Yy} + \hat{\mathbf{K}}\Phi_{Zy} \quad (3.167)$$

$$\mathbf{k} = \hat{\mathbf{I}}\Phi_{Xz} + \hat{\mathbf{J}}\Phi_{Yz} + \hat{\mathbf{K}}\Phi_{Zz} \quad (3.168)$$

Consideration of the space-fixed unit vectors in terms of the body-fixed ones provides an inverse set of relations

$$\hat{\mathbf{I}} = \hat{\mathbf{i}}\Phi_{Xx} + \hat{\mathbf{j}}\Phi_{Xy} + \hat{\mathbf{k}}\Phi_{Xz} \quad (3.169)$$

$$\hat{\mathbf{J}} = \hat{\mathbf{i}}\Phi_{Yx} + \hat{\mathbf{j}}\Phi_{Yy} + \hat{\mathbf{k}}\Phi_{Yz} \quad (3.170)$$

$$\hat{\mathbf{K}} = \hat{\mathbf{i}}\Phi_{Zx} + \hat{\mathbf{j}}\Phi_{Zy} + \hat{\mathbf{k}}\Phi_{Zz} \quad (3.171)$$

Having established the relationships among unit vectors it is now possible to develop the relationships between the two axis systems for a general vector, $\mathbf{G}$. In the space fixed axis system, $\mathbf{G} = G_X\hat{\mathbf{I}} + G_Y\hat{\mathbf{J}} + G_Z\hat{\mathbf{K}}$, while in the body fixed axis system, $\mathbf{G} = G_x\hat{\mathbf{i}} + G_y\hat{\mathbf{j}} + G_z\hat{\mathbf{k}}$. Since the vector $\mathbf{G}$ is independent of any particular reference axis system there exists a transformation which relates the components of $\mathbf{G}$ in the two axis systems. The general method for finding the transformation can be established by referring to Fig. 3.3. The components of $\mathbf{G}$ are related to $|\mathbf{G}|$ by the angles, θ, ϕ, and ψ

$$G_x = |\mathbf{G}| \cos \psi \quad (3.172)$$

$$G_y = |\mathbf{G}| \cos \theta \quad (3.173)$$

$$G_z = |\mathbf{G}| \cos \phi \quad (3.174)$$

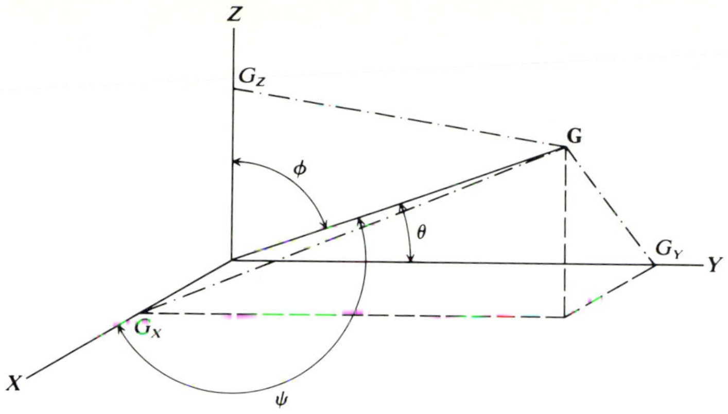

FIGURE 3.3
Relationships for a general vector.

From the definition of the scalar product, remembering that $|\hat{\mathbf{I}}| = |\hat{\mathbf{J}}| = |\hat{\mathbf{K}}| = 1$, it follows that

$$\mathbf{G} \cdot \hat{\mathbf{I}} = |\mathbf{G}||\hat{\mathbf{I}}| \cos \psi = |\mathbf{G}| \cos \psi \tag{3.175}$$

$$\mathbf{G} \cdot \hat{\mathbf{J}} = |\mathbf{G}||\hat{\mathbf{J}}| \cos \theta = |\mathbf{G}| \cos \theta \tag{3.176}$$

$$\mathbf{G} \cdot \hat{\mathbf{K}} = |\mathbf{G}||\hat{\mathbf{K}}| \cos \phi = |\mathbf{G}| \cos \phi \tag{3.177}$$

By pairwise comparison of Eqs. (3.172) to (3.174) with Eqs. (3.175) to (3.177) and incorporating Eqs. (3.169) to (3.171) one obtains

$$G_X = \mathbf{G} \cdot \mathbf{I} = \Phi_{Xx} G_x + \Phi_{Xy} G_y + \Phi_{Xz} G_z \tag{3.178}$$

$$G_Y = \mathbf{G} \cdot \mathbf{J} = \Phi_{Yx} G_x + \Phi_{Yy} G_y + \Phi_{Yz} G_z \tag{3.179}$$

$$G_Z = \mathbf{G} \cdot \mathbf{K} = \Phi_{Zx} G_x + \Phi_{Zy} G_y + \Phi_{Zz} G_z \tag{3.180}$$

Thus the direction cosines are the elements of the matrix which transforms a vector from one coordinate system to another with a common origin, but having undergone a rotation relative to the first. Using the symbol $\mathscr{F}$ to denote the space-fixed axis system, F to denote a space-fixed axis, $\mathscr{G}$ to denote the body-fixed axis system, and g to denote a body-fixed axis, the above equations can be written in a shorthand manner as

$$G_F = \sum_{g=x,y,z} \Phi_{Fg} G_g \qquad (F = X, Y, Z) \tag{3.181}$$

and the transformation can be written in vector-matrix terminology as

$$\mathbf{G}_{\mathscr{F}} = \underline{\mathbf{R}}\mathbf{G}_{\mathscr{G}} \tag{3.182}$$

or

$$\begin{pmatrix} G_X \\ G_Y \\ G_Z \end{pmatrix} = \begin{pmatrix} \Phi_{Xx} & \Phi_{Xy} & \Phi_{Xz} \\ \Phi_{Yx} & \Phi_{Yy} & \Phi_{Yz} \\ \Phi_{Zx} & \Phi_{Zy} & \Phi_{Zz} \end{pmatrix} \begin{pmatrix} G_x \\ G_y \\ G_z \end{pmatrix} \tag{3.183}$$

The inverse transformation, developed in an analogous manner from the relationships $G_x = \mathbf{G} \cdot \hat{\mathbf{i}}$, $G_y = \mathbf{G} \cdot \hat{\mathbf{j}}$, $G_z = \mathbf{G} \cdot \hat{\mathbf{k}}$, $\mathbf{G} = G_x\hat{\mathbf{i}} + G_y\hat{\mathbf{j}} + G_z\hat{\mathbf{k}}$, and Eqs. (3.166) to (3.168) is

$$G_g = \sum_{F=X,Y,Z} \Phi_{Fg} G_F \qquad (g = x, y, z) \tag{3.184}$$

Since it is known that only three coordinates are needed to specify the rotation of an object there must be some interrelationships among the nine direction cosines. Such relationships can be established from the orthogonality properties of the unit vectors. For example, the normalization condition for **I** gives

$$\hat{\mathbf{I}} \cdot \hat{\mathbf{I}} = \Phi_{Xx}^2 + \Phi_{Xy}^2 + \Phi_{Xz}^2 = 1 \tag{3.185}$$

and the orthogonality condition for **I** and **J** gives

$$\hat{\mathbf{I}} \cdot \hat{\mathbf{J}} = \Phi_{Xx}\Phi_{Yx} + \Phi_{Xy}\Phi_{Yy} + \Phi_{Xz}\Phi_{Yz} = 0 \tag{3.186}$$

Repeated application of such relations for other combinations of unit vectors gives a set of six such relationships which can be summarized as

$$\Phi_{Fx}\Phi_{F'x} + \Phi_{Fy}\Phi_{F'y} + \Phi_{Fz}\Phi_{F'z} = \delta_{FF'} \tag{3.187}$$

where $\delta_{FF'}$ is the Kronecker delta.

The discussion of molecular rotation and the development of the commutation rules for the direction cosines are facilitated by the use of the Euler angles. The Euler angles define a set of three rotations which describe the orientation of a body. They are defined by the following sequence of rotations. Beginning with a body-fixed axis system x, y, z and a space-fixed axis system X, Y, Z, in

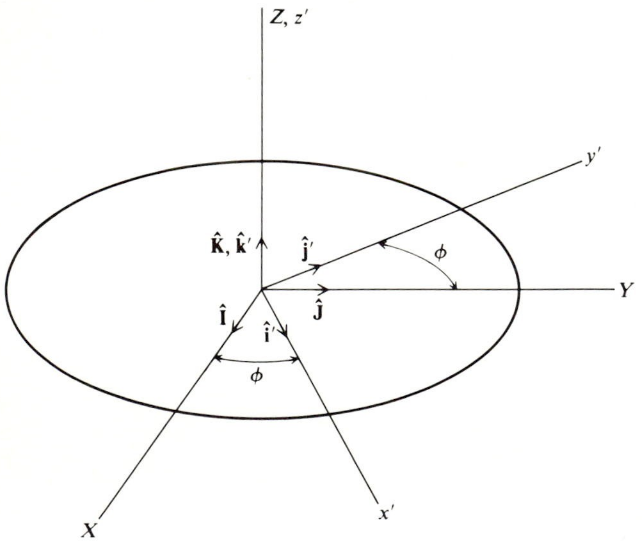

FIGURE 3.4
Euler angle ϕ.

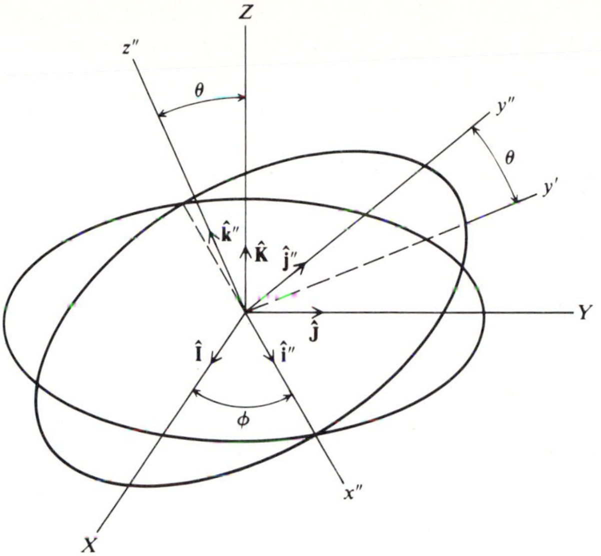

FIGURE 3.5
Euler angles ϕ and θ.

coincidence the body-fixed system is rotated about the common z, Z axis by an angle ϕ. This leads to the configuration shown in Fig. 3.4. Next the body-fixed system is rotated about the x' axis by an angle θ giving the configuration shown in Fig. 3.5. Finally the body-fixed system is rotated about the z axis (equivalent to the z'' axis) by an angle ψ giving rise to the final relationship shown in Fig. 3.6.

Examination of the rotation illustrated in Fig. 3.4 shows that a unit vector $\hat{\mathbf{i}}$ initially lying along the x, X axes will be transformed into a new vector $\hat{\mathbf{i}}'$ lying

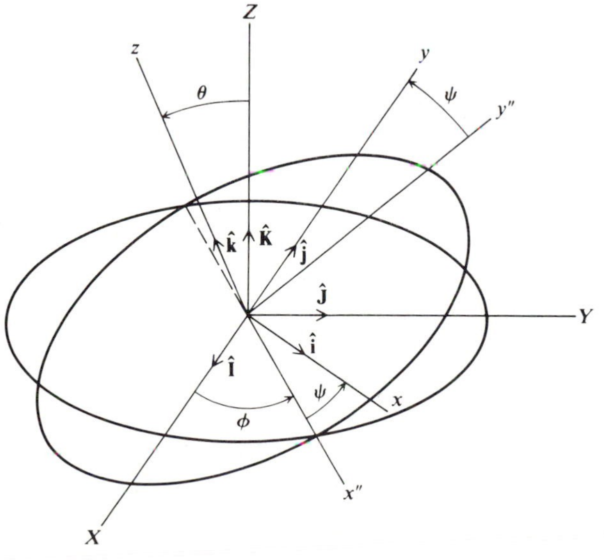

FIGURE 3.6
Euler angles ϕ, θ, and ψ.

along the x' axis. The vector $\hat{\mathbf{i}}'$ will be given by $\hat{\mathbf{i}}' = \hat{\mathbf{I}} \cos \phi + \hat{\mathbf{J}} \sin \phi$. Similarly a vector $\hat{\mathbf{j}}$ lying along the y, Y axes will be transformed to a new vector $\hat{\mathbf{j}}' = -\hat{\mathbf{I}} \sin \phi + \hat{\mathbf{J}} \cos \phi$, lying along the y' axis. A vector $\mathbf{k}$ lying along the z, Z axes will remain unchanged. This transformation of a vector, $\mathbf{R} = \hat{\mathbf{I}} + \hat{\mathbf{J}} + \hat{\mathbf{K}}$, is written in matrix terminology as

$$\mathbf{r} = \underline{\mathbf{S}}_\phi \mathbf{R} \tag{3.188}$$

or

$$\begin{pmatrix} \hat{\mathbf{i}}' \\ \hat{\mathbf{j}}' \\ \hat{\mathbf{k}}' \end{pmatrix} = \begin{pmatrix} \cos \phi & \sin \phi & 0 \\ -\sin \phi & \cos \phi & 0 \\ 0 & 0 & 1 \end{pmatrix} \begin{pmatrix} \hat{\mathbf{I}} \\ \hat{\mathbf{J}} \\ \hat{\mathbf{K}} \end{pmatrix} \tag{3.189}$$

The second rotation, through the angle θ, results in a similar transformation

$$\mathbf{r}'' = \underline{\mathbf{S}}_\theta \mathbf{r}' \tag{3.190}$$

or

$$\begin{pmatrix} \hat{\mathbf{i}}'' \\ \hat{\mathbf{j}}'' \\ \hat{\mathbf{k}}'' \end{pmatrix} = \begin{pmatrix} 1 & 0 & 0 \\ 0 & \cos \theta & \sin \theta \\ 0 & -\sin \theta & \cos \theta \end{pmatrix} \begin{pmatrix} \hat{\mathbf{i}}' \\ \hat{\mathbf{j}}' \\ \hat{\mathbf{k}}' \end{pmatrix} \tag{3.191}$$

The final rotation, through the angle ψ, gives the transformation

$$\mathbf{r} = \underline{\mathbf{S}}_\psi \mathbf{r}'' \tag{3.192}$$

or

$$\begin{pmatrix} \hat{\mathbf{i}} \\ \hat{\mathbf{j}} \\ \hat{\mathbf{k}} \end{pmatrix} = \begin{pmatrix} \cos \psi & \sin \psi & 0 \\ -\sin \psi & \cos \psi & 0 \\ 0 & 0 & 1 \end{pmatrix} \begin{pmatrix} \hat{\mathbf{i}}'' \\ \hat{\mathbf{j}}'' \\ \hat{\mathbf{k}}'' \end{pmatrix} \tag{3.193}$$

The overall rotational transformation can be expressed as

$$\mathbf{r} = \underline{\mathbf{S}}_\psi \underline{\mathbf{S}}_\theta \underline{\mathbf{S}}_\phi \mathbf{R} = \underline{\mathbf{S}}\mathbf{R} \tag{3.194}$$

where

$$\underline{\mathbf{S}} = \begin{pmatrix} \cos \psi \cos \phi - \sin \psi \cos \theta \sin \phi & \cos \psi \sin \phi + \sin \psi \cos \theta \cos \phi & \sin \psi \sin \theta \\ -\sin \psi \cos \phi - \cos \psi \cos \theta \sin \phi & -\sin \psi \sin \phi + \cos \psi \cos \theta \cos \phi & \cos \psi \sin \theta \\ \sin \theta \sin \phi & -\sin \theta \cos \phi & \cos \theta \end{pmatrix} \tag{3.195}$$

The inverse of this transformation is found by left-hand multiplication of Eq. (3.194) by the reciprocal matrix $\underline{\mathbf{S}}^{-1}$

$$\underline{\mathbf{S}}^{-1}\mathbf{r} = \underline{\mathbf{S}}^{-1}\underline{\mathbf{S}}\mathbf{R} = \mathbf{R} \tag{3.196}$$

where the matrix $\underline{\mathbf{S}}^{-1}$ as found by conventional matrix algebra is

$$\underline{\mathbf{S}}^{-1} = \begin{pmatrix} \cos\psi\cos\phi & -\sin\psi\cos\phi & \\ \quad -\cos\theta\sin\phi\sin\psi & \quad -\cos\theta\sin\phi\cos\psi & \sin\theta\sin\phi \\ \cos\psi\sin\phi & -\sin\psi\sin\phi & \\ \quad +\cos\theta\cos\phi\sin\psi & \quad +\cos\theta\cos\phi\cos\psi & -\sin\theta\cos\phi \\ \sin\theta\sin\psi & \sin\theta\cos\psi & \cos\theta \end{pmatrix} \tag{3.197}$$

The elements of the $\underline{\mathbf{S}}$ matrix are the coefficients which relate the unit vectors $\hat{\mathbf{I}}, \hat{\mathbf{J}}, \hat{\mathbf{K}}$ to the unit vectors $\hat{\mathbf{i}}, \hat{\mathbf{j}}, \hat{\mathbf{k}}$ and are the direction cosines expressed in terms of the Euler angles. The manner in which the $\underline{\mathbf{S}}$ matrix is written is such that the rows are indexed by x, y, z and the columns by X, Y, Z. The selection of the Euler angles is not entirely unique hence the form of the transformation matrix may be found to be different in other discussions. Other conventions which have been used are: (1) use of left-handed coordinate system, (2) measurement of ψ from the j' axis, (3) interchange of the use of the symbols ψ and ϕ, (4) use of χ and ϕ rather than ϕ and ψ, and (5) counterclockwise rotations about Z.

The body-fixed components of the angular momentum in terms of the Euler position and momenta coordinates as derived in App. H are

$$P_y = \frac{\cos\psi}{\sin\theta}(-p_\psi\cos\theta + p_\phi) - p_\theta\sin\psi \tag{3.198}$$

$$P_x = \frac{\sin\psi}{\sin\theta}(-p_\psi\cos\theta + p_\phi) + p_\theta\cos\psi \tag{3.199}$$

$$P_z = p_\psi \tag{3.200}$$

Commutation relationships between different Φ_{Fg} terms and between $\mathbf{P}_g$ and $\mathbf{\Phi}_{Fg}$ terms are formulated by using the Dirac commutator. As an illustration we will evaluate $[\mathbf{P}_y, \mathbf{\Phi}_{Zz}]$

$$[\mathbf{P}_y, \mathbf{\Phi}_{Zz}] = i\hbar\left[\frac{\partial\mathbf{P}_y}{\partial\phi}\frac{\partial\Phi_{Zz}}{\partial p_\phi} - \frac{\partial\mathbf{P}_y}{\partial p_\phi}\frac{\partial\Phi_{Zz}}{\partial\phi} + \frac{\partial\mathbf{P}_y}{\partial\theta}\frac{\partial\Phi_{Zz}}{\partial p_\theta} - \frac{\partial\mathbf{P}_y}{\partial P_\theta}\frac{\partial\Phi_{Zz}}{\partial\theta} + \frac{\partial\mathbf{P}_y}{\partial\psi}\frac{\partial\Phi_{Zz}}{\partial P_\psi} - \frac{\partial\mathbf{P}_y}{\partial P_\psi}\frac{\partial\Phi_{Zz}}{\partial\psi}\right] \tag{3.201}$$

Since all $\mathbf{\Phi}_{Fg}$'s are independent of momenta components the positive (1st, 3rd, 5th) terms on the right-hand side of Eq. (3.201) are zero. Taking the derivatives of the angular momentum components with respect to the Euler momentum coordinates gives

$$\frac{\partial P_y}{\partial P_\phi} = \frac{\cos\psi}{\sin\theta} \tag{3.202}$$

$$\frac{\partial P_y}{\partial P_\theta} = -\sin\psi \tag{3.203}$$

$$\frac{\partial P_y}{\partial P_\psi} = \frac{\cos\theta\cos\psi}{\sin\theta} \tag{3.204}$$

Since $\phi_{Zz} = \cos\theta$ the derivatives with respect to position coordinates are $\partial\phi_{Zz}/\partial\phi = 0$, $\partial\phi_{Zz}/\partial\theta = -\sin\theta$, and $\partial\phi_{Zz}/\partial\psi = 0$. Introducing these partial derivatives into Eq. (3.201) gives

$$\mathbf{P}_y\mathbf{\Phi}_{Zz} - \mathbf{\Phi}_{Zz}\mathbf{P}_y = i\hbar(\sin\psi\sin\theta) = -i\hbar\Phi_{Zx} \tag{3.205}$$

Repeated application of this procedure to other pairs of $\mathbf{\Phi}_{Fg}$ and P_g leads to a set of commutation relations

$$[\mathbf{\Phi}_{Fg}, \mathbf{\Phi}_{F'g'}] = 0 \tag{3.206}$$

$$[\mathbf{P}_g, \mathbf{\Phi}_{Fg'}] = -i\hbar\mathbf{\Phi}_{Fg''} \tag{3.207}$$

$$[\mathbf{P}_g, \mathbf{\Phi}_{Fg}] = 0 \tag{3.208}$$

where $F = X$, Y, or Z and $g, g', g' = x, y, z$ taken in cyclic order so that each of the last two equations represents three separate relations.

Since $\mathbf{P}_F \neq \mathbf{P}_g$ $(F, g = X, x;\ Y, y;\ Z, z)$ when both sets of components are expressed in terms of the Euler angles, the previously stated relations between the $\mathbf{P}_g$ and the Euler angles, Eqs. (3.198) to (3.200), cannot be used to evaluate relations of the form $[\mathbf{P}_F, \mathbf{\Phi}_{Fg}]$. They can be evaluated, however, by use of the transformation relationships

$$G_F = \sum_g \phi_{Fg} G_g \tag{3.209}$$

To illustrate the method consider the evaluation of $[\mathbf{P}_Y, \mathbf{\Phi}_{Zz}]$. Using this relationship $\mathbf{P}_Y$ is expressed as

$$\mathbf{P}_Y = \mathbf{\Phi}_{Yx}\mathbf{P}_x + \mathbf{\Phi}_{Yy}\mathbf{P}_y + \mathbf{\Phi}_{Yz}\mathbf{P}_z \tag{3.210}$$

and the commutator becomes

$$\begin{aligned}[\mathbf{P}_Y, \mathbf{\Phi}_{Zz}] &= \mathbf{\Phi}_{Yx}\mathbf{P}_x\mathbf{\Phi}_{Zz} + \mathbf{\Phi}_{Yy}\mathbf{P}_y\mathbf{\Phi}_{Zz} + \mathbf{\Phi}_{Yz}\mathbf{P}_z\mathbf{\Phi}_{Zz}\\ &\quad - \mathbf{\Phi}_{Zz}\mathbf{\Phi}_{Yx}\mathbf{P}_x - \mathbf{\Phi}_{Zz}\mathbf{\Phi}_{Yy}\mathbf{P}_y - \mathbf{\Phi}_{Zz}\mathbf{\Phi}_{Yz}\mathbf{P}_z\end{aligned} \tag{3.211}$$

By using the commutators for $\mathbf{P}_g$ and $\mathbf{\Phi}_{Fg}$, the first three terms on the right rearrange to $\mathbf{\Phi}_{Yx}(i\hbar\mathbf{\Phi}_{Zx} + \mathbf{\Phi}_{Zz}\mathbf{P}_x) + \mathbf{\Phi}_{Yy}(-i\mathbf{\Phi}_{Zx} + \mathbf{\Phi}_{Zz}\mathbf{P}_x) + \mathbf{\Phi}_{Yz}\mathbf{\Phi}_{Zz}\mathbf{P}_z$ which, when combined with the last three terms, leaves

$$[\mathbf{P}_Y, \mathbf{\Phi}_{Zz}] = i\hbar(\mathbf{\Phi}_{Zy}\mathbf{\Phi}_{Yz} - \mathbf{\Phi}_{Yy}\mathbf{\Phi}_{Zx}) \tag{3.212}$$

Since $\hat{\mathbf{I}}$, $\hat{\mathbf{J}}$, and $\hat{\mathbf{K}}$ are a set of orthogonal unit vectors, $\hat{\mathbf{J}} \times \hat{\mathbf{K}} = \hat{\mathbf{I}}$. Using the expressions for these unit vectors given by Eqs. (3.169) to (3.171) and taking this vector product gives

$$\begin{aligned}&(\mathbf{\Phi}_{Yy}\mathbf{\Phi}_{Zz} - \mathbf{\Phi}_{Yz}\mathbf{\Phi}_{Zy})\hat{\mathbf{i}} + (\mathbf{\Phi}_{Yz}\mathbf{\Phi}_{Zx} - \mathbf{\Phi}_{Yx}\mathbf{\Phi}_{Zz})\hat{\mathbf{j}} + (\mathbf{\Phi}_{Yx}\mathbf{\Phi}_{Zy} - \mathbf{\Phi}_{Yz}\mathbf{\Phi}_{Zx})\hat{\mathbf{k}}\\ &\quad = \mathbf{\Phi}_{Xx}\hat{\mathbf{i}} + \mathbf{\Phi}_{Xy}\hat{\mathbf{j}} + \mathbf{\Phi}_{Xz}\hat{\mathbf{k}}\end{aligned} \tag{3.213}$$

Equating the coefficients of like unit vector gives

$$[\mathbf{P}_y, \Phi_{Zz}] = i\hbar\Phi_{Xz} \tag{3.214}$$

Using a similar procedure the additional commutation relations are developed

$$[\mathbf{P}_F, \mathbf{\Phi}_{F'g}] = i\hbar\mathbf{\Phi}_{F''g} \tag{3.215}$$

$$[\mathbf{P}_F, \mathbf{\Phi}_{Fg}] = 0 \tag{3.216}$$

where $g = x$, y, or z and $F, F', F'' = X, Y, Z$ taken in cyclic order. For both sets of commutation relationships, if the cyclic order for the subscripts is not maintained, a change in signs results. For example

$$[\mathbf{P}_X, \mathbf{\Phi}_{Yg}] = -[\mathbf{P}_Y, \mathbf{\Phi}_{Xg}] = i\hbar\mathbf{\Phi}_{Zg} \tag{3.217}$$

or

$$[\mathbf{P}_x, \mathbf{\Phi}_{Fy}] = -[\mathbf{P}_y, \mathbf{\Phi}_{Fx}] = -i\hbar\mathbf{\Phi}_{Fz} \tag{3.218}$$

There is a difference in sign between the angular momentum commutators taken relative to a space-fixed or a body-fixed axis system. The occurrence of the negative sign in the commutators of the components taken relative to the body-fixed axis system can be demonstrated by the evaluation of terms of the form $[P_g, \mathbf{P}_{g'}]$. This is done by using Eqs. (3.198) to (3.200) to obtain the necessary derivative terms and substituting these into the general Dirac commutator. For example

$$[\mathbf{P}_z, \mathbf{P}_x] = i\hbar \sum_{q=\phi,\theta,\psi} \left[\frac{\partial \mathbf{P}_z}{\partial q}\frac{\partial \mathbf{P}_x}{\partial P_q} - \frac{\partial \mathbf{P}_x}{\partial q}\frac{\partial \mathbf{P}_z}{\partial P_q}\right] = -i\hbar\mathbf{P}_y \tag{3.219}$$

The sign difference may also be shown by direct substitution of $\mathbf{P}_F = \sum_g \mathbf{\Phi}_{Fg}\mathbf{P}_g$ into the $[\mathbf{P}_F, \mathbf{P}_{F'}] = i\hbar\mathbf{P}_F$ commutator, expansion of the resulting products and simplification by means of the $\mathbf{P}_F$ commutators [7].

3.8 ANGULAR MOMENTUM MATRIX ELEMENTS

In Sec. 3.6 the eigenvalues of the square of the angular momentum and the Z component of the angular momentum were found to be $J(J+1)\hbar^2$ and $M\hbar$ respectively. By using the differential operators for $\mathbf{P}^2$ and $\mathbf{P}_Z$ and the rigid rotor wavefunctions we established that the only nonzero matrix elements were those diagonal in J and M. In this section this fact will be confirmed and the matrix elements of $\mathbf{P}_X$ and $\mathbf{P}_Y$ will also be determined. The matrix elements for $\mathbf{P}_X$ and $\mathbf{P}_Y$ can best be evaluated by using the methods of matrix mechanics.

It was established in Secs. 3.6 and 3.7 that the angular momentum components could be related to either a space-fixed or a body-fixed axis system and that the commutators of the two differ only in sign. The development of the matrix elements will be for those related to a space-fixed system with subsequent conversion to body-fixed elements.

It frequently occurs that a set of eigenvectors are simultaneous eigenvectors of two or more operators. If two operators $\underline{\mathbf{A}}$ and $\underline{\mathbf{B}}$ commute, $\underline{\mathbf{AB}} = \underline{\mathbf{BA}}$, then it is possible to simultaneously measure values for the properties associated with the operators. Of particular interest will be a collection of operators which are classified as a completed set of commuting operators. Such a set is composed of operators, $\underline{\mathbf{A}}, \underline{\mathbf{B}}, \underline{\mathbf{C}}, \ldots$, which satisfy the conditions: (1) all operators commute with each other, (2) there exists a set of eigenstates, each described by an eigenvector, which are common to all operators, (3) the eigenstates have a unique set of eigenvalues, $a, b, c, \ldots$, (4) the set of eigenvectors forms a basis for a Hilbert space, and (5) there are no other operators which simultaneously commute with all members of the set. Such a state will contain the maximum information allowable and the quantum numbers associated with the eigenvalues are good quantum numbers.

It can be shown [8] that the Hamiltonian of a system in the absence of external fields commutes with both $\mathbf{P}_Z$ and $\mathbf{P}^2$, hence there exists a set of eigenstates, denoted by $\psi_{\alpha JM} = |\alpha, J, M\rangle$ which are simultaneous eigenstates of $\mathcal{H}$, $\mathbf{P}^2$, and $\mathbf{P}_Z$. In this notation the α, J, and M are quantum numbers which specify the energy, the total angular momentum, and a space-fixed component of angular momentum, respectively. These simultaneous eigenstates of $\mathcal{H}$, $\mathbf{P}^2$, and $\mathbf{P}_Z$ are defined by the eigenvalue equations

$$\mathcal{H}|\alpha, J, M\rangle = E|\alpha, J, M\rangle \qquad (3.220)$$

$$\mathbf{P}^2|\alpha, J, M\rangle = \mathcal{J}|\alpha, J, M\rangle \qquad (3.221)$$

$$\mathbf{P}_Z|\alpha, J, M\rangle = \mathcal{M}|\alpha, J, M\rangle \qquad (3.222)$$

where E, $\mathcal{J}$ and $\mathcal{M}$ are the eigenvalues of $\mathcal{H}$, $\mathbf{P}^2$, and $\mathbf{P}_Z$ respectively, α, J, and M are the accompanying quantum numbers (*Note*: α may be a set of quantum numbers), and the eigenstates are represented as $|\alpha, J, M\rangle$. These equations provide the starting point for the evaluation of $\mathcal{J}$ and $\mathcal{M}$. In order to proceed it is necessary to define two new operators

$$P_+ = P_X + iP_Y \qquad (3.223)$$

$$P_- = P_X - iP_Y \qquad (3.224)$$

where $i = \sqrt{-1}$, and to develop their commutation relationships with $\mathbf{P}_Z$ and with one another. For $\mathbf{P}_+$

$$[\mathbf{P}_Z, \mathbf{P}_+] = \mathbf{P}_Z\mathbf{P}_+ - \mathbf{P}_+\mathbf{P}_Z = \mathbf{P}_Z\mathbf{P}_X - \mathbf{P}_X\mathbf{P}_Z + i(\mathbf{P}_Z\mathbf{P}_Y - \mathbf{P}_Y\mathbf{P}_Z) \qquad (3.225)$$

Incorporating the commutation relationships given by Eqs. (3.145) to (3.147), this becomes

$$\mathbf{P}_Z\mathbf{P}_+ - \mathbf{P}_+\mathbf{P}_Z = i\hbar\mathbf{P}_Y + \hbar\mathbf{P}_X = \hbar\mathbf{P}_+ \qquad (3.226)$$

Using the same technique the additional relationships

$$\mathbf{P}_z\mathbf{P}_- - \mathbf{P}_-\mathbf{P}_z = -\hbar\mathbf{P}_- \qquad (3.227)$$

and

$$\mathbf{P}_-\mathbf{P}_+ - \mathbf{P}_+\mathbf{P}_- = -2\hbar\mathbf{P}_Z \qquad (3.228)$$

can be derived. For the present the operators $\mathbf{P}_+$ and $\mathbf{P}_-$ are simply mathematical entities, but it will be shown later that they have a useful physical significance.

The sequence of eigenvalues of $\mathbf{P}_Z$ is established by beginning with the eigenvalue equation (3.222), and left-multiplying by $\mathbf{P}_+$:

$$\mathbf{P}_+\mathbf{P}_Z|\alpha, J, M\rangle = \mathbf{P}_+\mathcal{M}|\alpha, J, M\rangle \qquad (3.229)$$

$\mathcal{M}$ is a constant and commutes with $\mathbf{P}_+$, so $\mathbf{P}_+\mathcal{M} = \mathcal{M}P_+$. Rearrangement of Eq. (3.226) to

$$\mathbf{P}_+\mathbf{P}_Z = (\mathbf{P}_Z - \hbar)\mathbf{P}_+ \qquad (3.230)$$

and substitution of Eq. (3.229) gives

$$(\mathbf{P}_Z - \hbar)\mathbf{P}_+|\alpha, J, M\rangle = \mathcal{M}\mathbf{P}_+|\alpha, J, M\rangle \qquad (3.231)$$

or

$$\mathbf{P}_Z\mathbf{P}_+|\alpha, J, M\rangle = (\mathcal{M} + \hbar)\mathbf{P}_+|\alpha, J, M\rangle \qquad (3.232)$$

Therefore, $\mathbf{P}_+|\alpha, J, M\rangle$ is an eigenstate of $\mathbf{P}_Z$ with eigenvalue $(\mathcal{M} + \hbar)$. Once $\mathcal{M}$ has been evaluated it will be possible to determine the precise effect of operating on the state, $|\alpha, J, M\rangle$ with the $\mathbf{P}_+$ or $\mathbf{P}_-$ operators. Left-multiplication of the eigenvalue equation of $\mathbf{P}_-$ followed by an analogous procedure will give

$$\mathbf{P}_Z\mathbf{P}_-|\alpha, J, M\rangle = (\mathcal{M} - \hbar)\mathbf{P}_-|\alpha, J, M\rangle \qquad (3.233)$$

showing the $\mathbf{P}_-|\alpha, J, M\rangle$ is an eigenstate of $\mathbf{P}_Z$ with eigenvalue $(\mathcal{M} - \hbar)$. This process can be repeated. For example, left-multiplication of Eq. (3.232) by $\mathbf{P}_+$ and employment of the commutation relationships give

$$\mathbf{P}_Z\mathbf{P}_+^2|\alpha, J, M\rangle = (\mathcal{M} + 2\hbar)\mathbf{P}_+^2|\alpha, J, M\rangle \qquad (3.234)$$

Repetition of this procedure will establish the sequence of eigenvalues of $\mathbf{P}_Z$ to be

State	Eigenvalue
...	...
$\mathbf{P}_+^n|\alpha, J, M\rangle$	$(\mathcal{M} + n\hbar)$
$\vdots$	$\vdots$
$\mathbf{P}_+^2|\alpha, J, M\rangle$	$(\mathcal{M} + 2\hbar)$
$\mathbf{P}_+|\alpha, J, M\rangle$	$(\mathcal{M} + \hbar)$
$|\alpha, J, M\rangle$	$(\mathcal{M})$
$\mathbf{P}_-|\alpha, J, M\rangle$	$(\mathcal{M} - \hbar)$
$\mathbf{P}_-^2|\alpha, J, M\rangle$	$(\mathcal{M} - 2\hbar)$
$\vdots$	$\vdots$
$\mathbf{P}_-^n|\alpha, J, M\rangle$	$(\mathcal{M} - n\hbar)$
...	...

Since the individual components of the angular momentum commute with the Hamiltonian

$$[\mathbf{P}_X, \mathscr{H}] = [\mathbf{P}_Y, \mathscr{H}] = [\mathbf{P}_Z, \mathscr{H}] = 0 \tag{3.235}$$

it then follows that $\mathbf{P}_+$ and $\mathbf{P}_-$ will commute with $\mathscr{H}$ and also with $\mathbf{P}^2$ since the latter two commute. Since the commutation of two operators signifies the occurrence of simultaneous eigenstates of the operators this signifies that the states $\mathbf{P}_+|\alpha, J, M\rangle$ and $\mathbf{P}_-|\alpha, J, M\rangle$ are simultaneous eigenstates of $\mathscr{H}$ and $\mathbf{P}^2$ having the same eigenvalues of $\mathscr{H}$ and $\mathbf{P}^2$ as $|\alpha JM\rangle$. However, $\mathbf{P}_Z$ does not commute with $\mathbf{P}_X$ or $\mathbf{P}_Y$ so the states $\mathbf{P}_+|\alpha, J, M\rangle$ and $\mathbf{P}_-|\alpha, J, M\rangle$ will not have the same eigenvalues as $\mathbf{P}_Z$. That is,

$$\mathscr{H}\mathbf{P}_+|\alpha, J, M\rangle = E\mathbf{P}_+|\alpha, J, M\rangle \tag{3.236}$$

$$\mathbf{P}^2\mathbf{P}_+|\alpha, J, M\rangle = \mathscr{J}\mathbf{P}_+|\alpha, J, M\rangle \tag{3.237}$$

but

$$\mathbf{P}_Z\mathbf{P}_+|\alpha, J, M\rangle \neq \mathscr{M}\mathbf{P}_+|\alpha, J, M\rangle \tag{3.238}$$

Since $\mathbf{P}_Z$ is a component of the angular momentum in a space-fixed Z direction its magnitude will be limited by the value of the total angular momentum P. Expanding the $\mathbf{P}_-\mathbf{P}_+$ product

$$\mathbf{P}_-\mathbf{P}_+ = (\mathbf{P}_X - i\mathbf{P}_Y)(\mathbf{P}_X + i\mathbf{P}_Y) = \mathbf{P}_X^2 + \mathbf{P}_Y^2 + i(\mathbf{P}_X\mathbf{P}_Y - \mathbf{P}_Y\mathbf{P}_X) \tag{3.239}$$

adding and subtracting $\mathbf{P}_Z^2$, and using the commutation relationships gives

$$\mathbf{P}_-\mathbf{P}_+ = \mathbf{P}^2 - \mathbf{P}_Z^2 - \hbar\mathbf{P}_Z \tag{3.240}$$

Therefore any eigenfunction of $\mathbf{P}^2$ and $\mathbf{P}_Z$ will also be an eigenfunction of $\mathbf{P}_-\mathbf{P}_+$. Using the eigenvalue equations (3.220) to (3.222)

$$(\mathbf{P}^2 - \mathbf{P}_Z^2 - \hbar\mathbf{P}_Z)|\alpha, J, M\rangle = (\mathscr{J} - \mathscr{M}^2 - \mathscr{M}\hbar)|\alpha, J, M\rangle \tag{3.241}$$

or

$$\mathbf{P}_-\mathbf{P}_+|\alpha, J, M\rangle = (\mathscr{J} - \mathscr{M}^2 - \mathscr{M}\hbar)|\alpha, J, M\rangle \tag{3.242}$$

Now look at the $\mathbf{P}_-\mathbf{P}_+$ operator from a different viewpoint. Since $\mathbf{P}_-\mathbf{P}_+$ is a quantum mechanical operator it can be represented by a matrix having elements

$$\langle \alpha, J, M|\mathbf{P}_-\mathbf{P}_+|\alpha', J', M'\rangle \equiv \int \psi^*_{\alpha JM}\mathbf{P}_-\mathbf{P}_+\psi_{\alpha' J' M'}\, dv \tag{3.243}$$

By examining the relationship of $\mathbf{P}_-\mathbf{P}_+$ to the Hamiltonian and angular momentum operators one can show that the $\mathbf{P}_-\mathbf{P}_+$ matrix must be diagonal. The Hamiltonian or energy matrix of any system is, or in principle can be, diagonalized, therefore the elements $\langle \alpha, J, M|\mathscr{H}|\alpha, J', M'\rangle$ are the energy eigenvalues. Since $\mathbf{P}^2$ and $\mathbf{P}_Z$ both commute with $\mathscr{H}$ their matrices are also diagonal with eigenvalues $\langle \alpha, J, M|\mathbf{P}^2|\alpha, J, M\rangle$ and $\langle \alpha, J, M|\mathbf{P}_Z|\alpha, J, M\rangle$. Since $\mathbf{P}_-\mathbf{P}_+ = \mathbf{P}^2 - \mathbf{P}_Z^2 - \hbar\mathbf{P}_Z$ it follows that the elements $\langle \alpha, J, M|\mathbf{P}_-\mathbf{P}_+|\alpha, J, M\rangle$, will be the eigenvalues of $\mathbf{P}_-\mathbf{P}_+$. It is to be noted, however, that since $\mathbf{P}_X$ and $\mathbf{P}_Y$ do not commute with both $\mathbf{P}^2$ and $\mathbf{P}_Z$ their matrices will not necessarily be diagonal. The diagonal elements of the $\mathbf{P}_-\mathbf{P}_+$ matrix can be written

$$\langle \alpha, J, M|\mathbf{P}_-\mathbf{P}_+|\alpha, J, M\rangle = \langle \alpha, J, M|(\mathbf{P}_X - i\mathbf{P}_Y)(\mathbf{P}_X + i\mathbf{P}_Y)|(\alpha, J, M\rangle \quad (3.244)$$

The matrices for $\mathbf{P}_X, \mathbf{P}_Y$ and $\mathbf{P}_Z$ are Hermitian, that is,

$$\langle \alpha, J, M|\mathbf{P}_F|\alpha', J', M'\rangle = \langle \alpha', J', M'|\mathbf{P}_F|\alpha, J, M\rangle^* \qquad (F = X, Y, Z) \quad (3.245)$$

Therefore

$$\langle \alpha, J, M|\mathbf{P}_X \pm i\mathbf{P}_Y|\alpha', J', M'\rangle = \langle \alpha', J', M'|\mathbf{P}_X \mp i\mathbf{P}_Y|\alpha, J, M\rangle \quad (3.246)$$

and $\mathbf{P}_X \pm i\mathbf{P}_Y$ is not Hermitian. The diagonal elements of the $\mathbf{P}_-\mathbf{P}_+$ matrix become

$$\begin{aligned}\langle \alpha, J, M|\mathbf{P}_-\mathbf{P}_+|\alpha JM\rangle &= \sum \langle \alpha, J, M|\mathbf{P}_-|\alpha', J', M'\rangle\langle \alpha', J', M'|\mathbf{P}_+|\alpha, J, M\rangle \\ &= \sum \langle \alpha, J, M|\mathbf{P}_+|\alpha, J, M\rangle^2 \end{aligned} \quad (3.247)$$

The square of the absolute value of any number must be real and positive, therefore the eigenvalues of $\mathbf{P}_-\mathbf{P}_+$ are real and positive. Hence, $\mathcal{J} - \mathcal{M}^2 - \mathcal{M}\hbar \leq 0$. However,

$$\mathcal{J} = \langle \alpha, J, M|\mathbf{P}^2|\alpha, J, M\rangle = \sum_{F=XYZ} \langle \alpha, J, M|\mathbf{P}_F^2|\alpha, J, M\rangle \quad (3.248)$$

and $\mathbf{P}_X$, $\mathbf{P}_Y$, $\mathbf{P}_Z$ are Hermitian with real positive eigenvalues. Therefore $\mathcal{J}$ is also real and positive so

$$\mathcal{J} \geqq \mathcal{M}^2 + \mathcal{M}\hbar \quad (3.249)$$

For any given eigenvalue of P^2 there will be an upper limit of the eigenvalues of $\mathbf{P}_Z$. Denoting the upper limiting eigenvalue of $\mathbf{P}_Z$ by $\mathcal{M}_{\max}$, with quantum number $M_{\max}$, and realizing that any state $|\alpha, J, M'\rangle$ with M' greater than $M_{\max}$ does not exist, this upper limit is defined by

$$\mathbf{P}_+|\alpha, J, M_{\max}\rangle = 0 \quad (3.250)$$

Therefore

$$\mathbf{P}_-\mathbf{P}_+|\alpha, J, M_{\max}\rangle = (\mathcal{J} - \mathcal{M}_{\max}^2 - \mathcal{M}_{\max}\hbar)|\alpha, J, M_{\max}\rangle \quad (3.251)$$

Since the state $|\alpha, J, M_{\max}\rangle$ exists it follows that

$$\mathcal{J} - \mathcal{M}_{\max}^2 - \mathcal{M}_{\max}\hbar = 0 \quad (3.252)$$

or

$$\mathcal{J} = \mathcal{M}_{\max}(\mathcal{M}_{\max} - \hbar) \quad (3.253)$$

Using the reverse operator $\mathbf{P}_+\mathbf{P}_-$ it can be shown, in an analogous manner, that there is a lower limit to the eigenvalues of $\mathbf{P}_Z$ such that

$$\mathcal{J} = \mathcal{M}_{\min}(\mathcal{M}_{\min} + \hbar) \quad (3.254)$$

Since the eigenvalues have been shown to be symmetrical about the origin

$$\mathcal{M}_{\max} = -\mathcal{M}_{\min} \tag{3.255}$$

The eigenvalues of $\mathbf{P}_Z$ are multiples of the fundamental unit of momentum $\hbar$, so we let $\mathcal{M}_{\max} = J$ where J is an integer, and $\mathcal{M} = M\hbar$ where M is an integer.

The properties of angular momentum operators are independent of the Hamiltonian of the system, so there is no need to consider the energy eigenvalue equation of energy eigenstates any further at this time, and a simplified notation for the eigenstates of the angular momentum will be adopted. The eigenvalue equations will be written

$$\mathbf{P}^2|J, M\rangle = J(J+1)\hbar^2|J, M\rangle \tag{3.256}$$

and

$$\mathbf{P}_Z|J, M\rangle = M\hbar|J, M\rangle \tag{3.257}$$

Looking back, Eq. (3.232) becomes

$$P_Z P_+|J, M\rangle = (M+1)\hbar\mathbf{P}_+|J, M\rangle \tag{3.258}$$

Examination of this equation shows that $\mathbf{P}_+|J, M\rangle$ is an eigenstate of $\mathbf{P}_Z$ with eigenvalue $(M+1)\hbar$ or

$$\mathbf{P}_+|J, M) = \hbar[J(J-1) - M(M+1)]^{1/2}|J, M+1\rangle \tag{3.259}$$

The effect of operation on a state with the $\mathbf{P}_+$ operator is to raise it to the next higher state. The effect of $\mathbf{P}_-$ is to lower a state to the next lower one

$$\mathbf{P}_-|J, M) = \hbar[J(J+1) - M(M-1)]^{1/2}|J, M-1\rangle \tag{3.260}$$

$\mathbf{P}_+$ and $\mathbf{P}_-$ are called *stepping* or *raising and lowering operators.*

The matrix elements of the stepping operators are used to evaluate the matrix element for $\mathbf{P}_X$ and $\mathbf{P}_Y$. Operating on the state $|J, M\rangle$ with $\mathbf{P}_-\mathbf{P}_+$ gives

$$\mathbf{P}_-\mathbf{P}_+|J, M\rangle = (P^2 - P_Z^2 - \hbar\mathbf{P}_Z)|J, M\rangle = [J(J+1) - M(M+1)]\hbar^2|J, M\rangle \tag{3.261}$$

Left-multiplying by the complex conjugate $\langle J, M|$ gives the diagonal elements

$$\langle J, M|\mathbf{P}_-\mathbf{P}_+|J, M\rangle = [J(J+1) - M(M+1)]\hbar^2\langle J, M|J, M\rangle \tag{3.262}$$

Expanding these diagonal elements as a matrix product yields

$$\langle J, M|\mathbf{P}_-\mathbf{P}_+|J, M\rangle = \sum_{M'} \langle J, M|\mathbf{P}_-|J, M'\rangle\langle J, M'|\mathbf{P}_+|J, M\rangle \tag{3.263}$$

It is to be noted that the sum is only over M' and not J. This occurs because $\mathbf{P}_+$ and $\mathbf{P}_-$ commute with $\mathbf{P}^2$, thus the matrix elements of $\mathbf{P}_+$ and $\mathbf{P}_-$ are diagonal in J since $\mathbf{P}^2$ is diagonal in J. The stepping properties of $\mathbf{P}_+$ and $\mathbf{P}_-$ dictate that

$$\langle J, M|\mathbf{P}_+|JM'\rangle = 0 \qquad M' \neq m-1 \tag{3.264}$$

$$\langle J, M|\mathbf{P}_-|JM'\rangle = 0 \qquad M' \neq M+1 \tag{3.265}$$

That is, states with different values of M are orthogonal

$$\langle J, M | J, M' \rangle = \delta_{MM'} \tag{3.266}$$

The summation in Eq. (3.263) reduces to

$$\langle J, M | \mathbf{P}_- \mathbf{P}_+ | J, M \rangle = \langle JM | \mathbf{P}_- | J, M+1 \rangle \langle J, M+1) | \mathbf{P}_+ | J, M \rangle \tag{3.267}$$

Taking the complex conjugate of $\langle J, M+1 | \mathbf{P}_+ | J, M \rangle$ and employing the Hermitian character of $\langle J, M | \mathbf{P}_F | J, M' \rangle$,

$$\begin{aligned} \langle J, M+1 | \mathbf{P}_+ | J, M \rangle^* &= \langle J, M+1) | \mathbf{P}_X | J, M \rangle^* + \langle J, M+1 | i\mathbf{P}_Y (J, M \rangle^* \\ &= \langle J, M | \mathbf{P}_X | J, M+1 \rangle + \langle J, M | -iP_X | J, M+1 \rangle \\ &= \langle J, M | \mathbf{P}_- | J, M+1 \rangle \end{aligned} \tag{3.268}$$

and using Eq. (3.262), Eq. (3.263) reduces to

$$\langle J, M | \mathbf{P}_- \mathbf{P}_+ | JM \rangle = \langle J, M+1 | \mathbf{P}_+ | J, M \rangle^2 = [J(J+1) - M(M+1)]\hbar^2 \tag{3.269}$$

or

$$\langle J, M+1 | \mathbf{P}_+ | J, M \rangle = [J(J+1) - M(M+1)]^{1/2}\hbar \tag{3.270}$$

Beginning with $\mathbf{P}_+\mathbf{P}_-$ one can derive the analogous expression

$$\langle J, M | \mathbf{P}_- | J, M+1 \rangle = [J(J+1) - M(M+1)]^{1/2}\hbar \tag{3.271}$$

The form of Eqs. (3.270) and (3.271) follows from the Condon choice of phase [9].[2] Recognizing the raising and lowering properties of $\mathbf{P}_+$ and $\mathbf{P}_-$ and the orthogonality properties of states one finds that

$$\langle J, M | \mathbf{P}_+ | J, M+1 \rangle = 0 \tag{3.272}$$

$$\langle J, M+1 | \mathbf{P}_- | J, M \rangle = 0 \tag{3.273}$$

The matrix elements of $\mathbf{P}_X$ are now found by addition of Eqs. (3.270) and (3.273)

$$\begin{aligned} \langle J, M+1 | \mathbf{P}_X + i\mathbf{P}_Y | J, M \rangle &+ \langle J, M+1 | \mathbf{P}_X - i\mathbf{P}_Y | J, M \rangle \\ &= [J(J+1) - M(M+1)]^{1/2}\hbar \end{aligned} \tag{3.274}$$

or

$$\langle J, M+1 | \mathbf{P}_X | J, M \rangle = \frac{\hbar}{2}[J(J+1) - M(M+1)]^{1/2} \tag{3.275}$$

[2] The general nature of the raising and lowering operators is expressed as $\mathbf{P}_\pm | J, M \rangle = N_\pm | J, M \pm 1 \rangle$ where $N_\pm$ are the normalizing factors for the $| J, M \pm 1 \rangle$ states when $| J, M \rangle$ is normalized such that $\langle J, M | J, M \rangle = 1$. Since any two states are not distinct unless they are linearly independent the factors $N_\pm$ will contain an arbitrary phase factor, exp $(i\delta)$, such that $N_\pm = \hbar[(J \mp M)(J \pm M + 1)]^{1/2} \exp(i\delta)$ and δ is an arbitrary real number. Any results of physical significance will not be affected by this arbitrary choice of phase so any convenient explicit choice can be made. The elements defined by Eqs. (3.270) and (3.271) for $P_\pm$ result from the choice of $\delta = 0$.

Those for $\mathbf{P}_Y$ result from subtraction of the same equations

$$\langle J, M+1|P_Y|J, M\rangle = \frac{-i\hbar}{2}[J(J+1)-M(M+1)]^{1/2} \tag{3.276}$$

Similarly the addition of Eqs. (3.271) and (3.272) gives

$$\langle J, M|\mathbf{P}_X|J, M+1\rangle = \frac{\hbar}{2}[J(J+1)-M(M+1)]^{1/2} \tag{3.277}$$

and subtraction gives

$$\langle J, M|\mathbf{P}_Y|J, M+1\rangle = \frac{i\hbar}{2}[J(J+1)-M(M+1)]^{1/2} \tag{3.278}$$

A comparison of Eq. (3.276) with (3.278) and of Eq. (3.275) with (3.277) illustrates the Hermitian properties of the matrix elements

$$\langle J, M|\mathbf{P}_F|J, M+1\rangle = \langle J, M+1|\mathbf{P}_F|J, M\rangle^* = \langle J, M+1|P_F^*|J, M\rangle \tag{3.279}$$

It was previously found that angular momentum may be referred to a set of body-fixed axes as well as those fixed in space. The commutation rules for the body-fixed momentum components P_g $(g = x, y, z)$ differ from those for the space-fixed components by a sign [Eqs. (3.152) to (3.154)]. Using these commutation relationships and letting K be the quantum number representing the component of angular momentum along the body-fixed z axis, the nonzero matrix elements relative to the body-fixed axes are derived in a manner analogous to those for the space-fixed axes.[3] The results of such derivations, along with a summary of the previous discussion regarding the elements relative to the space-fixed axes, are presented in Table 3.1. In this summary the notation has been expanded to include both the K and M quantum numbers. When using the values in this table it must be recognized that elements of the form $\langle J, K, M|\mathbf{P}_Y|J, K, M-1\rangle$ may be found by letting $M \to M-1$ in Eq. (3.276), giving $\langle J, K, M|\mathbf{P}_Y|J, K, M-1\rangle = (-i\hbar/2)\,[J(J+1)-M(M-1)]^{1/2}$.

A comparison of Table 3.1 with the comparable tables in other sources shows that sometimes the role of X and Y are interchanged. This is equivalent to making a choice of different phase factors for the matrix elements. It is important to be consistent within any calculation even though most practical problems involve squares of angular momentum components and the physical results are not affected.

---

[3] Due to the sign reversal between the commutators of space and body fixed components of angular momenta the operator $P_+ = P_x + iP_y$ will be a lowering operator while $P'_+ = P_y + iP_x$ is a raising operator.

TABLE 3.1
Angular momentum matrix elements

$\langle J, K, M \vert \mathbf{P}_Z \vert J, K, M\rangle = M\hbar$	$\langle J, K, M \vert \mathbf{P}_z \vert J, K, M\rangle = K\hbar$
$\langle J, K, M \vert \mathbf{P}_Z \vert J, K, M'\rangle = 0 \qquad (M \neq M')$	$\langle J, K, M \vert \mathbf{P}_z \vert J, K', M\rangle = 0 \qquad (K \neq K')$
$\langle J, K, M \vert \mathbf{P}_X \vert J, K, M \pm 1\rangle = \frac{\hbar}{2}[J(J+1) - M(M \pm 1)]^{1/2}$	$\langle J, K, M \vert \mathbf{P}_x \vert J, K \pm 1, M\rangle = \frac{\hbar}{2}[J(J+1) - K(K \pm 1)]^{1/2}$
$\langle J, K, M \pm 1 \vert \mathbf{P}_X \vert J, K, M\rangle = \frac{\hbar}{2}[J(J+1) - M(M \pm 1)]^{1/2}$	$\langle J, K \pm 1, M \vert \mathbf{P}_x \vert J, K, M\rangle = \frac{\hbar}{2}[J(J+1) - K(K \pm 1)]^{1/2}$
$\langle J, K, M \vert \mathbf{P}_Y \vert J, K, M \pm 1\rangle = \frac{\pm\hbar i}{2}[J(J+1) - M(M \pm 1)]^{1/2}$	$\langle J, K, M \vert \mathbf{P}_y \vert J, K \pm 1, M\rangle = \frac{\mp\hbar i}{2}[J(J+1) - K(K \pm 1)]^{1/2}$
$\langle J, K, M \pm 1 \vert \mathbf{P}_Y \vert J, K, M\rangle = \frac{\mp\hbar i}{2}[J(J+1) - M(M \pm 1)]^{1/2}$	$\langle J, K \pm 1, M \vert \mathbf{P}_y \vert J, K, M\rangle = \frac{\pm\hbar i}{2}[J(J+1) - K(K \pm 1)]^{1/2}$
$\langle J, K, M \vert \mathbf{P}_X \vert J, K, M\rangle = 0$	$\langle J, K, M \vert \mathbf{P}_x \vert J, K, M\rangle = 0$
$\langle J, K, M \vert \mathbf{P}_Y \vert J, K, M\rangle = 0$	$\langle J, K, M \vert \mathbf{P}_y \vert J, K, M\rangle = 0$
$\langle J, K, M \vert \mathbf{P}_- \vert J, K, M+1\rangle = [J(J+1) - M(M+1)]^{1/2}\hbar$	$\langle J, K, M \vert \mathbf{P}_- \vert J, K+1, M\rangle = [J(J+1) - K(K+1)]^{1/2}\hbar$
$\langle J, K, M+1 \vert \mathbf{P}_+ \vert J, K, M\rangle = [J(J+1) - M(M+1)]^{1/2}\hbar$	$\langle J, K+1, M \vert \mathbf{P}_+ \vert J, K, M\rangle = [J(J+1) - K(K+1)]^{1/2}\hbar$
$\langle J, K, M+1 \vert \mathbf{P}_- \vert J, K, M\rangle = 0$	$\langle J, K+1, M \vert \mathbf{P}_- \vert J, K, M\rangle = 0$
$\langle J, K, M \vert \mathbf{P}_+ \vert J, K, M+1\rangle = 0$	$\langle J, K, M \vert \mathbf{P}_+ \vert J, K+1, M\rangle = 0$

$$\langle JKM \vert \mathbf{P}^2 \vert JKM\rangle = J(J+1)\hbar^2$$

3.9 DIRECTION COSINE MATRIX ELEMENTS

In this section we present an introduction to the determination of direction cosine matrix elements. While this is a very informative section for the student who wishes to develop a more complete understanding of matrix mechanics its omission will not deter the reader from having sufficient background to follow the applications of these elements in later chapters.

Having established the commutation relationships for the direction cosines in Sec. 3.7, their matrix elements are found by considering them as vectors and using the methods of noncommuting vector algebra. When considering the direction cosine matrix elements it must be kept in mind that they relate to axis systems and will be designated by the set of three quantum numbers J, K, and M. The development of these matrix elements is a lengthy procedure but one which not only provides a substantial exercise in matrix manipulation but also establishes the basis of a number of concepts which are seldom encountered other than as a table of results.

As an example of the method the evaluation of the K dependency of the $\boldsymbol{\phi}_{Fx}$, $\boldsymbol{\phi}_{Fy}$, and $\boldsymbol{\phi}_{Fz}$ elements will be used.

As with the evaluation of the angular momentum matrix elements it is expedient to employ stepping operators. To determine the dependency of $\langle J, K, M|\boldsymbol{\phi}_{Fg}|J'K'M'\rangle$ on K the two operators[4] $\mathbf{P}'_- = \mathbf{P}_y - i\mathbf{P}_x$ and $\boldsymbol{\Phi}_- = \boldsymbol{\Phi}_{Fy} - i\boldsymbol{\Phi}_{Fx}$ are employed. By use of the commutator relations for $\boldsymbol{\Phi}_{Fg}$ and $\mathbf{P}_g$ [Eqs. (3.206), (3.207), and (3.208)], it is found that $\mathbf{P}'_-$ and $\boldsymbol{\Phi}_-$ commute.

The nonvanishing matrix elements of $\mathbf{P}'$ are

$$\langle J, K-1, M|\mathbf{P}'_-|J, K, M\rangle = -i\hbar[J(J+1) - K(K-1)]^{1/2} \tag{3.280}$$

Equating the Kth element of $\mathbf{P}'_-\boldsymbol{\Phi}_- = \boldsymbol{\Phi}_-\mathbf{P}'_-$

$$\begin{aligned}\langle J, K-1, M|\mathbf{P}'_-|J', K, M'\rangle\langle J', K, M'|\boldsymbol{\Phi}_-|J, K+1, M\rangle \\ = \langle J, K-1, M|\boldsymbol{\Phi}_-|J', K, M'\rangle\langle J', K, M'|\mathbf{P}'_-|J, K+1, M\rangle\end{aligned} \tag{3.281}$$

and introducing Eq. (3.280), the elements of $\boldsymbol{\Phi}_-$ can be related:

$$\begin{aligned}[J(J+1) - K(K+1)]^{1/2}\langle J, K, M|\boldsymbol{\Phi}_-|J', K+1, M'\rangle \\ = \langle J, K-1, M|\boldsymbol{\Phi}_-|J', K, M'\rangle[J'(J'+1) - K(K+1)]^{1/2}\end{aligned} \tag{3.282}$$

We must now digress briefly from the main object of finding the K dependency to that of finding the allowed values of J'. This is done in order to limit the evaluation of the elements to those with values of J' which do not lead to zero values for the elements. The procedure involves the evaluation of the second commutator of $\mathbf{P}^2$ and $\boldsymbol{\Phi}$

$$[\mathbf{P}^2, [\mathbf{P}^2, \boldsymbol{\Phi}]] = \mathbf{P}^2(\mathbf{P}^2\boldsymbol{\Phi} - \boldsymbol{\Phi}\mathbf{P}^2) - (\mathbf{P}^2\boldsymbol{\Phi} - \boldsymbol{\Phi}\mathbf{P}^2)\mathbf{P}^2 \tag{3.283}$$

where $\boldsymbol{\Phi} = \boldsymbol{\Phi}_{Fx}\hat{\mathbf{i}} + \boldsymbol{\Phi}_{Fy}\hat{\mathbf{i}} + \boldsymbol{\Phi}_{Fz}\hat{\mathbf{k}}$. The first commutator of $\mathbf{P}^2$ and $\boldsymbol{\Phi}_{Fz}$ is

$$\mathbf{P}^2\boldsymbol{\Phi}_{Fz} - \boldsymbol{\Phi}_{Fz}\mathbf{P}^2 = (\mathbf{P}_x^2 + \mathbf{P}_y^2 + \mathbf{P}_z^2)\boldsymbol{\Phi}_{Fz} - \boldsymbol{\Phi}_{Fz}(\mathbf{P}_x^2 + \mathbf{P}_y^2 + \mathbf{P}_z^2) \tag{3.284}$$

Since $\mathbf{P}_z\boldsymbol{\Phi}_{Fz} = \boldsymbol{\Phi}_{Fz}\mathbf{P}_z$ it follows that

$$\mathbf{P}^2\boldsymbol{\Phi}_{Fz} - \boldsymbol{\Phi}_{Fz}\mathbf{P}^2 = (\mathbf{P}_x^2\boldsymbol{\Phi}_{Fz} - \boldsymbol{\Phi}_{Fz}\mathbf{P}_x^2) + (\mathbf{P}_y^2\boldsymbol{\Phi}_{Fz} - \boldsymbol{\Phi}_{Fz}\mathbf{P}_y^2) \tag{3.285}$$

Evaluating the right-hand terms by employing commutation relations gives

$$\begin{aligned}\mathbf{P}_x^2\boldsymbol{\Phi}_{Fz} - \boldsymbol{\Phi}_{Fz}\mathbf{P}_x^2 &= \mathbf{P}_x(i\hbar\boldsymbol{\Phi}_{Fy} + \boldsymbol{\Phi}_{Fz}\mathbf{P}_x) - (\mathbf{P}_x\boldsymbol{\Phi}_{Fy} - i\hbar\boldsymbol{\Phi}_{Fy})\mathbf{P}_x \\ &= i\hbar(\mathbf{P}_x\boldsymbol{\Phi}_{Fy} + \boldsymbol{\Phi}_{Fy}\mathbf{P}_x)\end{aligned} \tag{3.286}$$

Likewise

$$\mathbf{P}_y^2\boldsymbol{\Phi}_{Fz} - \boldsymbol{\Phi}_{Fz}\mathbf{P}_y^2 = -i\hbar(\mathbf{P}_y\boldsymbol{\Phi}_{Fx} + \boldsymbol{\Phi}_{Fx}\mathbf{P}_y) \tag{3.287}$$

---

[4] Note that in the expressions for $\mathbf{P}'_-$ and $\boldsymbol{\Phi}_-$ the phase choice results in an interchange of x and y relative to the prior development. This will eventually lead to an interchange in Table 3.2 relative to the tables in some presentations. Since we generally are interested in the squares of terms this is of insignificant consequence.

Combining these latter two quantities into Eq. (3.285) yields

$$\mathbf{P}^2\mathbf{\Phi}_{Fz} - \mathbf{\Phi}_{Fz}\mathbf{P}^2 = i\hbar[(\mathbf{P}_x\mathbf{\Phi}_{Fy} + \mathbf{\Phi}_{Fy}\mathbf{P}_x) - (\mathbf{P}_y\mathbf{\Phi}_{Fx} + \mathbf{\Phi}_{Fx}\mathbf{P}_y)] \tag{3.288}$$

An analogous treatment of the x and y components of $\mathbf{\Phi}_{Fg}$ leads to similar relationships which, along with that just derived, can be designated in vector terminology by

$$\mathbf{P}^2\mathbf{\Phi} - \mathbf{\Phi}\mathbf{P}^2 = i\hbar(\mathbf{P}X\mathbf{\Phi} - \mathbf{\Phi}X\mathbf{P}) \tag{3.289}$$

where $\mathbf{P} = \mathbf{P}_x\mathbf{i} + \mathbf{P}_y\mathbf{j} + \mathbf{P}_z\mathbf{k}$.

The second commutator can then be reduced by repeated applications of the commutation rules for $\mathbf{P}_g$ and $\mathbf{\Phi}_{Fg}$ to

$$\begin{aligned}[\mathbf{P}^2, [\mathbf{P}^2, \mathbf{\Phi}]] &= \mathbf{P}^4\mathbf{\Phi} - 2\mathbf{P}^2\mathbf{\Phi}\mathbf{P}^2 + \mathbf{\Phi}\mathbf{P}^4 \\ &= 2\hbar^2(\mathbf{P}^2\mathbf{\Phi} + \mathbf{\Phi}\mathbf{P}^2) - 4\hbar^2\mathbf{P}(\mathbf{P}\cdot\mathbf{\Phi})\end{aligned} \tag{3.290}$$

Examining the J, K, M; J', K', M' element of the second commutator it is found that

$$\begin{aligned}&\langle J, K, M|\mathbf{P}^4\mathbf{\Phi} - 2\mathbf{P}^2\mathbf{\Phi}\mathbf{P}^2 + \mathbf{\Phi}\mathbf{P}^4|J', K', M'\rangle \\ &\quad = 2\hbar^2\langle J, K, M|\mathbf{P}^2\mathbf{\Phi} + \mathbf{\Phi}\mathbf{P}^2|J'K'M'\rangle - 4\hbar^2\langle J, K, M|\mathbf{P}(\mathbf{P}\cdot\mathbf{\Phi})|J', K', M'\rangle\end{aligned} \tag{3.291}$$

Since the matrix elements for angular momentum are known, this equation can be separated into a known and an unknown part. The expression is kept simple by remembering that the only nonzero elements of $\mathbf{P}^2$ are those with $J, K, M = J'K'M'$

$$\begin{aligned}&[\langle J, K, M|\mathbf{P}^4|J, K, M\rangle\langle J, K, M|\mathbf{\Phi}|J', K', M'\rangle - 2\langle J, K, M|\mathbf{P}^2|J, K, M\rangle \\ &\quad \times\langle J, K, M|\mathbf{\Phi}|J', K', M'\rangle\langle J', K', M'|\mathbf{P}^2|J', K', M'\rangle \\ &\quad + \langle J, K, M|\mathbf{\Phi}|J', K', M'\rangle\langle J', K', M'|\mathbf{P}^2|J'K'M'\rangle] \\ &= [2\hbar^2\langle J, K, M|P^2|J, K, M\rangle\langle JKM|\mathbf{\Phi}|J'K'M'\rangle + \langle J, K, M|\mathbf{\Phi}|J', K', M'\rangle \\ &\quad \times\langle J', K', M'|\mathbf{P}^2|J', K', M'\rangle - 4\hbar^2\langle J, K, M|\mathbf{P}(\mathbf{P}\cdot\mathbf{\Phi})|J', K', M'\rangle]\end{aligned} \tag{3.292}$$

Since $\langle J, K, M|\mathbf{P}^2|J, K, M\rangle = J(J+1)\hbar^2$ this equation reduces to

$$\langle J, K, M|\mathbf{\Phi}|J, K', M'\rangle = \frac{\langle J, K, M|\mathbf{P}(\mathbf{P}\cdot\mathbf{\Phi})|J, K', M'\rangle}{\hbar^2 f(J, J')} \tag{3.293}$$

where $f(J, J') = J^2(J+1)^2 - 2J(J+1)J'(J'+1) + J'^2(J'+1)^2 - 2J(J+1) - 2J'(J'+1)$.

The right-hand numerator is evaluated by developing the commutators of $\mathbf{P}_z$ and $\mathbf{P}^2$ with $\mathbf{P}\cdot\mathbf{\Phi}$. Since $\mathbf{P}\cdot\mathbf{\Phi} = \mathbf{P}_x\mathbf{\Phi}_{Fx} + \mathbf{P}_y\mathbf{\Phi}_{Fy} + \mathbf{P}_z\mathbf{\Phi}_{Fz}$ the $\mathbf{P}_z$ commutator becomes

$$[\mathbf{P}_z, (\mathbf{P} \cdot \mathbf{\Phi})] = \mathbf{P}_z(\mathbf{P}_x\mathbf{\Phi}_{Fx} + \mathbf{P}_y\mathbf{\Phi}_{Fy} + \mathbf{P}_z\mathbf{\Phi}_{Fz}) - (\mathbf{P}_x\mathbf{\Phi}_{Fx} + \mathbf{P}_y\mathbf{\Phi}_{Fy} + \mathbf{P}_z\mathbf{\Phi}_{Fz})\mathbf{P}_z \tag{3.294a}$$

Introducing the $\mathbf{P}_g$, $\mathbf{\Phi}_{Fg}$ commutation rules shows this to be zero. A similar treatment of the $\mathbf{P}^2$, $\mathbf{\Phi}_{Fg}$ commutator gives an analogous result

$$[\mathbf{P}_z, (\mathbf{P} \cdot \mathbf{\Phi})] = [\mathbf{P}^2, (\mathbf{P} \cdot \mathbf{\Phi})] = 0 \tag{3.294b}$$

Therefore, remembering that the only nonzero element of $\langle J, K, M | \mathbf{P}_z | J', K', M' \rangle$ is for $J, K, M = J', K', M'$

$$\begin{aligned}\langle J, K, M | \mathbf{P}_z(\mathbf{P} \cdot \mathbf{\Phi}) - (\mathbf{P} \cdot \mathbf{\Phi})\mathbf{P}_z | J', K', M' \rangle \\ = [\langle J, K, M | \mathbf{P}_z | J, K, M \rangle \langle J, K, M | \mathbf{P} \cdot \mathbf{\Phi} | J', K', M' \rangle \\ - \langle J, K, M | \mathbf{P} \cdot \mathbf{\Phi} | J', K', M' \rangle \langle J', K', M' | P_z | J', K', M' \rangle] = 0\end{aligned} \tag{3.295}$$

Since $\langle J, K, M | \mathbf{P}_z | J, K, M \rangle = K\hbar$ it follows that

$$(K - K')\langle J, K, M | \mathbf{P} \cdot \mathbf{\Phi} | J', K', M' \rangle = 0 \tag{3.296}$$

For this equality to be true the matrix element must be zero unless $K = K'$. Evaluating the $[\mathbf{P}^2, (\mathbf{P} \cdot \mathbf{\Phi})]$ commutator gives

$$\begin{aligned}\langle JKM | \mathbf{P}^2(\mathbf{P} \cdot \mathbf{\Phi}) - (\mathbf{P} \cdot \mathbf{\Phi})\mathbf{P}^2 | J'K'M' \rangle \\ = \langle J, K, M | \mathbf{P}^2 | J, K, M \rangle \langle J, K, M \rangle \langle J, K, M | (\mathbf{P} \cdot \mathbf{\Phi}) | J', K', M' \rangle \\ - \langle J, K, M | (\mathbf{P} \cdot \mathbf{\Phi}) | J', K', M' \rangle \langle J', K', M' | \mathbf{P}^2 | J', K', M' \rangle = 0\end{aligned} \tag{3.297}$$

which, since $\langle J, K, M | \mathbf{P}^2 | J, K, M \rangle = J(J+1)\hbar^2$, reduces to

$$[J(J+1) - J'(J'+1)]\langle J, K, M | (\mathbf{P} \cdot \mathbf{\Phi}) | J', K', M' \rangle \tag{3.298}$$

Therefore $\langle J, K, M | \mathbf{P} \cdot \mathbf{\Phi} | J', K', M' \rangle = 0$ unless $J = J'$.

While the determination of allowed elements between states designated with J, J', and K, K' are being investigated it is convenient to extend the discussion to the quantum number M as well. The nonzero elements for M, M' are found by examination of the commutator $[\mathbf{P}_Z, \boldsymbol{\psi}_-]$, where $\boldsymbol{\psi}_- = \mathbf{\Phi}_{Yg} - i\mathbf{\Phi}_{Xg}$. This commutator can be written as

$$[\mathbf{P}_Z, \boldsymbol{\psi}_-] = [\mathbf{P}_Z, (\mathbf{\Phi}_{Yg} - i\mathbf{\Phi}_{Xg})] = [\mathbf{P}_Z, \mathbf{\Phi}_{Yg}] - i[\mathbf{P}_Z, \mathbf{\Phi}_{Xg}] \tag{3.299}$$

which, upon introduction of the commutation relations [Eqs. (3.215) and (3.216)], becomes

$$\mathbf{P}_Z\boldsymbol{\psi}_- - \boldsymbol{\psi}_-\mathbf{P}_Z = -i\hbar\mathbf{\Phi}_{Xg} + \hbar\mathbf{\Phi}_{Yg} = \hbar\boldsymbol{\psi}_- \tag{3.300}$$

Remembering that $\langle J, K, M | \mathbf{P}_Z | J, K, M \rangle = \hbar M$ and looking at the $J, K, M; J', K', M'$ matrix element of $[\mathbf{P}_Z, \boldsymbol{\psi}_-]$ the following condition emerges

$$(M - M' - 1)\langle J, K, M | \boldsymbol{\psi}_- | J', K', M' \rangle = 0 \tag{3.301}$$

It follows that $\langle J, K, M | \boldsymbol{\psi}_- | J'K'M' \rangle = 0$ unless $M' = M - 1$. By repetition of this derivation with ψ_-^* it is found that $\langle J, K, M|\psi_-|J'K'M'\rangle = 0$ unless $M' = M + 1$. Therefore the only nonvanishing elements of $\mathbf{\Phi}_{Yg}$ and $\mathbf{\Phi}_{Xg}$ are those with $M' = M \pm 1$. Since $\mathbf{\Phi}_{Zg}$ commutes with $\mathbf{P}_Z$, the only nonvanishing component $\langle J, K, M|\mathbf{\Phi}_{Zg}|J', K', M'\rangle$ is for $M = M'$.

If the arguments just presented are applied in an analogous manner only starting with P_z and $\mathbf{\Phi}_- = \mathbf{\Phi}_{Fy} - i\mathbf{\Phi}_{Fx}$ it is found that the only nonvanishing elements for $\langle J, K, M|\mathbf{\Phi}_{Fx}|J', K', M'\rangle$ and $\langle J, K, M|\mathbf{\Phi}_{Fy}|J', K', M'\rangle$ are for $K' = K \pm 1$ and that the only nonvanishing element for $\langle J, K, M|\mathbf{\Phi}_{Fz}|J', K', M'\rangle$ is for $K' = K$.

Having established the primary limitations on the relationship of J, K, M to J', K', M' the discussion of the original problem, that of determining the values of the direction cosine matrix elements, can be continued. Returning to Eq. (3.282) consideration will be given independently to the three cases, $J' - J = 0, 1, -1$.

For the case of $J' = J$ Eq. (3.282) becomes

$$\frac{\langle J, K, M|\mathbf{\Phi}_-|J, K+1, M'\rangle}{[J(J+1) - K(K+1)]^{1/2}} = \frac{\langle J, K-1, M|\mathbf{\Phi}_-|J, K, M'\rangle}{[J(J+1) - K(K-1)]^{1/2}} \tag{3.302}$$

When establishing this relationship no restrictions were placed on the value of K, hence these ratios must be independent of K. Using the symbol $\langle J, M|\mathbf{\Phi}_-|J, M'\rangle$ to denote the right-hand K independent ratio the dependence of the matrix elements of $\mathbf{\Phi}_-$ on K may be written

$$\langle J, K, M|\mathbf{\Phi}_-|J, K+1, M'\rangle = [J(J+1) - K(K+1)]^{1/2}\langle J, M|\mathbf{\Phi}_-|J, M'\rangle \tag{3.303}$$

Allowing $J' = J + 1$ in Eq. (3.282) and rearranging the bracketed terms gives

$$[(J+K)(J-K+1)]^{1/2}\langle J, K, M|\mathbf{\Phi}_-|J+1, K+1, M'\rangle$$
$$= [(J+K+2)(J-K+1)]^{1/2}\langle J, K-1, M|\mathbf{\Phi}_-|J+1, K, M'\rangle \tag{3.304}$$

Multiplying this by $[(J+K+1)/(J-K+1)]^{1/2}$ and rearranging yields

$$\frac{\langle J, K, M|\mathbf{\Phi}_-|J+1, K+1, M'\rangle}{[(J+K+2)(J+K+1)]^{1/2}} = \frac{\langle J, K-1, M|\mathbf{\Phi}_-|J+1, K, M'\rangle}{[(J+K)(J+K+1)]^{1/2}} \tag{3.305}$$

Again there are two ratios which are independent of K, so by representing the right-hand ratio with $-\langle J, M|\mathbf{\Phi}_-|J+1, M'\rangle$ the K dependency of $\mathbf{\Phi}_-$ for the case of $J' = J + 1$ is written

$$\langle J, K, M|\mathbf{\Phi}_-|J+1, K+1, M'\rangle$$
$$= -[(J+K+2)(J+K+1)]^{1/2}(JM|\mathbf{\Phi}_-|J+1, M') \tag{3.306}$$

Consideration of the $J' = J - 1$ case by an analogous method, where the necessary multiplication factor is $[(J-K)/(J+K)]^{1/2}$, leads to

$$\langle J, K, M|\mathbf{\Phi}_-|J-1, K+1, M'\rangle = [(J-K)(J-K-1)]^{1/2}\langle J, M|\mathbf{\Phi}_-|J-1, M'\rangle \tag{3.307}$$

where $\langle J, M|\mathbf{\Phi}_-|J-1, M'\rangle$ is a K independent ratio.

By developing analogous relationships for the elements of the $\mathbf{\Phi}_+ = \mathbf{\Phi}_{Fy} + i\mathbf{\Phi}_{Fx}$ matrix the elements for $\mathbf{\Phi}_{Fy}$ and $\mathbf{\Phi}_{Fx}$ may be independently evaluated. These can be developed simultaneously while investigating the dependence of $\mathbf{\Phi}_{Fz}$ on K. Beginning with the commutator $[\mathbf{P}'_+, \mathbf{\Phi}_-]$, where $\mathbf{P}'_+ = \mathbf{P}_y + i\mathbf{P}_x$, expanding and introducing the commutation relations [Eqs. (3.206), (3.207), and (3.208)], it is found that

$$[\mathbf{P}'_+, \mathbf{\Phi}_-] = [\mathbf{P}'_-, \mathbf{\Phi}_-] + 2i[\mathbf{P}_x, \mathbf{\Phi}_-] = 2\hbar\mathbf{\Phi}_{Fz} \tag{3.308}$$

The nonzero matrix elements of $\mathbf{P}_+$ are $\langle J, K, M|\mathbf{P}'_+|J, K-1, M\rangle = \hbar[J(J+1) - K(K-1)]^{1/2}$. Using these elements, along with the known elements for $\mathbf{\Phi}_-$ [Eqs. (3.303), (3.306), and (3.307)], the elements for $\mathbf{\Phi}_{Fz}$ are evaluated directly. Using the case of $J' = J$ and recalling that $\langle J, K, M|\mathbf{\Phi}_{Fz}|J', K', M'\rangle = 0$ for $K \neq K'$; the method of evaluation is illustrated. Using Eq. (3.308)

$$\begin{aligned}2\hbar\langle J, K, M|\mathbf{\Phi}_{Fz}|J, K, M'\rangle &= \langle J, K, M|\mathbf{P}'_+\mathbf{\Phi}_-|J, K, M'\rangle - \langle J, K, M|\mathbf{\Phi}_-\mathbf{P}'_+|J, K, M'\rangle \\ &= \langle J, K, M|\mathbf{P}'_+|J, K-1, M'\rangle\langle J, K-1, M|\mathbf{\Phi}_-|J, K, M'\rangle \\ &\quad - \langle J, K, M|\mathbf{\Phi}_-|J, K+1, M'\rangle\langle J, K+1, M|\mathbf{P}'_+|J, K, M'\rangle\end{aligned} \tag{3.309}$$

Substitution of Eq. (3.303) into this expression along with the relation $\langle J, K+1|\mathbf{P}_y + i\mathbf{P}_x|J, K\rangle = \hbar[J(J+1) - K(K+1)]^{1/2}$ gives

$$2\hbar\langle J, K, M|\mathbf{\Phi}_{Fz}|J, K, M'\rangle = 2\hbar K\langle JM|\mathbf{\Phi}_-|J, M'\rangle \tag{3.310}$$

or

$$\langle J, K, M|\mathbf{\Phi}_{Fz}|J, K, M'\rangle = K\langle J, M|\mathbf{\Phi}_-|J, M'\rangle \tag{3.311}$$

By expansion of the equality

$$-2\hbar\langle J, K, M|\mathbf{\Phi}_{Fz}|J \pm 1, K, M'\rangle = \langle J, K, M|\mathbf{P}'_+\mathbf{\Phi}_- - \mathbf{\Phi}_-\mathbf{P}'_+|J \pm 1, K, M'\rangle \tag{3.312}$$

the analogous relations for $J' = J \pm 1$ are found:

$$\langle J, K, M|\mathbf{\Phi}_{Fz}|J+1, K, M'\rangle = [(J+1)^2 - K^2]\langle J, M|\mathbf{\Phi}_-|J+1, M'\rangle \tag{3.313}$$

$$\langle J, K, M|\mathbf{\Phi}_{Fz}|J-1, K, M'\rangle = [J^2 - K^2]\langle J, M|\mathbf{\Phi}_-|J-1, M'\rangle \tag{3.314}$$

The quantity $\mathbf{\Phi}_{Fz}$ is real, therefore $\langle J, K, M|\mathbf{\Phi}_{Fz}|J, K, M'\rangle = \langle J, K, M'|\mathbf{\Phi}_{Fz}|J, K, M\rangle^*$. Also, since K is real,

$$\langle J, M|\mathbf{\Phi}_-|J, M'\rangle = \langle J, M'|\mathbf{\Phi}_-|J, M\rangle^* = \langle J, M'|\mathbf{\Phi}_+|J, M\rangle \tag{3.315}$$

Likewise using Eqs. (3.313) and (3.314) it is found that

$$\langle J, M|\mathbf{\Phi}_-|J \pm 1, M'\rangle = \langle J \pm 1, M'|\mathbf{\Phi}_+|J, M\rangle \tag{3.316}$$

This means that the $\langle J, M|\mathbf{\Phi}_-|J'M'\rangle$ matrices are Hermitian, and the elements for $\mathbf{\Phi}_+$ can be obtained directly from those for $\mathbf{\Phi}_-$. For example, since

$$\langle J, K, M|\mathbf{\Phi}_-|J+1, K+1, M'\rangle$$
$$= -[(J+K+1)(J+K+2)]^{1/2}[\langle J, M|\mathbf{\Phi}_-|J+1, M'\rangle] \quad (3.317)$$

it follows that

$$\langle J+1, K+1, M'|\mathbf{\Phi}_+|J, K, M\rangle$$
$$= \langle J, K, M|\mathbf{\Phi}_-|J+1, K+1, M'\rangle^*$$
$$= -[(J+K+2)(J+K+1)]^{1/2}\langle J+1, M'|\mathbf{\Phi}_-|J, M\rangle \quad (3.318a)$$

Likewise

$$\langle J-1, K+1, M'|\mathbf{\Phi}_+|J, K, M\rangle$$
$$= \langle J, K, M|\mathbf{\Phi}_-|J-1, K+1, M'\rangle^*$$
$$= [(J-K)(J-K-1)]^{1/2}\langle J-1, M'|\mathbf{\Phi}_-|J, M\rangle \quad (3.318b)$$

By application of the reasoning developed between Sec. 3.9 and here to the evaluation of the matrix elements for the $\mathbf{\Phi}_{Xg}$, $\mathbf{\Phi}_{Yg}$, and Φ_{Zg} it is found, by employing the matrix elements for the angular momentum relative to the space-fixed axis system $\mathbf{P}_Y \pm i\mathbf{P}_X$ and the operators $\boldsymbol{\psi}_\pm = \mathbf{\Phi}_{Yg} \pm i\mathbf{\Phi}_{Xg}$, that an M dependency having the following nonzero elements emerges

$$\langle J, K, M|\mathbf{\Phi}_{Zg}|J, K', M\rangle = M\langle J, K|\boldsymbol{\psi}_-|J, K'\rangle \quad (3.319)$$

$$\langle J, K, M|\mathbf{\Phi}_{Zg}|J\pm 1, K', M\rangle = [(J+\tfrac{1}{2}\pm\tfrac{1}{2})^2 - M^2]^{1/2}\langle J, K|\boldsymbol{\psi}_-|J\pm 1, K'\rangle \quad (3.320)$$

The terms $\langle J, K|\boldsymbol{\psi}_-|J, K'\rangle$ and $\langle J, K|\boldsymbol{\psi}_-|J\pm 1, K'\rangle$ are independent of M. The additional relations

$$\langle J, K|\boldsymbol{\psi}_-|J, K'\rangle = \langle J, K'|\boldsymbol{\psi}_-|J, K\rangle^* = \langle J, K'|\boldsymbol{\psi}_+|J, K\rangle \quad (3.321)$$

$$\langle J, K|\boldsymbol{\psi}_-|J\pm 1, K'\rangle = \langle J\pm 1, K'|\boldsymbol{\psi}_-|J, K\rangle^* = \langle J\pm 1, K'|\boldsymbol{\psi}_+|J, K\rangle \quad (3.322)$$

$$\langle J+1, K', M+1|\boldsymbol{\psi}_+|J, K, M\rangle = \langle J, K, M|\boldsymbol{\psi}_-|J+1, K', M+1\rangle^*$$
$$= [(J+M+2)(J+M+1)]^{1/2}$$
$$\times\langle J+1, K'|\boldsymbol{\psi}_-|J, K\rangle \quad (3.323)$$

$$\langle J-1, K', M+1|\boldsymbol{\psi}_+|J, K, M\rangle = \langle J, K, M|\boldsymbol{\psi}_-|J-1, K', M+1\rangle^*$$
$$= [(J-M)(J-M-1)]^{1/2}$$
$$\times\langle J-1, K'|\boldsymbol{\psi}_-|J, K\rangle \quad (3.324)$$

are formulated by consideration of the Hermitian properties of the $\langle J, K, M|\boldsymbol{\psi}_+|J', K', M'\rangle$ elements.

The dependency of the $\mathbf{\Phi}_{Fg}$ elements on K and M can be found from the elements of $\boldsymbol{\psi}_+$ and $\boldsymbol{\phi}_+$ in a manner analogous to that used in Sec. 3.8 for obtaining

$\mathbf{P}_x$ and $\mathbf{P}_y$ from $\mathbf{P}_+$ and $\mathbf{P}_-$. It has been previously established that $\langle J, K, M|\psi_\pm|J'K'M'\rangle = 0$ unless $M' = M \pm 1$ and $K' = K \pm 1$; therefore all elements of the form $\langle J, K, M|\psi_+|J'K, M\rangle$ will be zero and it follows that the only nonzero elements which are diagonal ($K = K'$, $M = M'$) in K and M will be those for $\boldsymbol{\Phi}_{Fz}$ and $\boldsymbol{\Phi}_{Zg}$. There dependencies have already been shown in Eqs. (3.311) and (3.319). An example will illustrate how the elements for the $\boldsymbol{\Phi}_{Fx}$, $\boldsymbol{\Phi}_{Fy}$, $\boldsymbol{\Phi}_{Xg}$, and $\boldsymbol{\Phi}_{Yg}$ terms are formed. Beginning with

$$\langle J, K, M|\boldsymbol{\Phi}_{Fy} - i\boldsymbol{\Phi}_{Fx}|J+1, K\pm 1, M'\rangle$$
$$= -[(J\pm K+1)(J\pm K+2)]^{1/2}\langle J, M|\boldsymbol{\Phi}_-|J+1, M'\rangle \quad (3.325)$$

and

$$\langle J, K, M|\boldsymbol{\Phi}_{Fy} + i\boldsymbol{\Phi}_{Fx}|J+1, K\pm 1, M'\rangle = 0 \quad (3.326)$$

addition yields

$$\langle J, K, M|\boldsymbol{\Phi}_{Fy}|J+1, K\pm 1, M'\rangle$$
$$= \pm\tfrac{1}{2}[(J\pm K+1)(J\pm K+2)]^{1/2}\langle J, M|\boldsymbol{\Phi}_-|J+1, M'\rangle \quad (3.327)$$

Subtraction gives a similar equation

$$\langle J, K, M|\boldsymbol{\Phi}_{Fx}|J+1, K\pm 1, M'\rangle$$
$$= \pm\frac{i}{2}[(J\pm K+1)(J\pm K+2)]^{1/2}\langle J, M|\boldsymbol{\Phi}_-|J+1, M'\rangle \quad (3.328)$$

Denoting the K dependency by $(J, K|\boldsymbol{\Phi}_{Fg}|J', K')$ and remembering that $\langle J, M|\boldsymbol{\Phi}_-|J', M'\rangle$ is independent of K leads to

$$\langle J, K|\boldsymbol{\Phi}_{Fy}|J+1, K\pm 1\rangle = \pm i\langle J, K|\boldsymbol{\Phi}_{Fx}|J+1, K\pm 1\rangle$$
$$= \mp\tfrac{1}{2}[(J\pm K+1)(J\pm K+2)]^{1/2} \quad (3.329)$$

In a similar manner the following elements are developed

$$\langle J, K|\boldsymbol{\Phi}_{Fy}|J-1, K\pm 1\rangle = \pm i\langle J, K|\boldsymbol{\Phi}_{Fx}|J-1, K\pm 1\rangle$$
$$= \mp\tfrac{1}{2}[(J\mp K)(J\mp K-1)]^{1/2} \quad (3.330)$$

$$\langle J, K|\boldsymbol{\Phi}_{Fy}|J, K\pm 1\rangle = \pm i\langle J, K|\boldsymbol{\Phi}_{Fx}|J, K\pm 1\rangle$$
$$= \tfrac{1}{2}[(J\mp K)(J\pm K+1)]^{1/2} \quad (3.331)$$

The M dependency is summarized by an analogous set of elements where $\boldsymbol{\Phi}_{Fy}$ is replaced by $\boldsymbol{\Phi}_{Yg}$, etc., and K is replaced by M. These are summarized in Table 3.2 following the evaluation of the J dependency.

Elements of the form $\langle J, K, M|\boldsymbol{\Phi}_{Fg}|J', K', M'\rangle$ are dependent on the quantum numbers J, K, and M. In the preceding discussions the nonzero elements and the K and M dependency have been established. Writing such elements as a product of three factors which are individually dependent on the numbers J,

K, and M,

$$\langle J, K, M|\mathbf{\Phi}_{Fg}|J', K', M'\rangle = \langle J|\mathbf{\Phi}_{Fg}|J'\rangle\langle J, K|\mathbf{\Phi}_{Fg}|J', K'\rangle \times \langle J, M|\mathbf{\Phi}_{Fg}|J'M'\rangle \tag{3.332}$$

the only factor remaining to be evaluated is $\langle J|\mathbf{\Phi}_{Fg}|J'\rangle$. Since the quantum number J is characteristic of the total angular momentum rather than any individual space or body-fixed component, the elements $\langle J|\mathbf{\Phi}_{Fg}|J'\rangle$ will be the same for all combinations of F and g. The evaluation of the $\langle J|\mathbf{\Phi}_{Fg}|J'\rangle$ matrix for $J' = J$ will be considered as an example. The algebra involved is rather lengthy so only an outline of the procedure will be given. Beginning with the relationship between angular momentum components, $\mathbf{P}_F = \sum_{g=xyz} \mathbf{\Phi}_{Fg}\mathbf{P}_g$, and letting $F = Z$, the diagonal elements are equated to give

$$\langle J, K, M|\mathbf{P}_Z|J, K, M\rangle = \langle J, K, M|\mathbf{\Phi}_{Zx}\mathbf{P}_x|J, K, M\rangle + \langle J, K, M|\mathbf{\Phi}_{Zy}\mathbf{P}_y|J, K, M\rangle + \langle J, K, M|\mathbf{\Phi}_{Zz}\mathbf{P}_z|J, K, M\rangle \tag{3.333}$$

The left-hand side has previously been found to equal $M\hbar$. The last term on the right side will be used to illustrate the method for evaluation of these terms.

Using the rules for matrix multiplication and recalling that the only nonzero elements for $\langle J, K, M|\mathbf{P}_z|J', K', M'\rangle$ are for $J' = J$, $K' = K$, and $M' = M$, one has

$$\begin{aligned}\langle J, K, M|\mathbf{\Phi}_{Zz}\mathbf{P}_z|J, K, M\rangle &= \langle J, K, M|\mathbf{\Phi}_{Zz}|J, K, M\rangle\langle J, K, M|P_z|J, K, M\rangle \\ &= K\hbar\langle J, K, M|\mathbf{\Phi}_{Zz}|J, K, M\rangle\end{aligned} \tag{3.334}$$

Introduction of Eq. (3.332) leads to

$$\langle J, K, M|\mathbf{\Phi}_{Zz}\mathbf{P}_z|J, K, M\rangle = K\hbar\langle J|\mathbf{\Phi}_{Zz}|J)(J, K|\mathbf{\Phi}_{Zz}\rangle\langle J, M|\mathbf{\Phi}_{Zz}|J, M\rangle \tag{3.335}$$

which, upon substitution of the appropriate K and M dependency, becomes

$$\langle J, K, M|\mathbf{\Phi}_{Zz}\mathbf{P}_z|J, K, M\rangle = 4K^2M\hbar\langle J|\mathbf{\Phi}_{Zz}|J\rangle \tag{3.336}$$

Evaluation of the first two right-hand terms in Eq. (3.333) proceeds along a similar but algebraically more complicated path, and yields as an end result

$$\begin{aligned}&M\hbar[(J-K)(J+K+1)]\langle J|\mathbf{\Phi}_{Zx}|J\rangle \\ &\quad + M\hbar[(J+K)(J-K+1)]\langle J|\mathbf{\Phi}_{Zx}|J\rangle + M\hbar[(J-K)(J+K+1)]\langle J|\mathbf{\Phi}_{Zy}|J\rangle \\ &\quad + M\hbar[(J+K)(J-K+1)]\langle J|\mathbf{\Phi}_{Zy}|J\rangle + 4K^2M\hbar\langle J|\mathbf{\Phi}_{Zz}|J\rangle = M\hbar\end{aligned} \tag{3.337}$$

Since $\langle J|\mathbf{\Phi}_{Fg}|J\rangle$ is the same for all Fg, this expression reduces to

$$\langle J|\mathbf{\Phi}_{Fg}|J\rangle = [4J(J+1)]^{-1} \tag{3.338}$$

The elements $\langle J|\mathbf{\Phi}_{Fg}|J \pm 1\rangle$ are evaluated in an analogous manner and are given in Table 3.2 along with a complete summary of the direction cosine matrix elements.

TABLE 3.2
Direction cosine matrix elements

	J'		
Factor	$J+1$	J	$J-1$
$\langle J \vert \Phi_{Fg} \vert J' \rangle$	$[4(J+1)\{(2J+1)(2J+3)\}^{1/2}]^{-1}$	$[4J(J+1)]^{-1}$	$[4J(4J^2-1)^{1/2}]^{-1}$
$\langle J, K \vert \mathbf{\Phi}_{Fz} \vert J', K \rangle$	$2[(J+1)^2 - K^2]^{1/2}$	$2K$	$-2(J^2-K^2)^{1/2}$
$\langle J, K \vert \mathbf{\Phi}_{Fy} \vert J', K \pm 1 \rangle = \pm i \langle J, K \vert \mathbf{\Phi}_{Fx} \vert J', K \pm 1 \rangle$	$\mp [(J \pm K + 1)(J \pm K + 2)]^{1/2}$	$[(J + K)(J \pm K + 1)]^{1/2}$	$\mp [(J \mp K)(J \mp K - 1)]^{1/2}$
$\langle J, M \vert \mathbf{\Phi}_{Zg} \vert J', M \rangle$	$2[(J+1)^2 - M^2]^{1/2}$	$2M$	$-2(J^2 - M^2)^{1/2}$
$\langle J, M \vert \mathbf{\Phi}_{Yg} \vert J', M \pm 1 \rangle = \pm i \langle J, M \vert \mathbf{\Phi}_{Xg} \vert J', M \pm 1 \rangle$	$\mp [(J \pm M + 1)(J \pm M + 2)]^{1/2}$	$[(J \pm M)(J \pm M + 1)]^{1/2}$	$\mp [(J \mp M)(J \mp M - 1)]^{1/2}$

$$\langle J, K, M \vert \mathbf{\Phi}_{Fg} \vert J', K', M' \rangle = \langle J \vert \mathbf{\Phi}_{Fg} \vert J' \rangle \langle J, K \vert \mathbf{\Phi}_{Fg} \vert J', K' \rangle \langle J, M \vert \mathbf{\Phi}_{Fg} \vert J', M' \rangle$$

3.10 TWO SHORT EXAMPLES

Extensive use will be made of the angular momentum and direction cosine matrix elements in the development of relationships in later chapters, but two brief examples will help to illustrate the ideas just presented. The general utility in the use of these matrix elements lies in the fact that if the energy of a system can be written as a function of the angular momentum or its components, then only matrix algebra is necessary to obtain the energy matrix which may then be diagonalized, if not already so, to get the energy states.

The classical three-dimensional rigid, diatomic rotor has an energy given by

$$E = \frac{P^2}{2I} \tag{3.339}$$

The elements of $\underline{\mathcal{H}}$, the energy matrix, will be given by

$$\langle J, K, M|\mathcal{H}|J'K'M'\rangle = \frac{\langle J, K, M|\mathbf{P}^2|J, K, M\rangle}{2I} \tag{3.340}$$

Reference to Table 3.1 shows that only nonzero elements in $\underline{\mathbf{P}}^2$ are

$$\langle J, K, M|\mathbf{P}^2|J, K, M\rangle = J(J+1)\hbar^2 \tag{3.341}$$

Therefore the only nonzero elements of $\underline{\mathcal{H}}$ will be the diagonal terms

$$\langle J, K, M|\mathcal{H}|J, K, M\rangle = \frac{\hbar^2}{2I}J(J+1) \tag{3.342}$$

and the energy of the quantized diatomic rotor will be

$$E_J = \frac{\hbar^2}{2I}J(J+1) \tag{3.343}$$

in agreement with the value derived using wave mechanics.

Another example is the determination of the energy of the rigid diatomic rotor in an electric field. For simplicity constrain the rotor to the XY plane and apply the electric field along the Y direction. In this case there will be, in addition to the energy of free rotation, a contribution due to the electric dipole moment of the molecule $\boldsymbol{\mu}$, interacting with the electric field $\boldsymbol{\mathcal{E}}$. The total classical energy of the system is given by

$$E = \frac{P^2}{2I} - \boldsymbol{\mu}\cdot\boldsymbol{\mathcal{E}} = \frac{P^2}{2I} - \mu\mathcal{E}\mathbf{\Phi}_{Yz} \tag{3.344}$$

where $\mathbf{\Phi}_{Yz}$ is the cosine of the angle between the applied electric field and the electric dipole (bond) direction. The matrix elements for the first term are given by Eq. (3.342) and for the second term

$$\mathbf{E}_s = -\mu\mathcal{E}\mathbf{\Phi}_{Yz} \tag{3.345}$$

the matrix elements will be

$$\langle J, K, M|\mathbf{E}_s|J', K', M'\rangle = -\mu\mathcal{E}\langle J, K, M|\mathbf{\Phi}_{Yz}|J', K', M'\rangle \tag{3.346}$$

Referring to Table 3.2 the nonzero elements are found to be, considering that $K = 0$ for a diatomic rotor since there is no component of angular momentum about the internuclear axis (z axis),

$$\langle J, M|\mathbf{E}_s|J+1, M\pm 1\rangle = \frac{\pm[(J\pm M+1)(J\pm M+2)]^{1/2}(J+1)\mu\mathscr{E}}{4(J+1)[(2J+1)(2J+3)]^{1/2}} \tag{3.347}$$

$$\langle J, M|\mathbf{E}_s|J-1, M\pm 1\rangle = \frac{\pm[(J\mp M)(J\mp M-1)]^{1/2}2J\mu\mathscr{E}}{4J[4J^2-1]^{1/2}} \tag{3.348}$$

The foremost observation at this point is that the nonzero elements lie off the diagonal, and a matrix diagonalization is necessary to find the energies. Since the purpose at this point was just to illustrate the general application of matrices further discussion of this problem will be deferred.

REFERENCES

A. Specific

1. Heisenberg, W., *Zeit. for Physik*, **33**, 879 (1925).
2. Schrödinger, E., *Ann. der Physik*, **79**, 361, 489 (1926).
3. Dirac, P. A. M., *Quantum Mechanics*, Oxford University Press, London, England, 1958, p. 84.
4. Eckhart, C., *Phys. Rev.*, **28**, 711 (1926). (Equivalence of methods.)
5. Schrödinger, E., *Ann. der Physik*, **79**, 734 (1926).
6. Boas, M. L., *Mathematic Methods in the Physical Sciences*, John Wiley & Sons, New York, NY, 1966, chap. 6.
7. Van Vleck, J. H., *Rev. Mod. Ph.*, **23**, 213 (1951).
8. Kauzmann, W., *Quantum Chemistry*, Academic Press, New York, NY, 1957, p. 256.
9. Condon, E. V., and G. H. Shortley, *Theory of Atomic Spectra*, Cambridge University Press, Cambridge, England, 1935, p. 48.

B. General references

Allen, H. C., Jr., and P. C. Cross, *Molecular Vibrators*, John Wiley and Sons, New York, NY, 1963.

Edmons, A. R., *Angular Momentum in Quantum Mechanics*, Princeton University Press, Princeton, NJ, 1957. (Intermediate Level.)

Feenberg, E., and G. E. Pake, *Quantum Theory of Angular Momentum*, Addison Wesley Publishing Co., Reading, MA, 1953.

Green, H. S., *Matrix Methods in Quantum Mechanics*, Barnes & Noble, Inc., New York, NY, 1965.

Heisenberg, W., *The Physical Principles of the Quantum Theory*, Dover Publications, Inc., New York, NY, 1930.

Rojansky, V., *Introductory Quantum Mechanics*, Prentice-Hall, Inc., Englewood Cliffs, NJ, 1938.

Schiff, L. I., *Quantum Mechanics*, McGraw Hill Book Co., New York, NY, 1955.

Temple, G., *The General Principles of Quantum Theory*, Methuen and Co. Ltd., London, G.B., 1953.

Also see references following Chap. 2.

PROBLEMS

3.1. Determine the Heisenberg matrix elements for P_{01}^2, P_{11}^2, P_{01}, P_{11} for the planar rotor by using the Schrödinger wavefunctions and the differential operator.

3.2. Show that Eqs. (3.33) to (3.35) follow from Eqs. (3.20) to (3.22) by use of matrix algebra.

3.3. Using the classical Poisson bracket evaluate the following commutators:
(*a*) $[\mathbf{H}, \mathbf{x}]$ —one-dimensional harmonic oscillator
(*b*) $[\mathbf{P}_x, \mathbf{x}]$ —one dimensional harmonic oscillator
(*c*) $[\mathbf{P}^2, \mathbf{P}_z]$ —planar rotor

3.4. Show that the $m = n \pm 2$ elements of the $\underline{\mathbf{E}}$ matrix for the harmonic oscillator are zero.

3.5. Diagonalize to terms in ξ the matrix

$$\underline{\mathbf{A}} = \begin{pmatrix} 3+6\xi & 2\xi \\ 3\xi & 2+\xi^2 \end{pmatrix}$$

Find the transformation matrix $\underline{\mathbf{T}}$ such that $\underline{\mathbf{B}} = \underline{\mathbf{T}}^{-1}\underline{\mathbf{AT}}$ where $\mathbf{B}$ is diagonal to terms in ξ.

3.6. Show that

$$y\frac{\partial}{\partial z} - z\frac{\partial}{\partial y} = \sin\phi\frac{\partial}{\partial\theta} + \cot\theta\cos\phi\frac{\partial}{\partial\phi}$$

3.7. Using the Dirac general commutation relationship show that

$$\mathbf{P}_z\mathbf{P}_x - \mathbf{P}_x\mathbf{P}_z = -i\hbar\mathbf{P}_y$$

3.8. Show that the commutator of $\mathbf{P}_+$ and $\mathbf{P}_-$ is $2\hbar\mathbf{P}_z$.

3.9. Find the direction cosines for the M—X bonds relative to the Cartesian axis system for the following:
(*a*) A tetrahedral molecule, MX_4, oriented such the X atoms are on alternate corners of a cube and the Cartesian system is through the cube center and parallel to the edges:
(*b*) A tetrahedral molecule oriented such that one M—X bond is along the negative z direction and one M—X bond is in the xz plane.

3.10. Determine the direction cosines between a fixed-axis system XYZ and one denoted by xyz that has been rotated such that the Euler angles are $\phi = 45°$, $\theta = 30°$, and $\psi = 30°$.

3.11. Using differential operators show that for the rotational wavefunction, $\psi_{2,-2}$

$$\mathbf{P}_+\psi_{2,-2} = 2\hbar\psi_{2,-1}$$

$$\mathbf{P}_-\psi_{2,-2} = 0$$

3.12. Show for the harmonic oscillator that by equating the $(n, n+1)$th elements of

$$m\ddot{x} = -kx$$

one can prove that

$$E_n = nh\nu_0 + E_0$$

3.13. Find the eigenvalues of

$$\begin{pmatrix} 1 & \varepsilon & 4\varepsilon \\ \varepsilon & 2 & \varepsilon \\ 4\varepsilon & \varepsilon & 3 \end{pmatrix}$$

correct to terms in ε^2.

3.14. Compute the Hamiltonian matrix of the harmonic oscillator by matrix algebra starting with the position and momentum matrices.

3.15. Beginning with the state $|J, M\rangle$ and the operator $\mathbf{P}_+\mathbf{P}_-$ show that

$$\langle J, M|\mathbf{P}_-|J, M+1\rangle = [J(J+1) - M(M+1)]^{1/2}\hbar$$

3.16. Show that $\langle J, M+1|\mathbf{P}_-|J, M\rangle = 0$.

3.17. Using the Dirac general commutation relationship show that

(*a*) $[\mathbf{P}_z, \mathbf{\Phi}_{Xz}] = 0$

(*b*) $[\mathbf{P}_z, \mathbf{\Phi}_{Xx}] = -i\hbar\Phi_{Xy}$

3.18. Show that

$$\mathbf{P}^2\mathbf{\Phi}_{Fx} - \mathbf{\Phi}_{Fx}\mathbf{P}^2 = i\hbar[(\mathbf{P}_y\mathbf{\Phi}_{Fz} + \mathbf{\Phi}_{Fz}\mathbf{P}_y) - (\mathbf{P}_z\mathbf{\Phi}_{Fy} + \mathbf{\Phi}_{Fy}\mathbf{P}_z)]$$

3.19. Show that the commutator of $\mathbf{P}_+$ and $\mathbf{\Phi}_-$ is $2\hbar\mathbf{\Phi}_{Fz}$.

CHAPTER 4

MOLECULAR SYMMETRY AND GROUP THEORY

4.1 INTRODUCTION

Central to the description of molecular structure and spectroscopy are the basic concepts of molecular symmetry. Most courses in physical chemistry provide an introduction to the elements of symmetry and the use of symmetry in discussing molecular structure. There are several excellent books available dealing exclusively with the subject, some of which are listed at the end of the chapter. The reader is referred to these for an in-depth study of the subject. Most of these sources place primary emphasis on point group symmetry described by the notation due to Schoenflies. This notation is adequate for discussion of the symmetry of individual molecules, but must undergo considerable extensions in order to be applicable to the symmetry of solid crystalline systems. The subject of crystal symmetry as developed by crystallographers over the past half-century employs a different symbolism, the Hermann-Mauguin notation which, in addition to having some fundamental differences in notation for describing point groups, also reflects the extension of symmetry concepts to a three-dimensional, periodic array of molecules. A short summary of concepts related to symmetry and point groups is presented for review purposes.

The spectroscopist is not limited to studies of the gaseous state but frequently examines the solid state and needs to be familiar with the methods used to describe molecules in the crystalline form. Although this book does not include detailed descriptions of crystallography and crystal structure the need for this familiarity is sufficient that a parallel presentation of the two notations is given. The mathematical method for dealing with the application of symmetry concepts to spectroscopy is group theory. A summary of the essential features of group theory needed for the latter chapters is presented.

4.2 MOLECULAR SYMMETRY—ELEMENTS AND OPERATIONS

The possession of symmetrical attributes appears to be one of the fundamental properties of nature. Nature abounds in a variety of objects that exhibit varying degrees of symmetry. With the exception of minor imperfections, most animals of higher order possess a plane of symmetry which divides the external body into a right- and left-hand side. In many cases, such as the diatoms of the ocean floor, or the snowflakes in winter, even higher elements of symmetry persist. It is not at all surprising then that individual molecules and crystals composed of molecules possess a variety of symmetry elements.

In the discussion of symmetry there are two related terms between which a distinction must be made: *symmetry element* and *symmetry operation.* A symmetry element is a geometrical figure, such as a line or a plane, with respect to which the object being considered displays a certain equivalence. If the figure is a plane, then the equivalence is that of considering it to be a mirror, placing half of the object adjacent to the mirror, and observing the resulting figure to be identical to the original. If the figure is a line, then moving the object around the line by $360°/n$ (n = integer) results in the encountering of equivalent points in the object. A symmetry operation is the motion of an object with respect to a symmetry element such that the new configuration is indistinguishable from the original.

There are two fundamental types of symmetry elements and associated operations. The first of these leaves at least one point in the object unchanged. Various combinations of these elements lead to the formation of point groups and are used for the description of single finite molecules. If one considers an extensive array of molecules then one can have additional symmetry elements which lead to the translation of an object in space. These latter type elements must be considered when discussion crystals and, in combination with the point groups, leads to the formation of space groups. The use of the term "group" implies a collection of symmetry elements and associated operations which obey the formal laws of group theory.

Restricting the discussion to those symmetry elements and operations which leave at least one point in the object unchanged (point symmetry) there are five operations and associated symmetry elements.

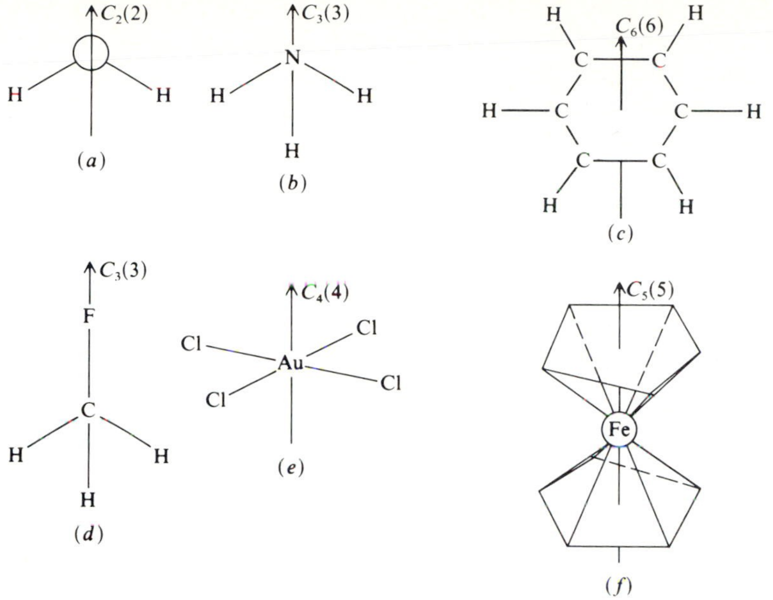

FIGURE 4.1
Molecules having symmetry axes.

1. If rotation of an object about an axis by $360°/n$ produces an identical configuration of the object, then the object possesses an n-fold proper rotational axis. Such an axis is denoted by a C_n in the Schoenflies notation and by an n in the Hermann-Mauguin notation. For the discussions and illustrations in this section the Schoenflies notation will be followed by the Hermann-Mauguin notation in parentheses. Figure 4.1 shows several molecular examples illustrating symmetry axes.
2. If the appearance of an object can be reconstructed by slicing the object in half and placing one half adjacent to a mirror, then it possesses a plane or mirror plane of symmetry. In the Schoenflies notation, such a plane is denoted by σ_v, σ_d, or σ_h while in the Hermann-Mauguin notation an m is used. The subscript on the σ is used to differentiate between different types of symmetry planes. σ_v and σ_d planes are parallel to the highest-fold symmetry axis in a molecule, while a σ_h plane is perpendicular to the highest-fold symmetry axis. Several examples of symmetry planes in molecules are shown in Fig. 4.2.
3. All objects will possess a $C_1(1)$ axis. The symmetry operation associated with this element is the identity operation, or the operation of leaving the object unchanged. The Schoenflies notation for the identity operation is $E \equiv C_1$ while the Hermann-Mauguin notation is just 1.
4. If the substitution of $(-x, -y, -z)$ for all points (x, y, z) in an object leads to an indistinguishable configuration, the object has a center of inversion. In the Schoenflies notation such an element is designated by i while in the Hermann-

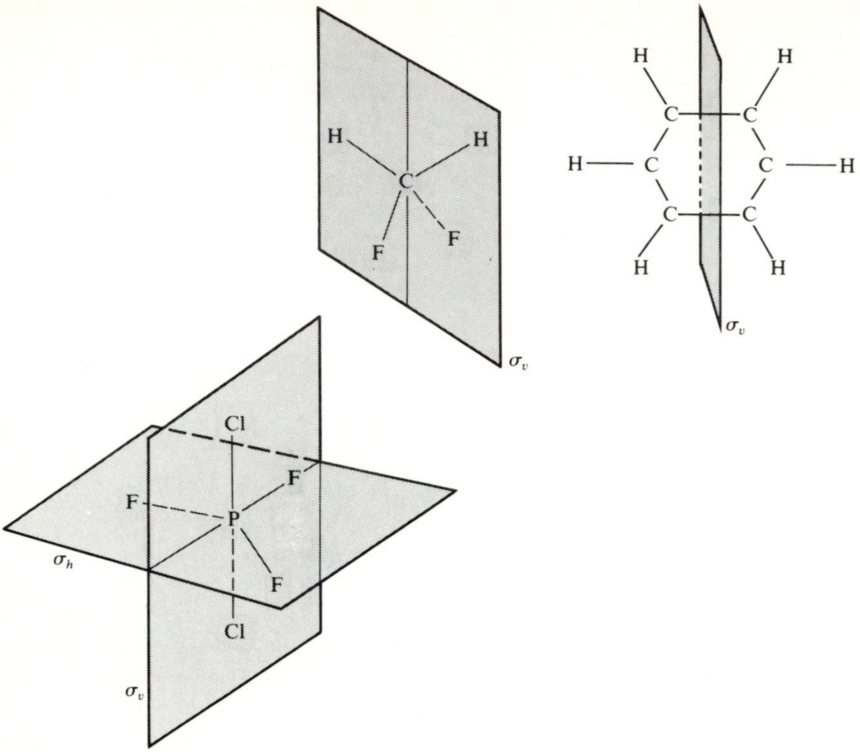

FIGURE 4.2
Symmetry planes in molecules.

Mauguin notation a $\bar{1}$ is used. Examples of molecules having a center of symmetry are shown in Fig. 4.3.

5. The last point symmetry element to be considered is the improper axis. The approach to describing the operation associated with such an axis differs between the Schoenflies and the Hermann-Mauguin notations, not only with regard to symbols but also with regard to the physical performance of the operations involved. The operation associated with an n-fold improper axis in the Schoenflies system is a rotary reflection designated by S_n. This involves the rotation about the axis of $360°/n$ followed by a reflection in a plane perpendicular to the axis. Such an axis is depicted for some simple molecules

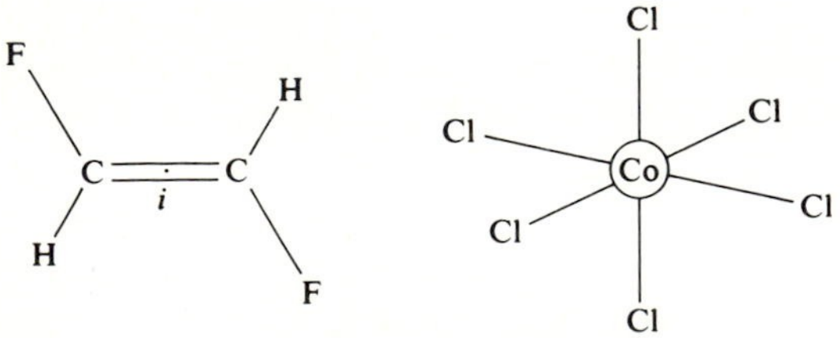

FIGURE 4.3
Molecules with centers of symmetry.

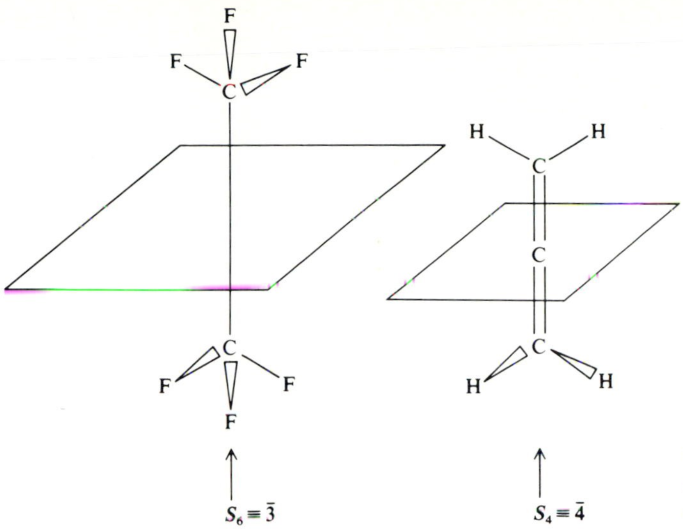

FIGURE 4.4
Improper axes and rotary reflection operations.

in Fig. 4.4. In the Hermann-Mauguin notation, one has a rotary inversion operation which consists of a $360°/n$ rotation about the n-fold improper axis, followed by an inversion of all points through the origin. Such an operation is denoted by $\bar{n}$ and is not necessarily equivalent to an S_n operation. Figure 4.4 also illustrates the occurrence of $\bar{n}$ axes in molecules. The relationship of the S_n and $\bar{n}$ operations is illustrated in more detail in Figs. 4.5 to 4.7.

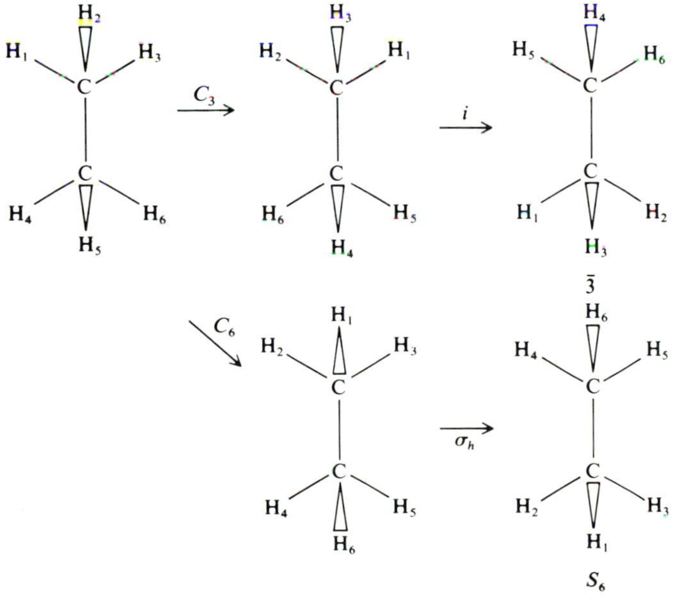

FIGURE 4.5
S_{2n} and $\bar{n}$ operations applied to the transform of ethane.

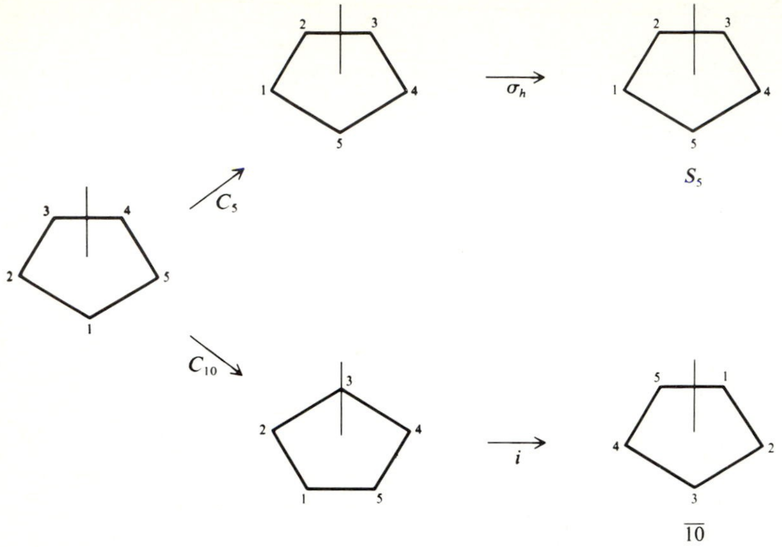

FIGURE 4.6
S_n and $2\bar{n}$ operations applied to cyclopentene. (Rotation axis perpendicular to plane of ring.)

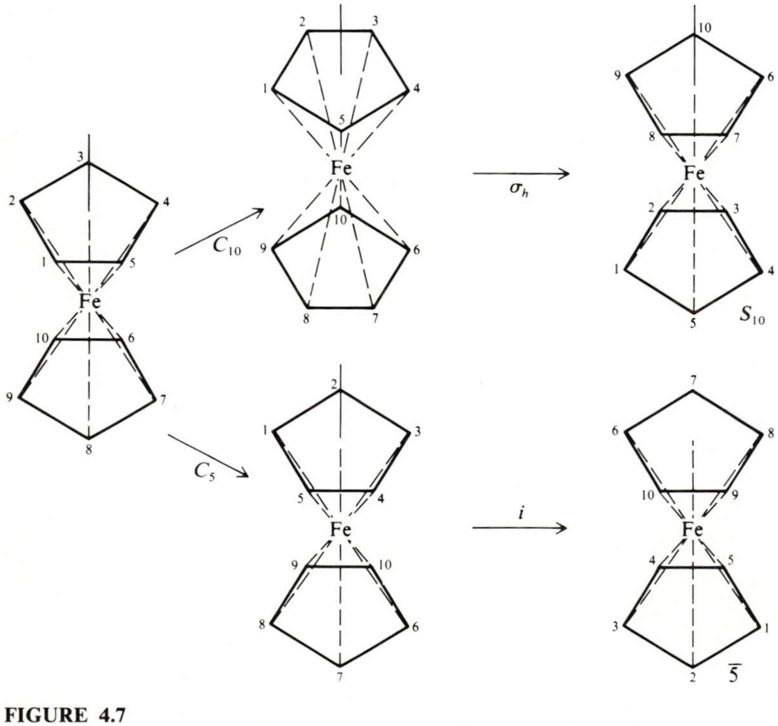

FIGURE 4.7
S_n and $2\bar{n}$ operations applied to ferrocene. (Rotation axis perpendicular to plane of ring.)

Consideration of these examples shows that $S_6 \equiv \bar{3}$, and $S_5 \equiv \overline{10}$. These observations can be generalized to show $S_{2n} \equiv \bar{n}$. The lone exception to this generalization is the case where $n = 4$, in which case $S_4 = \bar{4}$. Also note that $S_2 \equiv \bar{1} \equiv i$.

4.3 MOLECULAR SYMMETRY—POINT GROUPS

Inspection of the examples shown in the earlier figures shows that there are often more symmetry elements in a molecule than those that were specifically mentioned for illustrative purposes. For example, the H_2O molecule in Fig. 4.1 has, in addition to the C_2 axis, a pair of mirror planes parallel to the C_2 axis. The collection of symmetry operations belonging to an object of a particular symmetry is called a point group, and if they are three-dimensional, they constitute a three-dimensional point group. Although only the latter are of use for discussing the symmetry of real molecules, initial consideration of the simple plane point groups aids in understanding the concept of three-dimensional point groups. The set of symmetry operations comprising a point group satisfy the properties of a mathematical group.

In a two-dimensional world there are only two symmetry elements to be considered: the line, the two-dimensional analog of the plane; and the point, the two-dimensional analog of the proper axis. The associated operations consist of a reflection across the line and a rotation about the point. If we begin with the least symmetric planar configuration and progressively add symmetry elements, it is possible to construct a maximum of 12 plane point groups provided the highest symmetry is limited to a 6-fold rotation. Although higher symmetries are encountered in individual molecules, this is sufficient to illustrate the concept. Figure 4.8 shows examples of objects belonging to 13 plane point groups, 12 based on the points through 6-fold and the ∞-fold point. In this figure, the n-fold point is denoted n, and the symmetry lines an σ. Each configuration is captioned with either an $n(n)$ or $n(p)$, depending on whether it possesses only an n-fold point of symmetry or both an n-fold point and planes.

The three-dimensional point groups are developed by combination of the C_n, σ, S_n and i symmetry operations. The occurrence of certain combinations of elements often precludes the occurrence of some other types, or demands the presence of still others. The incorporation of point groups into three-dimensional structures (crystals) leads to further restrictions which should be recognized at this point although this aspect of symmetry will not be discussed in detail.

The nomenclature of a point group generally reflects the highest symmetry present in the system. The symbols used and how the elements are combined in the two different notations are summarized below.

Schoenflies Notation

1. The principal symbol used denotes the axis of highest symmetry:
   C—highest symmetry axis is a proper axis

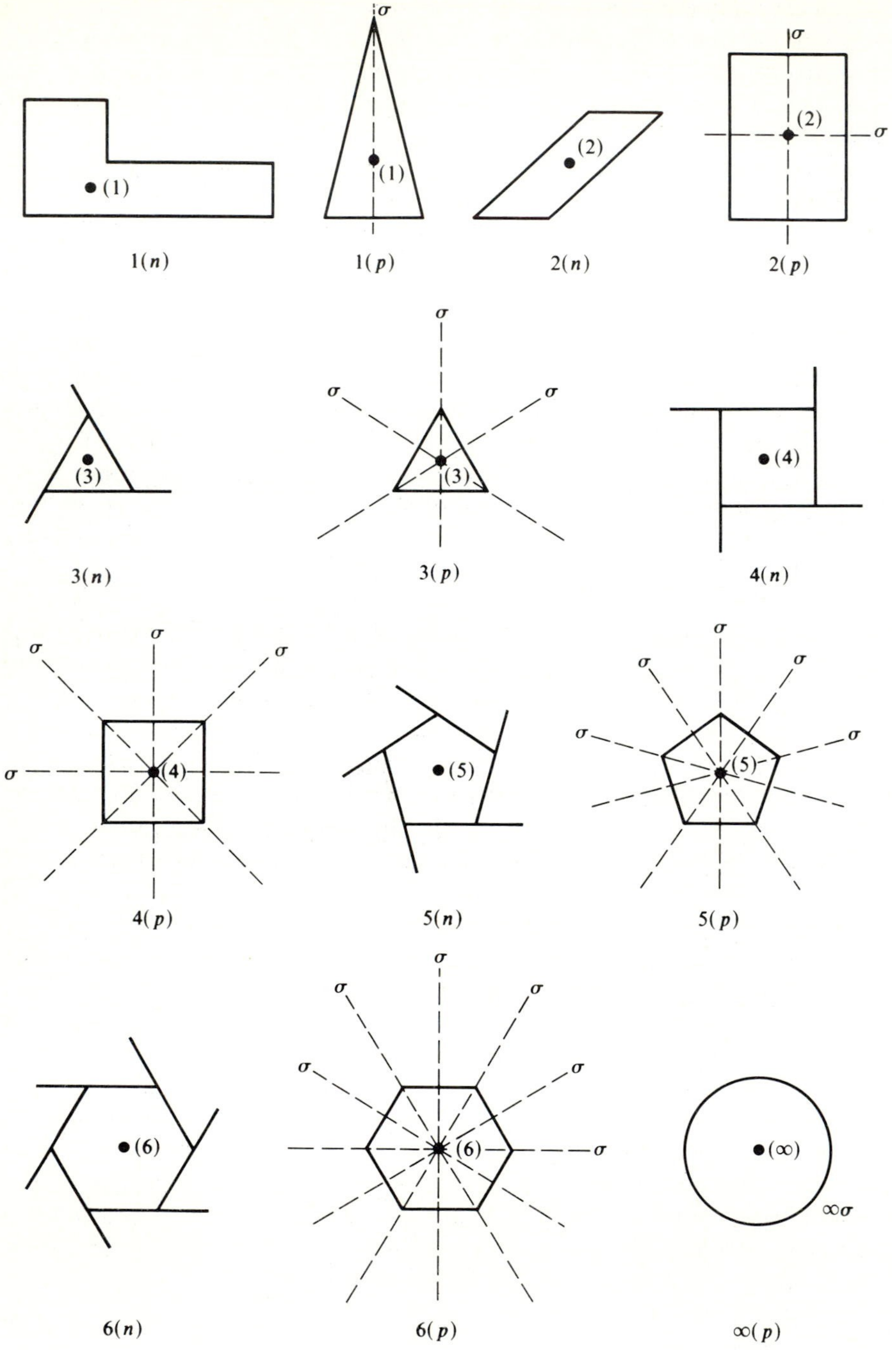

FIGURE 4.8.
Plane point groups.

S—highest symmetry axis is an improper axis
E—highest symmetry element is a proper axis with n 2-fold axes perpendicular to the principle axis
T, O—special symbols for the highly symmetric cubic and octahedral configurations

2. A numerical subscript is used to indicate the order of the axis of highest symmetry.
3. An alphabetic subscript is used to indicate the presence of symmetry planes.
   v, d—planes parallel to the major symmetry axis
   h—plane perpendicular to the major symmetry axis
4. The subscript i denotes the presence of a center of symmetry in the absence of any symmetry planes.
5. The occurrence of either an inversion point or a mirror plane as the only symmetry elements is denoted by C_i or C_s, respectively.
6. Linear molecules belong to the special groups, C_{∞_v} and D_{∞_h}, depending on the absence or presence of a center of symmetry.

Hermann-Mauguin Notation

1. In general the symbol contains three parts, each of which refers to a different direction.
2. The order of axes are denoted by a number n for a proper axis, and $\bar{n}$ for an improper axis.
3. A mirror plane is denoted by an m. The position of the m in the symbol indicates the direction perpendicular to the plane. For example, the symbol, $mm2 = C_{2v}$, denotes a 2-fold axis in the z direction and two mirror planes with their normals in the x and y directions, the planes then being the yz and xz planes.
4. The presence of a horizontal plane of symmetry (σ_h) is denoted by n/m (n over m) where n is the order of the axis perpendicular to the plane. For example, the symbol $4/mmm \equiv D_{4h}$ denotes a 4-fold axis with a perpendicular plane of symmetry and two other planes of symmetry. Note that the entry n/m constitutes a single component of the total symbol.
5. For the less symmetric systems, the symbol does not always contain three parts. For example, if there is only one symmetry axis, then a single number denotes the group.
6. Since this notation is used primarily for crystalline solids the ordering of the symbols is dependent on the basic symmetry of the crystal. For completeness, these rules will be included.

Triclinic. Only a single axis is specified and it is either 1 or $\bar{1}$.

Monoclinic. Only a single unique axis is present.

Orthorhombic. The first component is the 4 or $\bar{4}$ axis and is taken as the z direction. The second component refers to the x and y axes which are mutually perpendicular. The last component refers to the directions of the bisectors of the angles between the x and y axes.

Trigonal-Hexagonal. The first component, either a 3, $\bar{3}$ or 6, $\bar{6}$ refers to the major symmetry axis. The second component refers to the set of equivalent directions. The third component refers to the bisectors of the angles formed by the equivalent directions.

Cubic. The first component refers to the axes of the cube. The second refers to the body diagonal and is always a 3. The third refers to the direction of the face diagonals.

Certain symmetry point groups are forbidden in crystals due to the fact that they cannot be combined with translational elements to produce repetitive structures. Groups with 5-fold and ∞-fold axes are of this type. Such point groups, however, are necessary for discussing individual molecular symmetry. Since the Hermann-Mauguin system was developed for crystallography, it is not commonly used to specify such point groups.

The development of a point group for some of the molecules shown in Fig. 4.1 will illustrate the use of these rules. In addition to the $C_2(2)$ axis shown for H_2O in Fig. 4.1*a* there are two planes of symmetry parallel to the 2-fold axis and perpendicular to one another. These three elements, along with the identity element, constitutes the $C_{2v}(mm2)$ point group. For $AuCl_4^-$ shown in Fig. 4.1*e* there will be in addition to the $C_4(4)$ axis, a $C_2(2)$ axis parallel to $C_4(4)$, four planes parallel to $C_4(4)$, one plane perpendicular to $C_4(4)$, an $S_4(\bar{4})$ axis, four $C_2(2)$ axes perpendicular to $C_4(4)$, and an inversion center giving rise to 16 operations, E, $2C_4(4)$, $C_2(2)$, $2C_2'(2)$, $2C_2''(2)$, $2S_4(\bar{4})$, $\sigma_h(4/m)$, $2\sigma_v(m)$, $2\sigma_d(m)$ and $i(\bar{1})$. Due to their equivalence both $i(\bar{1})$ and $S_4(\bar{4})$ are not included in the group. This set comprises a formal group and is denoted $D_{4h}(4/mmm)$.

A listing of the common point groups along with the symmetry operations (total number $\equiv$ order of group), notations and examples are shown in App. I. The first 32 groups are those common to both free molecules and crystals, and for information have been listed by crystal class. The remaining nine groups occur only in free molecules.

4.4 GROUP THEORY

This section summarizes the important concepts, definitions, and rules relating to the restricted groups that are commonly used in the discussion of molecular structure and spectroscopy. This more restricted definition considers a group to be a set of elements, each of which represents a geometrical operation and all of which obey the following set of mathematical relationships.

1. The group contains, in addition to the elements representing the geometrical operations, the identity element E whose operation leaves the system unchanged.

2. The group also contains
   (*a*) The products of all pairs of elements
   (*b*) The squares of all elements
   (*c*) The reciprocals of all elements
3. All elements obey the associative law of multiplication.
4. All elements commute with the identity operator, but do not necessarily commute with each other.

The total number of elements in a group is called the order of the group. These points are illustrated by the C_{4v} point group having elements

E	identity operation
$C_2 \equiv C_4^2$	rotation by 180° about a four-fold axis
C_4	rotation by 90° about a four-fold axis
C_4^3	rotation by −90° about a four-fold axis
σ_v, σ'_v	reflection in a plane parallel to and including the four-fold axis
σ_d, σ'_d	reflection in a plane parallel to and including the four-fold axis but at 45° to the σ_v planes

and the multiplication table

	E	C_2	C_4	C_4^3	σ_v	σ'_v	σ_d	σ'_d
E	E	C_2	C_4	C_4^3	σ_v	σ'_v	σ_d	σ'_d
C_2	C_2	E	C_4^3	C_4	σ'_v	σ_v	σ'_d	σ_d
C_4	C_4	C_4^3	C_2	E	σ_d	σ'_d	σ'_v	σ_v
C_4^3	C_4^3	C_4	E	C_2	σ'_d	σ_d	σ_v	σ'_v
σ_v	σ_v	σ_v	σ'_v	σ_d	E	C_2	C_4^3	C_4
σ'_v	σ'_v	σ_v	σ_d	σ'_d	C_2	E	C_4	C_4^3
σ_d	σ_d	σ'_d	σ'_v	σ_v	C_4	C_4^3	E	C_2
σ'_d	σ'_d	σ_d	σ_v	σ'_v	C_4^3	C_4	C_2	E

The operations of a group can be divided into classes. A class consists of a complete set of elements that are conjugate with each other. Two elements of a group, A and B, are considered to be conjugate if $A = X^{-1}BX$ and $B = XAX^{-1}$ where X is also a group element. The basic commutation relationships are:

1. Every element is conjugate with itself. That is, there exists at least one element S such that $R = S^{-1}RS$. The sufficient condition for this is (*a*) that $S \equiv E$, or (*b*) that R and S commute, that is, $RS = SR$.
2. If R is conjugate with R' then R' is conjugate with R. In terms of similarity transformation this means that if $R = S^{-1}\ R'S$ then there exists an element T such that $R' = T^{-1}RT$.

Looking at the C_{4v} multiplication table one observes that there are five classes, two of order one and three of order two.

$$E$$

$$C_2$$

$$C_4, C_4^3$$

$$\sigma_v, \sigma_v'$$

$$\sigma_d, \sigma_d'$$

Since the groups of interest are composed of geometrical operations they can be represented by sets of matrices. Such matrix represents a single operation and will transform a vector in accordance with that operation. For the C_{4v} group the matrices are

$$E = \begin{pmatrix} 1 & 0 & 0 \\ 0 & 1 & 0 \\ 0 & 0 & 1 \end{pmatrix} \qquad C_2 = \begin{pmatrix} -1 & 0 & 0 \\ 0 & -1 & 0 \\ 0 & 0 & 0 \end{pmatrix} \tag{4.1}$$

$$C_4 = \begin{pmatrix} 0 & 1 & 0 \\ -1 & 0 & 0 \\ 0 & 0 & 1 \end{pmatrix} \qquad C_4^3 = \begin{pmatrix} 0 & -1 & 0 \\ 1 & 1 & 0 \\ 0 & 0 & 1 \end{pmatrix} \tag{4.2}$$

$$\sigma_v = \begin{pmatrix} -1 & 0 & 0 \\ 0 & 1 & 0 \\ 0 & 0 & 1 \end{pmatrix} \qquad \sigma_v' = \begin{pmatrix} 1 & 0 & 0 \\ 0 & -1 & 0 \\ 0 & 0 & 1 \end{pmatrix} \tag{4.3}$$

$$\sigma_d = \begin{pmatrix} 0 & -1 & 0 \\ -1 & 0 & 0 \\ 0 & 0 & 1 \end{pmatrix} \qquad \sigma_d' = \begin{pmatrix} 0 & 1 & 0 \\ 1 & 0 & 0 \\ 0 & 0 & 1 \end{pmatrix} \tag{4.4}$$

For any group it is possible to devise an unlimited number of representations. If, for a representation there exists a similarity transformation that will transform all matrices of the representation into ones consisting of comparable diagonal block structures, then that representation is reducible. If no such transformation exists for a given representation then it is irreducible.

The important group theoretical relationships necessary for use in molecular structure and spectroscopy are derived from the orthogonality theorem.

$$\sum_R \Gamma_j^*(R)_{nm}\Gamma_k(R)_{n'm'} = \frac{h}{l_j l_k}\,\delta_{jk}\delta_{nn'}\delta_{mm'} \tag{4.5}$$

where R = group operation

$\Gamma_j(R)_{nm}$ = nmth matrix element of the matrix representing the jth irreducible representation

l_j = dimension of the jth representation

h = order of the group

δ_{ij} = Kronecker delta

and the sum is over all operations of the group. From this theorem the following useful concepts can be derived:

1. The number of irreducible representations in a group is equal to the number of classes in the group.
2. The order of the group is related to the dimension of the irreducible representation by

$$h = \sum_i l_i^2 \tag{4.6}$$

3. The order of a group is related to the characters (traces of the matrices) or any irreducible representation by

$$h = \sum_R [\chi_i(R)]^2 \tag{4.7}$$

where $\chi_i(R)$ is the character (trace) of the matrix representing the operation R in the ith irreducible representation and the summation is over all operations of the group.
4. The characters of all matrices representing operators belonging to the same class are equal.
5. Considering that the characters of the matrices for a given representation constitute the components of a vector then the vectors whose components are the characters of two different irreducible representations are orthogonal

$$\sum_R \chi_i(R)\chi_j(R) = 0 \qquad i \neq j \tag{4.8}$$

The properties of a point group are summarized by means of a character table. The character table contains a listing of the irreducible representations of the group and the characters of the irreducible matrices associated with each class of operations of the group. Character tables often contain auxiliary information relating to the transformation of algebraic quantities. As an illustration the C_{4v} character table is presented and analyzed.

I	II					III	IV
C_{4v}	E	$2C_4$	C_2	$2\sigma_v$	$2\sigma_d$		
A_1	1	1	1	1	1	z	x^2+y^2, z^2
A_2	1	1	1	−1	−1	R_z	
B_1	1	−1	1	1	−1		x^2-y^2
B_2	1	−1	1	−1	1		xy
E	2	0	−2	0	0	$(x, y)(R_x, R_y)$	(xy, yz)

The table consists of four sections

I. A listing of the conventional symbols for the irreducible representations. The general symbol used in previous discussions is Γ_i. The conventions used for the symbols are:

a. *A*, *B* One-dimensional
   E Two-dimensional
   T(*F*) Three-dimensional

b. *A* Symmetric $[\chi(C_n) = 1]$ with respect to rotation of $2\pi/n$ about the unique C_n axis
   B Antisymmetric $[\chi(C_n) = -1]$ with respect to a rotation of $2\pi/n$ about the unique C_n axis

c. Subscripts on *A* and *B*
   1 Symmetric with respect to a C_2 axis perpendicular to C_n or with respect to σ_v if no C_n exists
   2 Antisymmetric with respect to a C_2 axis perpendicular to C_n or with respect to σ_v if no C_n exists
   g Symmetric with respect to inversion
   u Antisymmetric with respect to inversion

d. Primes on *A* and *B*
   N' Symmetric with respect to σ_h
   N'' Antisymmetric with respect to σ_h

II. The characters for each irreducible representation. Note that the character for the identity operator *E* is always equal to the dimension of the irreducible representation.

III. Designates the irreducible representation associated with translations *x*, *y*, *z* and rotations R_x, R_y, R_z.

IV. Designates the irreducible representations associated with algebraic combinations of the coordinates.

Appendix J contains character tables for some common point groups. Several other important relations will be given without proof.

1. Any arbitrary representation can be written as a sum of irreducible representations with the characters being related by

$$\chi_{\text{red}}(R) = \sum_j a_j \chi_j(R) \tag{4.9}$$

   where a_j is the number of times the *j*th irreducible representation occurs in the reducible one.

2. The coefficients a_j are given by

$$a_j = \frac{1}{h} \sum_R \chi_{\text{red}}(R) \chi_j(R) \tag{4.10}$$

   where the sum is over *all* operations of the group.

3. In a Cartesian framework, vectors can be considered as the basis of representations. Sets of orthogonal functions, such as encountered in quantum mechanics, can be basis functions for representations. Just as

an operation on a vector by a group operator gives a new vector that can be expressed in terms of components along the other members, the operation on a basis function by a group operator gives a new function that is a linear combination of the other basis functions

$$\boldsymbol{R}A_i = \sum_{k=1}^{N} a_{ki}A_k \qquad (4.11)$$

where A_i is one basis function of a set of N functions and a_{ki} is a scalar coefficient.

4. A function $A_j^m B_l^m$, defined by

$$\boldsymbol{R}A_j^m B_l^m = \sum_{k=1}^{N} \sum_{i=1}^{N} a_{kj}b_{il}A_k^m B_i^m \qquad (4.12)$$

is a direct product representation of A_j^m and B_l^m, where the latter are members of a basis set for the representation Γ_m and R is any operation of the group.

5. The characters of the representations of a direct product are equal to the products of the characters of the representations based on the individual sets of functions.

$$\chi(RR') = \sum_{i=1}^{N} \sum_{k=1}^{N} [\Gamma_j(R)_{ii}][\Gamma_l(R')_{kk}] = \chi(R)\chi(R') \qquad (4.13)$$

4.5 PROJECTION OPERATORS

A frequently used method for establishing permissible quantum states for a molecule involves the use of projection operators. Their use frequently enables one to employ the molecular symmetry to obtain state functions of the proper symmetry by beginning with an unsymmetrized set of functions. If one has a set of basis functions for the ith irreducible representation of a group, projection operators based on the group representations may be used to formulate appropriate linear combinations which have the desired symmetry properties.

The details regarding the formulation of projection operators from group theoretical principles can be found in most current textbooks on group theory. Only the final form of these operators will be presented. The projection operator for the ith irreducible representation of dimension l_i belonging to a group of order h is defined as

$$P^i = \frac{l_i}{h}\sum_R \chi_i(R)R \qquad (4.14)$$

It is to be noted that the sum is over all individual operations and not just over classes. The first use of this operator will be given in Chap. 5 when spin functions are discussed.

REFERENCES

General

1. Cotton, F. A., *Chemical Applications of Group Theory*, 2d ed., Wiley Interscience, New York, NY, 1971.
2. Douglas, B. E., and C. A. Hollingsworth, *Symmetry and Bonding in Spectra*, Academic Press Inc., Orlando, FL, 1985.
3. Ferraro, J. R., and J. S. Ziomek, *Introductory Group Theory and Its Application to Molecular Structure*, Plenum Press, New York, NY, 1969.
4. Flurry, R. L., Jr., *Symmetry Groups: Theory and Chemical Applications*, Prentice-Hall, Inc., Englewood Cliffs, NJ, 1980.
5. Hall, L. H., *Group Theory and Symmetry in Chemistry*, McGraw-Hill Book Co., New York, NY, 1969.
6. Hochstrasser, R. M., *Molecular Aspects of Symmetry*, W. A. Benjamin Inc., New York, NY, 1966.
7. Tinkham, M., *Group Theory and Quantum Mechanics*, McGraw-Hill, New York, NY, 1964.

PROBLEMS

4.1. Identify the symmetry elements in each of the following molecules or ions.

(*a*) $Co(NH_3)_4Cl_2^+$ (all possible isomers)
(*b*) H_2S
(*c*) O_3
(*d*) NSF
(*e*) $POCl_3$
(*f*) SF_5I
(*g*) Gauch, 1,2-dichloroethane
(*h*) Propane
(*i*) Isopropanol
(*j*) Tetrafluoroethylene
(*k*) Tin tetrachloride
(*l*) 1,2-Dicyanoethylene (both isomers)
(*m*) Trimethylamine
(*n*) Cyanoacetylene
(*o*) Formaldehyde
(*p*) Acetone
(*q*) Benz(o)anthracene
(*r*) Chrysene
(*s*) Pyrene
(*t*) Benzo(c)phenanthrene
(*u*) Difluoroacetylene
(*v*) ClF_3
(*w*) Dibenzene chromium

4.2. Determine the point group for each of the following molecules or ions. Give both the Schoenflies and Hermann-Mauguin notation.

(*a*) Cyclohexane (chair form)
(*b*) Cyclohexane (boat form)
(*c*) H_3BO_3
(*d*) SF_5NSF_2
(*e*) N_2O
(*f*) NO_2
(*g*) cis-$[COCl_2(NH_3)_4]^+$
(*h*) trans-$[COCl_2(NH_3)_4]^+$
(*i*) S_4H_4
(*j*) 1,3-Difluoroallene
(*k*) Facial-$[CrCl_3(NH_3)_3]$
(*l*) NO_2^+
(*m*) SF_4
(*n*) B_2H_6
(*o*) XeF_4
(*p*) Spiropentane
(*q*) I_3^-
(*r*) IF_7
(*s*) $Ni(C_4H_4)_2$
(*t*) P_4O_6
(*u*) P_4O_{10}
(*v*) $Mo_6Cl_8^{4+}$
(*w*) $W_2Cl_9^{3-}$
(*x*) CO_3^{2-}

4.3. The molecule HFClC–CHFCl in a staggered confirmation can exist as four different geometric isomer. Determine the point group for each isomer.

4.4. Determine the point group for each of the following geometric forms.

(*a*) Baseball (including seams)
(*b*) Football (ignore lacing)

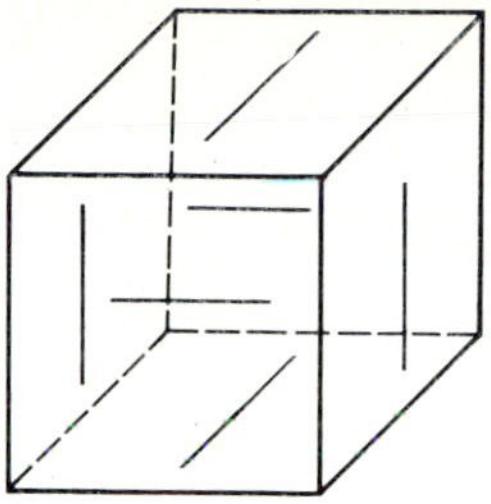

FIGURE P4.4c

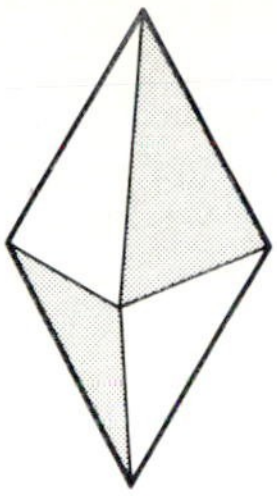

FIGURE P4.4d

(c) Cube with line faces
(d) Octahedron with 4 colored faces
(e) Moebius band
(f) Klein bottle
(g) Trigonal bipyramid

4.5. Determine which point groups are subgroups of the following.
(a) O_h (d) D_{3h}
(b) T_d (e) C_{4v}
(c) D_{2d} (f) D_{3d}

4.6. Point groups are frequently related by the addition or deletion of one symmetry element.
(a) What point group results for the following additions?
(1) $O + i$
(2) $D_3 + i$
(3) $D_4 + S_8$
(b) What point group results for the following deletions?
(1) $C_{4h} - i$
(2) $D_{4h} - S_4$
(3) $C_{3h} - S_3$

4.7. Determine the irreducible representations to which the Cartesian coordinates belong for each of the following point groups.
(a) C_{3v} (d) D_{6h}
(b) C_{4h} (e) T_d
(c) D_2 (f) O_h

4.8. Determine the irreducible representations contained in the following direct products. (Point group in parenthesis).
(a) $B_1XB_2(D_2)$ (d) $T_1XT_2(T_d)$
(b) $B_2XE(C_{4v})$ (e) $\Pi_bX\Delta_g(D_{\infty_h})$
(c) $E_{1u}XE_{2u}(D_{6h})$ (f) $E_{1g}XE_2(D_{\infty h})$

4.9. Set up projection operators for the following irreducible representations. (Point group in parenthesis).
(a) $B_2(C_{4v})$ (d) $T_2(T_d)$
(b) $E_g(D_{3d})$ (e) $E_1'(C_{5h})$
(c) $E(C_4)$ (f) $E_u(O_h)$

CHAPTER 5

ELECTRIC AND MAGNETIC PROPERTIES OF MATTER

5.1 INTRODUCTION

The first four chapters have reviewed the theoretical background necessary for the study of the quantum mechanics of spectroscopic methods. We now direct our attention to the nature and description of those microscopic properties of matter which are related to the physical interaction of matter with radiation and which are responsible for both electric and magnetic macroscopic properties. The ensuing discussion will examine concepts regarding electric and magnetic moments, which are obtainable from spectroscopic studies, and develop relationships which relate the microscopic properties to the bulk electric and magnetic susceptibilities of the material. We will also use this forum to introduce the systems of units used to characterize electric and magnetic properties. The important relationships of classical electrostatics and electromagnetics are reviewed but not derived. The quantum mechanics of angular momentum is extended to spin systems and the development of symmetrized spin functions reviewed.

5.2 FUNDAMENTAL ELECTRICAL QUANTITIES AND THE LAWS THAT DESCRIBE THEIR BEHAVIOR

The fundamental particles of matter, insofar as molecular structure is concerned, are the neutron (which possesses no electric charge) and the proton and electron (which possess equal, but opposite, charges). The unit of charge added to those of length, time, and mass, form the fundamental units of measurement. In the cgs system of units the proton charge is 4.80298×10^{-10} esu (electrostatic units) or 1.60219×10^{-2} emu (electromagnetic units) while in the SI system it is 1.60219×10^{-19} C (coulomb). A summary of units is given in App. K. The electrostatic attractive and repulsive interactions among the protons and electrons is one of the contributions to the internal energies of molecules.

The force of interaction between two particles of charge q_1 and q_2 separated by a distance r and in free space of permittivity ε_0 is expressed by Coulomb's law

$$\mathbf{F} = \frac{q_1 q_2 \hat{\mathbf{r}}}{4\pi\varepsilon_0 r^2} \tag{5.1}$$

In the cgs system $4\pi\varepsilon_0 = 1$ while in the SI system $4\pi\varepsilon_0 = 1.112650 \times 10^{-10}\ \mathrm{C^2\ J^{-1}\ m^{-1}}$.

The region of space influenced by an electric charge is denoted as the electric field. The intensity of the free space field $\boldsymbol{\mathscr{E}}$ at a distance r from a single charge q will be

$$\boldsymbol{\mathscr{E}} = \frac{q\mathbf{r}}{4\pi\varepsilon_0 r^2} \tag{5.2}$$

The units of the electric field will be statvolt cm^{-2} in the cgs system and newtons/coulomb (N/C) in the SI system.

The work necessary to move a unit charge from infinity to a distance r from a charge q is the electrostatic potential of the charge and is given by

$$V = \frac{q}{4\pi\varepsilon_0 r} \tag{5.3}$$

The potential and field are related by

$$\boldsymbol{\mathscr{E}} = \frac{dV}{dr}\hat{\mathbf{r}} \tag{5.4}$$

A property of the electric field which will be of importance when discussing nuclear quadrupole resonance is the electric field gradient (EFG) or the rate of change of the field in a particular direction. Before examining the detailed formulation of this concept some further remarks regarding field properties are in order. Although the electric potential is a scalar quantity its derivative with respect to $\mathbf{r}$, which is a vector is also a vector. Thus the electric field may be written

$$\boldsymbol{\mathscr{E}} = \mathscr{E}_x\hat{\mathbf{i}} + \mathscr{E}_y\hat{\mathbf{j}} + \mathscr{E}_z\hat{\mathbf{k}} \tag{5.5}$$

where

$$\mathscr{E}_g = -\left(\frac{\partial V}{\partial g}\right)_{g'g''} \quad (g, g', g'' = x, y, z \text{ cycled}) \tag{5.6}$$

For example, the x component of the field due to a single proton of charge e is

$$\mathscr{E}_x = -\left(\frac{\partial V}{\partial x}\right)_{y,z} = -\frac{e}{4\pi\varepsilon_0}\left(\frac{\partial}{\partial x}\frac{1}{r}\right)_{y,z} \tag{5.7}$$

Since $\mathbf{r} = x\hat{\mathbf{i}} + y\hat{\mathbf{j}} + z\hat{\mathbf{k}}$ and $r^2 = x^2 + y^2 + z^2$ it follows, introducing spherical coordinates, that

$$\mathscr{E}_x = \frac{ex}{4\pi\varepsilon_0 r^3} = \frac{e}{4\pi\varepsilon_0}\left(\frac{\sin\theta\cos\phi}{r^3}\right) \tag{5.8}$$

Similarly,

$$\mathscr{E}_y = \frac{ey}{4\pi\varepsilon_0 r^3} = \frac{e}{4\pi\varepsilon_0}\left(\frac{\sin\theta\sin\phi}{r^3}\right) \tag{5.9a}$$

$$\mathscr{E}_z = \frac{ez}{4\pi\varepsilon_0 r^3} = \frac{e}{4\pi\varepsilon_0}\left(\frac{\cos\theta}{r^3}\right) \tag{5.9b}$$

The electric field gradient is defined as

$$\underline{\boldsymbol{\nabla}\boldsymbol{\mathscr{E}}} = \frac{d\boldsymbol{\mathscr{E}}}{d\mathbf{r}} = -\frac{d^2V}{dr^2} \tag{5.10}$$

and

$$\boldsymbol{\nabla} = \frac{\partial}{\partial x}\hat{\mathbf{i}} + \frac{\partial}{\partial y}\hat{\mathbf{j}} + \frac{\partial}{\partial z}\hat{\mathbf{k}} \tag{5.11}$$

Combining Eqs. (5.5) and (5.11) with (5.10) shows that

$$\begin{aligned}\underline{\boldsymbol{\nabla}\boldsymbol{\mathscr{E}}} = {} & \nabla\mathscr{E}_{xx}\hat{\mathbf{i}}\hat{\mathbf{i}} + \nabla\mathscr{E}_{xy}\hat{\mathbf{i}}\hat{\mathbf{j}} + \nabla\mathscr{E}_{xz}\hat{\mathbf{i}}\hat{\mathbf{k}} \\ & + \nabla\mathscr{E}_{yx}\hat{\mathbf{j}}\hat{\mathbf{i}} + \nabla\mathscr{E}_{yy}\hat{\mathbf{j}}\hat{\mathbf{j}} + \nabla\mathscr{E}_{yz}\hat{\mathbf{j}}\hat{\mathbf{k}} \\ & + \nabla\mathscr{E}_{zx}\hat{\mathbf{k}}\hat{\mathbf{i}} + \nabla\mathscr{E}_{zy}\hat{\mathbf{k}}\hat{\mathbf{j}} + \nabla\mathscr{E}_{zz}\hat{\mathbf{k}}\hat{\mathbf{k}}\end{aligned} \tag{5.12}$$

where

$$\nabla\mathscr{E}_{gg'} = \frac{\partial\mathscr{E}_g}{\partial g'} = -\frac{\partial^2 V}{\partial g\,\partial g'} \quad (g, g' = x, y, z) \tag{5.13}$$

Again referring to a single proton two examples are

$$\nabla\mathscr{E}_{zz} = \frac{\partial\mathscr{E}_z}{\partial z} = -\frac{\partial^2 V}{\partial z^2} = -\frac{\partial^2}{\partial z^2}\left(\frac{e}{4\pi\varepsilon_0 r}\right) = -\frac{e}{4\pi\varepsilon_0}\left(\frac{3\cos^2\theta - 1}{r^3}\right) \tag{5.14}$$

and

$$\nabla\mathscr{E}_{xy} = \frac{\partial\mathscr{E}_x}{\partial y} = -\frac{\partial^2 V}{\partial x\,\partial y} = -\frac{\partial^2}{\partial x\,\partial y}\left(\frac{e}{4\pi\varepsilon_0 r}\right) = -\frac{e}{4\pi\varepsilon_0}\left(\frac{3\sin^2\theta\sin\phi\cos\phi}{r^3}\right) \tag{5.15}$$

The first of these gives the gradient or rate of change of the z component of the field as one moves in the z direction while the second gives the gradient of the x component of the field as one moves in the y direction. The *EFG*, $\underline{\nabla \mathscr{E}}$ is a second-rank tensor having nine components (see App. L).

5.3 DIPOLE MOMENTS

The primary mode of interaction between molecules and electromagnetic radiation in the microwave through the ultraviolet region of the electromagnetic spectrum is due to the interaction of the electric field of the radiation with a changing electric dipole, resulting from some change or motion of the system. The detailed formulation of this interaction requires an understanding of the nature of permanent and induced molecular electric dipole moments.

An electric dipole moment is present in a molecule when there is a lack of coincidence between the centers of positive and negative charge. The electric dipole may be either permanent, due to electronegativity differences among atoms in the molecule, or induced (temporary), due to momentary distortions of the molecule due to either interactions with other molecules or to intramolecular motions. This latter phenomenon is referred to as *polarization.*

For a system comprised of two equal, but opposite, charges $\pm q$ separated by a distance $\mathbf{r}$, the electric dipole moment $\boldsymbol{\mu}$ is given by $\boldsymbol{\mu} = q\mathbf{r}$. Note that the electric dipole moment is a vector quantity. The SI unit for the electric dipole moment is the coulomb meter, Cm, and the cgs unit is the debye, D, where $1\,D = 10^{-18}$ esu $= 3.338 \times 10^{-30}$ Cm. For the example of a proton and an electron having a 0.1 nm separation the electric dipole moment will be $\mu = 1.602 \times 10^{-19}\,\text{C} \times 10^{-10}\,\text{m} = 1.602 \times 10^{-29}\,\text{Cm} = 4.8\,D$.

It is often useful when predicting the nature of molecular spectra to have estimates of molecular electric dipole moments. For diatomic molecules a crude estimate can be made by using a Pauling [1] type relationship to determine the ionic character of the molecule

$$I = 1 - \exp - [0.25(X_A - X_B)^2]$$

where X_A and X_B are the electronegatives. The value of I can then be used to estimate the partial charge on each atom and hence the electric dipole moment. For example HCl with $X_{\text{H}} = 2.1$ and $X_{\text{Cl}} = 3.0$ has $I = 0.18$ which, with a bond length of 0.136 nm estimated from covalent radii, gives $\mu = 2.87 \times 10^{-30}$ Cm. The experimental value is 3.60×10^{-30} Cm so we observe that this is at best an order-of-magnitude calculation. The process is even more unreliable for molecules containing larger atoms which are more easily polarized.

A convenient method to estimate moments for complex molecules is by use of "bond moments". By using an oversimplified model of a molecule being composed of point atoms having partial charges due to electronegativity differences and being held in a rigid structure, the net electric dipole moment can be considered as a vector sum of the bond moments of the system. For example, if we consider the hypothetical pyramidical molecule $AXYZ$ shown in Fig. 5.1 then

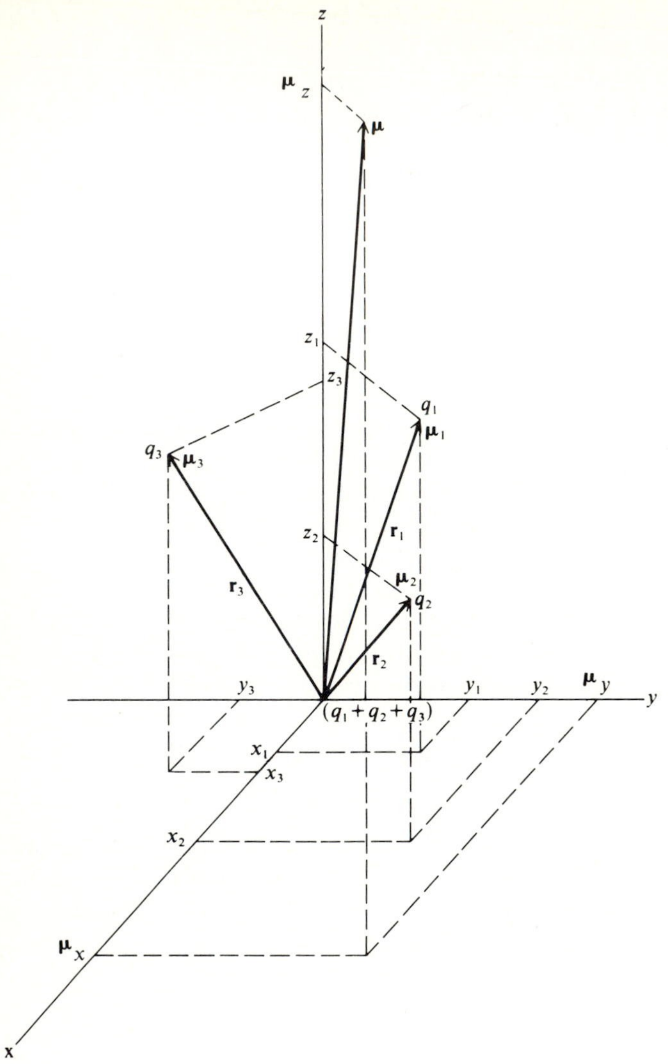

FIGURE 5.1
Concept of bond moments.

the net electric dipole moment is given by

$$\boldsymbol{\mu} = \boldsymbol{\mu}_1 + \boldsymbol{\mu}_2 + \boldsymbol{\mu}_3 \tag{5.16}$$

where

$$\boldsymbol{\mu}_1 = q_1\mathbf{r}_1 \tag{5.17}$$

$$\boldsymbol{\mu}_2 = q_2\mathbf{r}_2 \tag{5.18}$$

$$\boldsymbol{\mu}_3 = q_3\mathbf{r}_3 \tag{5.19}$$

TABLE 5.1
Estimated bond moments for common bonds

Bond†	Moment Cm × 10^{30}	Bond	Moment Cm × 10^{30}
Single bonds (covalent)		Multiple bonds	
F—As	6.7	O=C	7.7
F—Br	4.3	O=N	6.7
F—C	4.7	N=C	3.0
F—Cl	3.0	N≡C	11.7
F—H	6.3	S=C	8.7
F—N	0.7		
O—C	2.3		
O—Cl	2.3		
O—H	5.0		
O—N	1.0		
Cl—As	5.3		
Cl—Br	2.0		
Cl—C	4.7		
Cl—H	3.7	Single bonds (coordinate)	
Cl—I	3.3	O←As	14.0
Cl—P	2.7	O→B	12.0
Cl—S	2.3	O←N	14.3
Cl—Sb	8.7	O←P	9.0
N—C	0.7	O←S	9.3
N—H	4.3	O←Se	10.3
Br—As	4.3	O←Te	7.7
Br—C	4.7	N→B	13.0
Br—H	2.7	S→B	12.7
Br—I	4.0	S→C	16.7
Br—P	1.3	S←Sb	15.0
Br—Sb	4.0	P→B	14.7
C—H	1.3	P→S	10.3
C—I	4.0	P→Se	10.7
C—S	3.0		
C—Se	2.7		
C—Te	2.0		
I—As	2.7		
I—C	4.0		
I—H	1.3		
I—P	0.0		
I—Sb	2.7		
H—As	0.3		
H—P	1.3		
H—Sb	0.3		

† Bonds are listed with the most electronegative element first.

or, looking at the Cartesian components,

$$\boldsymbol{\mu} = (q_1x_1 + q_2x_2 + q_3x_3)\hat{\mathbf{i}} + (q_1y_1 + q_2z_2 + q_3y_3)\hat{\mathbf{j}} + (q_1z_1 + q_2z_2 + q_3z_3)\hat{\mathbf{k}} \tag{5.20}$$

Although this approach is oversimplified it nevertheless constitutes a relatively good method for making initial estimates of dipole moments for preliminary spectroscopic calculations. Table 5.1 summarizes bond moments for some of the more common bonds. In using such data to estimate dipole moments one must keep in mind that when particular pairs of atoms are incorporated into a molecule their charge distributions may be modified from a simple diatomic combination, and that often there are lone pairs of electrons which can influence the calculated moment. For example, using the N—H and N—F bond moments from Table 5.1, and taking the bond angle for the two pyramidical molecules NH_3 and NF_3 to be 105°, their electric dipole moments are estimated to be 4.34×10^{-30} Cm and 1.23×10^{-29} Cm, respectively. The experimental values are 5.67×10^{-30} Cm and 2.33×10^{-30} Cm respectively, showing that the lone pairs of electrons tend to enhance the magnitude of the moment for NH_3 and to diminish that for NF_3 relative to the calculated values.

For a charge $+q$ surrounded by a continuous electron charge distribution which can be represented by a charge density function $\rho(x, y, z)$, the electric dipole moment is given by

$$\boldsymbol{\mu} = \int \rho(xyz)\mathbf{r}\, dv \tag{5.21}$$

where

$$\int \rho(xyz)\, dv = -q \tag{5.22}$$

A more sophisticated method for the estimation of dipole moments based on molecular quantum mechanics has been developed by Jaffe and coworkers [2]. Anyone wishing to make more accurate estimates will find this method useful.

5.4 DIELECTRIC PROPERTIES—INDUCED POLARIZATION

The occurrence of a dipole moment in a molecule not only provides the physical mechanism for the interaction of the free molecule with electromagnetic radiation but also contributes to the bulk electrical properties of the material in all states. Before we proceed to the consideration of spectroscopic phenomenon which involve the former type of interaction we will examine the nature of the bulk electrical properties of matter in order to establish their relationship to the microscopic electric dipole moments.

When a nonconducting (dielectric) material is placed between two parallel conducting plates which are at a potential difference due to the presence of charges on the plates the electric field between the plates is a function of the electrical nature of the material. In this case the modification of the field can be

FIGURE 5.2
Parallel plate capacitor.

related to the real dielectric permittivity ε of the substance. The factor $\varepsilon_r = \varepsilon/\varepsilon_0$ is the relative dielectric constant of the material and is the ratio of its real dielectric permittivity to that of free space[1].

This concept is perhaps best understood in terms of electrical capacitance C, which is the ratio of charge stored to the electrostatic potential between two parallel charged plates. For this configuration, illustrated in Fig. 5.2, the magnitude of the electric field is related to the potential by

$$\mathscr{E} = \frac{V\hat{\mathbf{r}}}{r} = \frac{q\hat{\mathbf{r}}}{\varepsilon_0 A} \tag{5.23}$$

In terms of surface charge density $\sigma = q\mathbf{r}/A$ the magnitude of the free space electric field is given by

$$\mathscr{E} = \frac{\boldsymbol{\sigma}}{\varepsilon_0} \tag{5.24}$$

It then follows that the free space capacitance is

$$C_0 = \frac{q}{V} = \frac{\varepsilon_0 A}{r} \tag{5.25}$$

If the space between the plates is filled with a dielectric material the capacitance will be changed to

$$C = \frac{\varepsilon A}{r} \tag{5.26}$$

One will observe that a measurement of the capacitance of the parallel plates for free space C_0, and with a material of dielectric constant ε, allows one to obtain the relative permittivity from

$$\varepsilon_r = \frac{C}{C_0} \tag{5.27}$$

---

[1] The electric permittivity of a material is a complex quantity $\varepsilon + i\varepsilon'$, where ε is the real part or dielectric constant and ε' is the dielectric loss. The dielectric constant is a measure of energy stored and ε' is a measure of energy dissipated by the material.

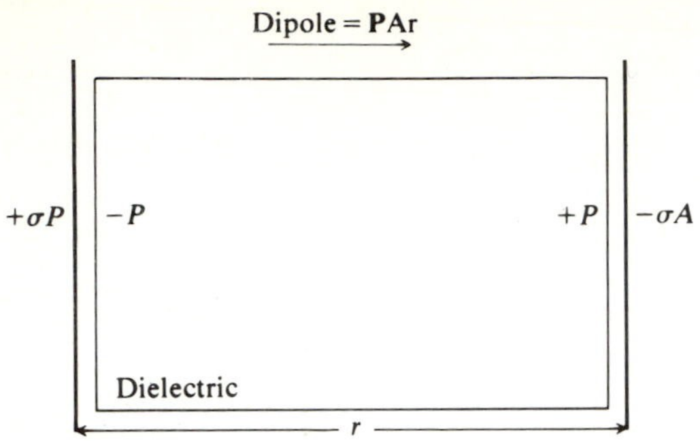

FIGURE 5.3
Electrical behavior of a dielectric material.

We next consider the problem of relating the macroscopic dielectric constant to the microscopic dipole moments. This will be approached by considering the bulk behavior of a dielectric medium placed between two oppositely charged plates. Considering that the dielectric medium is composed of permanent electric dipoles and electric dipoles induced by the field, and that these dipoles can be oriented by the field, the net effect will be to create a surface charge on the dielectric proportional to the charge on the plates. This is illustrated in Fig. 5.3.

This surface charge created on the dielectric material will lower the effective electric field. Denoting the surface charge per unit area on the capacitor plate by $\boldsymbol{\sigma}$, the electric field in the dielectric medium $\mathscr{E}_e$, and the polarization $\mathbf{P}$, are related by

$$e_0\mathscr{E}_e + \mathbf{P} = \boldsymbol{\sigma} \tag{5.28}$$

The polarization is the charge per unit area on the dielectric or the average dipole moment per unit volume. The factor $\varepsilon_0\mathscr{E}_e + \mathbf{P}$ is referred to as the dielectric displacement or electric flux density $\mathbf{D}$. Hence

$$\varepsilon_0\mathscr{E}_e + \mathbf{P} = \mathbf{D} \tag{5.29}$$

In a dielectric

$$\mathscr{E}_e = \frac{\boldsymbol{\sigma}}{\varepsilon} \tag{5.30}$$

Note that for free space, where $\varepsilon_r = 1$, the electric field and the displacement are equivalent or the polarization is zero. Equation (5.29) upon substitution of Eq. (5.31) and rearrangement yields

$$\mathscr{E}_e = \frac{\mathbf{P}}{(\varepsilon_r - 1)\varepsilon_0} \tag{5.31}$$

The electric field in a dielectric medium is thus a function of both the polarization and the dielectric constant of the medium.

The polarization of a dielectric will depend on both the orientation of the permanent electric dipoles $\mathbf{P}_0$ and the induced electric dipoles $\mathbf{P}_i$ in the system and is a sum of contributions from both

$$\mathbf{P} = \mathbf{P}_0 + \mathbf{P}_i \tag{5.32}$$

We will examine the two contributions separately, beginning with that of the induced dipoles, since they will be present in all materials. When an atom or a molecule is subjected to an external electric field the charges will be distorted from their normal configuration in such a manner as to produce a change in the positions of the centers of positive and negative charge density. For a system which has spherical symmetry this will result in the creation of an electric dipole. If the molecule possesses a permanent electric dipole moment the resulting separation of the charge density centers will be different from that of the field free molecule, and the total dipole moment will then have both a permanent and an induced component, the former generally being the larger of the two. The magnitude of the induced moment $\boldsymbol{\mu}_i$ will be proportional to the electric field at the molecular site $\mathscr{E}_e$. Hence

$$\boldsymbol{\mu}_i = \alpha \mathscr{E}_e \tag{5.33}$$

where the proportionality constant α is called the *polarizability*, and is the electric dipole induced in the medium by a unit electric field. The relationship just given for $\boldsymbol{\mu}_i$ is only for an isotropic dielectric. It is applicable to gases, most liquids, and some solids, but is not valid for nonisotropic solids or oriented liquids, such as liquid crystals.

If the dielectric material contains N molecules per unit volume, then the induced dipole per unit volume, which is the induced polarization P_i, is given by

$$\mathbf{P}_i = N\boldsymbol{\mu}_i = \alpha N \mathscr{E}_e \tag{5.34}$$

The electric field in a dielectric is difficult to evaluate except for simple systems such as ideal gases and nonpolar liquids. For an ideal gas where the molecules are far apart and intermolecular interactions are negligible the effective electric field is equal to the applied external field. The induced electric dipole moment is, then, $\boldsymbol{\mu}_i = \alpha \mathscr{E}$. Expressing the molecular concentration as $N = N_0\rho/M$ where N_0 = Avagadro number, ρ = density, and M = molecular weight, the induced polarization is then

$$\mathbf{P}_i = \frac{\alpha N_0 \rho \mathscr{E}}{M} \tag{5.35}$$

Combining this equation with Eq. (5.31) gives

$$(\varepsilon_r - 1) = \frac{\alpha N_0 \rho}{M \varepsilon_0} \tag{5.36}$$

This can be rearranged to put the readily determined experimental parameters on the left-hand side, thus

$$\frac{(\varepsilon_r - 1)M}{\rho} = \frac{\alpha N_0}{\varepsilon_0} \tag{5.37}$$

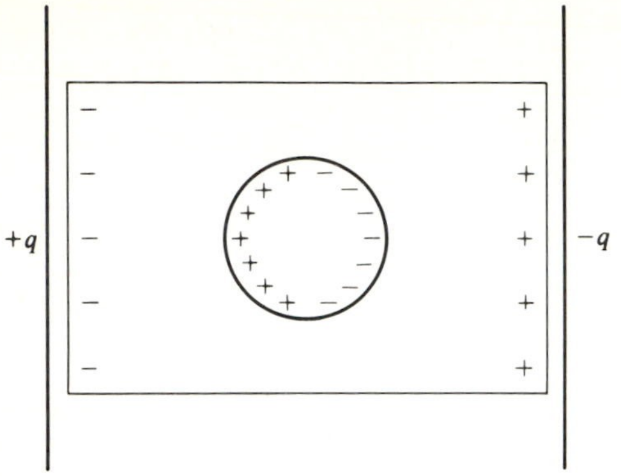

FIGURE 5.4
Contributing charges to the internal field of a dielectric.

For a nonpolar gas, where there are no permanent electric dipole moments, the polarizability, and hence the magnitude of the induced electric dipole moment for a given electric field, can be determined by measuring the dielectric constant and the density.

For a condensed phase it is not possible to exactly solve the problem of calculating a general internal electric field. For systems such as nonpolar liquids, high-density gases, or solutions of polar solutes in nonpolar solvents, it is possible to approximate a general solution. By use of an idealized model illustrated in Fig. 5.4 one can evaluate the local electric field in an isotropic nonpolar dielectric material. The local internal field within a spherical cavity of molecular dimensions will be due to three contributions

$$\mathscr{E}_{\text{loc}} = \frac{\boldsymbol{\sigma}}{\varepsilon_0} - \frac{\mathbf{P}}{\varepsilon_0} + \mathscr{E}_s \tag{5.38}$$

where $\mathscr{E}_s$ is the contribution of the charges bound to the walls of the cavity.

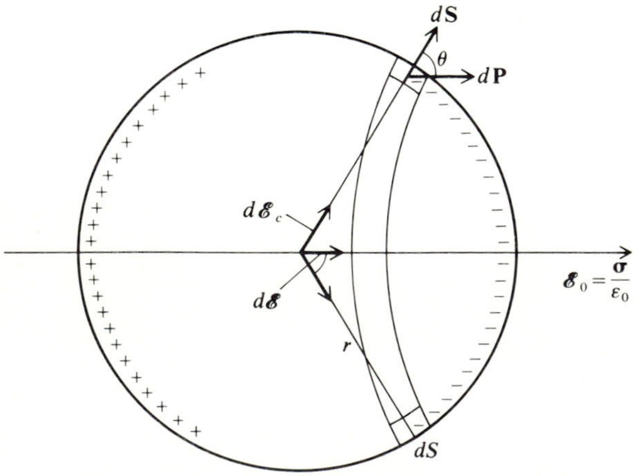

FIGURE 5.5
Coordinates for the evaluations at internal electric fields in a dielectric.

Referring to Fig. 5.5 for the designation of the necessary parameters one can express an incremental field at the center of the cavity as

$$d\mathscr{E}_c = \frac{\sigma_s d\mathbf{S}}{4\pi\varepsilon_0 r^2} \tag{5.39}$$

where σ_s is the cavity surface charge density and will be related to the polarization by $\sigma_s = P\cos\theta$. The incremental field at a point within the cavity will be given by $d\mathscr{E}_s = d\mathscr{E}_c \cos\theta$, hence

$$d\mathscr{E}_s = \frac{\mathbf{P}\cos^2\theta}{4\pi\varepsilon_0 r^2}\,dS \tag{5.40}$$

For a ring the differential element of area is $dS = 2\pi r^2 \sin\theta\, d\theta$ and

$$d\mathscr{E}_s = \frac{P\cos^2\theta\sin\theta}{2\varepsilon_0}\,d\theta \tag{5.41}$$

Integration then gives

$$\mathscr{E}_S = \frac{1}{2\varepsilon_0}\int_0^{\pi} P\cos^2\theta\sin\theta\, d\theta = \frac{\mathbf{P}}{3\varepsilon_0} \tag{5.42}$$

The local electric field inside the dielectric is then

$$\mathscr{E}_{\text{loc}} = \frac{\boldsymbol{\sigma}}{\varepsilon_0} - \frac{\mathbf{P}}{\varepsilon_0} + \frac{\mathbf{P}}{3\varepsilon_0} = \mathscr{E}_e + \frac{\mathbf{P}}{3\varepsilon_0} \tag{5.43}$$

Since $P = \varepsilon_e(\varepsilon_r - 1)\varepsilon_0$ the induced electric dipole moment will be given by

$$\boldsymbol{\mu}_i = \alpha\mathscr{E}_{\text{loc}} = \alpha\left[\mathscr{E}_e + \frac{\mathscr{E}_e(\varepsilon_r - 1)\varepsilon_0}{3\varepsilon_0}\right] \tag{5.44}$$

and the induced polarization becomes

$$\mathbf{P}_i = \frac{N_0}{M}\boldsymbol{\mu}_i = \frac{\mathscr{E}_e N_0\rho\alpha}{M}\left[1 - \frac{(\varepsilon_r - 1)}{3}\right] \tag{5.45}$$

Since we are considering only induced polarization it is equal to the total polarization

$$\mathbf{P} = \mathscr{E}_e(\varepsilon_r - 1)\varepsilon_0 = \frac{\mathscr{E}_e N_0\rho\alpha}{M}\left[1 + \frac{(\varepsilon_r - 1)}{3}\right] \tag{5.46}$$

which rearranges to the Clausius-Mossotti equation [3, 4]

$$\frac{M(\varepsilon_r - 1)}{\rho(\varepsilon_r + 2)} = \frac{N_0\alpha}{3\varepsilon_0} \tag{5.47}$$

where we again have the experimentally determinable parameters on the left-hand side. The product of terms $P_M = N_0\alpha/3\varepsilon_0$ is referred to as the molar polarization.

5.5 DIELECTRIC PROPERTIES—ORIENTATION POLARIZATION

Having examined the contribution of induced electric dipole moments to the total polarization of a dielectric we next turn our attention to the contributions of permanent electric dipoles. In this instance we must examine the balance between the electrical orienting effect of the field and the disordering effect of thermal motion in the substance.

The energy of a dipole moment $\boldsymbol{\mu}$, in an internal electric field $\boldsymbol{\mathscr{E}}_{\text{loc}}$, is given by

$$E = -\boldsymbol{\mu} \cdot \boldsymbol{\mathscr{E}}_{\text{loc}} = -|\mu||\mathscr{E}_{\text{loc}}|\cos\theta \tag{5.48}$$

The tendency of the dipoles to align in the field due to this energy of interaction is opposed by the thermal agitation of the molecules. The number of dipoles aligned at an angle between θ and $\theta + d\theta$ is given by the Boltzmann distribution as

$$dN = Ae^{\mu\mathscr{E}_{\text{loc}}\cos\theta/kT}\sin\theta\, d\theta\, d\phi \tag{5.49}$$

The distribution is symmetric with respect to ϕ so

$$dN = 2\pi Ae^{\mu\mathscr{E}_{\text{loc}}\cos\theta/kT}\sin\theta\, d\theta \tag{5.50}$$

The contribution of the molecules lying within the solid angle $2\pi\sin\theta\, d\theta$ will be $\mu\cos\theta\, dN$ or

$$\mu\cos\theta\, dN = 2\pi Ae^{\mu\mathscr{E}_{\text{loc}}\cos\theta/kT}\mu\cos\theta\sin\theta\, d\theta \tag{5.51}$$

The average value of the dipolar contribution by a single molecule $\bar{\mu}$ will be the ratio of the total dipolar contribution to the total number of molecules

$$\bar{\mu} = \frac{\int_0^{2\pi} 2\pi A\mu e^{\mu\mathscr{E}_{\text{loc}}\cos\theta/kT}\cos\theta\sin\theta\, d\theta}{\int_0^{2\pi} 2\pi Ae^{\mu\mathscr{E}_{\text{loc}}\cos\theta/kT}\sin\theta\, d\theta} \tag{5.52}$$

Letting $b = \mu\mathscr{E}_e/kT$, $y = \cos\theta$ and $dy = -\sin\theta\, d\theta$ this expression becomes

$$\bar{\mu} = \frac{\mu\int_{-1}^{1} ye^{by}\, dy}{\int_{-1}^{1} e^{by}\, dy} \tag{5.53}$$

which integrates to give

$$\bar{\mu} = \mu\left(\coth b - \frac{1}{b}\right) \tag{5.54}$$

For a typical molecule, $\mu \approx 10^{-20}$ Cm in a 100-volt electric field at room temperature, the value of b will be approximately 10^{-5}. For $b \ll 1$ $\coth b$ can be expanded as a power series

$$\coth b = \frac{1}{\tanh b} = \frac{1}{[b - (b^3/3) - \cdots]} \approx \frac{1}{b}\left[1 - \frac{b^2}{3}\right]^{-1} \tag{5.55}$$

Invoking the binomial expansion for $(1 - b^2/3)^{-1}$ gives

$$\bar{\mu} = \mu\left[\frac{1}{b}\left\{1 + \frac{b^2}{3}\right\} - \frac{1}{b}\right] = \frac{\mu b}{3} = \frac{\mu^2 \mathscr{E}_{\text{loc}}}{3kT} \tag{5.56}$$

The total polarization in the system is then

$$\mathbf{P} = \mathbf{P}_i + \mathbf{P}_0 = N\boldsymbol{\mu}_i + N\boldsymbol{\mu} = \frac{N_0\rho}{M}\left[\frac{\alpha\mathscr{E}_e(\varepsilon + 2)}{3} + \frac{\mu^2}{3kT}\left\{\mathscr{E}_e + \frac{4\pi}{3}\mathbf{P}\right\}\right] \tag{5.57}$$

Rearrangement of Eq. (5.31) to

$$\mathbf{P} = \mathscr{E}_e(\varepsilon_r - 1)\varepsilon_0 \tag{5.58}$$

followed by substitution into Eq. (5.57) and equating the result to Eq. (5.58) gives

$$\mathscr{E}_e(\varepsilon_r - 1)\varepsilon_0 = \frac{N_0\rho}{M}\left[\frac{\alpha\mathscr{E}_e}{3} + \frac{\mu^2\mathscr{E}_e}{9kT}\right](\varepsilon + 2) \tag{5.59}$$

The molar polarization is given as

$$P_M = \frac{M}{\rho}\frac{(\varepsilon_r - 1)}{(\varepsilon_r + 2)} = \frac{N_0}{\varepsilon_0}\left[\frac{\alpha}{3} + \frac{\mu^2}{9kT}\right] \tag{5.60}$$

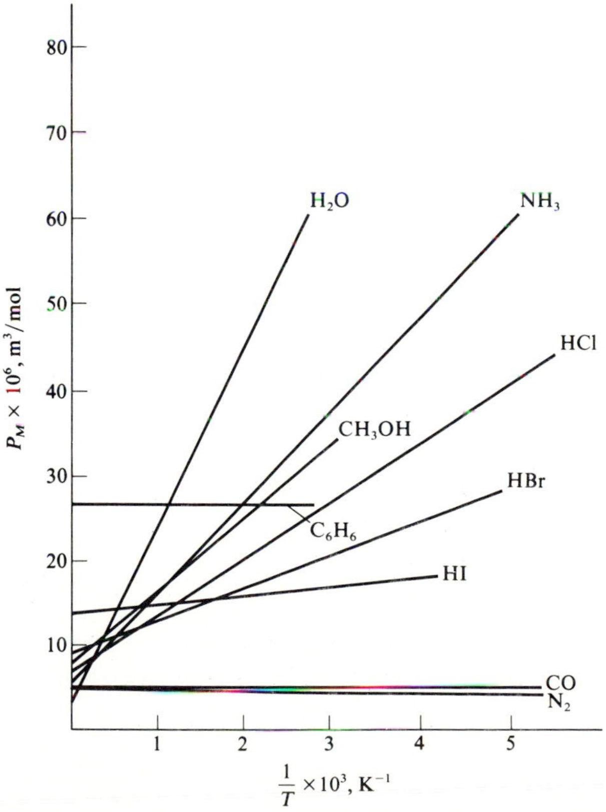

FIGURE 5.6
Qualitative behavior of molar polarization.

This equation, known as the Debye equation, is the basis for the evaluation of permanent dipole moments from the temperature dependence of the dielectric constant. This is illustrated qualitatively in Fig. 5.6 by curves for molecules with moments increasing from N_2 to H_2O. This method is relatively accurate for the determination of dipole moments of polar gases or polar solutes in nonpolar solvents where minimal intermolecular interactions occur.

5.6 BASIC MAGNETIC PROPERTIES OF MATTER

If a sample of any substance is placed in an inhomogeneous magnetic field it will find itself either repelled by the field or attracted by the field. Those materials which exhibit repulsion are said to be diamagnetic, while those that are attracted are termed paramagnetic or ferromagnetic, depending on whether the attraction is weak or very strong. The study of ferromagnetism is a subject to itself and since it has no direct bearing on the material to be presented in this book, it will be given no further consideration.

When a nonferromagnetic substance is placed in a magnetic field of strength **H** (tesla) (1 T = 10^4 gauss) the magnetic induction **B** (tesla) is given by

$$\mathbf{B} = \mu_0\mathbf{H} + \mu_0\mathbf{M} = \mu_0\mathbf{H} + \mu_0\chi\mathbf{H} \tag{5.61}$$

where μ_0 is the vacuum magnetic permeability ($\mu_0 = 4\pi \times 10^{-7} NA^{-2}$)[2] and **M**, called the magnetization, is the magnetic moment per unit volume. A bulk magnetic moment is the magnetic analogy to the electrostatic polarization moment. The magnitude of a unit magnetic moment is such that when placed in a field of 1 tesla it will experience a torque of 1 N-m. The magnetization is proportional to the applied field with the dimensionless proportionality constant χ referred to as the magnetic susceptibility. For a diamagnetic substance which is repelled by an inhomogeneous field the magnetic induction inside the substance will be less than that in the surrounding air, hence, χ will be negative for such materials. The opposite is true for paramagnetic substances, these having positive values of χ. The magnetic susceptibility is frequently expressed as the gram susceptibility χ_g, or the molar susceptibility χ_M, these being related by

$$\chi_M = M\chi_g = \frac{M\chi}{\rho} \tag{5.62}$$

where M = molecular weight and ρ = density.

Before exploring further aspects of magnetic behavior we will look at the microscopic magnetic nature of particles so that we may more completely integrate the two aspects of magnetism. One of the fundamental properties of electrons and some nuclei is the possession of an intrinsic angular momentum referred to

[2] Note: $\mu_0\varepsilon_0 = c^{-2}(c = 2.99793 \times 10^8\ \text{ms}^{-1})$.

as spin. The magnitude of the angular momentum possessed by any particular particle is related to a spin quantum number s for an electron or I for a nucleus.

Prior to the advent of modern quantum mechanics, the observation of an unexpected doubling of some lines in the atomic spectra of certain elements defied explanation. A reasonable explanation of this observed phenomenon was formulated by requiring that there be an angular momentum of $\frac{1}{2}\hbar$ associated with each electron in such systems. In Chap. 3 it was shown that the total angular momentum of a system is given by $[J(J+1)]^{1/2}\hbar$ where J is an integer. Since the orbital angular momentum quantum number of an electron was restricted to integer values, there had to be some new phenomenon to account for this observed momentum requirement. In 1925, Gouldsmit and Uhlenbeck postulated that the electron possessed an angular momentum due to its spinning about an axis. This spin angular momentum has a magnitude of $[s(s+1)]^{1/2}\hbar$. They further postulated that, since the spinning of the electron involved the motion of an electric charge, there would be a magnetic moment $\mu = -(ge/2mc)[s(s+1)]^{1/2}\hbar$, where e = electronic charge, m = electron mass, c = velocity of light, and g is a proportionality constant called the Landé g factor. Furthermore, in analogy to known properties of quantum angular momentum, the projections of $[s(s+\frac{1}{2})]^{1/2}\hbar$ on a body- or space-fixed axis will be $m_s\hbar$ where $-s < m_s < +s$. Using values of $g = 2$ and $s = \frac{1}{2}$ it was found that the aforementioned spectral doubling could be explained. The electron was therefore assigned a spin quantum number of $s = \frac{1}{2}$ with possible values of $m_s = \pm\frac{1}{2}$.

In 1928 [5] Dirac extended the Schrödinger equation for the hydrogen atom by incorporating relativistic mechanics and determined that the solution involved four quantum numbers rather than the three conventional ones. Furthermore, the fourth number was restricted to values of $\pm\frac{1}{2}$ and the factor $g = 2$ was also predicted. Although the Dirac theory gave rise to the same quantum number as the postulation of Gouldsmit and Uhlenbeck it involved no a priori assumption of electron spin. It was concluded that the spin angular momentum and the associated magnetic moment were fundamental properties of the electron just as are mass and charge, and are not necessarily associated with any physical rotation of the electron. For this reason this property is referred to as the intrinsic spin.

For nuclei there exists a similar situation but the intrinsic spin varies from one nuclei to another. The protons and neutrons which compose the nucleus have spin quantum numbers of $I = \frac{1}{2}$. Depending on the number and the manner in which these fundamental particles are combined to form a given nucleus the nucleus may have a spin quantum number of 0, positive integer, or half-integer value. The magnitude of the associated magnetic moment is given by

$$\mu_N = g_N \frac{e\hbar}{2m_p c}[I(I+1)]^{1/2} = g_N\beta_N[I(I+1)]^{1/2} \tag{5.63}$$

where m_p is the proton mass and g_N is a proportionality constant called the *nuclear g factor*, and is a fundamental property of the nucleus.

The magnetic moments of both particles are generally expressed in multiples of the Bohr magneton, $\beta = e\hbar/2mc = 0.92741 \times 10^{-20}$ erg gauss^{-1}, $= 9.724 \times 10^{-24}\,JT^{-1}$, or the nuclear magneton $\beta_N = e\hbar/2m_pc = 0.50508 \times 10^{-23}$ erg gauss^{-1} $= 0.50508 \times 1 - ^{-27}\,JT^{-1}$. While we have derived the moments by using scalar magnitudes they can also be expressed as vector quantities. The eigenvalue of I^2 and s^2 are $I(I+1)\hbar^2$ and $s(s+1)\hbar^2$, hence we can also express these last relationships as, $\boldsymbol{\mu} = -g\beta\mathbf{s}/\hbar$ and $\boldsymbol{\mu}_N = g_N\beta_N\mathbf{I}/\hbar$.

Before encountering the discussion of magnetic resonance we need to examine two preliminary concepts. The first of these is the relationship between the microscopic properties of spin systems and the macroscopic property of magnetic susceptibilities, and the second is the manner by which spins are considered by quantum mechanics.

5.7 MAGNETIC SUSCEPTIBILITY

In this section we will examine the fundamental relationships between the microscopic properties of individual atoms and molecules and the bulk magnetic properties of a system. We are interested at this point only in basic concepts which will be of use later, and not in the detailed study of the experimental methods for susceptibility measurements or the application of such measurements to particular chemical problems. The classical theory of diamagnetism was first extensively investigated by Langevin [6].

Diamagnetism is a basic property of all matter. Its nature can be illustrated by considering a simple one-electron atom placed in a magnetic field of increasing magnitude. The geometry of such a system is illustrated in Fig. 5.7. If we impose the increasing magnetic field **H** in the vertical direction it produces a circular electric field $\mathscr{E}$ perpendicular to the direction of the magnetic field. This field will have a tangential component which gives rise to an electric field of magnitude

$$\mathscr{E} = -\frac{R}{2c}\frac{d\mathbf{H}}{dt} \tag{5.64}$$

This field in turn produces a torque on the electron which tends to move it in a circular path. The torque $-e\mathscr{E}R$ is equal to the rate of change of the angular momentum **L** hence

$$\frac{d\mathbf{L}}{dt} = -e\mathscr{E}R = \frac{eR^2}{2c}\frac{d\mathbf{H}}{dt} \tag{5.65}$$

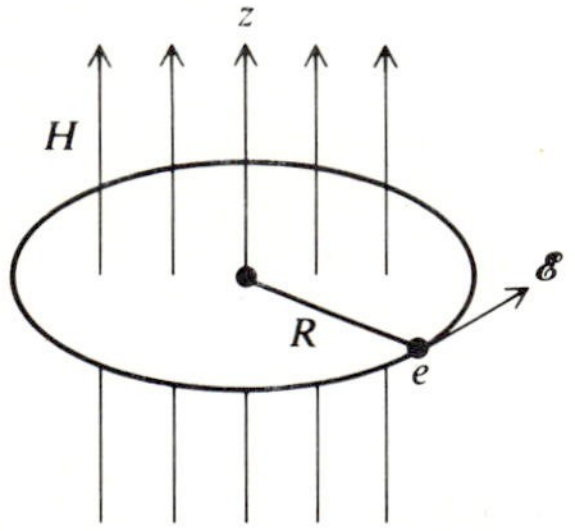

FIGURE 5.7
Parameters related to atomic diamagnetism.

If we now integrate from time zero where $\mathbf{L} = \mathbf{L}_0$ and $\mathbf{H} = O$ ($\mathbf{L}_0$ being the angular momentum in the absence of a field) to some later time where the magnetic field has reached a value $\mathbf{H}$, then

$$\Delta\mathbf{L} = \frac{eR^2\mathbf{H}}{2c} \tag{5.66}$$

The angular momentum of the electron has been increased by the application of the magnetic field. The magnetic moment associated with electron's angular momentum is given by $\boldsymbol{\mu} = (-e/2mc)\mathbf{L}$ and the resulting change in magnetic moment will be

$$\Delta\boldsymbol{\mu} = -\frac{e^2R^2\mathbf{H}}{4mc^2} \tag{5.67}$$

The net effect is to set up a magnetic field in opposition to the applied magnetic field. We now realize the reason for the tendency of a diamagnetic material to be repelled by a magnetic field.

In any real atomic system, we do not have all of the atoms oriented so that their orbital planes are perpendicular to an arbitrarily applied magnetic field, but rather they will have random orientations. For a randomly oriented atom, the distance R will be given by $R = (x^2 + y^2)^{1/2}$. This is shown in Fig. 5.8. The average value of $x^2 + y^2$ is related to the true radial coordinate r by $(x^2 + y^2)_{av} = \frac{2}{3}(r^2)_{av}$. The change in magnetic moment is thus

$$\Delta\boldsymbol{\mu} = -\frac{e^2\mathbf{H}}{6mc^2}(r^2)_{av} \tag{5.68}$$

At this point we can recognize the possibility of relating the diamagnetism of a substance to its electron distribution through the calculation of average values based on particular wave functions for the electrons. It is also to be noted that since $\Delta\boldsymbol{\mu}$ depends directly on r^2 the outer or valence electrons will have a greater effect on the diamagnetic susceptibility.

It is conventional to discuss diamagnetism in terms of the susceptibility rather than the change in dipole moment per atom. If we multiply Eq. (5.68) by

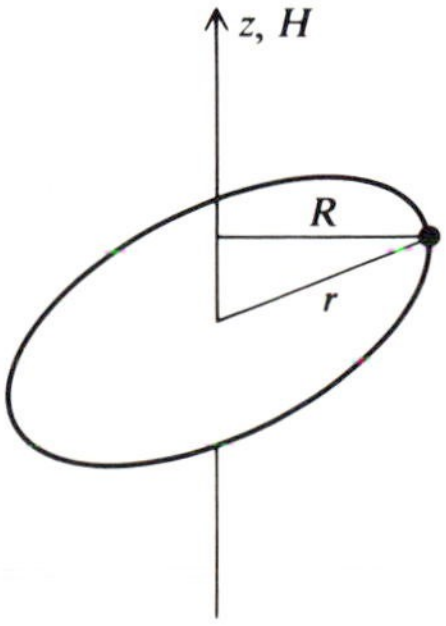

FIGURE 5.8
Effect of a magnetic field on a randomly oriented atom.

N_0, the Avagadro number, we have the total magnetic moment per mole, $\mathbf{M} = N_0\Delta\boldsymbol{\mu} = \chi_M\mathbf{H}$. Hence the molar diamagnetic susceptibility is

$$\chi_M^d = -\frac{N_0e^2}{6mc^2}(r^2)_{\text{av}} = -2.832 \times 10^{10}(r^2)_{\text{av}} \qquad (5.69)$$

The negative sign which occurs in this expression indicates that the diamagnetic susceptibility results in a decrease in the magnetic induction through the sample and produces a repulsion of the material by the magnetic field. Diamagnetism is a fundamental property of all matter irrespective of whether it possesses a permanent magnetic moment.

The quantum mechanical theory of susceptibilities was first presented by Van Vleck [7]. We will not attempt to derive or even condense the work of Van Vleck but will state the final results following a brief discussion of paramagnetism.

The classical theory of paramagnetism as developed by Langevin [6] considers each atom or molecule as a small magnet, having a magnetic moment $\boldsymbol{\mu}$, and undergoing thermal agitation proportional to the absolute temperature. The magnetic moment may be due to either orbital motion of an electron or to an unpaired electron spin. Considering that the thermal agitation of the system serves to disorder the system, the molar susceptibility, due to the permanent moments, was found to be

$$\chi_M^p = \frac{N_0\mu^2}{3kT} = \frac{C}{T} \qquad (5.70)$$

We note that C will be a property of the substance. This relationship had been established experimentally by P. Curie [8] prior to the advent of the theory. Since all materials possess diamagnetic behavior the total molar susceptibility of a substance will be given by the sum of Eqs. (5.69) and (5.70). In practice, the diamagnetic contribution is only a small fraction of the paramagnetic one when the latter is present so, except for accurate measurements or for large molecules, it is often neglected. The development of Eq. (5.70) also assumes that the system is rigid and unchanged by any external fields. In practice, an external field will distort (polarize) the system and the paramagnetic molar susceptibility is more correctly given by

$$\chi_m^p = \frac{N_0\mu^2}{3kT} + N_0\alpha \qquad (5.71)$$

where α is a measure of this distortion.

The magnetic moment due to electron spin can be written as

$$\mu = -g\beta[s(s+1)]^{1/2} \qquad (5.72)$$

By analogy the moment due to an orbital electron specified by quantum number l would be

$$\mu = -g\beta[l(l+1)]^{1/2} \qquad (5.73)$$

If we let μ_{eff} be the number of Bohr magnetons making up the moment of a single atom or ion of a particular system $\mu = \mu_{\text{eff}}\beta$ then, neglecting the last term in Eq. (5.71), we have

$$\mu_{\text{eff}} = \left[\frac{\chi_M^p}{N_0}\frac{3kT}{\beta}\right]^{1/2} \tag{5.74}$$

It is to be noted that μ_{eff} can be found without neglecting $N_0\alpha$ if one measures χ_M as a function of temperature and uses the slope of the resulting curve. From Eqs. (5.72) and (5.73) we can write

$$\mu_{\text{eff}} = g[s(s+1)]^{1/2} \tag{5.75}$$

$$\mu_{\text{eff}} = g[l(l+1)]^{1/2} \tag{5.76}$$

If a system has several unpaired electrons or orbital electrons to be considered then the total momentum quantum numbers S and L may be substituted for s and l. In practice, the experimentally determined vaue of μ_{eff} for a system does not indicate an integer number of electron spins or orbital electrons, but reflects some admixture of the two. The determination of μ_{eff}, however, can often serve as a guide for more detailed electron paramagnetic resonance (EPR) studies and can often give an indication of the electronic structure of the system. Further discussion of these points may be found in the general references presented at the end of the chapter.

Before turning our attention to the quantum theory of spin we will state, without any derivation, the results of the quantum theory of magnetic susceptibility as formulated by Van Vleck. This is being done primarily to illustrate the parallelism with the classical expression rather than to provide working relationships. The total molar susceptibility of a polyatomic molecule is given by

$$\chi_m = \frac{N_0\mu^2}{3kT} + \frac{2N_0}{3}\sum_{n'\neq n}\frac{(n|\mathbf{M}^\circ|n')^2}{(E_{n'}-E_n)} - \frac{N_0e^2}{6mc^2}\sum_{n'\neq n}\bar{r}_n^2 \tag{5.77}$$

where (1) the second term involves the sum over the electronic states of the system and gives the polarization term, and (2) the third term is the diamagnetic correction and also involves a sum over the electron states.

5.8 QUANTUM MECHANICS OF SPIN

The early work of Dirac lead to the establishment of the wave function of the electron as being the product of a function of the spatial coordinates, $\psi_{nlm}(r\theta\phi)$, and a spin function ψ_s. These functions can be considered independently and the eigenstates of ψ_s obtained without recourse to the spatial function. Since the "spin" of a particle manifests itself as an angular momentum, its behavior will be analogous to that for angular momentum as discussed previously. At this point we will consider a system of electrons and later expand these concepts to include nuclei. If the total spin momentum is denoted by $\mathbf{S}$ then use of the angular momentum relationships in Sec. 3.6 shows that $\mathbf{S}_X\mathbf{S}_Y - \mathbf{S}_Y\mathbf{S}_X = i\hbar\mathbf{S}_Z$ (two additional relationships are obtained by cycling x, y, and z), $\mathbf{S}^2 = \mathbf{S}_X^2 + \mathbf{S}_Y^2 + \mathbf{S}_Z^2$

and $(\mathbf{S}^2\mathbf{S}_F - \mathbf{S}_F\mathbf{S}^2) = 0$ $(F = X, Y, Z)$. Furthermore, the eigenvalue of $\mathbf{S}^2$ is $S(S+1)\hbar^2$ and that of $\mathbf{S}_Z$ is $M_S\hbar$ where $M_S = S, S-1, S-2, \ldots, -S+1, -S$. This means that operations on the spin states ψ_s by $\mathbf{S}^2$ and $\mathbf{S}_Z$ give

$$\mathbf{S}^2\psi_S = S(S+1)\hbar^2\psi_S \tag{5.78}$$

$$\mathbf{S}_Z\psi_S = M_S\hbar\psi_S \tag{5.79}$$

In general the matrix elements for the various components and their combinations of $\mathbf{S}$ can be obtained by referring to Table 3.1 and by substituting $\mathbf{S}$ for $\mathbf{P}$, S for J, and M_s for M. The same holds true for nuclei only $\mathbf{P} \rightarrow \mathbf{I}$, $J \rightarrow I$, and $M \rightarrow M_I$.

Examination of the case of a single electron having $S = s = \frac{1}{2}$ and $M_s = m_s = \pm\frac{1}{2}$ will provide further insight into the nature of the spin functions. Since the proton has a nuclear spin, $I = \frac{1}{2}$, the concepts examined regarding the electron hold equally well for a proton. Since $S = \frac{1}{2}$, M_s is limited to $\pm\frac{1}{2}$, and the electron has only two spin states which are designated as α and β so that

$$\mathbf{S}_Z\alpha = \tfrac{1}{2}\hbar\alpha \qquad \text{or} \qquad \mathbf{S}_Z|\alpha\rangle = \tfrac{1}{2}\hbar|\alpha\rangle \tag{5.80}$$

$$\mathbf{S}_Z\beta = -\tfrac{1}{2}\hbar\beta \qquad \text{or} \qquad \mathbf{S}_Z|\beta\rangle = -\tfrac{1}{2}\hbar|\beta\rangle \tag{5.81}$$

Use of Eqs. (3.230) and (3.240) and the preceding arguments show

$$\mathbf{S}_Z\mathbf{S}_\pm = \mathbf{S}_\pm(\mathbf{S}_Z \pm \hbar) \tag{5.82}$$

$$\mathbf{S}^2 - \mathbf{S}_Z^2 = \mathbf{S}_-\mathbf{S}_+ + \hbar\mathbf{S}_Z = \mathbf{S}_+\mathbf{S}_- - \hbar\mathbf{S}_Z \tag{5.83}$$

where $\mathbf{S}_\pm = \mathbf{S}_\mathbf{X} \pm i\mathbf{S}_\mathbf{Y}$. We will now find the eigenvalues of the spin state $\mathbf{S}_+\beta$. Using Eqs. (5.82) and (5.81) gives

$$\mathbf{S}_Z\mathbf{S}_+|\beta\rangle = \mathbf{S}_+(\mathbf{S}_Z + \hbar)|\beta\rangle \tag{5.84}$$

and

$$\mathbf{S}_+\mathbf{S}_Z|\beta\rangle = \mathbf{S}_+[\tfrac{1}{2}\hbar|\beta\rangle] = \tfrac{1}{2}\hbar\mathbf{S}_+|\beta\rangle \tag{5.85}$$

Therefore

$$\mathbf{S}_+|\beta\rangle = |\alpha\rangle \tag{5.86}$$

and we find that the operators $\mathbf{S}_\pm$ behave in the same manner as the $\mathbf{P}_\pm$ operators of Chap. 3. Conventional orthogonality relationships are also valid for spin states

$$\langle\alpha|\alpha\rangle = \langle\beta|\beta\rangle = 1 \tag{5.87}$$

$$\langle\alpha|\beta\rangle = \langle\beta|\alpha\rangle = 0 \tag{5.88}$$

A problem frequently encountered in manipulation of spin operators is that of a physical interaction of particles being represented by a vector sum or by the scalar product of a pair of spin operators. When one has a pair of commuting spin operators such as $\mathbf{I}$ and $\mathbf{S}$, for example, one mode of interaction is for them to combine vectorially to give a resultant $\mathbf{F} = \mathbf{I} + \mathbf{S}$. The spin states of the coupled system are the eigenstates of $\mathbf{F}^2$ and $\mathbf{F}_Z$ such that

$$\mathbf{F}^2\psi_S = F(F+1)\hbar^2\psi_S \tag{5.89}$$

$$\mathbf{F}_Z\psi_S = M_F\hbar\psi_S \tag{5.90}$$

and the possible values of F, found by looking at the possible vector combinations by $\mathbf{I}$ and $\mathbf{S}$, will be $F = (I + S), (I + S - 1) \cdots |I - S|$. In some instances a physical interaction involves the scalar product of two spin operators, such as nuclear spin-electron spin interactions $\mathbf{I} \cdot \mathbf{S}$. Since $(\mathbf{I} + \mathbf{S})^2 = \mathbf{I}^2 + 2\mathbf{I} \cdot \mathbf{S} + \mathbf{S}^2$ it follows that

$$\mathbf{I} \cdot \mathbf{S} = \tfrac{1}{2}[(\mathbf{I} + \mathbf{S})^2 - \mathbf{I}^2 - \mathbf{S}^2] = \tfrac{1}{2}[\mathbf{F}^2 - \mathbf{I}^2 - \mathbf{S}^2] \tag{5.91}$$

and if $\psi_{S,I}$ is an eigenstate of F^2 then

$$\mathbf{I} \cdot \mathbf{S}\psi_{S,I} = \frac{\hbar^2}{2}[F(F+1) - I(I+1) - S(S+1)]\psi_{S,I} \tag{5.92}$$

Having examined the quantum mechanical properties of a single spin operator and its interaction with a second commuting operator we will now expand these concepts by looking at multiple spin systems where all of the spin quantum numbers are $\frac{1}{2}$. This treatment is applicable to nuclei with $I = \frac{1}{2}$ as well as electrons.

For an electron i with $s(i) = \frac{1}{2}$ there will be two spin states, α_i and β_i, having $m_s(i) = \frac{1}{2}$ and $m_s(i) = -\frac{1}{2}$. For two electrons whose moments are vectorially coupled the spin wave function is a product of the individual functions and the total spin operator is a sum of the individual operators $\mathbf{S} = \mathbf{s}(1) + \mathbf{s}(2)$. The magnitude of S will be limited to values of 0 or 1 since $s(i) = \frac{1}{2}$. For $S = 0$ and $M_S = 0$ there exists a single spin state. Such a state is called a singlet state. For $S = 1$, $M_S = 0, \pm 1$ and there are three spin states. This state is called the triplet state. We can systematically evaluate the possible spin states by beginning with the $\alpha_1\alpha_2$ state having both electrons with $M_S(i) = \frac{1}{2}$.

Looking at the eigenvalues of $\alpha_1\alpha_2$ we find, using $\mathbf{S}^2 = (\mathbf{s}(1) + \mathbf{s}(2))^2 = [\mathbf{S}_-\mathbf{S}_+ + \mathbf{S}_Z(\mathbf{S}_Z + \hbar)]$, $\mathbf{S}_\pm = \mathbf{s}_\pm(1) + \mathbf{s}_\pm(2)$ and $\mathbf{S}_Z = \mathbf{s}_Z(1) + \mathbf{s}_Z(2)$, that

$$\mathbf{S}^2\alpha_1\alpha_2 = [\mathbf{S}_-\mathbf{S}_+ + \mathbf{S}_Z(\mathbf{S}_Z + \hbar)]\alpha_1\alpha_2 \tag{5.93}$$

However,

$$\mathbf{S}_+\alpha_1\alpha_2 = [\mathbf{s}_+(1) + \mathbf{s}_+(2)]\alpha_1\alpha_2 = 0 \tag{5.94}$$

and

$$\mathbf{S}_Z\alpha_1\alpha_2 = [s_Z(1) + s_Z(2)]\alpha_1\alpha_2 = \hbar[m_s(1) + m_s(2)]\alpha_1\alpha_2 \tag{5.95}$$

For the α_i states $m_s(i) = +\frac{1}{2}$, and it follows that $\mathbf{S}_z\alpha_1\alpha_2 = \hbar\alpha_1\alpha_2$. Since $\alpha_1\alpha_2$ is the highest of the states and $M_S = S, S-1, \ldots, 0, \ldots, -S+1, -S$

$$\mathbf{S}^2\alpha_1\alpha_2 = S(S+1)\hbar^2\alpha_1\alpha_2 \tag{5.96}$$

and the eigenvalues of $\mathbf{S}^2$ are $S(S+1)\hbar^2$. Applying the $\mathbf{S}_-$ operator to the $\alpha_1\alpha_2$ state we have

$$\mathbf{S}_-\alpha_1\alpha_2 = [\mathbf{s}_-(1) + \mathbf{s}_-(2)]\alpha_1\alpha_2 = (\beta_1\alpha_2 + \alpha_1\beta_2)\hbar \tag{5.97}$$

A repeat gives

$$\mathbf{S}_-(\alpha_1\beta_2 + \beta_1\alpha_2) = \beta_1\beta_2\hbar \tag{5.98}$$

We have thus generated, except for normalization factors, the triplet spin states $\alpha_1\alpha_2$, $\alpha_1\beta_2 + \beta_1\alpha_2$, and $\beta_1\beta_2$. Since $m_s(i) = \frac{1}{2}$ is the eigenfunction of α_i and $m_s(i) = -\frac{1}{2}$ that of β_i, it follows that the eigenvalue is $S = 1$ and the M_S values of these three states are $1, 0, -1$, respectively.

Since one of the triplet functions combines $\alpha_1\beta_2$ and $\beta_1\alpha_2$, the remaining singlet function must do likewise. In order to conserve dimensionality of function space and retain normalization it must be $1/\sqrt{2}\,(\alpha_1\beta_2 - \beta_1\alpha_2)$ while the normalized form of the triplet function is $1/\sqrt{2}\ (\alpha, \beta_2 + \beta_1\alpha_2)$. Operation on this singlet function by $\mathbf{S}^2$ shows that $S = 0$. The total allowed states for a two-electron spin system is summarized as follows:

State	S	M_S
$\alpha_1\alpha_2$	1	1
$\frac{1}{\sqrt{2}}(\alpha_1\beta_2 + \beta_1\alpha_2)$	1	0
$\frac{1}{\sqrt{2}}(\alpha_1\beta_2 - \beta_1\alpha_2)$	0	0
$\beta_1\beta_2$	1	−1

It is appropriate at this point to inject a few explanatory notes regarding the occurrence of the spin function involving $\alpha_1\beta_2 \pm \beta_1\alpha_2$. Upon initial examination of a two-spin system one would conclude that the spin states would be $\alpha_1\alpha_2$, $\alpha_1\beta_2$, $\beta_1\alpha_2$, and $\beta_1\beta_2$. Since all wavefunctions must be either symmetric or antisymmetric with respect to the exchange of two particles, the functions $\alpha_1\beta_2$ and $\beta_1\alpha_2$ are not satisfactory functions for a pair of indistinguishable spins. Satisfactory functions, which have the necessary symmetry properties, can be constructed by taking linear combinations and appropriately normalizing them. It should be noted that when discussing nuclei with $I = \frac{1}{2}$ the $\alpha\beta$ notation for spin states is commonly used whereas for $I \geqq 1$ a more general symbolism, such as ψ_{m_I}, is employed.

A more convenient derivation of the spin functions $\alpha_1\beta_2 \pm \beta_1\alpha_2$ is provided by the use of group theory and projection operators which were discussed in Chap. 4. If we consider the two-spin system to belong to the C_{2v} point group then we can determine the irreducible representations to which the spin functions belong and then construct the functions themselves. For two spins symmetrically situated about the origin the effects of the C_{2v} group operations as they permute the spins are shown in Table 5.2. By examination of the number of changed and unchanged spins upon application of the permutations encountered by applying the group operations we can arrive at a general representation having the characters

	E	C_2	σ	σ'	
Γ	4	2	4	2	(5.99)

TABLE 5.2
Transformation of double-spin functions

	New function			
Initial function	E	C_2	σ	σ'
$\alpha_1\alpha_2$	$\alpha_1\alpha_2$	$\alpha_1\alpha_2$	$\alpha_1\alpha_2$	$\alpha_1\alpha_2$
$\alpha_1\beta_2$	$\alpha_1\beta_2$	$\beta_1\alpha_2$	$\alpha_1\beta_2$	$\beta_1\alpha_2$
$\beta_1\alpha_2$	$\beta_1\alpha_2$	$\alpha_1\beta_2$	$\beta_1\alpha_2$	$\alpha_1\beta_2$
$\beta_1\beta_2$	$\beta_1\beta_2$	$\beta_1\beta_2$	$\beta_1\beta_2$	$\beta_1\beta_2$

Using the C_{2v} character table, this general representation is shown to be composed of

$$\Gamma = 3A_1 + B_1 \qquad (5.100)$$

The necessary projection operators are found by use of Eq. (4.16) to be

$$\mathbf{P}^{A_1} = \tfrac{1}{4}(\mathbf{E} + \mathbf{C}_2 + \boldsymbol{\sigma} + \boldsymbol{\sigma}') \qquad (5.101)$$

$$\mathbf{P}^{B_1} = \tfrac{1}{4}(\mathbf{E} - \mathbf{C}_2 + \boldsymbol{\sigma} - \boldsymbol{\sigma}') \qquad (5.102)$$

Using these and Table 5.2 we can generate the basic functions

$$\mathbf{P}^{A_1}\alpha_1\alpha_2 = \alpha_1\alpha_2 \qquad (5.103)$$

$$\mathbf{P}^{A_1}\alpha_1\beta_2 = \tfrac{1}{2}(\alpha_1\beta_2 + \beta_1\alpha_2) \qquad (5.104)$$

$$\mathbf{P}^{A_1}\beta_1\beta_2 = \beta_1\beta_2 \qquad (5.105)$$

$$\mathbf{P}^{B_1}\alpha_1\beta_2 = \tfrac{1}{2}(\alpha_1\beta_2 - \beta_1\alpha_2) \qquad (5.106)$$

which, when normalized, give the previously derived set of spin functions.

As a second example of a multiple-spin system let us consider three equivalent interacting nuclei each with $I_i = \frac{1}{2}$. If the three nuclei were completely distinguishable from each other then one would expect a set of spin states as follows, where $M_F = \sum_{i=1}^{3} M_I(i)$:

Function	$M_I(1)$	$M_I(2)$	$M_I(3)$	M_F
$\alpha_1\alpha_2\alpha_3$	$\frac{1}{2}$	$\frac{1}{2}$	$\frac{1}{2}$	$\frac{3}{2}$
$\alpha_1\alpha_2\beta_3$	$\frac{1}{2}$	$\frac{1}{2}$	$-\frac{1}{2}$	$\frac{1}{2}$
$\alpha_1\beta_2\alpha_3$	$\frac{1}{2}$	$-\frac{1}{2}$	$\frac{1}{2}$	$\frac{1}{2}$
$\beta_1\alpha_2\alpha_3$	$-\frac{1}{2}$	$\frac{1}{2}$	$\frac{1}{2}$	$\frac{1}{2}$
$\alpha_1\beta_2\beta_3$	$\frac{1}{2}$	$-\frac{1}{2}$	$-\frac{1}{2}$	$-\frac{1}{2}$
$\beta_1\alpha_2\beta_3$	$-\frac{1}{2}$	$\frac{1}{2}$	$-\frac{1}{2}$	$-\frac{1}{2}$
$\beta_1\beta_2\alpha_3$	$-\frac{1}{2}$	$-\frac{1}{2}$	$\frac{1}{2}$	$-\frac{1}{2}$
$\beta_1\beta_2\beta_3$	$-\frac{1}{2}$	$-\frac{1}{2}$	$-\frac{1}{2}$	$-\frac{3}{2}$

If the spins are indistinguishable then only the first and last of these spin states are suitable functions and the others must be synthesized into correct linear

TABLE 5.3
Transformation of triple-spin functions

Initial function	New function E	C_3	C_3^2	σ_1	σ_2	σ_3
$\alpha_1\alpha_2\alpha_3$	$\alpha_1\alpha_2\alpha_3$	$\alpha_1\alpha_2\alpha_3$	$\alpha_1\alpha_2\alpha_3$	$\alpha_1\alpha_2\alpha_3$	$\alpha_1\alpha_2\alpha_3$	$\alpha_1\alpha_2\alpha_3$
$\alpha_1\alpha_2\beta_3$	$\alpha_1\alpha_2\beta_3$	$\beta_1\alpha_2\alpha_3$	$\alpha_1\beta_2\alpha_3$	$\beta_1\alpha_2\alpha_3$	$\alpha_1\alpha_2\beta_3$	$\alpha_1\beta_2\alpha_3$
$\alpha_1\beta_2\alpha_3$	$\alpha_1\beta_2\alpha_3$	$\alpha_1\alpha_2\beta_3$	$\beta_1\alpha_2\alpha_3$	$\alpha_1\beta_2\alpha_3$	$\beta_1\alpha_2\alpha_3$	$\alpha_1\alpha_2\beta_3$
$\beta_1\alpha_2\alpha_3$	$\beta_1\alpha_2\alpha_3$	$\alpha_1\beta_2\alpha_3$	$\alpha_1\alpha_2\beta_3$	$\alpha_1\alpha_2\beta_3$	$\alpha_1\beta_2\alpha_3$	$\beta_1\alpha_2\alpha_3$
$\alpha_1\beta_2\beta_3$	$\alpha_1\beta_2\beta_3$	$\beta_1\alpha_2\beta_3$	$\beta_1\beta_2\alpha_3$	$\beta_1\beta_2\alpha_3$	$\beta_1\alpha_2\beta_3$	$\alpha_1\beta_2\beta_3$
$\beta_1\alpha_2\beta_3$	$\beta_1\alpha_2\beta_3$	$\beta_1\beta_2\alpha_3$	$\alpha_1\beta_2\beta_3$	$\beta_1\alpha_2\beta_3$	$\alpha_1\beta_2\beta_3$	$\beta_1\beta_2\alpha_3$
$\beta_1\beta_2\alpha_3$	$\beta_1\beta_2\alpha_3$	$\alpha_1\beta_2\beta_3$	$\beta_1\alpha_2\beta_3$	$\alpha_1\beta_2\beta_3$	$\beta_1\beta_2\alpha_3$	$\beta_1\alpha_2\beta_3$
$\beta_1\beta_2\beta_3$	$\beta_1\beta_2\beta_3$	$\beta_1\beta_2\beta_3$	$\beta_1\beta_2\beta_3$	$\beta_1\beta_2\beta_3$	$\beta_1\beta_2\beta_3$	$\beta_1\beta_2\beta_3$

combinations. The nature of these linear combinations can be determined by use of group theory. Since there are three identical nuclei let us consider the geometry where they are ordered so as they have C_{3v} symmetry and construct a transformation table for the behavior of the component functions. Looking at the numbers of changed and unchanged functions given in Table 5.3 we can write, for the characters of the general representation of these functions,

$$\begin{array}{cccc} & E & C_3 & \sigma \\ \Gamma & 8 & 2 & 4 \end{array} \tag{5.107}$$

This representation is found to be composed of

$$\Gamma = 4A_1 + 2E \tag{5.108}$$

The appropriate projection operators are

$$\mathbf{P}^{A_1} = \tfrac{1}{6}(\mathbf{E} + \mathbf{C}_3 + \mathbf{C}_3^2 + \boldsymbol{\sigma}_1 + \boldsymbol{\sigma}_2 + \boldsymbol{\sigma}_3) \tag{5.109}$$

$$\mathbf{P}^{E} = \tfrac{1}{3}(2\mathbf{E} - \mathbf{C}_3 - \mathbf{C}_3^2) \tag{5.110}$$

Applying the $\mathbf{P}^{A_1}$ operator to the initial functions we obtain

$$\mathbf{P}^{A_1}\alpha_1\alpha_2\alpha_3 = \alpha_1\alpha_2\alpha_3 \tag{5.111}$$

$$\mathbf{P}^{A_1}\alpha_1\alpha_2\alpha_3 = \mathbf{P}^{A_1}\alpha_1\beta_2\alpha_3 = \mathbf{P}^{A_1}\beta_1\alpha_2\alpha_3 = \tfrac{1}{3}(\alpha_1\alpha_2\beta_3 + \alpha_1\beta_2\alpha_3 + \beta_1\alpha_2\alpha_3) \tag{5.112}$$

$$\mathbf{P}^{A_1}\alpha_1\beta_2\beta_3 = \mathbf{P}^{A_1}\beta_1\alpha_2\beta_3 = \mathbf{P}^{A_1}\beta_1\beta_2\alpha_3 = \tfrac{1}{3}(\alpha_1\beta_2\beta_3 + \beta_1\alpha_2\beta_3 + \beta_1\beta_2\alpha_3) \tag{5.113}$$

$$\mathbf{P}^{A_1}\beta_1\beta_2\beta_3 = \beta_1\beta_2\beta_3 \tag{5.114}$$

as the A_1 type spin functions. The situation regarding the remaining functions is a little more complicated due to the degeneracy of the E representation. The straightforward application of the projection operator $\mathbf{P}^E$ to the first three mixed spin functions gives

$$\mathbf{P}^E\alpha_1\alpha_2\beta_3 = \tfrac{1}{3}(2\alpha_1\alpha_2\beta_3 - \beta_1\alpha_2\alpha_3 - \alpha_1\beta_2\alpha_3) \tag{5.115}$$

$$\mathbf{P}^E\alpha_1\beta_2\alpha_3 = \tfrac{1}{3}(2\alpha_1\beta_2\alpha_3 - \alpha_1\alpha_2\beta_3 - \beta_1\alpha_2\alpha_3) \tag{5.116}$$

$$\mathbf{P}^E\beta_1\alpha_2\alpha_3 = \tfrac{1}{3}(2\beta_1\alpha_2\alpha_3 - \alpha_1\beta_2\alpha_3 - \alpha_1\alpha_2\beta_3) \tag{5.117}$$

TABLE 5.4
Basic spin functions for three equivalent $I = \frac{1}{2}$ nuclei

Spin function	C_{3v} symmetry designation	M_F
$\alpha_1\alpha_2\alpha_3$	A_1	$\frac{3}{2}$
$\frac{1}{\sqrt{3}}(\alpha_1\alpha_2\beta_3 + \alpha_1\beta_2\alpha_3 + \beta_1\alpha_2\alpha_3)$	A_1	$\frac{1}{2}$
$\frac{1}{\sqrt{6}}(2\alpha_1\alpha_2\beta_3 - \beta_1\alpha_2\alpha_3 - \alpha_1\beta_2\alpha_3)$	E	$\frac{1}{2}$
$\frac{1}{\sqrt{2}}(\alpha_1\beta_2\alpha_3 - \beta_1\alpha_2\alpha_3)$	E	$\frac{1}{2}$
$\frac{1}{\sqrt{2}}(\beta_1\alpha_2\beta_3 + \alpha_1\beta_2\beta_3)$	E	$-\frac{1}{2}$
$\frac{1}{\sqrt{6}}(2\beta_1\beta_2\alpha_3 + \alpha_1\beta_2\beta_3 + \beta_1\alpha_2\beta_3)$	E	$-\frac{1}{2}$
$\frac{1}{\sqrt{3}}(\alpha_1\beta_2\beta_3 + \beta_1\alpha_2\beta_3 + \beta_1\beta_2\alpha_3)$	A_1	$-\frac{1}{2}$
$\beta_1\beta_2\beta_3$	A_1	$-\frac{3}{2}$

We have here one more function than is necessary, and even when normalized they are not orthogonal. Since they are eigenfunctions of the system, linear combinations are also eigenfunctions. One finds that the difference of any pair is orthogonal to the third, hence one pair of suitable unnormalized functions would be $(2\alpha_1\alpha_2\beta_3 - \beta_1\alpha_2\alpha_3 - \alpha_1\beta_2\alpha_3)$ and $(\alpha_1\beta_2\alpha_3 - \beta_1\alpha_2\alpha_3)$. The complete set of normalized spin functions along with their M_F values and symmetry designations are given in Table 5.4.

When one has a system with $I \geqq \frac{1}{2}$, then a multiplicity of spin functions is accessible to the system. If, for example, $I = \frac{5}{2}$, then there are $2I + 1 = 6$ possible values of m_I ranging from $-\frac{5}{2}$ to $\frac{5}{2}$ and 6 spin functions represented by m_I. The major research effort in NMR has involved spin $I = \frac{1}{2}$ nuclei, while that in pure nuclear quadrupole resonance (NQR) spectroscopy has of necessity involved higher-spin nuclei.

Before continuing with the discussion of interactions of spins and external fields the Pauli matrices are introduced. If one examines the nature of the angular momentum matrix elements (Table 3.1) for the case of $I = \frac{1}{2}$, $M_I = \pm\frac{1}{2}$ it is observed that the matrices for the spin components may be written

$$\underline{\mathbf{I}}_X = \frac{\hbar}{2}\begin{pmatrix} 0 & 1 \\ 1 & 0 \end{pmatrix} \tag{5.118}$$

$$\underline{\mathbf{I}}_Y = i\frac{\hbar}{2}\begin{pmatrix} 0 & -1 \\ 1 & 0 \end{pmatrix} \tag{5.119}$$

$$\underline{\mathbf{I}}_Z = \frac{\hbar}{2}\begin{pmatrix} 1 & 0 \\ 0 & -1 \end{pmatrix} \tag{5.120}$$

A further convention is that for any operator **0** the elements $(\alpha|\mathbf{0}|\alpha)$, $(\alpha|\mathbf{0}|\beta)$, $(\beta|\mathbf{0}|\alpha)$ and $(\beta|\mathbf{0}|\beta)$ are the upper-left, upper-right, lower-left and lower-right elements respectively.

The material reviewed in this chapter provides background material for most of the following chapters. Magnetic and electric moments of bulk materials can be experimentally determined, but such methods lie outside the immediate area of spectroscopy. This presentation has been made to provide background information to aid in the spectroscopic discussions which follow and not to provide a thorough discussion of the nonspectroscopic methods which are reviewed in several of the general references.

REFERENCES

A. Specific

1. Pauling, L., *The Nature of the Chemical Bond*, 3d ed., Cornell University Press, Ithaca, 1960, p. 98.
2. Hinze, J., M. A. Whitehead, and H. H. Jaffe, *J. Am. Chem. Soc.*, **85**, 148 (1963).
3. Mossotti, O. F., *Mem. di Mathem. e di fisica in Modena*, **24**, II, 49 (1850).
4. Clausius, R., *Die Mechansiche Warmetheorie*, vol. II, Vieweg, 1879, p. 62.
5. Dirac, P. A. M., *The Principles of Quantum Mechanics*, 4th ed., Clarendon Press, Oxford, 1958, ch. XI, A.
6. Langevin, P. J., *Phys.*, **4**, 468, 678 (1905).
7. Van Vleck, J. H., *The Theory of Electric and Magnetic Susceptibilities*, Oxford University Press, Oxford, 1932.
8. Curie, P., *Ann. Chim. et Phys.*, **5**, 289 (1895).

B. General

Atkins, P. W., *Molecular Quantum Mechanics*, 2d ed., Oxford University Press, Oxford, 1983, part 4.
Bottcher, C. J. F., and P. Bordewijk, *Theory of Electric Polarization*, Elsevier, Amsterdam, 1978.
Davies, D. W., *The Theory of Electric and Magnetic Properties of Molecules*, Wiley, New York, 1967.
Debye, P., *Polar Molecules*, Dover Publications, New York, 1929.
Guillory, W. A., *Introduction to Molecular Structure and Spectroscopy*, Allyn and Bacon, Boston, 1977, chs. 1, 8.
LeFevre, R. J., *Dipole Moments*, Wiley, New York, 1953.
Nussbaum, A., *Electronic and Magnetic Behavior of Materials*, Prentice-Hall, Englewood Cliffs, 1967.
Van Vleck, J. H., *The Theory of Electric and Magnetic Susceptibilities*, Oxford University Press, Oxford, 1932.

PROBLEMS

5.1. A charge of 4.81×10^{-19} C is placed at a position given by $(x, y, z) = (0.1$ nm, 0.2 nm, 0.3 nm$)$ relative to the origin of a coordinate system. For this system calculate the magnitude and direction of each of the following quantities at the origin of the system.
(*a*) Electric field components, $\mathscr{E}_x$, $\mathscr{E}_y$, $\mathscr{E}_z$
(*b*) Electric field, $\boldsymbol{\mathscr{E}}$
(*c*) Electric potential, V
(*d*) The components of the electric field gradient tensor
(*e*) The components of the electric field gradient tensor in the axis system where it will be diagonal.

5.2. Using Eq. (5.16) and common Pauling electronegativity values estimate the dipole moment of the following molecules and compare them to the experimental values given in parentheses. Assume an ideal geometry for each molecule and use average bond lengths.

(*a*) CF_2Cl_2 (0.5 *D*) (*d*) CF_3NF_2 (1.4 *D*)
(*b*) $CHClF_2$ (1.4 *D*) (*e*) CH_3CHF_2 (2.2 *D*)
(*c*) $CHCl_2F$ (1.3 *D*) (*f*) $CF_2 = CH_2$ (1.4 *D*)

5.3. Using the data in Table 5.1 estimate the dipole moments of the molecules given in Prob. 5.2.

5.4. Using the data in Table 5.1 estimate the direction of the dipole moment in the following molecules.

(*a*) O = C = S (*d*) ClF_3
(*b*) CH_3OH (*e*) Phenol
(*c*) C_2H_5Cl (*f*) SO_2F_2

5.5. Since the refractive index, η, of a substance is related to the relative dielectric constant by $\eta^2 = e_r$ the molar distortion polarization can be determined by measuring the former. Given the following data:

Compound	*M*	ρ(g/cm^3)	η
CCl_2	153.82	1.594	1.4664
$SnCl_4$	260.50	2.226	1.512

calculate P_M and α for each compound.

5.6. Given the following information calculate α and μ for each compound:

Compound	P_M, m^3/mole	dP_m/d, $1/T$ m^3/mole K	*T*, K
CH_3OH	33×10^{-6}	8.5×10^{-3}	333
H_2O	60.5×10^{-6}	2.072×10^{-2}	360

5.7. The average value of r^2 for a hydrogen-like atom can be estimated from

$$\langle r^2 \rangle_{\text{av}} = \frac{a_0^2 n^4}{Z_e^2} 1 + \frac{3}{2}\left\{1 - \frac{l(l+1) - 0.333}{n^2}\right\}$$

Estimate the molar diamagnetic susceptibility of atomic H, Li, Na and K. Assume $Z_e = 1$ for all alkali metals.

5.8. For the paramagnetic molecule NO, the molar magnetic susceptibilities at 290 and 145 K are 1.455×10^{-3} and 2.307×10^{-3} respectively. Calculate μ using only these two points and calculate μ_{eff} at both temperatures assuming $\alpha = 0$.

5.9. The Gouy method for the determination of bulk magnetic susceptibility involves determining the change in weight of a sample when it is field-free and when one end experiences a field H_0 while the other end is at zero field. Calculate the force (in grams) that will be exerted by a 10,000 gauss field on a 0.5-cm diameter sample of K_3CoF_6 (μ = 4.26 Bohr magnetons) if polarization is neglected, $T = 25°C$, and the force is given by

$$F = \tfrac{1}{2}H^2(\chi_v - \chi_v')A$$

when $\chi' = 0.629 \times 10^{-6}$, the volume susceptibility of air.

5.10. Referring to a handbook for values of g_N and I calculate the magnetic moments of

(1) ^1H; (2) ^2H; (3) ^{13}C; (4) ^{35}Cl; (5) ^{14}N; (6) ^{127}I

5.11. Develop the symmetrized spin functions and classify them according to their irreducible representations for sets of spin $I = \frac{1}{2}$ nuclei having the following geometries.

(*a*) Three nuclei located at the corners of an isosceles triangle (C_{2v} symmetry)

(*b*) Four equivalent nuclei having T_d symmetry

(*c*) Four equivalent nuclei having the configurations

(1) C_{4v}; (2) C_{4h}; (3) D_{4d}; (4) D_{4h}

CHAPTER 6

THE ELECTRONIC STRUCTURE OF DIATOMIC MOLECULES—AN OVERVIEW

6.1 INTRODUCTION

For a real molecule it is impossible to separate the effects of different types of motion or interactions so that each can independently be described mathematically and discussed as separate entities. The orderly presentation of the various aspects of spectroscopy, without encountering the need to introduce as yet undiscussed concepts, poses a challenge. For example, (1) the description of the vibration of a diatomic molecule requires a prior knowledge of the potential function which in turn is related to the electronic structure of the molecule; (2) the discussion of hyperfine effects in the rotational spectra of diatomic molecules necessitates a prior understanding of the electronic states and of electrical and magnetic interactions of the molecule.

In the first five chapters mathematical and conceptual tools of quantum mechanics and group theory, and basic electric and magnetic properties have been summarized. In addition, the less familiar concepts of matrix methods were examined in some detail. In this chapter concepts which are basic to a general understanding of molecular structure and energies will be reviewed. These are, for the most part, related to the quantum mechanical characterization of molecules, and full development of the details of these concepts is outside the immediate concerns of this treatment of spectroscopy.

Following this review of molecular quantum concepts the organization of the presentation is structured to prevent, insofar as possible, having to introduce complex concepts regarding various interactions prior to a detailed discussion of their nature. This will be accomplished by (1) examining the general interactions of matter and radiation, (2) discussing the intramolecular interactions of electrons and nuclei, and (3) proceeding to the discussions of the different areas of molecular spectroscopy.

6.2 MOLECULAR ENERGIES

The total energy of a molecule is composed of contributions due to the electronic configuration, vibration, rotation, translation, and various electron-electron, nuclear-nuclear, and nuclear-electron interactions. For real molecules it is not possible to view these contributing factors as being totally independent of each other. However, in practice these various interactions are often considered separately to provide a simpler introduction to the phenomenon. The key to being able to view the different energy contributions separately lies in the fact that the energies associated with different interactions often differ by several orders of magnitude. This can allow one to consider one type of interaction as a perturbation relative to a larger one. This is particularly true for energies due to spin-spin interactions relative to those due to electronic, vibrational, and rotational effects.

The dominant terms in the Hamiltonian for a molecule in the absence of external fields will be

$$\mathcal{H} = \mathbf{T}_e + \mathbf{T}_t + \mathbf{T}_v + \mathbf{T}_r + \mathbf{V}_{en} + \mathbf{V}_{ee} + \mathbf{V}_{nn} \tag{6.1}$$

where $\mathbf{T}_e$, $\mathbf{T}_t$, $\mathbf{T}_v$, and $\mathbf{T}_r$ are kinetic energy operators representing electronic, translational, vibrational, and rotational motions and $\mathbf{V}_{en}$, $\mathbf{V}_{ee}$, and $\mathbf{V}_{nn}$ are potential energy operators representing interactions among the nuclei and electrons. In 1927, Born and Oppenheimer [1] considered the nature and magnitude of these energy terms and concluded, at least as a first approximation, that since the nuclear motion is slow relative to that of the electrons the energies due to these motions are separable or the nuclei may be considered to move under the influence of an average electronic potential. The electronic Hamiltonian may be written as

$$\mathcal{H}_e = \mathbf{T}_e + \mathbf{V}_{en} + \mathbf{V}_{ee} + \mathbf{V}_{nn} \tag{6.2}$$

and the nuclear Hamiltonian as

$$\mathcal{H}_n = \mathbf{T}_t + \mathbf{T}_v + \mathbf{T}_r \tag{6.3}$$

For a diatomic molecule the average electronic potential is a function only of a single variable, the internuclear separation, and is represented by the conventional potential energy curve.

The Born-Oppenheimer separation considers the total wavefunction ψ_{en}, which is a function of both electronic and nuclear coordinates, to be written as a product function $\psi_{en}(r_e, r_n) = \psi(r_e, r_n)\psi_n(r_n)$. Substitution of this product function into the complete Schrödinger equation

$$\mathcal{H}\psi_{en} = (\mathcal{H}_e + \mathcal{H}_n)\psi_{en} = E\psi_{en} \tag{6.4}$$

considering that the Schrödinger equations for the electronic energy is

$$\mathcal{H}_e\psi_e = E_e\psi_e \tag{6.5}$$

where the electronic eigenfunctions E_e have an explicit dependency on the electron coordinates and a parametric dependency on nuclear coordinates, leads to the Schrödinger equation for the nuclear motion

$$(\mathcal{H}_n + E_e)\psi_n = E\psi_n \tag{6.6}$$

By choosing the zero reference of energy to be at the minimum of the electronic state. The energy of the nuclear motion E_n is the solution of this Schrödinger equation

$$(\mathcal{H}_n + E_e)\psi_n = E_n\psi_n \tag{6.7}$$

where E_e contains the nuclear coordinates as a parameter, and E_n encompasses the translational, rotational, and vibrational energies.

The Born-Oppenheimer approximation has been found to be a valid method to use for the calculation of energies of molecules with zero order electronic wavefunctions. This is the case for many systems of spectroscopic interest. Further consideration of the separability of translational, vibrational, and rotational motions will be examined in Chap. 11 following discussions of the nature of molecular electronic wavefunctions and energies.

6.3 POTENTIAL ENERGY FUNCTIONS OF DIATOMIC MOLECULES

In order to solve the Schrödinger equation to obtain the rotational and vibrational energies of a molecule it is convenient to have an analytical expression for the electronic potential function. An oversimplification of this process was illustrated in Chap. 2 where it was assumed that rotational and vibrational motion were completely independent. In this case the molecule was considered rigid insofar as the rotational motion was concerned and the potential function for the solution of the vibrational problem was taken to be the classical harmonic oscillator potential. In practice there are severe limitations associated with these views of molecular motion and a more realistic treatment requires the use of a potential function which can more closely describe the real behavior of molecules. For polyatomic molecules where numerous internuclear distances are needed to describe the interactions, the potential energy function is a $3N-6$-dimensional surface whose characterization and use will not be discussed. For a diatomic molecule the potential function depends only on a single internuclear distance, hence $V = V(r)$.

Theoretically, an analytical expression for $V(r)$ can be found by solving the electronic wave equation for the molecule. Such solutions are limited to the lighter molecules though progress has been made on the problem of solving heavier molecules [2, 23]. Because of the relatively small amount of information available from quantum mechanical calculations it is common practice to use an empirical relationship for $V(r)$. From experimental studies of the electronic spectra of diatomic molecules and from quantum mechanical calculations the general form of $V(r)$ is known to be that shown by the solid curve in Fig. 6.1*a*. At large internuclear separations the interactions between the atoms are negligible and the energy is that of the individual atoms. This condition is arbitrarily taken to be the reference level $V(\infty) = 0$. Considering only electrostatic interactions one can conclude that small internuclear separations of the electron clouds of

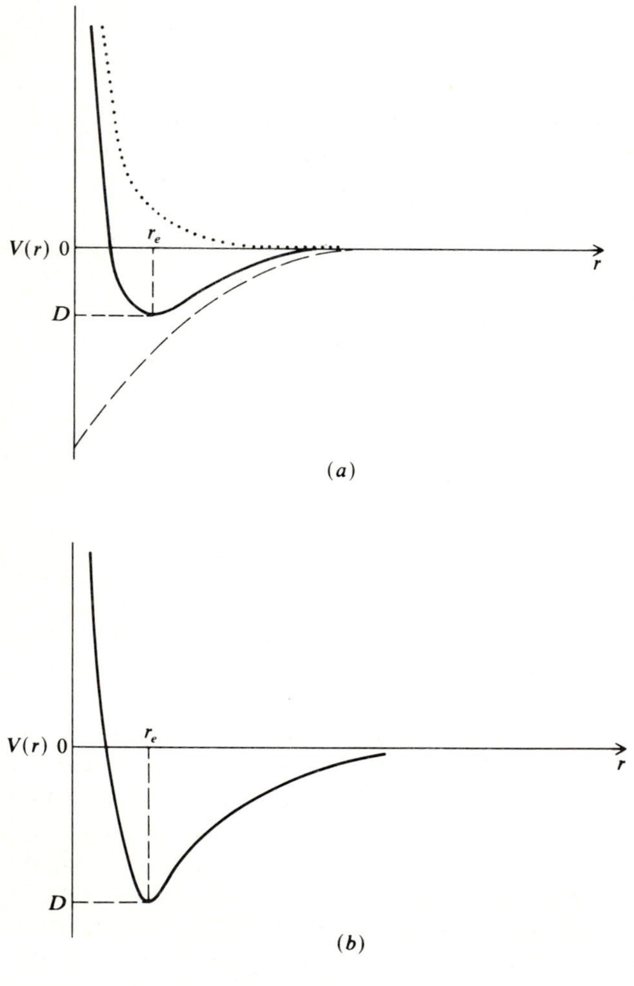

FIGURE 6.1
(*a*) Variation of potential energy $V(r)$ with internuclear separation for diatomic molecule; (*b*) Typical potential functions for a stable diatomic molecule.

the two atoms become sufficiently close together that they interpenetrate and the mutual repulsion increases. Also, at small internuclear distances nuclear-nuclear repulsion is effective. The total electrostatic repulsive potential of nuclei for nuclei and electrons for electrons is illustrated by the dotted line in Fig. 6.1*a*. As two atoms, initially at a very large separation, are brought together, the attractive forces resulting from the opposite charged particles will give rise to an attractive potential illustrated by the dashed line in Fig. 6.1*a*. The sum of these two curves gives the total electrostatic potential illustrated by the solid curve. This indicates that at one value of r, r_e, the potential energy of the system is at a minimum and the system is in a stable configuration. This discussion has been based on simplified electrostatic interactions and hence presents only a qualitative picture of the process of molecule formation and the nature of the resulting potential function. A complete quantum mechanical treatment [3*a*, 3*b*] for moderately stable molecules gives functions like that in Fig. 6.1*b*.

Over the past fifty years a number of empirical potential functions have been proposed. Following a discussion of the commonly used Morse [4] potential function several others will be presented. In Chap. 11 the use of potential functions for the solution of the rotational-vibrational Schrödinger equation will be examined in detail.

The Morse potential function, which closely approximates the behavior of real diatomic molecules over a wide range of r, is

$$V(r) = D[1 - e^{-a(r-r_e)}]^2 \tag{6.8}$$

where

a = a constant for a given electronic state
D = dissociation energy of the molecule
r_e = equilibrium internuclear separation

The constant a is a measure of the curvature or narrowness of the function and the D is the depth of the potential well. The Morse function, while being quite large at $r = 0$, does not become as large as a real potential. This is of little consequence, however, because in this region the vibrational wavefunction is found to be very small.

The Morse potential function is not the only potential function which can be used. Reflection on the discussion of the Morse potential leads to the conclusion that a number of different functions can be used. It is necessary that as a minimum such functions fulfill the following three conditions: (1) The potential $V(r)$ exhibits a minimum at $r = r_e$, and the derivative $(dV(r)/dr)_{r=r_e}$ is zero. (2) As $r \to \infty$ the potential asymptotically approaches a constant value such that $V(\infty) - V(r_e) = D$. The value $V(\infty)$ is often arbitrarily taken as zero, hence the potential of the stable system is negative and $V(r_e) = -D$. [This is not reflected in the form of Eq. (6.8).] (3) When $r \to 0$ the potential goes to infinity. This reflects the impossibility of total superposition of two nuclei. In actual practice a function that goes to a very large value rather than infinity is adequate. The Morse function satisfies these conditions:

$$(1)\ r = r_e \qquad \left(\frac{dV(r)}{dr}\right)_{r=r} = 0 \tag{6.9}$$

$$(2)\ r = \infty \qquad V(\infty) - V(r_e) = D \tag{6.10}$$

$$(3)\ r = 0 \qquad V(0) = De^{2ar_e} \qquad (\text{since } e^{2ar}e \gg 2e^{ar_e}) \tag{6.11}$$

If $V(\infty) = 0$ then $V(r_e) = -D$ and condition 2 is satisfied. An order of magnitude calculation with $D = 6.5\ \text{eV} = 5 \times 10^4\ \text{cm}^{-1}$ and $a = 1.2 \times 10^8\ \text{cm}^{-1}$ gives $V(0) = 120\ D$.

The most general diatomic potential function $V(r)$ is a Taylor series in $(r - r_e)$ about the point r_e

$$V(r) = V(r_e) + (r - r_e)\left[\frac{dV(r)}{dr}\right]_{r_e} + \frac{(r - r_e)^2}{2}\left[\frac{d^2V(r)}{dr^2}\right]_{r_e} + \frac{(r - r_e)^3}{3}\left[\frac{d^3V(r)}{dr^3}\right]_{r_e} + \cdots \tag{6.12}$$

If the potential is measured relative to $V(\infty) = 0$ then the first term is a constant $V(r_e) = -D$. The second term is zero since the potential has a minimum at $r = r_e$. Since $d^2V(r_e)/dr^2$ is a constant the third term is a harmonic potential term corresponding to the harmonic oscillator potential. This constant is the harmonic force constant for infinitesimal vibrations about r_e. If the series is terminated after the third term the potential is that used in Chap. 2 to find the harmonic oscillator wavefunctions. The fourth and higher terms in the series characterize the anharmonicity of the oscillator.

The Morse and Taylor series potential functions are not the only useful ones. The common characteristics of all empirical potential functions is that they satisfy the conditions stated by Eqs. (6.9) to (6.11). Some of the more commonly encountered potential functions are those due to Rydberg [5]

$$V(r) = -D[1 + b(r - r_e)]e^{-b(r-r_e)} \tag{6.13}$$

Lippincott [6]

$$V(r) = D\left[1 - \exp\frac{-n(r - r_e)^2}{2r}\right] \tag{6.14}$$

Rosen and Morse [7]

$$V(r) = A \tanh\frac{r}{d} - C \operatorname{sech}^2\frac{r}{d} \tag{6.15}$$

Frost and Musulin [8]

$$V(r) = \mathrm{e}^{-ar}\left(\frac{1}{r} - b\right) \tag{6.16}$$

and Varshni [9]

$$V(r) = D\left[1 - \frac{r}{r_e}e^{-a(r-r_e)}\right]^2 \tag{6.17}$$

One of the best empirical-type potential functions is the five-parameter function of Hulburt and Hirschfelder [10]

$$V(r) = D[(1 - e^{-x})^2 + cx^3 e^{-2x}(1 + bx)] \tag{6.18}$$

where $x = a(r - r_e)$. In this function a, r_e, and D have the same physical significance as in the Morse potential function and b and c are two additional constants.

As will be discussed in Chap. 11, a rigorous separation of vibrational and rotational motion is not possible and interactions must be allowed for when any potential function is employed. A diatomic potential function which includes the effect of rotation and can be employed with the WKB method [11] to give the energy levels of a diatomic vibrotor is that developed by Dunham [12]

$$V(\mathscr{L}) = a_0\mathscr{L}(1 + a_1\mathscr{L} + a_2\mathscr{L}^2 + \cdots) + B_eJ(J+1)(1 - 2\mathscr{L} + 3\mathscr{L}^2 - 4\mathscr{L}^3 + \cdots) \tag{6.19}$$

where $\mathscr{L} = (r - r_e)/r_e$, $B_e = h/8\pi^2\mu r_e^2$ and the a_i are arbitrary constants which are related to the coefficients in Eq. (6.12). The first sum of terms describes the nature of the potential as a function of nuclear displacement while the second sum introduces the effect of molecular rotation on the potential function of the molecule.

Not only is the potential energy function of a molecule dependent on the internuclear separation but also on the electronic state. For example, the electronic state might be represented by a particular configuration of molecular orbitals. When an electron is excited into a molecular orbital configuration which is

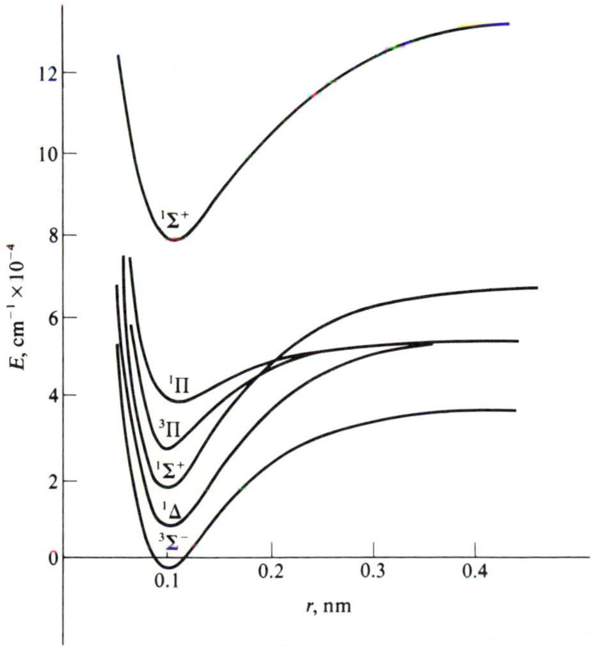

FIGURE 6.2
Potential energy functions for the ground and five stable excited states of the NH molecule.

energetically higher than that of the ground state the excited state molecule will be represented by a different potential function. If the molecule remains in a bonding state the new function will have the same general shape as the ground state function. However, the position of the minimum ($r = r_e$) will be displaced upward in energy and in all probability to a different value of r_e. This is illustrated in Fig. 6.2 which shows the ground and several excited state potential functions for NH.

6.4 ELECTRONIC WAVEFUNCTIONS

The development of electronic wavefunctions for diatomic molecules is in the domain of quantum mechanics and valence theory. We will only review those results necessary to clarify discussions of spectroscopy. Of particular interest are the concepts related to the characterization of electronic states with appropriate quantum numbers, the nature of angular momenta of electronic states, and the symmetry of the electronic wavefunctions. These are all necessary for the development of selection rules and analysis of observed spectra.

The lack of high-speed computers during the early years of quantum mechanics led to the development of two general approaches to the description of bonding in molecules, the valence bond method [13] and the molecular orbital method [14]. While they differ in their basic tenets they both lead to a set of molecular wavefunctions which describe the electronic behavior of molecules. Both methods employ varying degrees of approximation and would, in the limit where exact solutions could be found, yield the same wavefunctions.

From the viewpoint of the spectroscopist, who is generally familiar with the concepts of atomic spectroscopy and the description of atomic states using atomic wavefunctions, the molecular orbital method is preferable. The nature of the molecular orbital method is illustrated by reviewing the results for the hydrogen molecule ion and the hydrogen molecule, and extending the concepts to larger molecules. The basic concept of the molecular orbital (MO) theory is the consideration of the electrons in a molecule as occupying a set of orbitals analogous to those in an atom. When two atoms bond, the space which is accessible to an electron in one atom overlaps with that accessible to an electron in the second atom. This results in an appreciable change in the geometry of the electron distribution in the region between the two atoms but causes little change near the individual nuclei. One technique for determining the wavefunctions (molecular orbitals) which approximately describe the electron distribution in a molecule is to employ a linear combination of the atomic orbitals (LCAO) of the two combining atoms.

For example, the hydrogen molecule ion can be treated using the variational method and an approximate LCAO wavefunction of the form

$$\psi = C_A \psi_A^{1s} + C_B \psi_B^{1s} \tag{6.20}$$

Here ψ_A^{1s} and ψ_B^{1s} are the $1s$ atomic wavefunctions for atoms A and B respectively. As a first approximation, when constructing a set of molecular orbitals by use of the LCAO method there are three criteria to be observed: (1) the energies of the atomic orbitals used must be comparable, (2) the atomic orbitals from the bonding atoms should overlap to the maximum extent, and (3) the combined atomic orbitals must have a symmetry consistent with that of the molecule. As will be shown in Chap. 18, MOs for complex molecules frequently have contributions from AOs of differing energies. They also exhibit relative AO contributions of widely different magnitudes. Thus these three criteria are only guidelines for the formation of conventional strongly bonding or strongly antibonding MOs.

Since only $1s$ atomic orbitals are employed in the formation of H_2^+ molecular orbitals these three criteria are satisfied. The solutions of the variational integral for the H_2^+ molecule ion gives, to a first approximation [15],

$$E_\pm = \frac{H_{AA} \pm H_{AB}}{I \pm S_{AB}} \tag{6.21}$$

where

$$H_{IJ} = \langle \psi_I(1s) \mathscr{H} \psi_J(1s) \rangle \tag{6.22}$$

and

$$\mathscr{H} = -\frac{\hbar^2}{2m}\nabla^2 - \frac{e^2}{r_A} - \frac{e^2}{r_B} + \frac{e^2}{r_{AB}} \tag{6.23}$$

The normalized MO wavefunctions are

$$\psi_\pm = \frac{1}{\sqrt{2}}[\psi_A(1s) \pm \psi_B(1s)] \tag{6.24}$$

There are two important observations to be made regarding this solution: (1) The LCAO method applied to a pair of atomic orbitals gives a pair of molecular orbitals. There is a conservation of the number of orbitals. (2) Associated with the pair of molecular orbitals are two energy states, one lower and one higher than the initial atomic orbital states.

Examination of the symmetry, with respect to inversion of coordinates, of the two molecular orbitals, shows that ψ_+ is even with respect to inversion about the midpoint of the bond axis while ψ_- is odd with respect to the same operation. This $\pm$ character of wavefunctions is referred to as parity. Furthermore, ψ_- contains a midplane node perpendicular to the bond axis, a fact indicative of it being an antibonding function. The state with energy E_+ will be described by a potential function as shown by curve (a) in Fig. 6.3 while that with E_- will follow curve (b). These functions, while not exact, due to the approximation method used for their derivation, nevertheless are qualitatively correct.

The relationship between the potential function of a diatomic molecule and the quantum mechanical description of the system becomes apparent at this point. The Hamiltonian contains the internuclear distance R as a variable, hence, insofar as the analytical expression for E_+ is exact, it is a useful potential function. For molecules having many electrons, and even for H_2 the Hamiltonian will contain a large number of interaction terms, and it will not be possible to obtain exact solutions or an explicit formulation of the potential function. It is due to this

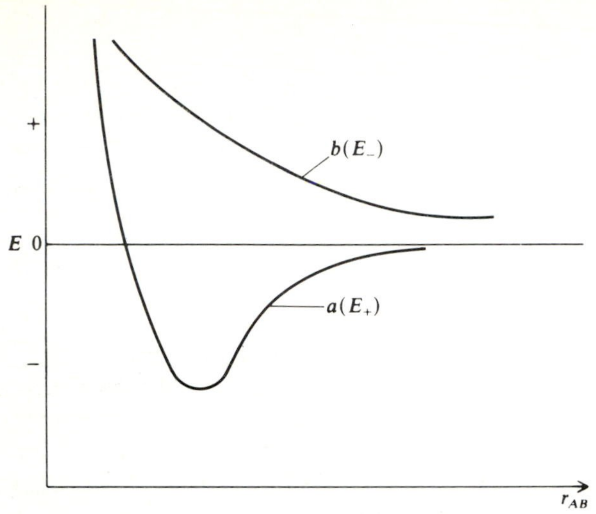

FIGURE 6.3
Energy curves for H_2^+ molecule ion.

limitation on obtaining exact analytical solutions for the state energies that empirical potential functions are employed.

In addition to the even-odd inversion symmetry shown by ψ_+ and ψ_- both functions have cylindrical symmetry relative to rotation about the internuclear axis. Molecular orbitals of these forms are denoted as $\sigma_g 1s$ and $\sigma_u^* 1s$ respectively. The g (gerade) and u (ungerade) denotes the inversion symmetry, σ the axial symmetry, * the antibonding character, and $1s$ the nature of the atomic orbitals from which the molecular orbitals are constructed.

Since the H_2^+ molecule ion contains only one electron the spin of the electron did not have to be considered in the formulation of the MO wavefunctions. When molecules which contain two or more electrons are considered it is necessary to allow for the proper inclusion of the spin wavefunctions. Furthermore, just as is the case for polyelectronic atoms, it is necessary in the formulation of the wavefunctions to allow for electron indistinguishability. To review the nature of these two factors we will summarize the treatment of the H_2 molecule [15].

For the hydrogen molecule which contains two electrons the ground state wavefunction can be formulated as the Slater determinant [15]

$$\psi_0 = N \begin{vmatrix} \psi_1(1)\alpha(1) & \psi_1(1)\beta(1) \\ \psi_1(2)\alpha(2) & \psi_1(2)\beta(2) \end{vmatrix} \tag{6.25}$$

where ψ_1 is an MO (less spin), generally of an LCAO type. Recalling the Pauli exclusion principle, which states that the total wavefunction of any fermion must be antisymmetric, we recognize that for the molecular MOs to be correct they must combine spatial functions with antisymmetric spin functions and vice versa. Equation (6.25) gives just this form (Recall discussion of the two-spin system in Chap. 5)

$$\psi_0 = N\psi_1(1)\psi_1(2)\{\alpha(1)\beta(2) - \beta(1)\alpha(2)\} \tag{6.26}$$

The LCAO form of ψ_1, limiting the AOs to the $n = 1$ levels, will be

$$\psi_1 = N'(\psi_A^{1s} + \psi_B^{1s}) \tag{6.27}$$

Combining Eqs. (6.26) and (6.27) gives as the ground state wavefunction

$$\psi_0 = N[\psi_A^{1s}(1)\psi_A^{1s}(2) + \psi_B^{1s}(1)\psi_B^{1s}(2) + \psi_A^{1s}(1)\psi_B^{1s}(2) + \psi_B^{1s}(1)\psi_A^{1s}(2)] \cdot (\alpha(1)\beta(2) - \beta(1)\alpha(2)) \tag{6.28}$$

The energy of the state is found from

$$E_0 = \langle\psi_0|\mathscr{H}|\psi_0\rangle \tag{6.29}$$

Writing the Hamiltonian operator $\mathscr{H}$ as a function of hydrogen one-electron operators $\mathbf{H}_i$

$$\mathscr{H} = -\frac{\hbar^2}{2m}\nabla_1^2 - \frac{\hbar 2}{2m}\nabla_2^2 - \frac{e^2}{r_{A1}} - \frac{e^2}{r_{A2}} - \frac{e^2}{r_{B1}} - \frac{e^2}{r_{B2}} + \frac{e^2}{r_{12}} + \frac{e^2}{r_{AB}}$$
$$= \mathbf{H}_1 + \mathbf{H}_2 + \frac{e^2}{r_{12}} + \frac{e^2}{r_{AB}} \tag{6.30}$$

it is possible to find more explicit expressions for the energy. Since the Hamiltonian does not depend on the spin the evaluation of the energy will depend only on the spatial part of the wavefunction.

Before examining the energy further let us look at the physical implications of the separate parts of the wavefunction, Eq. (6.28). The first two terms describe the system when the configuration is ionic, $H_A^-H_B^+$ or $H_A^+H_B^-$ while the latter two are covalent terms. The equal weighting of ionic and covalent forms for homonuclear molecules in this limited formulation of MO theory is a limitation in its use, especially for determination of energies. The omission of the ionic terms leads to the valence bond wavefunctions and a somewhat more satisfactory calculation of the energy.

Despite its disadvantages for calculation of energies the MO method has an inherent simplicity in its formulation that makes it appealing. The disadvantages of the equal weighting of ionic and covalent contributions can be minimized by incorporating the technique of forming the MOs from an LCAO of not only the lowest atomic orbitals but also inclusions of higher ones. For example, the MO for H_2 can be taken to have the form (less spin)

$$\psi = C^{1s}(\psi_A^{1s} + \psi_B^{1s}) + C^{2s}(\psi_A^{2s} + \psi_B^{2s}) + C^{2p}(\psi_A^{2p} + \psi_B^{2p}) + \cdots \tag{6.31}$$

Using the Roothaan-Hartree-Fock method [15] the C^{AO} coefficients and the minimum energy can be obtained.

Another method for improvement of MO results is to include configuration interaction. Any wavefunction which describes only a single configuration, even when developed using self-consistent field techniques, is still not a true wavefunction of the Hamiltonian because it does not adequately compensate for electron correlation. This correlation can be recognized by "mixing into" the wavefunction of the state of interest contributions from higher states of comparable symmetry.

By "mixing in" a sufficiently large number of excited state configurations the wavefunction will converge to the true wavefunction. The advent of high-speed computers which can evaluate thousands of integrals in the "blink of an eye" allows for the incorporation of these techniques and provides the mechanism for making accurate calculations of molecular electronic energies and associated properties. Although improvements in the speed and accuracy of such calculations within recent years have provided much in the way of numerical data for correlation with spectroscopic results the concept of orbitals and their useful pictorial and spectroscopic uses tend to become diffused.

It will, therefore, be to our advantage to continue to use the concept of one-electron MOs which relate to the bonding in molecules and to which we can assign energy levels and correlate spectroscopic data. This will be particularly true in the discussion of the electronic states and spectra of polyatomic molecules. We will then build the hierarchy of energy levels for diatomic molecules by considering the MOs to be formed from pairs of atomic orbitals. This method will enable us to develop the symmetry relationship and selection rules necessary to make correct spectral assignments. The exact spectral transition frequencies may then be correlated with the results of more sophisticated quantum mechanical calculations.

For larger molecules it is necessary to determine the nature and ordering of molecular orbitals which can be constructed from higher atomic orbitals. The qualitative nature of the molecular orbitals formed from *s* and *p* atomic orbitals is illustrated in Fig. 6.4 and the energy level scheme is qualitatively illustrated in Fig. 6.5 for the second row (Li_2–Ne_2) diatomic molecules. The notation can

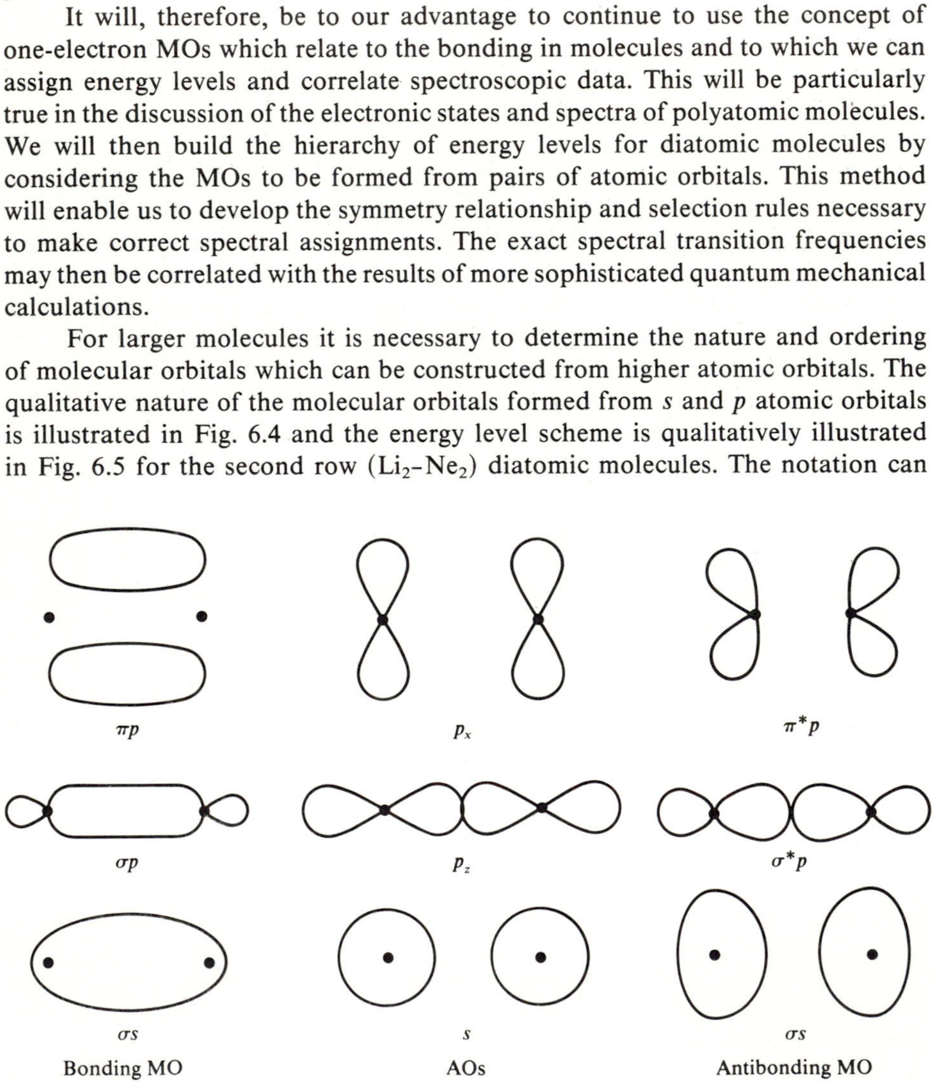

FIGURE 6.4
Angular distribution of MOs in homonuclear diatomic molecules.

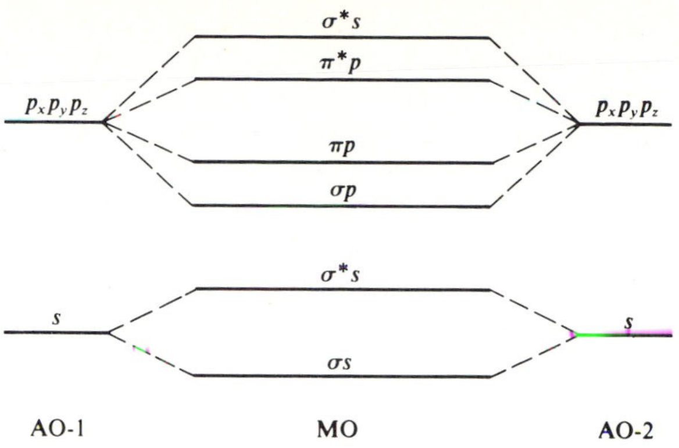

FIGURE 6.5
Energy levels for homonuclear molecular orbitals.

be expanded such that π denotes molecular orbitals which have their maximum electron density in either the *xz* or *yz* plane, are antisymmetric with respect to the plane perpendicular to that of maximum density but containing the *z* axis, and contain electrons having one unit of angular momentum. The ordering of the energy levels in this diagram is idealized, and real molecules frequently deviate from this arrangement. It should be noted that: (1) there is a conservation of the number of orbitals in going from the atomic orbitals to the molecular orbitals, (2) two equivalent atomic orbitals give rise to a symmetric and antisymmetric molecular orbital pair. The electronic configuration of a homonuclear diatomic molecule is found by filling the molecular orbitals in their proper order of energy with the available electrons in accordance with the Aufbau principle. A determination of the ordering of the MO energy levels for a molecule in both ground and excited states frequently can be inferred from the analysis of its electronic spectrum.

The development of molecular orbital schemes for heteronuclear diatomic molecules is somewhat more complex and will be examined further following the introduction of the classification and characterization of energy states and the correlation of energy states to idealized "united atom" and "separated atom" constructs. As has been pointed out, although modern quantum mechanics has progressed to the point where even geometric parameters of simple molecules can be calculated rather precisely by ab initio methods [16] this subject will not be pursued further since it would not add appreciably to our discussion of the assignment and utilization of diatomic electronic spectra.

6.5 ANGULAR MOMENTUM COUPLING AND TERM SYMBOLS FOR DIATOMIC MOLECULES

In order to classify and discuss the electronic spectra of diatomic molecules it is expedient to use a shorthand notation which is directly related to the momenta of the electrons. The electrons in a molecule will possess angular momenta due to both their orbital motions and their intrinsic spins. The coupling of these two types of momenta can be represented by a vector model analogous to that used for atoms. The coupling in atoms is summarized in App. M.

There is a primary difference between the nuclear environment experienced by the electrons in an atom and in a diatomic molecule. As a first approximation, by ignoring electron-electron interactions, the electron in an atom experiences a spherically symmetric central force field, the total angular momentum is a constant of motion (i.e., constant magnitude) and the angular momentum quantum number L is a good quantum number so long as there is insignificant interaction with electron spin. In a diatomic molecule the electron is subjected to a strong electric field located axially between the nuclei. The behavior of the electron is then determined by its interaction with this field.

The introduction of this directed electrostatic field produces an effect similar to the Stark effect in atoms in which the orbital angular momentum precesses about the field direction with quantized projections of magnitude $M_L\hbar$ ($M_L = L, L-1, \ldots, -L+1, -L$) along the field direction. Concurrent with the establishment of this precession the angular momentum L and the spin momentum **S** are uncoupled, and the spin momentum also precesses about the internuclear axis. The velocity of the precession of the electron about the internuclear axis will in general be so large that its magnitude is undefined. As the internuclear electrostatic field is increased the precession of **L** increases further and it becomes even less of a constant of motion. However, as with the Stark effect in atoms, the projection of **L** on the internuclear axis remains a constant of motion (it is conserved) and M_L is a good quantum number.

The electronic motion is then considered relative to the molecule fixed-axis system with the z axis lying in the internuclear direction. Later, the model will be expanded to combine the coupling of molecular rotational angular momenta and electronic angular momenta.

For molecules having a magnitude of the orbital angular momentum which is nonzero there will exist multiple energy levels corresponding to the different allowed values of M_L. However, as in the case of the Stark effect in atoms, the energies of these levels depend on $|M_L|$ and hence, with the exception of the $M_L = 0$ level, are doubly degenerate in the quantum number M_L. It is conventional to denote the absolute value of M_L with the quantum number Λ such that $\Lambda = |M_L| = 0, 1, \ldots, L$. This degeneracy can be viewed in simple terms as being due to states of equal energy in which the direction of the electron precession is reversed. Analogous to the designations for atomic angular momenta the states of diatomic molecules are denoted by term symbols Σ, Π, Δ, Φ, etc., for $\Lambda = 0, 1, 2, 3, \ldots$.

The magnitude of Λ can be correlated with the symmetry of the electron cloud just as that of the electron orbital quantum number l could be correlated with the symmetry of atomic orbitals. Thus, for $\Lambda = 0$ the electron cloud possesses cylindrical symmetry about the internuclear axis, for $\Lambda = 1$ it is symmetric with respect to a plane parallel to the internuclear axis, etc. For a given molecule there will generally be a large difference in energy between states possessing different values of Λ.

The previous discussion has provided a qualitative view of the interactions in diatomic molecules but is not a quantitative quantum mechanical view. A rigorous quantum mechanical treatment shows that the effect of the axial field is to mix states of angular momentum denoted by L with those denoted by $L \pm 1$, thereby limiting the usefulness of L as a good quantum number.

In addition to the orbital angular momentum characterized by Λ the angular momenta due to the intrinsic spins of the electrons must be considered. Each electron in a molecule will have an intrinsic spin angular momentum of magnitude $\frac{1}{2}\hbar$. These individual spin momenta can couple to produce a total resultant spin momentum **S**. Associated with this total spin momentum will be a magnetic moment which is capable of interacting with other magnetic fields in the molecule. This magnetic moment cannot interact with the axial electric field of the nuclei. However, the orbital motion of the electrons (unless $\Lambda = 0$) will produce an axial magnetic field with which the spin magnetic moment can interact. The result of this interaction will be an independent precession of the total spin momentum **S** about the z axis. The total spin momentum component along the z axis is denoted by the symbol S, analogous to M_S for atoms, and can have values of $\Sigma = S, S-1, \ldots, -S+1, -S$. Care must be taken not to confuse the use of Σ to denote spin components and to represent the molecular state with $\Lambda = 0$. The relationships between **L** and Λ and between **S** and Σ are illustrated in Fig. 6.6.

Although the rapid precession of **L** dictates that the only defined components lie along the z axis the situation for **S** is not the same. For $\Lambda \geqq 1$ such that there exists a resulting magnetic field about which **S** can precess it can have well-defined directions relative to the z axis and hence will have $2S+1$ possible components in the z direction. The quantity $2S+1$ is referred to as the multiplicity of the state, and for a given orbital angular momentum state is denoted with a left superscript on the term symbol. Figure 6.7 illustrates the vector relationships for molecules having $\Lambda = 1$, $S = 1$and $\Lambda = 2$, $S = \frac{1}{2}$. In addition to Λ and Σ a total electronic angular momentum $\Omega = |\Lambda + \Sigma|$ is defined. Its magnitude is often denoted by a right subscript on the term symbol. The total angular momentum is also shown for the examples in Fig. 6.7. For $\Lambda \geqq S$ there will be $2S+1$ different values of Ω for each value of Λ. Due to the interaction of **S** with the magnetic field produced by the orbital motion of the electrons the electronic levels corresponding to different values of Ω will have different energies. If $\Lambda = 0$, due to the absence of any orienting magnetic field, both Σ and Ω are undefined. For this situation no splitting of energy levels occurs and the spin states are single in the absence of any molecular rotation or external fields.

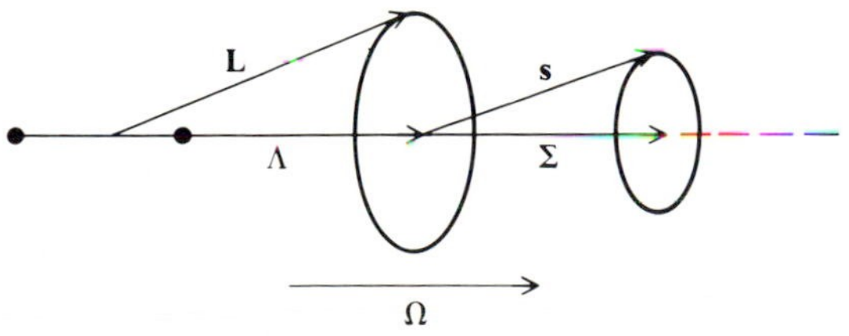

FIGURE 6.6
Momenta relationships for a diatomic molecule.

Quantum numbers				Vector relationships	Term symbol
Λ	Σ	Ω	$2s+1$		
1	1	2	3	$\Lambda = 1$ $\Sigma = 1$	$^3\Pi_2$
1	0	1	3	$\Lambda = 1$ $\Sigma = 0$	$^3\Pi_1$
1	-1	0	3	$\Lambda = 1$ $\Sigma = -1$	$^3\Pi_0$
2	$\frac{1}{2}$	$\frac{5}{2}$	2	$\Lambda = 2$ $\Sigma = \frac{1}{2}$	$^2\Delta_{5/2}$
2	$-\frac{1}{2}$	$\frac{3}{2}$	2	$\Lambda = 2$ $\Sigma = -\frac{1}{2}$	$^2\Delta_{3/2}$

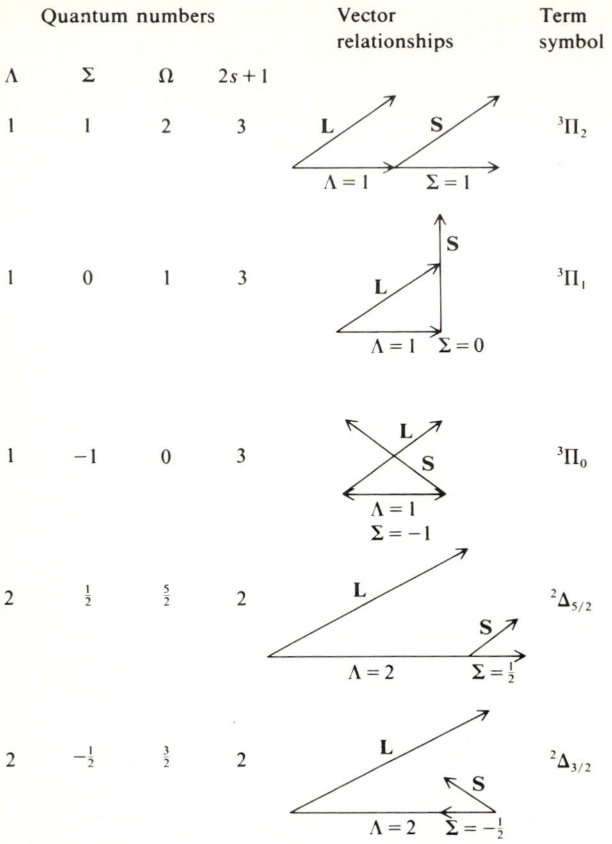

FIGURE 6.7
Momenta and designations for diatomic electronic states.

   If the electronic energy of a molecule having a given value of Λ, and in the absence of spin, is denoted by T_0 then the electronic energies of the multiplets can be qualitatively determined by examining the interactions which occur due to the spin. The internal magnetic field H_i due to the orbital electronic momentum will be proportional to that moment

$$\mathbf{H}_i = \frac{g_L \beta \Lambda}{\hbar} \tag{6.32}$$

The spin magnetic moment of a set of coupled electrons is

$$\boldsymbol{\mu}_s = \frac{-g\beta \mathbf{S}}{\hbar} \tag{6.33}$$

The magnetic interaction energy will be

$$T_\mu = -\boldsymbol{\mu}_s \cdot \mathbf{H}_i \tag{6.34}$$

The electronic energies of the multiplet terms will then be given by

$$T_e = T_0 + A\mathbf{\Lambda} \cdot \mathbf{S} = T_0 + A\Lambda\Sigma \tag{6.35}$$

where A is a constant.

6.6 "UNITED ATOM" AND "SEPARATED ATOMS" CONSTRUCTS

The understanding of the ordering of diatomic molecular orbital energy levels can be further enhanced by examination of the "*united atom*" and the "*separated atoms*" constructs. In the first case the molecular orbitals of a diatomic molecule and the corresponding atomic orbitals of the atom formed by allowing the internuclear separation to go to zero are correlated. This is in contrast to the previous concept of a molecular orbital being related to two atomic orbitals of combining atoms, the "separated atom" approach.

The basis for the "united atom" construct is shown by considering the nature of the wavefunctions for a single electron in a diatomic molecule. The simplest example of such a system is the H_2^+ molecule ion (an electron constrained to move in the potential of two nuclei separated by a distance R) which is illustrated in Fig. 6.8 and is characterized by the Schrödinger equation

$$\left[\frac{\hbar^2}{2m_e}\nabla^2 - \frac{e^2}{r_A} - \frac{e^2}{r_B} + \frac{e^2}{R}\right]\psi = E\psi \tag{6.36}$$

A transformation to elliptical coordinates allows this equation to be solved [17] giving as the wavefunctions for the electron

$$\psi_e = M(\mu)N(\nu)\Phi(\phi) \tag{6.37}$$

where ϕ is the angle about the z axis measured from the xz plane and μ and ν are elliptical coordinates related to the nuclear-electron distances r_A and r_B, and the internuclear distance R by

$$\mu = \frac{r_A + r_B}{R} \tag{6.38}$$

$$\nu = \frac{r_A - r_B}{R} \tag{6.39}$$

The angular part of this function originates from the solution of an angular equation analogous in form to that for the three-dimensional rigid rotor. It is

$$\Phi(\phi) = \frac{1}{\sqrt{2\pi}} e^{-i\lambda\phi} \tag{6.40}$$

where λ is an integer. The integer λ is a quantum number designating the

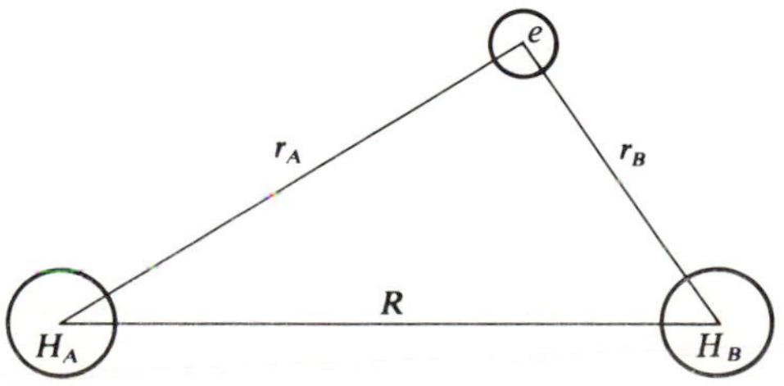

FIGURE 6.8
Coordinates for the H_2^+ molecule ion.

component of electronic angular momentum about the z axis. The previous discussion of angular moments showed that the magnitude of the angular momentum about the z axis is $\lambda\hbar$ where $\lambda = 0, \pm 1, \pm 2, \ldots$. There are several important points related to the solution of Eq. (6.36):

1. There will be two wavefunctions analogous in symmetry properties to the molecular orbitals given by Eq. (6.24).
2. There will be two energy states whose magnitudes will vary with internuclear separation in a manner analogous to the states shown in Fig. 6.2.
3. The $M(\mu)$ and $N(\nu)$ functions will contain a dependency on λ^2 so that the energy of a state is independent of the sign of λ.
4. If the internuclear distance R is allowed to approach zero and the nuclear repulsion term e^2/R in Eq. (6.36) is dropped, then the system becomes an He^+ ion and the two functions revert to He^+ atomic orbitals.
5. The quantum number λ is a good quantum number for all values of R so it will correlate with the m_l atomic orbital value of the "united atom" and the molecular orbitals can be identified with the n or l quantum numbers of the united atom atomic orbitals.
6. For small values of R the ordering of the energies of the molecular orbitals follows that of the "united atom" atomic orbitals.

6.7 CORRELATION DIAGRAMS FOR DIATOMIC MOLECULES

The ordering of the energies of the homonuclear diatomic molecular orbitals can be represented in a semiquantitative manner by the use of a Mulliken [18] type correlation diagram where the levels are viewed as intermediate between a pair of "separated atom" atomic orbitals and a "united atom" atomic orbital. These two extremes for the H_2 molecule will be a pair of H atoms and a He atom. The correlation diagram for the lower molecular orbitals of a homonuclear molecule is shown in Fig. 6.9. There are four features of this figure which are of particular importance:

1. In the "united atom" limit the state symbol for the molecular orbital lists the n and l quantum numbers (s, p, d, f, etc., for l) for the atomic orbital followed by a symbol (σ, π, δ, ϕ) which indicates the value of $|\lambda|$, (0, 1, 2, 3) for the molecular orbital.
2. For molecules with small values of R the ordering of the states for the molecular orbitals based on the "united atom" atomic orbitals will be

$$1s\sigma < 2s\sigma < 2p\sigma < 2p\pi < 3s\sigma < 3p\sigma < 3p\pi < 3d\sigma < 3d\pi < 3d\delta < \cdots$$

3. In the "separated atoms" limit the state symbol consists of the designation for $|\lambda|$ followed by the values of n and l for the limiting separated atomic orbitals.

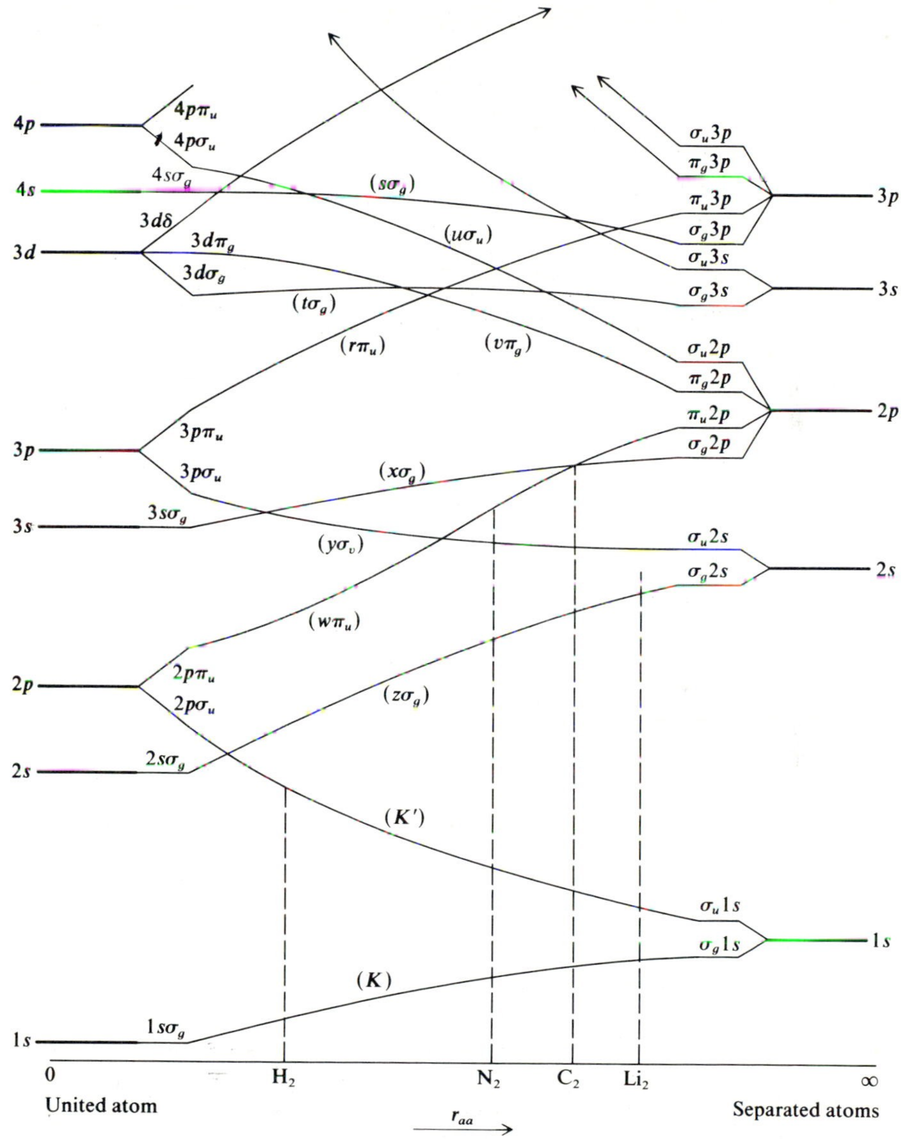

FIGURE 6.9
Qualitative correlation diagram for homonuclear diatomic molecules (not to scale).

4. In the limit of large internuclear distances the energy levels for the molecular orbitals based on the "separated atoms" atomic orbitals will be in the order

$$\sigma_g 1s < \sigma_u^* 1s < \sigma_g 2s < \sigma_u^* 2s < \sigma_g 2p < \pi_u 2p < \pi_g^* 2p < \sigma_u^* 2p < \cdots$$

For many homonuclear diatomic molecules neither of these limiting cases predicts the correct order for the electronic energy levels. The manner in which the energy levels of a molecule deviate from these idealized limits can be qualitatively discussed by correlating the levels of the two limiting cases. If the exact form of the $M(\mu)$ and $N(\nu)$ functions for each electron in a molecule were known then there would be available a quantitative analytical function to link pairs of levels. Fortunately, although these functions are not readily obtainable for molecules more complex than H_2, a qualitative but still useful correlation can be developed. The $\Phi(\phi)$ part of the electronic wavefunction is independent of R, hence λ is a good quantum number and must correlate in the two extreme cases of a given molecular orbital. The formation of a molecular orbital beginning from either extreme must lead to no change in total orbital symmetry. Therefore, only levels of the same symmetry (g or u) can correlate. Furthermore, from an examination of the symmetry of the normal atomic orbitals it can be established that the symmetries of the "united atom" atomic orbitals alternate $g, u, g, u, \ldots$ for $\sigma, \pi, \delta, \phi, \ldots$. Using these basic concepts and beginning with the two lowest limiting levels the correlations shown in Fig. 6.9 were established. Note that in the correlation of levels the nl designations for the two limiting cases for each molecular orbital are not necessarily the same. Another important relationship illustrated in Fig. 6.9 is the noncrossing rule of Von Neumann and Wigner [19] which states that as the internuclear distance changes in a progression of electronic energy levels there can be no crossing of levels belonging to the same total symmetry including $u - g$ symmetry and parity ($\pm$).

The concept of bonding and antibonding molecular orbitals is inherent in the correlation diagram also. This can be illustrated by reference to the behavior of the molecular orbitals which originate from the correlation of a pair of "separated atoms" $1s$ atomic orbitals and the $1s$ and $2p$ "united atom" atomic orbitals. The order of the energy levels for atomic orbitals is always $1s < 2s < 2p$. When a molecule is transformed from the "separated atoms" limit to the "united atom" limit the increase in nuclear charge will result in the energies of the "united atoms" ns atomic orbitals lying below those of the "separated atoms" ns atomic orbitals. Also the energies of the "united atoms" np atomic orbitals will lie above those of either the "united atoms" ns atomic orbitals or the "separated atoms" $(n-1)s$ atomic orbitals. This is illustrated in Fig. 6.9 where it is observed that $E(ua - 1s) < E(sa - 1s)$, $E(ua - 2p) > E(sa - 1s)$, etc. If one considers the Li_2 molecule (six electrons) shown at the right of Fig. 6.9 then it is observed that when the two valence electrons are located in the $z\sigma_g$ molecular orbital they will be at a lower energy level than if they were in the "separated atoms". Hence a more stable or bonding configuration results. According to the Aufbau principle only two electrons may occupy the $z\sigma_g$ molecular orbital, so for the Li_2^- molecule with three valence electrons one must be located in the $y\sigma_u$ molecular orbital. This results in the electron being at a higher energy than when in the "separated atoms" and the molecular orbital is less stable or antibonding. On the correlation diagram those molecular orbitals which are bonding have correlation lines with a positive slope.

6.8 ELECTRONIC ENERGY STATES OF HETERONUCLEAR DIATOMIC MOLECULES

The same general concepts inherent in the development of the correlation diagram for homonuclear diatomic molecules can be applied to heteronuclear ones. The primary difference is the loss of the inversion symmetry for the heteronuclear case and the accompanying relaxing of symmetry requirements when correlating "united atom" and "separated atoms" atomic orbitals. Figure 6.10 illustrates the heteronuclear case. In this diagram, as in Fig. 6.9, it must be remembered that the correlation is depicted in a qualitative manner and no explicit curvature is indicated by the correlation lines.

The two constructs of the "united atom" and the "separated atoms" can serve as an aid for the determination of the allowed electronic states for heteronuclear diatomic molecules. Consider two different "separate atoms" whose individual electron angular and spin momenta are LS-coupled and whose total angular and spin momenta are designated by L_1, S_1, L_2, and S_2. When the atoms are brought together adiabatically the spherically symmetric electrostatic fields of the atoms are converted to a cylindrically symmetric field and the individual angular momenta $\mathbf{L}_1$ and $\mathbf{L}_2$ precess about the internuclear axis with components M_{L_1} and M_{L_2} in the z direction. For this situation the resulting angular momentum along the z axis will be

$$\Lambda = |M_{L_1} + M_{L_2}| \tag{6.41}$$

If L_1 and L_2 are nonzero then there will be $2L_i + 1$ values of M_{L_i} leading to several possible values of Λ. As the separation between the two atoms is decreased until it is equal to the equilibrium internuclear separation of the diatomic molecule, the individual angular momenta of the atoms merge to form the molecular angular momentum. As this occurs the precession of $\mathbf{L}$ about the z direction becomes rapid and, as stated previously, the magnitude is not well defined. However, in this hypothetical adiabatic process of molecule formation there must be a conservation of the z component of angular momentum so the orbital angular momentum Λ can be correlated with that of the atoms M_{L_1} and M_{L_2}. This concept is illustrated in Fig. 6.11 which shows two examples: (1) the combination of two atoms in P states ($L = 1$, $M_L = 0, \pm 1$) and (2) the combination of one atom in an S state ($L_1 = 0$, $M_{L_1} = 0$) and one in a D state ($L_2 = 2$, $M_{L_2} = 0, \pm 1, \pm 2$). Table 6.1 enumerates the characteristics of the five possible angular momentum states resulting from the combination of two atoms in P states and the three resulting from the combination of an atom in an S state with one in a

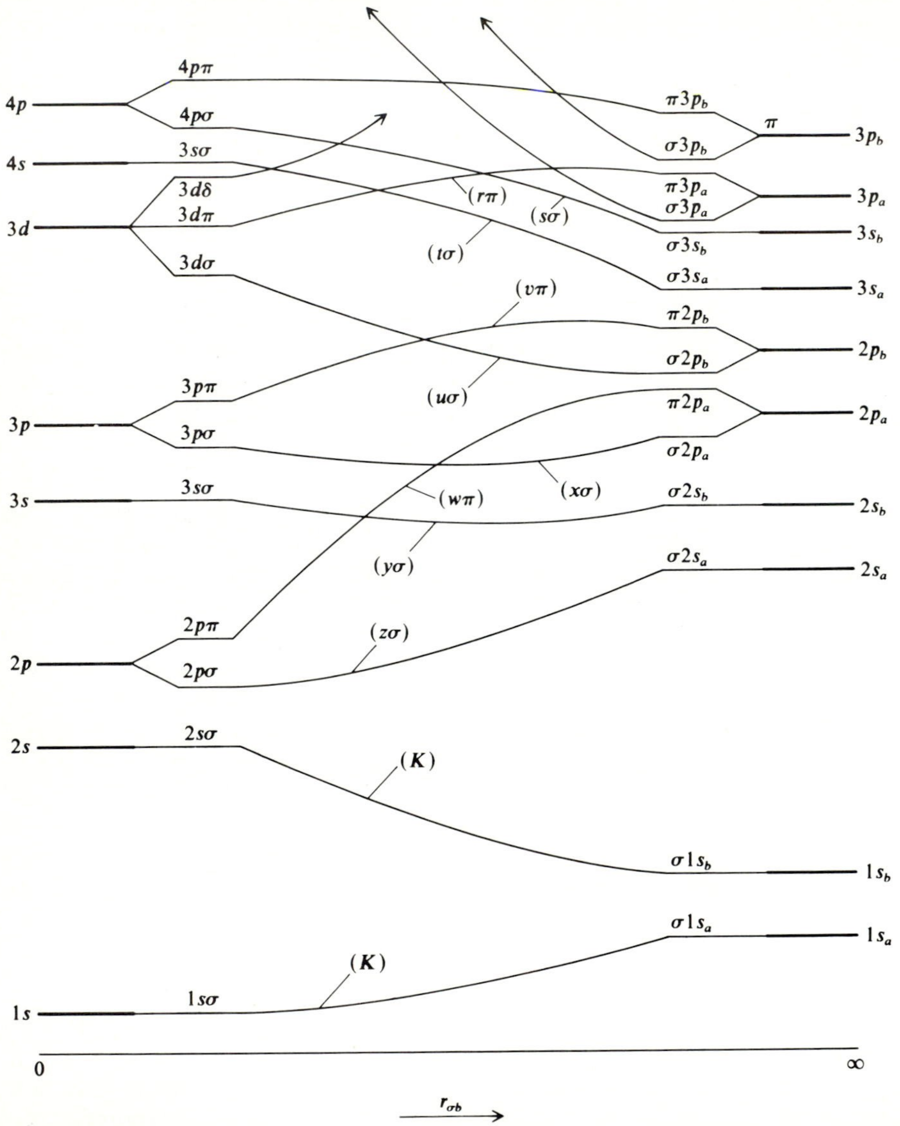

FIGURE 6.10
Qualitative correlation diagram for heteronuclear diatomic molecules (not to scale) (alternate notations given in parentheses).

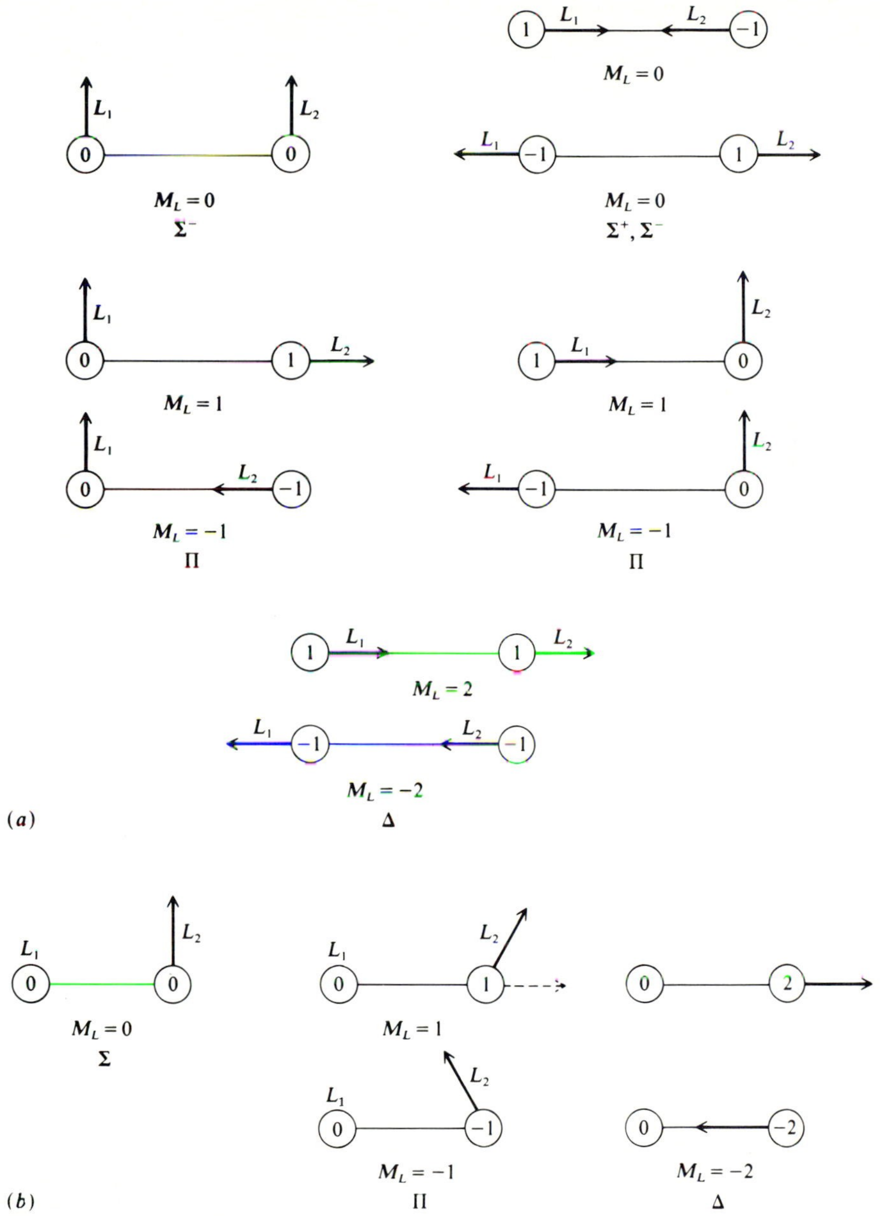

FIGURE 6.11
Allowed angular momenta components for two separated atoms in (*a*) two *P* states and (*b*) *S* and *D* states.

TABLE 6.1
Angular momentum states for heteronuclear diatomic molecules formed from atoms in (*a*) two *P* states of opposite parity and (*b*) *S* and *D* states of like parity

No.	M_{L1}	M_{L2}	M_L	Λ	Term symbol
		(*a*) Two *P* states			
1	0	±1	±1	1	Π
2	0	0	0	0	Σ^-
3	±1	±1	±2	2	Δ
4	±1	±1	0	0	Σ^+, Σ^-
5	±1	0	±1	1	Π
		(*b*) *S* and *P* states			
1	0	0	0	0	Σ^+
2	0	±1	±1	1	Π
3	0	±2	±2	2	Δ

D state. All states having $\Lambda > 0$ will be doubly degenerate but the pair of Σ states arising from the combination of $\pm M_{L_1}$ and $\pm M_{L_2}$ when $M_{L_1} = M_{L_2}$ are not. The symmetry of the Σ state arising from the $M_{L_1} = M_{L_2} = 0$ combination can be determined from the parity of $L_1 + L_2 + \sum_i l_{1_i} + \sum_i l_{2_i}$ where the sums are over all electrons in the incomplete electron shells of the atoms. If this sum is even then the state is Σ^+. Since $\sum_i l_{N_i}$ determines the parity of the atomic state for atom *N* atom *N* the parity of the molecular state depends on those of the two "separated atoms."

The coupling of spin momenta is not affected by the strong internuclear field and the total spin momentum of the system is found by conventional vector addition to give $S = S_1 + S_2, S_1 + S_2 - 1, \ldots, |S_1 - S_2|$. The resulting multiplicity of a given state is then $2S + 1$. Referring to the examples in Table 6.1 and combining the angular and spin momenta leads to a manifold of states characterized in part by Table 6.2. For each individual term in Table 6.1 the manifold of spin states enumerated in Table 6.2 will be repeated. For example, the combination of two atoms in ^3P and ^2P states to give a molecule with $\Lambda = 2$ will have as possible states $^2\Delta$ and $^4\Delta$. Likewise, a molecule resulting from the combination of two atoms in ^1S and ^3D states and with $\Lambda = 2$ will have as the only possible state a $^3\Delta$.

The previous discussion has shown that for each molecular angular momentum state arising from a specific pair of atomic states and characterized by a given value of Λ there will be $2S + 1$ states due to the spin momentum. For example, the molecule formed by the combination of ^3P and ^3D atomic states of opposite parity will have possible angular momentum components of $\Lambda = 0$, 1, 2, 3. When these states are combined with the possible spin states there results a total of 27 possible terms ranging from a $^1\Sigma^-$ to a $^5\Phi$ with each state having a different energy. Each of these substates will be split into multiplets of closer but

TABLE 6.2
Total electronic states for heteronuclear diatomic molecules found from atoms in (*a*) two *P* states of opposite parity and (*b*) *S* and *D* states of like parity

No. refers to corresponding entry in Table 6.1.

No.	Λ	Atomic states 1	2	M_{S_1}	M_{S_2}	M_S	$2S+1$	Possible terms
				(*a*) Two *P* states				
1	1	^1P	^1P	0	0	0	1	$^2\Pi$
1	1	^1P	^2P	0	$\frac{1}{2}$	$\frac{1}{2}$	2	$^2\Pi$
1	1	^1P	^3P	0	1	1	3	$^3\Pi$
1	1	^2P	^1P	$\frac{1}{2}$	0	$\frac{1}{2}$	2	$^2\Pi$
1	1	^2P	^2P	$\frac{1}{2}$	$\frac{1}{2}$	0, 1	1, 3	$^1\Pi$, $^3\Pi$
1	1	^2P	^3P	$\frac{1}{2}$	1	$\frac{1}{2}$, $\frac{3}{2}$	2, 4	$^2\Pi$, $^4\Pi$
1	1	^3P	^1P	1	0	1	3	$^3\Pi$
1	1	^3P	^2P	1	$\frac{1}{2}$	$\frac{1}{2}$, $\frac{3}{2}$	2, 4	$^2\Pi$, $^4\Pi$
1	1	^3P	^3P	1	1	0, 1, 2	1, 3, 5	$^1\Pi$, $^3\Pi$, $^5\Pi$
				(*b*) *S* and *D* states				
3	2	^1S	^1D	0	0	0	1	$^1\Delta$
3	2	^1S	^2D	0	$\frac{1}{2}$	$\frac{1}{2}$	2	$^2\Delta$
3	2	^1S	^3D	0	1	1	3	$^3\Delta$
3	2	^2S	^1D	$\frac{1}{2}$	0	$\frac{1}{2}$	2	$^2\Delta$
3	2	^2S	^2D	$\frac{1}{2}$	$\frac{1}{2}$	0, 1	1, 3	$^1\Delta$, $^3\Delta$
3	2	^2S	^3D	$\frac{1}{2}$	1	$\frac{1}{2}$, $\frac{3}{2}$	2, 4	$^2\Delta$, $^4\Delta$
3	2	^3S	^1D	1	0	1	3	$^3\Delta$
3	2	^3S	^2D	1	$\frac{1}{2}$	$\frac{1}{2}$, $\frac{3}{2}$	2, 4	$^2\Delta$, $^4\Delta$
3	2	^3S	^3D	1	1	0, 1, 2	1, 3, 5	$^1\Delta$, $^3\Delta$, $^5\Delta$

yet different energies, there being a total of 81 for the 27 terms of this example. Although it appears that the problem of characterizing molecular electronic states is hopelessly complicated it is possible to express the necessary information in short tabular form as shown in Tables 6.3 and 6.4. In Table 6.4 the parity of the atomic state is denoted by a subscripted *o* or *e* for odd or even.

The development of the possible terms for a diatomic molecule can also proceed via the hypothetical separation of a "united atom" into the two component atoms of the molecule. In this process the spherical electrostatic field of the "united atom" is distorted into a strong axially symmetric one, the *LS*-type coupling of the "united atom" is broken down and **L** precesses rapidly about the *z* axis leading to quantized *z* components designated by $\Lambda = |M_L| = L, L-1, \ldots, 0$. The states with $\Lambda > 0$ will be doubly degenerate and the symmetry of the Σ states ($\Lambda = 0$) are found from the parity of the "united atom". If $L + \sum_i l_i$ is even then the state is Σ^+. The spin **S** of the molecular state will be the same as that of the "united atom" state from which it originates. An example of this would be the separation of a carbon atom (six electrons) into a BH molecule as shown by Fig. 6.12.

TABLE 6.3
Allowed multiplicities of molecular electronic states given in terms of those of the separated atoms

Separated atom				Molecule	
M_{S_1}	M_{S_2}	$2S_1 + 1$	$2S_2 + 1$	M_S	$2S + 1$
0	0	1	1	0	1
0	$\frac{1}{2}$	1	2	$\frac{1}{2}$	2
0	1	1	3	1	3
0	$\frac{3}{2}$	1	4	$\frac{3}{2}$	4
$\frac{1}{2}$	$\frac{1}{2}$	2	2	1, 0	3, 1
$\frac{1}{2}$	1	2	3	$\frac{3}{2}, \frac{1}{2}$	4, 2
$\frac{1}{2}$	$\frac{3}{2}$	2	4	2, 1	5, 3
1	1	3	3	2, 1, 0	5, 3, 1
1	$\frac{3}{2}$	3	4	$\frac{5}{2}, \frac{3}{2}, \frac{1}{2}$	6, 4, 2
$\frac{3}{2}$	$\frac{3}{2}$	4	4	3, 2, 1, 0	7, 5, 3, 1

TABLE 6.4
Molecular electronic angular momentum states for heteronuclear diatomic molecules

The total number of allowed states for a given value of Λ is shown in parentheses and alternate atomic state pairs which will give the same molecular states is shown in brackets

States of "separated atoms" Atom 1	Atom 2	Possible molecular states
$S_e[S_o]$	$S_e[S_o]$	Σ^+
$S_e[S_o]$	$S_o[S_e]$	Σ^-
$S_e[S_o]$	$P_e[P_o]$	Σ^-, Π
$S_e[S_o]$	$P_o[P_e]$	Σ^+, Π
$S_e[S_o]$	$D_e[D_o]$	Σ^+, Π, Δ
$S_e[S_o]$	$D_o[D_e]$	Σ^-, Π, Δ
$S_e[S_o]$	$F_e[F_o]$	$\Sigma^-, \Pi, \Delta, \Phi$
$S_e[S_o]$	$F_o[F_e]$	$\Sigma^+, \Pi, \Delta, \Phi$
$P_e[P_o]$	$P_e[P_o]$	$\Sigma^+(2), \Sigma^-, \Pi(2), \Delta$
$P_e[P_o]$	$P_o[P_e]$	$\Sigma^+, \Sigma^-(2), \Pi(2), \Delta$
$P_e[P_o]$	$D_e[D_o]$	$\Sigma^+, \Sigma^-(2), \Pi(3), \Delta(2), \Phi$
$P_e[P_o]$	$D_o[D_e]$	$\Sigma^+(2), \Sigma^-, \Pi(3), \Delta(2), \Phi$
$P_e[P_o]$	$F_e[F_o]$	$\Sigma^+(2), \Sigma^-, \Pi(3), \Delta(3), \Phi(2), \Gamma$
$P_e[P_o]$	$F_o[F_e]$	$\Sigma^+, \Sigma^-(2), \Pi(3), \Delta(3), \Phi(2), \Gamma$
$D_e[D_o]$	$D_e[D_o]$	$\Sigma^+(3), \Sigma^-(2), \Pi(4), \Delta(3), \Phi(2), \Gamma$
$D_e[D_o]$	$D_o[D_e]$	$\Sigma^+(2), \Sigma^-(3), \Pi(4), \Delta(3), \Phi(2), \Gamma$
$D_e[D_o]$	$F_e[F_o]$	$\Sigma^+(2), \Sigma^-(3), \Pi(5), \Delta(4), \Phi(3), \Gamma(2), H$
$D_e[D_o]$	$F_o[F_e]$	$\Sigma^+(3), \Sigma^-(2), \Pi(5), \Delta(4), \Phi(3), \Gamma(2), H$
$F_e[F_o]$	$F_e[F_o]$	$\Sigma^+(4), \Sigma^-(3), \Pi(6), \Delta(5), \Phi(4), \Gamma(3), H(2), I$
$F_e[F_o]$	$F_o[F_e]$	$\Sigma^+(3), \Sigma^-(4), \Pi(6), \Delta(5), \Phi(4), \Gamma(3), H(2), I$

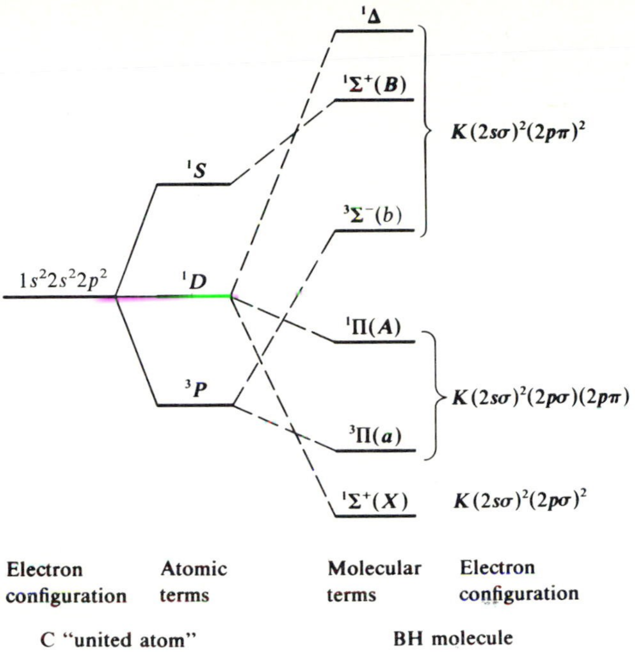

FIGURE 6.12
Relationships among the energy states of the BH molecule and the C atom. (The ordering of the atomic terms is not in accordance with Hund's rules for atoms but agrees with the values reported by Herzberg [20].)

A second and somewhat more complex example, that of the Si atom and the BF molecule, is illustrated in Fig. 6.13.

The features of both of these constructs are combined in Fig. 6.14 which depicts the energy states of the NH molecule relative to those of the "united" oxygen atoms and the "separated" nitrogen and hydrogen atoms.

6.9. ELECTRONIC ENERGY STATES OF HOMONUCLEAR DIATOMIC MOLECULES

The concepts discussed in Sec. 6.8 for heteronuclear diatomic molecules are also applicable to homonuclear species. In some respects the situation is less complex since the energies of the two "separated atoms" are identical. However, the occurrence of equivalent atoms leads to the necessity for further symmetry classification of the states. Also the level separations between the *nsA-nsB* and *npA-npB* pairs shown in Fig. 6.10 disappear and the separations of equivalent levels shown on the right-hand side of Fig. 6.9 are present only for labeling purposes.

Using the "separated atoms" construct, the angular momentum states for a homonuclear diatomic molecule are found to be the same as those found for

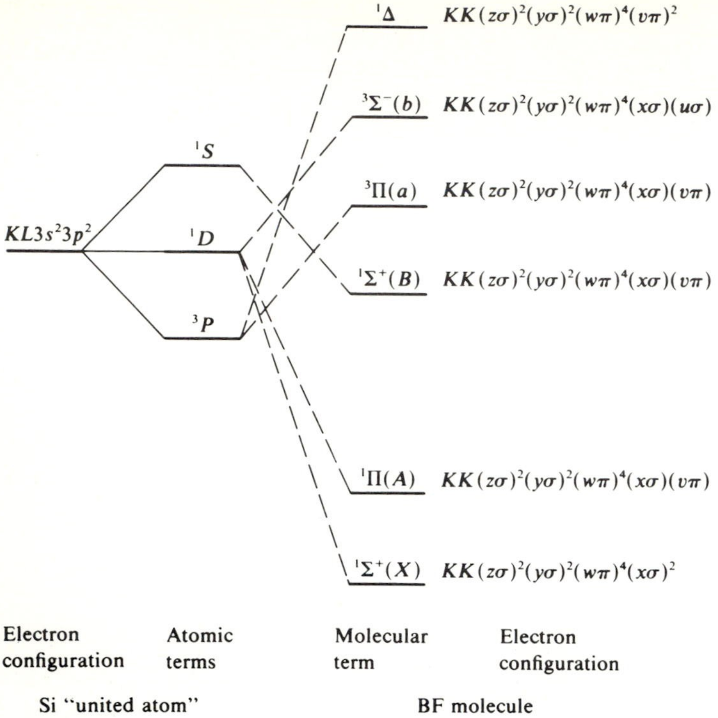

FIGURE 6.13
Relationships among the energy states of the BF molecule and the Si atom.

the heteronuclear case with the restriction that only pairs of equivalent atomic terms can be combined. Hence, Table 6.1*a* is equally applicable to a homonuclear molecule but Table 6.1*b* has no meaning. If the combining states are absolutely identical, including multiplicity, then the total number of possible "separated atoms" state combinations as illustrated by Table 6.2*a* is further restricted, and only the first, fifth, and ninth entries are applicable.

Due to the homonuclear molecule having a center of symmetry the molecular wavefunctions for the resulting states will be either symmetric (g = gerade) or antisymmetric (u = ungerade) with respect to inversion and the term symbol will have an added subscript of g or u to denote this property. The inversion symmetry of the homonuclear molecular orbital will be the same as that of the two "separated atoms" orbitals and that of the "united atom" orbital which correlate with the molecular orbital. These symmetry relationships are illustrated in Fig. 6.15 for several orbitals.

The use of group theory to determine orbital symmetry is based on the fact that the molecular orbitals of a molecule constitute a set of basis functions, each of which belongs to some irreducible representation of the molecule. The determination of the symmetries of the molecular orbitals best done by the use of the general example of a second period homonuclear molecule whose "separated

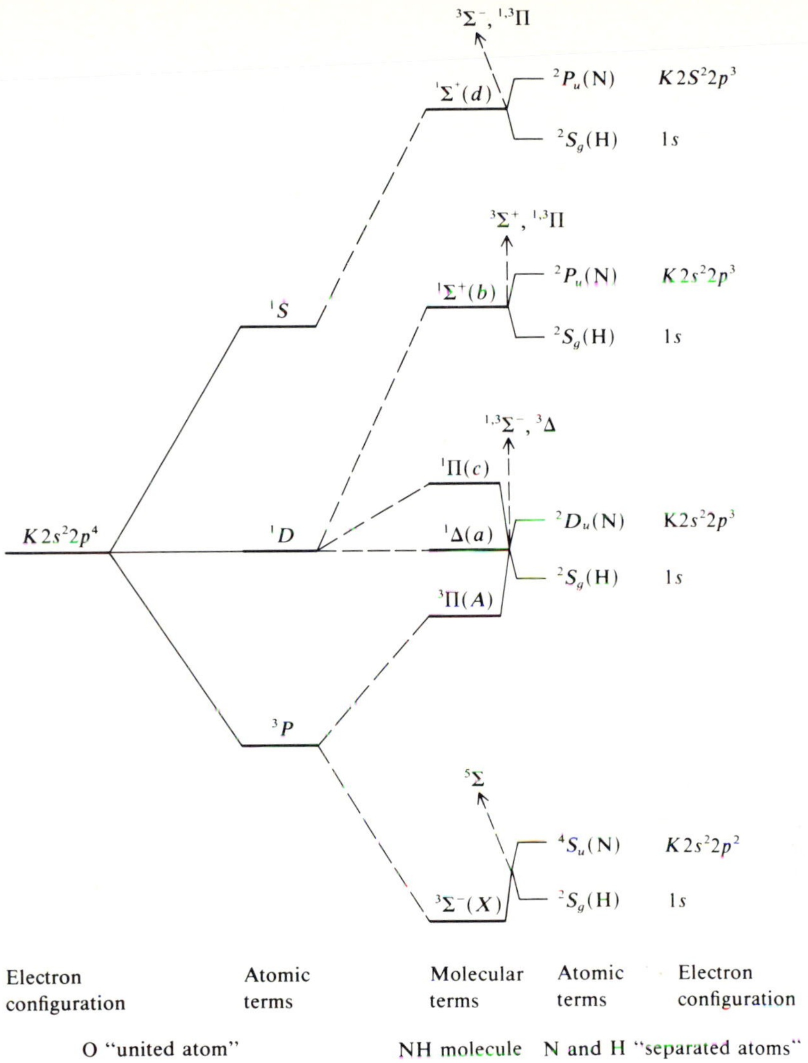

FIGURE 6.14
Relationships among the energy levels of the NH molecule, "united atom"-"separated atoms" correlation. (Dashed arrows indicate states which correlate to higher "united atom" states.)

atoms" have $1s$, $2s$, and $2p$ atomic orbitals available. The combination of the $1s$ atomic orbitals will be analogous to the case of the H_2^+ molecule ion discussed in Sec. 6.4 and is qualitatively illustrated in Fig. 6.16.

Homonuclear molecules belong to the $D_{\infty h}$ group which contains the E, C_∞^ϕ, σ_v, i, S_∞^ϕ and C_2 operations (see App. J). Examination of the effect of these operations on a pair of $1s$ atomic orbitals shows that E, C_∞^ϕ, and σ_v leave the system unchanged while i, S_∞^ϕ and C_2 interchange the atomic orbitals. Thus the

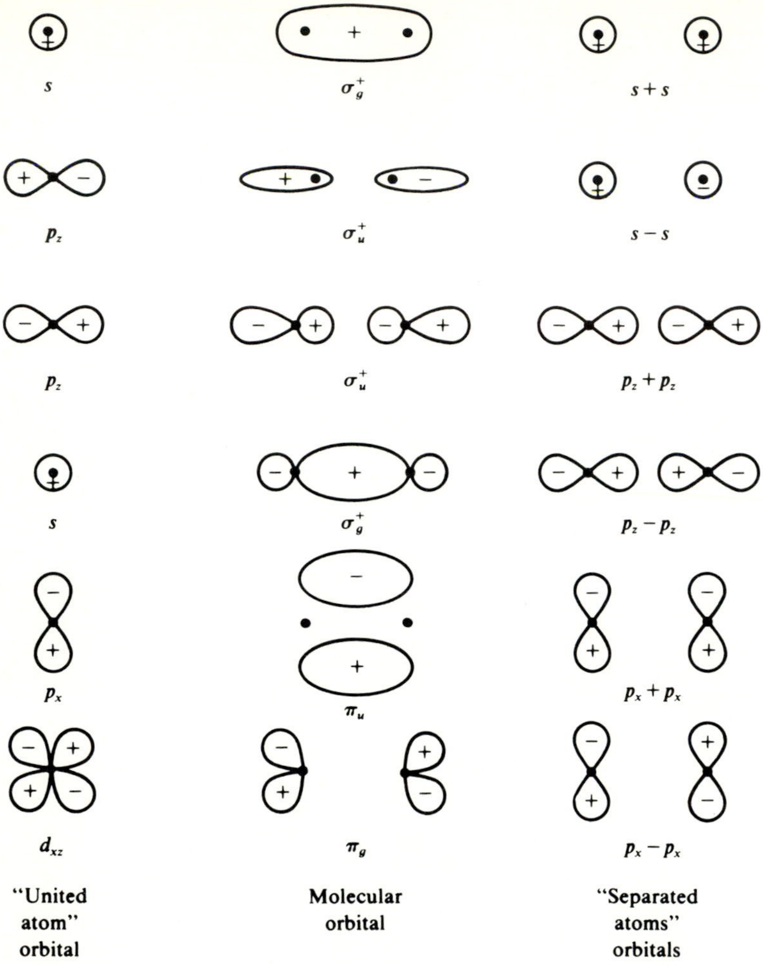

FIGURE 6.15
Symmetry relationships between the atomic orbitals of a "united atom" and "separated atoms," and molecular orbitals of a homonuclear diatomic molecule.

characters for the reducible representation to which this pair of atomic orbitals belongs are 2, 2, 2, 0, 0, 0. By reference to the $D_{\infty h}$ character table and employing Eq. (4.9) it is found that the two molecular orbitals belong to the σ_g^+ and σ_u^+ irreducible representations. (*Note:* In order to avoid confusing symbols lowercase letters are used to refer to entries in the character table and the uppercase ones will be reserved for molecular terms.) The inversion symmetries of the bonding and antibonding states, $\sigma 1s$ and $\sigma^* 1s$, will then be g and u, respectively.

The treatment of the set of eight orbitals for $n = 2$ proceeds by determining the effect of all of the symmetry operations in the atomic orbitals. Using Fig. 6.17 as a basis the results are summarized in Table 6.5. The characters of the

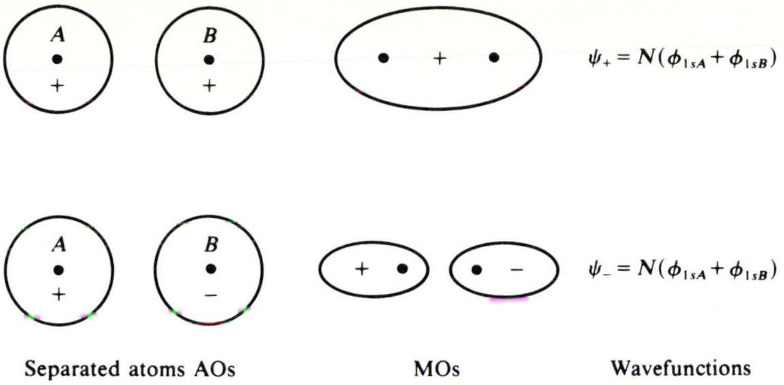

FIGURE 6.16
"Separated atom" atomic orbitals, homonuclear orbitals, and wavefunctions for molecular orbitals formed from $1s$ atomic orbitals.

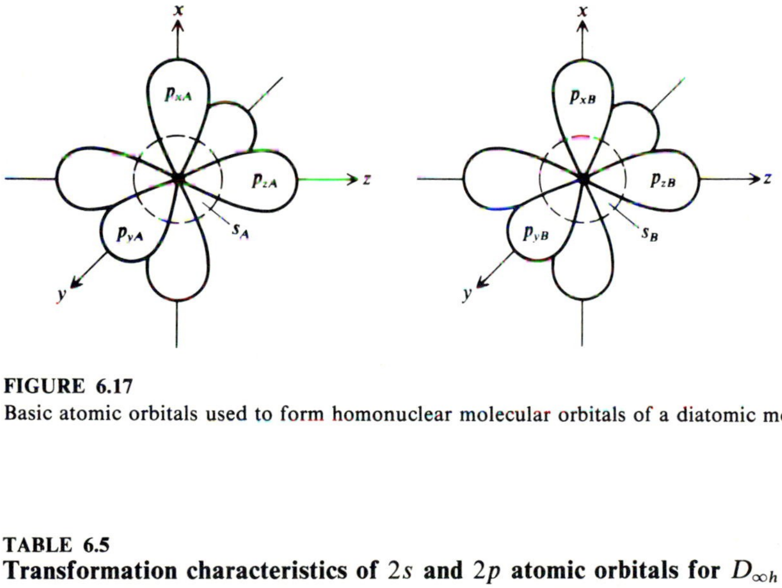

FIGURE 6.17
Basic atomic orbitals used to form homonuclear molecular orbitals of a diatomic molecule.

TABLE 6.5
Transformation characteristics of $2s$ and $2p$ atomic orbitals for $D_{\infty h}$ group

	Operation					
Atomic orbital	E	C_∞^ϕ	σ_v	i	S_∞	C_2
s	s	s	s	$-s$	$-s$	$-s$
p_x	p_x	$p_x \cos\phi + p_y \sin\phi$	p_x	$-p_x$	$-p_x$	$-p_x$
p_y	p_y	$-p_x \sin\phi + p_y \cos\phi$	p_y	$-p_y$	$-p_y$	$-p_y$
p_z	p_z	p_z	p_z	$-p_z$	$-p_z$	$-p_z$

reducible representation to which the eight resulting molecular orbitals belong are found from these transformations to be 8, $4 + 4\cos\phi$, 4, 0, 0, 0. Again use of Eq. (4.9) and the character table shows that the irreducible representations will be $2\sigma_g^+$, $2\sigma_u^-$, π_g and π_u. Thus the resulting molecular orbitals will be $2s\sigma_g$, $2s\sigma_u$, $2p\sigma_g$, $2p\sigma_u$, $2p\pi_g$ and $2p\pi_u$ with the latter two being doubly degenerate.

Having established the symmetry of each of the molecular orbitals the next step is to find the appropriate symmetry of the molecular terms. Initially it is instructive to look at the case of two $1s$ atomic orbitals combining to form the H_2 molecule. The Pauli principle requires that the total electronic wavefunction must be antisymmetric with respect to the inversion of any pair of electrons. The total electronic wavefunction ψ_e consists of the product of an orbital function ψ_o and a spin function ψ_s. The two electrons in the H_2 molecule will be in the $(\sigma 1s)$ molecular orbital, $\psi_o = \sigma 1s(1) \times \sigma 1s(2)$. The $\sigma 1s$ molecular orbital belongs to the totally symmetric representation, therefore ψ_o will be symmetric. A pair of electrons, each of which can have m_s values of $\pm\frac{1}{2}$ (denoted by α or β) can combine to form three symmetric spin states, $\alpha\alpha$, $\beta\beta$, $\frac{1}{2}(\alpha\beta + \beta\alpha)$ and one antisymmetric one, $\frac{1}{2}(\alpha\beta - \beta\alpha)$. In the symmetric states the spins are considered to be unpaired and in the antisymmetric one they are paired. In order for ψ_e to be antisymmetric, ψ_o must be combined with the antisymmetric spin state. Because the multiplicity of the antisymmetric spin state is 1 ($S = 0$) and the symmetry of the electronic orbital state ψ_o is gerade, it follows that the molecular term for the ground state of H_2 will be ${}^1\Sigma_g^+$. In view of the fact that the "separated atoms" for H_2 are H atoms having $L = 0$ and $S = \frac{1}{2}$ it can be concluded that one possible term arising from the combination of two "separated atoms" in ${}^2S$ states will be ${}^1\Sigma_g^+$.

Symmetry considerations, along with the spin combinations given in Table 6.3, when applied to other possible molecular orbitals arising from the combinations of like "separated atoms," give rise to the terms given in Table 6.6. For these states the parity ($\pm$) of the Σ states has been determined by Wigner and Witmer [21] and will not be discussed further. The relationship between the

TABLE 6.6
Molecular electronic angular momentum states for homonuclear diatomic molecules

States of "separated atoms" Atom 1	Atom 2	Possible molecular states
${}^1S$	${}^1S$	${}^1\Sigma_g^+$
${}^2S$	${}^2S$	${}^1\Sigma_g^+, {}^3\Sigma_u^+$
${}^3S$	${}^3S$	${}^1\Sigma_g^+, {}^3\Sigma_u^+, {}^5\Sigma_g^+$
${}^4S$	${}^4S$	${}^1\Sigma_g^+, {}^3\Sigma_u^+, {}^5\Sigma_g^+, {}^7\Sigma_u^+$
${}^1P$	${}^1P$	${}^1\Sigma_g^+(2), {}^1\Sigma_u^-, {}^1\Pi_g, {}^1\Pi_u, {}^1\Delta_g$
${}^2P$	${}^2P$	${}^1\Sigma_g^+(2), {}^1\Sigma_u^-, {}^1\Pi_u, {}^1\Pi_g, {}^1\Delta_g, {}^3\Sigma_u^+(2), {}^3\Sigma_g^-, {}^3\pi_u, {}^3\Pi_g, {}^3\Delta_g$
${}^3P$	${}^3P$	${}^1\Sigma_g^+(2), {}^1\Sigma_u^-, {}^1\Pi_u, {}^1\Pi_g, {}^1\Delta_g, {}^3\Sigma_u^+(2), {}^3\Sigma_g^-, {}^3\Pi_u, {}^3\Pi_g, {}^3\Delta_g, {}^5\Sigma_g^+(2), {}^3\Sigma_u^-, {}^5\Pi_u, {}^5\Delta_g$
${}^1D$	${}^1D$	${}^1\Sigma_g^+(3), {}^1\Sigma_u^-(2), {}^1\Pi_g(2), {}^1\Pi_u(2), {}^1\Delta_g(2), {}^1\Delta_u, {}^1\Phi_g, {}^1\Phi_u, {}^1\Gamma_g$

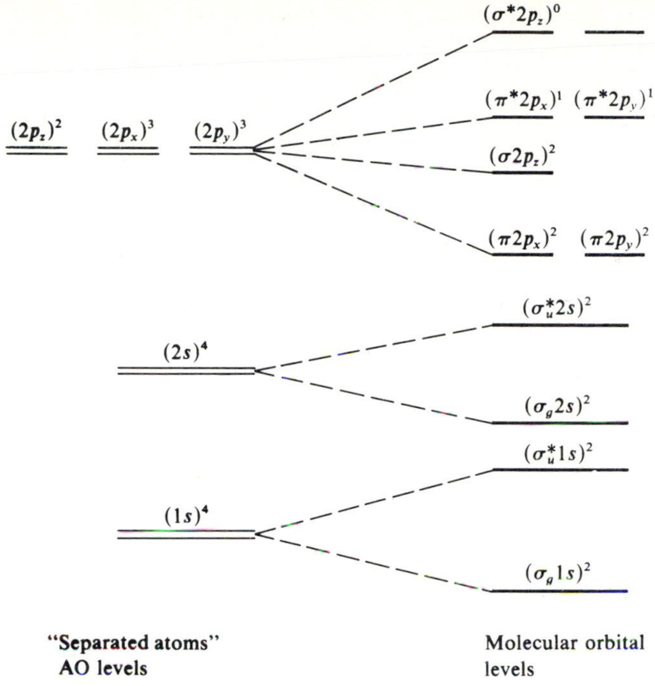

FIGURE 6.18
Diagram of the qualitative relationships between the "separated atoms" states of two oxygen atoms and the molecular states of the O_2 molecule. (Superscript outside parenthesis denotes the number of electrons.)

"separated atoms" states of two oxygen atoms and the states of the O_2 molecule are illustrated in Fig. 6.18.

For excited states the resulting terms do not necessarily relate to equivalent "separated atoms" levels and Table 6.6 is not usable. For example, the first excited state of H_2 will have the molecular orbital configuration $(\sigma 1s)(\sigma^* 1s)$. In this case the orbital function $\psi_o = \sigma 1s(1)\sigma^* 1s(2)$ is neither symmetric or antisymmetric and appropriate wavefunctions must be generated by taking the linear combinations

$$\psi_o(\text{sym}) = \frac{1}{\sqrt{2}}[\sigma 1s(1)\sigma^* 1s(2) + \sigma 1s(2)\sigma^* 1s(1)] \tag{6.42}$$

$$\psi_o(\text{antisym}) = \frac{1}{\sqrt{2}}[\sigma 1s(1)\sigma^* 1s(2) - \sigma 1s(2)\sigma^* 1s(1)] \tag{6.43}$$

Appropriate total electronic wavefunctions are formed by combining $\psi_o(\text{sym})$ with $\psi_s(\text{antisym})$ and $\psi_o(\text{antisym})$ with one of the three $\psi_s(\text{sym})$. The symmetry species of these two states can be determined by recognizing that, since of $\sigma 1s$ belongs to the same symmetry species as an s atomic orbital, and $\sigma^* 1s$ belongs to the same symmetry species as a p_z atomic orbital, $\psi_o(\text{sym})$ will belong to the same representation as a p_z atomic orbital while ψ_o(antisym) will belong to the same as an s atomic orbital. Therefore, the excited state gives rise to two molecular terms $^1\Sigma_g^+$ and $^3\Sigma_u^+$ and two additional molecular terms are possible from the combination of two "separated atoms" in ^2S configurations.

The states obtained by the division of a "united atom" into two equal parts of a molecule are the same as those obtained by the consideration of the "separated atoms" construct. Wigner and Witmer [21] have shown that the parity of the molecular state depends on the even or odd character of $L + \sum_i 1_i$. Furthermore, the total spin of the molecule will be equal to that of the "united atom" and the inversion symmetry (g or u) of the molecular terms will correspond to the "united atom" term from which they originated. This approach is illustrated in Fig. 6.19 for the case of the C_2 molecule formed from an Mg atom.

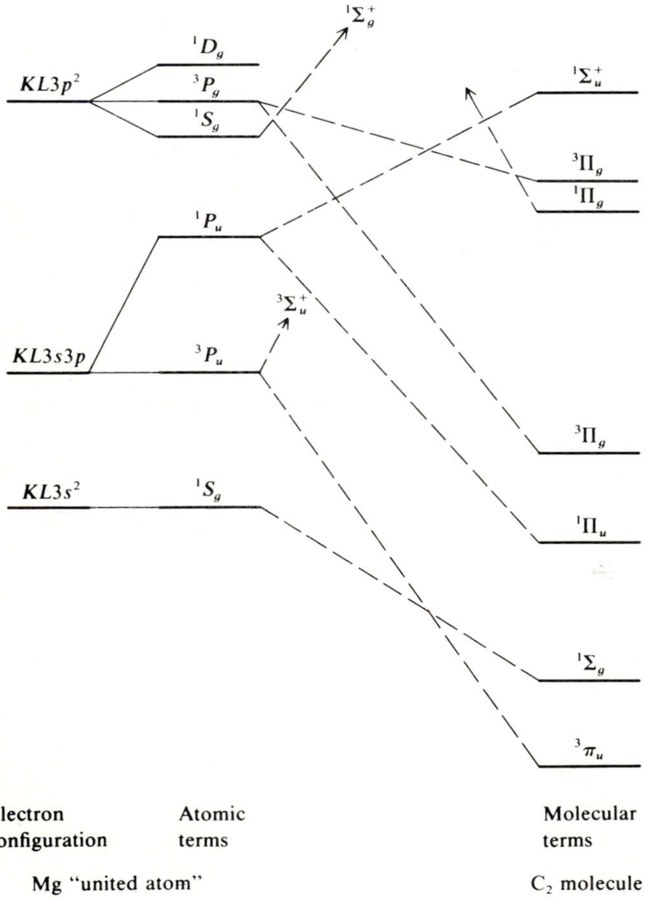

FIGURE 6.19
Qualitative relationship between the "united atom" Mg, and the C_2 molecule energy levels.

6.10 DETERMINATION OF MOLECULAR TERMS FROM THE ELECTRON CONFIGURATION

The consideration of the combination of "separated atoms" states, which lead to the manifold of terms listed in Tables 6.4 and 6.6, placed no restrictions on the electron occupancy of a state designated by a given term. In order to determine the allowed states for a particular molecule one must apply the Pauli principle just as is done for electron occupancy in single atoms. This, of course, dictates that for the electrons in a given molecule no two may have the same set of quantum numbers.

Because of the continuous transition of molecular properties as one proceeds from a "united atom" through a molecule to the "separated atoms" we can select, as a point for consideration, a position where the internuclear separation is small and the quantum numbers n and 1 are well characterized. In this case the axial component of angular momentum λ, for a single electron, will be $\lambda = |M_1|$, and it follows that there can be only two electrons with a given n and $1 = 0$ those with $m_1 = 0$ and $m_s = \pm\frac{1}{2}$. This means that a σ molecular orbital is filled with two electrons. For molecular orbitals containing electrons with $\lambda \neq 0$, $m_1 = \pm\lambda$, $m_s = \pm\frac{1}{2}$ so π, δ, ϕ molecular orbitals will have a maximum occupancy of four electrons. The consequences of the Pauli principle are summarized in Table 6.7 where the column labeled MO gives the "united atom" level designation and that labeled N is the maximum electron occupancy.

If one begins with the "separated atoms" limit and applies the Pauli principle to the molecular orbitals which are formed from the available atomic orbitals $\sigma 1s_A$, $\sigma 1s_B$, $\sigma 2s_A, \ldots, \pi 2p_B, \ldots$, etc., the same conclusions regarding electron occupancy can be drawn. Although it would appear that with two equivalent atoms there would be four electrons for a σ state, this is not the case since as the atoms are merged to form a molecule two separate molecular orbitals relate back to different atomic orbitals on the "united atom".

TABLE 6.7
Electron occupancy and allowed quantum numbers for molecular orbitals near the "united atom" limit

MO	n	1	λ	m_1	m_s	N
$1s\sigma$	1	0	0	0	$\pm\frac{1}{2}$	2
$2s\sigma$	2	0	0	0	$\pm\frac{1}{2}$	2
$2p\sigma$	2	1	0	0	$\pm\frac{1}{2}$	1
$2p\pi$	2	1	1	± 1	$\pm\frac{1}{2}$	4
$3s\sigma$	3	0	0	0	$\pm\frac{1}{2}$	2
$3p\sigma$	3	1	0	0	$\pm\frac{1}{2}$	2
$3p\pi$	3	1	1	± 1	$\pm\frac{1}{2}$	4
$3d\sigma$	3	2	0	0	$\pm\frac{1}{2}$	2
$3d\pi$	3	2	1	± 1	$\pm\frac{1}{2}$	4
$3d\delta$	3	2	2	± 2	$\pm\frac{1}{2}$	4

In molecules, as in atoms, there will be coupling between the orbital and spin angular momenta. The nature of this coupling will determine the allowed molecular terms for a given electronic configuration in a molecule. For the lighter atoms it is well established [22] that the most common type of coupling is Russell-Saunders. In this situation the orbital angular momenta of a single electron is more strongly coupled to that of the other electrons than it is to its own spin angular momentum. The same type interaction also holds for the individual spin momenta. If it is assumed that Russel-Saunders coupling is equally valid, at least for the lighter diatomic molecules, then the total molecular angular momentum about the internuclear axis and the total molecular spin momenta can be written as

$$\Lambda = \sum_i m_i \tag{6.44}$$

and

$$S = \sum_i \mathbf{s}_i \tag{6.45}$$

The coupling for nonequivalent electrons is less subtle so it will be considered first. For this case the values of the n quantum numbers of the electrons will differ so any combination of the 1_i, m_{1_i} and m_{s_i} will be allowed. There will, however, be restrictions arising from the fact that $\boldsymbol{\Lambda}$ can lie only along the axial direction.

For a single electron in a σ molecular orbital $\Lambda = \lambda_1 = 0$ and $m_{1_1} = 0$. The total spin will be $S = s_1 = \frac{1}{2}$. Hence the only possible state will be a $^2\Sigma^+$. The $+$ designation arises from the fact that $L + \sum_i 1_i$ for the "united atom" limiting state is even. Although Table 6.4 indicates the possibility of both Σ^+ and Σ^- states for s molecular orbitals formed from atomic s orbitals only the Σ^+ will in fact be observed. For a single electron in a π molecular orbital $\Lambda = 1$, $s = \frac{1}{2}$, and there will be a $^2\Pi$ term.

These same conclusions can be reached by examination of the symmetry of the σ and π molecular orbitals and determining the irreducible representation of the $C_{\infty v}$ group to which they belong. The σ molecular orbital has cylindrical symmetry and will transform according to the Σ^+ representation while the π molecular orbital will transform according to the Π representation.

The extension of this methodology to several electrons is aided by Fig. 6.20 illustrating the vector addition of the individual orbital momenta for (a) two electrons having $\lambda_1 = 1$ and $\lambda_2 = 2$, ($\pi\delta$) and (b) three electrons having $\lambda_1 = \lambda_2 = 1$ and $\lambda_3 = 2$ ($\pi\pi\delta$). For the $\pi\delta$ configuration it is observed that there will be two energy states having Λ values of 1 and 3. Since the electrons are nonequivalent their spins can be either parallel or antiparallel and there will result a total of four doubly degenerate molecular terms, $^1\Pi$, $^3\Pi$, $^1\Phi$, and $^3\Phi$. Reference to Table 6.4 indicates that the number of allowed states has been reduced to two from the total of nine determined by combining "separated atom" P and D terms. If the electrons in these four states were totally independent the states would all have the same energy. However, due to the interaction between the electrons the singlet and triplet states will differ in energy.

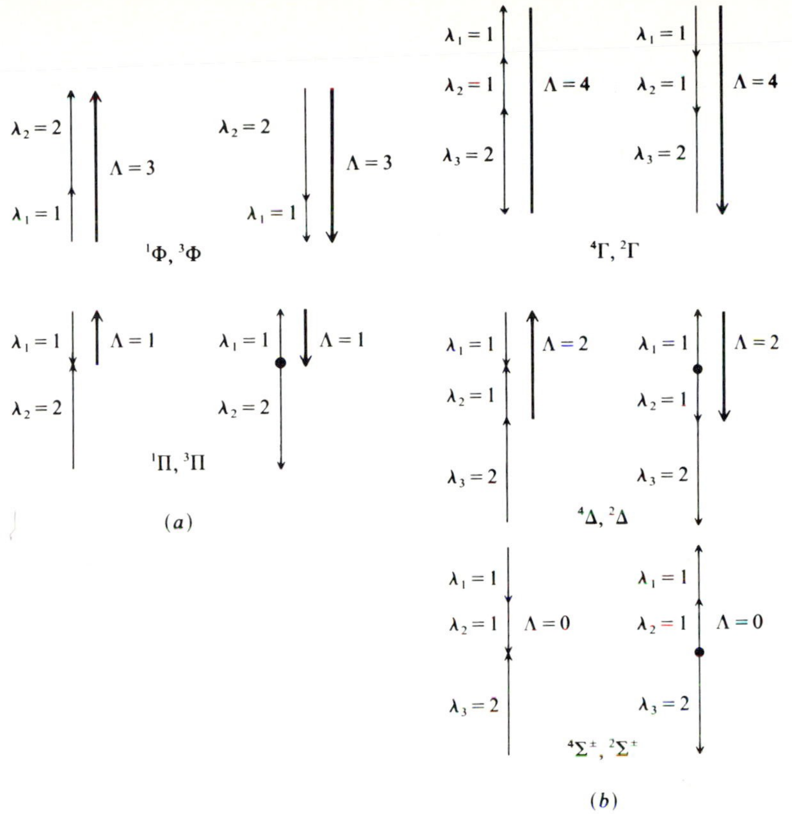

FIGURE 6.20
Vector relationships for nonequivalent electrons with (*a*) $\lambda_1 = 1$, $\lambda_2 = 2$ and (*b*) $\lambda_1 = \lambda_2 = 1$, $\lambda_3 = 2$.

Group theory can also be used to determine the allowed terms but in this case the direct product of Π and Δ must be employed. By either determining the irreducible representations contained in $\Pi \times \Delta$ using conventional methods of group theory or by consulting a table of direct products (App. N) for the $C_{\infty v}$ group it is found that $\Pi \times \Delta = \Pi + \Phi$.

The $\pi\pi\delta$ configuration will have possible values of $\Lambda = 0, 2, 4$ and $S = \frac{3}{2}, \frac{1}{2}$. This dictates that there will be Σ, Δ, and Γ states of both doublet and quartet multiplicity. The application of the methods of group theory shows that the direct product is

$$\Pi \times \Pi \times \Delta = \Sigma^+ + \Sigma^- + 2\Delta + \Gamma \tag{6.46}$$

The advantage of the use of the direct product is the introduction of the $\pm$ designation on the Σ states. The allowed terms and multiplicities of several states of nonequivalent electrons are listed in Table 6.8.

For equivalent electrons the n and 1 quantum numbers are the same so they must be differentiated by the m_l and m_s values. The vector relationships for

TABLE 6.8
Terms and multiplicities for nonequivalent electrons in molecular orbitals

Configuration	Allowed terms	Allowed multiplicities
σ	Σ^+	2
π	Π	2
δ	Δ	2
$\sigma\sigma$	Σ^+	1, 3
$\sigma\pi$	Π	1, 3
$\sigma\delta$	Δ	1, 3
$\pi\pi$	$\Sigma^+, \Sigma^-, \Delta$	1, 3
$\pi\delta$	Π, Φ	1, 3
$\delta\delta$	$\Sigma^+, \Sigma^-, \Gamma$	1, 3
$\sigma\sigma\sigma$	Σ^+	2, 4
$\sigma\sigma\pi$	Π	2, 4
$\sigma\sigma\delta$	Δ	2, 4
$\sigma\pi\pi$	$\Sigma^+, \Sigma^-, \Delta$	2, 4
$\sigma\pi\delta$	Π, Φ	2, 4
$\pi\pi\pi$	Π, Φ	2, 4
$\pi\pi\delta$	$\Sigma^+, \Sigma^-, \Delta, \Gamma$	2, 4
$\pi\delta\delta$	Π, Φ, H	2, 4
$\delta\delta\delta$	Δ, I	2, 4

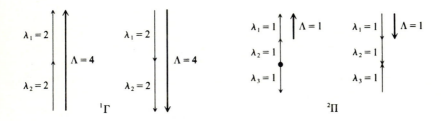

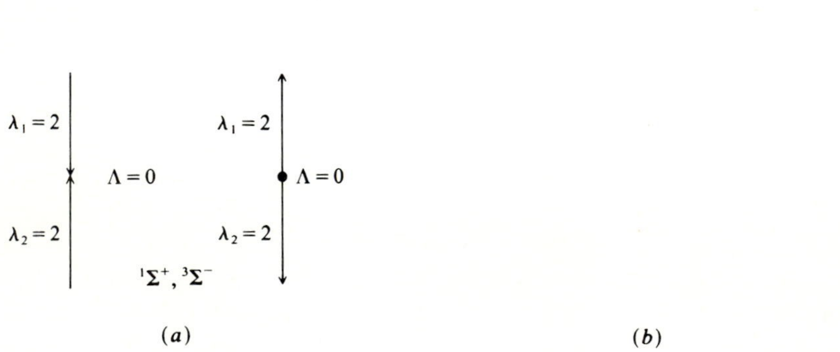

FIGURE 6.21
Vector relationship for equivalent electrons with (a) $\lambda_1 = \lambda_2 = 2$ and (b) $\lambda_1 = \lambda_2 = \lambda_3 = 1$.

TABLE 6.9
Terms and multiplicities for equivalent electrons in molecular orbitals

Configuration	Allowed terms
σ^2	$^1\Sigma^+$
π^2	$^1\Sigma^+, {}^3\Sigma^-, {}^1\Delta$
δ^2	$^1\Sigma^+, {}^3\Sigma^-, {}^1\Gamma$
π^3	$^2\Pi$
δ^3	$^2\Delta$
π^4	$^1\Sigma^+$
δ^4	$^1\Sigma^+$

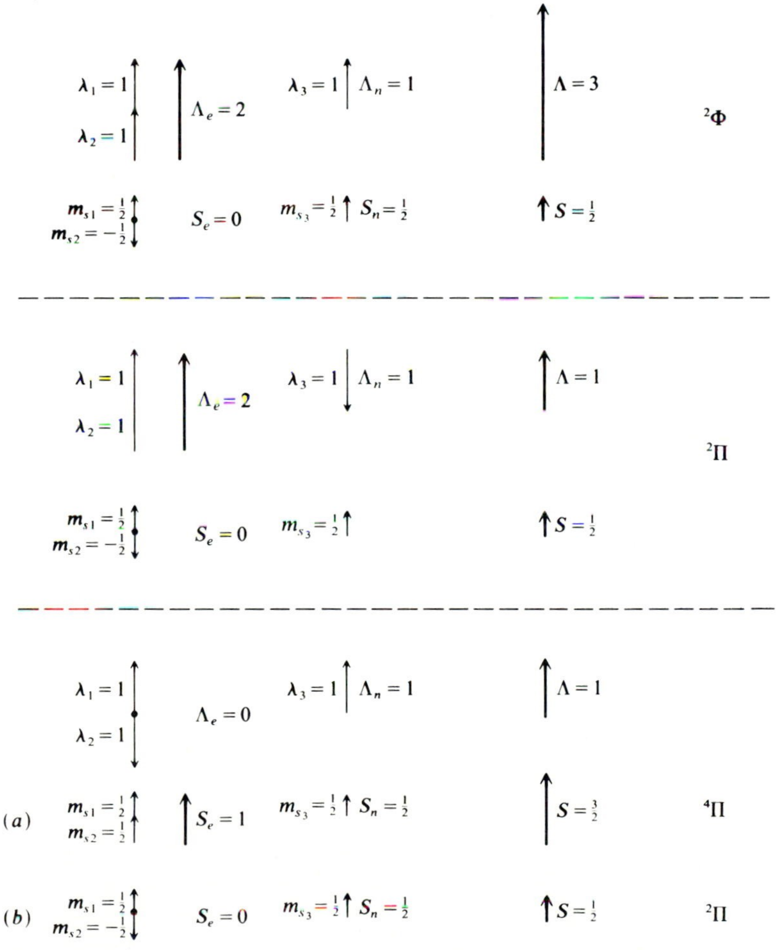

FIGURE 6.22
Vector relationships for two equivalent and one nonequivalent electrons.

TABLE 6.10
Terms for mixed electron configurations

Configuration (equiv) (nonequiv)	Allowed terms
$(\sigma^2)(\pi)$	$^2\Pi$
$(\sigma^2)(\delta)$	$^2\Delta$
$(\pi^2)(\sigma)$	$^2\Sigma^+, ^2\Sigma^-, ^2\Delta, ^4\Sigma^-$
$(\pi^2)(\pi)$	$^2\Pi, ^2\Phi, ^4\Pi$
$(\pi^2)(\delta)$	$^2\Sigma^+, ^2\Sigma^-, ^2\Delta, ^2\Gamma, ^4\Delta$
$(\pi^2)(\sigma\sigma)$	$^1\Sigma^+, ^1\Sigma^-, ^1\Delta, ^3\Sigma^+, ^3\Sigma^-, ^3\Delta, ^5\Sigma^-$
$(\pi^2)(\sigma\pi)$	$^1\Pi, ^1\Phi, ^3\Pi, ^3\Phi, ^5\Pi$
$(\pi^2)(\sigma\delta)$	$^1\Sigma^+, ^1\Sigma^-, ^2\Delta, ^1\Gamma, ^3\Sigma^+, ^3\Sigma^-, ^3\Delta, ^3\Gamma, ^5\Delta$
$(\pi^2)(\pi\pi)$	$^1\Sigma^+, ^1\Sigma^-, ^1\Delta, ^1\Gamma, ^3\Sigma^+, ^3\Sigma^-, ^3\Delta, ^3\Gamma, ^5\Sigma^+, ^5\Sigma^-, ^5\Delta$
$(\pi^2)(\pi^2)$	$^1\Sigma^+, ^1\Sigma^-, ^1\Delta, ^1\Gamma, ^3\Sigma^+, ^3\Sigma^-, ^3\Delta, ^5\Sigma^+$
$(\pi^3)(\sigma)$	$^1\Pi, ^3\Pi$
$(\pi^3)(\pi)$	$^1\Sigma^+, ^1\Sigma^-, ^1\Delta, ^3\Sigma^+, ^3\Sigma^-, ^3\Delta$
$(\pi^3)(\delta)$	$^1\Pi, ^1\Phi, ^3\Pi, ^3\Phi$

the λ_i are shown in Fig. 6.21 for the δ^2 and π^3 cases. For the δ^2 case there will be three possible states. If λ_1 and λ_2 are parallel then $m_{s_1} \neq m_{s_2}$ and one has a $^1\Gamma$. If λ_1 and λ_2 are antiparallel then m_{s_1} can equal m_{s_2} and both $^1\Sigma$ and $^3\Sigma$ states occur. For the π^3 case there will be only one state, since the occurrence of two parallel λ_i dictates that those two electrons must have opposite spins. This lone state will be a $^2\Pi$. Table 6.9 summarizes the allowed terms and multiplicities for equivalent electrons.

Not all electron configurations fall clearly into one of the two categories just discussed but often involve a combination of equivalent and nonequivalent electrons. When this occurs the procedure employed to determine the allowed terms is to determine the values of Λ and S separately for the equivalent and nonequivalent electrons and then combine these separate values. As an example the case of two equivalent and one nonequivalent π electrons is illustrated in Fig. 6.22. A summary of the allowed terms for a few mixed configurations is given in Table 6.10.

6.11 EXAMPLES OF DIATOMIC MOLECULE ELECTRONIC ENERGY STATES

Although it has not been possible to provide more than a brief introduction to electronic energy states the prior discussions, along with a few additional specific examples, should be sufficient to provide the necessary background to understand the nature and analysis of electronic spectra. For a more in-depth treatment of electronic energy states the reader can consult the classic treatise by Herzberg [20]. This discussion of selected examples will begin by looking at the electron occupancy and ordering of the molecular orbitals in representative molecules and the relationship of the energy levels of these molecules to one another via correlation diagrams. This will be followed by examining the energy level schemes of a set of molecules having the same total number of electrons. Also given is a tabular summary of properties of common diatomic molecules.

It has been shown in Figs. 6.9 and 6.10 that the molecular energy states may be indexed either in terms of the "united atom" atomic orbitals or in terms of the "separated atoms" atomic orbitals. In the latter case the designation is different for homonuclear and heteronuclear molecules because the principal quantum number of the combining orbitals will not necessarily be the same for a heteronuclear molecule. The use of different terminologies by various authors is a feature which must be recognized, and care should be taken to maintain clarity in the mind of the reader. For example, the electronic occupancy of the three-molecular orbitals in the Li_2 molecule will be found to be written in various ways such as: $(\sigma 1s)^2(\sigma 1s^*)^2(\sigma 2s)^2$; $(1s\sigma)^2(2p\sigma)^2(2s\sigma)^2$; $(\sigma_g 1s)^2(\sigma_u 1s)^2(\sigma_g 2s)^2$; $KK(\sigma 2s)^2$; $KK(\sigma_g 2s)^2$; $(1\sigma_g^+)^2(1\sigma_u^+)^2(2\sigma_g^+)^2$; $(\sigma_g 1s)^2(\sigma_u^* 1s)^2(\sigma_g 2s)^2$; $KK(z\sigma_g)^2$. For a heteronuclear molecule such as LiH the representations can be: $(1s)^2(\sigma 2s)^2$; $(1s\sigma)^2(2s\sigma)^2$; $K(\sigma 2s)^2$; $(1\sigma^+)^2(2\sigma^+)^2$; $K(z\sigma)^2$. In this molecule it can be considered that the $1s$ atomic orbital of the Li atom remains virtually unchanged ($1s \equiv 1s\sigma \equiv K$) and there is a single molecular orbital ($z\sigma$) formed from the Li $2s$ atomic orbital and the H $1s$ atomic orbital.

The electronic configurations and resulting terms for excited state molecules can be determined by establishing the order of the molecular orbital energies and their occupancies and then examining the possibilities for electron excitation. As examples of homonuclear molecules F_2, N_2, and B_2 will be used.

The F_2 molecule contains 18 electrons and, due to its longer bond length, will lie toward the right side of the correlation diagram, Fig. 6.9. The energy levels will be in the "normal" order for the "separated atoms" limit and the electronic configuration of the ground state is

$$KK(z\sigma_g)^2(y\sigma_u)^2(x\sigma_g)^2(w\pi_u)^4(v\pi_g)^4$$

Each of the molecular orbitals is fully occupied with spin-paired electrons. Reference to Table 6.9 indicates that this configuration containing σ^2 and π^4 equivalent electron sets will be in a ${}^1\Sigma_g^+$ state. If sufficient energy is absorbed by one of the $v\pi_g$ electrons to promote it to the antibonding $u\sigma$ molecular orbital the configuration becomes a $\pi^3\sigma$ state which has two allowed terms, ${}^3\Pi$ and ${}^1\Pi$.

The electronic configuration of nitrogen, N_2 will have a different ordering of the molecular orbital energies than that for F_2 since N_2 has a shorter bond length and lies further to the left on the correlation diagram. It will be

$$KK(z\sigma_g)^2(y\sigma_u)^2(w\pi_u)^4(x\sigma_g)^2$$

The ground state of N_2 will also be a ${}^1\Sigma_g^+$ state. The next empty molecular orbital in N_2 will be the $v\pi_g$ so the first excited state will be

$$KK(z\sigma_g)^2(y\sigma_u)^2(w\pi_u)^4(x\sigma_g)(v\pi_g)$$

having ${}^3\Pi_u$ and ${}^1\Pi_g$ terms. This is consistent with experiment where in agreement with Hunds' rules these states are found at 50,206 cm^{-1} and 69,290 cm^{-1} above the ground state.

The electronic configuration of B_2 will depend on the position of the intersection of the $x\sigma_g$ and the $w\pi_u$ orbital curves on the correlation diagram and will be either

$$KK(z\sigma_g)^2(y\sigma_u)^2(x\sigma_g)^2$$

or

$$KK(z\sigma_g)^2(y\sigma_u)^2(w\pi_u)^2$$

The first configuration will give rise to only a $^1\Sigma_g^+$ term while the second will give rise to $^1\Sigma_u^+$, $^3\Sigma_g^-$, and $^1\Delta$ terms. Excitation from the upper level of the first configuration would give

$$KK(z\sigma_g)^2(y\sigma_u)^2(x\sigma_g)(w\pi_u)$$

and from the second would give

$$KK(z\sigma_g)^2(y\sigma_u)^2(w\pi_u)(x\sigma_g)$$

These two excited states are identical, involve a $\sigma\pi$ nonequivalent electron pair, and would give rise to $^1\Pi$ and $^3\Pi$ terms. Experimentally the lowest transition is from a $^3\Sigma_g^-$ ground state to a $^3\Sigma_u^-$ excited state indicating that the observed excited state may be

$$KK(z\sigma_g)^2(y\sigma_u)(w\pi_u)^2(x\sigma_g)$$

It is important to note that the scale for internuclear separation in the correlation diagrams is neither linear nor regular. Stated another way, the relative ordering of the energies of the molecular orbitals will differ from one molecule to another. Table 6.11, which lists the observed configurations for a number of homonuclear diatomic molecules, serves to illustrate this point. Both N_2^+ and O_2^+ species have identical internuclear separations but the order of the $w\pi_u$ and $x\sigma_g$ molecular orbitals is reversed. Hence it is impossible to represent both by a single vertical line in Fig. 6.8. Also C_2 and O_2 both have internuclear separations greater than N_2^+ but have reversed ordering of the $w\pi_u$ and $x\sigma_g$ orbitals. One must conclude that correlation diagrams are at best qualitative in nature and do not present a quantitative description of the bonding. The behavior which occurs regarding the molecular orbital energies as more electrons are added is analogous to the progressive decrease in energy of the atomic orbitals as more electrons are added to the atom [23]. Mulliken [18] has constructed curves which illustrate this behavior.

For heteronuclear diatomic molecules between atoms in the same period the electronic configurations are generally the same as those of the isoelectronic homonuclear molecules. For example the configurations of CN^+, BN and BeO will be the same as that of C_2 while those of CO, NO^+, CF^+, and BF will be the the same as that of N_2. Due to different nuclear charges there will not be an exact correspondence of energy levels. Excited state configurations can be deter-

TABLE 6.11
Ground state electronic configurations, equilibrium bond lengths and molecular terms of common homonuclear diatomic molecules and molecule ions

Molecule	Number of electrons	Configuration	$r_e(A)$†	Term
H_2^+	1	$\sigma_g 1s$	1.05	$^2\Sigma_g^+$
H_2	2	$(\sigma_g 1s)^2 \equiv K$	0.74	$^1\Sigma_g^+$
He_2^+	3	$K(\sigma_u 1s)$	1.08	$^2\Sigma_u^+$
He_2	4	$(\sigma_g 1s)^2(\sigma_u 1s)^2 \equiv KK$	2.97	$^1\Sigma_g^+$
Li_2	6	$KK(\sigma_g 2s)^2$	2.67	$^1\Sigma_g^+$
Be_2	8	$KK(\sigma_g 2s)^2(\sigma_u 2s)^2 \equiv HH$	....	$^1\Sigma_g^+$
B_2	10	$HH(\pi_u 2p)^2$	1.59	$^3\Sigma_g^-$
C_2	12	$HH(\pi_u 2p)^3(\sigma_g 2p)$	1.24	$^1\Sigma_g^+$
N_2^+	13	$HH(\pi_u 2p)^4(\sigma_g 2p)$	1.12	$^2\Sigma_g^+$
N_2	14	$HH(\pi_u 2p)^4(\sigma_g 2p)^2$	1.10	$^1\Sigma_g^+$
O_2^+	15	$HH(\sigma_g 2p)^2(\pi_u 2p)^4(\pi_g 2p)$	1.12	$^2\Pi_g$
O_2	16	$HH(\sigma_g 2p)^2(\pi_u 2p)^4(\pi_g 2p)^2$	1.21	$^3\Sigma_g^-$
F_2	18	$HH(\sigma_g 2p)^2(\pi_u 2p)^4(\pi_g 2p)^4$	1.41	$^1\Sigma_g^+$
Ne_2	20	$HH(\sigma_g 2p)^2(\pi_u 2p)^4(\pi_g 2p)^4(\sigma_u 2p)^2$	3.15	$^1\Sigma_g^+$

† Huber, K. P., and G. Herzberg, *Molecular Spectra and Molecular Structure, IV. Constants of Diatomic Molecules*, Van Nostrand Reinhold Co., New York, NY, 1979.

mined in the same manner as was done for homonuclear molecules. Figure 6.23 illustrates the relationships of the energy levels and electronic configurations of a series of 13 electron molecules.

When the atoms forming a heteronuclear diatomic molecule are from different periods the situation is changed and one must refer to the correlation diagram, Fig. 6.10. One common type of heteronuclear molecule which is frequently encountered is the hydride, MH. A typical example will be the CH species which contain a total of seven electrons. Since the $1s$ atomic orbital of carbon can be considered not to be involved in the bonding and there will be only one electron available from hydrogen the electronic configuration is found by looking at the left-hand side of Fig. 6.10 to be

$$K(2s\sigma)^2(2p\sigma)^2(p\pi)$$

which, by reference to Table 6.8, is a $^2\Pi$ term. The configuration of the first excited state of CH will depend on the relative spacings between the members of the $2p\sigma$-$2p\pi$ and $2p\pi$-$3s\sigma$ energy level pairs. The "united atom" in this case will be the nitrogen atom whose energy levels are such that it will take less energy to promote an electron from $2p_z$ to $2p_x$ than to promote one from $2p_x$ to $3s$. Hence the excited state configurations will be

$$K(2s\sigma)^2(2p\sigma)(2p\pi)^2$$

and

$$K(2s\sigma)^2(2p\pi)^3$$

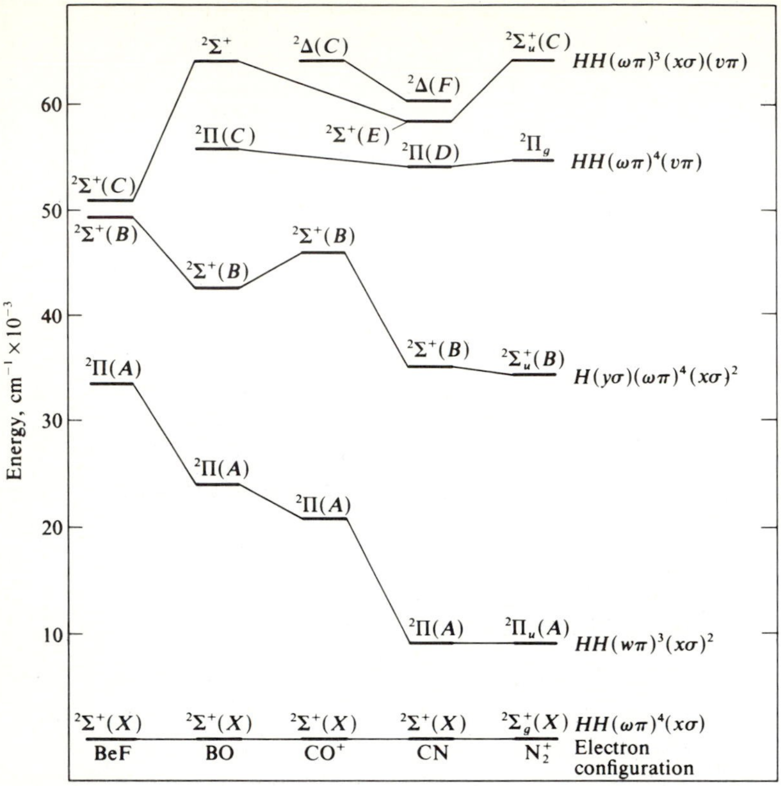

FIGURE 6.23
Energy levels, electronic configurations, and molecular terms of 13 electron diatomic molecules (*Based on data from K. P. Huber and G. Herzberg, "Molecular Spectra and Molecular structure, IV, Constants of Diatomic Molecules," Van Nostrand Reinhold Co., New York, NY,* 1979.)

which have allowed terms of $^4\Sigma^-$, $^2\Delta$, $^2\Sigma^+$, and $^2\Sigma^-$ for the former and $^2\Pi$ for the latter.

The basic concepts of this chapter can be correlated once more by use of the CH molecule as an example. The "united atom" for CH will be nitrogen with an s^2p^3 electronic configuration. Reference to App. M gives the possible atomic terms which in turn give rise to the following molecular terms:

$$^2P \rightarrow {}^2\Sigma^+, {}^2\Pi$$

$$^2D \rightarrow {}^2\Delta, {}^2\Pi, {}^2\Sigma^-$$

$$^4S \rightarrow {}^4\Sigma^-$$

Approached from the "separated atoms" limit the CH molecule is synthesized from a H atom, which has a $1s$ electronic configuration and a ^2S atomic term, and a C atom, which has a $1s^22s^22p^2$ configuration with ^1S, ^1D and ^3P

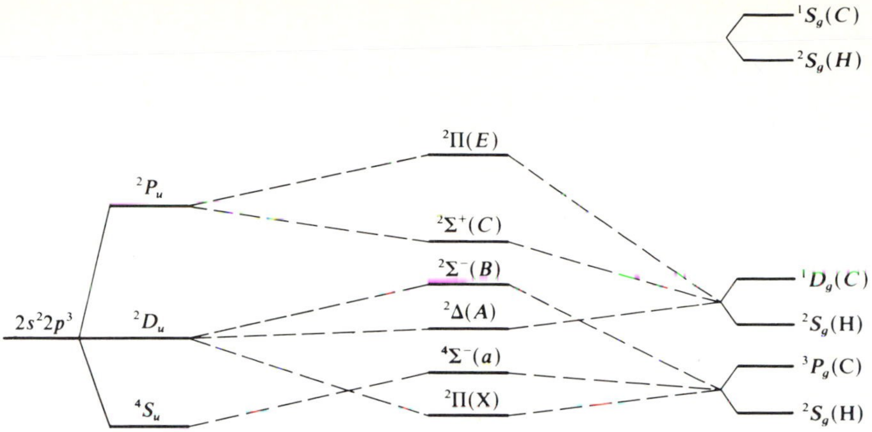

FIGURE 6.24
Diagrammatic relationships of energy levels and terms for the CH molecule and its limiting "united atom" and "separated atoms."

atomic terms, which give rise to the following molecular terms:

$$^2S + {}^1S \rightarrow {}^2\Sigma^+$$

$$^2S + {}^1D \rightarrow {}^2\Delta, {}^2\Pi, {}^2\Sigma^+$$

$$^2S + {}^3P \rightarrow {}^4\Sigma^-, {}^2\Sigma^-, {}^2\Pi, {}^4\Pi$$

If the energy levels and terms of the two limiting cases are correlated with those of the molecule, keeping in mind the noncrossing rule, the diagram shown in Fig. 6.24 emerges. It should be noted again that this diagram, which is also a correlation diagram, differs from that of Fig. 6.10 in that the latter is for a series of one-electron energy states.

REFERENCES

A. Specific

1. Born, M., and J. R. Oppenheimer, *Ann. Physik*, **84**, 457 (1927).
2. Prosser, F., and H. Shull, *Ann. Rev. Phys. Chem.*, **17**, 37 (1966).
3*a*. Mulliken, R. S., *Diatomic Molecules*, Academic Press, Orlando, FL, 1977.
3*b*. Schaefer, H. F., III, *The Electronic Structure of Atoms and Molecules*, Addison-Wesley, Reading, MA, 1972, ch. 3.
4. Morse, P. M., *Phys. Rev.*, **34**, 57 (1929).
5. Rydberg, R., *Z. fur Physik*, **73**, 376 (1931).
6. Lippincott, E. R., *J. Chem. Phys.*, **21**, 2070 (1953).
7. Rosen, N., and P. M. Morse, *Phys. Rev.*, **42**, 210 (1932).
8. Frost, A. A., and B. Musulin, *J. Chem. Phys.*, **22**, 1017 (1954).
9. Varshni, Y. P., *Rev. Med. Phys.*, **29**, 664 (1957).
10. Hulburt, H. M., and J. O. Hirschfelder, *J. Chem. Phys.*, **9**, 611 (1941); ibid, **35**, 1901 (1961).
11. Pauling, L., and E. B. Wilson, Jr., *Introduction to Quantum Mechanics*, McGraw-Hill Book Co., New York, NY, 1935. ch. 7.

12. Dunham, J. L., *Phys. Rev.*, **41**, 721 (1932).
13. Heitler, H., and F. London, *Zeit, fur Phys.*, **44**, 455 (1927).
14. Mulliken, R. S., *Phys. Rev.*, **4**, 1 (1932).
15. Weissbluth, M., *Atoms and Molecules*, Academic Press, New York, NY, 1978, ch. 26.
16. Loc. cit., Schaefer.
17. James, H., and A. Coolidge, *J. Chem. Phys.*, **1**, 825 (1933).
18. Mulliken, R. S., *Rev. Mod. Phys.*, **2**, 60 (1930); ibid **3**, 89 (1931); ibid **4**, 1 (1932).
19. Neumann, J. V., and E. Wigner, *Physik Z.*, **30**, 467 (1929).
20. Herzberg, G., *Molecular Spectra and Molecular Structure: I. Spectra of Diatomic Molecules*, 2d ed., Von Nostrand Reinhold Co., New York, NY 1950, p. 510.
21. Wigner, E., and E. E. Witmer, *Zeit for Physik*, **51**, 859 (1928).
22. Herzberg, G., *Atomic Spectra and Atomic Structure*, Dover Publications, New York, NY, 1944, p. 128.
23. Offenhartz, P. O'D., *Atomic and Molecular Orbital Theory*, McGraw-Hill Book Co., New York, NY, 1970.

B. General

See 3a, 3b, 20 and 22b of Special References.

Coulson, C. A., *Valence*, Oxford University Press, London, GB, 1952.

Hollas, M. J., *High Resolution Spectroscopy*, Butterworths, London, GB, 1982.

Huber, R. P., and G. Herzberg, *Molecular Spectra and Molecular Structure: IV. Constants of Diatomic Molecules*, Von Nostrand Reinhold Co., New York, NY, 1979.

King, G. W., *Spectroscopy and Molecular Structure*, Holt Reinhart and Winston, Inc., New York, NY, 1964.

PROBLEMS

6.1. For a hypothetical diatomic molecule having a vibrational frequency $\omega_e = 2000\ \text{cm}^{-1}$, a dissociation energy $D_o = 80\ \text{kcal/mol}$, a reduced mass $\mu = 20$ amu, and an equilibrium internuclear separation $r_e = 2$ A.

(*a*) Evaluate the **a** of the Morse potential function and construct the Morse curve.

(*b*) Evaluate the **b** of the Rydberg potential functions and construct the Rydberg curve.

(*c*) Evaluate the **a** of the Varshni potential function and construct the Varshni curve.

(*d*) Evaluate **b** and **c** for the Hulburt-Hirshfelder potential function and construct the Hulburt-Hirshfelder curve.

(*e*) Compare the curves just constructed.

6.2. Show that the Rydberg and Varshni potential functions have a minimum value at $r = r_e$.

6.3. What molecular terms, including multiplicities, can be obtained by combining the following?

(*a*) Two atoms both in ^2D states

(*b*) Two atoms both in ^6S states

(*c*) Two atoms in ^2P and ^1P states

(*d*) Two atoms in ^1DE and ^1S states

6.4. What molecular terms, including multiplicities, can arrive from the ground state of the following "united atom"?

(*a*) 0

(*b*) Be

(*c*) V

(*d*) Br

6.5. Determine the ground state term for the following species.
(*a*) NO^+ (*b*) O_2^- (*c*) NH^+ (*d*) P_2^-

6.6. Determine the ground state terms for the hydrides of the elements from Li through F.

6.7. Construct "united atom"-"separated atom" correlation diagrams analogous to Fig. 6.13 for the following species. For each case the term symbols for molecular electronic levels and their energies (T_e in cm^{-1}) are given.

(*a*) MgO: $^1\Sigma^+(0)$, $^3\Pi(2400)$, $^1\Pi(3563)$, $^1\Sigma^+(19984)$, $^3\Sigma^+(28300)$, $^3\Delta(29300)$, $^1\Delta(29851)$, $^3\Sigma^-(<31250)$, $^1\Sigma^-(30080)$, $^1\Sigma^+(37722)$, $^1\Pi(37922)$.

(*b*) BeH: $^2\Sigma^+(0)$, $^2\Pi(20033)$, $^2\Pi(50882)$, ($^2\Delta$, $^2\Pi$, $^2\Sigma^+$) (overlapping at 54000), $^2\Sigma^+(54134)$, $^2\Sigma^+(56606)$, $^2\Pi(58711)$.

(*c*) AlF: $^1\Sigma^+(0)$, $^3\Pi(27241)$, $^1\Pi(43949)$, $^3\Sigma^+(44813)$, $^1\Sigma^+(54251)$, $^3\Sigma^+(54957)$, $^1\Sigma^+(57688)$, $^1\Delta(61229)$, $^3\Delta(63203)$, $^1\Pi(63689)$, $^3\Sigma^+(65010)$, $^1\Pi(65796)$.

(*d*) CF: $^2\Pi(0)$, $^4\Sigma^-(22000)$, $^2\Sigma^+(42693)$, $^2\Delta(49399)$, $^2\Pi(52272)$, $^2\Sigma^+(>53000)$.

(*e*) Na_2: $^1\Sigma_g^+(0)$, $^3\Pi(<14680)$, $^1\Sigma_u^+(14681)$, $^1\Pi_u(20320)$, $^1\Pi_u(29382)$, $^1\Sigma_g^+(33000)$, $^1\Pi_u(33487)$, $^1\Pi_u(35557)$.

(*f*) P_2: $^1\Sigma_g^+(0)$, $^3\Sigma^+(18795)$, $^3\Sigma_u^-(28503)$, $^3\Pi_g(28330, 28197, 28069)$, $^1\Pi_g(34515)$, $^1\Sigma_u^+(46941)$, $^3\Pi_u(47177, 47159, 47139)$, $^1\Pi_u(50846)$, $^1\Pi_u(59446)$, $^1\Sigma_u^+(66313)$.

CHAPTER 7

RADIATIVE TRANSITIONS

7.1 INTRODUCTION

The initial discussions of the rigid rotor and the harmonic oscillator have shown that the energy levels of such systems are quantized and that the energy differences between the levels in these systems is of the order of magnitude of electromagnetic radiation in the microwave and infrared regions respectively. These examples are only two of the many areas of spectroscopy which may encompass the interaction of quantized energy states with electromagnetic radiation from the radiofrequency region through the X-ray region. However, the occurrence of quantized energy levels in a system is not a sufficient condition for electromagnetic radiation, having an energy content equivalent to the energy difference of a pair of the levels (that is, $\Delta E = h\nu$), to be adsorbed or emitted. Whether a particular transition between two quantized energy levels can occur is determined by the nature of the interaction between the system and the radiation. Both the probability of a transition occurring and the amount of radiation adsorbed or emitted are controlled by this interaction. We will first look at the physical nature of the phenomenon involved and then explore the more rigorous quantum mechanical approach to the interaction.

7.2 THE INTERACTION OF RADIATION AND MATTER—CLASSICAL APPROACH

If a molecule is to interact with electromagnetic radiation and adsorb or emit energy, then there must be a mechanism for the interaction. Consider the case of classical rotation. If one has a flywheel and wishes to increase its speed of rotation (i.e., increase its energy) then the simplest method is to couple it to a driving motor via a gear or belt system and hence transfer energy from the motor to the flywheel. The complexity of the mechanism for energy transfer can be increased, and the analogy to the molecular rotor more closely approached, if we were to mount a bar magnet along the diameter of the flywheel and mount a second such magnet on the motor shaft. By proper placement of the two magnets the flywheel could be caused to rotate as the motor supplied energy without any direct mechanical linkage. This type of interaction between magnetic fields originating either from permanent magnets or from an electric current in a coil of wire constitutes the basic interaction in electric motors and is a well-known principle of electromagnetic theory.

The transfer of these concepts to the molecular scale is relatively straightforward. For molecules the physical interaction may be either magnetic or electrostatic in nature. The source of energy, the electromagnetic radiation, has associated with it an oscillating electric and magnetic field. Hence, it occupies the role of the motor-driven rotating magnet. If the molecule possesses either a magnetic dipole moment or an electric dipole moment which can interact with one of the oscillating fields then it occupies the role of the magnetic flywheel and the analogy is complete. Before proceeding to the more formal development of the nature of this interaction let us examine some specific cases in a little more detail.

For any molecule that does not have a center of electric charge symmetry there will exist a permanent electric dipole moment. If we consider such a dipole moment to be located at a fixed point in space and at that point there is an alternating electric field due to the electromagnetic radiation then the dipole will experience an instantaneous electrostatic interaction that will exert a torque on the molecule. Considered in a classical mechanical sense, if the frequency of rotation of the molecular electric dipole and the frequency of the oscillating electromagnetic field are the same then there will be an "in-phase" reinforcement of the interaction and the dipole will absorb energy from the radiation and its rotational energy will increase. The fact that the rotation of an electrostatic dipole is a periodically varying function is illustrated in Fig. 7.1; the dipole is considered to be rotating in the xy plane and the y component is plotted vs. time.

Any phenomenon which leads to a time-variant magnetic or electric dipole can give rise to a similar interaction. Since the electric dipole moment of a molecule is a function of its geometry, a change in structure during the course of internal vibration often will lead to a periodic change in the dipole moment. If we consider a linear XY_2 molecule where there is a net negative charge on the terminal Y atoms and a net positive charge on the central X one observes that there will be no permanent electric dipole for this symmetrical linear molecule, but harmonic bending of the molecule will produce a periodically varying dipole moment as is shown in Fig. 7.2. The mechanism for the interaction in the case of vibrational spectroscopy will be the electrostatic interaction between the periodically changing molecular dipole and the electric field of the electromagnetic radiation.

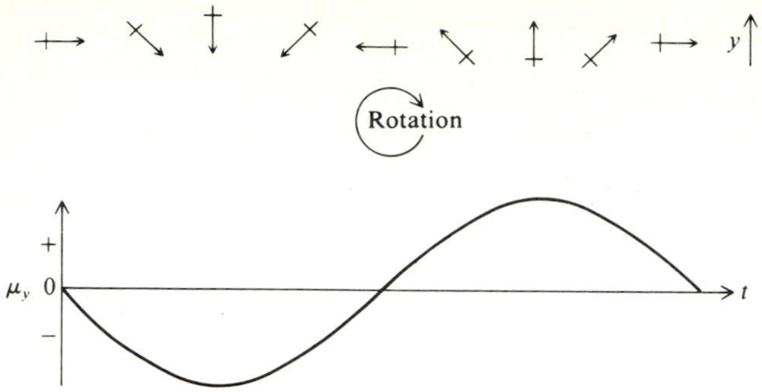

FIGURE 7.1
Variation of the y component of a rotating electrostatic dipole with time.

In the areas of nuclear magnetic resonance, electron spin resonance, and nuclear quadrupole resonance the interaction is between a permanent magnetic moment of a nucleus or an electron and the magnetic field of the electromagnetic radiation. In electronic spectroscopy the motion of the electron relative to the positive nucleus provides a changing electric dipole which can interact with the electric field of the radiation.

The previous discussions have shown that one classical mechanism for the occurrence of an interaction between a molecular system and electromagnetic radiation involves either the existence of a permanent electric or magnetic dipole or the instantaneous creation of an electric dipole due to internal motions. The magnitude of the interactions, which will be reflected in the probability of a given system undergoing a particular transition with the accompanying emission or absorption of radiation, will depend on the size of the molecular dipole moment.

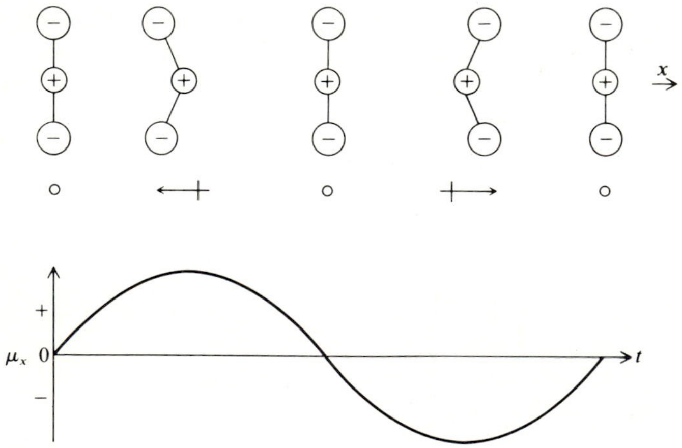

FIGURE 7.2
Creation of a periodically varying electric dipole moment in XY_2 due to a bending motion.

7.3 THE TIME-DEPENDENT WAVE EQUATION

The next problem associated with the study of the absorption of radiation by molecules is to quantitatively relate the magnitude of the absorption to molecular parameters. Since the principles involved are common to all areas of spectroscopy the general concepts will be formulated and these will be used in later discussions of specific systems. There are two topics which will be considered, the quantitative aspects of the basic physical phenomenon and the effect of quantum mechanical selection rules.

The methods and examples considered up to this point have been primarily stationary state situations and have involved time-independent wave equations. The interaction of atoms or molecules with electromagnetic radiation necessitates consideration of time dependency. The general time-dependent Schrödinger equation for a single particle is

$$\mathscr{H}\Psi(x, y, z, t) = -\frac{\hbar^2}{2m}\nabla^2\Psi(x, y, z, t) + V(x, y, z)\Psi(x, y, z, t) = -\frac{\hbar}{i}\frac{\partial\Psi(x, y, z, t)}{\partial t} \tag{7.1}$$

In the treatment which follows, in order to avoid unnecessary complexities when writing expressions, the symbols Ψ and ψ will be used to denote the time-dependent wave function $\Psi\ (x, y, z, t)$ and the time-independent wave function $\psi\ (x, y, z)$.

If the potential energy of a system is time-independent, the time-dependent equation can be simplified by introducing a separation of variables and writing the time-dependent function as a product of the time-independent function ψ, and a function of time alone $\phi = \phi(t)$

$$\Psi = \psi\phi \tag{7.2}$$

This separation, while being a mathematical technique to simplify the solution, is acceptable on physical grounds since it produces a solution which can predict properties of systems which are in agreement with experiment. Introducing this product function into Eq. (7.1), recognizing that ∇^2 and $V(x, y, z)$ depend only on coordinates, that $\partial/\partial t$ depends only on the time, and dividing by $\psi\phi$ one obtains

$$-\frac{\hbar^2}{2m}\frac{1}{\psi}\nabla^2\psi + V = -\frac{\hbar}{i}\frac{1}{\phi}\frac{\partial\phi}{\partial t} \tag{7.3}$$

A solution of an equation of this type is possible only if each side is independently equal to a constant, which we shall call E. Therefore

$$-\frac{\hbar^2}{2m}\nabla^2\psi + V\psi = E\psi \tag{7.4}$$

and

$$-\frac{\hbar}{i}\frac{d\phi}{dt} = E\phi \tag{7.5}$$

Equation 7.4 is the time-independent Schrödinger equation whose solution for a particular system provides the wavefunctions from which physical properties can be deduced. Furthermore, the constant E is the energy of the system which, being quantized, is more appropriately designated as E_n where n is the quantum number or set of numbers which characterizes the state.

The time-dependent part of the wave equation is a first-order differential equation whose solution is

$$\phi(t) = e^{-iE_n t/\hbar} \tag{7.6}$$

One very important concept evolves at this point; the fact that the time-dependent factor in the solution of the Schrödinger equation has the same form regardless of the system being considered. This was not at all the case for the time-independent solution since the form of the Hamiltonian operator depends on the mechanics of the system. The complete wavefunction for a single particle can be written as

$$\Psi_n = \psi_n e^{-iE_n t/\hbar} \tag{7.7}$$

The product $\psi_n^* \psi_n$ is a probability distribution function which gives information as to the spatial distribution of a system. The product $\Psi_n^* \Psi_n$ is then a probability distribution function which conveys information as to the distribution of the system in space as a function of both coordinates and time. Looking at this product function

$$\Psi_n^* \Psi_n = \psi_n^* \psi_n e^{iE_n t/\hbar_e - iE_n t/\hbar} = \psi_n^* \psi_n \tag{7.8}$$

one sees that the time-dependent terms drops out. Systems which are not being subjected to electromagnetic radiation are of this type and are described by the time-independent function ψ_n.

When a molecule absorbs radiation there is a change in the system from one stationary state ψ_n to a higher state ψ_m. Since the irradiation process requires a small but nevertheless finite time span, the process of absorption, or emission, of radiation is a time-dependent phenomenon and must be treated as such.

7.4 ABSORPTION OF RADIATION

The absorption of radiation by a quantized system involves allowing a certain incident radiative energy, having some definite spectral distribution, and denoted by $P_0(\nu)$ to pass into a sample which adsorbs a narrow band of frequencies centered at $\nu_{\sigma\sigma'}$ where σ and σ' represent sets of quantum numbers which specify the two states of the system between which the transition occurs. If we consider that the number of molecules per unit volume C remains constant then the decrease in power $dP(\nu)$ for passage of the radiation through a length dx of the sample is given by Beer's law as

$$dP(\nu) = -a(\nu_{\sigma\sigma'})CP(\nu)\,dx \tag{7.9}$$

where $a(\nu_{\sigma\sigma'})$ is a frequency-dependent proportionality constant and is referred to as the absorption coefficient. Integration of this relation over a sample of thickness x gives

$$\ln P(\nu) = -a(\nu_{\sigma\sigma'})Cx + k' \tag{7.10}$$

When $x = 0 P(\nu) = P_0(\nu)$ so $k' = \ln P_0(\nu)$ and

$$P(\nu) = P_0(\nu)e^{-a(\nu_{\sigma\sigma'})Cx} \tag{7.11}$$

or

$$\ln \frac{P(\nu)}{P_0(\nu)} = -a(\nu_{\sigma\sigma'})Cx \tag{7.12}$$

Using the terminology of solution spectrophotometry and considering the incident radiation to be a single frequency this expression becomes the Beer-Lambert relation as discussed in most elementary physical chemistry texts

$$\log \frac{I}{I_0} = -aCx \tag{7.13}$$

where a is the absorbancy index or extinction coefficient for a particular frequency and I is the intensity (energy per unit time per unit area) of the radiation.

For the cases of absorption of radiation due to transitions between low-lying and closely spaced energy levels, such as rotational and vibrational transitions, the concentration factor is not the molar concentration or even the total number of molecules, but rather depends on both the concentration of molecules in the lower of the two states and the difference between the populations of the two states involved in the transition. This is also true for absorptions in the visible and ultraviolet regions that are commonly used for solution spectrophotometry. However, the transitions involved in these regions are usually electronic transitions and for most molecules essentially all are in the ground electronic state at ordinary temperatures. The state concentration in these cases is the molar concentration of the adsorbing species.

Since in general the concentration factor refers to the population of the state specified by the set of quantum numbers σ, it is common procedure to include this factor in a new absorption coefficient $\alpha(\nu_{\sigma\sigma'})$ defined by

$$\alpha \equiv \alpha(\nu_{\sigma\sigma'}) = C_\sigma a(\nu_{\sigma\sigma'}) \tag{7.14}$$

where C_σ is the concentration of molecules in the state denoted by σ. Rearrangement of Eq. (7.9) gives for the absorption coefficient

$$\alpha = \frac{-1}{P(\nu)} \frac{dP(\nu)}{dx} \tag{7.15}$$

which upon multiplication of the right side by dt/dt becomes

$$\alpha = \frac{-1}{P(\nu)} \frac{dP(\nu)}{dt} \frac{dt}{dx} \tag{7.16}$$

Noting that the last derivative term is just the reciprocal of the velocity of the radiation, i.e., the velocity of light c, this becomes

$$\alpha = \frac{-1}{cP(\nu)} \frac{dP(\nu)}{dt} \tag{7.17}$$

The term $dP(\nu)/dt$ is the energy absorbed from the radiation beam per unit time per unit frequency range and will be proportional to the number of transitions occurring per unit time. It can be considered to be the product of the energy absorbed per transition times the number of transitions per unit time. This latter quantity depends on the radiation density $\rho(\nu)$, which is the energy per unit volume per unit frequency range in the radiation beam and is related to the radiation power by $P(\nu) = k\rho(\nu)$ so

$$\frac{dP(\nu)}{dt} = k \frac{d\rho(\nu)}{dt} \tag{7.18}$$

7.5 INDUCED TRANSITIONS

Before one can enumerate the factors which affect the intensities of transitions between stationary states of a system it is necessary to consider the process whereby electromagnetic radiation interacts with the system. In general a quantum mechanical system will have a ladder of energy values E_0, E_1, E_2, etc., corresponding to a set of states Ψ_0, Ψ_1, Ψ_2, etc. In order to simplify our discussion we will consider a system which is restricted to two nondegenerate states Ψ_l and Ψ_m. This restriction results in no loss of generality of the results. For such a system the time-dependent Schrödinger equation is

$$-\frac{\hbar^2}{2m} \nabla^2 \Psi_n + V\Psi_n = -\frac{\hbar}{i} \frac{\partial \Psi_n}{\partial t} \tag{7.19}$$

and there will be only two solutions:

$$\Psi_l = \psi_l e^{-iE_l t/\hbar} \tag{7.20}$$

$$\Psi_m = \psi_m e^{-iE_m t/\hbar} \tag{7.21}$$

We next examine a wavefunction of the general form

$$\Psi = c_l(t)\Psi_l + c_m(t)\Psi_m \tag{7.22}$$

to see if it can be an acceptable solution to Eq. (7.19). The constants $c_n(t)$ are weighting coefficients which are independent of the coordinates but can be functions of time. The two stationary states are described by this general function with $c_l(t) = 1$, $c_m(t) = 0$ for Ψ_l and $c_m(t) = 1$, $c_l(t) = 0$ for Ψ_m. During an absorptive transition from one state to the other the coefficients will be a function of time changing from $c_l = 1 \rightarrow 0$ and $c_m = 0 \rightarrow 1$ if Ψ_l is the lower state. The parameter of time will enter the treatment via these coefficients and Eq. (7.22) is the general solution for a system in transition between two states.

When a system is exposed to an electromagnetic radiation field the physical effect is that of imposing on the system an additional time-dependent potential.

The general Schrödinger equation during irradiation becomes

$$-\frac{\hbar^2}{2m}\nabla^2\Psi + (V + \mathscr{V})\Psi = -\frac{\hbar}{i}\frac{\partial\Psi}{\partial t} \tag{7.23}$$

where $\mathscr{V} = \mathscr{V}(x, y, z, t)$, the added potential due to the radiation field, will be small compared to the molecular potential $V(x, y, z)$ and can be considered to be a perturbation.

If we assume that our system is initially in a state Ψ_l then it will remain there so long as $\mathscr{V} = 0$. Writing the unperturbed time-dependent wave equation as

$$\mathscr{H}^\circ\Psi = -\frac{\hbar}{i}\frac{\partial\Psi}{\partial t} \tag{7.24}$$

it then becomes, during the irradiation process,

$$(\mathscr{H}^\circ + \mathscr{V})\Psi = -\frac{\hbar}{i}\frac{\partial\Psi}{\partial t} \tag{7.25}$$

Substituting the wavefunction with undetermined coefficients into Eq. (7.25), using c_i for $c_i(t)$ for simplicity and expanding, gives

$$c_l\mathscr{H}^\circ\Psi_l + c_m\mathscr{H}^\circ\Psi_m + c_l\mathscr{V}\Psi_l + c_m\mathscr{V}\Psi_m$$

$$= -\frac{\hbar}{i}\Psi_l\frac{\partial c_l}{\partial t} - \frac{\hbar}{i}\Psi_m\frac{\partial c_m}{\partial t} - \frac{\hbar}{i}c_l\frac{\partial\Psi_l}{\partial t} - \frac{\hbar}{i}c_m\frac{\partial\Psi_m}{\partial t} \tag{7.26}$$

Since Ψ_l and Ψ_m are solutions of the unperturbed wave equation, the first two left-hand terms cancel the last two right-hand terms leaving

$$c_l\mathscr{V}\Psi_l + c_m\mathscr{V}\Psi_m = -\frac{\hbar}{i}\left[\Psi_l\frac{\partial c_l}{\partial t} + \Psi_m\frac{\partial c_m}{\partial t}\right] \tag{7.27}$$

This expression can be simplified by multiplying both sides by Ψ_m^* and integrating over all space, keeping in mind that c_l and c_m are independent of the spatial coordinates.

Thus

$$c_l\int\Psi_m^*\mathscr{V}\Psi_l\,dv + c_m\int\Psi_m^*\mathscr{V}\Psi_m\,dv = -\frac{\hbar}{i}\left[\frac{\partial c_l}{\partial t}\int\Psi_m^*\Psi_l\,dv + \frac{\partial c_m}{\partial t}\int\Psi_m^*\Psi_m\,dv\right] \tag{7.28}$$

Using the orthonormal properties of Ψ_l and Ψ_m

$$\int\Psi_i^*\Psi_j\,dv = \delta_{ij} \tag{7.29}$$

and substituting Eqs. (7.20) and (7.21) into Eq. (7.28) this reduces to

$$c_m\int\psi_m^*\mathscr{V}\psi_m\,dv + c_l[e^{-i(E_l-E_m)t/\hbar}]\int\psi_m^*\mathscr{V}\psi_l\,dv = -\frac{\hbar}{i}\frac{dc_m}{dt} \tag{7.30}$$

Rearrangement shows that the rate of increase of c_m, which is a measure of the population of the state Ψ_m or the rate of absorption of energy or the rate of transition from one state to another, is given by

$$\frac{dc_m}{dt} = \frac{-i}{\hbar}\left[c_m \int \psi_m^* \mathscr{V} \psi_m \, dv + c_l(e^{-i(E_l - E_m)t/\hbar}) \int \psi_m^* \mathscr{V} \psi_l \, dv \right] \tag{7.31}$$

Since $\Psi = c_l \Psi_l + c_m \Psi_m$ this expression can be written as

$$\frac{dc_m}{dt} = \frac{-i}{\hbar} \int \Psi_m^* \mathscr{V} \Psi \, dv \tag{7.32}$$

While Eq. (7.31) is necessary to determine the intensity of a transition, the possibility of the occurrence of a transition (selection rule) can be determined by use of a simpler form. Considering that initially $c_l = 1$ and $c_m = 0$ the initial rate of growth of population of state Ψ_m is given by substitution of these conditions into Eq. (7.31)

$$\frac{dc_m}{dt} = \frac{-i}{\hbar}[e^{-i(E_l - E_m)t/\hbar}] \int \psi_m^* \mathscr{V} \psi_l \, dv \tag{7.33}$$

If the value of the integral is zero then the initial transition rate is zero and the transition is forbidden. To proceed further we must investigate the nature of the perturbing potential $\mathscr{V}$ and its mode of interaction with a system.

7.6 INTERACTION OF RADIATION WITH MATTER

Electromagnetic radiation has both an electric and magnetic component and it is possible for two types of interaction to occur, that of the radiation with an electric dipole of the system or with a magnetic dipole of the system. The same general treatment can be applied to both cases. For simplicity we will restrict the treatment to an electric dipole constrained to move in one dimension interacting with plane polarized radiation in the same dimension. Extension to the three-dimensional case will follow this development. Furthermore, it is assumed that the wavelength of the radiation is large compared to the size of the adsorbing molecular system, thereby immersing the entire molecule in a homogeneous radiation field over the volume of the molecule. This assumption is justified when one considers that electronic, vibrational, and rotational transitions correspond to wavelengths of the order of magnitude of 10^{-9}, 10^{-7}, and 10^{-4} m respectively while molecular dimensions are of the order of 10^{-10} m.

The amplitude of the electric field $\mathscr{E}_x$ as a function of time for plane polarized electromagnetic radiation of frequency ν is given by

$$\mathscr{E}_x = 2\mathscr{E}_x^\circ \cos 2\pi\nu t = \mathscr{E}_x^\circ(e^{i2\pi\nu t} + e^{-i2\pi\nu t}) \tag{7.34}$$

where $2\mathscr{E}_x^\circ$ is the maximum amplitude of the electric field. Since the electric dipole of the molecular system results from the displacement of an electric charge by a distance x from its equilibrium position, the system can be considered to have a dipole moment of magnitude $\mu_x = ex$. The classical interaction between a dipole moment $\boldsymbol{\mu}$ and electric field $\mathscr{E}$ gives the perturbing potential $\mathscr{V}$

$$\mathscr{V} = -\boldsymbol{\mu} \cdot \boldsymbol{\mathscr{E}} = -\mu\mathscr{E} \cos \theta \tag{7.35}$$

where θ is the angle between the dipole and the field. Since μ_x and $\mathscr{E}_x$ are parallel, $\cos \theta = 1$ and

$$\mathscr{V} = -\mu_x\mathscr{E}_x = -ex\mathscr{E}_x \tag{7.36}$$

Expressing $\mathscr{E}_x$ in the complex form and introducing this relation for $\mathscr{V}$ into Eq. (7.33) the initial rate of change of c_m of the upper level becomes

$$\frac{dc_m}{dt} = \left\{\frac{-i\mathscr{E}_x^\circ}{\hbar} \int \psi_m^* \boldsymbol{\mu}_x \psi_l \, dv\right\} \{e^{i(E_m - E_l + h\nu)t/\hbar} + e^{i(E_m - E_l - h\nu)t/\hbar}\} \tag{7.37}$$

The value of c_m as a function of time is found by integration of Eq. (7.37) to be

$$c_m = (m|\mu_x|l)\mathscr{E}_x^\circ \left\{\frac{1 - e^{i(E_m - E_l + h\nu)t/h}}{E_m - E_l + h\nu} + \frac{1 - e^{i(E_m - E_l - h\nu)t/h}}{E_m - E_l - h\nu}\right\} \tag{7.38}$$

where the integral $\int \psi_m^* \boldsymbol{\mu}_x \psi_l \, dv$ is represented by $\langle m | \mu_x | l \rangle$ and is designated as the dipole moment integral or dipole matrix element. An examination of the terms in this expression, in view of the physical process being considered, will enable further simplification. The energy difference between the two levels involved in the transition is given by $(E_m - E_l)$ and will be positive for the case where l denotes the lower energy state since our initial condition was that the transition was from state Ψ_l to state Ψ_m. This is the case of absorption. If the state Ψ_l has the higher energy then $(E_m - E_l)$ will be negative and an emission process is occurring. We will consider the absorption case in more detail.

For the case of absorption we see that as the frequency of the radiation field ν approaches the value where $h\nu = E_m - E_l$, the denominator of the second term becomes very small and this term becomes important. For the case of emission the first term is the important one since its denominator then becomes small. Since the magnitudes of $(m | \mu_x | l)$ and $\mathscr{E}_x^\circ$ are small, as will be shown later, it is also clear that if the frequency of the radiation is such that both terms in the denominators are large then the entire expression in the brackets is very small and the rate of increase of the population of state Ψ_m is very small. This means that to induce absorption in a molecular system the frequency of the radiation field must correspond to the difference in energy between energy levels of the systems. For the absorption process we have

$$c_m = \langle m | \mu_x | l \rangle \mathscr{E}_x^\circ \left\{\frac{1 - e^{i(E_m - E_l - h\nu)t/\hbar}}{E_m - E_l - h\nu}\right\} \tag{7.39}$$

The probability of a system existing in a given state having a wavefunction ψ_n is proportional to $\psi_n^*\psi_n$. For the system being considered the probability of its being in a state $\Psi = c_l\Psi_l + c_m\Psi_m$ at some time following irradiation is $\Psi^*\Psi \, dx$.

Expressing this probability in terms of the general function gives

$$\int \Psi^*\Psi \, dx = c_l^* c_l \int \Psi_l^*\Psi_l \, dx + c_m^* c_l \int \Psi_m^*\Psi_l \, dx$$

$$+ c_l^* c_m \int \Psi_l^*\Psi_m \, dx + c_m^* c_m \int \Psi_m^*\Psi_m \, dx \tag{7.40}$$

Use of the orthonormal properties of Ψ_l and Ψ_m reduces this to

$$\int \Psi^*\Psi \, dx = c_l^* c_l + c_m^* c_m \tag{7.41}$$

Thus the contribution of c_m to the total probability of finding the system in a state Ψ is $c_m^* c_m$. This product can be found by multiplying Eq. (7.39) by its complex conjugate giving

$$c_m^* c_m = \langle m|\mu_x|l\rangle^2 (\mathscr{E}_x^\circ)^2 \left\{ \frac{2 - e^{-i2A} - e^{i2A}}{(E_m - E_l - h\nu)^2} \right\} \tag{7.42}$$

where $A = \pi t(E_m - E_l - h\nu)/\hbar$. Using the Euler relationship $e^{\pm iy} = \cos y \pm i \sin y$, this reduces to

$$c_m^* c_m = \langle m|\mu_x|l\rangle^2 (\mathscr{E}_x^\circ)^2 \left\{ \frac{2 - 2\cos 2A}{(E_m - E_l - h\nu)^2} \right\} \tag{7.43}$$

and, since $\cos 2x = 1 - 2\sin^2 x$, further reduces to

$$c_m^* c_m = \langle m|\mu_x|l\rangle^2 (\mathscr{E}_x^\circ)^2 \left\{ \frac{4\sin^2 A}{(E_m - E_l - h\nu)^2} \right\} \tag{7.44}$$

This expression applies only when the radiation is monochromatic but one can obtain the final desired result for the total transition probability by integration over all frequencies.

$$c_m^* c_m = 4\langle m|\mu_x|l\rangle^2 (\mathscr{E}_x^\circ)^2 \int \frac{\sin^2 A}{(Ah/\pi t)^2} \, d\nu \tag{7.45}$$

In so doing $\mathscr{E}_x^\circ$ is considered to be a constant, although in reality it is seldom truly constant but is a function of the frequency. In actual experiments one generally uses the narrowest band of frequencies which is obtainable. Taking the variable of integration to be $\pi t\,(E_m - E_l - h\nu)/\hbar$, and integrating from $-\infty$ to $+\infty$ (this being permissible since the integral will have an appreciable value only when $(E_m - E_l - h\nu)$ is near zero) by using the definite integral

$$\int_{-\infty}^{\infty} \frac{\sin^2 x}{x^2} \, dx = \pi \tag{7.46}$$

one obtains the relation

$$c_m^* c_m = \frac{1}{\hbar^2} \langle m|\mu_x|l\rangle^2 (\mathscr{E}_x^\circ)^2 t \tag{7.47}$$

Hence, the probability of a system in state Ψ_l undergoing a transition to state Ψ_m is proportional to the dipole moment integral, the intensity of the incident radiation, and the time of irradiation.

Since it is more convenient to compare the energy density ρ to experimental results of spectroscopic experiments this can be substituted for the electric field strength $\mathscr{E}_x^\circ$ by use of the classical electrostatic relationship $\rho = (2\pi)^{-1}(\mathscr{E}_x^\circ)^2$.

Differentiation of Eq. (7.47) with respect to time then gives

$$\frac{d[c_m^* c_m]}{dt} = \frac{8\pi^3}{h^2}\langle m|\mu_x|l\rangle^2\rho(\nu) \tag{7.48}$$

as the expression for the transition probability per unit time from state Ψ_l to state Ψ_m under the influence of radiation which is polarized in the x direction. A repetition of this procedure, considering radiation polarized in the y or z direction, gives similar results. For nonpolarized isotropic radiation the transition probability per unit time, which is related to the Einstein transition probability coefficient of induced absorption B_{lm}, is given as

$$\frac{d[c_m^* c_m]}{dt} = \frac{8\pi^3}{3h^2}\langle m|\mu|l\rangle^2\rho(\nu) = B_{lm}\rho(\nu) \tag{7.49}$$

where

$$\langle m|\mu|l\rangle^2 = \sum_{i=x,y,z}\langle m|\mu_i|l\rangle^2 \tag{7.50}$$

This expression shows that the two principal factors upon which a transition depends are the square of the dipole moment integral and the energy density of the radiation.

If one begins with the conditions $c_l(0) = 0$ and $c_m(0) = 1$ then it is possible to derive an expression for the induced emission of radiation from state Ψ_m to Ψ_l. The expression for induced emission is identical to that for induced absorption, that is $B_{lm} = B_{ml}$. For a two-level system where the state has degeneracies g_l and g_m the relationship becomes $B_{lm}g_l = B_{ml}g_m$.

The treatment given so far has been only for induced radiation changes and has not considered the possibility of spontaneous changes from the upper state Ψ_m to the lower state Ψ_l. The change in population of the upper state with time due to both induced and spontaneous emission is given by

$$\left(\frac{dN_m}{dt}\right)_e = N_{m\to l} = -N_m B_{ml}\rho(\nu) - N_m A_{ml} \tag{7.51}$$

where A_{ml} is the Einstein coefficient for spontaneous emission. In the presence of radiation there will also be a change in the population of the upper state due to induced absorption by the lower state

$$\left(\frac{dN_m}{dt}\right)_a = N_{l\to m} = N_l B_{lm}\rho(\nu) \tag{7.52}$$

The net change of the upper-state population will be given by

$$\frac{dN_m}{dt} = N_{m\to l} + N_{l\to m} = (N_l - N_m)B_{lm}\rho(\nu) - N_m A_{ml} \tag{7.53}$$

The coefficients A_{ml} and B_{lm} can be related by use of the Boltzmann relationship for two nondegenerate levels

$$N_m = N_l e^{-h\nu_{lm}/kT} \tag{7.54}$$

and considering that the system is a collection of blackbody radiators where the radiative density is given by the Planck radiation law

$$\rho(\nu) = \frac{8\pi h\nu_{lm}^3}{c^3(e^{-h\nu_{lm}/kT} - 1)} \tag{7.55}$$

For a system in equilibrium $dN_m/dt = 0$ and

$$(N_l - N_m)B_{lm}\rho(\nu) - N_m A_{ml} = 0 \tag{7.56}$$

Incorporating Eqs. (7.54) and (7.55) into Eq. (7.56) gives

$$A_{ml} = \frac{8\pi h\nu_{lm}^3 B_{lm}}{c^3} = 6.2 \times 10^{-58}\,(\mathrm{Jm^{-3}\,s^4})\,\nu_{lm}^3 B_{lm} \tag{7.57}$$

It is obvious that for rotational and vibrational transitions having 10^9 Hz $< \nu <$ 10^{12} Hz the magnitude of A_{ml} relative to B_{lm} is insignificant and can be ignored. As the frequency increases the importance of spontaneous emissions increases and for the processes occurring in optical lasers they are of importance. Referring to Eqs. (7.49) now shows that

$$A_{ml} = \frac{64\pi^2\nu_{ml}^3}{3hc^3}\langle m|\mu|l\rangle^2 \tag{7.58}$$

7.7 ABSORPTION COEFFICIENTS

The absorption coefficient for a transition is expressed as

$$\alpha_{lm} = -\frac{1}{c\rho}\frac{d\rho}{dt} \tag{7.59}$$

where the term $d\rho/dt$ is proportional to the energy absorbed per transition per unit frequency range times the number of transitions per unit time. From the last section we have the transition probability per unit time as $B_{lm}\rho(\nu)$ for either induced absorption or emission. The total number of molecules undergoing absorption per unit volume per unit time $N_{l\to m}$ was given by Eq. (7.52) while the total number of molecules undergoing emission per unit volume per unit time was given by Eq. (7.51). Having observed that for radiation falling below 10^{12} Hz, A_{ml} can be neglected, B_{ml} will be the predominant term in Eq. (7.51) and the net rate of absorption of energy will be proportional to the difference

$$N_{l\to m} - N_{m\to l} = (N_l B_{lm} - N_m B_{ml})\rho(\nu) \tag{7.60}$$

Since the coefficients for spontaneous emission and spontaneous absorption are equal one has

$$N_{l\to m} - N_{m\to l} = (N_l - N_m)B_{lm}\rho(\nu) \tag{7.61}$$

Assuming thermal equilibrium, the net absorption of energy per unit volume per unit time per unit frequency is the product of the net rate of absorption times the energy per transition $\Delta E_{l\to m} = E_m - E_l$ divided by the bandwidth $\Delta\nu$ of the incident radiation or

$$\Delta\nu \frac{d\rho(\nu)}{dt} = -\Delta E_{l\to m}(N_l - N_m)B_{lm}\rho(\nu) \tag{7.62}$$

For a system at thermal equilibrium where the populations of the two states are related by the Boltzmann distribution, Eq. (7.54), it follows that

$$N_l - N_m = N_l(1 - e^{-\Delta E_{l\to m}/kT}) \tag{7.63}$$

and

$$\frac{1}{\rho(\nu)}\frac{d\rho(\nu)}{dt} = \frac{-N_l \Delta E_{l\to m} B_{lm}}{\Delta\nu}(1 - e^{-\Delta E_{l\to m}/kT}) \tag{7.64}$$

Combining Eqs. (7.59) and (7.64) and using $\Delta E_{l\to m} = h\nu_{lm}$ the absorption coefficient for a unit bandwidth becomes

$$\alpha_{lm} = \frac{h\nu_{lm}}{c\Delta\nu} N_l B_{lm}(1 - e^{-h\nu_{lm}/kT}) \tag{7.65}$$

It is possible to simplify this expression further for the case where ν_{lm} is less than 10^{12} Hz. At room temperature 300 K, and $\nu_{lm} = 10^{12}$ Hz, the value of $h\nu_{lm}/kT = 0.15$. Expanding $\exp(-h\nu_{lm}/kT)$ in a power series one has

$$e^{-h\nu_{lm}/kT} = 1 - \frac{h\nu_{lm}}{kT} + 2\left(\frac{h\nu_{lm}}{kT}\right)^2 - \frac{1}{6}\left(\frac{h\nu_{lm}}{kT}\right)^3 + \cdots \tag{7.66}$$

which for $h\nu_{lm}/kT \ll 1$ reduces to

$$e^{-h\nu_{lm}/kT} \approx 1 - \frac{h\nu_{lm}}{kT} \tag{7.67}$$

and the absorption coefficient becomes

$$\alpha_{lm} = \frac{h^2\nu_{lm}^2}{ckT\Delta\nu} N_l B_{lm} \tag{7.68}$$

This expression is satisfactory for discussion of spectra where transition frequencies are generally less than 10^{12} Hz.

This expression for the absorption coefficient was based on the assumption that the energy being absorbed was a single frequency. Actually the absorption observed may be a band of frequencies which is quite narrow for a single rotational transition or very broad for an unresolved infrared vibrational-rotational band.

Some comments regarding the effect of frequency distributions will be reviewed in the next section. Any discussion of the evaluation of state populations and dipole moment integrals is best deferred until additional details regarding the nature of the wavefunctions and energy states for specific systems are discussed.

7.8 TRANSITION LINE SHAPES

In the previous development of the absorption coefficient the spectroscopic transition was considered to be an infinitesimally narrow line of frequency ν_{lm} and the intensity of the radiation field was considered to be constant for all frequencies. Experimentally it is generally possible to obtain a reasonable approximation to the latter condition by appropriate control of the radiation source of proper manipulation of the radiation beam to give a narrow band of frequencies. However, the assumption of an infinitesimally narrow transition cannot be realized in practice and it would be misleading to leave this impression. The absorption coefficient is modified to allow for the finite spectral line width by inclusion of a multiplicative term. One commonly used term is the Van Vleck-Weisskopf line shape function $S(\nu_{lm}, \nu)$ [1]. Figure 7.3 illustrates the parameters associated with the discussions of line shapes and widths.

In the gas phase there are five factors which contribute to the width of a spectral line, but due to differences in their magnitudes not all five significantly affect the shape factor simultaneously. The two factors which determine the lower limit of the line width in the absence of any more dominant effect are the natural line width and the Doppler effect. The three factors which usually determine the shape function at lower frequencies are intermolecular collisions, wall collisions, and application of a saturation level of radiation. The effects of these five factors will be enumerated and the physical consequences considered, but no details regarding the origin or derivation of analytical relationships will be presented. In condensed phases intermolecular interactions primarily determine the line widths and will not be discussed further.

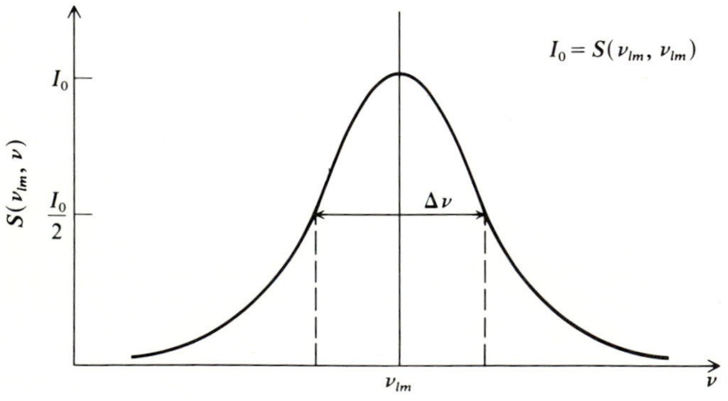

FIGURE 7.3
Schematic illustration $S(\nu_{lm}, \nu)$ and $\Delta\nu$.

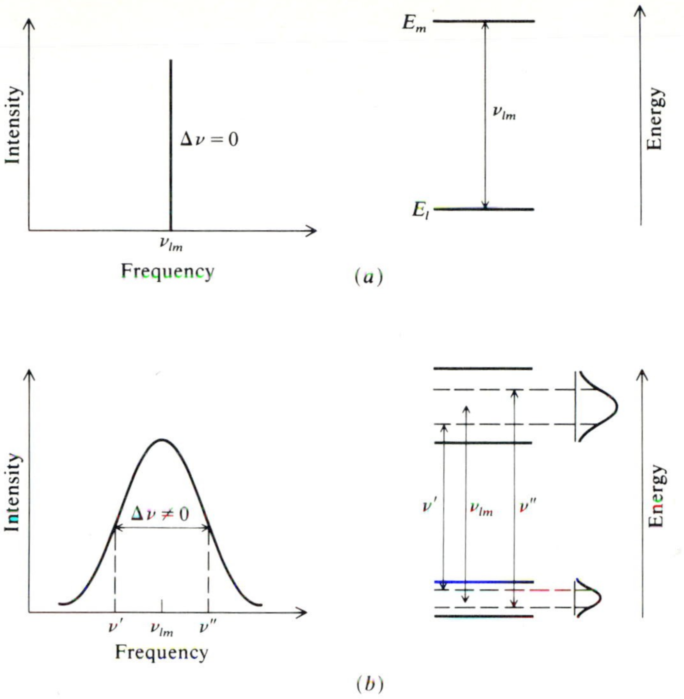

FIGURE 7.4
Schematic illustration of the origin of the natural line widths. (*a*) ΔE_l and ΔE_m assumed to be zero; (*b*) $\Delta E_l \neq 0$, $\Delta E_m \neq 0$.

The natural line width of a transition is a consequence of the *uncertainty principle.* In terms of energy and time the uncertainty principle is expressed as $\Delta E \Delta t \approx \hbar$. This indicates that within the time necessary for a transition to occur, Δt, there will be an uncertainty ΔE in the energy of the participating levels, and a resulting variation in the energy difference or transition frequency between the levels. This results in the total absorption over any finite period consisting of a band of frequencies centered at ν_{lm}. This phenomenon is illustrated in Fig. 7.4.

For electronic transitions in atoms this phenomenon can be viewed classically to be a damped harmonic oscillator and the shape function for the normalized natural line width of a spontaneous emission has been shown to be [2]

$$S_N(\nu_{lm}, \nu) = \frac{1}{2\pi^2}\left[\frac{\Delta\nu_N}{(\nu_{lm} - \nu)^2 + (\Delta\nu)^2}\right] \tag{7.69}$$

where

$$\Delta\nu_N = \frac{A_{lm}}{2\pi} = \frac{1}{2\pi\tau_m} = \frac{32\pi\nu_{lm}^3}{3hc^3}\langle m|\mu|l\rangle^2 \tag{7.70}$$

and τ_m is the lifetime of the upper state. These relationships have assumed that the transition ended at the ground state and the uncertainty resided solely in the upper state. In general one must consider that the lifetimes of both states contribute to the uncertainty, that is, $\Delta\nu = (1/2\pi)(1/\tau_m + 1/\tau_l)$, as shown in Fig. 7.4*b*. For real transitions the possibility of nonradiative relaxation from the upper state must be considered, in which case Eq. (7.70) is only approximate. In all of these cases a Lorentzian line shape is valid but the relationship of $\Delta\nu$ to other parameters will differ. At temperatures above absolute zero molecules are subjected to random electromagnetic fields due to blackbody radiation and the magnitude of the line width at zero Kelvin, Eq. (7.70), is multiplied by a factor of $2kT/h\nu_0$ ($kT \gg h\nu_0$). This is necessary because at ordinary temperatures thermal (blackbody) radiation is much more intense than the zero-point radiation, and each mode of oscillation has a mean energy equal to kT rather than $\frac{1}{2}h\nu$.

A couple of examples will serve to establish the magnitude of the natural line width. For the vibrational transition in HCl at 2989 cm^{-1} the magnitude of $(0|\mu|1)^2$ is approximately 10^{36} D^2 so $\Delta\nu_N \approx 140$ Hz (4.7×10^{-9} cm^{-1}). This is well below observed line widths of 1-2 cm^{-1}. For a rotational transition at 30 GHz with $(0|\mu|1)^2 \approx 10^{-36}$ D^2, $\Delta\nu \approx 10^{-7}$ Hz, which again is well below observed values of a few kilohertz. The two examples just cited were for absolute zero where the radiation field is due to zero-point vibrations only. For a temperature of 300 K the half-widths determined using Eq. (7.70) are multiplied by the temperature-dependent factors $2kT/h\nu = 0.14$ and 420 respectively for the two cases considered, giving $\Delta\nu$ values of 20 Hz for the vibrational transition and 420 Hz for the rotational one. We see that even considering the radiation fields present from blackbody radiation at 300 K the natural line widths are less than the generally observed values by several orders of magnitude.

The Doppler effect occurs when a molecule is moving so that a component of its velocity is parallel to the direction of propagation of the radiation field with which it is interacting. It is important only in gases where molecular velocities are large and times between collisions are long. The shift in frequency is given by

$$\varepsilon = \pm\nu\frac{\bar{v}}{v_p} \tag{7.71}$$

where ν is the transition frequency, $\bar{v}$ is the average molecular velocity, and v_p is the phase velocity of the radiation. Except for the special case of the sample cell being a metal waveguide operating at a radiation frequency near its cutoff, the phase velocity can be taken to equal the velocity of free space propagation of radiation, c. The probability of a molecule of a gas having an average velocity $\bar{v}$ is proportional to $\exp[-m\bar{v}^2/2kT]$, where m is the molecular mass [4]. The absorption line intensity as a function of ε, the change from the resonant frequency, is given by

$$I = \exp - \left[\frac{mc^2}{2kT}\left(\frac{\varepsilon}{\nu}\right)^2\right] \tag{7.72}$$

The half-width at half maximum intensity is found by using the conditions,

$$I = I_{\max} = 1 \qquad \text{at} \qquad \varepsilon = 0 \tag{7.73}$$

$$I = \frac{I_{\max}}{2} = \exp - \left[\frac{mc^2\varepsilon_{1/2}^2}{2kT\nu^2}\right] \quad \text{at} \quad \varepsilon = \varepsilon_{1/2} \tag{7.74}$$

solving both expressions for $I_{\max}$, equating, and solving for $\varepsilon_{1/2} = \Delta\nu_0$

$$\Delta\nu_0 = \frac{\nu}{c}\left(\frac{2kT\ln 2}{m}\right)^{1/2} = 3.6 \times 10^{-7}\left(\frac{T}{M}\right)^{1/2} \nu \quad \text{(Hz)} \tag{7.75}$$

where M is the molecular weight. Looking at the previous two examples shows that for the vibrational transition of HCl ($\nu = 10^{14}$ Hz) at 300 K, $\Delta\nu_D = 10^8$ Hz (0.003 cm^{-1}) and for the rotational transition ($\nu = 3 \times 10^9$ Hz) of a molecule with $M = 50$ amu, $\Delta\nu_D = 25$ kHz. Since $\Delta\nu_D$ depends directly on the transition frequency it will be the principal contributor to the line width for electronic transitions in the optical region of the spectrum. It is noted that the Doppler effect can be minimized by lowering the temperature or by velocity selection of the molecules being observed. If the Doppler effect is the limiting factor in determining the line shape then Eq. (7.72) shows that it will have a Gaussian rather than a Lorentzian profile.

For spectroscopic transitions occurring in the infrared and lower-frequency regions of the electromagnetic spectrum both the natural line width and the Doppler line width are small, and the observed line widths are primarily dependent on the other three factors. The latter two factors, wall collisions and radiation saturation, may be experimentally minimized, so the most important contribution to the line widths will be the effect of intermolecular collisions. The effect of intermolecular collisions on the absorption coefficient has been considered from a classical basis by Lorentz [5] giving

$$\alpha = \frac{Ne^2}{mc}\left(\frac{\nu}{\nu_0}\right) \frac{\left(\frac{1}{2\pi\tau}\right)}{(\nu - \nu_0)^2 + \left(\frac{1}{2\pi\tau}\right)^2} - \frac{\left(\frac{1}{2\pi\tau}\right)}{(\nu + \nu_0)^2 + \left(\frac{1}{2\pi\tau}\right)^2} \tag{7.76}$$

where τ is the mean time between collisions. By substitution of $8\pi^2\nu_{\sigma\sigma}(\sigma|\mu|\sigma')^2/3h$ for e^2/m, assuming a Boltzmann distribution of oscillators, and summing over all possible transitions [6], the absorption coefficient becomes

$$\alpha = \frac{8\pi^3 N}{3hc} \frac{\sum_{\sigma}\sum_{\sigma'} \nu_{\sigma\sigma'}\langle\sigma|\mu|\sigma'\rangle^2 S(\nu, \nu_{\sigma\sigma})e^{-E_{\sigma'}/kT}}{\sum_{\sigma'} e_{-E_{\sigma'}/kT}} \tag{7.77}$$

where

$$S(\nu_{\sigma\sigma'}, \nu) = \frac{1}{\pi}\left[\frac{\Delta\nu_c}{(\nu_{\sigma\sigma'} - \nu)^2 + (\Delta\nu_c)^2} - \frac{\Delta\nu_c}{(\nu_{\sigma\sigma'} + \nu)^2 + (\Delta\nu_c)^2}\right] \tag{7.78}$$

and $\Delta\nu_c = (2\pi\tau)^{-1}$. These same relationships have been formulated by Wigner and Weisskopf [7] using a formal quantum mechanical treatment.

Since $\Delta\nu$ is related to the mean time between collisions it can be estimated from a knowledge of the kinetic properties of the sample being investigated. The mean time between collisions for an ideal gas is given by $\tau = 1/\pi\rho^2\bar{v}N$, where ρ is the collision diameter, $\bar{v}$ the average velocity, and N the concentration. For a molecule with $\rho = 0.2$ nm, $\bar{v} = 5 \times 10^2$ m s^{-1}, and at a pressure of 0.1 Torr, $\Delta\nu \approx 1$ GHz. Thus for rotational frequencies of the order of 25 GHz the line width is an appreciable fraction of the line frequency, and if there are numerous closely spaced lines in a spectrum the resolution will be poor. The line width can be reduced by lowering the temperature to decrease the average molecular velocity and by lowering the pressures to reduce the concentration, hence increasing τ. Except at very low pressures the pressure broadening will exceed the broadening due to the Doppler effect. If they are of the same order of magnitude then the line width is given approximately by $\Delta\nu = [\Delta\nu_D^2 + \Delta\nu_C^2]^{1/2}$.

Broadening of lines due to collisions with the walls of the containing vessel is analogous to that produced by intermolecular collision broadening, but if the cell dimensions are sufficiently large the number of intermolecular collisions far exceeds that with the walls, and the latter may be neglected. A detailed treatment of wall broadening has been presented by Danos and Geschwind [8], but since the magnitude may be experimentally reduced in many experiments by proper sample cell design it will be considered no further.

The last phenomenon which affects the line shape is *power saturation*, which can occur when the applied radiation field is sufficiently large to induce absorption at a faster rate than that at which the upper energy level population can decay. As a result the normal excess population of the lower state is depleted. Experimentally this effect can be minimized by proper attenuation of the input power to the sample. Since it does provide further insight into the absorption phenomenon a short discussion of saturation is presented. In the vicinity of $\nu_{\sigma\sigma'}$ the factor $(\nu - \nu_{\sigma\sigma'})$ will be much smaller than $\nu + \nu_{\sigma\sigma'}$, and the second term in the shape function, Eq. (7.78), can be neglected and the absorption coefficient becomes

$$\alpha(\nu_{\sigma\sigma'}) = \frac{8\pi^2}{3ch}(N_\sigma - N_{\sigma'})\langle\sigma|\mu|\sigma'\rangle^2\left[\frac{\nu\Delta\nu}{(\nu - \nu_\tau)^2 + (\Delta\nu)^2}\right] \tag{7.79}$$

If no radiation is applied the molecules are maintained in equilibrium by collisions such that

$$N_{\sigma'} = N_\sigma e^{-h\nu_{\sigma\sigma'}/kT} \tag{7.80}$$

Letting $1/t_{\sigma\sigma'}$ and $1/t_{\sigma'\sigma}$ be the probabilities per unit time that a molecule is transformed from $\psi_\sigma \to \psi_{\sigma'}$ and from $\psi_{\sigma'} \to \psi_\sigma$ by collisions, then at equilibrium, with no applied radiation, the number of absorption transitions equals the number of emission transitions and

$$\frac{N_\sigma}{t_{\sigma\sigma'}} = \frac{N_{\sigma'}}{t_{\sigma'\sigma}} \tag{7.81}$$

When radiation of intensity I (quanta $s^{-1}\,m^{-2}$) is absorbed a steady state is established

$$\frac{N_\sigma}{t_{\sigma\sigma'}} + I\alpha(\nu_{\sigma\sigma'}) = \frac{N_{\sigma'}}{t_{\sigma\sigma'}} \tag{7.82}$$

Since saturation effects are most frequently encountered in the microwave and lower frequency regions where rotational and magnetic resonance transitions occur the factor $h\nu_{\sigma\sigma'}/kT$ will be much less than 1 and Eq. (7.80) becomes

$$N_{\sigma'} = N_\sigma\left[1 - \frac{h\nu_{\sigma\sigma'}}{kT}\right] \tag{7.83}$$

For equilibrium during irradiation

$$N_\sigma - N_{\sigma'} = N_\sigma \frac{h\nu_{\sigma\sigma'}}{kT} - I\alpha(\nu_{\sigma\sigma'})t \tag{7.84}$$

where $t_{\sigma\sigma'} = t$. Substitution of this result into Eq. (7.79) gives

$$\alpha(\nu_{\sigma\sigma'}) = \frac{8\pi^2}{3hc}\langle\sigma|\mu|\sigma'\rangle^2\left[\frac{\nu\Delta\nu}{(\nu - \nu_{\sigma\sigma'})^2 + (\Delta\nu)^2}\right]\left(N_\sigma\frac{h\nu_{\sigma\sigma'}}{kT} - I\alpha(\nu_{\sigma\sigma'})t\right) \tag{7.85}$$

Solving this expression for $\alpha(\nu_{\sigma\sigma'})$ gives

$$\alpha(\nu_{\sigma\sigma'}) = \frac{8\pi^2 N_\sigma}{3ckT}\langle\sigma|\mu|\sigma'\rangle^2\nu^2\left[\frac{\Delta\nu}{(\nu - \nu_{\sigma\sigma'})^2 + (\Delta\nu)^2 + (8\pi^2 t/3ch)(\sigma|\mu|\sigma')^2\nu I(\Delta\nu)}\right] \tag{7.86}$$

If I is small then the third term in the denominator approaches zero and the absorption coefficient is normal. When I becomes large the denominator increases and the absorption is diminished.

REFERENCES

A. Specific

1. Van Vleck, J. H., and V. F. Weisskopf, *Rev. Mod. Ph.*, **17**, 227 (1945).
2. Demtroder, W., *Laser Spectroscopy*, Springer-Verlag, New York, N.Y., 1981, ch. 3.
3. Townes, C. H., and A. L. Schawlow, *Microwave Spectroscopy*, McGraw-Hill Co. Inc., New York, N.Y., 1955, p. 336.
4. Golden, S., *Elements of the Theory of Gases*, Addison Wesley Publishing Co. Inc., Reading, MA, 1964, ch. 4.
5. Lorentz, H. A., *Proc. Amst. Akad. Sci.*, **8**, 591 (1906).
6. Heitler, W., *The Quantum Theory of Radiation*, Oxford University Press, London, England, 1954, pp. 40, 108.
7. Wigner, E., and V. Weisskopf, *Z. for Physik*, **63**, 54 (1930), ibid., **65**, 18 (1930), ibid., **75**, 287 (1932).
8. Danos, M., and S. Geschwind, *Phys. Rev.*, **91**, 1159 (1953).

B. General

Debye, P., *Polar Molecules*, Chemical Catalog Co., Inc., New York, N.Y., 1929, ch. 5.
Gordy, W., and R. L. Cooke, *Microwave Molecular Spectra*, 2d ed., Wiley-Interscience Publishers, New York, N.Y., 1985.
Townes, C. H., and A. L. Schawlow, *Microwave Spectroscopy*, McGraw-Hill Book Co., New York, N.Y., 1955.
Steinfeld, J. I., *An Introduction to Modern Spectroscopy*, 2d ed., MIT Press, Cambridge, MA, 1985, ch. 1.

PROBLEMS

7.1. For an absorption coefficient $\alpha(\nu) = 10^{-18}\ \text{cm}^{-1}$, calculate the percent of power absorbed from a $10\ \text{mW/cm}^2$ beam of radiation in passing through a 3 m sample path.

7.2. Consider the oversimplified, nonexistent case of molecular rotation with only two states $J = 1$ and $J = 2$. For a diatomic molecule with $B = 4000$ GHz and $\mu_0 =$ 2 debye, make a plot of $C_m^* C_m$ vs. t for the case where the irradiation field has $\varepsilon_x^0 = 100$ mV/cm. (Use $\mu = 1.0D$.)

7.3. Using the rigid rotor and harmonic oscillator wavefunctions calculate the Einstein coefficient for induced absorption for the $J = 1 \leftarrow 0$ and the $n = 1 \leftarrow 0$ transitions and account for the factors that cause them to differ. [Use $\mu = 1.0D$ and $(d\mu/dx)_0 = 0.01D\ \text{cm}^{-1}$.]

7.4. Using the results of Prob. 7.3, a rotational transition of $\nu_{1\leftarrow 0} = 15{,}000$ GHz, and a vibrational transition of $\bar{\nu}_{1\leftarrow 0} = 2000\ \text{cm}^{-1}$, calculate the Einstein coefficients for spontaneous emission for the two transitions.

7.5. Derive an expression for $d[C_l^* C_l]/dt$ and show that $B_{l\to m} = B_{m\to l}$.

7.6. Evaluate the dipole moment integral connecting the $n = 1$ and $n = 2$ states of an electron in a one-dimensional box of 10 A lengths.

7.7. Calculate the magnitude of the Einstein coefficient for the $n = 1 \to 2$ transition for the electron in a one-dimensional box of 10 A length.

7.8. Calculate the ratio, at 25°C, of the populations of two energy states separated by (1) 60 MHz, (2) 2 GHz, (3) $200\ \text{cm}^{-1}$, and (4) $4000\ \text{cm}^{-1}$.

7.9. For a system of one mole of particle calculate the ratio of $\alpha(\nu_{lm})$, using the complete expression for $\alpha(\nu_{lm})$, and with $h\nu/kT \ll 1$, for the case where $E_n - E_l$ is given by (1) 60 MHz, (2) 2 GHz, (3) $200\ \text{cm}^{-1}$, (4) $4000\ \text{cm}^{-1}$.

7.10. Calculate $\Delta\nu\alpha(\nu_{12})$ for one mole of electrons in a one-dimensional box of 10 A length at 25°C. Consider the distribution of electrons to be governed by the Boltzmann distribution so that

$$N_1 = N_0 \frac{e^{-E_1/kT}}{(2\pi mkTa^2/h^2)^{1/2}}$$

where N_0 = Avogadro number.

7.11. Plot the shape functions $S(\nu, \nu_{\sigma\sigma'})$ for the case of
(1) $\nu_{\sigma\sigma'} = 15$ GHz $\quad \Delta\nu = 100$ kHz
(2) $\nu_{\sigma\sigma'} = 15$ GHz $\quad \Delta\nu = 500$ kHz
(3) $\nu_{\sigma\sigma'} = 15$ GHz $\quad \Delta\nu = 2$ MHz
How will line width and absorption maxima be related?

7.12. Calculate the absorption coefficients $\Delta\nu\alpha(\nu_{lm})$ [Eq. (7.68)] for the $J = 1 \leftarrow 0$, $5 \leftarrow 4$, $10 \leftarrow 9$, and $15 \leftarrow 14$ transitions for the following molecules at 100 K, 300 K, 600 K (assume to be gases at all temperatures). Use Eq. (11.73) to evaluate the dipole matrix element and take $N_j = (2J + 1)(hB/kT) \exp[-hBJ(J + 1)/kT]$.

a. $H^{35}Cl$	$r_e = 1.27455$ A	$\mu_0 = 1.07D$
b. $^{205}Tl^{127}I$	$r_e = 2.8135$ A	$\mu_0 = 2.6D$
		$\mu_0 = 4.60D$

CHAPTER 8

MAGNETIC RESONANCE

8.1 INTRODUCTION

The domain of magnetic resonance has become one of the widest and the most diversely employed of any of the modern techniques used for the investigation of structure and bonding. Although it has limited use for the direct determination of interatomic distances and bond angles its use as a tool for qualitative structure determinations and as a method for measuring electron densities in molecules is very extensive. The phenomenon of magnetic resonance is due to the interaction of an external magnetic field with the magnetic moment associated with the intrinsic spin of a nucleus of an electron. Although the fundamental behavior of these two types of particles is very much alike, the experimental apparatus used to study their interactions with magnetic fields and the types of information which can be gained from the studies differ. These differences have lead to a separation of the subject into two broad categories: (1) nuclear magnetic resonance (NMR), and (2) electron paramagnetic resonance (EPR) or electron spin resonance (ESR). In practice the term EPR is frequently used when the discussion centers about the behavior of ions containing one or more unpaired electrons (paramagnetic ions) while the term ESR more frequently denotes the study of uncharged species containing one or more unpaired electrons (free radicals).

In Chap. 5 a review of the general quantum mechanical problem of the coupling of two or more spins was presented. In this section we will discuss the interaction of a single nuclear spin with an external magnetic field from both the quantum mechanical and classical viewpoints. This will be followed by discussions of the analysis of high-resolution NMR spectra and the use of NMR to investigate the structure of solids. Although these two areas of NMR are of paramount importance they are only two of many applications. They have been chosen to illustrate the nature and utility of NMR. The subject of electron resonance will be presented in Chap. 9.

8.2 INTERACTION OF A SINGLE SPIN WITH AN EXTERNAL MAGNETIC FIELD

In the absence of an external field the energy of a spin system, denoted by either S or I, is degenerate in the quantum numbers M_S or M_I. In this case there is only a single energy level and there is no possibility for observation of a spectroscopic transition. The situations of interest arise when a spin system interacts with an external magnetic field to remove this degeneracy and to establish energy levels between which one can observe the emission or absorption of radiation. In this section we will consider this type interaction for a single-spin system.

For a nuclear magnetic moment $\boldsymbol{\mu}$, the interaction energy with a magnetic field $\mathbf{H}$ is given by

$$E = -\boldsymbol{\mu} \cdot \mathbf{H} \tag{8.1}$$

Since the nuclear magnetic moment is proportional to the nuclear spin angular momentum $\boldsymbol{\mu} = g_N \beta_N \mathbf{I}/\hbar$, the quantum mechanical energies of such a system are the eigenvalues of the diagonal energy matrix whose elements are

$$E_{I,M_I} = \frac{-1}{\hbar} \langle I, M_I \,|\, g_N \beta_N \mathbf{I} \cdot \mathbf{H} \,|\, I, M_I \rangle \tag{8.2}$$

The eigenvalues of $\mathbf{I}$ and $\mathbf{S}$ contain the factor $\hbar$, so we will consider that this factor is implicitly contained in the operators and from this point on write our energy expressions with an appropriate prefactor. Since $\mathbf{I}$ and $\mathbf{H}$ are vector quantities, they may be written

$$\mathbf{I} = I_X \hat{\mathbf{i}} + I_Y \hat{\mathbf{j}} + I_Z \hat{\mathbf{k}} \tag{8.3}$$

$$\mathbf{H} = H_X \hat{\mathbf{i}} + H_Y \hat{\mathbf{j}} + H_Z \hat{\mathbf{k}} \tag{8.4}$$

If we consider a unidirectional external magnetic field and arbitrarily let its direction be along the Z axis then $H_X = H_Y = 0$, $\mathbf{I} \cdot \mathbf{H} = (I_X \hat{\mathbf{i}} + I_Y \hat{\mathbf{j}} + I_Z \hat{\mathbf{k}}) \cdot H_Z \hat{\mathbf{k}} = I_Z H_Z$ and

$$E_{I,M_I} = \frac{-1}{\hbar} \langle I, M_I \,|\, g_N \beta_N H_Z I_Z \,|\, I, M_I \rangle \tag{8.5}$$

Since the only nonzero matrix elements of I_z are the diagonal ones having values of $M_I \hbar$ the energy of a single spin in an external field is given by

$$E_{M_I} = -g_N \beta_N H_Z M_I \tag{8.6}$$

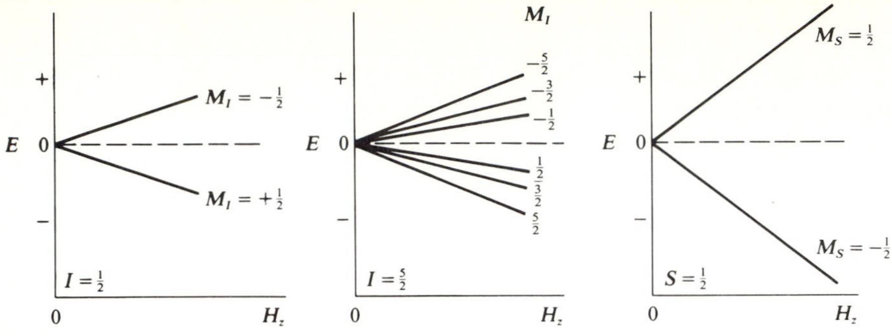

FIGURE 8.1
Spin energy levels in an external magnetic field.

The nature of the energy levels for $I = \frac{1}{2}$ and $I = \frac{5}{2}$ as well as the analogous case of the electron where

$$E_{M_I} = g\beta H_Z M_S \tag{8.7}$$

is illustrated in Fig. 8.1. Note the inversion of the energy levels relative to the sign of M between $I = \frac{1}{2}$ and $S = \frac{1}{2}$. This arises from the opposite charges on protons and electrons. It is also to be noted that the energy scale for the $I = \frac{1}{2}$, $\frac{5}{2}$ systems and that of the $S = \frac{1}{2}$ system are different since at any particular field the separation of the levels for the electron, $S = \frac{1}{2}$, will be greater than those for a proton, $I = \frac{1}{2}$, by a factor of approximately 2000. When discussing NMR, the concept of angular frequency, $\omega = 2\pi\nu$, is frequently used. The separation between two energy levels, for the spin $I = \frac{1}{2}$ case, is given by

$$\Delta E = h\nu = E_{-1/2} - E_{1/2} = g_N\beta_N H_Z \tag{8.8a}$$

or

$$\omega = 2\pi\nu = \frac{g_N\beta_N}{\hbar} H_Z = \gamma H_Z \tag{8.8b}$$

where $\gamma = g_N\beta_N/\hbar$. The factor γ, called the magnetogyric ratio, is generally used to replace the more cumbersome combination of factors to which it is equal. If we assume that a transition between the states $E_{-1/2}$ and $E_{1/2}$ is allowed, a point which will be demonstrated later, we can look at the magnitude of the effect for several representative systems. A comparison of resonance frequencies and magnetic fields for several common $I = \frac{1}{2}$ nuclei are shown in Table 8.1. By contrast, the analogous parameters for the electron are shown in Table 8.2.

Having looked at the quantum mechanical nature of magnetic resonance and the magnitudes of possible resonance effects we will now examine in more detail the general selection rules and the detailed nature of magnetic resonance spectra. It will be helpful at this point to introduce a classical description of the phenomenon.

TABLE 8.1
Magnetic resonance parameter for some common spin $I = \frac{1}{2}$ nuclei

Nucleus	μ, nuclear magnetons	$\gamma \times 10^{-7}$ rad S^{-1} T^{-1}	ν, MHz			H, T		
			H = 1.5 T	3.0 T	8.0 T	ν = 90 MHz	300 MHz	500 MHz
^{1}H	2.79268	26.751	63.9	127.7	340.6	2.1	7.0	11.7
^{13}C	0.0700220	6.726	16.1	32.2	85.9	8.4	28.0	46.7
^{19}F	2.6273	25.167	60.1	120.2	320.4	2.2	7.5	12.5
^{31}P	1.1305	10.830	25.8	51.7	137.9	5.3	17.4	29.0
^{119}Sn	1.0409	8.369	20.0	39.9	106.4	6.8	22.5	37.5

TABLE 8.2
Magnetic resonance parameters for the electron

ν (MHz)	20	100	1000	10,000	25,000	50,000
H (T)	0.0007	0.004	0.04	0.36	0.9	1.8

8.3 STATE POPULATIONS AND SELECTION RULES

The problem of determining the possibility and intensity of transitions involves two factors. The first is the nature of the formal quantum mechanical selection rules while the second relates to the population of the energy states involved in the transition.

The property of the nucleus which enables it to interact with an external electromagnetic field is the nuclear magnetic moment, so any interaction will be with the magnetic vector of the applied field. The Z component of an applied electromagnetic field will only periodically change the energy levels of the system and will not serve to induce transitions between levels. A component perpendicular to Z will, however, be able to interact with the moment. This point is illustrated by use of a classical description of the NMR phenomenon as applied to protons. Considering the nucleus to be a classical magnetic dipole $\boldsymbol{\mu}$, oriented at an angle θ to an applied magnetic field $\mathbf{H}$, we have the condition necessary for a precessional motion to occur. Since the field will apply a torque $\mathbf{T}$ to the dipole it will precess about the field at some angular frequency which is designated by $\boldsymbol{\omega}$. Also associated with the nucleus is an angular momentum $\mathbf{I}$. These various vector quantities are illustrated in Fig. 8.2. The torque is defined by classical mechanics as

$$\mathbf{T} = \boldsymbol{\omega} \times \mathbf{I} \tag{8.9}$$

The physical origin of the torque is the interaction of $\boldsymbol{\mu}$ and $\mathbf{H}$ so

$$\mathbf{T}' = \mathbf{H} \times \boldsymbol{\mu} \tag{8.10}$$

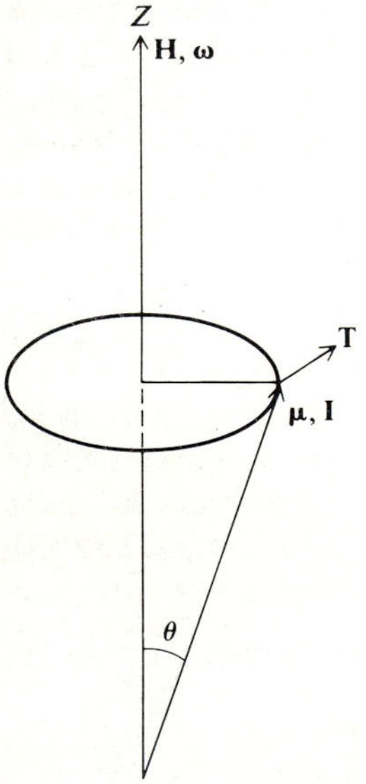

FIGURE 8.2
Classical picture of magnetic interactions.

Since the torque is the same in both cases, $\mathbf{T} = \mathbf{T}'$, it follows that

$$\boldsymbol{\omega} \times \mathbf{I} = \mathbf{H} \times \boldsymbol{\mu} \tag{8.11}$$

If the exceptionally simple, but questionable, physical concept that the magnetic moment is due to the rotation of a nucleus of charge e and radius r is invoked, then we can write

$$\boldsymbol{\mu} = \mathbf{i}A \tag{8.12}$$

where the equivalent current is $\mathbf{i} = e\mathbf{v}/2\pi cr$, $A = \pi r^2$ and $\mathbf{v}$ is the tangential velocity. Therefore

$$\boldsymbol{\mu} = \frac{e\mathbf{v}r}{2c} \tag{8.13}$$

Since the angular momentum can be expressed in terms of the linear momentum

$$\mathbf{I} = mr\mathbf{v} \tag{8.14}$$

it follows that

$$\boldsymbol{\mu} = \frac{e}{2mc}\mathbf{I} = \gamma_H \mathbf{I} \tag{8.15}$$

Substituting this expression back into Eq. (8.11) we get

$$\boldsymbol{\omega} \times \mathbf{I} = \gamma_H \mathbf{H} \times \mathbf{I} \tag{8.16}$$

or

$$\boldsymbol{\omega} = \gamma_H \mathbf{H} \tag{8.17}$$

and the precession frequency is proportional to the magnetic field experienced by the nucleus. This expression is equivalent to Eq. (8.8), thereby showing the relationship between the classical precession frequency and the quantum mechanical transition frequency.

If we further examine this classical picture we can verify that the physical interaction between the nuclear magnetic moment and the applied radiation is with a magnetic field component perpendicular to H_Z. Given a small oscillating electromagnetic field along the X axis

$$\mathbf{H}_1 = \mathbf{H}_X \cos \omega' t \tag{8.18}$$

it is equivalent to the sum of the two components

$$\mathbf{H}_1' = \tfrac{1}{2}\mathbf{H}_X \cos \omega' t + \tfrac{1}{2}\mathbf{H}_X \sin \omega' t \tag{8.19}$$

$$\mathbf{H}_1'' = \tfrac{1}{2}\mathbf{H}_X \cos \omega' t - \tfrac{1}{2}\mathbf{H}_X \sin \omega' t \tag{8.20}$$

These components constitute circularly polarized radiation and can be represented by a pair of rotating vectors of magnitude H_1. This is illustrated with relation to H_Z and $\boldsymbol{\mu}$ in Fig. 8.3.

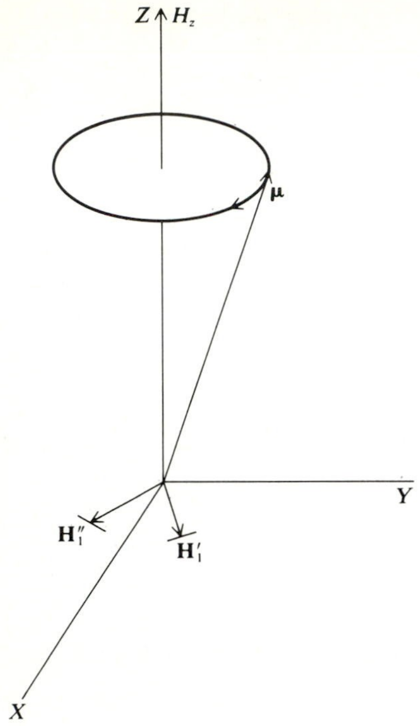

FIGURE 8.3
Field orientations for NMR experiment.

If the frequency ω' of the applied electromagnetic field H_1 is equal to the precession frequency ω of $\boldsymbol{\mu}$, then $\mathbf{H}''$ and $\boldsymbol{\mu}$ will process together and there will be a perturbation, $\boldsymbol{\mu} \cdot \mathbf{H}_1''$ which will tend to change the orientation of μ or in quantum mechanical terms will induce a transition. The other component, $\mathbf{H}_1'$, will be at the precession frequency also, but it will be rotating in the opposite direction relative to the precession and will not spend sufficient time oriented in a plane with $\boldsymbol{\mu}$ in order to interact so as to produce a torque. On the other hand, if $\mathbf{H}_1$ were oriented along the z axis it would not be able to correctly interact with $\boldsymbol{\mu}$ in order to produce a transition.

The interacting field is then a periodic electromagnetic field of the form $\mathbf{H}_1 = \mathbf{H}_X \cos \omega t$. Using this as the perturbing term one can apply the concepts of Chap. 7 in order to evaluate the transition probability. The perturbing potential [see Eq. (7.24)] will be of the form

$$v(t) = -\boldsymbol{\mu} \cdot \mathbf{H}_1 = -\mu_X H_X \cos wt \tag{8.21}$$

since $\mathbf{H}_1$ has only an X component. Following the procedure outlined in Sec. 7.3 we evaluate the transition probability [see Eq. (7.48)] to be

$$c_m^* c_m = P_{nm} = \frac{\langle n | \boldsymbol{\mu}_X | m \rangle^2}{\hbar^2} H_X^2 \tag{8.22}$$

Since $\mu = g_N \beta_N I / \hbar$, it follows that the evaluation of P_{nm} reduces to an evaluation of $\langle n | \mathbf{I}_X | m \rangle$. Looking at Table 3.1 we see that the only nonzero values for

$\langle I, M_I | \mathbf{I}_X | I', M'_I \rangle$ occur for $I' = I$, $M'_I = M_I \pm 1$, hence we have evaluated the selection rule

$$\Delta M_I = \pm 1 \tag{8.23}$$

An analogous procedure yields, for an ESR transition,

$$\Delta M_S = \pm 1 \tag{8.24}$$

Since the transition probability is the same for both the induced emission and induced absorption there must be a net difference between the populations of the states involved for there to be an observable absorption of radiation. Restricting our consideration to spin $\frac{1}{2}$ systems and assuming a Boltzmann distribution we find that, for the proton,

$$\frac{N_{-1/2}}{N_{+1/2}} = e^{-\gamma_N \hbar H/kT} \tag{8.25}$$

and, for the electron,

$$\frac{N_{+1/2}}{N_{-1/2}} = e^{-g\beta H/kT} \tag{8.26}$$

Inserting the appropriate values for the physical constants, letting $H = 1\,T$ and $T = 300$ K we find that, for the proton,

$$N_{-1/2}/N_{1/2} = 0.9999932 \tag{8.27}$$

while for the electron

$$N_{+1/2}/N_{-1/2} = 0.95 \tag{8.28}$$

For electrons the upper state is appreciably less populated than the lower, but for protons the population is nearly the same. This means that unless there is some mechanism whereby the protons, once excited to the upper state, can lose energy and return to the lower one quickly there will be rapid equalization of state populations and the system will exhibit no further net absorption. In this condition it is said to be saturated. The processes whereby the spin systems lose their excess energy are referred to as *relaxation processes*. If there is a suitable mechanism whereby the excited state spin system can relax then one can observe an absorption of radiation in compliance with the previously developed selection rules.

8.4 RELAXATION—AN INTRODUCTION

The subject of magnetic relaxation is quite extensive, and no attempt will be made to explore all of its aspects in this discussion. We will present sufficient background to provide an understanding of the methods of relaxation and the relationship of relaxation to spectra. The two primary mechanisms for relaxation are provided by interactions of the resonant nuclei with the lattice and with other nuclei. The former mechanism involves the loss of energy from the spin system by conversion into translation, rotational, or vibrational energy of the surrounding atoms and molecules, the lattice, and is called *spin-lattice relaxation.* The latter mechanism, referred to as *spin-spin relaxation* involves a transfer of energy between the nuclear spins. For a solid where the nuclei are fixed at a distance **r** from one nucleus to the other the Hamiltonian for the interaction of two nuclei is

$$\mathscr{H}_D = \frac{\boldsymbol{\mu}_1 \cdot \boldsymbol{\mu}_2}{r^3} - \frac{3(\boldsymbol{\mu}_1 \cdot \mathbf{r})(\boldsymbol{\mu}_2 \cdot \mathbf{r})}{r^5} \tag{8.29}$$

For fluid states $\mathscr{H}_D$ becomes a function of time due to the motion of the molecules, and the interaction becomes more complex [1].

Before looking in detail at these two relaxation mechanisms it is informative to define a new quantity called *spin temperature*, T_s. This parameter can be defined under certain conditions for a general spin system, but this discussion will be limited $I = \frac{1}{2}$ nuclei. Equation (8.25) defined the ratio of the state populations for a system as a function of magnetic field and temperature. If the magnetic field is constant then for any arbitrary distribution of spins between the states $M_I = \frac{1}{2}$ and $M_I = -\frac{1}{2}$ the ratio of the populations can be expressed as a Boltzmann distribution involving a temperature T_s, called the *spin temperature*,

$$\frac{N_{1/2}}{N_{-1/2}} = e^{\gamma_N hH/kTs} \tag{8.30}$$

If, for a given lattice temperature T, there is some mechanism which causes $(N_{1/2} - N_{-1/2})$ to be less than the equilibrium value of $(N_{1/2} - N_{-1/2})_e$ at the temperature T, then the system is said to have a spin temperature in excess of the lattice temperature. Looking at Eq. (8.30), we see that if there is some mechanism which causes a population inversion $N_{-1/2} > N_{1/2}$, then the exponential term must be negative, and the system is said to have a negative spin temperature.

If one observes the environment of a nucleus embedded in either a solid or liquid, it is found that it is surrounded by neighboring ions and molecules that are constantly undergoing rotation, vibration, and translation due to thermal excitation. This motion is entirely random and, since the ions and molecules are composed of electric charges and dipoles, gives rise to randomly fluctuating magnetic fields. Any arbitrary electromagnetic field is composed of an infinite number of individual components and can be formally expressed as a Fourier series of the form

$$H(t) = \sum_{n=1}^{\infty} [a_n \cos(n\omega t) + b_n \sin(n\omega t)] \tag{8.31}$$

Any such field will contain components at, or sufficiently near, the frequency of a magnetic resonance to interact with the nuclei. It is the depopulation of the higher ($M_I = -\frac{1}{2}$) level by interaction with these random fields that results in spin-lattice relaxation. The nature of the random fields is measured in terms of a correlation time τ_c, which is a measure of the time necessary for local fluctuations, such as molecular rotation or vibration, to occur. A more definitive treatment of the quantitative aspects of τ_c is given by Carrington and McLachlan [2].

Spin-lattice relaxation is generally characterized by a spin-lattice relaxation time T_1. A relationship between the transition probability and T_1 can be developed by considering a spin $I = \frac{1}{2}$ system which has been displaced from thermal equilibrium to the point of saturation by the application of an electromagnetic field of the correct frequency. At this point $N_{-1/2}$ (upper state) = $N_{1/2}$ (lower state). It has been shown that for spin systems the transition probability for induced emission and induced absorption are equal, hence there appears to be no way to depopulate the upper state relative to the lower one. If, however, the spins can interact with the lattice in the manner just described, then we can define two new probabilities which apply to the coupled spin-lattice system. Denoting the total probability per unit time of a spin undergoing a transition from the low to the high state by P_+ and the inverse by P_-, the number of transitions per unit time in either direction will be given by $P_\pm N_{\pm 1/2}$. At the point where thermal equilibrium is attained, that is, the condition whereby the spin-lattice interaction is depopulating the upper state at the rate it is being populated by induced adsorption in the lower state and vice versa, the total number of transitions in each direction are equal and

$$(N_{+1/2}P_+)_{\text{eq}} = (M_{-1/2}P_-)_{\text{eq}} \tag{8.32}$$

Rearrangement followed by introduction of a Boltzmann distribution factor gives

$$\frac{P_+}{P_-} = \left(\frac{N_{-1/2}}{N_{+1/2}}\right)_{\text{eq}} = e^{-\gamma_N Hh/kT} \tag{8.33}$$

If it is assumed that energy of the lattice motion is only oscillatory and given by $E_n = nh\nu_0$ where $\nu_0 = \omega/2\pi$(ω = NMR frequency), then the population of a given energy state is given by

$$N_n = N_0 e^{-E_n/kT} \tag{8.34}$$

At the point of thermal equilibrium it is required that for every upward transition in the spin system a downward transition must occur in the lattice. If both transitions are equally probable then P_+ is proportional to the number of lattice energy states capable of making a downward transition. Since no downward transition is possible from the lowest level

$$P_+ = A \sum_{n=1}^{\infty} N_n = AN_0 \sum_{n=1}^{\infty} e^{-nh\nu_0/kT} \tag{8.35}$$

Since upward transition can occur from all levels

$$P_- = A \sum_{n=0}^{\infty} N_n = AN_0 \sum_{n=0}^{\infty} e^{-nh\nu_0/kT} \tag{8.36}$$

Combining these gives

$$\frac{P_+}{P_-} = e^{-nh\nu_0/kT} \tag{8.37}$$

This is analogous to Eq. (8.33) and demonstrates that the probability of upward and downward transitions due to lattice interactions are not equal as they are for the spin system. Since $h\nu_0/kT$ is small Eq. (8.37) becomes

$$\frac{P_+}{P_-} = 1 - \frac{h\nu_0}{kT} \tag{8.38}$$

It is seen that $P_+ < P_-$ and the net effect of the spin-lattice interaction will be to depopulate the higher level.

The rate at which the upper level depopulates can be expressed in a quantitative manner in terms of the spin-lattice relaxation time T_1 by considering a system which has been displaced from equilibrium and determining the time necessary for thermal equilibrium to be restored via lattice relaxation. Letting $N = N_{+1/2} + N_{-1/2}$ the equilibrium populations are given by

$$N_{\pm 1/2} = \frac{N}{2} e^{\pm \gamma_N hH/2kT} \approx \frac{N}{2}\left(1 \pm \frac{\gamma_N hH}{2kT}\right) \tag{8.39}$$

with the second term being applicable for temperatures above 10 K. We now define a mean transition probability P by

$$P_\pm = Pe^{\mp \gamma_N hH/2kT} \approx P\left(1 \mp \frac{\gamma_N hH}{2kT}\right) \tag{8.40}$$

When the system is displaced from equilibrium $N_{+1/2}P_+$ will no longer equal $N_{-1/2}P_-$, but the rate of approach to equilibrium will be given by

$$\frac{dN_{+1/2}}{dt} = -\frac{dN_{-1/2}}{dt} = N_{-1/2}P_- - N_{+1/2}P_+ \tag{8.41}$$

It is really the population difference $N' = N_{+1/2} - N_{-1/2}$ that is of interest. Since N' changes by 2 for each transition which occurs

$$\frac{dN'}{dt} = 2(N_{-1/2}P_- - N_{+1/2}P_+) \tag{8.42}$$

or substituting Eqs. (8.33) and (8.40)

$$\frac{dN'}{dt} = 2P(N'_{\text{eq}} - N') \tag{8.43}$$

and

$$N'_{\text{eq}} = (N_{+1/2} - N_{-1/2})\frac{\gamma_N hH}{2kT} \tag{8.44}$$

Equation (8.43) is the form of a first-order kinetic rate equation and upon integration gives

$$(N'_{\text{eq}} - N') = (N'_{\text{eq}} - N'_0)e^{-2Pt} \tag{8.45}$$

This is a typical exponential decay law with a decay constant of $2P$ or a relaxation time $T_1 = 0.5\ P^{-1}$. In this discussion we have envisioned the relaxing forces as being a general random field. The advanced monographs listed in the references explore the particular origins and effects of these fields in more detail.

A second type of relaxation mechanism exists in molecules that contain coupled spins. When one has near-neighbor magnetic dipoles in a substance they can produce small local magnetic fields at an adjacent resonant nucleus. The accumulated effect of these small local fields due to many dipoles at random but slowly fluctuating positions is to cause the effective magnetic field, and hence the precession frequencies (transition frequencies), to have a spread of values. This will produce an absorption line of appreciable width, $\Delta\nu$. In view of the uncertainty principle which relates uncertainty in energy and time we can express this line width in terms of a time factor $T_2 = \frac{1}{2}\pi\Delta\nu$ which is called the *spin-spin relaxation time.*

8.5 CLASSICAL TREATMENT OF RESONANCE AND RELAXATION

The understanding of a number of modern experimental methods in NMR requires a more detailed view of the mechanisms of relaxation. To enhance this understanding it is necessary to introduce the use of rotating coordinate systems and to examine the classical relationships between the bulk magnetization of a sample and the relaxation times T_1 and T_2.

In Sec. 8.3 an initial discussion of the interaction of an oscillating magnetic field and a magnetic moment associated with the intrinsic spin of a nucleus lead to the model of a synchronously precessing magnetic moment and an electromagnetic field component. The classical equation of motion for this gyrating nucleus with angular momentum $\mathbf{I}$ will be

$$\frac{d\mathbf{I}}{dt} = \boldsymbol{\mu} \times \mathbf{H} \tag{8.46}$$

Substituting $\boldsymbol{\mu} = \gamma\mathbf{I}$, multiplying both sides by N_0, and recalling that the bulk magnetization $\mathbf{M}$ is given by $\mathbf{M} = N_0\boldsymbol{\mu}$, leads to

$$\frac{d\mathbf{M}}{dt} = \gamma\mathbf{M} \times \mathbf{H} \tag{8.47}$$

Expansion of this vector equation into its components shows that the changes in magnetization along the axes of a Cartesian system can be expressed as

$$\frac{dM_X}{dt} = \gamma(M_Y H_Z - H_Y M_Z) \tag{8.48}$$

$$\frac{dM_Y}{dt} = \gamma(M_Z H_X - H_Z M_X) \tag{8.49}$$

$$\frac{dM_Z}{dt} = \gamma(M_X H_Y - H_X M_Y) \tag{8.50}$$

For the case of free precession with a large applied magnetic field along the Z direction ($H_Z = H_0$) and no applied electromagnetic radiation ($H_X = H_Y = 0$) these equations will reduce to

$$\frac{dM_X}{dt} = \gamma H_0 M_Y = \omega_0 M_Y \tag{8.51}$$

$$\frac{dM_Y}{dt} = -\gamma H_0 M_X = -\omega_0 M_X \tag{8.52}$$

$$\frac{dM_Z}{dt} = 0 \tag{8.53}$$

The transverse components of magnetization M_X and M_Y and the longitudinal component M_Z will be changed by any mechanisms which tends to reorient the spins. Following any perturbation of a spin system, such as would be caused by the application of electromagnetic radiation having $H_X \neq 0$ and $H_Y \neq 0$, it will tend to return to an equilibrium value at a rate which is proportional to the displacement. This return to equilibrium, or relaxation, adds another contribution to the rate equations [Eqs. (8.48) to (8.50)] giving in general

$$\frac{dM_X}{dt} = \gamma(M_Y H_Z - M_Z H_Y) - \frac{M_X}{T_2} \tag{8.54}$$

$$\frac{dM_Y}{dt} = \gamma(M_Z H_X - M_X H_Z) - \frac{M_Y}{T_2} \tag{8.55}$$

$$\frac{dM_Z}{dt} = \gamma(M_X H_Y - M_Y H_X) - \frac{M_Z - M_0}{T_1} \tag{8.56}$$

or as a vector equation

$$\frac{d\mathbf{M}}{dt} = \gamma(\mathbf{M} \times \mathbf{H}_0 + \mathbf{M} \times \mathbf{H}_1) - \frac{(M_X\hat{\mathbf{i}} + M_Y\hat{\mathbf{j}})}{T_2} - \frac{(M_Z - M_0)\hat{\mathbf{k}}}{T_1} \tag{8.57}$$

For the case of $\mathbf{H}_0 = H_0\hat{\mathbf{k}}$ and no applied electromagnetic radiation Eqs. (8.54) to (8.56) reduce to the equivalent of Eqs. (8.51) to (8.53), where the transverse components of $\mathbf{M}$ decay to an equilibrium value of zero and M_Z goes to an equilibrium value of $M_Z = M_0$;

$$\frac{dM_X}{dt} = \omega_0 M_Y - \frac{M_X}{T_2} \tag{8.58}$$

$$\frac{dM_Y}{dt} = -\omega_0 M_X - \frac{M_Y}{T_2} \tag{8.59}$$

$$\frac{dM_Z}{dt} = \frac{M_0 - M_Z}{T_1} \tag{8.60}$$

and T_1 and T_2 are referred to as the longitudinal and transverse relaxation times respectively.

If we now include the contributions from the magnetic components of the applied electromagnetic radiation field

$$\mathbf{H}_1 = H_X\hat{\mathbf{i}} + H_Y\hat{\mathbf{j}} = (H_1 \cos \omega t)\mathbf{i} - (H_1 \sin \omega t)\mathbf{j} \tag{8.61}$$

retain the static field in the Z direction ($H_Z = H_0$), and incorporate the concepts used to develop Eqs. (8.58) to (8.60), Eqs. (8.54) to (8.56) give the following equations of motion

$$\frac{dM_X}{dt} = \gamma M_Z H_1 \sin \omega t + \omega_0 M_Y - \frac{M_X}{T_2} \tag{8.62}$$

$$\frac{dM_Y}{dt} = \gamma M_Z H_1 \cos \omega t - \omega_0 M_X - \frac{M_Y}{T_2} \tag{8.63}$$

$$\frac{dM_Z}{dt} = -\gamma M_X H_1 \sin \omega t - \gamma M_Y H_1 \cos \omega t - \frac{M_Z - M_0}{T_1} \tag{8.64}$$

The solution of these equations is more readily obtained by transformation to a rotating coordinate system whose x' axis is parallel to $\mathbf{H}_1$ and rotating in the XY plane at angular frequency ω. Employing the classical mechanics of rotating bodies the relationship between the rates of change of magnetization relative to the laboratory (X, Y, Z) and rotating (x', y', z') frames, as illustrated in Fig. 8.4, can be found. In the rotating coordinate systems the vectors $\mathbf{M}$, $\mathbf{H}_1$, and $\boldsymbol{\omega}$ become

$$\mathbf{M}' = M'_x\hat{\mathbf{i}}' + M'_y\hat{\mathbf{j}}' + M'_z\hat{\mathbf{k}}' \tag{8.65}$$

$$\mathbf{H}'_1 = \hat{\mathbf{i}}' H_1 \tag{8.66}$$

$$\boldsymbol{\omega}' = -\omega\hat{\mathbf{k}}' \tag{8.67}$$

where M'_x and M'_y are the transverse components of the magnetization in the

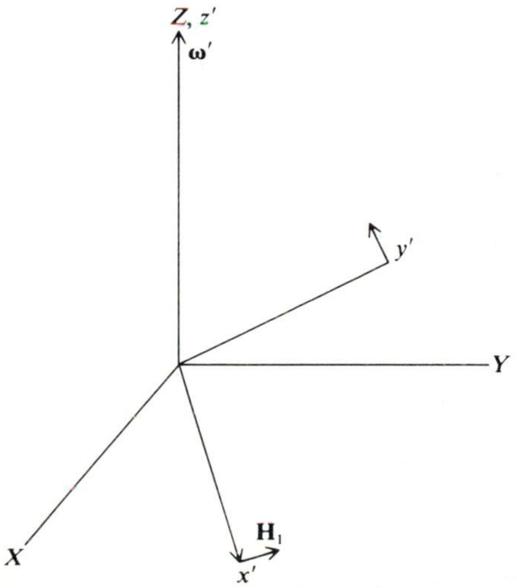

FIGURE 8.4
Rotating coordinate system.

$x'y'$ axis system. For an observer located in the rotating frame the unit vector $\hat{\mathbf{i}}'$, $\hat{\mathbf{j}}'$, and $\hat{\mathbf{k}}'$ will appear to be stationary and the apparent variation of the total magnetization can be expressed as

$$\frac{d\mathbf{M}'}{dt} = \frac{dM'_x}{dt}\hat{\mathbf{i}}' + \frac{dM'_y}{dt}\hat{\mathbf{j}}' + \frac{dM'_z}{dt}\hat{\mathbf{k}}' \tag{8.68}$$

For an observer located in the stationary axis system the unit vector in the rotating systems will change at a rate given by

$$\frac{d\hat{\mathbf{i}}'}{dt} = \boldsymbol{\omega}' X \hat{\mathbf{i}}' \tag{8.69}$$

$$\frac{d\hat{\mathbf{j}}'}{dt} = \boldsymbol{\omega}' X \hat{\mathbf{j}}' \tag{8.70}$$

$$\frac{d\hat{\mathbf{k}}'}{dt} = \boldsymbol{\omega}' X \hat{\mathbf{k}}' \tag{8.71}$$

The total magnetization $\mathbf{M}$ is independent of the choice of coordinate system chosen to describe its motion, so $\mathbf{M} = \mathbf{M}'$ and, using Eq. (8.65),

$$\frac{d\mathbf{M}}{dt} = \frac{dM'_x}{dt}\hat{\mathbf{i}}' + \frac{dM'_y}{dt}\hat{\mathbf{j}}' + \frac{dM'_z}{dt}\hat{\mathbf{k}}' M'_x \frac{d\hat{\mathbf{i}}'}{dt} + M'_y \frac{d\hat{\mathbf{j}}'}{dt} + M'_z \frac{d\hat{\mathbf{k}}'}{dt} \tag{8.72}$$

Incorporating Eqs. (8.65) to (8.67) this becomes

$$\frac{d\mathbf{M}}{dt} = \frac{d\mathbf{M}'}{dt} + \boldsymbol{\omega}' \times \mathbf{M}' \tag{8.73}$$

and substituting into Eq. (8.57) gives

$$\frac{d\mathbf{M}'}{dt} = \gamma\left(M' X \frac{\boldsymbol{\omega}}{\gamma} + \mathbf{M} \times \mathbf{H}_0 + \mathbf{M} \times \mathbf{H}_1\right) - \frac{(M_X \hat{\mathbf{i}} + M_Y \hat{\mathbf{j}})}{T_2} - \frac{(M_Z - M_0)\hat{\mathbf{k}}}{T_1} \tag{8.74}$$

The sum $\mathbf{H}_0 + \boldsymbol{\omega}/\gamma = \mathbf{H}_e$ is the effective magnetic field as observed from the rotating system. These relationships are illustrated in Fig. 8.5.

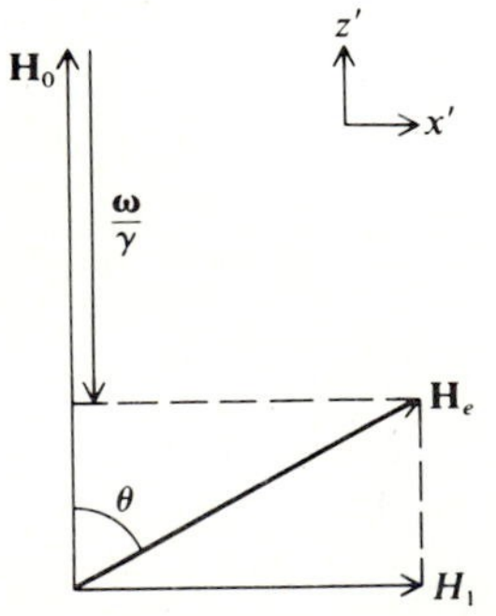

FIGURE 8.5
Effective magnetic field.

At some point in time the two-axis systems will be coincidental and Eq. (8.74) can be written, in terms of the rotating coordinate system, as

$$\frac{d\mathbf{M}'}{dt} = \gamma\mathbf{M}' \times \mathbf{H}'_e + \gamma\mathbf{M}'\mathbf{X}\mathbf{H}'_1 - \frac{M'_x\hat{\mathbf{i}}' + M'_y\hat{\mathbf{j}}'}{T_2} - \frac{(M_z - M_0)\hat{\mathbf{k}}'}{T_1} \tag{8.75}$$

This is the vector form of the Bloch phenomenological equations [3] which are

$$\frac{dM'_x}{dt} = (\gamma H_0 - \omega)M'_y - \frac{M'_x}{T_2} \tag{8.76}$$

$$\frac{dM'_y}{dt} = (\gamma H_0 - \omega)M'_x + \gamma M'_z H_1 - \frac{M'_y}{T_2} \tag{8.77}$$

$$\frac{dM'_z}{dt} = \frac{dM_Z}{dt} = -\gamma M'_y H_1 - \frac{(M_Z - M_0)}{T} \tag{8.78}$$

in view of Eqs. (8.65) to (8.67), and since **k** is collinear with **k**′.

A unidirectional oscillating magnetic field $2H_1 \cos \omega t$ can be regarded as being composed of two counterrotating field components $H_1 \cos \omega t \pm H_1 \sin \omega t$, but only the component rotating in the same direction as the Larmor precession of the nuclei can be in resonance with the nuclear spins. Hence the solutions of Eqs. (8.76) to (8.78) can be obtained under the steady state condition which exists after the oscillating magnetic field has been on for a long time and is characterized by $dM'_x/dt = dM'_y/dt = dM'_z/dt = 0$. These solutions are:

$$M'_x = \frac{\gamma M_0 H_1 T_2^2(\omega_0 - \omega)}{1 + T_2^2(\omega_0 - \omega)^2 + \gamma^2 H_1^2 T_1 T_2} \tag{8.79}$$

$$M'_y = \frac{\gamma M_0 H_1 T_2}{1 + T_2^2(\omega_0 - \omega)^2 + \gamma^2 H_1^2 T_1 T_2} \tag{8.80}$$

$$M'_z = \frac{M_0(1 + T_2^2(\omega_0 - \omega)^2)}{1 + T_2^2(\omega_0 - \omega)^2 + \gamma^2 H_1^2 T_1 T_2} \tag{8.81}$$

where $\omega_0 = \gamma H_0$.

The classical behavior of a spin system in the vicinity of a resonance can be viewed in terms of the behavior of the magnetic susceptibility. The behavior of static magnetic susceptibility discussed in Chap. 5 can be expanded to oscillating fields. M_X and M_Y are components of the magnetization which oscillate respectively in and 90° out of phase with H_1. They are related to M'_x and M'_y by

$$M_X = M'_x \cos \omega t + M'_y \sin \omega t \tag{8.82}$$

$$M_Y = M'_x \sin \omega t - M'_y \cos \omega t \tag{8.83}$$

These phase relationships can be described by use of a complex magnetic susceptibility

$$\chi(\omega) = \chi'(\omega) + i\chi''(\omega) \tag{8.84}$$

Writing the X component of the magnetic field in complex notation

$$H_{1X} = 2H_1 e^{-i\omega t} \tag{8.85}$$

the component of the magnetization is the real part of

$$\chi(\omega)H_{1X} = H_1[\chi(\omega)e^{-i\omega t} + \chi^*(\omega)e^{i\omega t}] \tag{8.86}$$

which is

$$M_X = 2H_1[\chi'(\omega)\cos\omega t + \chi''(\omega)\sin\omega t] \tag{8.87}$$

A comparison of this relationship with Eq. (8.82), coupled with reference to Eqs. (8.79) and (8.81), then gives

$$\chi'(\omega) = \frac{M'_x}{2H_1} = \frac{\chi_0\omega_0}{2}\left\{\frac{T_2^2\Delta\omega}{1 + T_2^2(\Delta\omega)^2 + \gamma^2H_1^2T_1T_2}\right\} \tag{8.88}$$

$$\chi''(\omega) = \frac{M'_y}{2H_1} = \frac{\chi_0\omega_0}{2}\left\{\frac{T_2}{1 + T_2^2(\Delta\omega)^2 + \gamma^2H_1^2T_1T_2}\right\} \tag{8.89}$$

where $\chi_0\omega_0 = \chi_0\gamma H_0 = M_0\gamma$ and $\Delta\omega = \omega_0 - \omega$. The magnetic field H_1 will exert a torque on the spins causing them to move away from the z axis. This torque is proportional to χ'' and results in a change in the energy of the spin system.

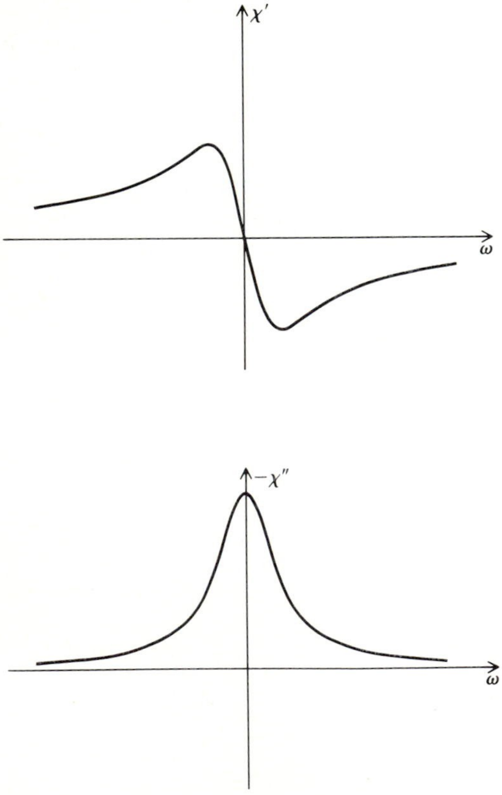

FIGURE 8.6
χ' and χ'' vs. ω.

Looking at the plot of Eq. (8.89) in Fig. 8.6 one observes that for $\omega \approx \omega_0$, χ'' becomes very large and the spin system will absorb energy from the oscillating electromagnetic field and a resonance will be observed. For weak fields $\gamma^2 H_1^2 T_1 T_2 \ll 1$; where there is no field saturation, the plot of χ'' vs. ω, and hence the resonance, will have a modified Lorentzian line shape.

In this section we have given an expanded discussion of the concepts introduced in Secs. 8.3 and 8.4. Since the observed resonance will be proportional to χ'', reference to Eq. (8.89) shows that the width of a resonance line, in the absence of saturation by a large H_1, will be proportional to T_2, the transverse or spin-spin relaxation time. The intensity of the absorption line as determined by the amount of saturation depends more on T_1, the longitudinal or spin-lattice relaxation.

Following the termination of the application of an H_1 field to a spin system the magnetization requires a finite time to return to its equilibrium value. During this time period the individual spins, which were precessing in a coherent manner, lose this phase coherence and there is a time-dependent decay of M_X and M_Y to zero and of M_Z to M_0. The measurement of this decay rate by several different types of experiments can result in the determination of T_1 and T_2. The measurement of this free-induction decay is the basis for the experimental techniques of *Fourier transform nuclear magnetic resonance.*

Having now established the basic relationships and defined the fundamental terms associated with magnetic resonance we will next turn our attention to problems associated with the relation of spectra to parameters related to molecular structure.

8.6 NMR IN CRYSTALLINE SOLIDS

Although the number of systems which are amenable to study in the non-motional solid state is very limited, a discussion of this technique provides a good introduction to dipole-dipole interactions in general, reviews a method which accounted for some early measurements of interproton separations in crystals, and provides the background for the discussion of the "magic angle" spinning technique.

In a crystalline solid the nuclei are located in a quite restricted environment, their change in position being only the slight motion associated with vibration of the lattice or molecules. Motion due to translation and, in most cases, rotation of the molecules, is not observed. In the presence of an externally applied magnetic field the nuclear dipoles will be thermally distributed among their allowed M_I values. Magnetic dipoles thus restricted can directly interact with each other via a dipole-dipole interaction. The net effect of this interaction is to produce at any given dipole an additional magnetic field, termed the local magnetic field, which is superimposed on the applied field. Depending on the crystalline environment of a particular nucleus, this effect can both broaden and split the observed resonance. The order of magnitude of this broadening is many times larger than the splittings observed in liquid systems. For a set of N magnetic moments located in a rigid lattice the Hamiltonian for the system is given by

$$\mathcal{H} = \sum_{i=1}^{N} \boldsymbol{\mu}_i \cdot \mathbf{H} + \sum_{i>j}^{N} \left[\frac{\boldsymbol{\mu}_i \cdot \boldsymbol{\mu}_j}{r^3} - \frac{3(\boldsymbol{\mu}_i \cdot \mathbf{r}_{ij})(\boldsymbol{\mu}_j \cdot \mathbf{r}_{ij})}{r^5} \right] \tag{8.90}$$

where the first term gives the magnetic energy levels $-g_{N_i}\beta_N I_{z_i} H_z$ and the second is the classical potential for a set of N magnetic dipoles.

For a system containing several magnetic nuclei one will have a rather complex Hamiltonian to consider as a general case, so let us restrict the discussion to a pair of identical nuclei with spin $I = \frac{1}{2}$ related by the coordinates shown in Fig. 8.7, where the two nuclei are at the origin and at x_2, y_2, z_2. The magnetic moment of a nucleus is a function of the spin $\boldsymbol{\mu}_N = g_N \beta_N \mathbf{I}/\hbar$. If the applied field is in the z direction then the Hamiltonian for this system is

$$\mathcal{H} = -\frac{g_N \beta_N \mathbf{I}_{Z_1} H_Z}{\hbar} - \frac{g_N \beta_N \mathbf{I}_{Z_2} H_Z}{\hbar} + \frac{g_N^2 \beta_N^2}{\hbar^2} \left[\frac{\mathbf{I}_1 \mathbf{I}_2}{r^3} - \frac{3(\mathbf{I}_1 \cdot \mathbf{r})(\mathbf{I}_2 \cdot \mathbf{r})}{r^5} \right] \tag{8.91}$$

The magnitude of the last term will be less than the first two, by at least two orders of magnitude, so it can be considered as a perturbation, $\mathcal{H}^{(1)}$. Using the vector relationships for $\mathbf{I}_1$, $\mathbf{I}_2$, and $\mathbf{r}$ and preforming the indicated scalar products lead to

$$\begin{aligned} \mathcal{H}^{(1)} = \frac{g_N^2 \beta_N^2}{\hbar^2 r^5} [&\mathbf{I}_{X_1} \mathbf{I}_{X_2} (r^2 - 3X^2) + \mathbf{I}_{Y_1} \mathbf{I}_{Y_2} (r^2 - 3Y^2) + \mathbf{I}_{Z_1} \mathbf{I}_{Z_2} (r^2 - 3Z^2) \\ &- 3(\mathbf{I}_{X_1} \mathbf{I}_{Y_2} + \mathbf{I}_{Y_1} \mathbf{I}_{X_2}) XY - 3(\mathbf{I}_{Y_1} \mathbf{I}_{Z_2} + \mathbf{I}_{Z_1} \mathbf{I}_{Y_2}) YZ \\ &- 3(\mathbf{I}_{Z_1} \mathbf{I}_{X_2} + \mathbf{I}_{X_1} \mathbf{I}_{Z_2}) XZ] \end{aligned} \tag{8.92}$$

This expansion can be expressed in a more compact notation by resorting to the use of tensor notation as outlined in App. L. Hence,

$$\mathcal{H}^{(1)} = \frac{g_N^2 \beta_N^2}{h^2} (\mathbf{I}_1 \cdot \underline{\mathbf{D}} \cdot \mathbf{I}_2) \tag{8.93}$$

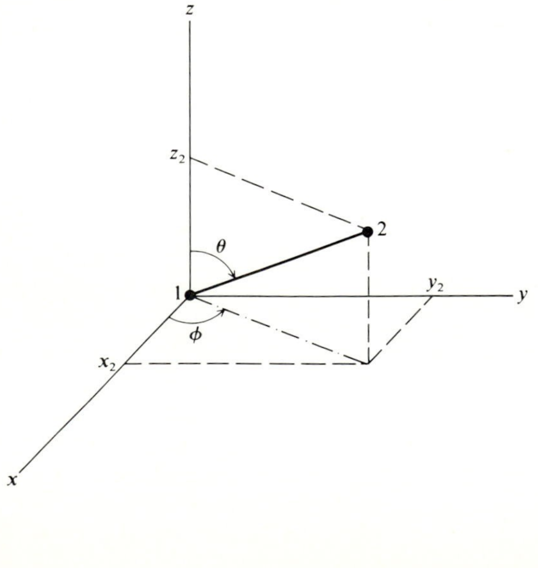

FIGURE 8.7
Coordinates for two interacting nuclei.

where $\underline{\mathbf{D}}$, the dipole coupling tensor has the components

$$\frac{1}{r^5}\begin{vmatrix} r^2 - 3X^2 & -3XY & -3XZ \\ -3XY & r^2 - 3Y^2 & -3YZ \\ -3XZ & -3YZ & r^2 - 3Z^2 \end{vmatrix} \tag{8.94}$$

It is to be noted that if the direction of $\mathbf{r}$ is chosen to be parallel to z, that is the solid is so oriented that the dipoles lie along the direction of the external field, then the off-diagonal terms vanish and the diagonal ones reduce to

$$\frac{1}{r^3}\begin{vmatrix} 1 & 0 & 1 \\ 0 & 1 & 0 \\ 0 & 0 & 1 \end{vmatrix} \tag{8.95}$$

The wavefunctions for such a system were found in Chap. 5 to be expressed as

$$\begin{aligned} \psi_{s_1} &= \alpha(1)\alpha(2) \\ \psi_{s_0} &= \frac{1}{\sqrt{2}}[\alpha(1)\beta(2) + \beta(1)\alpha(2)] \\ \psi_{s_a} &= \frac{1}{\sqrt{2}}[\alpha(1)\beta(2) - \beta(1)\alpha(2)] \\ \psi_{s_{-1}} &= \beta(1)\beta(2) \end{aligned} \tag{8.96}$$

In the absence of the perturbation they will be associated with the energies

$$E = -g_N\beta_N H_Z(M_{I_1} + M_{I_2}) \tag{8.97}$$

as shown in Fig. 8.8. The allowed transitions, $\Delta M_I = \pm 1$ are also shown in Fig. 8.8.

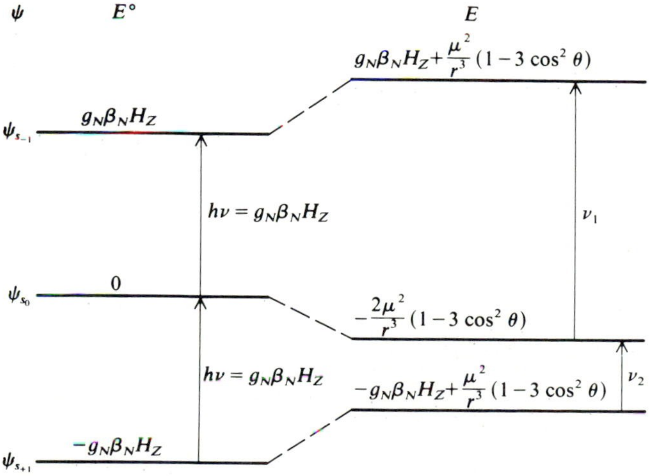

FIGURE 8.8
Energy levels of two-spin system.

To evaluate the perturbing Hamiltonian $\mathscr{H}^{(1)}$, it is best to transform to spherical coordinates and to employ the raising and lowering operators associated with the spin systems. Using the relationships

$$\mathbf{I}_X = \tfrac{1}{2}(\mathbf{I}_+ - \mathbf{I}_-) \tag{8.98}$$

$$\mathbf{I}_Y = \frac{1}{2i}(\mathbf{I}_+ - \mathbf{I}_-) \tag{8.99}$$

$$\cos\phi = \tfrac{1}{2}(e^{i\phi} + e^{-i\phi}) \tag{8.100}$$

$$\cos\phi = \frac{1}{2i}(e^{i\phi} - e^{-i\phi}) \tag{8.101}$$

substituting into Eq. (8.92), and rearranging, gives

$$\begin{aligned}\mathscr{H}^{(1)} = \frac{g_N^2\beta_N^2}{\hbar^2 r^3}[(1 - 3\cos^2\theta)\mathbf{I}_{Z_1}\mathbf{I}_{Z_2} - \tfrac{1}{4}(1 - 3\cos^2\theta)(\mathbf{I}_{+_1}\mathbf{I}_{-_2} + \mathbf{I}_{-_1}\mathbf{I}_{+_2})\\ - \tfrac{3}{2}\sin\theta\cos\theta e^{-i\phi}(\mathbf{I}_{Z_1}\mathbf{I}_{+_2} + \mathbf{I}_{+_1}\mathbf{I}_{Z_2}) - \tfrac{3}{2}\sin\theta\cos\theta e^{i\phi}(\mathbf{I}_{Z_1}\mathbf{I}_{-_2} + \mathbf{I}_{-_1}\mathbf{I}_{Z_2})\\ - \tfrac{3}{4}\sin^2 e^{-2i\phi}\mathbf{I}_{+_1}\mathbf{I}_{+_2} - \tfrac{3}{4}\sin^2\theta e^{2i\phi}\mathbf{I}_{-_1}\mathbf{I}_{-_2}]\end{aligned} \tag{8.102}$$

The interaction can be evaluated to first order by employing first-order perturbation theory and evaluating the terms in $\langle\psi_M|\mathscr{H}^{(1)}|\psi_M\rangle$ to get the perturbed energy $E^{(1)}$. This can be accomplished by examining the effect of the various operators in Eq. (8.102) on the spin functions. For example, looking at the effect of the individual terms $\langle\alpha(1)\alpha(2)|\mathscr{H}^{(1)}|\alpha(1)\alpha(2)\rangle$

$$\langle\alpha(1)\alpha(2)|\mathbf{I}_{Z_1}\mathbf{I}_{Z_2}|\alpha(1)\alpha(2)\rangle = \frac{\hbar^2}{4}\langle\alpha(1)\alpha(2)|\alpha(1)\alpha(2)\rangle = \frac{\hbar^2}{4} \tag{8.103}$$

$$\langle\alpha(1)\alpha(2)|\mathbf{I}_{+_1}\mathbf{I}_{-_2}|\alpha(1)\alpha(2)\rangle = 0 \tag{8.104}$$

$$\langle\alpha(1)\alpha(2)|\mathbf{I}_{-_1}\mathbf{I}_{+_2}|\alpha(1)\alpha(2)\rangle = 0 \tag{8.105}$$

$$\langle\alpha(1)\alpha(2)|\mathbf{I}_{Z_1}\mathbf{I}_{-_2}|\alpha(1)\alpha(2)\rangle = 0 \tag{8.106}$$

$$\langle\alpha(1)\alpha(2)|\mathbf{I}_{+_1}\mathbf{I}_{+_2}|\alpha(1)\alpha(2)\rangle = 0 \tag{8.107}$$

Hence,

$$\begin{aligned}E^{(1)}(\psi_a) = \langle\alpha(1)\alpha(2)|\mathscr{H}^{(1)}|\alpha(1)\alpha(2)\rangle = \frac{1}{4}\frac{g_N^2\beta_N^2}{r^3}(1 - 3\cos^2\theta)\\ = \frac{\mu^2}{4r^3}(1 - 3\cos^2\theta)\end{aligned} \tag{8.108}$$

The first-order perturbation energy for the three symmetric spin states are

$$E^{(1)}(\psi_{s_-}) = \frac{\mu^2}{r^3}(1 - 3\cos^2\theta) \tag{8.109}$$

$$E^{(1)}(\psi_{s_0}) = \frac{-2\mu^2}{r^3}(1 - 3\cos^2\theta) \tag{8.110}$$

$$E^{(1)}(\psi_{s_+}) = \frac{\mu^2}{r^3}(1 - 3\cos^2\theta) \tag{8.111}$$

The energy levels which exist in the presence of this dipole-dipole interaction are shown on the right-hand side of Fig. 8.8.

Using the selection rule $\Delta M_I = \pm 1$ the observable transitions are given by

$$\nu_1 = \frac{g_N\beta_N H_Z}{h} + \frac{3\mu^2}{hr^3}(1 - 3\cos^2\theta) \tag{8.112}$$

$$\nu_2 = \frac{g_N\beta_N H_Z}{h} + \frac{3\mu^2}{hr^3}(1 - 3\cos^2\theta) \tag{8.113}$$

An example of this interaction is reviewed in Sec. 8.9.

If the molecules in a system containing magnetic dipoles is subjected to thermal motion there is a reduction of the dipole-dipole interactions. In the limit of a liquid, where there is complete isotropic motion of the molecules, the dipole-dipole interactions average to zero and only the first summation in Eq. (8.90) remains. In this situation we can observe high-resolution NMR spectra as will be discussed in Sec. 8.7. If the two-spin system is a polycrystalline solid then there is a random distribution of θ value due to the random distribution of individual crystallites. For the ideal case of only two nonidentical spins one would observe a separate pair of lines each corresponding to the resonance of a different nucleus. In practice longer-distance interactions tend to broaden the lines, and the spectra of most powdered solids have little structure. A study of this phenomenon can yield information relative to molecular motions [4] but will not be pursued further.

Examination of Eqs. (8.109) to (8.111) or Eqs. (8.112) to (8.113) shows that if, by some means, one could experimentally average $(1 - 3\cos^2\theta)$ to zero then the dipole-dipole interactions would be quenched and the spectrum would be greatly simplified. This would be particularly true for systems with more than two interacting nuclei. Although this possibility was first noticed nearly 30 years ago it has only appeared as a viable experimental technique within the past decade. This technique, referred to as "*magic angle*" *spinning*, is illustrated by

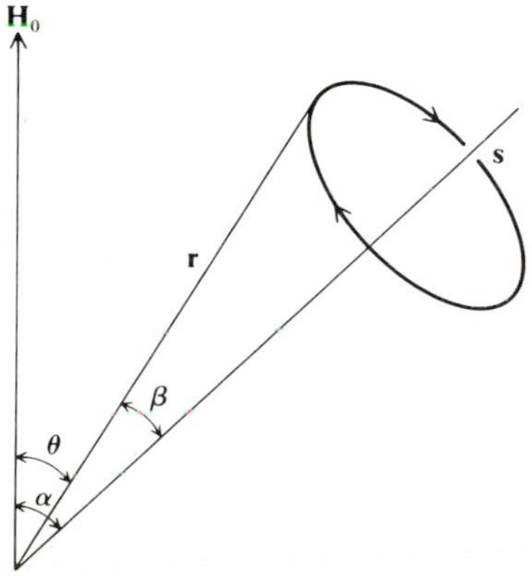

FIGURE 8.9
Geometry of "magic angle" spinning.

^{113}Cd in $CdCl_2$

These two spectra illustrate dramatically the effect of magic angle spinning (MAS). The powder pattern of the upper static spectrum collapses to a narrow line in the lower MAS spectrum, with spinning sidebands which reflect the chemical shift anisotropy of the original spectrum.

$CdCl_2$
spinning and nonspinning
XL-400

Observe:
 Frequency 88.73 MHz
 Spectral width 100 kHz
 Acq. time 10.2 ms
 Relaxation delay 500.0 s
 Pulse width 90°
 Ambient temperature
 No. repetitions 33
 Spin rate 3420 Hz
 Double precision acquisition
Data processing:
 Line broadening 200.0 Hz
 Ft size 2K
Total time 4 h 35.0 min

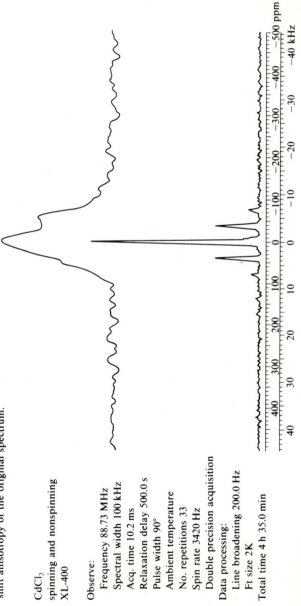

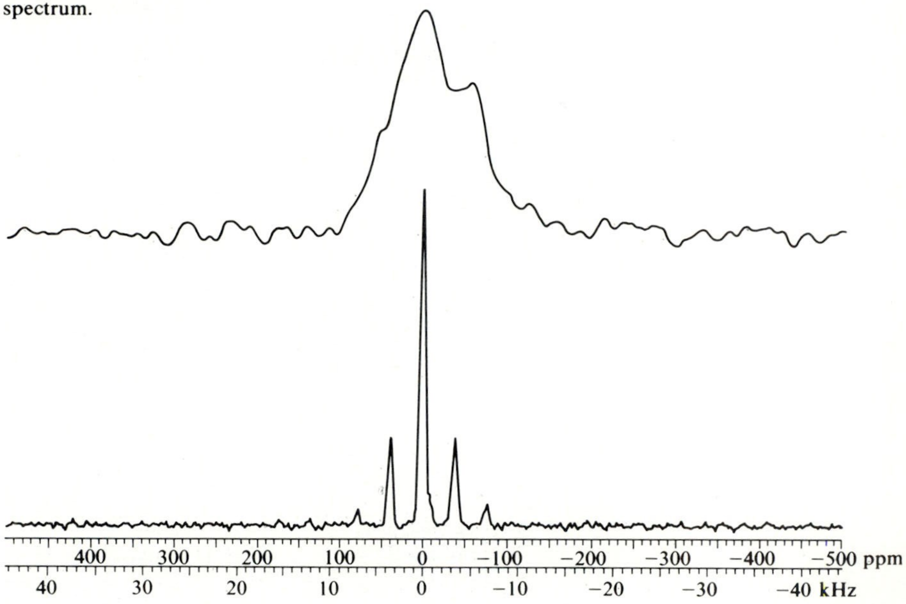

FIGURE 8.10
Spectra of $CdCl_2$: nonspinning and spinning polycrystalline sample ($\alpha = 54.736°$).

means of Fig. 8.9. In this technique a polycrystalline sample is rapidly rotated about an axis **s** at an angle α to the external magnetic field **H**. Any given internuclear vector $\mathbf{r}_{ij}$ will then sweep out a conic section at an angle β to **s** and will be at the angle θ_{ij} to the magnetic field. For a single crystal β will have a fixed value but it will be averaged for a polycrystalline sample. The problem is to find the average value of $(1 - 3\cos^2\theta_{ij})$ as $\mathbf{r}_{ij}$ rotates about **s** and show that there will be a value of α for which that average is zero. Harris [5] has shown that the average value of $(1 - 3\cos^2\theta)$ is $\frac{1}{2}(1 - 3\cos^2\alpha)(1 - 3\cos^2\beta)$. Thus if $\alpha = 54.736°$ the factor of $(1 - 3\cos^2\theta)$ vanishes, the dipole-dipole coupling averages to zero, and the spectrum resembles that of a liquid. The contrast between a nonspinning powder spectrum and a "magic angle" spectrum is shown in Fig. 8.10.

8.7 HIGH-RESOLUTION NMR SPECTRA: BASIC CONCEPTS

The use of high-resolution NMR spectra for structural characterization of molecules has become a powerful technique surpassing that of infrared spectroscopy. In this section we will look at the basic phenomenon giving rise to high-resolution NMR spectra and examine the quantum mechanical problems associated with the analysis of such spectra.

The principal emphasis will be on the use of matrix methods for evaluating the energy level and hence spectral schemes of multinuclear systems.

If the only interaction of a nucleus with a magnetic field was described by $\omega = \gamma H$ where H is the applied external magnetic field, then the technique would be of little use to chemists. It was discovered very early, however, that if the applied magnetic field was of sufficient homogeneity an NMR transition might be split into several components. Since the application of a magnetic field to any substance causes the electrons in the matter to react so as to produce a diamagnetic effect then, if different nuclei, protons for example, are located in different electronic environments within a molecule, the local field which opposes the applied field will be different for the different proton sites. Since this diamagnetic effect is field-dependent we can write the Hamiltonian for the interaction of a nucleus of spin **I** with a magnetic field as

$$\mathscr{H}^\circ = \frac{-g_N\beta_N}{\hbar}\mathbf{H}_{\text{loc}} \cdot \mathbf{I} = \frac{-g_N\beta_N}{\hbar}(H_Z - \sigma H_Z)\mathbf{I}_Z \tag{8.114}$$

assuming that the applied and induced fields are both in the Z direction.

The constant σ is called a shielding constant and is a measure of the shielding of the nuclei from the effect of the external field by its surrounding electrons. The shielding constant can be related to the electronic structure of the molecules but this topic will not be considered further.

The transition frequency for a system with an $I = \frac{1}{2}$ nucleus is

$$\nu = \frac{g_N\beta_N}{h}(1 - \sigma)H_Z = \nu_0(1 - \sigma) \tag{8.115}$$

In a continuous wave (cw) NMR experiment the frequency of the rf field is held constant and the external magnetic field H_Z is swept. This means that the resonance frequencies of protons which experience large shielding will occur at larger external magnetic fields than those which experience small shielding. This is easily rationalized by considering that at a fixed frequency for the rf field the local magnetic field necessary for resonance is the same at all nuclei

$$H_{\text{loc}} = H_Z - \sigma H_Z \tag{8.116}$$

Hence, as σ becomes large H_Z must increase to keep H_{loc} constant. The NMR spectra of several $I = \frac{1}{2}$ systems as they would appear under moderately high resolution are depicted in Fig. 8.11.

Due to the difficulty in making absolute measurements of large magnetic fields and of obtaining exact theoretical relationships between shielding constants and the electronic structure of a molecule it is convenient to express the chemical shifts for a given nucleus relative to a reference. For a fixed-frequency rf field and an external magnetic field H, the resonance of a nucleus of interest will be

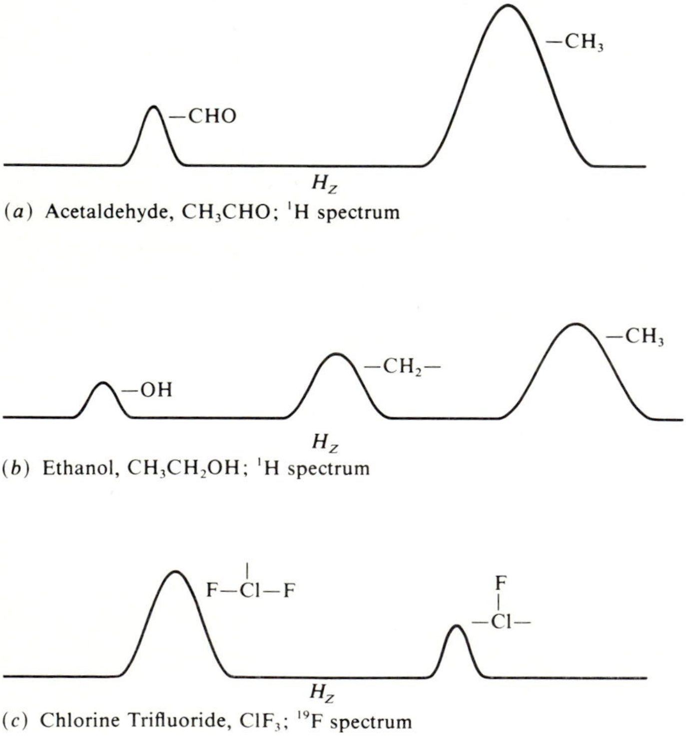

FIGURE 8.11
Qualitative NMR spectra of several $I = \frac{1}{2}$ systems depicting lines due only to chemical shifts.

given by

$$\nu = \frac{g_N \beta_N}{h}(1 - \sigma)H \tag{8.117}$$

while that for the same nucleus in a reference compound will be given by

$$\nu_r = \frac{g_N \beta_N}{h}(1 - \sigma_r)H_r \tag{8.118}$$

Since for a fixed frequency the local magnetic fields must be the same for resonance to occur in the two systems, we have $\nu = \nu_r$ or $(1 - \sigma)H = (1 - \sigma_r)H_r$. Upon rearrangement, since $H \approx H_r$ (this difference being of the order of a few parts per million), this gives

$$\delta = \sigma - \sigma_r = \frac{H - H_r}{H_r} \tag{8.119}$$

The quantity δ is called the chemical shift.

Although the chemical shift can be used extensively for qualitative structure characterization it is generally complicated by the simultaneous presence of further splitting due to additional interactions between nuclear spins. These interactions as well as the chemical shift are observed for liquids and gases using cw spectroscopy even though both are of much lesser magnitude than direct dipole-dipole interactions in solids. This results from the rapid tumbling of the nuclei averaging out the direct dipole-dipole interaction and results in the smaller effects being observed.

The further splittings beyond those due to chemical shifts are due to electron coupled spin-spin interactions which, unlike direct dipole-dipole interactions, do not average to zero due to molecular rotation. This effect was first discussed theoretically by Ramsay and Purcell [6] and can be considered as the interaction of the moment of one nucleus with that of another via a mutual interaction with the electrons in orbitals common to both nuclei. This interaction contributes energy terms which are proportional to the products of the involved spin operators. The proportionality constant is dependent on the system and independent of the applied magnetic field. This constant constitutes another measure of the nature of the electron distribution within the molecule. The discussion will proceed by considering a series of examples of increasing complexity. We will find that not only is there a splitting and displacement of energy levels due to spin-spin coupling but also, when the coupling is of the same order of magnitude as the chemical shift, there are anomalous changes in the spectral line intensities.

The general Hamiltonian (expressed in energy units) for the energy of a set of N nuclei will be given by

$$\mathcal{H} = -\sum_{i=1}^{N} \gamma_i(1 - \sigma_i)\mathbf{H} \cdot \mathbf{I}_i + \frac{h}{\hbar^2}\sum_{i>j} J_{ij}\mathbf{I}_i \cdot \mathbf{I}_j \tag{8.120}$$

The first summation contains the chemical shift terms and the second contains the spin-spin coupling terms. J_{ij} is the spin-spin coupling constant. It is customary to use a general designation to represent various types of spin systems. Nuclei having chemical shifts comparable in magnitude are denoted by alphabetic symbols that are adjacent or close while those with nonequivalent chemical shift parameters are denoted by letters at the extremes of the alphabet. For example, HF would be an AX type, H_2CO would be an A_2 type and 2-bromo-5-chlorothiophene an AB type, when only the H and F nuclei are considered.

The simplest case to analyze will be the spin $= \frac{1}{2}$ AX system. Taking the external field to be in the Z direction the Hamiltonian (expressed in hertz) reduces to

$$\mathcal{H} = \frac{-\gamma_A}{h}(1-\sigma_A)H_Z\mathbf{I}_{Z_A} - \frac{\gamma_X}{h}(1-\sigma_X)H_Z\mathbf{I}_{Z_X} + \frac{J_{AX}}{\hbar^2}\mathbf{I}_A \cdot \mathbf{I}_X \quad (8.121)$$

The energy levels of the system are found by evaluation of the matrix elements

$$E_i = \langle \psi_i | \mathcal{H} | \psi_i \rangle \quad (8.122)$$

For a two-spin system with nonidentical spins the wavefunctions are products of individual spin functions. Hence, there will be four energy states denoted as follows:

State	**Wavefunction**	I_{Z_A}	I_{Z_X}	$F_Z = I_{Z_A} + I_{Z_X}$
E_1	$\beta(A)\beta(X)$	$-\frac{1}{2}$	$-\frac{1}{2}$	-1
E_2	$\beta(A)\alpha(X)$	$-\frac{1}{2}$	$\frac{1}{2}$	0
E_3	$\alpha(A)\beta(X)$	$\frac{1}{2}$	$-\frac{1}{2}$	0
E_4	$\alpha(A)\alpha(X)$	$\frac{1}{2}$	$\frac{1}{2}$	1

If $J_{AX} = 0$ then the energies (in hertz) of the states are found, for example, as

$$E_4 = -\langle \alpha(A)\alpha(X)| \frac{\gamma_A}{h}(1-\sigma_A)H_Z\mathbf{I}_{Z_A} + \frac{\gamma_X}{h}(1-\sigma_X)H_Z\mathbf{I}_{Z_X}|\alpha(A)\alpha(X)\rangle \quad (8.123)$$

But

$$\mathbf{I}_{Z_A}\alpha(A)\alpha(X) = \frac{\hbar}{2}\alpha(A)\alpha(X) \quad (8.124)$$

$$\mathbf{I}_{Z_X}\alpha(A)\alpha(X) = \frac{\hbar}{2}\alpha(A)\alpha(X) \quad (8.125)$$

so

$$E_4 = -\frac{\gamma\!\!\!\!-_A}{2}(1-\sigma_A)H_z - \frac{\gamma\!\!\!\!-_X}{2}(1-\sigma_X)H_z \quad (8.126)$$

where

$$\gamma\!\!\!\!-_i = \frac{\gamma_i}{2\pi} \quad (8.127)$$

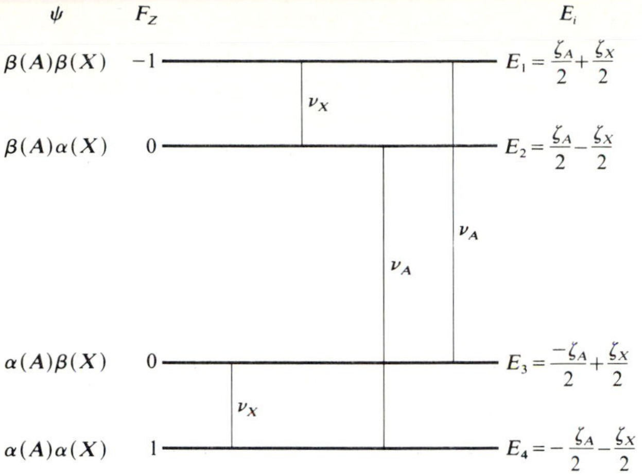

FIGURE 8.12
Magnetic energy levels of an *AX* system without spin-spin interactions.

Letting $\gamma_i(1 - \sigma_i)H_z = \zeta_i$ the energy levels for the case of $\gamma_A > \gamma_X$ are shown in Fig. 8.12.

Although one can rationalize the appropriate selection rules by use of the previously derived $\Delta M_I = \pm 1$ rule and by considering the two spins to function independently of one another there is a more general procedure which is applicable to the more complex cases. The transition probability for a single nucleus can be generalized to include several nuclei, hence

$$P_{nm} = \sum_i \frac{H_X^2}{h^2} \langle \psi_n | \boldsymbol{\mu}_{X_i} | \psi_m \rangle^2 g(\nu) = \frac{1}{h^2} \sum_i \gamma_i^2 H_X^2 \langle \psi_n | \mathbf{I}_{X_i} | \psi_m \rangle^2 g(\nu) \quad (8.128)$$

I_{X_i} can be written in terms of the stepping operators so

$$P_{nm} = \sum_i \frac{\gamma_i^2 H_x^2}{4\hbar^2} \langle \psi_n | \mathbf{I}_{+_i} + \mathbf{I}_{-_i} | \psi_m \rangle^2 g(\nu) \quad (8.129)$$

where $g(\nu)$ is a shape factor and will be ignored when dealing with $I = \frac{1}{2}$ nuclei where all transitions will be comparable in width. Often we will be interested in relative intensities of transitions due to like nuclei in which case the magnitudes of the matrix elements in Eq. (8.129) are needed. These elements can be further simplified by application of a knowledge of the properties of spin states and stepping operators. For a set of spin states, $\psi_1, \psi_2, \ldots, \psi_N$, the orthogonal nature of the states requires that $\langle \psi_i | \psi_j \rangle = \delta_{ij}$. If we are interested in the transition $\psi_n \to \psi_m$ then the element $\langle \psi_n | (\mathbf{I}_{+_i} + \mathbf{I}_{-_j}) | \psi_m \rangle$ must be nonzero or $(\mathbf{I}_{+_i} + \mathbf{I}_{-_j}) | \psi_m \rangle = | \psi_n \rangle$. If ψ_n is the lower of the two states then clearly $\mathbf{I}_{+_i} | \psi_m \rangle$ will give a function further removed and we can eliminate it from consideration. The intensity of the transition $\psi_n \to \psi_m (E_n < E_m)$ will then be proportional to $\langle \psi_n | \sum_i \mathbf{I}_{-_i} | \psi_m \rangle^2$. Likewise the probability of the transition $\psi_m \to \psi_n$ will be proportional to $\langle \psi_m | \sum_i \mathbf{I}_{+_i} | \psi_n \rangle^2$. In some presentations the discussion of the intensity problem is such that the states and operators are reversed so one must be on the alert to be sure that the conventions are understood.

For the AX system the selection rules will be determined by evaluation of the elements $\langle\psi_n|I_{-_A}+I_{-_X}|\psi_m\rangle$. For example, for the transition $\beta(A)\beta(X)\to\beta(A)\alpha(X)$ the transition probability will be proportional to

$$\langle\beta(A)\beta(X)|\mathbf{I}_{-_A}+\mathbf{I}_{-_X}|\beta(A)\alpha(X)\rangle=\langle\beta(A)\beta(X)|\beta(A)\beta(X)\rangle\hbar=\hbar \tag{8.130}$$

while for the transition $\beta(A)\beta(X)\to\alpha(A)\alpha(X)$ it will be proportional to

$$\begin{aligned}\langle\beta(A)\beta(X)|\mathbf{I}_{-_A}+\mathbf{I}_{-_X}|\alpha(A)\alpha(X)\rangle&=[\langle\beta(A)\beta(X)|\beta(A)\alpha(X)\rangle\\&\quad+\langle\beta(A)\beta(X)|\alpha(A)\beta(X)\rangle]\hbar=0\end{aligned} \tag{8.131}$$

The allowed transitions as evaluated in this manner are shown in Fig. 8.12. The observed spectrum in this case will consist of a pair of lines each characteristic of one particular nucleus and will be given by

$$\nu_A=\frac{\zeta_A}{2}=\gamma_A(1-\sigma_A)H_Z \tag{8.132}$$

$$\nu_X=\frac{\zeta_X}{2}=\gamma_X(1-\sigma_X)H_Z \tag{8.133}$$

Looking at the quantum number changes associated with each transition we find that first M_I changes for only one nucleus, which is expected if there is no coupling between the nucleus and $\Delta F_Z=\pm1$. This latter selection rule also is valid when spin-spin coupling occurs.

For the AX system when J_{AX} is not zero, but small compared to $\delta_{AX}H_Z$, the energy shifts due to spin-spin coupling are much less than those due to the chemical shifts and the energy matrix will contain additional terms of the form $\langle\psi_A|J_{AX}\mathbf{I}_A\cdot\mathbf{I}_X|\psi_m\rangle$. The scalar product $\mathbf{I}_A\cdot\mathbf{I}_X$ is equal to $\mathbf{I}_{X_A}\mathbf{I}_{X_X}+\mathbf{I}_{Y_A}\mathbf{I}_{Y_X}+\mathbf{I}_{Z_A}\mathbf{I}_{Z_X}$. Substituting this expression into the matrix element gives for the spin-spin coupling energy term (expressed in units of hertz), for example

$$E_3'=E'(\alpha(A)\beta(X))=\frac{J_{AX}}{\hbar^2}\langle\alpha(A)\beta(X)|\mathbf{I}_{X_A}\mathbf{I}_{X_X}+\mathbf{I}_{Y_A}\mathbf{I}_{Y_X}+\mathbf{I}_{Z_A}\mathbf{I}_{Z_X}|\alpha(A)\beta(X)\rangle \tag{8.134}$$

This term can be factored further giving

$$\begin{aligned}E_3'=\frac{J_{AX}}{\hbar^2}&[\langle\alpha(A)|\mathbf{I}_{X_A}|\alpha(A)\rangle\langle\beta(X)|\mathbf{I}_{X_X}|\beta(X)\rangle\\&+\langle\alpha(A)|\mathbf{I}_{Y_A}|\alpha(A)\rangle\langle\beta(X)|\mathbf{I}_{Y_X}\beta(X)\rangle\\&+\langle\alpha(A)|\mathbf{I}_{Z_A}|\alpha(A)\rangle\langle\beta(X)|\mathbf{I}_{Z_X}|\beta(X)\rangle]\end{aligned} \tag{8.135}$$

The individual terms in this expression are angular momentum matrix elements and can be evaluated by referring to Table 3.1. It is more convenient, however, to use the Pauli matrices given by Eqs. (5.18) to (5.20). For example,

$$\langle \alpha(A) | \mathbf{I}_{Z_A} \alpha | (A) \rangle = \frac{\hbar}{2} \tag{8.136}$$

$$\langle \beta(X) | \mathbf{I}_{X_X} | \beta(X) \rangle = 0 \tag{8.137}$$

The energy is thus evaluated to give $E_2' = -J_{AX}/4$. In an analogous manner the other energy terms are evaluated. The energy levels for the case with spin-spin coupling are shown in Fig. 8.13.

Since the state functions of the AX systems are the same with or without spin-spin coupling the selection rules will be the same as previously derived, and the allowed transitions are

$$\nu_X^{\pm} = \gamma_X(1 - \sigma_X)H_Z \pm \frac{J_{AX}}{2} \tag{8.138}$$

$$\nu_A^{\pm} = \gamma_A(1 - \sigma_A)H_Z \pm \frac{J_{AX}}{2} \tag{8.139}$$

The spectrum of the AX molecule will consist of two pairs of lines, each pair being centered at the positions of the non-spin-spin coupled transitions (the chemical shift) and each pair having an internal separation of J_{AX}. It is important to note that, unlike the magnitude of the chemical shift, that of the spin-spin splitting is independent of the applied magnetic field. Because of this fact the ratio $\delta H_Z/J_{AX}$ increases with increasing magnetic field and, for reasons which we will investigate soon, NMR spectra are often simplified by use of higher fields.

The form of the spectrum can also be easily rationalized on classical grounds. If we consider the magnetic field experienced by one nucleus it will be composed

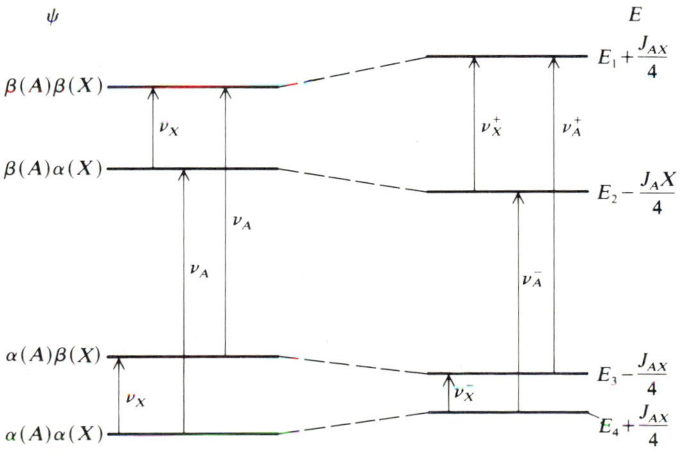

FIGURE 8.13
Energy levels for a spin-spin coupled AX system.

of three contributions. The major one will be the externally applied field. The two minor ones will be (1) the opposing field due to the diamagnetic effect of the electrons in the neighborhood of the nucleus and (2) a field due to the presence of the other nuclear moment. Although the interaction between the nuclei is through electron cloud coupling the classical effect is that of a weakened dipole-dipole interaction. The second nucleus has a nearly equal probability for two different orientations in the external field and, depending on the orientation, will produce a different contribution to the total field experienced by the first nucleus. For a large number of molecules there will be an almost equal division of the field experienced by the first nucleus, hence, one will observe two resonances at closely spaced field values. The same reasoning can be applied to the spectra of the second nucleus.

Although the procedure just described for evaluation of the energies of magnetic spin states is straightforward in approach there exists a simpler method based on the use of a permutation operator $\mathbf{P}_{ij}$ for evaluation of the spin-spin coupling energies. This operator is best defined by observation of its behavior when operating on spin states. The behavior is such that

$$\mathbf{P}_{12}|\alpha(1)\beta(2)\rangle = |\beta(1)\alpha(2)\rangle \tag{8.140}$$

or

$$\mathbf{P}_{12}|\alpha(1)\beta(2)\alpha(3)\rangle = |\beta(1)\alpha(2)\alpha(3)\rangle \tag{8.141}$$

In order to shorten the equations which follow, a shorthand notation that eliminates the numbers signifying particular nuclei will be used. Hence

$$\alpha(1)\beta(2)\beta(3) \rightarrow \alpha\beta\beta \tag{8.142}$$

or

$$\mathbf{P}_{12}|\alpha\beta\beta\rangle = |\beta\alpha\beta\rangle \tag{8.143}$$

We will not consider the details of the quantum mechanical derivation but Dirac [7] has shown that the scalar product of two spin operators for nuclei with $I = \frac{1}{2}$ is related to the permutation operator by $\mathbf{I}_i \cdot \mathbf{I}_j = \frac{1}{4}(2\mathbf{P}_{ij} - 1)$. The term in the Hamiltonian arising from the spin-spin coupling is then (in units of hertz)

$$\mathcal{H}_{ij} = \frac{J_{ij}}{\hbar^2}\mathbf{I}_i \cdot \mathbf{I}_j = \frac{J_{ij}}{4}(2\mathbf{P}_{ij} - 1) \tag{8.144}$$

In NMR spectroscopy it is more conventional to give energies in units of hertz, so from here on this procedure will be followed, hence the factor of h will be omitted from energy expressions including those involving $\mathbf{I}_F$ operators.

Examining the $\alpha(1)\beta(2)$ state of the AX system we find

$$E_3' = \langle\alpha\beta|\mathcal{H}_{AX}|\alpha\beta\rangle = \frac{J_{AX}}{4}\langle\alpha\beta|2\mathbf{P}_{AX} - 1|\alpha\beta\rangle = \frac{-J_{AX}}{4} \tag{8.145}$$

which is in agreement with the value previously found.

With the details of the AX system in hand we now turn our attention to the changes in energy levels and spectra of the system that is produced by (1) allowing the environment of the X nucleus to become similar to that of A nucleus, in which case the chemical shift is equivalent in magnitude to the spin-spin coupling $\delta_{AX}H_z \approx J_{AX}$ and (2) allowing the two nuclei to be chemically equivalent in which case $\sigma_A = \sigma_X$.

The discussion will be limited to $I = \frac{1}{2}$ nuclei as with the AX system. The primary difference encountered when transforming from an AX to an A_2 system is the fact that the two spins are identical and the state functions $\alpha\beta$ and $\beta\alpha$ are no longer good functions. In this case the spin state functions are given by $\alpha\alpha$, $(\alpha\beta + \beta\alpha)/\sqrt{2}$, $(\alpha\beta - \beta\alpha)/\sqrt{2}$, and $\beta\beta$. The Hamiltonian for the system is

$$\mathcal{H} = -2\gamma_A(1 - \sigma_A)H_Z\mathbf{I}_{Z_A} + J_{AA}\mathbf{I}_A \cdot \mathbf{I}_A \tag{8.146}$$

or

$$\mathcal{H} = -2\gamma_A(1 - \sigma_A)H_Z\mathbf{I}_{Z_A} + \frac{J_{AA}}{4}(2\mathbf{P}_{AA} - 1) \tag{8.147}$$

The energies of the four spin states are found by evaluating the elements $\langle\psi_m|\mathcal{H}|\psi_m\rangle$. For example the energy of the state $\psi_{s_0} = (\alpha\beta + \beta\alpha)/\sqrt{2}$ is

$$E_{s_0} = \frac{1}{2}\langle(\alpha\beta + \beta\alpha)|2\gamma_A(1 - \sigma_A)H_Z\mathbf{I}_{Z_A} + \frac{1}{2}J_{AA}(2\mathbf{P}_{AA} - 1)|(\alpha\beta + \beta\alpha)\rangle \tag{8.148}$$

$$\begin{aligned} E_{s_0} = \frac{1}{2}\gamma_A H(1 - \sigma_A)[&\langle\alpha\beta|\mathbf{I}_{Z_A}|\alpha\beta\rangle + \langle\alpha\beta|\mathbf{I}_{Z_A}|\beta\alpha\rangle + \langle\beta\alpha|\mathbf{I}_{Z_A}|\alpha\beta\rangle \\ &+ \langle\beta\alpha|\mathbf{I}_{Z_A}|\beta\alpha\rangle] + \frac{J_{AA}}{8}[\langle\alpha\beta|2\mathbf{P}_{AA} - 1|\alpha\beta\rangle + \langle\alpha\beta|2\mathbf{P}_{AA} - 1|\beta\alpha\rangle \\ &+ \langle\beta\alpha|2\mathbf{P}_{AA} - 1|\alpha\beta\rangle + \langle\beta\alpha|2\mathbf{P}_{AA} - 1|\beta\alpha\rangle] \end{aligned} \tag{8.149}$$

Considering that $I_{Z_A}|\alpha\rangle = \hbar/2|\alpha\rangle$ and $\mathbf{I}_{Z_A}|\beta\rangle = -\hbar/2|\beta\rangle$ for an $I = \frac{1}{2}$ system and recognizing the orthogonal properties of the spin states, that is, $\langle\alpha\beta|\alpha\beta\rangle = 1$, $\langle\alpha\beta|\beta\alpha\rangle = 0$ the sum of the first four terms is equal to zero. Since $\mathbf{P}_{AA}|\alpha\beta\rangle = |\beta\alpha\rangle$ the sum of the last four terms is $J_{AA}/4$ or $E_{s_0} = J_{AA}/4$. Repeating this procedure for the other three spin states leads to the energy level scheme tabulated in Table 8.3 and shown schematically in Fig. 8.14.

The selection rules for transition from $\psi_n \rightarrow \psi_m$ are found by evaluation of the elements $(\psi_n|\mathbf{I}_{+_A} + \mathbf{I}_{-_A}|\psi_m)$.

The effect of the operators $\mathbf{I}_\pm$ on the states of a system where mixed functions occur is as follows:

$$\begin{aligned} \mathbf{I}_+|\beta\beta\rangle &= |\alpha\beta\rangle + |\beta\alpha\rangle & \mathbf{I}_-|\beta\beta\rangle &= |00\rangle \\ \mathbf{I}_+|\alpha\beta\rangle &= |0\alpha\rangle + |\alpha\alpha\rangle & \mathbf{I}_-|\alpha\beta\rangle &= |\beta\beta\rangle + |\alpha 0\rangle \\ \mathbf{I}_+|\beta\alpha\rangle &= |\alpha\alpha\rangle + |\beta 0\rangle & \mathbf{I}_-|\beta\alpha\rangle &= |0\alpha\rangle + |\beta\beta\rangle \\ \mathbf{I}_+|\alpha\alpha\rangle &= |00\rangle & \mathbf{I}_-|\alpha\alpha\rangle &= |\alpha\beta\rangle + |\beta\alpha\rangle \end{aligned}$$

TABLE 8.3
Spin states and energy levels for the A_2 system

Notations	Wavefunction $-\psi_i$	F_z	$E_j(J_{AA}=0)$	$E_j(J_{AA}\neq 0)$	Symmetry of ψ_i
$\psi_{s_{-1}}$	$\beta\beta$	-1	$\gamma_A H_Z(1-\sigma_A)$	$\gamma_A H_Z(1-\sigma_A)+\frac{J_{AA}}{4}$	Symmetric
ψ_{s_0}	$\frac{\alpha\beta+\beta\alpha}{\sqrt{2}}$	0	0	$+\frac{J_{AA}}{4}$	Symmetric
ψ_{a_0}	$\frac{\alpha\beta-\beta\alpha}{\sqrt{2}}$	0	0	$-\frac{-3J_{AA}}{4}$	Antisymmetric
ψ_{s_1}	$\alpha\alpha$	-1	$-\gamma_A H_Z(1-\sigma_A)$	$-\gamma_A H_Z(1-\sigma_A)+\frac{J_{AA}}{4}$	Symmetric

Examination of the transition $\psi_{s_{-1}} \rightarrow \psi_{a_0}$, for example, shows that

$$\frac{1}{2}\langle\beta\beta|\mathbf{I}_{+_A}+\mathbf{I}_{-_A}|(\alpha\beta-\beta\alpha)\rangle=\frac{\hbar}{2}[\langle\beta\beta|\alpha\alpha)+(\beta\beta|\beta\beta\rangle-\langle\beta\beta|\alpha\alpha\rangle$$
$$-\langle\beta\beta|\beta\beta\rangle]=0 \qquad (8.150)$$

Using this method the allowed transitions are found to be $s_{-1} \rightarrow s_0$ and $s_0 \rightarrow s_1$ while those of the type $s \rightarrow a$ are forbidden. Therefore, although there is spin-spin coupling between like nuclei the magnitude of the level splittings and the selection rules combine so that all allowed transitions have the same frequency. This is experimentally confirmed by the observation of only a single proton resonance for molecules such as H_2O or CH_4.

Before investigating the AB system it will be informative to first examine the relative intensities of the transitions for an AX and an A_2 system. Since there is only a single resonance for the A_2 system all of the intensity will be in that line. The total intensity will depend on the several factors given in Eq. (8.128).

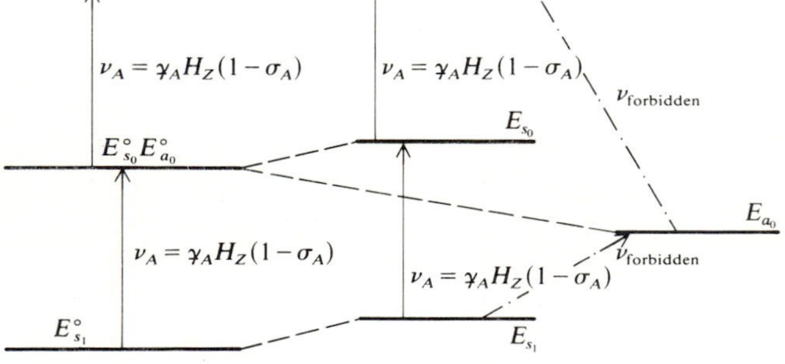

FIGURE 8.14
Energy levels and allowed transition for the A_2 system.

The four nonzero dipole terms $\langle \psi_n | I_{+_A} + I_{-_A} | \psi_m \rangle$ for the AX system we find to be equivalent in magnitude. Using Eq. (8.129) and considering that all nuclei in the system are being subjected to the same radiation field H_X, and the same external magnetic field H_Z, the relative intensities of the two pairs of lines are found to be in the ratio of the squares of the magnetogyric ratios of the nuclei. Thus, if there are two chemically inequivalent nuclei of the same atom the total integrated intensity of the two pairs of lines should be equivalent. This phenomenon is the basis for using the intensities of transitions as a measure of the relative number of chemically different nuclei in a system. An example of an AX spectrum is the proton spectrum of dichlorofluoromethane shown in Fig. 8.15 [8], where the spin-splitting is denoted by J_{HF}. This is only half of the total spectrum, the other half consisting of the F resonances that lie too far below the frequencies of the proton resonances to encompass both in a single sweep.

For an AB system one might suspect that the energy level and transition patterns would be intermediate between the AX and A_2 systems. Not only is this found to be the case but also there is an alteration of the relative intensities. The next problem will be to characterize the type of spin states which can account for such behavior. Since the two intermediate wavefunctions change from simple spin states of the form $\alpha\beta$ or $\beta\alpha$ to linear combinations containing equal weights of these two as one goes from the AX case to the A_2 case, the wavefunctions of the intermediate AB case will be taken to be

$$\psi_1 = \beta\beta \tag{8.151}$$

$$\psi_2 = a_2\alpha\beta + b_2\beta\alpha \tag{8.152}$$

$$\psi_3 = a_3\alpha\beta + b_3\beta\alpha \tag{8.153}$$

$$\psi_4 = \alpha\alpha \tag{8.154}$$

where the a_i and b_i variable coefficients which depend on the ratio $\delta_{AB}\nu_A/J_{AB}$.

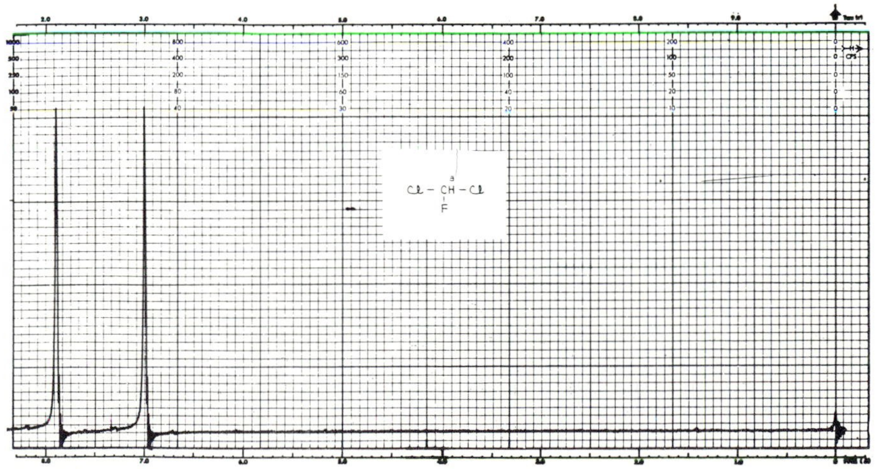

FIGURE 8.15
Proton NMR spectrum of dichlorofluoromethane[8]. © *Sadtler Research Laboratories, Division of Bio-Rad Laboratories, Inc. (1967).*

Using these arbitrary wavefunctions the variational method can be used to find the energy levels and the coefficients. The Hamiltonian for the system is

$$\mathcal{H} = -\gamma_A(1-\sigma_A)H_Z\mathbf{I}_{Z_A} - \gamma_B(1-\sigma_B)H_Z\mathbf{I}_{Z_B} + \mathbf{J}_{AB}\mathbf{I}_A\mathbf{I}_B \tag{8.155}$$

For the upper and lower states the energies are identical to those found for the AX system:

$$E_1 = \frac{1}{2}\gamma_A H_Z(1-\sigma_A) + \frac{1}{2}\gamma_B H_Z(1-\sigma_B) + \frac{J_{AB}}{4} \tag{8.156}$$

$$E_4 = -\frac{1}{2}\gamma_A H_Z(1-\sigma_A) - \frac{1}{2}\gamma_B H_Z(1-\sigma_B) + \frac{J_{AB}}{4} \tag{8.157}$$

Using as the mixed state ψ_2, the energies are found as follows. From the variational theorem

$$E = \frac{\langle\psi|\mathcal{H}|\psi\rangle}{\langle\psi|\psi\rangle} \tag{8.158}$$

or

$$E = \frac{\langle(a_2\alpha\beta + b_2\beta\alpha)|\mathcal{H}|(a_2\alpha\beta + b_2\beta\alpha)\rangle}{\langle(a_2\alpha\beta + b_2\beta\alpha)|(a_2\alpha\beta + b_2\beta\alpha)\rangle} \tag{8.159}$$

Multiplied out this becomes

$$(a_2^2 + b_2^2)E = a_2^2\langle\alpha\beta|\mathcal{H}|\alpha\beta\rangle + 2a_2b_2\langle\alpha\beta|\mathcal{H}|\beta\alpha\rangle + b_2^2\langle\beta\alpha|\mathcal{H}|\beta\alpha\rangle \tag{8.160}$$

Successively taking the derivative with respect to the coefficients and equating the resulting equations to zero leads to the familiar secular determinate, which has the form

$$\begin{vmatrix} H_{22} - E & H_{23} \\ H_{32} & H_{33} - E \end{vmatrix} = 0 \tag{8.161}$$

where

$$H_{22} = \langle\alpha\beta|\mathcal{H}|\alpha\beta\rangle \tag{8.162}$$

$$H_{23} = H_{32} = \langle\alpha\beta|\mathcal{H}|\beta\alpha\rangle = \langle\beta\alpha|\mathcal{H}|\alpha\beta\rangle \tag{8.163}$$

$$H_{33} = \langle\beta\alpha|\mathcal{H}|\beta\alpha\rangle \tag{8.164}$$

Solution of this determinate gives for the energy levels

$$E = \tfrac{1}{2}(H_{22} + H_{33}) \pm \tfrac{1}{2}[(H_{22} + H_{33})^2 - 4(H_{22}H_{33} - H_{23}^2)]^{1/2} \tag{8.165}$$

Substituting $J_{AB}(2\mathbf{P}_{AB} - 1)/4$ for $J_{AB}\mathbf{I}_A \cdot \mathbf{I}_B$ in the Hamiltonian the matrix elements H_{ij} in hertz become

$$H_{22} = \frac{1}{2}\gamma_A H_Z(1-\sigma_A) - \frac{1}{2}\gamma_B H_Z(1-\sigma_B) - \frac{J_{AB}}{4} \tag{8.166}$$

$$H_{33} = \frac{1}{2}\gamma_A H_Z(1-\sigma_A) - \frac{1}{2}\gamma_B H_Z(1-\sigma_B) - \frac{J_{AB}}{4} \tag{8.167}$$

$$H_{23} = \frac{J_{AB}}{2} \tag{8.168}$$

Considering that the two nuclei are of the same species $\gamma_A = \gamma_B = \gamma$, the energy levels become

$$E_2 = \frac{1}{2}\xi - \frac{J_{AB}}{4} \tag{8.169}$$

$$E_3 = -\frac{1}{2}\xi - \frac{J_{AB}}{4} \tag{8.170}$$

where

$$\xi = [\gamma^2 H_Z^2(\sigma_A - \sigma_B)^2 + J_{AB}^2]^{1/2} \tag{8.171}$$

The schematic presentation of the results is shown in Fig. 8.16 where the AB energy level scheme is depicted relative to those of both the AX and A_2 cases.

We can evaluate the selection rules in the same manner as used previously. Looking for example at the transition $\beta\beta \rightarrow a_3\alpha\beta + b_3\beta\alpha$, which was forbidden for the A_2, system the probability will be proportional to $P_{43} = \langle a_3\alpha\beta + b_3\beta\alpha | \mathbf{I}_+ + \mathbf{I}_- | \beta\beta\rangle^2 \gamma_A^2 H_X^2/2$. Since $\mathbf{I}_\pm|\beta\beta\rangle = |\alpha\beta + \beta\alpha\rangle$ and $a_i^2 + b_i^2 = 1$

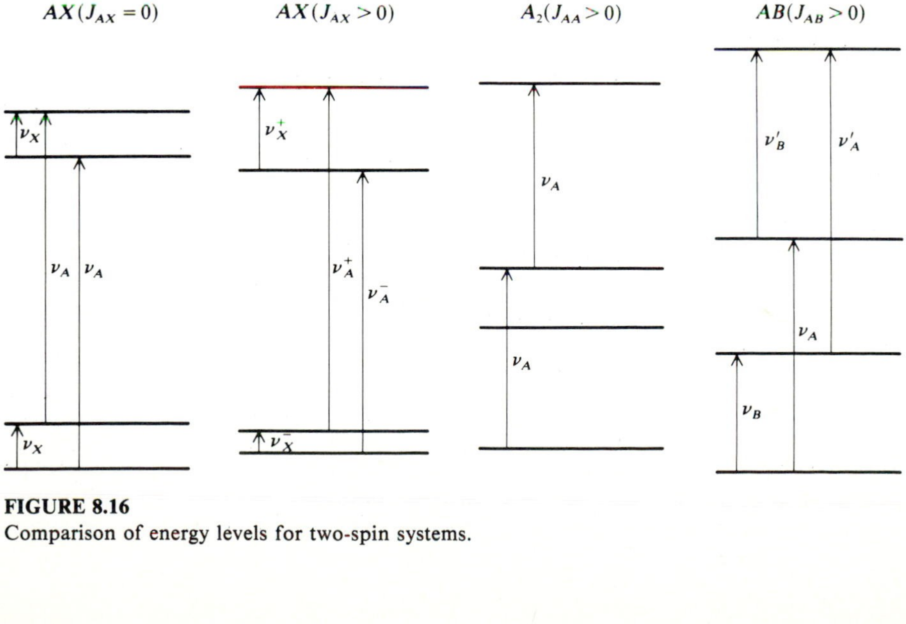

FIGURE 8.16
Comparison of energy levels for two-spin systems.

this term becomes $P_{43} = Y(1 - 2a_3b_3)^2$. A similar evaluation for the other transitions gives

$$P_{42} = (1 + 2a_2b_2)^2 Y \qquad P_{21} = (1 - 2a_2b_2)^2 Y$$
$$P_{31} = (1 + 2a_3b_3)^2 Y \qquad P_{41} = P_{32} = 0 \tag{8.172}$$

where

$$Y = \frac{\gamma_A H_X^2}{2} \tag{8.173}$$

It has been established that the relative intensities of the four transitions will depend on the coefficients in the wavefunctions. The allowed transitions are shown in Fig. 8.14 also.

By use of the secular equations the coefficients can be evaluated. These equations are

$$a_2(H_{22} - E) + b_2H_{23} = 0 \tag{8.174}$$

$$a_3H_{23} + b_3(H_{33} - E) = 0 \tag{8.175}$$

Therefore

$$\frac{a_2}{b_2} = \frac{-H_{23}}{H_{22} - E} = \frac{-J_{AB}}{\gamma H_Z(\sigma_A - \sigma_B) - [\gamma^2 H_Z^2(\sigma_A - \alpha_B)^2 + J_{AB}^2]^{1/2}} \tag{8.176}$$

$$\frac{a_3}{b_3} = \frac{-(H_{33} - E)}{H_{23}} = \frac{-\gamma H_Z(\sigma_A - \sigma_B) + [\gamma^2 H_Z^2(\sigma_A - \alpha_B)^2 + J_{AB}^2]^{1/2}}{J_{AB}} \tag{8.177}$$

There are two aspects of these relationships to note. First

$$\frac{a_2}{b_2} = -\frac{b_3}{a_3} \tag{8.178}$$

Knowing that $a_2^2 + b_2^2 = a_3^2 + b_3^2 = 1$ substitution into Eq. (8.178) shows that $a_2^2 + a_3^2 = b_2^2 + b_3^2 = 1$ also. The condition which simultaneously satisfies these equations is that $a_2 = -b_3$ and $a_3 = b_2$. The two intermediate spin state functions are then

$$\psi_2 = a\alpha\beta + b\beta\alpha \tag{8.179}$$

$$\psi_3 = b\alpha\beta - a\beta\alpha \tag{8.180}$$

Using Eqs. (8.176) and (8.177) along with these conditions we can individually solve for a and b in terms of the chemical shift and spin-spin coupling parameters. The second point involves the behavior of the coefficients in the two limiting cases. For the A_2 system $\sigma_A - \sigma_B \to 0$, $a/b \to 1$, and ψ_2 and ψ_3 reduce to the correct functions. For the AX system, $J_{AX} \ll \gamma H_z(\sigma_A - \sigma_B)$, $a_3/b_3 \to 0$ or $a_3 \to 0$, $a_2/b_2 \to \infty$ or $b_2 \to 0$, and the spin functions again reduce to the correct forms. To summarize the findings regarding two spin systems typical spectra for certain assumed relationships of the chemical shift and spin-spin splitting parameters are shown in Fig. 8.17. This figure illustrates the effect of changing the nuclear coupling on the numbers, spacings, and relative intensities of the transitions.

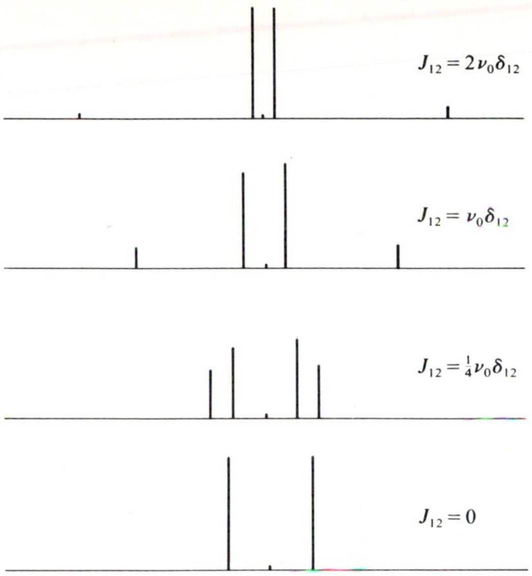

FIGURE 8.17
Typical two-spin NMR spectra as a function of $J_{12}/\delta_{12}\nu_0$.

8.8 HIGH-RESOLUTION NMR SPECTRA: GENERAL TREATMENT

One may raise the question at this point as to why the discussions have been limited to $I = \frac{1}{2}$ systems. Aside from the fact that several common nuclei of chemical significance have $I = \frac{1}{2}$ there is another reason that generally limits the application of high-resolution NMR to such systems. Nuclei having $I \geqq 1$ possess electric quadrupole moments. These moments can interact with the surrounding electrons to produce a mechanism whereby rapid relaxation of the higher spin states is enhanced. The decrease in the relaxation time in many cases leads to a broadening of the NMR spectra to the point where the line width for a particular nucleus exceeds any chemical shift splittings which may be present. An example of this is the behavior of the ^{35}Cl ($I = \frac{3}{2}$) nucleus where no chemical shift splittings of the NMR are observed for Cl bonded to carbon but some effect is noticeable for inorganic chlorides where the coupling of the nuclear electric quadrupole moment to the electron field is small and the relaxation time thus increased. Another example of this effect is provided by comparing the ^{14}N ($I = 1$) resonances in $(CH_3)_4N^+$, where the tetrahedral symmetry about the nitrogen atom produces a very small nuclear quadrupole coupling, and in the $(CH_3)_{4-n}NH_n^+$ ions where the coupling is much larger. These are illustrated in Fig. 8.18 [9]. Nuclear quadrupole coupling will be discussed in more detail in Chap. 10 and it will suffice for now to indicate that a short NMR relaxation time is associated with a large nuclear quadrupole coupling constant.

The effect of the quadrupolar interaction of one nucleus may also influence the spin-spin coupling with another, nonquadrupolar nucleus. This is illustrated in Fig. 8.19 [10] where we see that the ^1H spectrum of NH_3 shows the characteristic

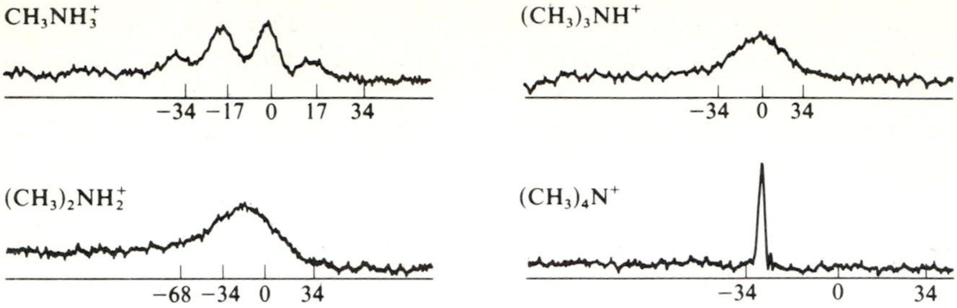

FIGURE 8.18
NMR spectra of ^{14}N in methylammonium chlorides in aqueous solution.

three-line pattern due to spin-spin coupling with a single $I = 1$ nucleus. The ^{14}N spectrum, however, does not show the expected quartet due to spin-spin coupling with a set of three identical $I = \frac{1}{2}$ nuclei. Instead it is quite broad due to the rapid relaxation. Looking at the ^{14}N spectrum of the NH_4^+ ion we again see good resolution and a pattern indicative of coupling of four equivalent protons. In this case the tetrahedral symmetry gives rise to a near-zero quadrupole coupling and a larger relaxation time.

One can readily imagine the number of possible systems that can occur for combinations of even a few nuclei. For a three-spin systems there will be 6; A_3, $A_2B \equiv AB_2$, ABC, $A_2X \equiv AX_2$, ABX, AMX. For more nuclei the number

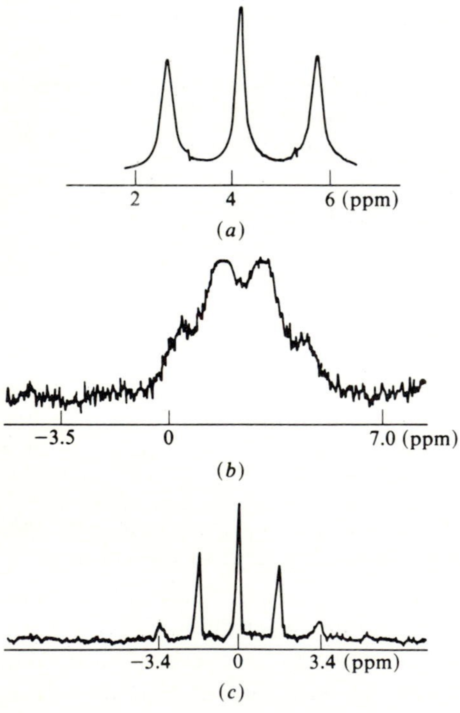

FIGURE 8.19
The proton spectrum of $^{14}NH_3$ and the ^{14}N spectrum in $^{14}NH_3$ and $^{14}NH_4^+$ ion. (*a*) ^1H spectrum in $^{14}NH_3$; (*b*) ^{14}N spectrum in $^{14}NH_3$; (*c*) ^{14}N spectrum in $^{14}NH_4^+$ ion.

increases rapidly. We cannot evaluate all of the individual cases but, having reviewed the necessary fundamentals, we can proceed to more advanced monographs and original literature.

The important principles involved in the previous discussions of the analysis of two-spin systems can be collected together to provide an explicit set of guidelines for the analysis of $I = \frac{1}{2}$ spin systems of varying complexity. After enumeration of these guidelines we will consider one further system as an example. To facilitate later reference to the rules they will be numbered.

1. The group theoretical methods regarding projection operators discussed in Chap. 5 are used to establish the spin functions of the system. The functions for the system are appropriate linear combinations of basic product functions of the type $\alpha(1)\beta(2)\alpha(3)\ldots\alpha(N) = \alpha\beta\alpha\ldots\alpha$ where N is the total number of nuclei involved. For example the general form of a wavefunction for a spin system of three nuclei would be

$$\psi_n = a_n\alpha\alpha\alpha + b_n\alpha\alpha\beta + c_n\alpha\beta\alpha + d_n\beta\alpha\alpha + e_n\alpha\beta\beta + f_n\beta\alpha\beta + g_n\beta\beta\alpha + h_n\beta\beta\beta \tag{8.181}$$

2. The total Hamiltonian consists of a sum of two parts, the first representing the chemical shift and external field interactions and the second representing the spin-spin coupling

$$\mathscr{H} = \mathscr{H}^\circ + \mathscr{H}' \tag{8.182}$$

The energy levels of the system are found by diagonalization of the matrix whose elements are $\langle\psi_n|\mathscr{H}|\psi_m\rangle$.

3. The matrix elements of the chemical shift Hamiltonian (in hertz) are given by

$$\langle\psi_n|\mathscr{H}^\circ|\psi_n\rangle = \sum_{i=1}^{N} \gamma_i(1-\sigma_i)H_Z(\mathbf{I}_{Z_i})_n \tag{8.183}$$

4. The matrix elements of the spin-spin Hamiltonian (in hertz) are given by

$$\langle\psi_n|\mathscr{H}'|\psi_n\rangle = \frac{1}{4}\sum_{i<j} J_{ij}T_{ij} \tag{8.184}$$

$$\langle\psi_n|\mathscr{H}'|\psi_m\rangle = \frac{1}{2}DJ_{ij} \tag{8.185}$$

where $T_{ij} = 1$ if spins i and j are parallel, $J_{ij} = -1$ if spins i and j are antiparallel, $D = 1$ if ψ_n and ψ_m differ by an interchange of spins i and j, and $D = 0$ if they differ otherwise.

5. No mixing occurs between spin functions with different values of the total spin component F_Z. This allows one to simplify the form of the matrix.

6. For a symmetrical molecule, one containing an A_n group of spins, there is no mixing between state functions of different symmetry type.

7. If there are several nuclear species present, that is, K, L, M, then, as a good approximation, rule 5 can be extended to state that there will be no mixing

between spin functions that differ in any of the total spin components $F_Z(K)$, $F_Z(L)$, $F_Z(M)$, etc. This same concept applies to the occurrence of several sets of nuclei when the chemical shifts among them are large compared to the spin-spin couplings.

8. For small applied radiofrequency fields the number of possible transitions are limited by the selection rule $\Delta F_Z = \pm 1$. Furthermore, for symmetrical systems transitions can occur only between levels of like symmetry.
9. For systems where the chemical shifts are large compared to the spin-spin splittings, transitions are allowed only between states that differ by ± 1 in only one of the total spin components $F_Z(K)$, $F_Z(L)$, etc.

8.9 DETERMINATION OF PROTON-PROTON DISTANCES IN A SINGLE CRYSTAL—AN EXAMPLE

It is difficult to precisely locate proton positions in crystals by means of X-ray diffraction. The technique of broad-line NMR can serve to compliment the X-ray method for simple systems where there are a limited number of protons. To illustrate this technique we will consider the determination of the H . . . H distance in gypsum, $CaSO_4 \cdot 2H_2O$, as investigated by Pake [11]. Gypsum forms a monoclinic unit cell with $a = 10.47A$, $b = 15.15A$, $C = 6.51A$ and $\beta = 151°33'$. The relationship of this unit cell to the parameters in Eqs. (8.129) to (8.130) and the experimentally determined parameters is shown in Fig. 8.20. The quantities $[hkl]$ denote the axes perpendicular to the respective (hkl) crystal faces. From a consideration of the symmetry of the crystal it can be concluded that the line joining the two protons in a water molecule and that joining the two sulfate oxygens are parallel, hence δ and ϕ_0 are known from the crystal structure and ϕ can be determined experimentally. Therefore θ, which is needed to get the interproton distance by use of Eqs. (8.112) and (8.113), can be calculated

$$\cos\theta = \cos\delta\cos(\phi - \phi_0) \qquad (8.186)$$

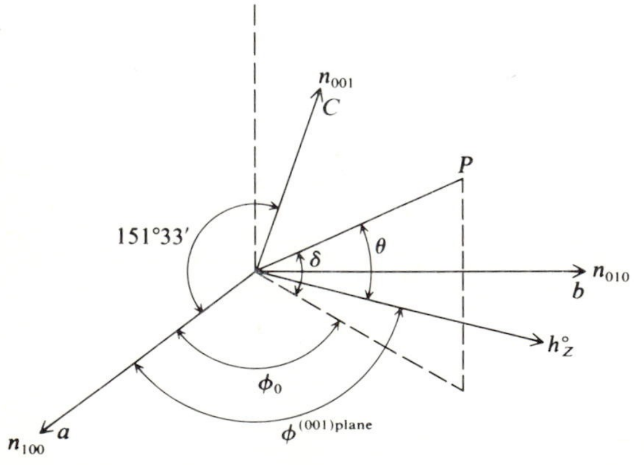

FIGURE 8.20
Relationships between crystal axes and magnetic field for gypsum crystal.

Since the gypsum crystal has two distinctly different water sites we would expect two pairs of resonances, assuming the two sites are sufficiently remote so as not to have any interaction between the pairs of protons. The $p - p$ direction in the second site will be related to that of the first by a reflection in the (010) plane. There will be two pairs of observed transitions derivable from Eqs. (8.112) and (8.113) and given, in terms of observed magnetic field rather than frequency, as

$$H_{\rm ob} = H_Z^\circ \pm \xi[3 \cos^2 \delta \cos^2 (\phi - \phi_0) - 1] \qquad (8.187)$$

where $H_Z^\circ = \nu_0/\gamma$ and $\xi = 3\mu/2r^3$. By cutting the crystal so it could be mounted to rotate with the H_Z° direction always lying in the (001) plane the splittings as a function of the angle ϕ was observed. The observed second derivative curves and the corresponding absorption curves are shown in Fig. 8.21. The two pairs

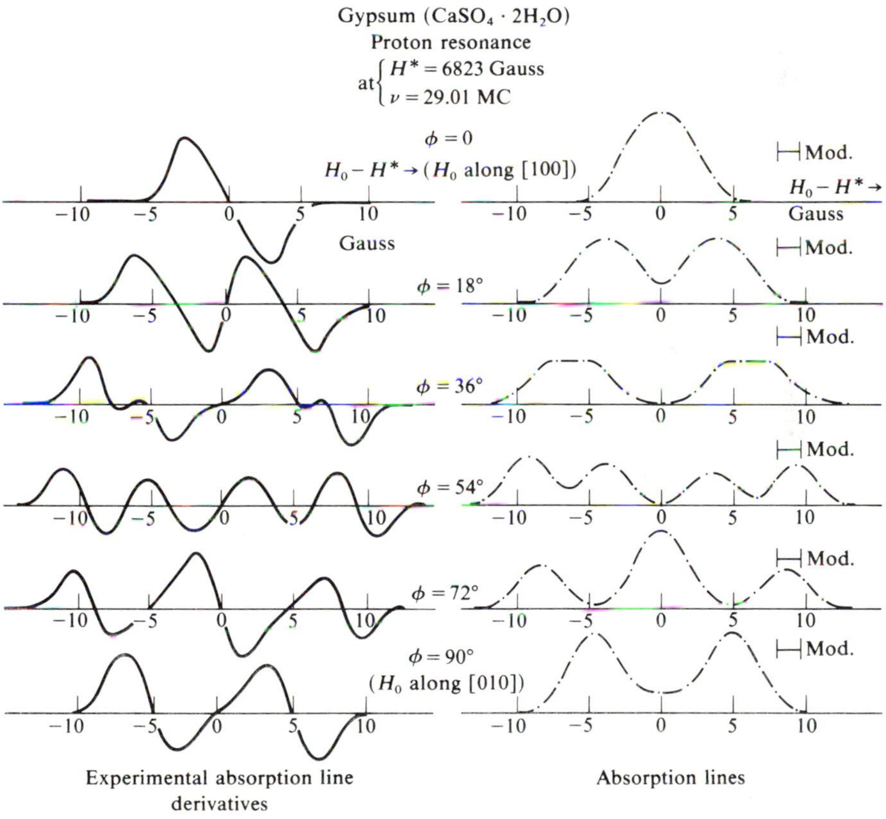

FIGURE 8.21
Representative experimental absorption curve derivatives and their integrals for various directions of the externally applied magnetic field H_0 in the (001) plane of a gypsum single crystal. The ordinates are measured in arbitrary units, and the region within which noise fluctuations occurred is indicated at the first maximum of each derivative. Peak-to-peak modulation sweep, indicated by the horizontal line near each integral, was 1.5 gauss for all curves.

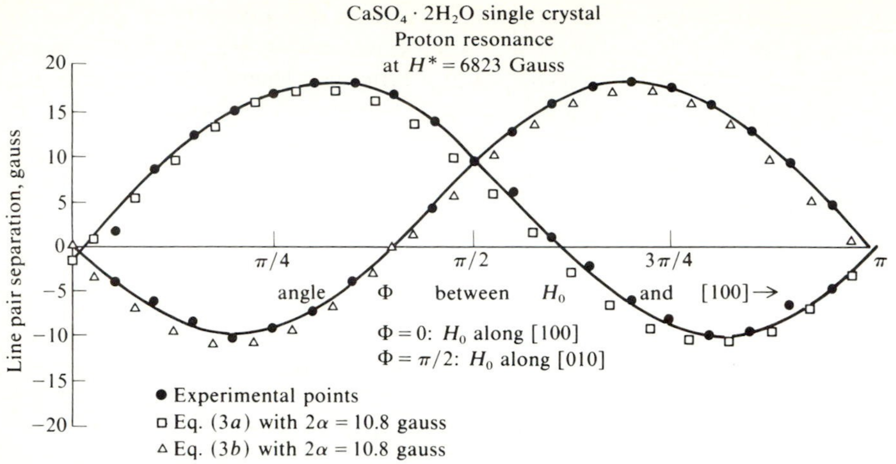

FIGURE 8.22
Line pair separation as a function of the angle ϕ between H_0 and [100]. Because two directions exist for proton-proton lines in a gypsum single crystal, there are two similar curves differing in phase.

of symmetrically displaced lines coalesce at some angles to give patterns other than a quadruplet. Using this data a plot of the separation of the members of each pair as a function of angle is constructed as shown in Fig. 8.22. Taking the crystal structure value of $\phi_0 = 54°34'$, the best fit of Eq. (8.187) to these curves is for $\xi = 5.4 \times 10^{-4} T$. From this[1]

$$r = \left(\frac{3\mu}{2\xi}\right)^{1/3} = \frac{3(2.79 \times 5.05 \times 10^{-27})\mathrm{J}\ T^{-1}}{2 \times 5.4 \times 10^{-4} T} = 0.158 \text{ nm} \tag{8.188}$$

Further discussions regarding the investigation of solids by NMR goes beyond the intended scope of this book. The technique is of value to the chemist for the investigation of molecular motions in solids as well as for the determination of proton-proton distances.

8.10 HIGH-RESOLUTION NMR SPECTRA ANALYSIS—EXAMPLES

In Sec. 8.8 it was found that the number of transitions, their separations, and their relative intensities are dependent on the ratio of chemical shifts to spin-spin coupling constants. Because of the interrelation of these factors in higher spin systems the analysis can become very complicated and soon reaches the point

[1] *Note:* $1\ \mathrm{J} = 10^{-7} T^2 m^3$.

where the most practical way to analyze spectra is to use a computer to simulate spectra as a function of the δ_i and J_{ij} parameters in the system. We will not attempt to develop the subject of computing simulated spectra but we will look at a couple of examples that can be reasonably analyzed without resorting to computer techniques.

The first case is the spectrum of 1,1,2-trichloroethane [12] shown in Fig. 8.23. Inspection of this spectrum indicates that it should be a three-proton A_2X type. The first problem is to establish the basis spin functions and to find the appropriate linear combinations where necessary. It will be helpful to start with the more specific case of an AMX system where there are no linear combinations of spin wavefunctions since the three nuclei are chemically well shifted apart. In this case, no application of projection operators is necessary, and we can evaluate the energy levels by use of Eqs. (8.182) to (8.184). The results of this are shown in Table 8.4, where $E_N^\circ = \gamma_N(1 - \sigma_N)H_Z(N = A, M, X)$.

If the parameters (δ_i and J_{ij}) are changed so that $AMX \rightarrow A_2X$ then the A_2 parts of the spin wavefunctions for the two state pairs $\psi_3 - \psi_4$ and $\psi_5 - \psi_6$ no longer have the appropriate forms. Considering the A_2 protons to belong to a C_{2v} group and following the procedure in Sec. 5.8 the appropriate linear combinations of the spin functions are developed. These are shown in Table 8.5, where again $E_i^\circ = \frac{1}{2}\gamma(1 - \sigma_i)H_Z$. The energy levels are determined by developing the energy matrix. Since $(\psi_m|\mathscr{H}^\circ|\psi_n) = 0$ for $n \neq m$ the E_i° terms occur only on the diagonal. For this set of spin functions the D in Eq. (8.185) will always be zero so the energy matrix will be diagonal and the energies will be as shown in Table 8.5. For example, the energy of ψ_3 is the sum of

$$\begin{aligned}
E^\circ &= \frac{1}{2}\langle(\alpha\beta + \beta\alpha)\alpha|\mathscr{H}^\circ|(\alpha\beta + \beta\alpha)\alpha\rangle \\
&= \frac{1}{2}\langle(\alpha\beta + \beta\alpha)\alpha|\zeta_A\mathbf{I}_{Z_{A_1}} + \zeta_A\mathbf{I}_{Z_{A_2}} + \zeta_X\mathbf{I}_{Z_X}|(\alpha\beta + \beta\alpha)\alpha\rangle \\
&= \frac{1}{2}\langle(\alpha\beta + \beta\alpha)\alpha|\zeta_A\mathbf{I}_{Z_{A_1}}|(\alpha\beta + \beta\alpha)\rangle\langle\alpha|\alpha\rangle \\
&\quad + \frac{1}{2}\langle(\alpha\beta + \beta\alpha)|\zeta_A\mathbf{I}_{Z_{A_2}}|(\alpha\beta + \beta\alpha)\rangle\langle\alpha|\alpha\rangle \\
&\quad + \frac{1}{2}\langle(\alpha\beta + \beta\alpha)|(\alpha\beta + \beta\alpha)\rangle\langle\alpha|\zeta_X\mathbf{I}_{Z_X}|\alpha\rangle \\
&= \frac{\zeta_{A_1}}{4}\langle(\alpha\beta + \beta\alpha)\alpha|(\alpha\beta + \beta\alpha)\rangle - \frac{\zeta_{A_2}}{4}\langle(\alpha\beta + \beta\alpha)|(\alpha\beta + \beta\alpha)\rangle + \frac{\zeta_X}{2}\langle\alpha|\alpha\rangle \\
&= \frac{\zeta_X}{2} = \frac{\gamma_X(1 - \sigma_X)H_Z}{2h} = \frac{1}{2}\gamma_X(1 - \sigma_X)H_z = E_X^\circ
\end{aligned} \tag{8.189}$$

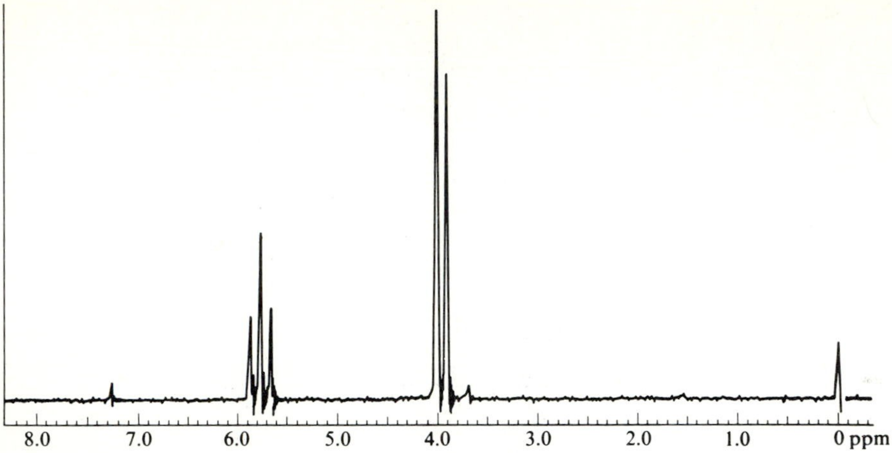

FIGURE 8.23
High-resolution spectrum of 1,1,2-trichloroethane[12].

TABLE 8.4
Spin wavefunctions and energy levels for the *AMX* system

State	Spin function	F_z	E_i, Hz
ψ_1	$\beta\beta\beta$	$\frac{3}{2}$	$E_A^\circ + E_M^\circ + E_X^\circ + \frac{1}{4}(J_{AM} + J_{AX} + J_{MX})$
ψ_2	$\beta\beta\alpha$	$\frac{1}{2}$	$E_A^\circ + E_M^\circ - E_X^\circ + \frac{1}{4}(J_{AM} - J_{AX} + J_{MX})$
ψ_3	$\beta\alpha\beta$	$\frac{1}{2}$	$E_A^\circ - E_M^\circ + E_X^\circ + \frac{1}{4}(-J_{AM} + J_{AX} - J_{MX})$
ψ_4	$\alpha\beta\beta$	$\frac{1}{2}$	$-E_A^\circ + E_M^\circ + E_X^\circ + \frac{1}{4}(-J_{AM} - J_{AX} + J_{MX})$
ψ_5	$\beta\alpha\alpha$	$-\frac{1}{2}$	$E_A^\circ - E_M^\circ - E_X^\circ + \frac{1}{4}(-J_{AM} - J_{AX} + J_{MX})$
ψ_6	$\alpha\beta\alpha$	$-\frac{1}{2}$	$-E_A^\circ + E_M^\circ - E_X^\circ + \frac{1}{4}(-J_{AM} + J_{AX} - J_{MX})$
ψ_7	$\alpha\alpha\beta$	$-\frac{1}{2}$	$-E_A^\circ - E_M^\circ + E_X^\circ + \frac{1}{4}(J_{AM} - J_{AX} - J_{MX})$
ψ_8	$\alpha\alpha\alpha$	$-\frac{3}{2}$	$-E_A^\circ - E_M^\circ - E_X^\circ + \frac{1}{4}(J_{AM} + J_{AX} + J_{MX})$

TABLE 8.5
Spin wavefunctions and energy levels for the A_2X system

State	Spin function	F_z	Desig-nation	E_i, Hz	Symmetry
ψ_1	$\beta\beta\beta$	$\frac{3}{2}$	$s_{-3/2}$	$2E_A^\circ + E_X^\circ + \frac{J_{AA}}{4} + \frac{J_{AX}}{2}$	Symmetric
ψ_2	$\beta\beta\alpha$	$\frac{1}{2}$	$2s_{-1/2}$	$2E_A^\circ - E_X^\circ + \frac{J_{AA}}{4} - \frac{J_{AX}}{2}$	Symmetric
ψ_3	$\frac{1}{\sqrt{2}}(\alpha\beta + \beta\alpha)\alpha$	$\frac{1}{2}$	$1s_{-1/2}$	$E_X^\circ + \frac{J_{AA}}{4}$	Symmetric
ψ_4	$\frac{1}{\sqrt{2}}(\alpha\beta - \beta\alpha)\alpha$	$\frac{1}{2}$	$a_{-1/2}$	$E_X^\circ - \frac{3J_{AA}}{4}$	Antisym-metric
ψ_5	$\frac{1}{\sqrt{2}}(\alpha\beta + \beta\alpha)\beta$	$-\frac{1}{2}$	$2s_{1/2}$	$-E_X^\circ + \frac{J_{AA}}{4}$	Symmetric
ψ_6	$\frac{1}{\sqrt{2}}(\alpha\beta - \beta\alpha)\beta$	$-\frac{1}{2}$	$a_{1/2}$	$-E_X^\circ - \frac{3J_{AA}}{4}$	Antisym-metric
‡$_7$	$\alpha\alpha\beta$	$-\frac{1}{2}$	$1s_{1/2}$	$-2E_A^\circ + E_X^\circ + \frac{J_{AA}}{4} - \frac{J_{AX}}{2}$	Symmetric
ψ_8	$\alpha\alpha\alpha$	$-\frac{3}{2}$	$s_{3/2}$	$-2E_A^\circ - E_X^\circ + \frac{J_{AA}}{4} + \frac{J_{AX}}{2}$	Symmetric

and

$$
\begin{aligned}
E' &= \left\langle \frac{1}{\sqrt{2}}(\alpha\beta + \beta\alpha)\alpha \,\middle|\, \mathscr{H}' \,\middle|\, \frac{1}{\sqrt{2}}(\alpha\beta + \beta\alpha)\alpha \right\rangle \\
&= \left\langle \frac{1}{\sqrt{2}}(\alpha\beta + \beta\alpha)\alpha \,\middle|\, \frac{J_{AA}}{4}(2\mathbf{P}_{12} - 1) + \frac{J_{AX}}{4}(2\mathbf{P}_{13} - 1) \right. \\
&\qquad \left. + \frac{J_{AX}}{4}(2\mathbf{P}_{23} - 1) \,\middle|\, \frac{1}{\sqrt{2}}(\alpha\beta + \beta\alpha)\alpha \right\rangle \\
&= \frac{J_{AA}}{4}
\end{aligned}
\tag{8.190}
$$

The symmetry of the spin functions, the F_z value, and the spectroscopic notation are also shown in Table 8.5. An energy level diagram for the A_2X system is shown in Fig. 8.24. The AMX system with no spin-spin coupling is shown for reference. The scale is conveniently set so $E_A^\circ + E_M^\circ + E_X^\circ$ for the AMX case is equal to $2E_A^\circ + E_X^\circ$ for the A_2X one.

The selection rules and the relative intensities of the transitions will be established by evaluation of the matrix elements [see Eq. (8.129)].

$$
P_{nm} \propto [\gamma_i \langle \psi_n | \mathbf{I}_{+_i} + \mathbf{I}_{-_i} | \psi_m \rangle]^2 \tag{8.191}
$$

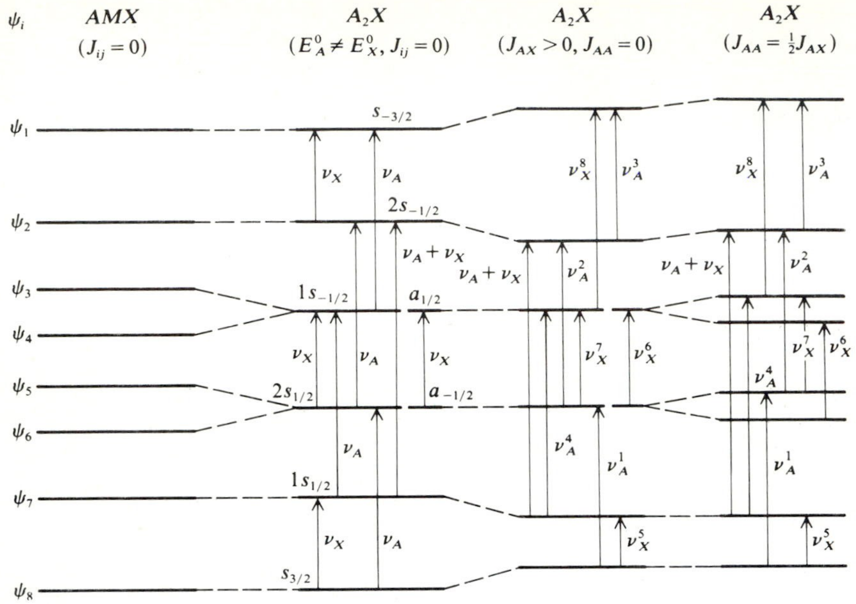

FIGURE 8.24
Energy levels of the A_2X and $AMX(J_{AM} = J_{AX} = J_{MX} = 0)$ systems.

where n and m denote the spin states and i denotes the nuclei. For example,

$$P_{78} \propto |\gamma_A\langle\beta\beta\alpha|\mathbf{I}_{+A_1} + \mathbf{I}_{-A_1}|\beta\beta\beta\rangle + \gamma_A\langle\beta\beta\alpha|\mathbf{I}_{+A_2} + \mathbf{I}_{-A_2}|\beta\beta\beta\rangle$$
$$+ \gamma_A\langle\beta\beta\alpha|\mathbf{I}_{+X} + \mathbf{I}_{-X}|\beta\beta\beta\rangle|^2$$
$$\propto \gamma_X^2 \tag{8.192}$$

The transition $\psi_7 \rightarrow \psi_8$ is allowed and furthermore from the character of the matrices in Eq. (8.192) we note that transition is associated only with the X nuclei.

For the $J_{ij} = 0$ case there are seven allowed transitions but they correspond to only two different frequencies. This spectrum is shown as Fig. 8.25a. If the spin-spin coupling is included the level diagrams shown to the right of Fig. 8.25 emerge. As with the A_2 case discussed previously the effect of J_{AA} is to raise or lower all states of identical symmetry by the same amount, hence it does not produce any observable splittings. The predicted spectrum, Fig. 8.25b, and that of 1,1,2-trichloroethylene shown in Fig. 8.23, are in agreement, and from the spectrum $J_{AX} = 6$ Hz and $\nu\delta_{AX} = 109.2$ Hz or $\delta_{AX} = 1.82$ ppm.

To conclude examples of high resolution NMR spectra we will look at some of the aspects of the spectra shown in Fig. 8.26. Both of the spectra shown are ABC types and exhibit the ultimate in mixing for a three-spin $I = \frac{1}{2}$ system. It is immediately obvious that the couplings in the two systems are not the same since the two patterns differ with respect to both intensities and line separations.

If the chemical shifts of an ABC system were large then it would be the AMX system discussed previously. Using the rules given in Sec. 8.6 the spin

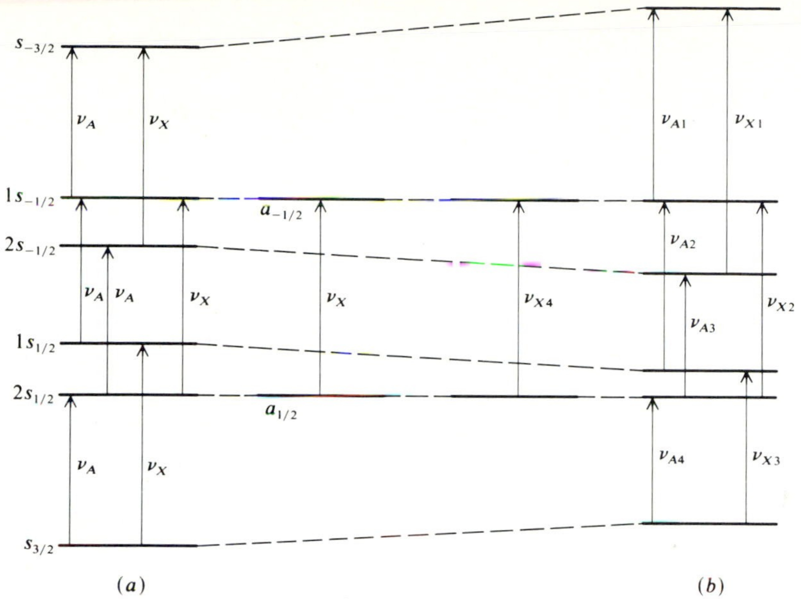

FIGURE 8.25
Spectrum of A_2X system. (*a*) $J_{ij} = 0$; (*b*) $J_{AX} > 0$.

functions and energy matrices can be determined. The basic set of spin wavefunctions is developed in a straightforward manner since no two of the nuclei are identical. The diagonal matrix elements are then found using Eqs. (8.183) and (8.184). Table 8.6 shows these terms; with $E_N^\circ = \frac{1}{2}\gamma_N(1 - \sigma_N)H_z$ ($N = A, B, C$).

Using Eq. (8.185) to obtain the off-diagonal terms the secular determinant becomes

$$\begin{vmatrix} E_1 - E & 0 & 0 & 0 & 0 & 0 & 0 & 0 \\ 0 & E_2 - E & \frac{J_{BC}}{2} & \frac{J_{AC}}{2} & 0 & 0 & 0 & 0 \\ 0 & \frac{J_{BC}}{2} & E_3 - E & \frac{J_{AB}}{2} & 0 & 0 & 0 & 0 \\ 0 & \frac{J_{AC}}{2} & \frac{J_{AB}}{2} & E_4 - E & 0 & 0 & 0 & 0 \\ 0 & 0 & 0 & 0 & E_5 - E & \frac{J_{AB}}{2} & \frac{J_{AC}}{2} & 0 \\ 0 & 0 & 0 & 0 & \frac{J_{AB}}{2} & E_6 - E & \frac{J_{BC}}{2} & 0 \\ 0 & 0 & 0 & 0 & \frac{J_{AC}}{2} & \frac{J_{BC}}{2} & E_7 - E & 0 \\ 0 & 0 & 0 & 0 & 0 & 0 & 0 & E_8 - E \end{vmatrix} = 0 \quad (8.193)$$

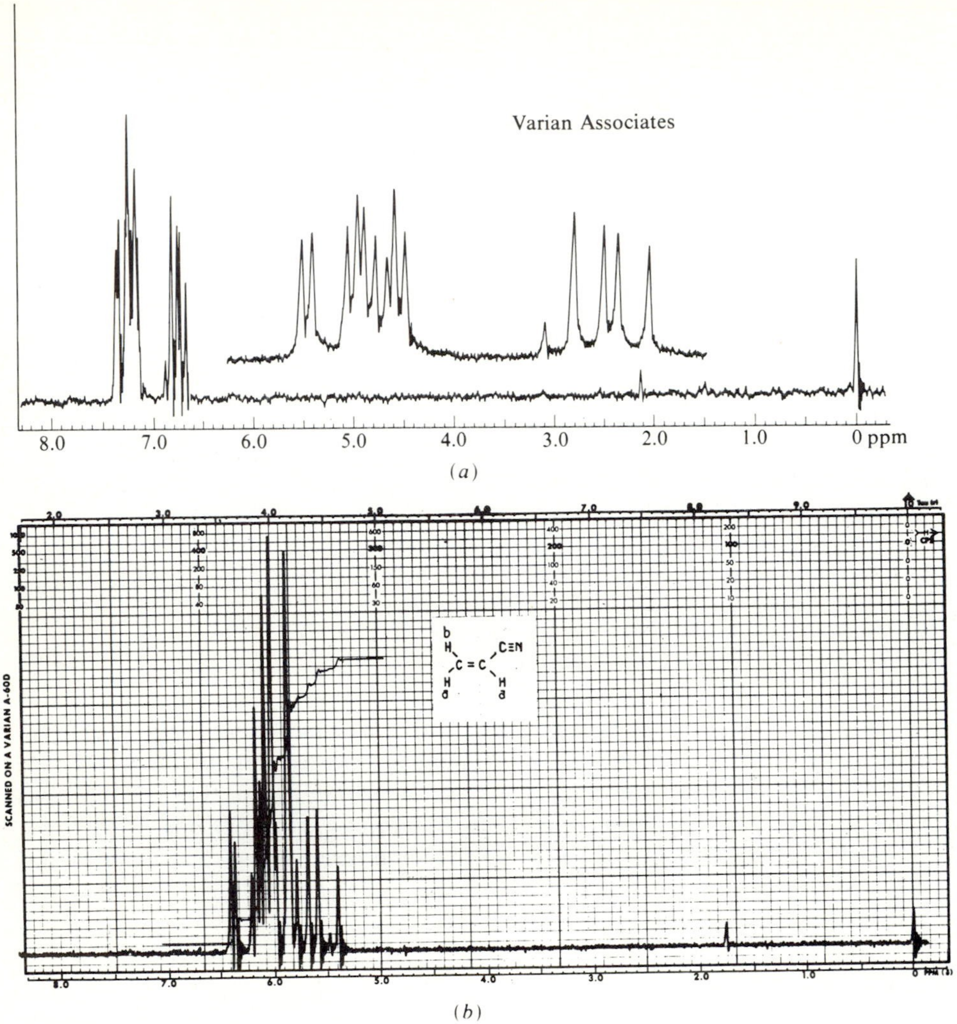

FIGURE 8.26
Experimental NMR spectra of (*a*) 2-iodothiophene[13] and (*b*) acrylonitrile[14]. © *Sadtler Research Laboratories, Division of Bio-Rad Laboratories, Inc. (1973).*

This results in a pair of cubic equations which must be solved to obtain the energies and the resulting transitions. At this point the usual procedure is to employ a high-speed computer and vary the σ_N and J_{NM} parameters to match the calculated spectra to the observed spectra.

The form of the energy matrix indicates that the wavefunctions for the system will involve mixtures of the two sets of states ψ_2, ψ_3, ψ_4, and ψ_5, ψ_6, ψ_7. Since there is a mixing of energy states one might expect that the true energy levels will not be exactly as shown by Eq. (8.193) but must be found by use of the variational procedure, which will also give the mixed spin functions. For states 2–4 the spin wavefunctions will be of the form $\psi_i = a_i\alpha\alpha\beta + b_i\alpha\beta\alpha + c_i\beta\alpha\alpha$

TABLE 8.6
Basic spin functions and diagonal matrix elements for the *ABC* system

State	Spin wavefunction	F_z	E_i, Hz
ψ_1	$\beta\beta\beta$	$-\frac{3}{2}$	$E_1 = E_A^\circ + E_B^\circ + E_C^\circ + \frac{1}{4}(J_{AB} + J_{BC} + J_{AC})$
ψ_2	$\beta\beta\alpha$	$-\frac{1}{2}$	$E_2 = E_A^\circ + E_B^\circ - E_C^\circ + \frac{1}{4}(J_{AB} - J_{BC} - J_{AC})$
ψ_3	$\beta\alpha\beta$	$-\frac{1}{2}$	$E_3 = E_A^\circ - E_B^\circ + E_C^\circ + \frac{1}{4}(J_{AB} - J_{BC} + J_{AC})$
ψ_4	$\alpha\beta\beta$	$-\frac{1}{2}$	$E_4 = -E_A^\circ + E_B^\circ + E_C^\circ + \frac{1}{4}(-J_{AB} + J_{BC} - J_{AC})$
ψ_5	$\beta\alpha\alpha$	$\frac{1}{2}$	$E_5 = E_A^\circ - E_B^\circ - E_C^\circ + \frac{1}{4}(-J_{AB} + J_{BC} - J_{AC})$
ψ_6	$\alpha\beta\alpha$	$\frac{1}{2}$	$E_6 = -E_A^\circ + E_B^\circ - E_C^\circ + \frac{1}{4}(-J_{AB} - J_{BC} + J_{AC})$
ψ_7	$\alpha\alpha\beta$	$\frac{1}{2}$	$E_7 = -E_A^\circ - E_B^\circ + E_C^\circ + \frac{1}{4}(J_{AB} - J_{BC} - J_{AC})$
ψ_8	$\alpha\alpha\alpha$	$\frac{3}{2}$	$E_8 = -E_A^\circ - E_B^\circ - E_C^\circ + \frac{1}{4}(J_{AB} + J_{BC} + J_{AC})$

and the corresponding matrix elements must be found by using these functions. Following the procedure used in Eqs. (8.158) through (8.164) one can develop the secular determinant for the ψ_2, ψ_3, ψ_4 set of states

$$\begin{vmatrix} (\alpha\alpha\beta|\mathcal{H}|\alpha\alpha\beta) - E & (\alpha\alpha\beta|\mathcal{H}|\alpha\beta\alpha) & (\alpha\alpha\beta|\mathcal{H}|\beta\alpha\alpha) \\ (\alpha\beta\alpha|\mathcal{H}|\alpha\alpha\beta) & (\alpha\beta\alpha|\mathcal{H}|\alpha\beta\alpha) - E & (\alpha\beta\alpha|\mathcal{H}|\beta\alpha\alpha) \\ (\beta\alpha\alpha|\mathcal{H}|\alpha\alpha\beta) & (\beta\alpha\alpha|\mathcal{H}|\alpha\beta\alpha) & (\beta\alpha\alpha|\mathcal{H}|\beta\alpha\alpha) - E \end{vmatrix} = 0 \qquad (8.194)$$

This is identical to the ψ_2, ψ_3, ψ_4 block of the secular determinants formed from the basic functions; thus for any system where the nuclei are all different the simple set of basic functions can be used to compute the energies. For a spin system where there are chemically alike nuclei this procedure will not be true and the basic functions must be developed by using the appropriate projection operators for the system as was done for the A_2X system.

The coefficients of the basic functions in the two-spin wavefunction are found by evaluation of the secular equations as was done for Eqs. (8.174) to (8.177). The selection rules and transition intensities will be determined from the P_{ij} elements as for the A_2X case. The selection rules are also stated by $\Delta F_Z = \pm 1$, but this alone does not give the relative intensities since the P_{nm} must be evaluated to get them. For example,

$$P_{27} \propto \langle\{a_2\alpha\alpha\beta + b_2\alpha\beta\alpha + c_2\beta\alpha\alpha\}|\gamma_A(\mathbf{I}_{+_A} + \mathbf{I}_{-_A}) + \gamma_B(\mathbf{I}_{+_B} + \mathbf{I}_{-_B}) + \gamma_C(\mathbf{I}_{+_C} + \mathbf{I}_{-_C})|\{a_7\alpha\beta\beta + b_7\beta\alpha\beta + c_7\beta\beta\alpha\}\rangle^2$$

$$P_{27} \propto \langle\{a_2\alpha\alpha\beta + b_2\alpha\beta\alpha + c_2\beta\alpha\alpha\}|\{\gamma_A(a_7\beta\beta\beta + b_7\alpha\alpha\beta + c_7\alpha\beta\alpha) + \gamma_B(a_7\alpha\alpha\beta + b_7\beta\beta\beta + c_7\beta\alpha\alpha) + \gamma_C(a_7\alpha\beta\alpha + b_7\beta\alpha\alpha + c_7\beta\beta\beta\}\rangle^2$$

$$P_{27} \propto (a_2b_7\gamma_A + a_2a_7\gamma_B + b_2c_7\gamma_A + b_2a_7\gamma_C + c_2c_7\gamma_B + c_2b_7\gamma_c)^2 \qquad (8.195)$$

We see that this is a mixed transition as contrasted to the pure X transition given by Eq. (8.192) for the A_2X case. Evaluation of all possible P_{ij} factors show that there will be a total of 15 allowed transitions where there were only 12 for the AMX case. Of these there will generally be three, those corresponding to combinations of states in the *AMX* case that are relatively weak and, depending on the coupling parameters, may escape detection.

Looking at the 2-iodothiophene spectrum we can see that its general appearance is that of three distorted quartets plus a couple of weak transitions. This is in good agreement with the application of the selection rules to the states given in Table 8.6. The situation with acrylonitrile is not as obvious but there are 13 discernible peaks. The single grouping of the spectrum relative to the dual groupings of the previous one indicates smaller chemical shifts or larger spin-spin couplings.

REFERENCES

A. Specific

1. Slichter, C. P., *Principles of Magnetic Resonance*, Harper and Row, New York, NY, 1963, chaps. 2 and 5.
2. Carrington, A., and A. D. McLachlan, *Introduction to Magnetic Resonance*, Harper and Row, New York, NY, 1967.
3. Bloch, F., *Phys. Rev.*, **70**, 460 (1946).
4. Fyfe, C. A., *Solid State NMR for Chemists*, C.F.C. Press, Guelph, Ont., 1983.
5. Harris, R. K., *Nuclear Magnetic Resonance Spectroscopy*, Pitman, Marshfield, MS, 1983, app. 4.
6. Ramsay, N. F., and E. M. Purcell, *Phys. Rev.*, **85**, 143 (1952).
7. Dirac, P. A. M., *The Principles of Quantum Mechanics*, 4th ed., Clarendon Press, Oxford, GB, 1967, chap. 9.
8. "Sadtler Standard Spectra", Sadtler Research Laboratories, Philadelphia, PA, No. 2429.
9. Ogg, R. A., Jr., and J. D. Ray, *J. Chem. Phys.*, **26**, 1339 (1957).
10. Ogg, R. A., Jr., and J. D. Ray, *J. Chem. Phys.*, **26**, 1516 (1957).
11. Pake, G. E., *J. Chem. Phys.*, **16**, 337 (1948).
12. Bhacca, N. S., L. F. Johnson, and J. N. Shoolery, *NMR Spectra Catalog*, Varian Associates, Palo Alto, CA, 1962, no. 2.
13. Bhacca, ibid, no. 49.
14. Sadtler, ibid, no. 14670.

General

Abragam, A., *The Principles of Nuclear Magnetism*, Clarendon Press, Oxford, England, 1961.

Abraham, R. J., and P. Loftus, *Proton and Carbon*-13 *NMR Spectroscopy*, Heydon & Sons Ltd., London, GB, 1978.

Becker, E. D., *High-Resolution NMR*, Academic Press, Orlando, FL, 1980.

Carrington, A., and A. D. MacLachlan, *Introduction to Magnetic Resonance*, Harper and Row, New York, NY, 1967.

Emsley, J. W., J. Feeney, and L. H. Sutcliffe, *High-Resolution NMR Spectroscopy*, vols. I, II, Pergamon Press, London, England, 1965, 1966.

Farrer, J. C., and E. D. Becker, *Pulse and Fourier Transform NMR*, Academic Press, Orlando, FL, 1971.

Fukushima, E., and S. B. W. Roeder, *Experimental Pulse NMR: A Nuts and Bolts Approach*, Addison-Wesley Publishing Co., 1981.

Griffith, J. S., *The Theory of Transition Metal Ions*, Cambridge University Press, Cambridge, England, 1961.

Gunther, H., *NMR Spectroscopy: An Introduction*, John Wiley and Sons, New York, NY, 1980.

Harris, R. K., *Nuclear Magnetic Resonance Spectroscopy*, Pitman, London, GB, 1983.

Hecht, H. G., *Magnetic Resonance Spectroscopy*, John Wiley and Sons, New York, NY, 1967.

Hill, H. A. O., and P. Day, *Physical Methods of Advanced Inorganic Chemistry*, Interscience Publishers, London, England, 1968.

Myers, R. J., *Molecular Magnetism and Magnetic Resonance Spectroscopy*, Prentice Hall, Englewood Cliffs, NJ, 1973.

Pople, J. A., W. G. Schneider, and H. J. Bernstein, *High-Resolution Nuclear Magnetic Resonance*, McGraw-Hill Book Co., New York, NY, 1959.

Slichter, C. P., *Principles of Magnetic Resonance*, Harper and Row, New York, NY, 1963.

PROBLEMS

8.1. Calculate the magnetic field necessary to observe resonances of the following nuclei in a 270 MHz spectrometer:
(*a*) ^1H (*d*) ^{31}P
(*b*) ^{19}F (*e*) ^{29}Si
(*c*) ^{13}C (*f*) ^{119}Sn

8.2. Calculate the relative populations, $N_{-1/2}/N_{1/2}$, in a field of 1.5 T at 300 K, for the following nuclei:
(*a*) ^{19}F (*c*) ^{31}P
(*b*) ^{13}C (*d*) ^{119}Sn

8.3. Since the state populations of spin systems follow the Boltzmann distribution they will be temperature-dependent. What temperature is necessary to produce a 10 percent difference between the spin state of (*a*) protons and (*b*) electrons if $H_0 =$ 1.5 T.

8.4. Calculate the relative transition probability, P_+/P_-, for the spin-lattice interaction of protons with a lattice at 80 K and in a field of 3 T.

8.5. Make a plot of $\nu_2 - \nu_1$ vs. θ for two protons located in a solid matrix at a separation of $2A$ and in a field of 2 T.

8.6. Show that $\mathbf{S}^2(\alpha_1\beta_1 + \beta_1\alpha_1) = 0$.

8.7. For the A_2 spin $I = \frac{1}{2}$ system show that the transitions $S_{-1} \rightarrow S_0$ and $S_0 \rightarrow S_1$ have finite values for their P_{nm} matrix elements.

8.8. Using the permutation operator evaluate the energies for the S_1 and A_0 states of an A_2 spin $I = \frac{1}{2}$ system. Verify the results by use of angular momentum matrix elements.

8.9. Calculate the relative intensities for the spin-spin multiplets in *HF*, an *AX* type system. Note this involves more than just the ratio of the dipole matrix elements.

8.10. For the NMR *AMX* system construct an energy level diagram, indicate the allowed transitions, and make a line drawing for a typical spectrum for $J_{AM} < J_{AX} < J_{MX}$.

8.11. Develop the spin functions and energy level expressions for an A_3X system.

8.12. Develop the symmetrized spin functions for an *ABX* proton system.

8.13. Evaluate all P_{ij} for the A_2X system.

8.14. Evaluate all P_{ij} for the *AMX* system.

8.15. Sketch an NMR line spectrum for the molecule H_3PBF_3, considering all nuclei present.

8.16. For the *AB* spin $I = \frac{1}{2}$ system show that $(\alpha\beta|H|\beta\alpha) = (\beta\alpha|H|\alpha\beta)$.

8.17. For the *AB* spin $I = \frac{1}{2}$ system confirm the transition probability factors P_{nm}.

8.18. Develop the symmetrized spin functions and classify them according to their irreducible representations for an A_2X_2 system with $I_A = \frac{1}{2}$ and $I_X = \frac{1}{2}$.

8.19. (a) At 270 MHz what magnetic field (in tesla) will be necessary to observe proton ($g_H = 5.5845$) resonances.

(b) What advantages, if any, are to be had by using a 270 MHz spectrometer rather than a 60 MHz spectrometer for the observation of the proton resonance(s) in the following. (Use chemical intuition to determine the relative magnitudes of $\nu_0\delta$ and J for the different systems.)

(i)
$$CH_3CCl_2CCl_2\overset{\overset{\displaystyle O}{\|}}{C}-OH$$

(ii)

(iii)
$$\begin{array}{c} H_2C-CH_2 \\ | \quad\ \ | \\ H_2C-CH_2 \end{array}$$

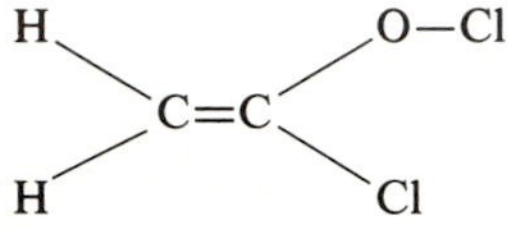

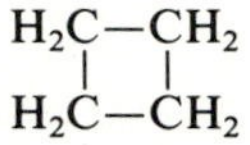

8.20. Sketch the expected proton ($I = \frac{1}{2}$) spectrum for 2,3,4,4-pentafluorocyclobutene given that

$$J_{HF} \text{ (crossing)} = 9.5 \text{ Hz}$$
$$J_{HF} \text{ (adjacent)} = 1.5 \text{ Hz}$$
$$J_{HF} \text{ (vinyl } F) = 6.9 \text{ Hz}$$
$$I(F) = \frac{1}{2}$$

8.21. At 60 MHz acetic acid gives to resonances, 602 Hz and 125 Hz downfield from TMS. These resonances belong to the carboxyl and methyl protons respectively.

(a) What conclusions can you draw concerning the chemical shielding of these two protons both relative to TMS and relative to each other.

(b) Calculate the chemical shift parameter δ for these two resonances.

8.22. A compound having two protons has an AB type spectrum with $\delta = 0.3$ ppm and $J_{AB} = 4$ Hz at 60 MHz. Construct a stick spectrum to indicate the relative positions and intensities of the lines in the spectrum of this compound at (a) 270 MHz and (b) 500 MHz.

CHAPTER 9

ELECTRON SPIN RESONANCE

9.1 INTRODUCTION

The phenomenon of magnetic resonance involving energy levels of an unpaired electron in a system is similar in many ways to that of NMR, but there are several additional types of interactions which must be considered. These interactions can be represented as additional terms in the Hamiltonian that describes the system. A complete discussion of the ESR (EPR) Hamiltonian and the determination of the resulting energy levels is beyond the scope of this book. We will, however, look at the more important terms in the case of some simple systems in order to provide a background from which the student can proceed to more detailed monographs.

For a single electron in an external magnetic field the energy levels and resulting transitions are given by an expression analogous to that for NMR

$$E = \frac{g\beta H_Z}{\hbar} \mathbf{S} \tag{9.1}$$

$$\nu = g\beta H_Z \tag{9.2}$$

where g is the electron g factor (= 2.002322 for a free electron) and β is the Bohr magneton. If all unpaired electrons in chemical systems behaved in this manner then, as for the case of NMR without a chemical shift, the subject would be of little use to the chemist since the only information that could be obtained would be the g factor. For the real situation the electron is located in an atom situated in a molecular and/or crystalline environment, and there are a variety of interactions that exist. We will first look at the physical nature of such interactions and then develop the total Hamiltonian. Following the presentation of the complete Hamiltonian, we will consider some simple systems where certain parts of the Hamiltonian can be ignored. It is possible to proceed in this manner since the predominant interactions are different for free radicals and paramagnetic ions in that one does not have the same interactions in solid and liquid systems.

9.2 THE TOTAL MAGNETIC HAMILTONIAN OF AN ELECTRON

The development of the total magnetic Hamiltonian will be restricted to the case of a single electron in a nonrotating, nonvibrating system. For a multielectron system, the primary difference will be the substitution of summations for single terms and the inclusion of electron-electron interactions in the potential. There are 11 major terms that must be included in the complete Hamiltonian of an electron located in a chemical system and subjected to an external magnetic field. Before examining these terms we will review some basic quantities and look at some properties of the chemical system.

The electron will be characterized by a spin angular momentum **S** and an orbital angular momentum **L**. Associated with these momenta will be magnetic moments $\boldsymbol{\mu}_S = -g\beta\mathbf{S}/\hbar$ and $\boldsymbol{\mu}_L = -g\beta\mathbf{L}/\hbar$. For each nucleus in the system, there will be an associated spin $\mathbf{I}_i$ and the accompanying magnetic moment $\boldsymbol{\mu}_{N_i} = g_{N_i}\beta_N\mathbf{I}_i/\hbar$.

Just as an electric field can be defined as the gradient of an electrostatic potential $\mathbf{E} = -\boldsymbol{\nabla} V$, the external magnetic field $\mathbf{H}_Z$ can be related to an electromagnetic potential $\mathbf{A}_Z$ by

$$\mathbf{H}_z = \boldsymbol{\nabla} \times \mathbf{A}_z \tag{9.3}$$

A nuclear moment μ_N will produce an effective magnetic field at the electron. The potential of this field will be

$$\mathbf{A}_N = \frac{\boldsymbol{\mu}_N \times \mathbf{r}}{r^3} \tag{9.4}$$

where r is the nuclear-electron distance. An atom or molecule can possess internal electric fields arising from the location and motion of its component charges. This field is denoted by **E**. The terms in the Hamiltonian follow. All terms are expressed in units of energy, and the momentum operators **I** and **S** contain the factor of $\hbar$, hence the $\hbar^{-1}$ factor is explicitly included in the equations.

1. *Electron kinetic energy*. This is the unperturbed term coming from the motion of the electron

$$-\frac{\hbar^2}{2m}\boldsymbol{\nabla}^2$$

2. *Internal potential.* This is the potential energy of the electron in the field of all of the other electrons and nuclei in the atom, ion, or molecule. It is denoted by V_0. Note that the sum of these first two terms is the Hamiltonian for a hydrogen-like atom.
3. *Crystal potential.* If the atom, ion, or molecule being observed is located in a solid environment there will be an electrostatic potential due to all of the charges surrounding the species of interest. This potential is represented by V_{xtal}. We can see that this term will be of no consequence for the study of free radicals in solution but could be quite appreciable in paramagnetic inorganic salts. The splittings of the electron energy levels due to such external potentials are referred to as *fine structure splittings.*
4. *Electron Zeeman energy.* This is the direct electron spin–magnetic field interaction. It is the direct interaction between the spin magnetic dipole of the electron and the external field,

$$E_{SH} = -\boldsymbol{\mu}_S \cdot \mathbf{H} = \frac{g\beta}{\hbar}\mathbf{S} \cdot \mathbf{H} \tag{9.5}$$

5. *Nuclear Zeeman energy.* This is the NMR Hamiltonian that has been discussed in Chap. 8. For a single nucleus it is

$$E_{IH} = \frac{g_N \beta_N (1 - \sigma)\mathbf{H} \cdot \mathbf{I}}{\hbar} \tag{9.6}$$

6. *Nuclear electric quadrupole energy.* For nuclei with $\mathbf{I} \geqq 1$ this term is of importance both as a perturbation of NMR and ESR spectra and as a pure absorption phenomenon. Since we will be concerning ourselves primarily with protons we will not discuss quadrupole perturbations in ESR. The subject of pure nuclear electric quadrupole resonance (NQR) will be treated in Chap. 10, and we will designate the term only as H_Q.

There are two types of interactions between nuclei and electrons that can give rise to small energy level splittings which in turn results in a hyperfine splitting of the observed spectra.

7. *Nuclear–electron spin coupling.* The first of these is the direct dipole-dipole coupling between the electron and nuclear moments. This interaction which is analogous to the classical coupling of two magnets and to the nuclear dipole-dipole coupling for NMR in solids is given by

$$E_{SI} = \frac{\boldsymbol{\mu}_S \cdot \boldsymbol{\mu}_N}{r^3} - \frac{3(\boldsymbol{\mu}_S \cdot \mathbf{r})(\mathbf{r} \cdot \boldsymbol{\mu}_N)}{r^5} = \frac{g\beta g_N \beta_N}{\hbar^2}\frac{\mathbf{S} \cdot \mathbf{I}}{r^3} - \frac{3(\mathbf{S} \cdot \mathbf{r})(\mathbf{r} \cdot \mathbf{I})}{r^5} \tag{9.7}$$

8. *Fermi contact term.* The second electron–nuclear interaction term is nonclassical in nature. It is the *Fermi contact interaction* and represents the energy of interaction between the nuclear moment and the magnetic field produced at the nucleus due to the electron spin. This interaction is limited to *s* electrons

since they are the only ones with a sufficient probability for being found within the nuclear volume. This interaction is given by

$$E_F = \left[\frac{8\pi}{3} g\beta g_N\beta_N |\psi(0)|^2\right]\frac{\mathbf{I}\cdot\mathbf{S}}{h^2} = \frac{a}{\hbar^2}\mathbf{I}\cdot\mathbf{S} \tag{9.8}$$

where $\psi(0)$ is the electronic wavefunction evaluated at the nucleus.

The remaining three terms involve interactions that are absent in a great many cases of interest but are of considerable importance in others, particularly in crystalline paramagnetic systems.

9. *Electron spin-orbit coupling.* The internal electric fields **E** of an atom, ion, or molecule can interact with the orbital angular momentum of an electron **L** to produce an internal Stark effect referred to as *orbital quenching.* The result of this interaction is to break down the coupling between the electron spin angular momentum and the electron orbital angular momentum. If this interaction between **L** and **S** is completely broken down, then the electron behaves as a free electron with $g = 2.002322$. This is the case for most organic free radicals and iron group salts. In many transition metal compounds, however, there is only partial quenching and the g factors range from 2 to 7 giving rise to complexities in the observed spectra. The quenching is a result of the application of the external magnetic field and is given by

$$E_{LS} = \frac{\beta}{mc\hbar}\mathbf{S}\cdot\left[\mathbf{E}X\left(\mathbf{L}+\frac{e}{c}\mathbf{A}_Z\right)\right] \tag{9.9}$$

10. *Nuclear moment-electron orbital coupling.* The electron orbital motion in the presence of the external magnetic field gives rise to an associated momentum defined by

$$\mathbf{P} = -i\hbar\nabla + \frac{e}{c}\mathbf{A}_Z \tag{9.10}$$

The magnetic moment associated with this momentum, $\mathbf{U} = \beta\mathbf{P}/\hbar$, couples with the magnetic field produced by the nuclear moment $\mathbf{A}_N$ to give an interaction

$$\mathbf{A}_N\cdot\mathbf{U} + \mathbf{U}\cdot\mathbf{A}_N = \frac{\beta}{\hbar}(\mathbf{A}_N\cdot\mathbf{P}\cdot\mathbf{A}_N) \tag{9.11}$$

11. *Electron orbital-magnetic field coupling.* The magnetic moment associated with the electron orbital angular momentum can interact with the external magnetic field to give a term

$$E_{LH} = \frac{\beta}{\hbar}(\mathbf{L}\cdot\mathbf{H}_Z + \mathbf{H}_Z\cdot\mathbf{L}) + \frac{\beta e}{\hbar c}\mathbf{A}_Z\cdot\mathbf{A}_Z \tag{9.12}$$

The Hamiltonian is the sum of these 11 terms. If the system is a molecule capable of undergoing rotation and vibrational also, the total Hamiltonian will include additional terms.

9.3 MAGNETIC INTERACTIONS IN ATOMS

This section will be concerned with magnetic interactions of atoms due to the application of an external magnetic field. Using the hydrogen atom as an example we will investigate the nature of some of the simpler terms in the Hamiltonian. For a free hydrogen atom, $V_{\text{xtal}} = 0$. Since $I = \frac{1}{2}$ for the proton, $H_Q = 0$. Terms 1 and 2 will be ignored since they relate to the field free electronic energy; we are concerned only with magnetic interactions. The electron being in an S orbital has no angular momentum so there will be no contribution from terms 9, 10, and 11. Since the dipole-dipole interaction (11) is dependent on the nuclear-electron distance this term must be averaged over the electron distribution. Due to the spherical symmetry of the s electron distribution, this term is zero for the hydrogen atom.

The remaining terms then give the magnetic Hamiltonian

$$\mathscr{H} = \frac{g\beta}{\hbar}\mathbf{H}_Z \cdot \mathbf{S} - \frac{g_N\beta_N}{\hbar}\mathbf{H}_Z \cdot \mathbf{I} + \frac{a}{\hbar^2}\mathbf{I} \cdot \mathbf{S} \tag{9.13}$$

Taking $\mathbf{H}$ to be in the Z direction, $\mathbf{H} \cdot \mathbf{S} = \mathbf{H}_Z\mathbf{S}_Z\hbar$ and $\mathbf{H} \cdot \mathbf{I} = \mathbf{H}_Z\mathbf{I}_Z\hbar$. For $a = 0$ there will be four energy levels as shown in Fig. 9.1. The transition probabilities will be given by

$$P_{lm} = \frac{\gamma^2 H_x^2}{h^2}\langle\psi_l|\mathbf{S}_X|\psi_m\rangle^2 S(\nu_{lm}, \nu) = \frac{\gamma^2 H_x^2}{4h^2}\langle\psi_l|\mathbf{S}_+ + \mathbf{S}_-|\psi_m\rangle^2 S(\nu_{lm}, \nu) \tag{9.14}$$

The nuclear spin term is omitted since the radiation frequency will be of a magnitude such that it cannot induce transitions between nuclear spin state and $\Delta M_I = 0$. The wavefunctions will be simple products of electron and nuclear spin functions. We have for $\beta_e\alpha_N \to \alpha_e\beta_N$

$$P_{13} = \frac{\gamma^2 H_x^2}{4h^2}\langle\beta_e\alpha_N|\mathbf{S}_+ + \mathbf{S}_-|\alpha_e\beta_N\rangle^2 S(\nu_{13}, \nu) \tag{9.15}$$

$$P_{13} = \frac{\gamma^2 H_x^2}{4h^2}\langle\beta_e|\mathbf{S}_-|\alpha_e\rangle^2\langle\alpha_N|\alpha_N\rangle^2 S(\nu_{13}, \nu) \tag{9.16}$$

$$P_{13} = \frac{\gamma^2 H_x^2}{4h^2} S(\nu_{13}, \nu) \tag{9.17}$$

The selection rules for the ESR transitions are then $\Delta M_S = \pm 1$, $\Delta M_I = 0$. Note that if the system we were considering had an $I = 0$ nucleus the NMR term in the Hamiltonian would vanish and we would have a two-level system. For the $a = 0$ case the two allowed transitions will have the same frequency, ν_0.

For the case where $a \neq 0$ the term $\mathscr{H}^{(1)} = (a/\hbar^2)\mathbf{I} \cdot \mathbf{S}$ is treated as a perturbation. This is justified by the fact that the observed hyperfine splitting in ESR spectra is of the order of a few gauss at total fields of several thousand gauss.

The first-order perturbation energies are found by averaging $\mathscr{H}^{(1)}$ over the ground state wavefunctions. For example, consider the E_2 level. The perturbing

Hamiltonian may be written as

$$\mathscr{H}^{(1)} = \frac{a}{\hbar^2}\mathbf{I}\cdot\mathbf{S} = \frac{a}{\hbar^2}(\mathbf{I}_X\mathbf{S}_X + \mathbf{I}_Y\mathbf{S}_Y + \mathbf{I}_Z\mathbf{S}_Z) \tag{9.18}$$

Transforming to stepping operators

$$\mathscr{H}^{(1)} = \frac{a}{\hbar^2}\left[\frac{(\mathbf{S}_+ + \mathbf{S}_-)(\mathbf{I}_+ + \mathbf{I}_-)}{4} + \frac{(\mathbf{S}_+ - \mathbf{S}_-)(\mathbf{I}_+ - \mathbf{I}_-)}{4} + \mathbf{I}_Z\mathbf{S}_Z\right] \tag{9.19}$$

the energy is found to be

$$\begin{aligned} E_2^{(1)} &= \frac{a}{\hbar^2}\langle e\beta_N|\,\frac{2\mathbf{S}_-\mathbf{I}_+ + 2\mathbf{S}_+\mathbf{I}_-}{4} + \mathbf{I}_Z\mathbf{S}_Z\,|\beta_e\beta_N\rangle \\ &= \frac{a}{\hbar^2}\left\{\frac{\langle\beta_e|\mathbf{S}_-|\beta_e\rangle\langle\beta_N|\mathbf{I}_+|\beta_N\rangle}{2} + \frac{\langle\beta_e|\mathbf{S}_+|\beta_e\rangle\langle\beta_N|\mathbf{I}_-|\beta_N\rangle}{2}\right. \\ &\quad \left. + \langle\beta_e|\mathbf{S}_Z|\beta_e\rangle\langle\beta_N|\mathbf{I}_Z|\beta_N\rangle\right\} \\ &= \frac{a}{\hbar^2}\left\{0 + 0 + \frac{\hbar^2}{4}\langle\beta_e|\beta_e\rangle\langle\beta_N|\beta_N\rangle\right\} = \frac{a}{4} \end{aligned} \tag{9.20}$$

The perturbation terms for the other three levels are found by an analogous

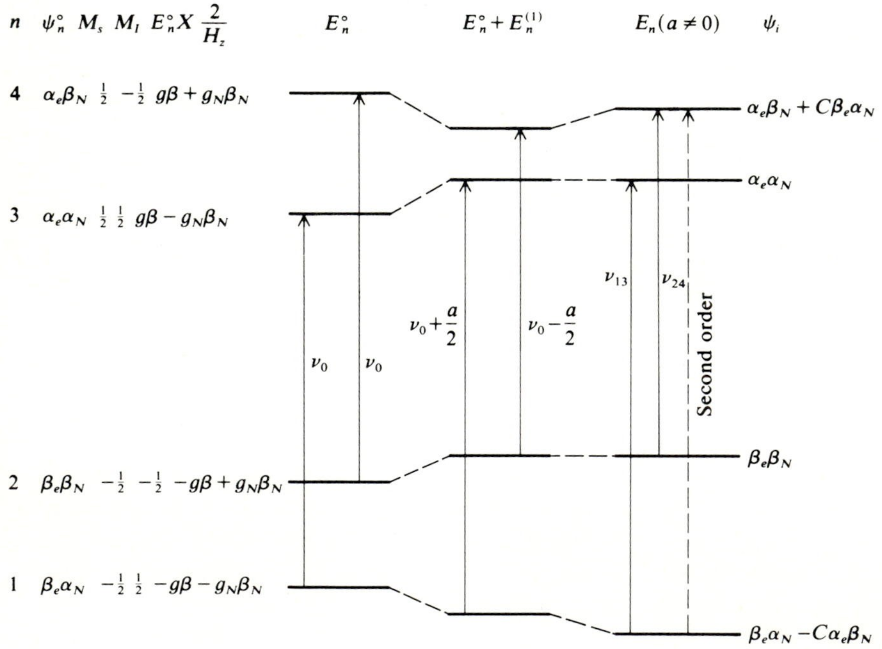

FIGURE 9.1
H atom magnetic energy levels.

calculation to be $E_1^{(1)} = -a/4$, $E_3^{(1)} = -a/4$, and $E_4^{(1)} = -a/4$. These new levels are also shown in Fig. 9.1.

Although the $I_XS_X + I_YS_Y$ term in Eq. (9.18) has no effect on the first-order perturbation energies they do cause a mixing of the ψ_1 and ψ_4 spin states and introduce a second-order energy correction. We next examine the values of the off-diagonal terms in the Hamiltonian matrix. For example

$$\begin{aligned}\langle \alpha_e\beta_N | \mathscr{H}^{(1)} | \beta_e\alpha_N\rangle &= \frac{a}{\hbar^2}\langle \alpha_e\beta_N | \frac{1}{2}(\mathbf{S}_+\mathbf{I}_- + \mathbf{S}_-\mathbf{I}_+) + \mathbf{I}_Z\mathbf{S}_Z | \beta_e\alpha_N\rangle \\ &= \frac{a}{2\hbar^2}\langle \alpha_e | \mathbf{S}_+ | \beta_e\rangle\langle \beta_N | \mathbf{I}_- | \alpha_N\rangle + \langle \alpha_e | \mathbf{S}_- | \beta_e\rangle\langle \beta_N | \mathbf{I}_+ | \alpha_N\rangle \\ &\quad + 2\langle \alpha_e | \mathbf{S}_Z | \beta_e\rangle\langle \beta_N | \mathbf{I}_Z | \alpha_N\rangle \\ &= \frac{a}{2\hbar^2}[\hbar^2 + 0 + 0] = \frac{a}{2}\end{aligned} \tag{9.21}$$

The $\langle \beta_e\alpha_N | \mathscr{H}^{(1)} | \alpha_e\beta_N\rangle$ term is equal to $a/2$ and all other off-diagonal ones are zero. The secular determinant then becomes

$$\begin{array}{c|cccc|}
 & \alpha_e\alpha_N & \alpha_e\beta_N & \beta_e\alpha_N & \beta_e\beta_N \\
\hline
\alpha_e\alpha_N & E_3^\circ + \frac{a}{4} - E & 0 & 0 & 0 \\
\alpha_e\beta_N & 0 & E_4^\circ - \frac{a}{4} - E & \frac{a}{2} & 0 \\
\beta_e\alpha_N & 0 & \frac{a}{2} & E_1^\circ - \frac{a}{4} - E & 0 \\
\beta_e\beta_N & 0 & 0 & 0 & E_2^\circ + \frac{a}{4} - E
\end{array} = 0 \tag{9.22}$$

where the spin functions are included for indexing. this secular equation may be solved directly or the second-order perturbation relation from Chap. 2 may be used to give

$$E_n = E_n^\circ + \langle \psi_n^\circ | \mathscr{H}^{(1)} | \psi_n^\circ\rangle - \sum_i{}' \frac{\langle \psi_n^\circ | \mathscr{H}^{(1)} | \psi_i^\circ\rangle\langle \psi_i^\circ | \mathscr{H}^{(1)} | \psi_n^\circ\rangle}{E_i^\circ - E_n^\circ} \tag{9.23}$$

We see that $E_2 = E_n^\circ + a/4$ and $E_3 = E_3^\circ + a/4$ are the same as for the first-order case but the other two levels are

$$E_1 = E_1^\circ - \frac{a}{4} - \frac{a^2}{4(E_4^\circ - E_1^\circ)} \tag{9.24}$$

$$E_4 = E_4^\circ - \frac{a}{4} + \frac{a^2}{4(E_4^\circ - E_1^\circ)} \tag{9.25}$$

The levels corrected to second-order are also shown in Fig. 9.1. Looking at all of the terms in the energy expressions we see that the energies of ψ_2 and ψ_3 are linear with respect to applied magnetic field while those for ψ_1 and ψ_4 contain a term that is inversely proportional to the field and thus are nonlinear. Keep in mind that the solution given by Eqs. (9.24) and (9.25) are valid only for $|E_n^\circ| \gg |a/4|$, and they will not reduce to the correct form at zero magnetic field where all of the E_i° approach zero.

To observe the situation at zero magnetic field one can set the $E_i^\circ = 0$ in the secular determinant and solve for the resulting energies

$$E_1(H_Z = 0) = -\frac{3a}{4} \tag{9.26}$$

$$E_2(H_Z = 0) = E_3(H_Z = 0) = E_4(H_Z = 0) = \frac{a}{4} \tag{9.27}$$

The mixing of states will have an effect on the selection rules. The wavefunctions for the mixed states are found by using perturbation theory

$$\psi_n = \psi_n^\circ - \sum_i{}' \frac{\langle \psi_i^\circ | \mathscr{H}^{(1)} | \psi_n^\circ \rangle}{E_i^\circ - E_n^\circ} \psi_i^\circ \tag{9.28}$$

Therefore

$$\psi_1 = \psi_1^\circ - \frac{a}{2(E_4^\circ - E_1^\circ)} \psi_4^\circ = \beta_e\alpha_N - c\alpha_e\beta_N \tag{9.29}$$

$$\psi_4 = \psi_4^\circ - \frac{a}{2(E_1^\circ - E_4^\circ)} \psi_1^\circ = \alpha_e\beta_N + c\beta_e\alpha_N \tag{9.30}$$

The probability for the transition $\psi_1 \rightarrow \psi_4$ for the rf field perpendicular to H_Z is

$$P_{14} = \frac{\gamma^2 H_x^2}{4h^2} [\langle(\beta_e\alpha_N - c\alpha_e\beta_N)|\mathbf{S}_+ + \mathbf{S}_-|(\alpha_e\beta_N + c\alpha_e\beta_N)\rangle]^2 S(\nu_{14}, \nu)$$
$$= \frac{\gamma^2 H_x^2}{4h^2} [\langle\beta_e\alpha_N|\mathbf{S}_-|\alpha_e\beta_N\rangle + \langle\beta_e\alpha_N|\mathbf{S}_+|c\beta_e\alpha_N\rangle - \langle c\alpha_e\beta_N|\mathbf{S}_-|\alpha_e\beta_N\rangle$$
$$- \langle c\alpha_e\beta_N|\mathbf{S}_+|\beta_e\alpha_N\rangle]^2 S(\nu_{14}, \nu) = 0 \tag{9.31}$$

However, if the radiation field has its magnetic vector parallel to the magnetic field H_z, then

$$P_{14} = \frac{\gamma^2 H_Z^2}{4h^2} [\langle(\beta_e\alpha_N - c\alpha_e\beta_N)|\mathbf{S}_Z|(\alpha_e\beta_N + c\beta_e\alpha_N)\rangle]^2 S(\nu_{14}, \nu)$$

$$P_{14} = \frac{\gamma^2 H_Z^2}{4h^2} \left[0 - \frac{c\hbar}{2} + 0 - \frac{ch}{2}\right]^2 S(\nu_{14}, \nu) = \frac{c^2\gamma^2 H_Z^2}{16\pi^2} S(\nu_{14}, \nu) \tag{9.32}$$

and there is an allowed second-order transition.

The general procedure just followed can be extended to more complicated systems where the size of the secular determinant is increased and additional terms in the Hamiltonian must be included. One particular point that needs to be mentioned before looking at a somewhat more complex system is the nature of the g factor. For a free electron the g factor is 2.002322 and for an isolated atom it is found from the Landé formula [1]. This reduces to $g = 2$

$$g = 1 + \frac{J(J+1) - L(L+1) + S(S+1)}{2J(J+1)} \tag{9.33}$$

for the free electron due to the fact that the Landé g factor does not contain a relative correction as does the factor $g = 2.002322$. For both of the cases just mentioned the systems have spherical symmetry and the g factor is isotropic. For a radical or paramagnetic ion in a solid this is not necessarily the case and the g factor will be anisotropic. In general, then, the g factor will be a second-rank tensor with nine components, analogous to the inertial tensor. (See App. L.) It will then be expressed as $\underline{\mathbf{g}}$ with components,

$$\begin{pmatrix} g_{XX} & g_{XY} & g_{XZ} \\ g_{YX} & g_{YY} & g_{YZ} \\ g_{ZX} & g_{ZY} & g_{ZZ} \end{pmatrix}$$

As with the inertial tensor, there will be a principal axis system in which the g tensor is diagonal. For an anisotropic g factor the basic electron spin Hamiltonian becomes

$$\mathscr{H}^0 = \frac{\beta}{\hbar} \mathbf{H} \cdot \underline{\mathbf{g}} \cdot \mathbf{S} \tag{9.34}$$

This same concept applies to nuclear magnetic resonance where the chemical shielding will in general be anisotropic and the basic Hamiltonian is[1]

$$\mathscr{H}_N^\circ = \frac{-g_N \beta_N}{h} \mathbf{H} \cdot (\underline{\mathbf{I}} - \underline{\boldsymbol{\sigma}}) \cdot \mathbf{I} \tag{9.35}$$

For high-resolution NMR in solution, one obtains an average scalar screening constant σ that is one-third of the trace of $\underline{\boldsymbol{\sigma}}$. This anisotropy in solids can be a useful phenomenon. By experimental location of the principal axis system of the g tensor, it is possible to obtain information pertaining to the orientation of the radical or ionic species in the solid.

For the hyperfine contributions to the energy levels the same concept is true. The Fermi contact parameter, since it involves only s electrons, is isotropic, but the magnetic dipole-dipole term is anisotropic. In the general case the two effects are given by

$$\mathscr{H}^{(1)} = \frac{1}{\hbar^2} \mathbf{S} \cdot \underline{\mathbf{T}} \cdot \mathbf{I} \tag{9.36}$$

where

$$\underline{\mathbf{T}} = \underline{\mathbf{I}} a - g_N \beta \beta_N \left[\frac{\mathbf{g}}{r^3} - \frac{3(\mathbf{g} \cdot \mathbf{r})(\mathbf{r} \cdot \mathbf{g})}{r^5} \right] \tag{9.37}$$

---

[1] Note that in Eq. 9.35 $\underline{\mathbf{I}}$ denotes a unit matrix while $\mathbf{I}$ is a spin operator.

9.4 FREE RADICAL ESR SPECTRA

The investigation of free radicals in solution is a useful technique for gaining insight about the mechanisms of the chemical reactions in which they are generated and for obtaining information regarding the electron distribution and hence, the structure of the radical. For radicals in solution, the magnetic part of the one electron Hamiltonian will contain only three parts

$$\mathscr{H} = \frac{g\beta}{\hbar}\mathbf{H}\cdot\mathbf{S} - \sum_i \frac{\gamma_i}{\hbar}(1-\sigma_i)\mathbf{H}\cdot\mathbf{I}_i + \sum_i \frac{a_i}{\hbar^2}\mathbf{I}_i\cdot\mathbf{S} \tag{9.38}$$

where the summation is over the nuclei in the system. The anisotropic part of the hyperfine tensor $\underline{\mathbf{T}}$, that is, the dipole-dipole interactions, will average to zero for liquids where the molecules are in rapid motion. The hydrogen atom just discussed was a special case of this Hamiltonian where $N = 1$. Since NMR transitions occur at frequencies lower than those due to ESR by a factor of approximately 2000 we will never be experimentally directly observing both type transitions simultaneously. There are, however, double-resonance type experiments where the system is irradiated at a frequency corresponding to one type of interaction and the observation made at that due to the other. Ignoring the NMR term the effective Hamiltonian becomes

$$\mathscr{H} = \frac{g\beta}{\hbar}\mathbf{H}\cdot\mathbf{S} + \sum_i \frac{a_i}{\hbar^2}\mathbf{I}_i\cdot\mathbf{S} \tag{9.39}$$

The observation of the ESR spectra of radicals can thus provide information, via the a_i factors, regarding the fraction of time the unpaired electron is associated with the various nuclei of the system.

Before examining some typical radical spectra we need to point out some additional considerations that are not implicit in the Hamiltonian. In addition to the averaging of the direct dipole-dipole interaction to a value much less than the hyperfine interaction, it is also necessary that there be a suitable relaxation mechanism present so that the lifetime of the upper state is large compared to the reciprocal of the half-width of the hyperfine transitions. If the relaxation is too rapid, then the line will be broadened to the point of obscuring the hyperfine structure. There are also two experiment conditions which must be satisfied for observation of hyperfine splittings of ESR spectra: (1) the magnitude of the magnetic component of the irradiation field H'_x must be much less than the hyperfine component separations; (2) the inhomogenities in the magnetic field $\mathbf{H}_Z$ must be appreciably less than the line widths of the hyperfine components.

It must be recognized that a complete description of the electron density distribution in a radical is generally not possible. This is illustrated by consideration of the radical ion $[(SO_3)_2NO]^{2-}$. Figure 9.2 [2] shows the spectrum and the ESR energy levels of this ion will be represented by

$$E = g\beta H_Z M_S + \sum_i \frac{a_i}{\hbar^2}\mathbf{I}_i\cdot\mathbf{S} \tag{9.40}$$

For ^{32}S and ^{16}O the nucleus spins are zero, hence the only coupling for which

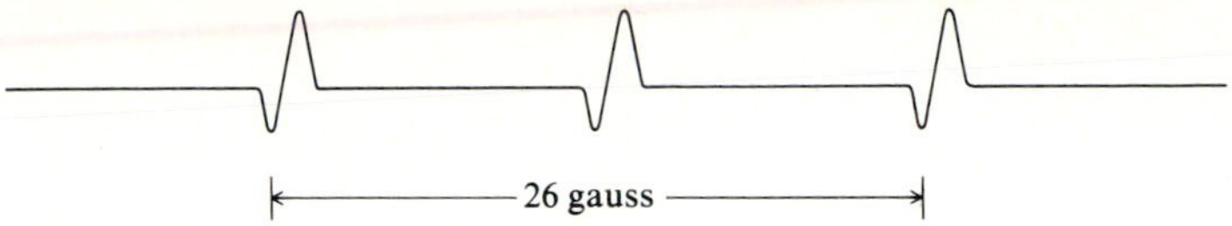

FIGURE 9.2
ESR spectrum of the $[(SO_3)_2NO]^{2-}$ ion.

there will be observable hyperfine spectra will be with the ^{14}N atom. Since the spin of ^{14}N is $I = 1$, $M_I = 0, \pm 1$, and M_S can have values of $\pm\frac{1}{2}$, the energy level diagrams and allowed transitions are as shown in Fig. 9.3. Actually the ESR spectrum in this case provides a minimum of information since the absence of a nuclear moment for sulfur and oxygen precludes any observation of the coupling to these atoms.

Another commonly observed type of system is one in which the unpaired electron density is distributed equally among several protons in the system. A simple example of a system of this type is the methyl radical [3] CH_3 or the benzene anion [4] $C_6H_6^-$. The simulated spectra of these two species are shown in Fig. 9.4. In both cases a very symmetrical spectrum exists. The set of energy levels that gives rise to the CH_3 spectrum is given by

$$E = g\beta H_Z M_S + \sum_{i=1}^{3} a_i M_S M_{I_i} = E^\circ_{\pm 1/2} \pm \frac{a}{4} \pm \frac{a}{4} \pm \frac{a}{4} \tag{9.41}$$

and are diagrammed in Fig. 9.5. The selection rules are $\Delta M_I = 0$ and $\Delta M_S = \pm 1$ so we find that there are four transitions allowed. Since two of the transitions arise from energy levels that are triply degenerate, they will have triple the intensity of the other two. The spectrum will then consist of a set of four equally spaced lines having an intensity ratio of 1:3:3:1. This is in agreement with the spectrum of CH_3 shown in Fig. 9.4.

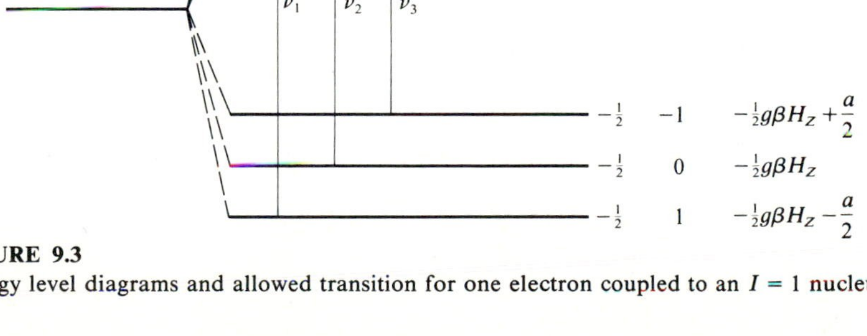

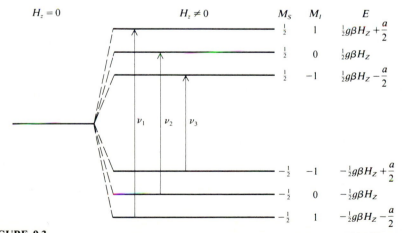

FIGURE 9.3
Energy level diagrams and allowed transition for one electron coupled to an $I = 1$ nucleus.

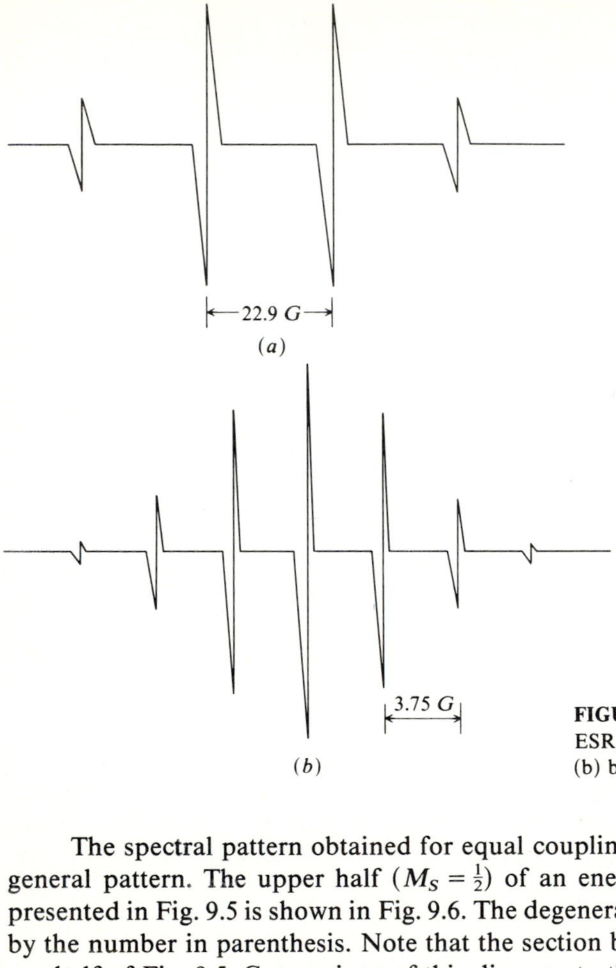

FIGURE 9.4
ESR spectra of (a) methyl radical and (b) benzene anion.

The spectral pattern obtained for equal coupling to three spins is part of a general pattern. The upper half ($M_S = \frac{1}{2}$) of an energy level diagram like that presented in Fig. 9.5 is shown in Fig. 9.6. The degeneracy of the levels is indicated by the number in parenthesis. Note that the section below I_3 corresponds to the top half of Fig. 9.5. Comparison of this diagram to the previous one shows that the relative intensities are numerically equal to the degeneracies of the levels. These numbers, incidentally, are the coefficients of the binomial series and can be obtained by a simple addition procedure

					1
				1	
				1	1
			1	1	
			1	2	1
		1	2	1	
		1	3	3	1
	1	3	3	1	
	1	4	6	4	1
1	4	6	4	1	
1	5	10	10	5	1

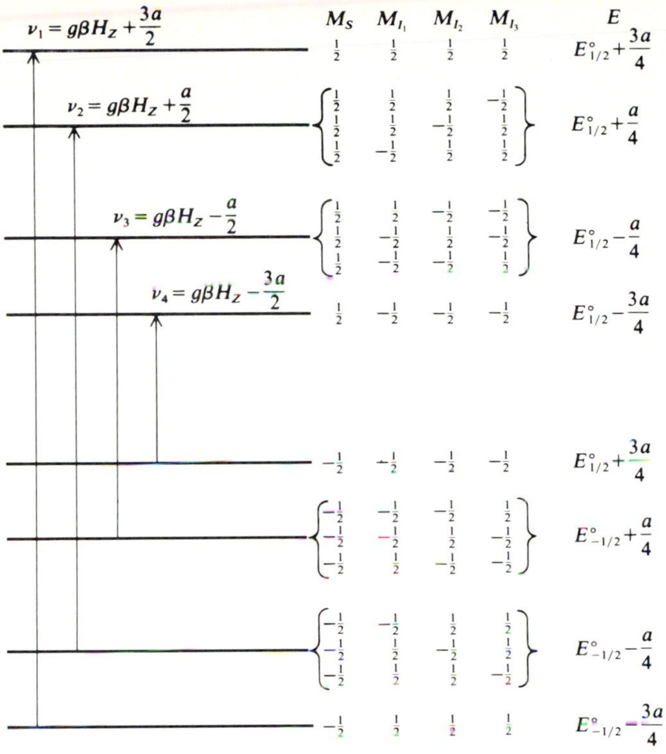

FIGURE 9.5
Energy levels and transitions for one electron coupled to three protons.

the benzene anion spectrum in Fig. 9.4 corresponds to the I_6 case in Fig. 9.6. The analysis of the spectra for these two cases shows that for CH_3, $a = 23$ gauss and for $C_6H_6^-$, $a = 3.76$ gauss.

The use of ESR for structure determination is illustrated by further consideration of the CH_3 radical. Although the isotropic coupling constants for the three protons have been obtained the electron density on the carbon is unknown since for ^{12}C, $I = 0$. An investigation of the ESR spectrum [5] of the $^{13}CH_3$ radical has found the coupling constant for the carbon to be $a = 41$ gauss. If the unpaired electron were to be in the carbon 2s orbital, which would result in a pyramidical configuration with the three hydrogen bonded through three p orbitals or through three sp^3 orbitals, then the coupling constant should be in excess of 1000. The comparatively low value of $a = 41$ gauss indicates that the unpaired electron is predominantly in a p orbital, hence the bonding is most likely sp^2 hybridization and the radical is planar.

The analysis of the $^{13}CH_3$ spectrum presents an example of equal coupling of an electron with the nuclei in a system. The energy levels of a four-spin system

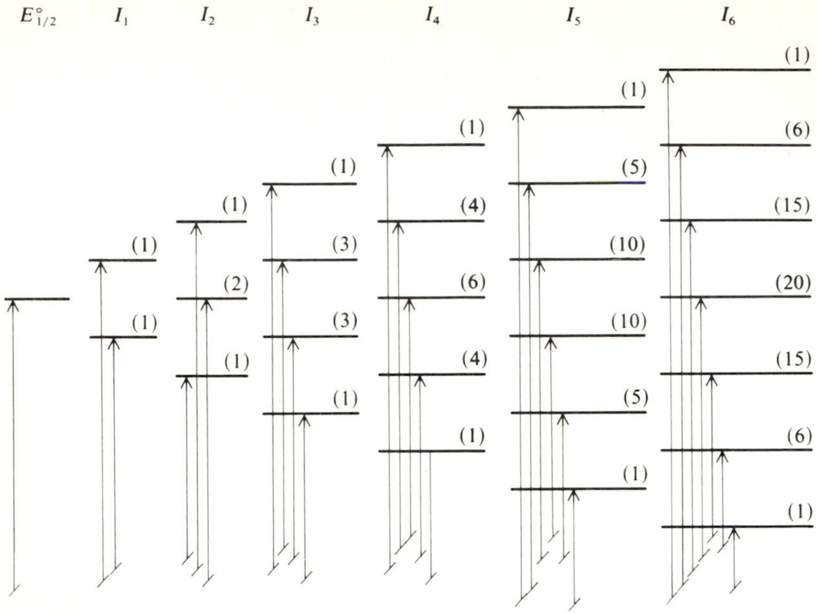

FIGURE 9.6
Energy levels for equal coupling of one electron to each member of a set of protons. (The diagram illustrates equal interaction of the electron with progressively more protons.)

where $I_1 = I_2 = I_3 = I_4 = \frac{1}{2}$ and $a = a_1 = a_2 = a_3 \neq a_4$ will be given by

$$E = E^\circ_{\pm 1/2} \pm \frac{a}{2} \pm \frac{a}{2} \pm \frac{a}{2} \pm \frac{a_4}{2} \tag{9.42}$$

and is shown in Fig. 9.7. Application of the normal selection rules shows that the resulting spectrum will be appreciably altered from that of a system with four equivalent coupling constants.

One final example depicting unequal coupling is afforded by the compound diphenylpicrylhydrazyl (DPPH), a stable radical. This example illustrates the case of both unequal coupling and $I > \frac{1}{2}$. The experimental spectrum of DPPH consists of a closely spaced set of five transitions having a relative intensity distribution of 1:2:3:2:1. Although this is a simple symmetrical pattern it does not have the correct intensity distribution to be an equal coupling to a set of four protons. The structure of DPPH is

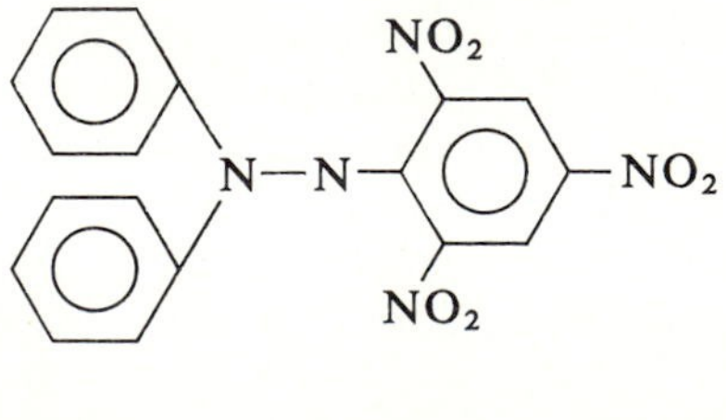

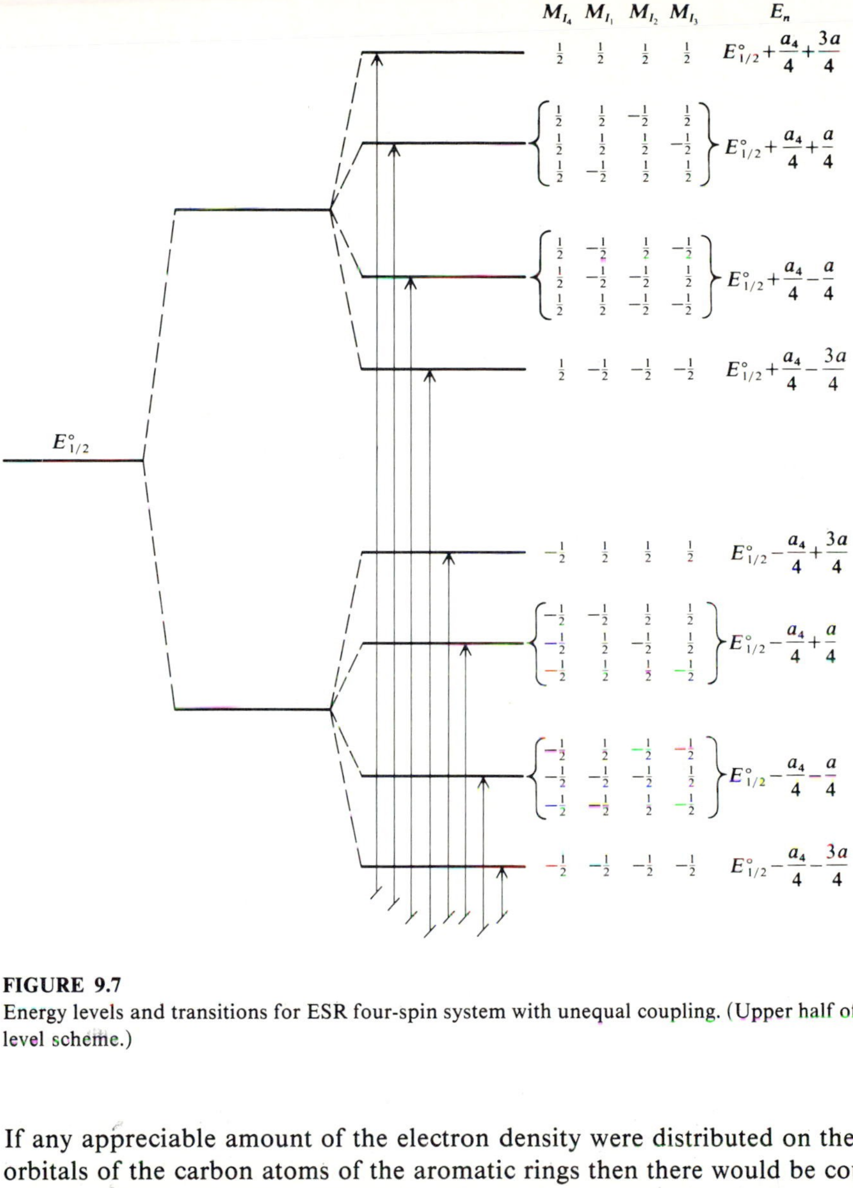

FIGURE 9.7
Energy levels and transitions for ESR four-spin system with unequal coupling. (Upper half of energy level scheme.)

If any appreciable amount of the electron density were distributed on the two *s* orbitals of the carbon atoms of the aromatic rings then there would be coupling to the ring protons also. The absence of any observed proton coupling indicates that the unpaired electron is either localized on the nitrogen and/or distributed through the π framework of the systems. If the electron spent any time in the π system of the picryl unit then it should exhibit some coupling to the NO_2 nitrogens. This would produce a rather more complicated spectrum than is observed. It is then possible to arrive at a preliminary conclusion that the electron is localized on the hydrazyl nitrogen or associated with the phenyl π systems. The latter interaction is not detectable due to the $I = 0$ of ^{12}C. For coupling with two

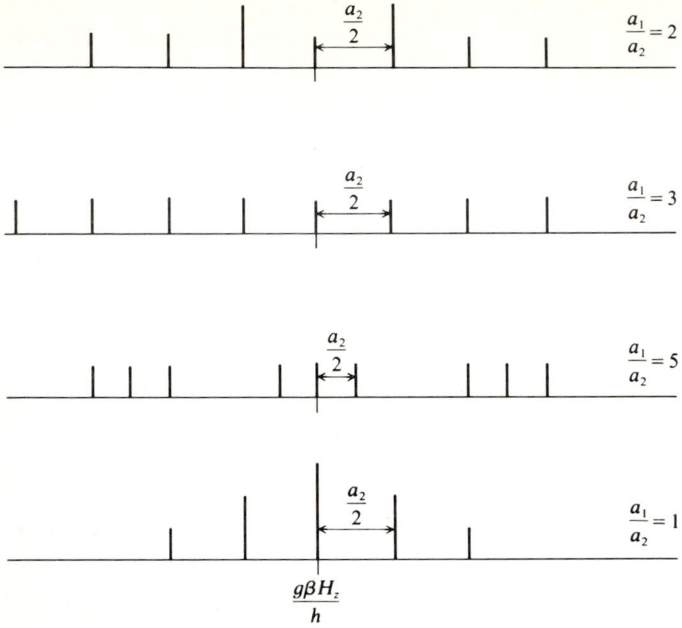

FIGURE 9.8
Spectra for coupling of one electron with two $I = 1$ nuclei.

nitrogens, $I = 1$, the energies are given by

$$E = E^{\circ}_{\pm 1/2} + \frac{a_1}{2} M_{I_2} + \frac{a_2}{2} M_{I_2} \tag{9.43}$$

For $\Delta M_I = 0$, the frequency of the transition will be given by

$$\nu = \frac{\Delta E}{h} = \frac{E^{\circ}}{h} + \frac{a_1}{2h} M_{I_1} + \frac{a_2}{2h} M_{I_2} \tag{9.44}$$

where $E^{\circ} = g\beta H_Z$. Expected spectral patterns of various ratios of a_1/a_2 are shown in Fig. 9.8. The relative intensities are shown by the height of the lines. The experimental spectrum agrees well with the case of equal coupling of the electron with the two hydrazyl nitrogens.

REFERENCES

A. Specific

1. Slater, J. C., *Quantum Theory of Atomic Structure*, McGraw-Hill, New York, NY, 1960, p. 249.
   Stratton, J. A., *Electromagnetic Theory*, McGraw-Hill, New York, NY, 1941, ch. I.

2. Townsend, J., S. I. Weissman, and G. E. Pake, *Phys. Rev.*, **89**, 606 (1953).
3. Jen, C. K., S. N. Foner, E. L. Cochran, and V. A. Brevera, *Phys. Rev.*, **112**, 1169 (1958).
4. Tuttle, T. R., and S. I. Weissman, *J. Am. Chem. Soc.*, **80**, 4549 (1958).
5. Cole, L. T., D. E. Pritchard, N. Davidson, and H. M. McConnell, *Mol. Phys.*, **1**, 406 (1958).

General

Carrington, A., *Microwave Spectroscopy of Free Radicals*, Academic Press, London, GB, 1974.
Carrington, A., and A. D. McLachlan, *Introduction to Magnetic Resonance*, Harper and Row, New York, NY, 1967.
McWeeny, R., *Spins in Chemistry*, Academic Press, Orlando, FL, 1970.
Pake, G. E., *Paramagnetic Resonance*, Benjamin, New York, NY, 1962.
Pool, C. P., Jr., *Electron Spin Resonance*, 2d ed., Wiley-Interscience, New York, NY, 1983.
Wertz, J. E., and J. R. Bolton, *Electron Spin Resonance: Elementary Theory and Practical Applications*, McGraw-Hill, New York, NY, 1972.

PROBLEMS

9.1. Show that $P_{24} \neq 0$ for the ESR levels of the hydrogen atom.

9.2. Evaluate the $E_4^{(1)}$ ESR energy level of the hydrogen atom when $a \neq 0$.

9.3. Evaluate the ESR matrix element $\langle \beta_e \alpha_N | \mathcal{H}^{(1)} | \alpha_e \beta_N \rangle$ for the hydrogen atom.

9.4. Show that for the hydrogen atom the ESR matrix element $\langle \alpha_e \alpha_n | \mathcal{H}^{(1)} | \alpha_e \beta_N \rangle$ is zero.

9.5. Plot the variation of the ESR energy levels vs. external magnetic field for the hydrogen atom. Plot from 00.2 T. Determine the tangents for the ψ_1 and ψ_4 curves and show that their intercepts on the $H_Z = 0$ axis is just the negative of the intercepts of the ψ_2 and ψ_3 curves. ($a/h = 1423$ MHz.)

9.6. Calculate the first-order energy levels (in megahertz) and transition frequencies for a hydrogen atom in a magnetic field of 1.5 T for the case of (*a*) $a = 0$ and (*b*) $a/h = 1423$ MHz.

9.7. When a liquid ethane is irridiated at 90 K a spectrum of 12 lines as shown schematically below is found. From this spectrum deduce the nature of the structure of the C_2H_5 radical and evaluate the hyperfine interaction constants [R. W. Fessenden and R. H. Schuler, *J. Chem. Phys.*, **34**, 2147 (1963)].

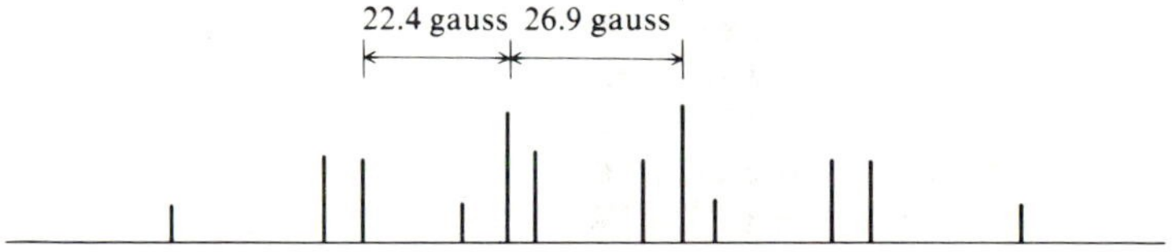

FIGURE P9.7

9.8. Construct the expected ESR spectrum for the $^{13}C^1H_3$ radical if $a_{^{13}C} = 41$ gauss and $a_H = 23$ gauss. For ^{13}C, $I = \frac{1}{2}$.

9.9. Below is given the ESR spectrum of a radical species formed by irridiation of CH_3CH_2Br with UV light. Predict the radical species formed and calculate the coupling constants.

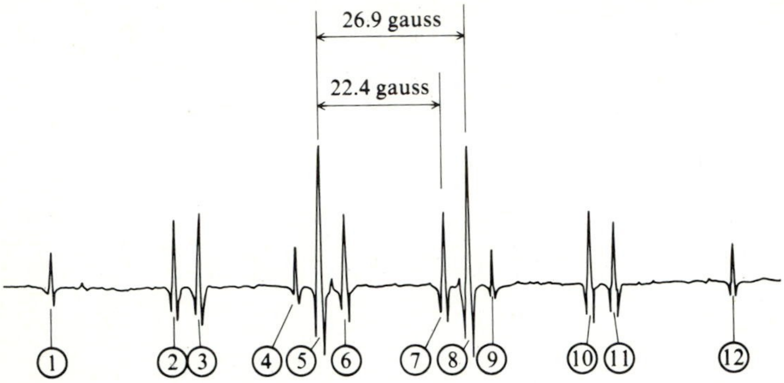

FIGURE P9.9

CHAPTER 10

NUCLEAR QUADRUPOLE INTERACTIONS

10.1 INTRODUCTION

The electrostatic interaction of a nuclear electric quadrupole moment and the electron charge cloud surrounding the nucleus can give rise to the observation of pure nuclear quadrupole resonance (NQR) spectra in solids, hyperfine splitting (HFS) of pure rotational spectra, and line broadening and splitting of nuclear magnetic resonance spectra. By use of information determined by these methods it is possible to obtain parameters which are related to the electron densities in molecules. In this chapter we will examine the basic interactions, develop suitable expressions for observed transitions, and then examine the areas of pure NQR of solids.

10.2 THE QUADRUPOLAR HAMILTONIAN

The interaction between the quadrupole moment of a nucleus and the surrounding electric field is different in nature from the direct charge-charge type interaction with which we are accustomed to dealing. For that reason we will first develop the classical Hamiltonian for such systems. This development can follow either of two methods, both of which involve the expression of the electrostatic interactions in terms of a multipole expansion. One method is to use an expansion in terms of Cartesian coordinates while the other uses an expansion based on the spherical harmonic functions. Although somewhat more complicated we will examine this latter method since the final form is more readily related to the quantum mechanical formulation.

The classical interaction of a set of nuclear charge points e_i, with a set of surrounding charge points e_j, is the product of the former multiplied by the potential due to the latter and summed over all pairs of points

$$H = \sum_i \sum_j e_i x \frac{e_j}{r_{ij}} \tag{10.1}$$

where r_{ij} is as shown in Fig. 10.1. In terms of the coordinates of the individual particles one can write

$$r_{ij}^2 = r_i^2 + r_j^2 - 2r_i r_j \cos \theta_{ij} \tag{10.2}$$

Therefore

$$H = \sum_i \sum_j e_i e_j [r_i^2 + r_j^2 - 2r_i r_j \cos \theta_{ij}]^{-1/2} \tag{10.3}$$

This is rearranged to give

$$H = \sum_i \sum_j \frac{e_i e_j}{r_j} \left[1 - 2\left(\frac{r_i}{r_j}\right) \cos \theta_{ij} + \left(\frac{r_i}{r_j}\right)^2 \right]^{-1/2} \tag{10.4}$$

The nuclear coordinate r_i will be approximately 10^{-14} m while the electron coordinate r_j will be approximately 10^{-10} m, therefore $(r_i/r_j) \ll 1$. For $y < 1$

$$\frac{1}{(1 - 2xy + y^2)^{1/2}} = \sum_{l=0}^{\infty} y^l P_l(x) \tag{10.5}$$

where $P_l(x)$ are the Legendre polynomials [1]. Hence,

$$H = \sum_i \sum_j \frac{e_i e_j}{r_j} \sum_{k=0}^{\infty} \left(\frac{r_i}{r_j}\right)^k P_k (\cos \theta_{ij}) \tag{10.6}$$

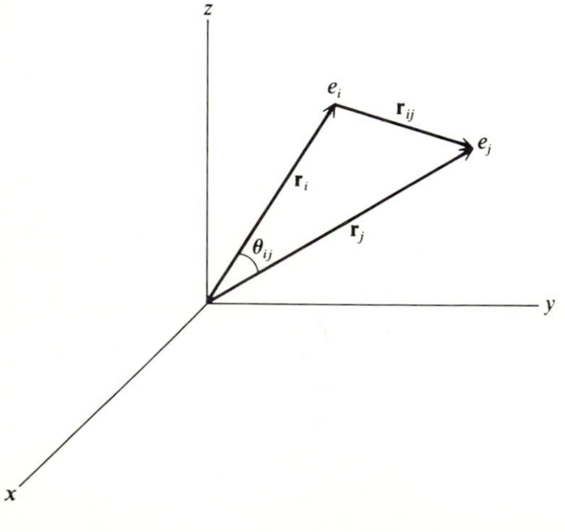

FIGURE 10.1
Nuclear-electron coordinates.

Using the properties of the Legendre polynomials given in App. F this becomes

$$H = \sum_i \sum_j \frac{e_i e_j}{r_j} \sum_{k=0}^{\infty} \left(\frac{r_i}{r_j}\right)^k \sum_{q=-k}^{k} (-1)^q \left(\frac{4\pi}{2k+1}\right) Y_k^{-q}(\theta_i\phi_i)\, Y_k^q(\theta_j\phi_j) \qquad (10.7)$$

where the $Y_k^{\pm q}$ are the spherical harmonics. Looking at the first few terms in this expansion one has

$$H = \sum_i \sum_j \frac{e_i e_j}{r_j} \left[P_0(\cos\theta_{ij}) + \left(\frac{r_i}{r_j}\right) P_1(\cos\theta_{ij}) + \left(\frac{r_i}{r_j}\right)^2 P_2(\cos\theta_{ij}) + \left(\frac{r_i}{r_j}\right)^3 P_3(\cos\theta_{ij}) + \cdots \right] \qquad (10.8)$$

We will now examine the individual terms in this expansion.

For $k = 0$

$$H_0 = \sum_i \sum_j \left(\frac{4\pi e_i e_j}{r_j}\right) Y_0^0(\theta_i\phi_i)\, Y_0^0(\theta_j\phi_j) = \sum_i \sum_j \frac{e_i e_j}{r_j} \qquad (10.9)$$

For a set of charges e_j the potential at a point is given by

$$V = \sum_j \frac{e_j}{r_j} \qquad (10.10)$$

and for a set of charges e_i the total charge is

$$Ze = \sum_i e_i \qquad (10.11)$$

hence

$$H_0 = ZeV \qquad (10.12)$$

which is the Coulombic interaction between a nucleus and the surrounding electrons. If one considers the nucleus to be composed of a continuous charge distribution denoted by $\rho(xyz)$ then

$$Ze = \int_{\text{nuc}} \rho(xyz)\, dx\, dy\, dz \qquad (10.13)$$

For $k = 1$

$$H_1 = -\frac{4\pi}{3} \sum_i \sum_j \frac{e_i e_j r_i}{r_j^2} [Y_1^1(i)\, Y_1^{-1}(j) + Y_1^0(i)\, Y_1^0(j) + Y_1^{-1}(i)\, Y_1^1(j)] \qquad (10.14)$$

where the (i) and (j) symbols following the $Y_k^{\pm q}$ denote $\theta_i\phi_i$ and $\theta_j\phi_j$, respectively. Employing the spherical harmonic from Table 11.1 this becomes

$$H_1 = \sum_i \sum_j \frac{e_i e_j r_i}{r_j^2} (\cos\theta_i \cos\theta_j + \sin\theta_i \cos\phi_i \sin\theta_j \sin\phi_j - \sin\theta_i \sin\phi_i \sin\phi_j \cos\phi_j) \qquad (10.15)$$

Looking at the sum over the nuclear charges we find that we must evaluate

$$\sum_i e_i r_i \cos\theta_i = \sum_i e_i z_i = \int_{\text{nuc}} \rho(xyz) z \, dx\, dy\, dz \tag{10.16}$$

$$\sum_i e_i r_i \sin\theta_i \cos\phi_i = \sum_i e_i x_i = \int_{\text{nuc}} \rho(xyz) x \, dx\, dy\, dz \tag{10.17}$$

$$\sum_i e_i r_i \sin\theta_i \sin\phi_i = \sum_i e_i y_i = \int_{\text{nuc}} \rho(xyz) y \, dx\, dy\, dz \tag{10.18}$$

These terms have the dimension of a dipole moment. The current state of nuclear theory indicates that the nucleus has cylindrical symmetry. In this case, the sums are all zero. Furthermore, we can see that any term which involves an integrand of the form $\rho(xyz)q^n(q = x, y, z,\ n = \text{odd integer})$ will be zero when integrated over symmetric limits. The zero character of this $k = 1$ term means physically that the nucleus does not possess an electric dipole moment. This is in keeping with current knowledge in that this phenomenon has never been observed.

For $k = 2$

$$H_2 = \frac{4\pi}{5} \sum_i \sum_j \frac{e_i e_j r_i^2}{r_j^3}\ Y_2^2(i)\,Y_2^{-2}(j) + Y_2^1(i)\,Y_2^{-1}(j) + Y_2^0(i)\,Y_2^0(j)$$
$$+ Y_2^{-1}(i)\,Y_2^1(j) + Y_2^{-2}(i)\,Y_2^2(j) \tag{10.19}$$

Before looking at the completely expanded version of this equation let us examine one term so as to introduce some physical significance. Selecting the middle term we have

$$\frac{4\pi}{5} \sum_i \sum_j \frac{e_i e_j r_i^2}{r_j^3}\ Y_2^0(i)\,Y_2^0(j) = \frac{1}{2}\sum_j e_j\left(\frac{3\cos^2\theta_j - 1}{r_j^3}\right) \times \frac{1}{2}\sum_i e_i r_i^2(3\cos^2\theta_i - 1) \tag{10.20}$$

The term

$$\frac{1}{2}\sum_j e_j\left(\frac{3\cos^2\theta_j - 1}{r_j^3}\right) = \frac{1}{2}\sum_j e_j\left(\frac{3z_j^2 - r_j^2}{r_j^5}\right) \tag{10.21}$$

is the expression for the classical electric field gradient tensor component in the z direction and is denoted by

$$\frac{1}{2}\sum_j e_j\left(\frac{3z_j^2 - r_j^2}{r_j^5}\right) = \frac{\partial E(xyz)}{\partial z} \equiv \frac{\partial^2 V(xyz)}{\partial z^2} \equiv \nabla E_0 \tag{10.22}$$

where $E(xyz)$ and $V(xyz)$ are the electric field and the electric potential due to the extranuclear charges. For a classical charge distribution denoted by $e(xyz)$

$$\nabla E_0 = \int_0^\infty \int_0^\infty \int_0^\infty e(xyz)\frac{3z^2 - r^2}{r^5}\, dx\, dy\, dz \tag{10.23}$$

If one has an atomic or molecular electronic charge distribution which is related to a wavefunction ψ, then the expectation value of ∇E_0 is

$$\langle \nabla E_0 \rangle = \iiint e\psi^* \left(\frac{3\cos^2\theta - 1}{r^3} \right) \psi \, dv \tag{10.24}$$

The term

$$\frac{1}{2} \sum_i e_i r_i^2 (3\cos^2\theta_i - 1) = \frac{1}{2} \sum_i e_i (3z_i^2 - r_i^2) = Q_0 \tag{10.25}$$

is a summation over the charges in the nucleus or, if one considers the nucleus to be a continuous charge distribution given by a density function $\rho(xyz)$, then

$$Q_0 = \frac{1}{2} \int_{\text{nuc}} \rho(xyz)[3z^2 - r^2] \, dv \tag{10.26}$$

Looking at this term we see that it will be nonzero if $3z^2 \neq r^2$, that is, if the nucleus is not spherically symmetric. Q_0 is a component of the nuclear quadrupole moment tensor and is a measure of the deviation of the nuclear charge distribution from spherical symmetry. If $I \geqq 1$ then $Q_0 \neq 0$. These points will be explored further a little later. The energy of interaction due to terms of the type $\nabla E_0 \times Q_0$ is the nuclear quadrupole interaction energy.

Before we look in more detail at the nuclear quadrupole interaction let us consider two other points. (1) If we look at the product $\nabla E_0 \times Q_0$ we can see that it has the dimensions of $[(e_j/r_j^3) \times e_i r_i^2]$. Since $e_i = e_j = 4.8 \times 10^{-10}$ esu, $r_i \approx 10^{-12}$ cm and $r_j \approx 10^{-8}$ cm, we find that the magnitude of $\nabla E_0 \times Q_0$ is about 40 MHz. If the spacings between energy levels are of this order of magnitude then the study of the absorption of energy due to this phenomenon will involve the use of radiofrequency spectroscopy. (2) We have previously shown that all terms with k = (odd integer) in the expansion for H are zero. A general restriction on the k = (even integer) terms is that terms for which $k > 2I$ vanish [2]. Looking at the form of the $k = 4$ term in Eq. (10.8) we see that it will have a magnitude of the order of $e^2 e r_i^4 / r_j^5 \approx 1$ Hz. This is of such a magnitude that its observation is rare using current techniques.

Beginning with Eq. (10.19) and employing the spherical harmonics given in Table 11.1 the quadrupole Hamiltonian $\mathcal{H}_2$ may be written as a product of two tensors $\underline{\nabla \mathbf{E}}$ and $\underline{\mathbf{Q}'}$;

$$\mathcal{H}_Q \equiv \mathcal{H}_2 = \underline{\nabla \mathbf{E}} : \underline{\mathbf{Q}'} = \sum_{m=-2}^{2} (\nabla E)_2^{-m} Q_2'^{m} \tag{10.27}$$

where

$$(\nabla E)_2^{-m} = \sqrt{\frac{4\pi}{5}} \sum_j \frac{e_j}{r_j^3} Y_2^{-m}(\theta_j \phi_j) \tag{10.28}$$

$$Q_2'^{m} = \sqrt{\frac{4\pi}{5}} \sum_i e_i r_i^2 Y_2^m(\theta_i \phi_i) \tag{10.29}$$

In terms of continuous charge distributions these become

$$(\nabla E)_2^{-m} = \sqrt{\frac{4\pi}{5}} \int_{\text{elec}} \frac{e(xyz)}{r^3} Y_2^{-m}(\theta\phi)\, dv \tag{10.30}$$

$$Q_2'^m = \sqrt{\frac{4\pi}{5}} \int_{\text{nuc}} \rho(xyz) r^2 Y_2^m(\theta\phi)\, dv \tag{10.31}$$

Equation (10.27) can also be written in terms of Cartesian coordinates

$$\underline{\mathbf{\nabla E}} = \frac{1}{2} \sum_{k=1}^{3} \sum_{l=1}^{3} \nabla E_{kl} \hat{\mathbf{e}}_k \hat{\mathbf{e}}_l \tag{10.32}$$

$$\underline{\mathbf{Q}'} = \sum_{k=1}^{3} \sum_{l=1}^{3} Q'_{kl} \hat{\mathbf{e}}_k \hat{\mathbf{e}}_l \tag{10.33}$$

which the $\hat{\mathbf{e}}_k$ are unit Cartesian vectors and

$$\mathscr{H}_Q = \underline{\mathbf{\nabla E}} : \underline{\mathbf{Q}'} = \frac{1}{2} \sum_{k=1}^{3} \sum_{l=1}^{3} \nabla E_{kl} Q_{kl} \tag{10.34}$$

Since we have previously expressed $\mathscr{H}_Q$ in terms of two tensors, each of which contains only five terms, it is apparent that all nine of the ∇E_{kl} and all nine of the Q_{kl} cannot be independent.

We will next determine the interrelationships of these tensor components and formulate the matrix elements for the interaction energy. Initial discussions will consider the electric field at the nucleus as being due to a static distribution of charge. The case of a rotating molecule where the nucleus experiences an interaction with a time average field will be discussed in Chap. 14. The minimum set of independent tensor components is referred to as the set of irreducible components. The general treatment that follows is equally valid whether we are considering a collection of point charges or a continuous charge distribution.

The irreducible components of $\underline{\mathbf{\nabla E}}$ in terms of the spherical harmonics are found from Eq. (10.28) to be

$$(\nabla E)_2^0 = \sqrt{\frac{4\pi}{5}} \sum_j \frac{e_j}{r_j^3} Y_2^0(\theta_j \phi_j) = \frac{1}{2} \sum_j \frac{e_j}{r_j^3} (3 \cos^2 \theta_j - 1) \tag{10.35a}$$

$$(\nabla E)_2^{\pm 1} = \sqrt{\frac{4\pi}{5}} \sum_j \frac{e_j}{r_j^3} Y_2^{\pm 1}(\theta_j \phi_j) = \mp \sqrt{\frac{3}{2}} \sum_j \frac{e_j}{r_j^3} \sin \theta_j \cos \theta_j\, e^{\pm i\phi_j} \tag{10.35b}$$

$$(\nabla E)_2^{\pm 2} = \sqrt{\frac{4\pi}{5}} \sum_j \frac{r_j}{r_j^3} Y_2^{\pm 2}(\theta_j \phi_j) = \sqrt{\frac{3}{8}} \sum \frac{e_j}{r_j^3} \sin^2 \theta_j\, e^{\pm 2i\phi_j} \tag{10.35c}$$

The Cartesian components ∇E_{kl} can be determined by a comparison of the form of H_Q when expressed in Cartesian coordinates and the $(\nabla E)_2^{\pm m}$ terms also expressed in Cartesian coordinates. Looking in Eq. (10.8) we can write the quadrupole term as

$$\mathscr{H}_Q = \sum_i \sum_j \frac{e_i e_j r_i^2}{r_j^3} P_2(\cos \theta_{ij}) = \frac{1}{2} \sum_k \sum_l \nabla E_{kl} Q'_{kl} \tag{10.36}$$

It is useful at this point to introduce a new tensor defined as

$$Q_{kl} = 3Q'_{kl} - \delta_{kl} \sum_e Q'_{ee} \tag{10.37}$$

where the sum is over the diagonal elements. The reason for this transformation will become apparent later when the quadrupole moment tensor is examined in more detail. This newly defined tensor, in addition to being symmetric, as was $\underline{\mathbf{Q}}'$, is also traceless. The Hamiltonian can now be written

$$\mathscr{H}_Q = \frac{1}{2} \sum_k \sum_l \frac{1}{3} (Q_{kl} + \delta_{kl} \sum_e Q'_{ee}) \nabla E_{kl} \tag{10.38a}$$

$$\mathscr{H}_Q = \frac{1}{6} \sum_k \sum_l Q_{kl} \nabla E_{kl} + \frac{1}{6} [\sum_k \nabla E_{kk}][\sum_e Q'_{ee}] \tag{10.38b}$$

Since the quadrupole moment is an inherent property of the nucleus and is independent of nuclear orientation, $\underline{\mathbf{Q}}'$ is an invariant tensor and its trace is a constant. Since all of the charge that contributes to $\underline{\nabla \mathbf{E}}$ lies outside the volume of the nucleus the components of $\underline{\nabla \mathbf{E}}$ satisfy Laplace's equation [3],

$$\nabla E_{xx} + \nabla E_{yy} + \nabla E_{zz} = 0 \tag{10.39}$$

The two factors in the second part of Eq. (10.38b) are thus independent of nuclear orientation and the expression for the quadrupole Hamiltonian becomes

$$\mathscr{H}_Q = \frac{1}{6} \sum_k \sum_l \nabla E_{kl} Q_{kl} \tag{10.40}$$

To reduce the number of summations and indices necessary for bookkeeping we will use the Hamiltonian as it applies to continuous charge distributions

$$\mathscr{H}_Q = \frac{1}{2} \int_{\text{elec}} \int_{\text{nuc}} \frac{e(xyz)}{r_e^3} \rho(xyz) r_n^2 (3 \cos^2 \theta - 1)\, dv_e\, dv_n \tag{10.41}$$

Using the law of cosines we have the relationship

$$r_e r_n \cos \theta = \sum_i x_{ei} x_{ni} \tag{10.42}$$

where the sum is over the three Cartesian coordinates. Substitution of this relationship into Eq. (10.41) leads to

$$\mathscr{H}_Q \int_{\text{elec}} \int_{\text{nuc}} \left[\frac{e(xyz)}{r_e^5}\right] \rho(xyz) \left[\frac{3}{2} \sum_k \sum_l x_{nk} x_{nl} x_{ek} x_{el} - \frac{1}{2} r_n^2 r_e^2\right] dv_e\, dv_n \tag{10.43}$$

This equation can be written as a sum of products of two terms, one involving the nuclear coordinates and one involving the electronic coordinates. Thus the

Hamiltonian has the same form as before

$$\mathscr{H}_Q = \frac{1}{6} \sum_{k=1}^{3} \sum_{l=1}^{3} Q_{kl} \nabla E_{kl} \tag{10.44}$$

where

$$Q_{kl} = \int_{\text{nuc}} \rho(xyz)[3x_{nk}x_{nl} - \delta_{kl} r_n^2]\, dv_n \tag{10.45}$$

$$\nabla E_{kl} = \int_{\text{elec}} \frac{e(xyz)}{r_e^5} [3x_{ek}x_{el} - \delta_{kl} r_e^2]\, dv_e \tag{10.46}$$

and both sets of coordinates are relative to the same Cartesian frame.

Dropping the e subscript for convenience and allowing the indices kl to relate to the Cartesian coordinates, that is, $x_{el} \rightarrow x$, $x_{e2} \rightarrow y$, and $x_{e3} \rightarrow z$, the components of $\underline{\mathbf{\nabla E}}$ become, for a single point charge at a distance r from the origin,

$$\underline{\mathbf{\nabla E}} = \frac{e}{r^5} \begin{pmatrix} 3x^2 - r^2 & 3xy & 3xz \\ 3yx & 3y^2 - r^2 & 3yz \\ 3zx & 3zy & 3z^2 - r^2 \end{pmatrix} \tag{10.47}$$

where the tensor components have been given in the form of a conventional array. It should be noted that the symbols V_{kl} and eq_{kl} are often used interchangeably with $-\nabla E_{kl}$ in the literature, and will be in the discussions which follow. They denote the nine Cartesian components of $\underline{\mathbf{\nabla E}}$. Using the conventional Cartesian-spherical coordinate relationships we can write $\underline{\mathbf{\nabla E}}$ in spherical coordinates as

$$\underline{\mathbf{\nabla E}} = \frac{e}{r^3} \begin{pmatrix} 3\sin^2\theta\cos^2\phi - 1 & 3\sin^2\theta\cos\phi\sin\phi & 3\sin\theta\cos\phi\cos\theta \\ 3\sin^2\theta\cos\phi\sin\phi & 3\sin^2\theta\sin^2\phi - 1 & 3\sin\theta\sin\phi\cos\theta \\ 3\sin\theta\cos\phi\cos\theta & 3\sin\theta\sin\phi\cos\theta & 3\cos^2\theta - 1 \end{pmatrix} \tag{10.48}$$

These components constitute the classical expressions for the EFG tensor and form the basis for the determination of quantum mechanical operators. For example, the expectation value of V_{zz} for an electron in an atomic orbital denoted by the wavefunction ψ would be

$$\langle V_{zz} \rangle = e \iiint \psi^* \left[\frac{3\cos^2\theta - 1}{r^3}\right] \psi\, dv \tag{10.49}$$

If we compare the components for $\underline{\mathbf{\nabla E}}$ given by Eqs. (10.35a) to (10.35c) with those in Eq. (10.48) we can enumerate the relationships among them. The relationship between $(\nabla E)_2^0$ and V_{zz} is

$$(\nabla E)_2^0 = \tfrac{1}{2} V_{zz} \tag{10.50}$$

The term $V_{xz} \pm iV_{yz}$ can be written as

$$V_{xz} \pm iV_{yz} = \frac{e}{r^3}(3\sin\theta\cos\theta\cos\phi \pm 3i\sin\phi\cos\theta\sin\phi) = \frac{e}{r^3}(3\sin\theta\cos\theta e^{\pm i\phi}) \tag{10.51}$$

Therefore

$$(\nabla E)_2^{\pm 1} = \mp \frac{1}{\sqrt{6}}(V_{xz} \pm iV_{yz}) \tag{10.52}$$

The term $V_{xx} - V_{yy} \pm 2iV_{xy}$ can be expressed as

$$V_{xx} - V_{yy} \pm i2V_{xy} = \frac{e}{r^3}(3\sin^2\theta\cos^2\phi - 3\sin^2\theta\cos 2\phi \pm i6\sin^2\theta\cos\phi\sin\phi) \tag{10.53}$$

$$V_{xx} - V_{yy} \pm i2V_{xy} = \frac{e}{r^3}[3\sin^2\theta\,(\cos\phi \pm i\sin\phi)^2] = \frac{e}{r^3}3\sin^2\theta e^{\pm i2\phi} \tag{10.54}$$

Therefore

$$(\nabla E)_2^{\pm 2} = \frac{1}{2\sqrt{6}}(V_{xx} - V_{yy} \pm i2V_{xy}) \tag{10.55}$$

While the expressions given by Eqs. (10.50), (10.52), and (10.55) show that there exists a relationship between the nine Cartesian components and the five irreducible components the derivations were made by use of a rather arbitrary method involving the introduction of certain quantities, that is, $V_{xz} \pm V_{yz}$, without any justification other than the final agreement produced. If we consider the situation of a single charge we can develop a more basic and general approach to developing these relations. This is done by taking the potential $V = e/r$ and evaluating the terms $\partial^2 V/\partial x_i \partial x_j$. One example will be considered and others left as an exercise for the student. Evaluating $\partial^2 V/\partial z^2$ gives

$$V_{zz} = \frac{\partial^2}{\partial z^2}\frac{e}{r} = e\left[\frac{3z^2 - r^2}{r^5}\right] = \frac{e}{r^3}(3\cos^2\theta - 1) \tag{10.56}$$

Reference to Eq. (10.35*a*) shows

$$(\nabla E)_2^0 = \tfrac{1}{2}V_{zz} \tag{10.57}$$

The theory of classical electrostatics allows us to make one further simplification of the EFG tensor components. For any arbitrary distribution of electric charge there will exist a principal axis system (X, Y, Z) in which the off-diagonal EFG tensor elements are zero. In this principal axis system the irreducible components of the EFG tensor become,[1]

$$(\nabla E)_2^0 = \frac{1}{2}V_{ZZ} \equiv \frac{1}{2}eq_{ZZ} \tag{10.58}$$

$$(\nabla E)_2^{\pm 1} = 0 \tag{10.59}$$

$$(\nabla E)_2^{\pm 2} = \frac{1}{2\sqrt{6}}(V_{XX} - V_{YY}) = \frac{1}{2\sqrt{6}}\eta e q_{ZZ} \tag{10.60}$$

[1] The previous convention has been to use X, Y, Z to denote a space-fixed axis system and x, y, z to denote a molecular fixed system. Since NQR is a solid state phenomenon the usage in this chapter is such that xyz denotes a general fixed system and XYZ a principal system.

where $\eta = (V_{XX} - V_{YY})/V_{ZZ}$ and the order of the axes is chosen such that $|V_{XX}| < |V_{YY}| < |V_{ZZ}|$. The parameter η is the asymmetry parameter and is a measure of the deviation of the electric field from cylindrical symmetry. The principal EFG tensor components may be evaluated from those in any arbitrary reference frame by a straightforward diagonalization procedure.

Having established the form of the EFG tensor components we next turn our attention to the nuclear quadrupole moment tensor. The quadrupole tensor elements

$$Q_{kl} = \int_{\text{nuc}} \rho(xyz)[3x_{nk}x_{nl} - \delta_{kl}r_n^2]\, dv_n \tag{10.61}$$

can all be expressed in terms of a single scalar quantity and the nuclear spin quantum numbers I and M_I. Since the components of the vector $\mathbf{r}_n$ commute among themselves, $x_{n1}x_{n2} = x_{n2}x_{n1}$, it follows that $Q_{kl} = Q_{lk}$ or $\underline{\mathbf{Q}}$ is symmetric. Furthermore we also observe that $\sum_{k=1}^{3} Q_{kk} = 0$, hence $\underline{\mathbf{Q}}$ is a symmetric traceless second-rank tensor. The relationships which we wish to establish involve the use of a theorem which states that the matrix elements of all traceless, second-rank, symmetric tensors are proportional. This theorem is developed in App. O. The dependency of a property on a set of states which in turn are specified by a set of quantum numbers is generally expressed as a set of average values or matrix elements involving the states, thus we will be interested in the form of the matrix elements $\langle I', M_I' | Q_{kl} | I, M_I \rangle$. If we assume that the nuclear charge precesses rapidly about I then the electric field interacts with an average nuclear charge. This is tantamount to I being a good quantum number. Since the quadrupolar interaction involves only a change in the relative nucleus–electric field direction and does not involve excitation of the nucleus to a higher spin state $\langle I', m_I' | \mathbf{Q}_{kl} | I, m_I \rangle$ will be diagonal in I. Due to the commutation of the components of $\mathbf{r}_n$ Eq. (10.61) may be written as

$$Q_{kl} = \int_{\text{nuc}} \rho(xyz)\left[3\left(\frac{x_{nk}x_{nl} + x_{nl}x_{nk}}{2}\right) - \delta_{kl}r_n^2\right] dv_n \tag{10.62}$$

Using the above-mentioned theorem and the procedure given in App. O we can construct an appropriate tensor from the components of $\mathbf{I}$ and have

$$\langle I, m' | \mathbf{Q}_{kl} | I, m \rangle = C \langle I, m' | \tfrac{3}{2}(\mathbf{I}_k\mathbf{I}_l + \mathbf{I}_l\mathbf{I}_k) - \delta_{kl}\mathbf{I}^2 | I, m \rangle \tag{10.63}$$

where C is the proportionality constant. We have dropped the subscript I from m_I as a matter of convenience since we will be restricting the discussions in this chapter to nuclear spin quantum numbers. The proportionality between $\underline{\mathbf{Q}}$ and $\underline{\mathbf{II}}$ is analogous to that shown in Chap. 5, where the nuclear magnetic moment was found to be proportional to the nuclear spin.

Regarding the z axis as that of spin quantization we next define a scalar nuclear electric quadrupole moment Q as the expectation value, in units of e, of Q_{zz} in the state $m = I$. This is the state where the component of I along z is a maximum. Hence

$$Q = \frac{1}{e}\langle I, I|\mathbf{Q}_{zz}|I, I\rangle \tag{10.64}$$

In Chap. 8 the factor of $\hbar$ was initially specifically written in expressions containing spin operators. In this and succeeding chapters it will be omitted, conforming to the more conventional method of presentation. Recalling the semiclassical concept of the nucleus being a rapidly precessing charge distribution we can conclude that the nondiagonal elements of $\underline{\mathbf{Q}}$ must be zero due to time-averaging, and that $Q_{xx} = Q_{yy}$ due to symmetry. Here $k = 3$ is taken as the unique or z direction. Since $\underline{\mathbf{Q}}$ is traceless, $Q_{xx} + Q_{yy} + Q_{zz} = 0$, it follows that

$$Q_{xx} = Q_{yy} = -\tfrac{1}{2}Q_{zz} \tag{10.65}$$

and all of the elements of $\underline{\mathbf{Q}}$ are related to Q_{zz}. It remains to obtain a more explicit definition for Q and evaluate the proportionality constant C. Using Eq. (10.45)

$$Q_{zz} = \int_{\text{nuc}} \rho_{II}(xyz)[3z^2 - r^2]\, dv_n \tag{10.66}$$

where all coordinates refer to the nucleus, and the nuclear charge distribution is that for $m = I$. Thus Q_{zz} is a property of the nucleus and is independent of any interactions.

Employing Eq. (10.63)

$$\langle I, I|\mathbf{Q}_{zz}|I, I\rangle = C\langle I, I|\tfrac{3}{2}[\mathbf{I}_z\mathbf{I}_z + \mathbf{I}_z\mathbf{I}_z] - \mathbf{I}^2|I, I\rangle \tag{10.67}$$

Using the angular momentum matrix elements in Table 3.1 with the exchange of quantum numbers $\langle JKM|\mathbf{P}_F|J'K'M'\rangle \rightarrow \langle Im|\mathbf{I}|I'm'\rangle$

$$\langle I, m|\mathbf{I}_z\mathbf{I}_z|I, m\rangle = \sum_{m'} \langle I, m|\mathbf{I}_z|I, m'\rangle\langle I, m'|\mathbf{I}_z|I, m\rangle \tag{10.68}$$

For $m = I$ there is a single term in the sum and $\langle I, I|\mathbf{I}_z^2|I, I\rangle = I^2$. Likewise, $\langle I, m|\mathbf{I}^2|I, m'\rangle = I(I+1)$ so $\langle I, I|\mathbf{I}^2|I, I\rangle = I(I+1)$. Therefore

$$\langle I, I|\mathbf{Q}_{zz}|I, I\rangle = C[\tfrac{3}{2}(I^2 + I^2) - I(I+1)] = CI(2I-1) \tag{10.69}$$

but $\langle I, I|\mathbf{Q}_{zz}|I, I\rangle = eQ$ therefore

$$C = \frac{eQ}{I(2I-1)} \tag{10.70}$$

and

$$Q_{kl} = \frac{eQ}{I(2I-1)}[\tfrac{3}{2}(\mathbf{I}_k\mathbf{I}_l + \mathbf{I}_l\mathbf{I}_k) - \delta_{kl}\mathbf{I}^2] \tag{10.71}$$

We can express the irreducible components, $Q_2^{\pm m}$, in terms of Q also. This is done by expressing Eq. (10.31) in terms of Cartesian coordinates and relating this to Eq. (10.71). The procedure is analogous to that employed to evaluate the

$(\nabla E)_2^{\pm m}$ elements. Thus

$$Q_2^0 = \frac{1}{2}Q_{zz} = \frac{eQ}{2I(2I-1)}[3\mathbf{I}_z^2 - \mathbf{I}^2] \qquad (10.72)$$

$$Q_2^{\pm 1} = \frac{1}{\sqrt{6}}(Q_{xz} \pm iQ_{yz}) = \left(\frac{eQ}{2I(2I-1)}\right)\frac{\sqrt{6}}{2}[\mathbf{I}_z(\mathbf{I}_x \pm i\mathbf{I}_y) + (\mathbf{I}_x \pm i\mathbf{I}_y)\mathbf{I}_z] \qquad (10.73)$$

$$Q_2^{\pm 2} = \frac{1}{2\sqrt{6}}(Q_{xx} - Q_{yy} \pm i2Q_{xy}) = \left(\frac{eQ}{2I(2I-1)}\right)\frac{\sqrt{6}}{2}[\mathbf{I}_x \pm i\mathbf{I}_y]^2 \qquad (10.74)$$

The matrix elements for the quadrupolar Hamiltonian may now be written in terms of either set of tensor components. The more commonly used form involves the use of the Cartesian components so we will limit our discussion to this case. Since

$$\mathscr{H}_Q = \frac{1}{6}\sum_{k=1}^{3}\sum_{l=1}^{3} Q_{kl}V_{kl} \qquad (10.75)$$

we have

$$\langle I, m'|\mathscr{H}_Q|I, m\rangle = \frac{eQ}{6I(2I-1)}\sum_{k=1}^{3}\sum_{l=1}^{3}\langle I, m'|\tfrac{3}{2}(\mathbf{I}_k\mathbf{I}_l + \mathbf{I}_l\mathbf{I}_k) - \delta_{kl}\mathbf{I}^2|I, m\rangle V_{kl} \qquad (10.76)$$

At this point the evaluation of quadrupole energies diverges depending on whether one considers a solid, where the interaction of $\underline{\mathbf{Q}}$ is with an EFG tensor determined by a set of static charges, or whether one considers a gaseous system, where $\underline{\nabla\mathbf{E}}$ must be averaged over the rotations of the molecule. These two situations will be considered separately, but first a few summary remarks on the nature of the nuclear quadrupole moments and EFG tensors.

Looking at the definition of Q

$$Q = \frac{1}{e}\int_{\text{nuc}} \rho(xyz)(3z^2 - r^2)\, dV \qquad (10.77)$$

we see that if $\rho(xyz)$ has an oblate distribution then Q is negative. On the other hand, if $\rho(xyz)$ has a prolate distribution then Q is positive. If $\rho(xyz)$ *has spherical symmetry, as for* $I = 0, \frac{1}{2}$, then $Q = 0$ and no quadrupolar interaction occurs. For an electric change distribution having spherical symmetry, $\underline{\nabla\mathbf{E}} = 0$, so no interaction can occur either. Physically, this denotes the fact that for spherical symmetry of either $\underline{\mathbf{Q}}$ or $\underline{\nabla\mathbf{E}}$ the interaction between a nucleus and an electron is independent of orientation and is composed only on the Coulombic term.

10.3 NUCLEAR QUADRUPOLE INTERACTIONS IN SOLIDS: ENERGY LEVELS FOR AXIALLY SYMMETRIC EFG TENSORS

In the case of a crystalline solid the components of $\underline{\nabla\mathbf{E}}$ are due to both the electron cloud surrounding the nucleus of interest and near neighbors. If it is assumed that the species constituting the solid are restricted relative to rotation the contributions from both of these effects are independent of any molecular rotation, hence the V_{kl} components may be considered as terms independent of rotational quantum numbers. Since, for a particular nucleus, I is constant and $\langle I, m'|H_Q|I'm\rangle$ is diagonal in I it is convenient to drop the I and write the Hamiltonian matrix elements as

$$\langle m'|\mathscr{H}_Q|m\rangle = A'' \sum_{k=1}^{3} \sum_{l=1}^{3} \langle m'|\tfrac{3}{2}(\mathbf{I}_k\mathbf{I}_l + \mathbf{I}_l\mathbf{I}_k) - \delta_{kl}\mathbf{I}^2|m\rangle V_{kl} \tag{10.78}$$

where $A'' = eQ/[6I(2I-1)]$. Expanding Eq. (10.78) where $k, l = x, y, z$ gives

$$\begin{aligned}(m'|\mathscr{H}_Q|m) = A''[&\langle m'|3\mathbf{I}_x^2 - \mathbf{I}^2|m\rangle V_{xx} + \tfrac{3}{2}\langle m'|\mathbf{I}_x\mathbf{I}_y + \mathbf{I}_y\mathbf{I}_x|m\rangle V_{xy} \\ &+ \tfrac{3}{2}\langle m'|\mathbf{I}_x\mathbf{I}_z + \mathbf{I}_z\mathbf{I}_x|m\rangle V_{xz} + \tfrac{3}{2}\langle m'|\mathbf{I}_y\mathbf{I}_x + \mathbf{I}_x\mathbf{I}_y|m\rangle V_{yx} \\ &+ \langle m'|3\mathbf{I}_y^2 - \mathbf{I}^2|m\rangle V_{yy} + \tfrac{3}{2}\langle m'|\mathbf{I}_y\mathbf{I}_z + \mathbf{I}_z\mathbf{I}_y|m\rangle V_{yz} \\ &+ \tfrac{3}{2}\langle m'|\mathbf{I}_z\mathbf{i}_x + \mathbf{I}_x\mathbf{I}_z|m\rangle V_{zx} + \tfrac{3}{2}\langle m'|\mathbf{I}_z\mathbf{I}_y + \mathbf{I}_y\mathbf{I}_z|m\rangle V_{zy} \\ &+ \langle m'|3\mathbf{I}_z^2 - \mathbf{I}^2|m\rangle V_{zz}]\end{aligned} \tag{10.79}$$

Using the alternate expression for H_Q

$$\langle m'|\mathscr{H}_Q|m\rangle = \sum_{k=-2}^{2} (\nabla E)_2^{\pm k} Q_2^{\pm k} \tag{10.80}$$

leads to

$$\begin{aligned}\langle m'|\mathscr{H}_Q|m\rangle = 2A_I'[&\langle m'|3\mathbf{I}_z^2 - \mathbf{I}^2|m\rangle(\nabla E)_2^0 \\ &+ \frac{\sqrt{6}}{2}\langle m'|\mathbf{I}+\mathbf{I}_z + \mathbf{I}_z\mathbf{I}_+|m\rangle(\nabla E)_2^{-1} + \frac{\sqrt{6}}{2}\langle m'|\mathbf{I}_-\mathbf{I}_z + \mathbf{I}_z\mathbf{I}_-|m\rangle(\nabla E)_2^1 \\ &+ \frac{\sqrt{6}}{2}\langle m'|\mathbf{I}_+^2|m\rangle(\nabla E)_2^{-2} + \frac{\sqrt{6}}{2}\langle m'|\mathbf{I}_-^2|m\rangle(\nabla E)_2^2]\end{aligned} \tag{10.81}$$

where $A_I' = eQ/4I(2I-1)$. Employing the symmetry of $\underline{\nabla \mathbf{E}}$ ($V_{xy} = V_{yx}$, etc.) and the angular momentum matrix elements

$$\langle I, m'|\mathbf{I}_z|I, m\rangle = \delta_{mn'}m \tag{10.82}$$

$$\langle I, m \pm 1|\mathbf{I}_\pm|I, m\rangle = [(I \mp m)(I \pm m + 1)]^{1/2} \tag{10.83}$$

$$\langle I, m|\mathbf{I}^2|I, m\rangle = I(I+1) \tag{10.84}$$

both of these forms for $(m'|\mathscr{H}_Q|m)$ will reduce, following some lengthy but straightforward manipulations to give the following nonzero elements

$$\langle m|\mathscr{H}_Q|m\rangle = 2A_I'[3m^2 - I(I+1)](\nabla E)_2^0 \tag{10.85}$$

$$\langle m \pm 1|\mathscr{H}_Q|m\rangle = A_I'(2m \pm 1)[(I \mp m)(I \pm m + 1)]^{1/2}(\nabla E)_2^{\pm 1} \tag{10.86}$$

$$\begin{aligned}\langle m \pm 2|\mathscr{H}_Q|m\rangle = \sqrt{6}A_I'[&(I \pm m)(I \mp m - 1)(I \pm m + 1) \\ &\times (I \pm m + 2)]^{1/2}(\nabla E)_2^{\pm 2}\end{aligned} \tag{10.87}$$

Consider the case where the electric field gradient tensor has axial symmetry, that is $\eta = 0$. In this case $(\nabla E)_2^0 = \frac{1}{2}eq_{ZZ}$ and $(\nabla E)_2^{\pm 2} = 0$ and the only remaining Hamiltonian matrix elements are the diagonal ones, which consequently specify the energy

$$E_m^Q = \langle m|\mathcal{H}_Q|m\rangle = \frac{e^2Qq_{ZZ}}{4I(2I-1)}[3m^2 - I(I+1)] = A_I[3m^2 - I(I+1)] \tag{10.88}$$

The product e^2Qq_{ZZ} is referred to as the nuclear quadrupole coupling constant. The energy levels are doubly degenerate in m except for $m = 0$ levels. For integer nuclear spins there will be $I + 1$ levels with one being singly degenerate, and for half-integer spins there will be $I + \frac{1}{2}$ doubly degenerate levels. For example if $I = \frac{5}{2}$ there will be three energy levels as shown in Fig. 10.2.

Although the energy levels of the system are determined by an electrostatic interaction, when a nucleus interacts with electromagnetic radiation the interaction is between the magnetic moment of the nucleus and the magnetic component of the external field. It is conceivable that the electric quadrupole moment of the nucleus could interact with the electric vector of an external field, but the required magnitude of the latter is too large for use of presently available instrumentation. The time-dependent Hamiltonian which gives the interaction is

$$\mathcal{H}'(t) = -\boldsymbol{\mu} \cdot \mathbf{H}(t) \tag{10.89a}$$

where $\boldsymbol{\mu}$ is the magnetic moment of the nucleus and is given in terms of the magnetogyric ratio as $\gamma\hbar\mathbf{I}$ and $\mathbf{H}(t)$ is the magnetic vector of the radiation field.

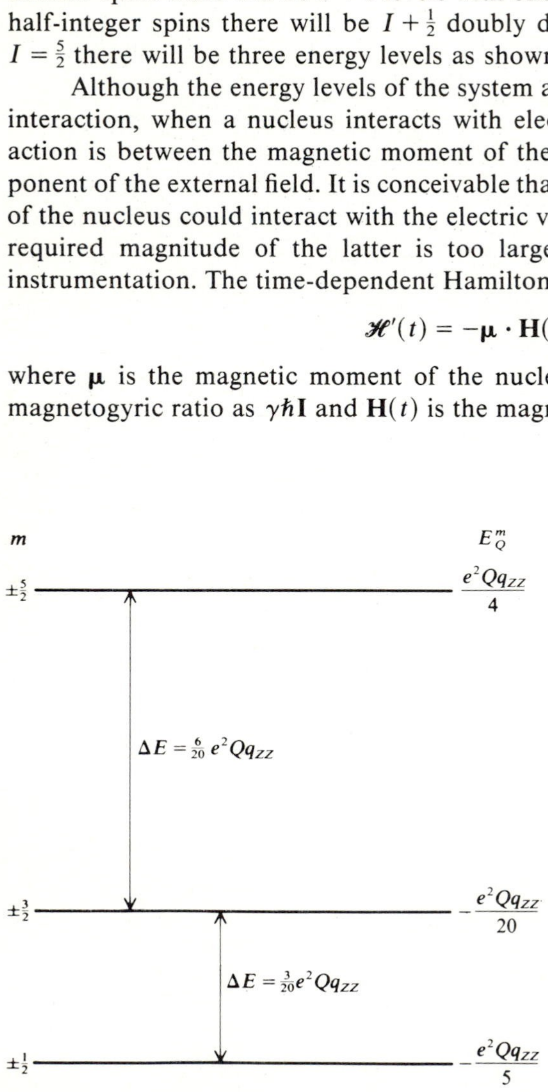

FIGURE 10.2
Quadrupole energy levels for an $I = \frac{5}{2}$ system.

In terms of the components of $\boldsymbol{\mu}$ and $\mathbf{H}(t)$

$$\mathscr{H}'(t) = -\gamma\hbar[H_x(t)\mathbf{I}_x + H_y(t)\mathbf{I}_y + H_z(t)\mathbf{I}_z] \tag{10.89b}$$

The intensity of a transition will be proportional to $\langle m'|\mathscr{H}'(t)|m\rangle^2$, hence the selection rules will be determined by the values of m and m' that lead to nonzero values for this element. Expanded, this element becomes

$$\langle m'|\mathscr{H}'(t)|m\rangle = -\gamma\hbar[\langle m'|\mathbf{I}_x|m\rangle H_x(t) + \langle m'|\mathbf{I}_y m\rangle H_y(t) + \langle m'|\mathbf{I}_z|m\rangle H_z(t)] \tag{10.90}$$

and we see that the selection rules will be determined from the nonzero values of the angular momentum matrix elements. For external fields with components in the x or y direction transitions with $\Delta m = \pm 1$ will be induced, and for those with components in the z direction $\Delta m = 0$ transitions will occur. The latter case is of no interest since it does not involve any change in energy for the system. The allowed transition for the $I = \frac{5}{2}$ case are also shown in Fig. 10.2.

10.4 NUCLEAR QUADRUPOLE INTERACTIONS IN SOLIDS: ENERGY LEVELS FOR NONAXIALLY SYMMETRIC EFG TENSORS

For systems having noncylindrical electric charge distribution such that $V_{XX} \neq V_{YY}$ the asymmetry is such that $\eta \neq 0$ and all of the off-diagonal Hamiltonian elements do not vanish. In the principal EFG tensor axis system $(\nabla E)_2^{\pm 1} = 0$, so the nonzero Hamiltonian elements will be those with $m' = m$ and $m' = m \pm 2$. The general form of the energy matrix for a spin I system will be

$$\begin{vmatrix} (I|\mathscr{H}_Q|I) & 0 & (I|\mathscr{H}_Q|I-2) & \cdots \\ 0 & (I-1|\mathscr{H}_Q|I-1) & 0 & \cdots \\ (I-2|\mathscr{H}_Q|I) & 0 & (I-2|\mathscr{H}_Q|I-2) & \cdots \\ 0 & (I-3|\mathscr{H}_Q|i-1) & 0 & \\ \vdots & \vdots & \vdots & (-I|\mathscr{H}_Q|-I) \end{vmatrix} \tag{10.91}$$

Using Eqs. (10.85) and (10.87) the individual matrix elements can be determined. For $I = \frac{3}{2}$ these are

$m' \backslash m$	$\frac{3}{2}$	$\frac{1}{2}$	$-\frac{1}{2}$	$\frac{3}{2}$
$\frac{3}{2}$	$3A_{3/2}$	0	$\sqrt{3}\eta A_{3/2}$	0
$\frac{1}{2}$	0	$-3A_{3/2}$	0	$\sqrt{3}\eta A_{3/2}$
$-\frac{1}{2}$	$\sqrt{3}\eta A_{3/2}$	0	$-3A_{3/2}$	0
$-\frac{3}{2}$	0	$\sqrt{3}\eta A_{3/2}$	0	$-3A_{3/2}$

(10.92)

where $A_{3/2} = e^2Qq_{ZZ}/12$. The evaluation of the eigenvalues of this equation leads to the secular equation

$$E^2 - A_{3/2}^2\eta^2 - 9A_{3/2}^2 = 0 \tag{10.93}$$

which has roots

$$E_{\pm 3/2} = 3A_{3/2}\left[1 + \frac{\eta^2}{3}\right]^{1/2} \tag{10.94}$$

$$E_{\pm 1/2} = -3A_{3/2}\left[1 + \frac{\eta^2}{3}\right]^{1/2} \tag{10.95}$$

Using the selection rule $\Delta m = \pm 1$, this system will have a single transition

$$\nu_Q = \frac{E_{\pm 3/2} - E_{\pm 1/2}}{h} = \frac{e^2 Q q_{zz}}{2h}\left[1 + \frac{\eta^2}{3}\right]^{1/2} \tag{10.96}$$

For $I = \frac{3}{2}$ there is only a single transition frequency, and it will be impossible to determine the nuclear quadrupole coupling constant e^2Qq_{ZZ} and the asymmetry parameters η simultaneously. This is a problem that is unique to an $I = \frac{3}{2}$ system. Section 10.6 will discuss how the use of Zeeman splittings can resolve this problem. This particular case is of considerable importance since there are a number of common nuclei with $I = \frac{3}{2}$. In an analogous manner the secular equations for other half-integral spin values are found to be

$$I = \tfrac{5}{2} \quad E^3 - 28A_{5/2}^2(3 + \eta^2)E - 160A_{5/2}^3(1 - \eta^2) = 0 \tag{10.97}$$

$$I = \tfrac{7}{2} \quad E^4 - 126A_{7/2}\left(1 + \frac{\eta^2}{3}\right)E^2 - 576A_{7/2}^2(1 - \eta^2)E + 8505A_{7/2}^4\left(1 + \frac{\eta^2}{3}\right)^2 = 0 \tag{10.98}$$

$$I = \tfrac{9}{2} \quad E^5 - 396A_{9/2}^2(3 + \eta^2)E^3 - 9504A_{9/2}^3(1 - \eta^2)E^2 + 19{,}008A_{9/2}^4(3 + \eta^2)^2E + 373{,}248A_{9/2}^5(3 + \eta^2)(1 - \eta^2) = 0 \tag{10.99}$$

Examination of these equations shows the formulation of exact solutions beyond $I = \frac{5}{2}$ will be difficult and will involve complicated expressions. In practice such systems are handled by either using series approximations to obtain solutions when η is small [4, 5] or by using numerical methods [6]. Both techniques are straightforward so the reader is directed to the original literature for further details. For half-integer spins $I > \frac{3}{2}$ there will always be two or more transitions, so both e^2Qq_{ZZ} and η can be independently determined.

Because of the chemical importance of nitrogen, $I = 1$, and the fact that the introduction of asymmetry removes the degeneracy in m for integer spin systems we will briefly consider an $I = 1$ system. Using Eqs. (10.85) and (10.87) the energy matrix is written as

$$\begin{vmatrix} A_1 - E & 0 & A_1\eta \\ 0 & -2A_1 - E & 0 \\ A_1\eta & 0 & A_1 - E \end{vmatrix} = 0 \tag{10.100}$$

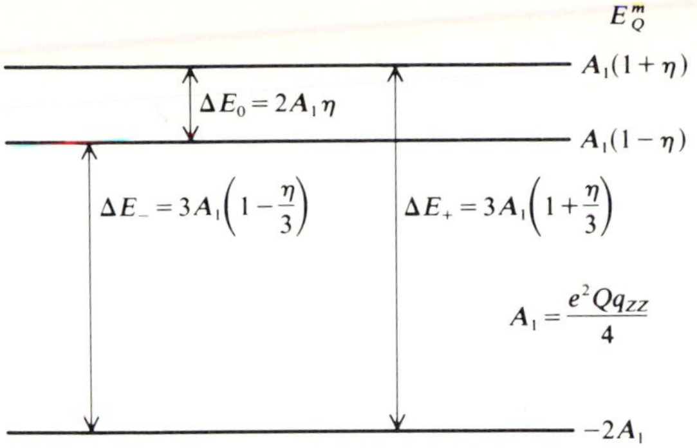

FIGURE 10.3
Energy level diagram for an $I = 1$ system.

The energy levels are then given by

$$E_0 = -2A_1 \tag{10.101}$$

$$E_{\pm 1} = A_1(1 \pm \eta) \tag{10.102}$$

For $\eta \neq 0$ the energy level diagram and transitions are as shown in Fig. 10.3. The observable transitions are given by

$$\nu_\pm = \frac{3}{4}\frac{e^2Qq_{zz}}{h}\left[1 \pm \frac{\eta}{3}\right] \tag{10.103}$$

$$\nu_0 = \frac{1}{2}\frac{e^2Qq_{zz}}{h}\eta \tag{10.104}$$

Since the coupling constant of nitrogen in its compounds is in the neighborhood of 4 MHz the ν_0 transition can only be observed for large values of η.

10.5 NUCLEAR QUADRUPOLE INTERACTIONS IN SOLIDS: STATES

The energy levels for systems having $\eta \neq 0$ have been indexed with the quantum number m. No mention has been made as to whether the states are the same as those for the $\eta = 0$ case. We can think of the EFG asymmetry as being a perturbation added to the systems. Perturbation theory may be used to take a more detailed look at state functions and intensities. Although the transition intensities are of importance in the study of single crystals they will not be discussed because of their limited use.

To illustrate the nature of the energy states we will consider the spin $I = \frac{3}{2}$ case. Equation (10.91) shows that the off-diagonal elements couple states which differ by $\Delta m = \pm 2$. Examination of the solution of Eq. (10.92) to give Eq. (10.93) shows that for half-integer spins the energy matrix divides into two submatrices of order $I + \frac{1}{2}$. These submatrices give rise to the same energies due to the degeneracy in m but each submatrix will have its own set of states. The general form for the mixed state functions will be

$$\Phi_m = C_{1/2,\,m}\psi_{1/2} + C_{-3/2,\,m}\psi_{-3/2} + C_{5/2,\,m}\psi_{5/2} + \cdots \tag{10.105}$$

$$\Phi_{-m} = C_{-1/2,\,m}\psi_{-1/2} + C_{3/2,\,m}\psi_{3/2} + C_{-5/2,\,m}\psi_{-5/2} + \cdots \tag{10.106}$$

where the ψ_m are the state functions for the $\eta = 0$ case and the $C_{m',\,m}$ are the mixing coefficients. The states $\Phi_{\pm m}$ are indexed by the level ψ_m to which they would reduce for zero asymmetry. For $I = \frac{3}{2}$ the state functions are

$$\Phi_{3/2} = C_{1/2,\,3/2}\psi_{1/2} + C_{-3/2,\,3/2}\psi_{-3/2} = a\psi_{1/2} + b\psi_{-3/2} \tag{10.107}$$

$$\Phi_{1/2} = C_{1/2,\,1/2}\psi_{1/2} + C_{-3/2,\,1/2}\psi_{-3/2} = c\psi_{1/2} + d\psi_{-3/2} \tag{10.108}$$

$$\Phi_{-1/2} = C_{-1/2,\,-1/2}\psi_{-1/2} + C_{3/2,\,-1/2}\psi_{3/2} = c\psi_{-1/2} + d\psi_{3/2} \tag{10.109}$$

$$\Phi_{-3/2} = C_{-1/2,\,-3/2}\psi_{-1/2} + C_{3/2,\,-3/2}\psi_{3/2} = a\psi_{-1/2} + b\psi_{3/2} \tag{10.110}$$

In the limit of $\eta = 0$ these will reduce as follows:

$$\Phi_{3/2} \to \psi_{-3/2},\ \Phi_{1/2} \to \psi_{1/2},\ \Phi_{-1/2} \to \psi_{-1/2},\ \text{and}\ \Phi_{-3/2} \to \psi_{3/2}$$

From these conditions and the symmetry of the problem we conclude that there is a relationship between pairs of coefficients such that $C_{1/2,\,m} = C_{-1/2,\,-m}$ and $C_{3/2,\,-m} = C_{-3/2,\,m}$. Thus the state functions can be written in a simpler fashion, as shown by the right-hand sections of Eqs. (10.107) to (10.110).

For a perturbed system the form of the secular equations is

$$\sum_k (H_{jk} - \delta_{jk}E_j)C_k = 0 \tag{10.111}$$

For the $I = \frac{3}{2}$ system the H_{jk} are the elements of the energy matrix, Eq. (10.92). The E_j are given by Eqs. (10.94) and (10.95), and the summation is over the pair of states which contribute to $\phi_{\pm m}$. We then have

$$(3A_{3/2} - E_{\pm m})a + 3A_{3/2}\eta b = 0 \tag{10.112}$$

$$\sqrt{3}A_{3/2}\eta c + (-3A_{3/2} - E_{\pm m})d = 0 \tag{10.113}$$

Using $E = -3A_{3/2}(1 + \eta^2/3)^{1/2}$, Eq. (10.112) becomes

$$\frac{b}{a} = \frac{\sqrt{3}\{(1 + \eta^2/3)^{1/2} + 1\}}{\eta} \tag{10.114}$$

Coupling this with the normalization condition $a^2 + b^2 = 1$ gives

$$a = \frac{\eta}{\{\eta^2 + 3[(1 + \eta^2/3)^{1/2} + 1]^2\}^{1/2}} \tag{10.115}$$

$$b = \frac{3\{1 + (1 + n^2/3)^{1/2}\}}{(\eta^2 + 3[(1 + \eta^2/3)^{1/2} + 1]^2)^{1/2}} \tag{10.116}$$

We see that as $\eta \to 0$ it follows that $a \to 0$, $b \to 1$, and $\phi_{3/2} \to \psi_{-3/2}$.

The transition intensities, and hence the selection rules, are determined by the nature of the interaction between the radiation field and the system as indicated by Eq. (10.90). Since for pure quadrupole transitions $h\nu \ll kT$ the intensities will be proportional to

$$\frac{8\pi^3 N_m \nu_{mm'}^2}{3hckT} \frac{1}{|\mathbf{H}'(t)|^2} [\mathbf{H}'(t) \cdot \langle m|\boldsymbol{\mu}|m'\rangle]^2 \tag{10.117}$$

In general the radiation field magnetic vector will have some arbitrary orientation relative to the principal EFG tensor axis system. Specifying the orientation of $\mathbf{H}'(t)$ relative to the XYZ axes with the spherical coordinates θ' and ϕ' we have

$$H'_x(t) = |\mathbf{H}'(t)| \sin\theta' \cos\phi' \tag{10.118}$$

$$H'_y(t) = |\mathbf{H}'(t)| \sin\theta' \sin\phi' \tag{10.119}$$

$$H'_z(t) = |\mathbf{H}'(t)| \cos\theta' \tag{10.120}$$

Substitution of these relations into Eq. (10.90) gives

$$\mathcal{H}'(t) = \gamma\hbar|\mathbf{H}'(t)| \cdot (\sin\theta' \cos\phi' \mathbf{I}_x + \sin\theta' \sin\phi' \mathbf{I}_y + \cos\theta' \mathbf{I}_z) \tag{10.121}$$

and the absorption coefficient is then proportional to

$$|\sin\theta' \cos\phi' \langle m'|\mathbf{I}_x|m\rangle + \sin\theta' \sin\phi' \langle m'|\mathbf{I}_y|m\rangle + \cos\theta' \langle m'|\mathbf{I}_z|m\rangle|^2 \tag{10.122}$$

A detailed review of the theory of transition intensities has been given by Cohen [6].

Although the absorption coefficients will depend on the conventional dipole moment matrix elements as would be expected, they also depend on the asymmetry parameter η. The effect of the introduction of asymmetry on the selection rules for NQR is analogous to the introduction of anharmonicity on the selection rules for the harmonic oscillator, and $\Delta m = \pm 2$ transitions are allowed.

10.6 NUCLEAR QUADRUPOLE INTERACTIONS IN SOLIDS: ZEEMAN EFFECT

For an $I = \frac{3}{2}$ system the single absorption frequency depends on both e^2Qq_{ZZ} and η, and it is impossible to obtain both quantities from a single measurement. This problem can be resolved by using the Zeeman effect. Although this technique is not necessary to independently determine e^2Qq_{ZZ} and η for spins other than $I = \frac{3}{2}$, it constitutes the most convenient method for locating the EFG tensor axis system relative to the crystal axis system. Although the relative orientation of the EFG tensor axis system and the radiation field axis system can be ascertained from intensity measurements, such measurements are experimentally difficult to make and are not as accurate as measurements involving frequencies of transitions.

If $\eta = 0$ there is no need to use the Zeeman method to get e^2Qq_{ZZ}. We will nevertheless examine this case first to provide some elementary background regarding the Zeeman effect. The discussion will be limited to the weak field Zeeman case where the magnitude of the interaction between the spin system and the external magnetic field is small compared to the magnitude of the

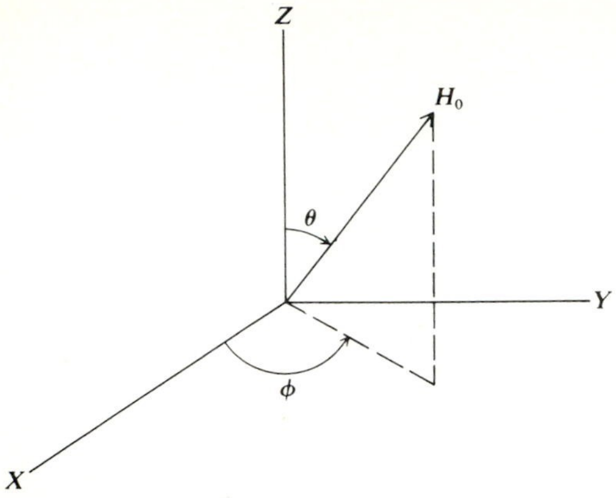

FIGURE 10.4
Orientation of state magnetic field and EFG tensor axes.

quadrupole energy. This permits first-order perturbation theory to be used to analyze the system.

For a nucleus of spin I located in an EFG tensor principal axis system X, Y, Z, and subjected to a static external magnetic field H_0, the relative orientations are depicted in Fig. 10.4. The total Hamiltonian will be

$$\mathscr{H} = \mathscr{H}_Q + \mathscr{H}_M \tag{10.123}$$

where H_Q is given by Eq. (10.88) for the $\eta = 0$ case, and

$$\mathscr{H}_M = -\mathbf{H}_0 \cdot \boldsymbol{\mu} = -\gamma\hbar\mathbf{H}_0 \cdot \mathbf{I} \tag{10.124}$$

or

$$\mathscr{H}_M = -\gamma\hbar H_0(\mathbf{I}_z \cos\theta + \mathbf{I}_y \sin\theta \sin\phi + \mathbf{I}_x \sin\theta \cos\phi) \tag{10.125}$$

The weak field case is defined by $\hbar\gamma H_0 \ll e^2Qq_{ZZ}$. We will again limit the discussion to an $I = \frac{3}{2}$ system and make some generalizations later regarding others. For this case, one with a half-integral spin, there will be $I + \frac{1}{2} = 2$ doubly degenerate energy levels; the degeneracy is removed by the application of the magnetic field giving $2I + 1 = 4$ discrete levels. The behavior of levels having $|m| \geqq \frac{3}{2}$ is somewhat different from those with $|m| = \frac{1}{2}$, there being in the latter case a mixing of states which cannot be neglected. Employing the general form shown by Eq. (10.91), but with $\eta = 0$, the energy matrix may be written

$$\begin{vmatrix} \langle\frac{3}{2}|\mathscr{H}_Q|\frac{3}{2}\rangle & 0 & 0 & 0 \\ 0 & \langle\frac{1}{2}|\mathscr{H}_Q|\frac{1}{2}\rangle & 0 & 0 \\ 0 & 0 & \langle-\frac{1}{2}|\mathscr{H}_Q|-\frac{1}{2}\rangle & 0 \\ 0 & 0 & 0 & \langle-\frac{3}{2}|\mathscr{H}_Q|-\frac{3}{2}\rangle \end{vmatrix} + \begin{vmatrix} \langle\frac{3}{2}|\mathscr{H}_M|\frac{3}{2}\rangle & \langle\frac{3}{2}|\mathscr{H}_M|\frac{1}{2}\rangle & \langle\frac{3}{2}|\mathscr{H}_M-\frac{1}{2}\rangle & \langle\frac{3}{2}|\mathscr{H}_M|-\frac{3}{2}\rangle \\ \langle\frac{1}{2}|\mathscr{H}_M|\frac{3}{2}\rangle & \langle\frac{1}{2}|\mathscr{H}_M|\frac{1}{2}\rangle & \langle\frac{1}{2}|\mathscr{H}_M|-\frac{1}{2}\rangle & \langle\frac{1}{2}|\mathscr{H}_M|-\frac{3}{2}\rangle \\ \langle-\frac{1}{2}|\mathscr{H}_M|\frac{3}{2}\rangle & \langle-\frac{1}{2}|\mathscr{H}_M|\frac{1}{2}\rangle & \langle-\frac{1}{2}|\mathscr{H}_M|-\frac{1}{2}\rangle & \langle-\frac{1}{2}|\mathscr{H}_M|-\frac{3}{2}\rangle \\ \langle-\frac{3}{2}|\mathscr{H}_M|\frac{3}{2}\rangle & \langle-\frac{3}{2}|\mathscr{H}_M|\frac{1}{2}\rangle & \langle-\frac{3}{2}|\mathscr{H}_M|-\frac{1}{2}\rangle & \langle-\frac{3}{2}|\mathscr{H}_M|-\frac{3}{2}\rangle \end{vmatrix} \tag{10.126}$$

Using the angular momentum matrix elements from Table 3.1 the nonzero elements for $\mathscr{H}$ become

$$\begin{vmatrix} 3A_{3/2}-\frac{3}{2}\gamma\hbar H_0\cos\theta & -\frac{\sqrt{3}}{2}\gamma\hbar H_0\sin\theta & 0 & 0 \\ \frac{\sqrt{3}}{2}\gamma\hbar H_0\sin\theta & -3A_{3/2}-\frac{1}{2}\gamma\hbar H_0\cos\theta & -\gamma\hbar H_0\sin\theta & 0 \\ 0 & \gamma\hbar H_0\sin\theta & -3A_{3/2}+\frac{1}{2}\gamma\hbar H_0\cos\theta & \frac{\sqrt{3}}{2}\gamma\hbar H_0\sin\theta \\ 0 & 0 & -\frac{\sqrt{3}}{2}\gamma\hbar H_0\sin\theta & 3A_{3/2}+\frac{3}{2}\gamma\hbar H_0\cos\theta \end{vmatrix} \tag{10.127}$$

Since $\eta = 0$ there will be no dependency on ϕ so we have arbitrarily set $\phi = 90°$. Solutions of this matrix for the eigenvalues results in a fourth-order equation for which there is no means of simplification without introducing some assumptions regarding the properties of the system. Letting $\hbar\gamma H_0 \cos\theta = \alpha$ and $\hbar\gamma H_0 \sin\theta = \beta$ the eigenvalue equation becomes

$$\begin{aligned} &(3A_{3/2}-\tfrac{3}{2}\alpha-E)(-3A_{3/2}-\tfrac{1}{2}\alpha-E)(-3A_{3/2}+\tfrac{1}{2}\alpha-E)(3A_{3/2}+\tfrac{3}{2}\alpha-E) \\ &\quad+\tfrac{3}{4}(3A_{3/2}-\tfrac{3}{2}\alpha-E)(-3A_{3/2}-\tfrac{1}{2}\alpha-E)\beta^2 \\ &\quad+(3A_{3/2}-\tfrac{3}{2}\alpha-E)(3A_{3/2}+\tfrac{3}{2}\alpha-E)\beta^2 \\ &\quad-\tfrac{3}{4}(-3A_{3/2}+\tfrac{1}{2}\alpha-E)(3A_{3/2}+\tfrac{3}{2}\alpha-E)\beta^2+\tfrac{9}{16}\beta^4=0 \end{aligned} \tag{10.128}$$

Taking ^{35}Cl as an example and comparing the magnitude of γH_0 to a e^2Qq_{ZZ} we find that for a 0.01 T field

$$\frac{\gamma H_0}{e^2Qq_{ZZ}} = \frac{4.26\times10^7\ \text{Hz T}^{-1}\times 0.01\ \text{T}}{80\times10^6\ \text{Hz}} = 5.3\times10^{-3} \tag{10.129}$$

Since first-order perturbation theory is being used we are justified in neglecting the off-diagonal terms with the exception of $\langle\frac{1}{2}|\mathscr{H}_M|-\frac{1}{2}\rangle$ and $(-\frac{1}{2}|\mathscr{H}_M|\frac{1}{2})$. In this case the energy matrix becomes

$$\begin{vmatrix} 3A_{3/2}-\frac{3}{2}\alpha-E & 0 & 0 & 0 \\ 0 & -3A_{3/2}-\frac{1}{2}\alpha-E & -\beta & 0 \\ 0 & \beta & -3A_{3/2}+\frac{1}{2}\alpha-E & 0 \\ 0 & 0 & 0 & 3A_{3/2}+\frac{3}{2}\alpha-E \end{vmatrix} = 0 \tag{10.130}$$

Solution of this matrix gives

$$E_{\pm3/2} = 3A_{3/2} \pm \tfrac{3}{2}\gamma\hbar H_0 \cos\theta \tag{10.131}$$

for the energies of the $m = \pm\frac{3}{2}$ levels and

$$\begin{vmatrix} -3A_{3/2}-\frac{1}{2}\alpha-E & -\beta \\ \beta & -3A_{3/2}+\frac{1}{2}\alpha-E \end{vmatrix} = 0 \tag{10.132}$$

for the remaining two. The presence of the off-diagonal terms leads to a mixing of the $\psi_{1/2}$ and $\psi_{-1/2}$ states to give a new pair of states, designated as $\psi_{\pm}$, having energies

$$E_{\pm} = -3A_{3/2} \mp \frac{f_{3/2}\hbar\gamma H_0}{2} \cos\theta \tag{10.133}$$

where $f_{3/2} = [1 + 4\tan^2\theta]^{1/2}$. For the general case of any half-integer spin the energies of this lower set of levels is given by

$$E_{\pm} = A_I[3/4 - I(I+1)] \mp \frac{f_I\gamma\hbar H_0}{2} \cos\theta \tag{10.134}$$

where $f_I = [1 + (I + \frac{1}{2})^2 \tan^2\theta]^{1/2}$.

The ordering of the resulting Zeeman energy levels for a system of spin $\frac{5}{2}$ is illustrated in Fig. 10.5. The selection rule $\Delta m = \pm 1$ strictly applies for transitions involving states of $m \geqq \frac{3}{2}$, thereby giving rise to a pair or pairs of transitions designated as $\nu_m^{\pm}$. For the general case these transitions is given by

$$\nu_m^{\pm} = \frac{3A_{3/2}}{h}(2|m| + 1) \pm \gamma H_0 \cos\theta \tag{10.135}$$

This pair, for the $\psi_{\pm 3/2} \rightarrow \psi_{\pm 5/2}$ transition, is shown in Fig. 10.5. The $\pm$ designation refers to the displacement from the original zero field transition. For $\psi_{\pm} \leftrightarrow \psi_{\pm 3/2}$ there will be four transitions. Using Eqs. (10.131) and (10.134) these four

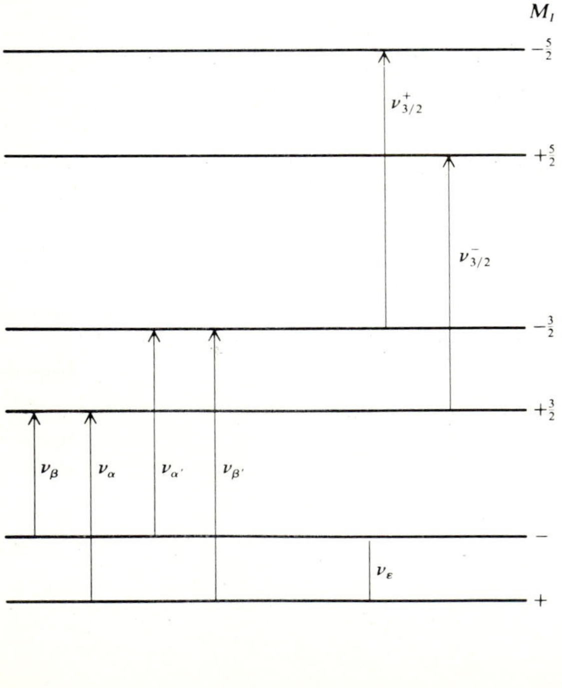

FIGURE 10.5
Zeeman energy levels for a spin $\frac{5}{2}$ system.

transitions are found to be

$$\nu_\alpha = \frac{6A_{3/2}}{h} - \frac{3-f}{2}\gamma H_0 \cos\theta \tag{10.136}$$

$$\nu_{\alpha'} = \frac{6A_{3/2}}{h} + \frac{3-f}{2}\gamma H_0 \cos\theta \tag{10.137}$$

$$\nu_\beta = \frac{6A_{3/2}}{h} - \frac{3+f}{2}\gamma H_0 \cos\theta \tag{10.138}$$

$$\nu_{\beta'} = \frac{6A_{3/2}}{h} + \frac{3+f}{2}\gamma H_0 \cos\theta \tag{10.139}$$

where

$$f = [1 + (I + \tfrac{1}{2}) \tan^2\theta]^{1/2} \tag{10.140}$$

These transitions are also shown in Fig. 10.5. The selection rules even allow for the transition $\psi_+ \to \psi_-$ but its frequency is too low to be observable for magnetic fields of the magnitude generally used for NQR Zeeman studies. The orientation of the principal EFG tensor axis system can be made by observing the behavior of the $\nu_+ \to \nu_{+3/2}$ transition as a function of the orientation of a crystal in an external magnetic field. This method becomes apparent by examination of Fig. 10.6 where the α and β pair transition frequencies have been plotted as a function of the polar angle θ. The two pairs are always symmetrically displaced about the position of the zero field line. Their separations as derived from Eqs. (10.136) to (10.139) and (10.140) are given by

$$\Delta\nu_\alpha = [3 + (1 + 4\tan^2\theta)^{1/2}]\gamma H_0 \cos\theta \tag{10.141}$$

$$\Delta\nu_\beta = [3 - (1 + 4\tan^2\theta)^{1/2}]\gamma H_0 \cos\theta \tag{10.142}$$

There are two unique positions on this plot: (1) the points where $\Delta\nu_\alpha = 0$ and (2) the points where $\Delta\nu_\alpha = \Delta\nu_\beta$. For the first situation we have $\sin^2\theta = \frac{2}{3}$. In this

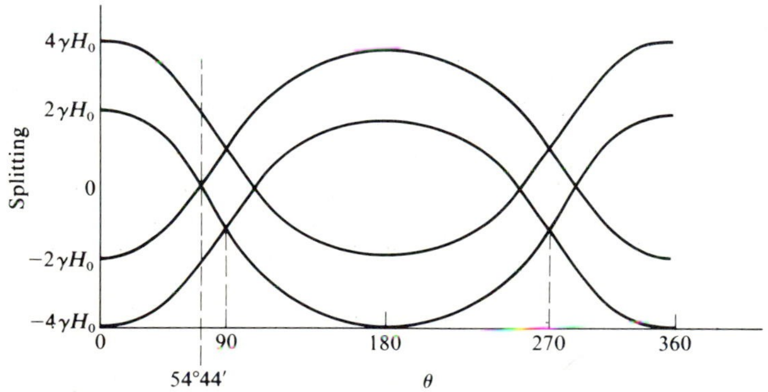

FIGURE 10.6
Variation of Zeeman components of $\psi_\pm \to \psi_{\pm 3/2}$ with θ.

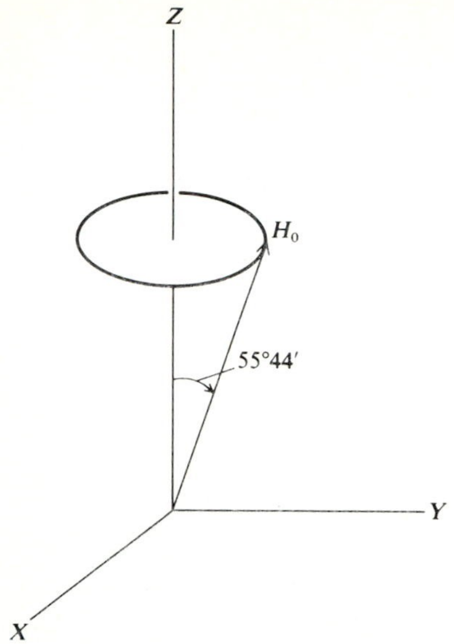

FIGURE 10.7
Zero-splitting locus for the Zeeman components of an $I = \frac{3}{2}$, $\eta = 0$ system.

case θ can have values of 54°44′, 125°16′, 234°44′, or 305°16′. For each of these values of θ there will be a 3-line pattern having the center line at the zero field position. These values of θ are said to define a locus of zero splitting as is illustrated in Fig. 10.7 for $\theta = 55°44'$. It is interesting to note that the loci for 55°44′ and 305°16′ superimpose, while those for 125°16′ and 234°44′ superimpose and have as their axis the $-Z$ direction.

If one experimentally determines the zero-splitting locus as a function of the positions of a crystal relative to an external magnetic field then the direction of the Z axis in the crystal is specified. The X and Y axes will be any arbitrary pair of perpendicular axes in the plane normal to the Z axis since for $\eta = 0$ they are equivalent. It is to be noted that the value of θ that defines the zero-splitting loci is a function of the spin quantum number. The second situation, where $\Delta\nu_\alpha = \Delta\nu_\beta$, occurs for $\theta = 90°, 270°$. The location of these splittings can often serve as a check on the information obtained from a determination of the zero-splitting loci.

The behavior of transitions between Zeeman levels of nuclei with half-integer values of spin and having $m > \frac{1}{2}$ is different. In this case there is only one pair of transitions whose separation is given by $\Delta\nu_{\pm m} = 2\gamma H_0 \cos\theta$. In this situation a particular value of θ will define a locus of constant splitting with the two-line pattern reducing to a single line when H_0 lies in the XY plane. This is illustrated in Fig. 10.8. The Zeeman effect for systems with integer values of spin is analogous to that for the $m > \frac{1}{2}$ case for nuclei with half-integer spins, with the exception that there is no degeneracy for the $m = 0$ state.

The background necessary to understand the Zeeman effect for a system having a nonaxial symmetric EFG tensor is embodied in the previous discussion. The procedure follows the same method except the matrix elements and zero

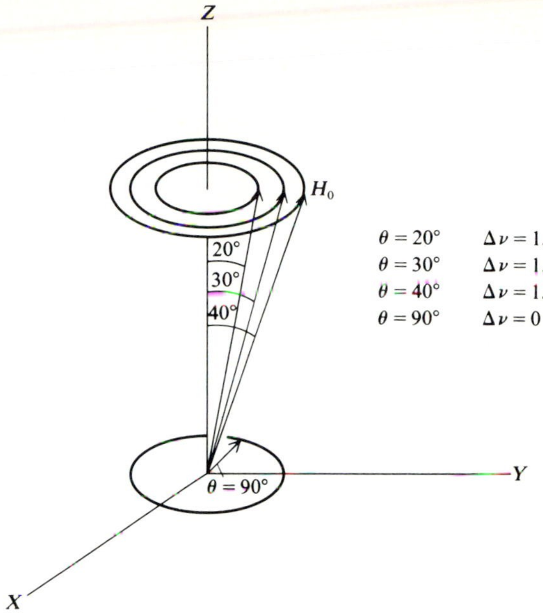

FIGURE 10.8
Loci of constant splittings for Zeeman components with $m > \frac{1}{2}$.

field energies are those of the $\eta \neq 0$ case presented in Sec. 10.5. The general references at the end of the chapter provides sources for the mathematical details, and only the results will be given.

Qualitatively the Zeeman splittings for an $\eta \neq 0$ system are like those for the $\eta = 0$ one. The two principal exceptions are the dependency of the zero field energies, and hence the Zeeman energy levels, on η and the dependence of the zero-splitting loci on the angle ϕ. The energy matrix for the Zeeman perturbation can be written

$$|\langle m'|\mathscr{H}_M|m\rangle - \delta_{mm'}E_m| = 0 \tag{10.143}$$

where for half-integer spin systems, the elements connect states given by Eqs. (10.105) and (10.106). The values of E_m found by solving these eigenvalue equations are added to the zero field energies. The only nonzero matrix elements will be

$$\langle m|\mathscr{H}_M|m\rangle = -\langle -m|\mathscr{H}_M|-m\rangle = \tfrac{1}{2}\hbar\gamma H_0 a_m \cos\theta \tag{10.144}$$

$$\langle m|\mathscr{H}_M|-m\rangle = \tfrac{1}{2}\hbar\gamma H_0 \sin\theta(b_m e^{-i\phi} + c_m e^{i\phi}) \tag{10.145}$$

where

$$a_m = c_{1/2,\,m}^2 - 3C_{3/2,\,m}^2 + 5C_{5/2,\,m}^2 - \cdots \tag{10.146}$$

$$b_m = (I + \tfrac{1}{2})C_{1/2,\,m}^2 + 2g_I(3/2)C_{3/2,\,m}c_{5/2,\,m} + 2g_I(7/2)C_{7/2,\,m}C_{7/2,\,m} + \cdots \tag{10.147}$$

$$c_m = 2g_I(1/2)C_{1/2,\,m}C_{3/2,\,m} + 2g_I(5/2)C_{5/2,\,m}C_{7/2,\,m} + \cdots \tag{10.148}$$

$$g_I(m) = [I(I+1) - m(m+1)]^{1/2} \tag{10.149}$$

and the $C_{m',m}$ are the coefficients in Eqs. (10.104) and (10.105). For a system of spin I the total magnetic energy matrix will be of the order $(2I + 1)$ and will produce $(I + \frac{1}{2})$ pairs of quadratic equations whose solutions are

$$E^M_{\pm m} = \pm \frac{\hbar \gamma H_0}{2} [a^2_m \cos^2 \theta + (b^2_m + c^2_m + 2b_m c_m \cos 2\phi) \sin^2 \theta]^{1/2} \quad (10.150)$$

For $I = \frac{3}{2}$ the coefficients become

$$a_{3/2} = -1 - 2\left(1 + \frac{\eta^2}{3}\right)^{-1/2} \quad (10.151)$$

$$a_{1/2} = -1 + 2\left(1 + \frac{\eta^2}{3}\right)^{-1/2} \quad (10.152)$$

$$b_{3/2} = 1 - \left(1 + \frac{\eta^2}{3}\right)^{-1/2} \quad (10.153)$$

$$b_{1/2} = 1 + \left(1 + \frac{\eta^2}{3}\right)^{-1/2} \quad (10.154)$$

$$c_{3/2} = -C_{1/2} = \eta\left(1 + \frac{\eta^2}{3}\right)^{-1/2} \quad (10.155)$$

and the transition frequencies are given by

$$\nu = \nu_0(m_i \rightarrow m_j) \pm \frac{\gamma H_0}{2} \{[d_i] \pm [d_j]\} \quad (10.156)$$

where

$$[d_i] = [a^2_{m_i} \cos^2 \theta + (b^2_{m_i} + c^2_{m_i} + 2b_{m_i} c_{m_i} \cos 2\phi) \sin^2 \theta]^{1/2} \quad (10.157)$$

This gives the four possible transitions, depending on the choice of sign combinations used. As with the $\eta = 0$ case it is conventional to refer to the outer pair as the β pair and the inner ones as the α pair. We find that $\Delta\nu_\alpha = \gamma H_0\{[d_i] - [d_j]\}$, so for $I = \frac{3}{2}$ the condition for zero splitting of the α pair becomes $[d_{1/2}] = [d_{3/2}]$. Substitution of Eq. (10.157) into the equality, and solving for $\sin^2 \theta$ yields

$$\sin^2 \theta = \frac{2}{3 - \eta(1 + \eta^2/3)^{-1/2} \cos 2\phi} \quad (10.158)$$

For small values of η (first-order in η) this reduces to

$$\sin^2 \theta = \frac{2}{3 - \eta \cos 2\phi} \quad (10.159)$$

Looking at these equations we see that for a given value of η the value of $\sin^2 \theta$, and hence that of θ itself, will have a maximum when $\phi = 0°$ or $180°$ and a minimum value $\phi = 90°$ or $270°$. The zero-splitting locus is an elliptical cone with its two-fold axis being the EFG tensor principal Z axis, and θ $(\phi = 0)$ and θ $(\phi = 90°)$ defining the directions of the Y and X axes, respectively. Using Eqs.

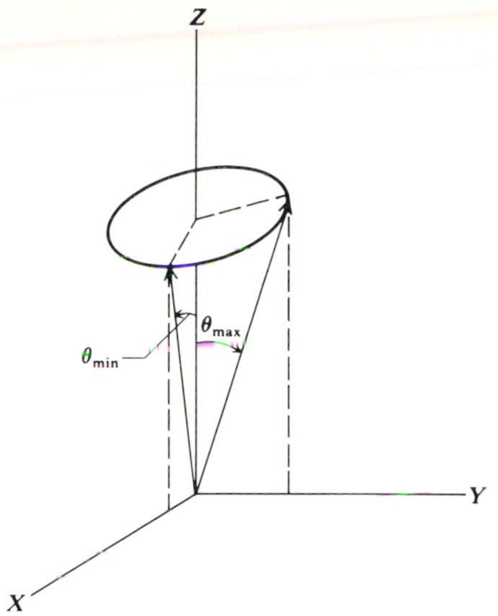

FIGURE 10.9
Zero splitting locus with $\eta \neq 0$.

(10.158) and (10.159) and solving for these two values of ϕ, $\cos 2\phi = 1$ for θ_{max} and $\cos 2\phi = -1$ for θ_{min} we have

$$\frac{3(\sin^2 \theta_{max} - \sin^2 \theta_{min})}{(\sin^2 \theta_{max} + \sin^2 \theta_{min})} = \frac{\eta}{(1 + \eta^2/3)^{1/2}} \tag{10.160}$$

or, to first order in η,

$$\frac{3(\sin^2 \theta_{max} - \sin^2 \theta_{min})}{(\sin^2 \theta_{max} + \sin^2 \theta_{min})} = \eta \tag{10.161}$$

Thus we see that the experimental determination of the zero-splitting locus for a nucleus will allow one to determine the direction of the EFG tensor principal axis system just as in the $\eta = 0$ case. The relations for the $\eta \neq 0$ case are illustrated in Fig. 10.9.

10.7 EXAMPLES

The analysis of polycrystalline NQR data is illustrated by considering the observed resonance of ^{59}Co and ^{35}Cl in bis(tetracarbonylcobalt)tin (IV)chloride. In this compound the ^{59}Co resonances are at 10.853, 21.243, and 31.926 MHz while the ^{35}Cl resonance is at 17.676 MHz. Since ^{59}Co has a nuclear spin of $I = \frac{7}{2}$ there are three transitions, and sufficient information is available to independently determine e^2Qq_{ZZ} and η. Looking at Fig. 10.2 one observes that for $\eta = 0$ the frequency of the $|\frac{3}{2}| \rightarrow |\frac{5}{2}|$ transition is twice that of the $|\frac{1}{2}| \rightarrow |\frac{3}{2}|$ transition. Comparing the lower two frequencies for ^{59}Co we see that this is almost the case, hence η will be small. To determine both e^2Qq_{ZZ} and η it is necessary to set up the secular determinate [Eq. 10.91)] for the $I = \frac{7}{2}$ case, solve for the resulting energies, and take the appropriate differences to obtain expressions for the frequencies. Since η is small this problem is simplified by neglecting higher powers of η. For $I = \frac{7}{2}$ the resulting frequency expressions are

$$\nu_{1/2\leftrightarrow 3/2} = \tfrac{1}{14}(e^2Qq_{ZZ})(1 + 50.865\eta^2) \tag{10.162}$$

$$\nu_{3/2\leftrightarrow 5/2} = \tfrac{2}{14}(c^2Qq_{ZZ})(1 - 15.867\eta^2) \tag{10.163}$$

$$\nu_{5/2\leftrightarrow 7/2} = \tfrac{3}{14}(e^2Qq_{ZZ})(1 - 2.8014\eta^2) \tag{10.164}$$

and the simultaneous solution of these gives $e^2Qq_{zz} = 149.1$ MHz and $\eta = 0.065$. For the ^{35}Cl nuclei having a spin $I = \frac{3}{2}$ only one transition is observed and one can only assume a zero value of η and use Eq. (10.96) to get $e^2Qq_{zz} = 35.352$ MHz.

A Zeeman study of a spin $I = \frac{3}{2}$ system is illustrated by the study of ^{137}Ba in $BaCl_2 \cdot 2H_20$. A stereographic plot of the observed zero-splitting loci (see App. P) for 76 data points is shown in Fig. 10.10. The two ellipses arise from two nonequivalent Ba positions in the unit cell. The two barium atoms are chemically equivalent but occupy different unit cell sites, hence the ellipses differ only in orientation. The positions of the major and minor axes of the ellipses, the parameters necessary to employ Eqs. (10.96) and (10.160), and the resulting NQR parameters, are given in Table 10.1.

In this chapter we have considered the phenomenon of the interaction between the electric quadrupole moment of a nucleus and the surrounding electric field. The emphasis has been directed toward gaining an understanding of the basic interaction in solids. Applications to gases will be considered in Chap. 14. There are two particular aspects of this subject which have been omitted but can be obtained from literature sources once the basic theory of this chapter is mastered.

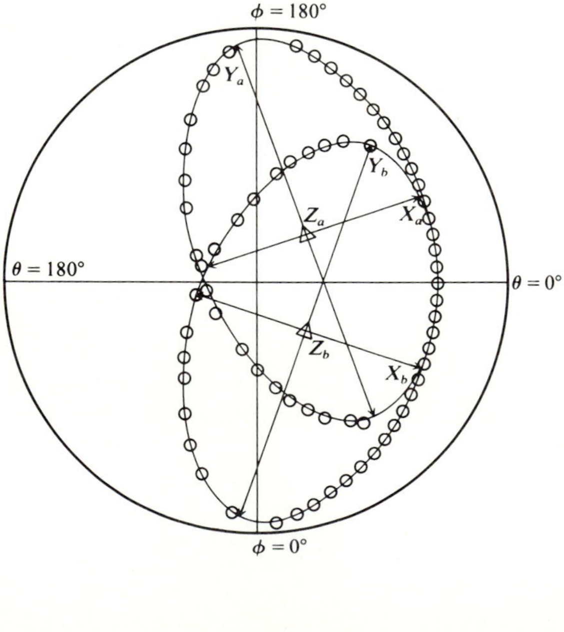

FIGURE 10.10
Zero-splitting loci for the ^{137}Ba NQR resonance in $BACl_2 \cdot 2H_2O$.

TABLE 10.1
NQR data and parameters for $BaCl_2 \cdot 2H_2O$

	Site A	Site B
Z	(0°, 0°)(68°, 110°)	(0°, 0°)(69°, 68°)
Y	(118°, 176°)(32°, 41°)	(121°, 5°)(30°, 138°)
X	(20°, 112°)(118°, 99°)	(22°, 68°)(121°, 79°)
θ_{max}	67.0° ± 2°	67.9° ± 2°
θ_{min}	46.3° ± 2°	4.65° ± 2°
η	0.78 ± 0.14	0.76 ± 0.14
e^2Qq_{zz}, MHz	29.75 ± 0.9	29.80 ± 0.9

1. The presence of a nuclear electric quadrupole moment can be a very important factor in the determination of the nature of the nuclear magnetic resonance spectrum of a compound. The NMR absorption lines for a quadrupole nuclei are broadened or split into several components by the presence of the quadrupole moment. The information which can be gained from such spectra is similar to that obtained from a pure NQR study. The method has a distinct advantage over pure NQR methods when the magnitude of e^2Qq is small, say less than 1 MHz, and the experimental observation of pure NQR becomes difficult or impossible. An early review of the theory of this effect was given by Cohen and Reif [7] and some monographs on NMR treat it to some extent. The presence of a quadrupole moment provides a means for nuclear relaxation thereby decreasing the lifetimes and broadening the resonance lines for the high-resolution NMR spectra of such nuclei. For this reason, it is difficult to study chemical shifts of nuclei such as N, Cl, or Br when they are in covalent environments where they exhibit large nuclear quadrupole coupling.

2. The second topic which has been omitted is the problem of evaluating q_{ZZ} for particular systems. This involves the use of average value calculations over electronic wavefunctions or some empirical method for correlation of results. The discussion of this problem is one of chemical bonding theories and its persual would lead us away from the main intent of this book. This aspect of NQR spectroscopy has been presented in several books and review articles given in the general references.

REFERENCES

A. Specific

1. Margenau, H., and G. M. Murphy, *The Mathematics of Physics and Chemistry*, Van Nostrand Co., Princeton, N.J., 1943, p. 96.
2. Ramsey, N. F., *Nuclear Moments*, John Wiley and Sons, New York, N.Y., 1953, p. 24.
3. Stratton, J. A., *Electromagnetic Theory*, McGraw-Hill, New York, N.Y., 1941, chap. III.
4. Bersohn, R., *J. Chem. Phys.*, **20**, 1505 (1952).
5. Dean, C., *Phys. Rev.*, **46**, 1053 (1954).

6. Cohen, M. H., *Phys. Rev.*, **96**, 1278 (1954).
7. Cohen, M. H., and F. Reif, "Nuclear Quadrupole Effects in NMR," *Solid State Physics*, **5**, 321 (1957).

B. General

Slichter, C. P., *Principles of Magnetic Resonance*, Harper and Row, New York, N.Y., 1963.
Townes, C. H., and A. L. Schawlow, *Microwave Spectroscopy*, McGraw-Hill, New York, N.Y., 1955, chs. 5, 6, 9.
Das, T. P., and E. L. Hahn, *NQR Spectroscopy*, Academic Press, New York, N.Y., 1958.
Lucken, E. A. C., *Nuclear Quadrupole Coupling Constants*, Academic Press, New York, N.Y., 1969.
Kubo, M., and D. Nakamura, *Adv. in Inorganic Chem. and Radiochem.*, **8**, 257 (1966).
O'Konski, C. T., "NQR Spectroscopy," in *Determination of Organic Structures by Physical Methods* (eds., F. C. Nachod and W. D. Phillip), Academic Press, New York, N.Y. (1962).
Ramsey, N. F., *Nuclear Moments*, John Wiley and Sons, New York, N.Y., 1953.
Segel, S. L., and R. G. Barnes, "Catalog of Nuclear Quadrupole Interaction and Resonance Frequencies in Solids," USAEC Research and Development Report, Ames Laboratory. Part I, "Elements and Inorganic Compounds," IS-520 (1962). Part II, "Halogen Measures in Organic Compounds," IS-1222 (1965).

PROBLEMS

10.1. Referring to Eq. 10.24 evaluate ∇E_0 for the 1s, $2p_z$ and $3dz_2$ hydrogen-like atomic orbitals allowing the atomic number Z to be a variable parameter.

10.2. Develop the matrix elements given by Eqs. (10.85) to (10.87) beginning with Eqs. (10.81).

10.3. For $I = \frac{5}{2}$ nuclei show that the matrix element $(M|H(t)|M+1)$ is nonzero if $H_x(t) \neq 0$ and $\eta = 0$.

10.4. Evaluate the energy levels for an iodine nucleus ($I = \frac{5}{2}$) in solid HI assuming axial symmetry.

10.5. Find the nonzero elements of the secular determinant for a nucleus with $I = \frac{5}{2}$ and $\eta = 0$.

10.6. Evaluate the Zeeman contribution to the quadrupole energy level $E_{3/2}$ for the following cases and determine if it is justified in using first-order perturbation theory in each case (i.e., is $E_{Zee} \ll E_Q$)
(*a*) ^{35}Cl in a 0.1 T field
(*b*) ^{63}Cu in a 1 T field
(*c*) ^{127}I in a 0.1 T field

10.7. Under conditions of optimum orientation ($f_I \cos\theta$ = maximum) what magnetic field would be needed to observe a 2 MHz value for the ν_ε Zeeman transition (that is, $E_+ \rightarrow E_-$) for (*a*) ^{35}C, (*b*) ^{63}Cu, (*c*) ^{127}I.

10.8. The unsplit resonance of ^{35}Cl in $(CH_3)_2SnCl_2$ is at 15.712 MHz. A Zeeman study of a single crystal established $\theta_{max} = 60°$ and $\theta_{min} = 51°$. Evaluate the quadrupole coupling constant and asymmetry parameter for the system considering the perturbation to be first-order in η.

CHAPTER

11

DIATOMIC VIBROTOR STATES AND SELECTION RULES

11.1 INTRODUCTION

Although the rigid rotor and the harmonic oscillator were used as examples to show the application of quantum mechanics to molecular systems, both of these were ideal examples in that they made the unreal assumptions that molecules are rigid and that vibrational motion is strictly harmonic. The application of these assumptions and the resulting solutions gave an approximation to the state of a real molecular system but such results are generally insufficiently accurate to account for the observed rotational-vibrational spectra of even diatomic molecules. The Born-Oppenheimer [1] approximation is reasonably effective in the separation of nuclear and electronic motions in molecular systems but, as will become evident from examples which follow, the complete separation of vibrational and rotational motions when analyzing high-resolution spectra is not possible. The gross aspects of either type of motion can be considered separately but their interactions do produce perturbations on one another that are well within the range of experimental measurements. We will first consider the combined rotational-vibrational motion of the diatomic vibrotor in order to point out the nature and magnitude of the interactions and later examine specific individual aspects characteristic of rotational and vibrational spectroscopy.

There are four principal interactions which will affect the energy of a diatomic molecular vibrotor as calculated by the rigid rotor or harmonic oscillator model.

1. The harmonic oscillator model indicates that a molecule always has at least a finite amount of vibrational energy, the zero point energy. This means that there can never be a rigid rotor in the true sense of the word and one must describe the molecule by an average or equilibrium internuclear separation.
2. Since the average bond length increases with increasing vibrational energy the effective moment of inertia and hence, the rotational energy, will depend on the vibrational energy state.
3. As the molecule goes to higher rotational states it is distorted by centrifugal force thereby perturbing the energies.
4. Even in the lowest vibrational state and especially in higher vibrational states the bond stretching motion does not obey Hook's law and the vibration is anharmonic in nature.

In order to develop an expression for the energy of a diatomic vibrotor in terms of experimentally measurable quantities the Born-Oppenheimer approximation as applied to the separation of nuclear and electronic motions will be accepted as valid. This means that the electronic motion in the system is so rapid compared to the time span of nuclear motion that the nuclei can be considered to be moving in an electric field whose magnitude is only a function of nuclear coordinates.

11.1 SEPARATION OF INTERNAL AND TRANSLATIONAL MOTION

A diatomic molecule consists of two nuclei located in an electronic charge cloud which, when combined with the internuclear repulsion, gives a potential denoted by $V(r)$, where r is the internuclear separation. The entire system is free to move in three-dimensional space. Specifying the location of the two nuclei by x_1, y_1, z_1, and x_2, y_2, z_2 in a Cartesian system the total classical energy of the system is written as

$$E = \frac{p_1^2}{2m_1} + \frac{p_2^2}{2m_2} + V(x_1, y_1, z_1, x_2, y_2, z_2) \tag{11.1}$$

where

$$p_i^2 = p_{x_i}^2 + p_{y_i}^2 + p_{z_i}^2 \tag{11.2}$$

and m_1 and m_2 are the masses of the two nuclei. Using the Schrödinger wave mechanical approach to the problem one establishes the wave equation to be

$$-\frac{\hbar^2}{2}\left\{\frac{1}{m_1}\nabla_1\Psi + \frac{1}{m_2}\nabla_2\Psi\right\} + V(x_1, y_1, z_1, x_2, y_2, z_2)\Psi = E_T\Psi \tag{11.3}$$

where Ψ is a function of the six coordinates x_1, y_1, z_1, x_2, y_2, z_2.

Since the present concern is with motion of one nucleus relative to the other and with the overall rotation of the molecule it is not necessary to solve Eq. (11.3) for a wavefunction which is dependent on all six coordinates since three will be used to specify the overall translational motion of the system. By analogy with the case of classical motion we will consider that the translational and internal (rotational and vibrational) energies are separable. The validity of this separation will be shown mathematically and is also confirmed by the applicability of the resulting energy expressions and their agreement with experiment. This separation can be achieved by specifying two new coordinate systems, one giving the location of nucleus-1 relative to nucleus-2, x, y, z, and one giving the location of the center of mass of the system, X, Y, Z, and transform to these systems. The relation of these systems to the Cartesian system of the two nuclei, x_i, y_i, z_i is shown in Fig. 11.1. From Fig. 11.1 one can observe that

$$x = x_2 - x_1 \tag{11.4}$$

$$y = y_2 - y_1 \tag{11.5}$$

$$z = z_2 - z_1 \tag{11.6}$$

and from the definition of the center of mass one has

$$X = \frac{m_1x_1 + m_2x_2}{m_1 + m_2} \tag{11.7}$$

$$Y = \frac{m_1y_1 + m_2y_2}{m_1 + m_2} \tag{11.8}$$

$$Z = \frac{m_1z_1 + m_2z_2}{m_1 + m_2} \tag{11.9}$$

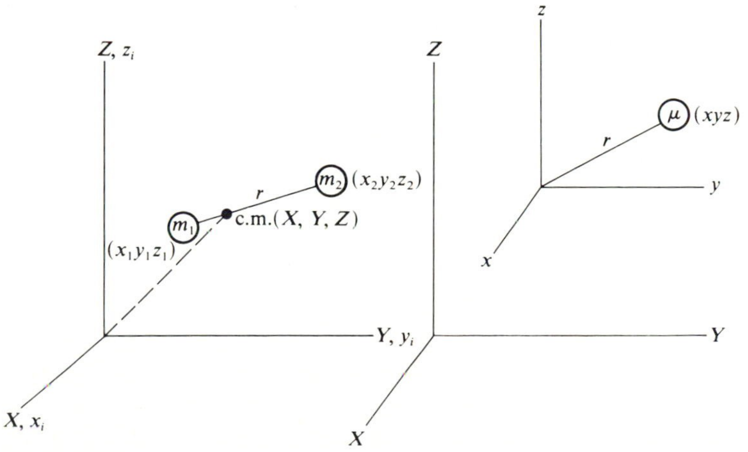

FIGURE 11.1
Relation of Cartesian axis system to center of mass and internal axis systems for a diatomic vibrotor.

The Laplacian operators, ∇_i^2 must be transformed to the new coordinate systems. This lengthy but straightforward procedure as outlined in App. Q gives

$$\frac{-\hbar^2}{2(m_1+m_2)}\nabla_F^2\Psi - \frac{\hbar^2}{2\mu}\nabla_g^2\Psi + V(g)\Psi = E_T\Psi \tag{11.10}$$

The assumption of separability of the translational and internal motion is expressed by writing the wavefunction $\Psi = \Psi\,(x_1, y_1, z_1, x_2, y_2, z_2)$ as a product of two functions, $F = F\,(X, Y, Z)$, which is a function of the center of mass coordinates only, and $\psi = \psi\,(x, y, z)$, which is a function of the internal coordinates only.

$$\Psi = F\Psi \tag{11.11}$$

Substituting $F\psi$ into Eq. (11.10) and recognizing that F is independent of x, y, z and that ψ is independent of X, Y, Z one obtains

$$\frac{-\hbar^2\Psi}{2(m_1+m_a)}\nabla_F^2 - \frac{\hbar^2 F}{2\mu}\nabla_g^2\Psi + V(g)F\Psi = E_T F\Psi \tag{11.12}$$

Dividing by the product function $F\psi$ gives

$$\frac{-\hbar^2}{2(m_1+m_a)F}\nabla_F^2 F - \frac{\hbar^2}{2\mu\Psi}\nabla_g^2\Psi - V(g) = E_T \tag{11.13}$$

which separates into two equations

$$\frac{-\hbar^2}{2(m_1+m_a)F}\nabla_F^2 F = E_{\mathrm{Tr}} \tag{11.14}$$

$$\frac{-\hbar^2}{2\mu\Psi}\nabla_g^2\Psi + V(g) = E \tag{11.15}$$

where E_{Tr} is the translational energy of the center of mass and E is the internal energy of the system. The equation for the translational motion is that for a particle in a three-dimensional box.

Examination of Fig. 11.1 shows that the coordinate system describing the internal motion is one having one of the mass points at the origin. This suggests that a change to spherical coordinates could simplify solution of Eq. (11.15). Using the appropriate quantum mechanical operators from Table 2.1 the wave equation for internal motion can be written

$$\frac{-\hbar^2}{2\mu}\left\{\frac{1}{r^2}\frac{\partial}{\partial r}r^2\frac{\partial\Psi}{\partial r} + \frac{1}{r^2\sin\theta}\frac{\partial}{\partial\theta}\sin\theta\,\frac{\partial\Psi}{\partial\theta} + \frac{1}{r^2\sin^2\theta}\frac{\partial^2\psi}{\partial\phi^2}\right\} + V(r)\psi = E\psi \tag{11.16}$$

If r is a constant, the rigid rotor approximation, then $V(r)$ is a constant and Eq. (11.16) reduces to the wave equation for the rigid rotor. In a real molecule, however, r and $V(r)$ are not constant. If $\psi = \psi(r, \theta, \phi)$ is written as a product function, $\psi = R(r)\Theta(\theta)\Phi(\phi)$, then the wave equation can be factored into three separate equations. Substituting $\psi = R\Theta\Phi$ into Eq. (11.16), remembering that R,

Θ, and Φ are functions of the variables r, θ, and ϕ respectively and dividing by $R\Theta\Phi$ leads to

$$\frac{1}{r^2R}\frac{\partial}{\partial r}r^2\frac{\partial R}{\partial r}+\frac{1}{r^2\sin\theta}\frac{\partial}{\partial\theta}\sin\theta\frac{\partial}{\partial\theta}+\frac{1}{\Phi r^2\sin^2\theta}\frac{\partial^2\Phi}{\partial\phi^2}+\frac{2\mu}{h^2}\{E-V(r)\}=0 \quad (11.17)$$

This is now of separable form. Multiplying by $r^2 \sin^2\theta$ and taking the separation constant as M^2 one has

$$\frac{\sin^2\theta}{R}\frac{\partial}{\partial r}r^2\frac{\partial R}{\partial r}+\frac{\sin\theta}{\Theta}\frac{\partial}{\partial\theta}\sin\theta\frac{\partial\Theta}{\partial\theta}+\frac{2\mu r^2\sin\theta}{\hbar^2}\{E-V(r)\}=M^2 \quad (11.18)$$

and

$$\frac{1}{\Phi}\frac{d^2\Phi}{d\phi^2}=-M^2 \quad (11.19)$$

Since the separation produces an equation in the variable ϕ only the partial derivative becomes an ordinary derivative. Substituting Eq. (11.19) back into Eq. (11.17), multiplying by r^2, and letting b be a second separation constant, we get

$$\frac{1}{\Theta\sin\theta}\frac{d}{d\theta}\frac{d\Theta}{d\theta}-\frac{M^2}{\sin^2\theta}=-b \quad (11.20)$$

and

$$\frac{1}{R}\frac{d}{dr}r^2\frac{dR}{dr}+\frac{2\mu r^2}{h^2}\{E-V(r)\}=b \quad (11.21)$$

The motion of the system is thus described by three second-order differential equations whose solutions, though somewhat complex, are obtainable. The solutions for the θ and Φ equations are related to those for the hydrogen atom, but the radial equation differs considerably. It is noted that the angular wave equations, which depend on the variables θ and ϕ, are independent of the potential function, $V(r)$. These equations are those found for the rigid rotor

$$\Phi_M=\frac{1}{\sqrt{2\pi}}e^{im\phi} \quad (11.22)$$

and

$$\Theta_{JM}=N_{JM}P_J^{|M|}(\cos\theta) \quad (11.23)$$

where

$$N_{JM}=\left[\frac{(2J+1)(J-|M|)!}{2(J+|M|)!}\right]^{1/2} \quad (11.24)$$

and

$$P_J^{|M|}(\cos\theta)=\frac{(-1)^J}{2^JJ!}\sin^{|M|}\theta\frac{d^{|M|}}{d(\cos\theta)^{|M|}}\frac{d^J(\cos^2\theta-1)^J}{d(\cos\theta)^J} \quad (11.25)$$

TABLE 11.1
Spherical harmonics

Y_0^0	$(\frac{1}{4}\pi)^{1/2}$
Y_1^0	$(\frac{3}{4}\pi)^{1/2}\cos\theta$
Y_1^1	$(\frac{3}{8}\pi)^{1/2}\sin\theta\, e^{i\phi}$
Y_1^{-1}	$(\frac{3}{8}\pi)^{1/2}\sin\theta\, e^{-i\phi}$
Y_2^0	$(\frac{5}{16}\pi)^{1/2}(3\cos^2\theta - 1)$
Y_2^1	$(\frac{15}{8}\pi)^{1/2}\sin\theta\cos\theta\, e^{i\phi}$
Y_2^{-1}	$(\frac{15}{8}\pi)^{1/2}\sin\theta\cos\theta\, e^{-i\phi}$
Y_2^2	$(\frac{15}{8}\pi)^{1/2}\sin^2\theta\, e^{2i\phi}$
Y_2^{-2}	$(\frac{15}{32}\pi)^{1/2}\sin^2\theta\, e^{-2i\phi}$

In order for Eqs. (11.19) and (11.20) to have acceptable solutions the necessary conditions are that $b = J(J + 1)$ where J is an integer and that M be limited to integer values of $M = J, J - 1, \ldots, 0, \ldots, -J + 1, -J$.

The nature of these results for the θ and ϕ functions, that is, their independence of any potential function, is a characteristic of any two-body problem in quantum mechanics. Because of this the two sets of functions are often combined to give a set known as the spherical harmonics:

$$Y_J^M(\theta, \phi) = \Theta_{JM}(\theta)\Phi_M(\phi) \tag{11.26}$$

Several of these functions, with normalization constants, are listed in Table 11.1 for lower values of J and M. These are recognized as being identical to the angular part of the hydrogen atom wavefunctions. In addition to the restrictions on J and M mentioned in the last paragraph an examination of Table 11.1 reveals that for each value of J there are $2J + 1$ different spherical harmonics having different values of M. Equation (11.25) shows that for M greater than J the associated Legendre function $P_J|M|(\cos\theta)$, will vanish.

Setting $b = J(J + 1)$ the radial equation, Eq. (11.21), becomes

$$\frac{1}{r^2}\frac{d}{dr}r^2\frac{dR}{dr} + \left\{\frac{2\mu}{\hbar^2}[E - V(r)] - \frac{J(J+1)}{r^2}\right\}R = 0 \tag{11.27}$$

The term $J(J + 1)/r^2$ may be viewed as an added potential term resulting from the centrifugal force due to the rotation of the molecule. The magnitude of this term increases approximately as J^2 and can become an appreciable fraction of the potential term for large values of J. This potential however, unlike that for the hydrogen atom where one has a central force field, is a function of the internuclear separation and the added centrifugal term. Since the solution of Eq. (11.27) will depend on the form of $V(r)$ we need to select an appropriate potential function.

11.3 ENERGY STATES

The energy states of the diatomic vibrotor are found by substituting an appropriate potential function into Eq. (11.27) and solving the resulting differential equation.

The solution of this equation is simplified by substitution of $S(r) = rR(r)$. The nature of this process is illustrated by making this substitution and incorporating the Morse potential function into Eq. (11.27) to give

$$\frac{d^2S}{dr^2} + \left\{\frac{-J(J+1)}{r^2} + \frac{2\mu}{\hbar^2}[E - D - De^{-2a(r-r_e)} + 2De^{-a(r-r_e)}]\right\}S = 0 \quad (11.28)$$

It should be recognized that since the Morse function involves only three parameters (r_e, a, D) the resulting solution will be limited in its applicability but nevertheless provides an energy expression which will be considerably more useful than the rigid rotor-harmonic oscillator approximations. By a proper change of variables and rearrangement Eq. (11.28) can be reduced to the Laguerre type of differential equation. The details of this reduction and the solution of the resulting equation [2] are given in App. R and a discussion of the Laguerre equation is in App. F.

The solution of Eq. (11.27), to terms second-order in v and J, the vibrational and rotational quantum numbers respectively, is

$$\frac{E_{vJ}}{h} = \omega_e(v + \tfrac{1}{2}) - \chi_e\omega_e(v + \tfrac{1}{2})^2 + B_eJ(J+1) - D_eJ^2(J+1)^2$$

$$- \alpha_e(v + \tfrac{1}{2})J(J+1) \quad (11.29)$$

where

$$\omega_e = \frac{a}{2\pi}\sqrt{\frac{2D}{\mu}} \quad (11.30)$$

$$\chi_e = \frac{h\omega_e}{4D} \quad (11.31)$$

$$B_e = \frac{\hbar}{4\pi I_e} \quad (11.32)$$

$$I_e = \mu r_e^2 \quad (11.33)$$

$$D_e = \frac{4B_e^3}{\omega_e^2} \quad (11.34)$$

$$\alpha_e = 6\sqrt{\frac{\chi_e B_e^3}{\omega_e}} - \frac{6B_3^2}{\omega_e} = \frac{3h^2\omega_e}{4\mu r_e^2 D}\left\{\frac{1}{ar_e} - \frac{1}{a^2r_e^2}\right\} \quad (11.35)$$

with ω_e, α_e, and B_e expressed in hertz.

Examination of this expression for the energy shows that the first term is that of the harmonic oscillator and the third term is that of a rigid diatomic rotor. The second term is due to anharmonicity in the oscillator motion which arises from the nonparabolic shape of the Morse potential and is analogous to the expression obtained by inclusion of a cubic term in the potential energy function of a simple harmonic oscillator. Higher-order terms involving the anharmonicity can be obtained if one retains higher-order terms in the solution of Eq. (11.27).

The fourth term is due to the centrifugal stretching of the molecule as it rotates. One notes that this centrifugal distortion is proportional to J^4 rather than J^2, as is the rotational energy, indicating that centrifugal distortion increases with larger values of J. Since the centrifugal distortion constant D_e is proportional to B_e^3 and inversely proportional to ω_e^2 one can conclude that this term will be larger for light molecules having small moments of inertia, hence large B_e values, and for molecules having low fundamental vibrational frequencies ω_e. The last term is a measure of the change in the average moment of inertia due to vibration and consequently contributes to the total rotational energy.

An analytical function which is generally more applicable can be obtained by using the Dunham potential function [Eq. (6.19)]. This expression is comprised of sequences of terms in increasing powers of the quantum numbers analogous to the form of the potential

$$\begin{aligned}\frac{E_{vJ}}{h} &= \omega_e(v+\tfrac{1}{2}) - \chi_e\omega_e(v+\tfrac{1}{2})^2 - y_e\omega_e(v+\tfrac{1}{2})^3 - z_e\omega_e(v+\tfrac{1}{2})^4 + \cdots \\ &\quad + [B_e - (v+\tfrac{1}{2})\alpha_e - (v+\tfrac{1}{2})^2\gamma_e - \cdots]J(J+1) \\ &\quad - D_eJ^2(J+1)^2 - H_eJ^3(J+1)^3 - \cdots \end{aligned} \tag{11.36}$$

The $y_e\omega_e$ and $z_e\omega_e$ are additional anharmonicity terms, the γ_e an additional rotation-vibration interaction constant and H_e is another centrifugal distortion constant.

11.4 WAVEFUNCTIONS OF THE HARMONIC OSCILLATOR

The discussion of wavefunctions in this section will be limited to consideration of systems in which the rotation and vibration can be considered separately, the vibrations are harmonic and the rotor is rigid. Since a knowledge of the wavefunction is used primarily to determine selection rules and is not involved in determination of the energies of the system this approach is satisfactory.

The wavefunctions for a one-dimensional harmonic oscillator were found to be related to the Hermite polynomials by $\psi_v = N_v e - \frac{1}{2}\eta^2 H_v(\eta)$ where $H_v(\eta)$ are the Hermite polynomials. The first six wavefunctions are given in Table 11.2

TABLE 11.2
Wavefunctions for a one-dimensional harmonic oscillator

$\psi_0 = \left(\frac{\alpha}{\pi}\right)^{1/4} e^{-(1/2)\eta^2}$	$\psi_3 = \left(\frac{\alpha}{\pi}\right)^{1/4} \frac{3}{12}(8\eta^3 - 12\eta)e^{-(1/2)\eta^2}$
$\psi_1 = \left(\frac{\alpha}{\pi}\right)^{1/4} 2\eta e^{-(1/2)\eta^2}$	$\psi_4 = \left(\frac{\alpha}{\pi}\right)^{1/4} \frac{6}{48}(16\eta^4 - 48\eta^2 + 12)e^{-(1/2)\eta^2}$
$\psi_2 = \left(\frac{\alpha}{\pi}\right)^{1/4} \frac{2}{4}(4\eta^2 - 2)e^{-(1/2)\eta^2}$	$\psi_5 = \left(\frac{\alpha}{\pi}\right)^{1/4} \frac{15}{240}(32\eta^5 - 160\eta^3 + 120\eta)e^{-(1/2)\eta^2}$

$$\eta = \sqrt{\alpha}\, q = (\sqrt[4]{\mu k/h^2})q = (\sqrt{2\pi\mu\nu_0/\hbar})q$$

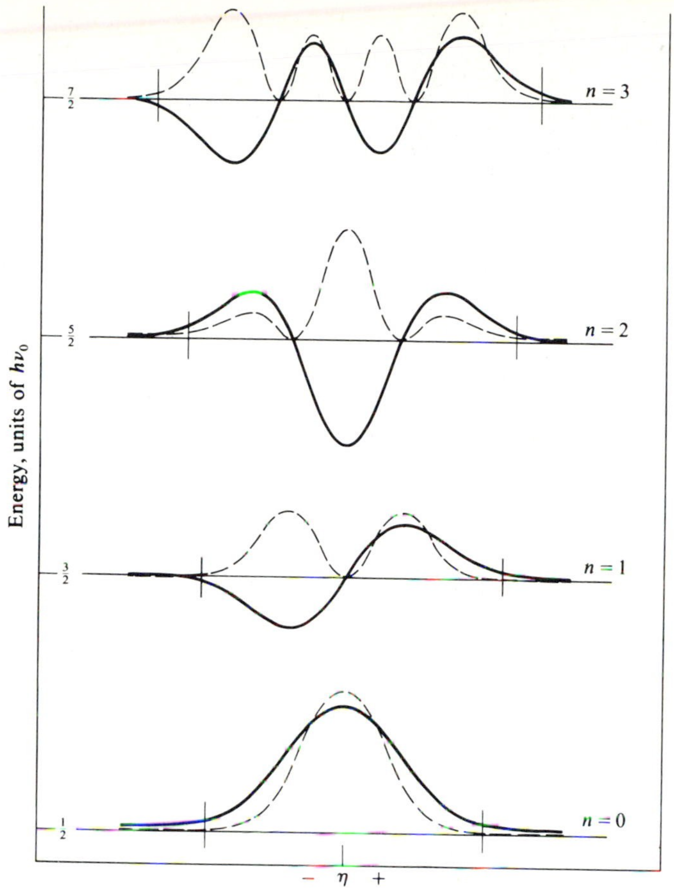

FIGURE 11.2
Schematic diagram of energy levels, wavefunctions (solid lines) and probability distribution functions (dotted lines) for a harmonic oscillator. The vertical marks on the η axes denote the position of $V(\eta_0)$. (These curves and that in Fig. 11.3 are computer-generated from the wavefunctions.)

and the first four are shown, along with the probability distribution functions and energy levels, in Fig. 11.2. In this discussion a new variable q, the displacement from the equilibrium value r_e, is introduced, that is, $q = r - r_e$.

There are several features to note with regard to the wavefunctions and probability amplitude functions.

1. The wavefunctions have a finite amplitude outside the bounds of the potential function, this being contrary to classical mechanics.
2. As the vibrational quantum number increases the portion of the wavefunction lying outside the bounds of the potential function decreases and the system approaches classical behavior.

3. For a classical harmonic oscillator the system would spend a longer proportion of its time at the extremes of oscillation than at the minimum of center of oscillation with a continuous distribution in between. The probability amplitude function for the quantum mechanical oscillator indicates a series of regions of maximum probability separated by nodes of zero probability.
4. In the limit of large values of the quantum number v, the points of maximum probability become very closely spaced and the system approaches classical behavior. This point is illustrated in Fig. 11.3. As was pointed out earlier, it is the approach to classical behavior in the limits of large values of the quantum number that formed part of the basis of the Heisenberg matrix formulation.

The first observation mentioned in the last paragraph warrants further consideration. If we consider the energies involved then penetration of the oscillator into the region beyond the potential boundary results in the oscillator having a total energy less than its potential energy, hence the situation is contrary to the law of conservation of energy. Such behavior on the part of the quantum mechanical oscillator is a direct result of the Heisenberg uncertainty principle which prohibits the simultaneous specification of its position and momentum (hence energy) beyond certain minimum limits. Another point-of-note which seems contrary to classical experience is the fact that the occurrence of nodes for the probability distribution function would indicate that the oscillator would be forever trapped in one region of space and since the probability of its being found at a node is zero, could not cross such a point. The uncertainty principle again is operative since, if we were to specify the position of the oscillator to be exactly at a node, then there would be an infinite uncertainty associated with the

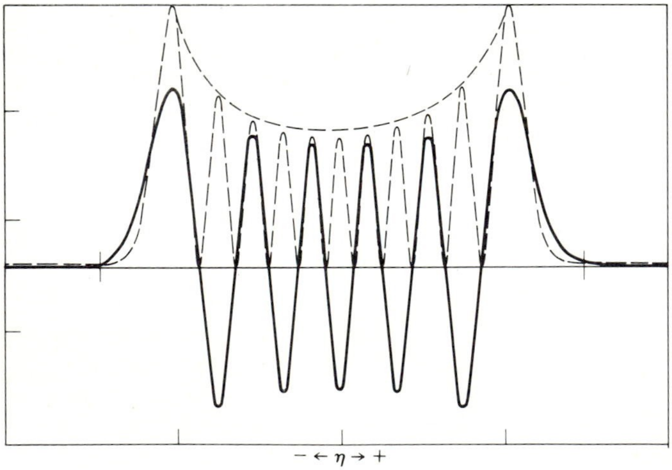

FIGURE 11.3
Schematic wavefunction (solid line) and probability distribution function (dotted line) of a harmonic oscillator with $v = 10$. Dashed line shows the approach of the average to classical behavior.

momentum. The simultaneous occurrence of a finite amount of uncertainty in both the position and momentum allows the system to pass through the nodal points. The amplitude of the oscillator is not constant as in the classical case but will vary with each oscillation. The average value of the amplitude will, however, be close to that of a classical oscillator with the same energy. These concepts regarding wavefunctions allow one to develop a better understanding of the quantum mechanical problem, and are in that sense quite useful. However, they must not be taken as providing an absolute physical picture of the system.

Examination of the wavefunctions given in Table 11.2 shows that if η is replaced with $-\eta$, that is, the coordinates are inverted, then those functions having an even or zero value of the quantum number v will remain unchanged while those with odd values of v will change sign. The former functions are classified as even and the latter as odd. These properties can greatly aid in the evaluation of the dipole moment integrals occurring in the expressions for the absorption coefficient.

11.5 SELECTION RULES FOR TRANSITIONS OF THE HARMONIC OSCILLATOR

The selection rules, which tell us whether or not transitions between two defined energy states (denoted by the quantum numbers v and v') is possible, are established by determining the zero or nonzero character of the absorption coefficient.

$$\alpha(\nu_{v,v'}) = \frac{8\pi^3\nu_{vv'}}{3hc} N_v \langle v|\boldsymbol{\mu}|v'\rangle^2 (1 - e^{-h\nu_{vv'}/kT}) S(\nu_{vv'})\nu \tag{11.37}$$

For a system at thermal equilibrium the absolute intensity of the transition will be determined by the magnitude of the absorption coefficient.

Examination of Eq. (11.37) shows that there are three factors which would lead to a zero value for the absorption coefficient: (1) $\nu_{vv'} = 0$, in which case $[1 - \exp(-h\nu_{vv}/kT)] = 0$, also (2) $N_v = 0$, and (3) $\langle v|\boldsymbol{\mu}|v'\rangle = 0$. If $\nu_{vv'} = 0$ then obviously there is no transition to measure and this case is of no further interest. Since the equilibrium populations of energy states are in accordance with a Boltzmann distribution, the value of N_v will be finite, so we can consider $N_v \neq 0$. We are then left with the problem of determining whether the dipole moment integral is zero or finite in order to ascertain the selection rules.

There are two factors which have to be considered in determining the value of the dipole moment integral, the nature of the dipole moment operator $\boldsymbol{\mu}$, and the symmetry of the dipole moment integral. Consideration of the former gives what are referred to as the *harmonic oscillator selection rules.* The nature of the dipole moment integral gives rise to the *symmetry selection rules.*

The dipole moment integral for the harmonic oscillator as previously derived had no qualifications on the nature of the dipole moment operator $\boldsymbol{\mu}$ other than to consider it as a product of charge and distance. In a vibrating diatomic molecule the dipole moment will be the sum of two components, a permanent dipole moment, μ_0, which is due to the partial electronic charge on each of the atoms when they are at the equilibrium separation, and a component that changes as the molecule undergoes vibration, $\mu(q)$,

$$\mu = \mu_0 + \mu(q) \tag{11.38}$$

and is a function of q, the displacement from equilibrium. The manner in which the dipole moment of a molecule varies with atomic displacements is not well-defined. Investigations of the electronic distributions in diatomic molecules lead to the conclusion that for any individual molecule the behavior can be as qualitatively illustrated by one of the curves in Fig. 11.4 [3]. This behavior is plausible since, if one considers the dipole moment μ to be the product of a partial charge on the atoms and the internuclear separation, then it should go to zero at small internuclear separations and go to zero as the atoms are separated and the charges flow back to give the separated neutral atoms. It is not possible however to definitely establish whether one is on the leading or trailing side of such a curve because, as we shall soon discover, the intensity of a transition depends only on the magnitude and not the sign of the slope of the curve at the $q = 0$ point. In either case the slope of the curve $d\mu/dq$ will, as a first approximation, be essentially constant over the amplitude of the oscillations, hence it is represented as $(d\mu/dq)_0$ which is the moment change per unit displacement from equilibrium, and the dipole moment term is written as

$$\mu = \mu_0 + \left(\frac{d\mu}{dq}\right)_0 q \tag{11.39}$$

The dipole moment integral is, since $\eta = \alpha q$,

$$\langle v|\boldsymbol{\mu}|v'\rangle = \int_{-\infty}^{\infty} \Psi_v^* \mu_0 \Psi_{v'}\, d\eta + \frac{1}{\sqrt{\alpha}} \int_{-\infty}^{\infty} \Psi_v^* \left(\frac{d\mu}{dq}\right)_0 \eta \Psi_{v'}\, d\eta \tag{11.40}$$

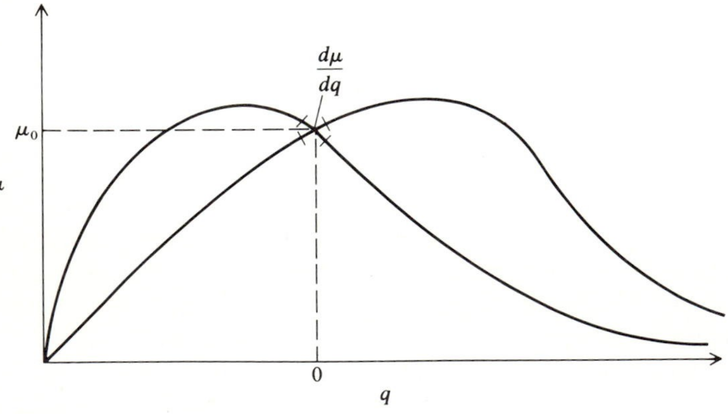

FIGURE 11.4
Qualitative behavior of the dipole moment of a diatomic molecule as a function of atomic displacement.

Since μ_0 and $(d\mu/dq)_0$ are constants they are removed from the integrals giving

$$\langle v|\boldsymbol{\mu}|v'\rangle = \mu_0 \int_{-\infty}^{\infty} \Psi_v^* \Psi_{v'}\, d\eta + \left(\frac{d\mu}{dq}\right)_0 \frac{1}{\sqrt{\alpha}} \int_{-\infty}^{\infty} \Psi_v^* \eta \Psi_{v'}\, d\eta \qquad (11.41)$$

Due to the orthogonality of the wavefunctions the first integral will be zero unless $v = v'$, in which case one has no transition. Thus neither the selection rules nor the intensities of vibrational transitions are determined by the magnitude of the permanent dipole moment of the molecule. We must now examine the conditions under which the second integral will be finite. A qualitative argument can first be advanced. Since the integral has symmetrical limits it will be finite only if the integrand is even. Since η is odd the product $\psi_v^* \psi_v$ must be odd. Our previous discussion of the harmonic oscillator wavefunctions showed that the symmetry of ψ_v followed the evenness or oddness of v. Therefore v' must differ from v by 1, 3, 5, ... (any odd integer) for their product to be odd. This then establishes a selection rule

$$v' - v = \Delta v = \pm 1, \pm 3, \pm 5, \ldots \qquad (11.42)$$

Further examination of the nature of the Hermite polynomials will restrict Δv even further. Using the recursions relationship given in App. C one can write

$$\eta \Psi_{v'} = N_{v'} H_{v'}(\eta) e^{-(1/2)\eta^2} = N_{v'} v' H_{v'-1}(\eta) e^{-(1/2)\eta^2} + \tfrac{1}{2} N_{v'} H_{v'+1}(\eta) e^{-(1/2)\eta^2} \qquad (11.43)$$

Therefore

$$\int_{-\infty}^{\infty} \Psi^* \eta \Psi_{v'}\, d\eta = N_v N_{v'} \left\{ v' \int_{-\infty}^{\infty} H_v(\eta) H_{v'-1}(\eta) e^{-\eta^2}\, d\eta + \frac{1}{2} \int_{-\infty}^{\infty} H_v(\eta) H_{v+1}(\eta) e^{-\eta^2}\, d\eta \right\} \qquad (11.44)$$

Due to the orthogonality of the Hermite functions the first integral is zero unless $v = v' + 1$ and the second is zero unless $v = v' - 1$. This condition further reduces the selection rule to

$$\Delta v = \pm 1 \qquad (11.45)$$

One must keep in mind that this selection rule is based on the dual assumption of harmonic motion and constant $(d\mu/dq)_0$ hence is valid only so long as those conditions are maintained. The relaxation of this selection rule and the subsequent occurrence of overtone vibrations will be discussed in Chap. 13.

The selection rule $\Delta v = \pm 1$ is based on the nature of the dipole moment and has ignored molecular symmetry. This will be considered during the discussion of vibronic spectra in Chap. 15. There are two types of diatomic molecules, homonuclear and heteronuclear. The former have $D_{\infty h}$ symmetry which is maintained during vibration, therefore $(d\mu/dq)_0$ is zero and the molecule has no vibrational spectrum. We find then that only heteronuclear diatomic molecules

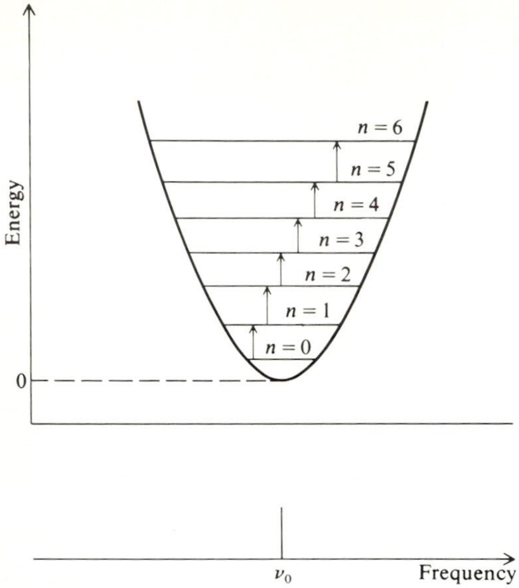

FIGURE 11.5
Energy levels, fundamental transitions, and spectrum of a heteronuclear diatomic molecule.

have vibrational spectra. This does not mean, however, that a homonuclear diatomic molecule is not undergoing vibration, but rather it has no means for exchanging energy with electromagnetic radiation via an oscillating electric dipole. We will find later that another mechanism does exist however.

In Fig. 11.5 are shown the spectrum, energy levels, and allowed transitions of a nonrotating heteronuclear diatomic harmonic oscillator. The frequency allowed by the $\Delta v = \pm 1$ selection rule is referred to as the *fundamental transition.*

11.6 WAVEFUNCTIONS OF THE RIGID ROTOR

The approach followed in this section will be that employed when discussing the wavefunction of the harmonic oscillator, namely the consideration of the vibrational and rotational motion separately. The problem of solving the Schrödinger equation for the rigid rotor was treated earlier. Examination of these results, as presented in Sec. 2.4, and those of the more complete treatment of the nonrigid rotor as given in Sec. 11.2, shows that the rotational wavefunctions are the same for both cases. The rotational wavefunctions are the spherical harmonics given in Table 11.1. Since the functions will be used to evaluate the dipole moment integrals a look at their properties is in order.

It was previously found that if a dipole moment integral whose integrand is a product of a coordinate and two state functions is to be nonzero, then the product of the two state functions must be odd to inversion. Another way of stating this is that transitions can occur only between states of opposite symmetry. When making this generalization, however, one must keep in mind that it is the total state function, which is a product of an electronic, a vibrational, and a rotational function which must be considered. If we are investigating a phenomenon such as rotation and keeping the system in a single electronic and vibrational state then the behavior of the rotational wave function alone will determine the probability of transitions. This general condition for selection rules is often represented as

$$+ \leftrightarrow - \qquad - \leftrightarrow + \qquad + \not\leftrightarrow + \qquad - \not\leftrightarrow -$$

The symmetry of the rotational wavefunctions can best be considered by converting to a set containing only real functions. Due to the *principle of superposition* we know that if Y_J^M and Y_J^{-M} are solutions to the rotational wave equation then the linear combinations

$$\Psi = \frac{1}{\sqrt{2}}(Y_J^M + Y_J^{-M}) \tag{11.46}$$

$$\Psi' = \frac{1}{i\sqrt{2}}(Y_J^M - Y_J^{-M}) \tag{11.47}$$

are also acceptable wavefunctions. Using the fact that $\exp(\pm iM\phi) = \cos M\phi \pm i \sin M\phi$ one can construct from the spherical harmonics a second set of acceptable wavefunctions, the first few of which are given in Table 11.3. Note

TABLE 11.3
Rotational wave functions

J	M	Function $[N_{JM} P_J^{\lvert M\rvert}(\cos\theta) f(\phi)]$
0	0	$\frac{1}{2\sqrt{\pi}}$
1	0	$\sqrt{\frac{3}{4\pi}}\cos\theta$
1	1	$\sqrt{\frac{3}{4\pi}}\sin\theta\cos\phi$
		$\sqrt{\frac{3}{4\pi}}\sin\theta\cos\phi$
2	0	$\sqrt{\frac{5}{16\pi}}(3\cos^2\theta - 1)$
2	1	$\sqrt{\frac{15}{4\pi}}\sin\theta\cos\theta\cos\phi$
		$\sqrt{\frac{15}{4\pi}}\sin\theta\cos\theta\sin\phi$
2	2	$\sqrt{\frac{15}{16\pi}}\sin^2\theta\cos^2\phi$
		$\sqrt{\frac{15}{16\pi}}\sin^2\theta\sin^2\phi$

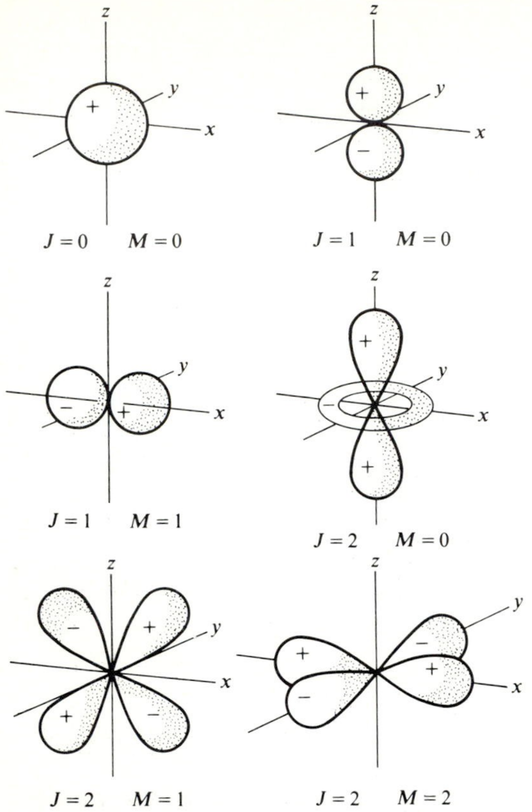

FIGURE 11.6
Pictorial representation of rotational wavefunctions.

that in this table the $|M|$ values are retained as an identifying feature. The geometries of these wavefunctions are illustrated in Fig. 11.6. The signs of the functions are indicated on this figure also. While it is difficult to attach much physical significance to the pictorial geometries of the functions in the absence of external fields they are nevertheless useful for presentation of symmetries. By looking at the spherical harmonic functions one can see that the probability of finding the rotor oriented in the direction specified by θ, ϕ is given by the probability distribution function

$$\Psi_{JM}^{*}\Psi_{JM} = N_{JM}^{2}[P_J^{|M|}(\cos\theta)]^2 f^2(\phi) \tag{11.48}$$

This distribution function is independent of ϕ for $M = 0$. The resulting functions are shown in Fig. 11.7. We see that the probability distributions are rotationally symmetric about the z axis. All orientations in space are equally probable for a rotor with $J = 0$, but as J increases certain preferred orientations develop. As J becomes large the distribution function for the state $J = |M|$ becomes a symmetrical set of closely spaced lobes, and the condition of classical rotation is suggested.

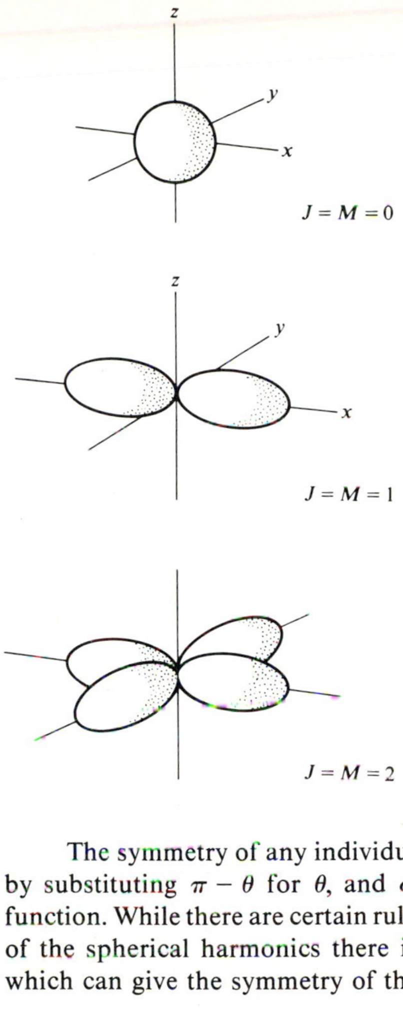

FIGURE 11.7
Probability distribution functions for the diatomic rotor.

The symmetry of any individual rotational wavefunction can be determined by substituting $\pi - \theta$ for θ, and $\phi + \pi$ for ϕ, and noting the behavior of the function. While there are certain rules relating to symmetrical integrals of products of the spherical harmonics there is no simple relationship involving J and M which can give the symmetry of the function.

11.7 SELECTION RULES FOR TRANSITIONS OF THE DIATOMIC ROTOR

As with the case of the harmonic oscillator, if we consider that a transition has a finite frequency $\nu_{J' \leftarrow J} \neq 0$ and the population of the lower of two states is nonzero, $N_J \neq 0$, then the selection rules are determined by the zero or nonzero character of the dipole moment integral or matrix element. Whereas in the discussion of vibrational selection rules we considered only the integrals which are dependent on vibrational functions and the vibrational quantum numbers we will now consider only those dependent on rotational functions and the rotational quantum numbers.

In addition to the fact that the square of the dipole matrix element $(J|\mu|J')^2$ is summed over the three orthogonal space-fixed coordinates X, Y, Z it is also summed over the degenerate states of the system. In the absence of an external field the M dependency of the rotation is degenerate, hence each state is $(2J+1)$-fold degenerate. The dipole matrix element necessary for the evaluation of rotational selection rules is then

$$\langle J|\boldsymbol{\mu}|J'\rangle^2 = \sum_{M=-J}^{J} \sum_{M'=-J}^{J} \sum_{F=X,Y,Z} \langle J, M|\boldsymbol{\mu}_F|J', M'\rangle^2 \tag{11.49}$$

It will be instructive at this point to consider two methods for evaluation of $\langle J|\boldsymbol{\mu}|J'\rangle$. The first method follows the wave-mechanical treatment of the problem, and while it is simple and gives the correct results for a diatomic molecule, it is not as easily applied to more complex molecules. The second method employs the matrix formulation and is the better method for discussing more complex molecules, and will be used in future discussions.

Since the permanent dipole moment of a diatomic molecule μ_0 lies along the bond axis the space-fixed components can be written

$$\mu_X = \mu_0 \sin\theta \cos\phi \tag{11.50}$$

$$\mu_Y = \mu_0 \sin\theta \cos\phi \tag{11.51}$$

$$\mu_Z = \mu_0 \cos\theta \tag{11.52}$$

where θ and ϕ are the spherical angles relating μ to the spaced-fixed X, Y, Z-axis system. The three-component dipole integrals are

$$\langle J, M|\boldsymbol{\mu}_X|J'M'\rangle = \mu_0 \int_0^{\pi}\int_0^{2\pi} \Psi_{JM}^* \sin\theta \cos\phi \Psi_{J'M'} \sin\theta \, d\theta \, d\phi \tag{11.53}$$

$$\langle J, M|\boldsymbol{\mu}_Y|J'M'\rangle = \mu_0 \int_0^{\pi}\int_0^{2\pi} \Psi_{JM}^* \sin\theta \sin\phi \Psi_{J'M'} \sin\theta \, d\theta \, d\phi \tag{11.54}$$

$$\langle J, M|\boldsymbol{\mu}_Z|J'M'\rangle = \mu_0 \int_0^{\pi}\int_0^{2\pi} \Psi_{JM}^* \cos\theta \Psi_{J'M'} \sin\theta \, d\theta \, d\phi \tag{11.55}$$

We are interested in the selection rules involving M as well as J. By using the wavefunctions written in terms of the spherical harmonics, the Z component can be written as

$$\langle J, M|\boldsymbol{\mu}_Z|J', M'\rangle = \frac{\mu_0 N_{JM} N_{J'M'}}{2\pi} \int_0^{\pi} P_J^{|M|}(\cos\theta) \cos\theta \, P_{J'}^{|M'|}(\cos\theta) \sin\theta \, d\theta \times \int_0^{2\pi} e^{i(M-M')\phi} \, d\phi \tag{11.56}$$

where

$$N_{JM} = \left[\frac{(2J+1)(J-|M|)!}{2(J+|M|)!}\right]^{1/2} \tag{11.57}$$

If the second integral has $M = M'$ then it reduces to a value of 2π. If $M \neq M'$

it goes to zero, hence the dipole integral becomes

$$\langle J, M|\boldsymbol{\mu}_Z|J'M'\rangle = \mu_0 N_{JM}N_{J'M'}\int_0^{\pi} P_J^{|M|}(\cos\theta)P_{J'}^{|M|}(\cos\theta)\cos\theta\sin\theta\,d\theta \tag{11.58}$$

and one selection rule is $\Delta M = 0$. Reflection on this point shows that it is a physically reasonable conclusion. If the applied radiation field is in the Z direction and the dipole moment is oriented in the same direction then the field cannot apply a torque to the molecule through its dipole and no change in orientation would result. Using the Legendre function recursion relations given in App. F one can express the μ_Z element as

$$\langle J, M|\boldsymbol{\mu}_Z|J', M'\rangle = \mu_0 N_{JM}N_{J'M}\left[\frac{(J+|M|)}{(2J+1)}\int_0^{\pi} P_{J-1}^{|M|}(\cos\theta)\sin\theta\,d\theta\right.$$
$$+\frac{(J-|M|+1)}{(2J+1)}$$
$$\left.\times\int_0^{\pi} P_{J+1}^{|M|}(\cos\theta)P_{J'}^{|M|}(\cos\theta)\sin\theta\,d\theta\right] \tag{11.59}$$

Using the properties of the Legendre polynomials it can be shown that the first of these integrals is zero unless $J' = J - 1$ and that the second is zero unless $J' = J + 1$. Hence the selection rule is $J' = J \pm 1$ or $\Delta J = \pm 1$. If we take J as the lower state then $J' = J + 1$, the first integral is zero and

$$\langle J, M||\boldsymbol{\mu}_Z|J+1, M\rangle = \mu_0\frac{N_{J,M}}{N_{J+1,M}}\frac{(J-|M|+1)}{(2J+1)}$$
$$\times\int_0^{\pi} N_{J+1,M}^2[P_{J+1}^{|M|}(\cos\theta)]^2\sin\theta\,d\theta \tag{11.60}$$

The integral in this expression is just the normalization integral for the angular wavefunctions $\psi_{J+1,M} = N_{J+1,M}P_{J+1}^{|M|}(\cos\theta)$ and is unity. Substituting the values for the N_{JM} into this expression gives the Z component of the dipole moment integral

$$\langle J, M|\boldsymbol{\mu}_Z|J+1, M\rangle = \mu_0\left(\frac{(J+1)^2 - M^2}{(2J+1)(2J+3)}\right)^{1/2} \tag{11.61}$$

The $\langle J, M|\boldsymbol{\mu}_X|J', M'\rangle$ and $\langle J, M|\boldsymbol{\mu}_Y|J', M'\rangle$ elements will be the products of pairs of integrals similar to those given by Eq. (11.56) but with the second integrals given by

$$\int_0^{2\pi} e^{i(M'-M)\phi}\cos\phi\,d\delta \tag{11.62}$$

$$\int_0^{2\pi} e^{i(M'-M)\phi}\sin\phi\,d\phi \tag{11.63}$$

respectively. Using

$$\cos\phi = \frac{1}{2}[e^{i\phi} + e^{-i\phi}] \tag{11.64}$$

$$\sin\phi = \frac{i}{2}[e^{i\phi} - e^{-i\phi}] \tag{11.65}$$

they become

$$\frac{1}{2}\int_0^{2\pi} e^{i(M+1-M')\phi}\,d\phi + \frac{1}{2}\int_0^{2\pi} e^{i(M-1-M')\phi}\,d\phi \tag{11.66}$$

and

$$\frac{i}{2}\int_0^{2\pi} e^{i(M+1-M')\phi}\,d\phi - \frac{i}{2}\int_0^{2\pi} e^{i(M-1-M')}\,d\phi \tag{11.67}$$

These integrals vanish except when $M' = M \pm 1$. By use of the recursion relations in App. F, and by following a procedure analogous to that used for the evaluation of $(J, M|\boldsymbol{\mu}_Z|J', M')$, one can verify the selection rule $\Delta M = \pm 1$ and evaluate the other two Cartesian components of the dipole matrix element

$$\begin{aligned}\langle J, M|\boldsymbol{\mu}_X|J+1, M+1\rangle &= -i\langle J, M|\boldsymbol{\mu}_Y|J+1, M+1\rangle \\ &= -\frac{\mu_0}{2}\left(\frac{(J+M+2)(J+M+1)}{(2J+1)(2J+3)}\right)^{1/2}\end{aligned} \tag{11.68}$$

$$\begin{aligned}\langle J, M|\boldsymbol{\mu}_X|J+1, M-1\rangle &= i\langle J, M|\boldsymbol{\mu}_Y|J+1, M-1\rangle \\ &= \frac{\mu_0}{2}\left(\frac{(J-M-1)(J-M+2)}{(2J+1)(2J+3)}\right)^{1/2}\end{aligned} \tag{11.69}$$

The dipole components thus determined are for the transition $J+1 \leftarrow J$.[1] The dipole components for the transitions $J \leftarrow J-1$ can be found by setting $J = J-1$ in Eqs. (11.61), (11.68), and (11.69).

The total squared dipole element is found by squaring the component terms and summing over the coordinates and the degenerate states

$$\langle J|\boldsymbol{\mu}|J+1\rangle^2 = \sum_{M'=-J}^{J}\ \sum_{F=X,Y,Z} \langle J, M|\boldsymbol{\mu}_F|J+1, M'\rangle^2 \tag{11.70}$$

or

$$\begin{aligned}\langle J|\boldsymbol{\mu}|J+1\rangle^2 &= \sum_{M+1=-J}^{J} \frac{\mu_0^2}{4}\left\{\left[\frac{(J+M+2)(J+M+1)}{(2J+1)(2J+3)}\right] - \left[\frac{(J+M+2)(J+M+1)}{(2J+1)(2J+3)}\right]\right\} \\ &+ \sum_{M-1=J}^{J} \frac{\mu_0^2}{4}\left\{\left[\frac{(J-M+1)(J-M+2)}{(2J+1)(2J+3)}\right] - \left[\frac{(J-M+1)(J-M+2)}{(2J+1)(2J+3)}\right]\right\} \\ &+ \sum_{M=-J}^{J} \mu_0^2\left\{\frac{(J+1)^2 - M^2}{(2J+1)(2J+3)}\right\}\end{aligned} \tag{11.71}$$

---

[1] The common convention is to write the quantum number of the highest state first and use the arrow to indicate an emission or absorption process.

The occurrence of the i in Eqs. (11.68) and (11.69) causes sign reversals and the cancellations of the first two pairs of summations leaving

$$\langle J|\boldsymbol{\mu}|J+1\rangle^2 = \mu_0^2 \sum_{M=-J}^{J} \frac{(J+1)-M^2}{(2J+1)(2J+3)} \tag{11.72}$$

Summing over M the dipole term becomes

$$\langle J|\boldsymbol{\mu}|J+1\rangle^2 = \mu_0^2\frac{(J+1)}{(2J+1)} \tag{11.73}$$

for the transition $J+1 \leftarrow J$. An analogous treatment for the $J \rightarrow J-1$ transition gives

$$\langle J|\boldsymbol{\mu}|J-1\rangle^2 = \mu_0^2\frac{J}{(2J+1)} \tag{11.74}$$

While the dipole element is nonzero for both the absorption transition, $J+1 \leftarrow J$, and the emission transition, $J \rightarrow J-1$, the magnitude is larger for the former. This is necessary in order for the system to maintain thermal equilibrium when transitions occur since there are $2J-1$, $2J+1$, and $2J+3$ angular momentum states for the $J-1$, J, and $J+1$ levels, respectively.

A second method for evaluation of the dipole integrals or matrix elements is to employ the angular momentum and direction cosine matrix elements. For the general case of a permanent molecular dipole moment μ_0 which does not lie along a unique molecular axis the dipole moment can be written in terms of components along the molecule-fixed axis system x, y, z

$$\boldsymbol{\mu}_0 = \mu_x\hat{\mathbf{i}} + \mu_y\hat{\mathbf{j}} + \mu_z\hat{\mathbf{k}} \tag{11.75}$$

The components along the molecular-fixed axis system are related to components along a space-fixed axis system X, Y, Z by a set of three equations

$$\mu_F = \mu_x \cos\alpha_{Fx} + \mu_y \cos\alpha_{Fy} + \mu_z \cos\alpha_{Fz} \qquad F = x, y, z \tag{11.76}$$

This is illustrated for the μ_Z component in Fig. 11.8. Since the diatomic molecule possesses a unique axis colinear with the dipole moment it is chosen as the molecular z axis, hence $\mu_z = \mu_0$ and $\mu_x = \mu_y = 0$, and the three-component dipole matrix elements which make up the total dipole matrix element given by Eq. (11.49) are

$$\langle J, M|\boldsymbol{\mu}_X|J', M'\rangle = \mu_0\langle J, M|\boldsymbol{\Phi}_{Xz}|J', M'\rangle \tag{11.77}$$

$$\langle J, M|\boldsymbol{\mu}_Y|J', M'\rangle = \mu_0\langle J, M|\boldsymbol{\Phi}_{Yz}|J', M'\rangle \tag{11.78}$$

$$\langle J, M|\boldsymbol{\mu}_Z|J', M'_-\rangle = \mu_0\langle J, M|\boldsymbol{\Phi}_{Zz}|J', M'\rangle \tag{11.79}$$

where $\Phi_{Fg} = \cos\alpha_{Fg}$ and the $\langle J, M|\Phi_{Fg}|J'M'\rangle$ terms are the direction cosine matrix elements. These terms can be evaluated by using Table 3.2. The entries in this table comprise the only nonzero direction cosine matrix elements. Hence the selection rules are established from the table as $\Delta J = 0, \pm 1$, and $\Delta M = 0$, ± 1. In formulating the direction cosine elements the quantum number K relates

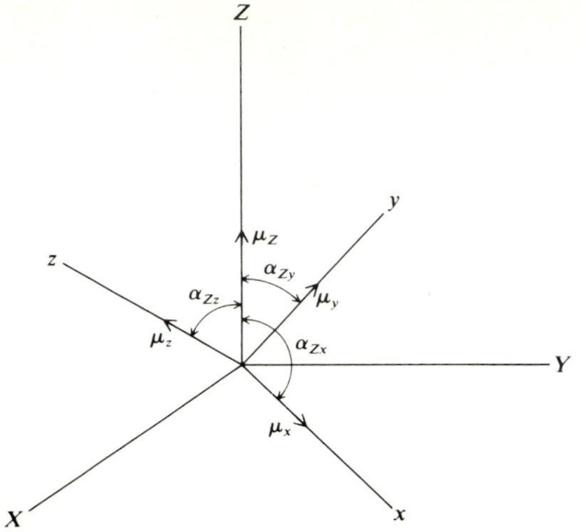

FIGURE 11.8
Angular relationship of molecule-fixed and space-fixed dipole moment components.

to the component of angular momentum about the unique molecular axis. For $^1\Sigma$ diatomic molecules which have no mass lying off the unique axis and no electronic, hence no angular, momentum about it, $K = 0$. Taking as an example the Zz component, one has as the only allowed terms

$$\mu_0\langle J, M|\mathbf{\Phi}_{Zz}|J, M\rangle = \mu_0\{[4J(J+1)]^{-1}X0X2M\} = 0 \tag{11.80}$$

$$\begin{aligned}\mu_0\langle J, M|\mathbf{\Phi}_{Zz}|J+1, M\rangle &= \mu_0\{[4J(J+1)(4J^2+8J+3)^{1/2}]^{-1} \times [2(J+1)] \\ &\quad \times 2[(J+1)^2 - M^2]^{1/2}\} \\ &= \left[\frac{(J+1)^2 - M^2}{(2J+1)(2J+3)}\right]^{1/2} \mu_0\end{aligned} \tag{11.81}$$

$$\begin{aligned}\mu_0\langle J, M|\mathbf{\Phi}_{Zz}|J-1, M\rangle &= \mu_0\{[4J(4J-1)^{1/2}]^{-1} \times [-2J] \times [-2(J^2 - M^2)^{1/2}]\} \\ &= \left[\frac{J^2 - M^2}{(4J^2-1)}\right]^{1/2} \mu_0\end{aligned} \tag{11.82}$$

The value obtained for $\mu_0(J, M|\mathbf{\Phi}_{Zz}|J+1, M)$ is observed to be the same as that given by Eq. (11.61). Substitution of $J-1$ in Eq. (11.61) gives the result shown in Eq. (11.82). Considering the Hermitian character of the direction cosine matrix elements, that is, $\langle J, M|\mathbf{\Phi}_{Fg}|J', M'\rangle = \langle J', M'|\mathbf{\Phi}_{Fg}|J, M\rangle^*$, and the fact that Φ_{Fg} is real, these results are in agreement. Since $\langle J, M|\mathbf{\Phi}_{Zz}|J, M\rangle = 0$ for a diatomic molecule $\Delta J = 0$ is not a valid selection rule. Analogous use of Table 3.2 to expand the expressions for the X and Y components gives terms which are the same as those given by Eqs. (11.68) and (11.69). The three-component terms are squared and combined as before to give the identical final results, Eqs. (11.73) and (11.74).

The great utility of the matrix method is realized when one considers that once the nonzero direction cosine matrix elements have been established they can then be used to determine selection rules and, as we will see later, other properties of a more complex molecular rotor. The use of the matrix formulation for the establishment of selection rules is in general simpler than the use of integrals involving wavefunctions.

Having established that in the absence of external fields the energy of a rigid rotor is degenerate in M and the selection rule is $\Delta J = \pm 1$, the use of the value derived for the energy of the rigid diatomic rotor

$$E = \frac{\hbar^2}{2I} J(J+1) = hBJ(J+1) \tag{11.83}$$

gives the energy level scheme shown in Fig. 11.9 and the spectrum shown in Fig. 11.10. The commonly used convention when discussing rotational energies is to express the energies in units of hertz, hence

$$E(\text{Hz}) = \frac{E}{h} = BJ(J+1) \tag{11.84}$$

where B is a pseudo reciprocal moment referred to as the rotational constant. B is generally expressed in units of megahertz or cm^{-1}. The moment of inertia is normally determined in units of kg-m^2 or amu-A^2, in which case the conversion factors are

$$B(\text{MHz}) = \frac{(8.37091 \pm 0.0005) \times 10^{-42}}{I(\text{kg-m}^2)} \tag{11.85}$$

and

$$B(\text{MHz}) = \frac{(5.05375 \pm 0.003) \times 10^{5}}{I(\text{amu-A}^2)} \tag{11.86}$$

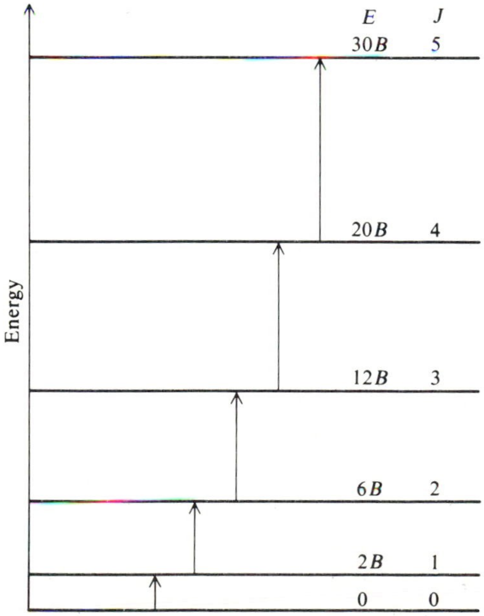

FIGURE 11.9
Energy level diagram for a rigid diatomic rotor.

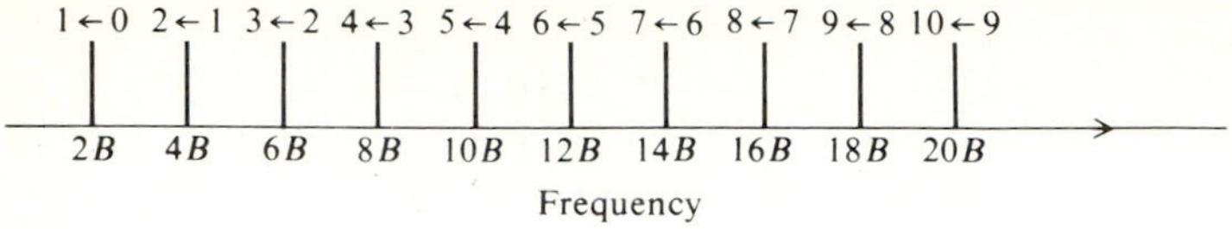

FIGURE 11.10
Rotational spectrum of the rigid diatomic rotor.

For a transition $J + 1 \leftarrow J$ the frequency is given by

$$\nu = 2B(J + 1) \tag{11.87}$$

In contrast to the diatomic harmonic oscillator spectrum which consists of a single transition the spectrum of the rigid diatomic rotor is a set of equally spaced frequencies separated by $2B$.

11.8 STATE POPULATIONS AND DIPOLE MATRIX ELEMENTS

Having established, for the diatomic harmonic oscillator and rigid rotor respectively, the conditions under which the dipole moment integrals are nonzero, $\Delta v = \pm 1$, $\Delta J = \pm 1$ and the expressions for the magnitude of the integrals

$$\langle v|\boldsymbol{\mu}|v+1\rangle = \left(\frac{d\mu}{dq}\right)_0 \frac{1}{\sqrt{2}} \int \Psi_v^* \eta \Psi_{v+1}\, d\eta \tag{11.88}$$

$$\langle J|\boldsymbol{\mu}|J+1\rangle = \mu_0 \left(\frac{[J+1]}{[2J+1]}\right)^{1/2} \tag{11.89}$$

we now direct our attention to the determination of state populations in order to complete the study of absorption coefficients and to be able to determine relative or absolute absorption coefficients.

Examination of the defining relationship for the absorption coefficient

$$\alpha(\nu_{\sigma\sigma'}) = \frac{8\pi^3 \nu_{\sigma\sigma'}}{3hc} N_\sigma \langle\sigma|\mu|\sigma'\rangle^2 (1 - e^{-h\nu_{\sigma\sigma'}/kT}) S(\nu, \nu_{\sigma\sigma'}) \tag{11.90}$$

shows that the absorption coefficient is proportional to the population of the lower state. We shall consider the phenomenon of vibration and rotation separately since the intensity factors are essentially additive when the two types of transitions occur simultaneously.

Prior to considering state populations let us digress for a moment to consider the magnitude of the dipole moment matrix element. If we wish to examine the relative intensities of transitions between vibrational levels in a one-type molecule the $(d\mu/dq)_0$ factor will be common to all levels and the relative value of the dipole matrix elements of the two transitions $\langle v|\mu|v'\rangle^2/\langle v'|\mu|v'\rangle^2$ will be given by

$$R\left(\frac{v' \leftarrow v}{v'' \leftarrow v'}\right) = \frac{\left[\int \Psi_v^8 \frac{\eta}{\sqrt{2}} \Psi_{v'}\, d\eta\right]^2}{\left[\int \Psi_{v'}^* \frac{\eta}{\sqrt{2}} \Psi_{v''}\, d\eta\right]^2} \tag{11.91}$$

Examining the magnitude of the $(v|\mu|v')$ and $R(v' \leftarrow v/v'' \leftarrow v')$ terms for some lower transitions will indicate their contributions to the intensities. For $n = 0 - 4$ the dipole elements are

$$\langle 0|\boldsymbol{\mu}|1\rangle = \nabla\mu_0 \int_{-\infty}^{\infty} \Psi_0^* \frac{\eta}{\sqrt{\alpha}} \Psi_1\, d\eta = \frac{\nabla\mu_0}{\sqrt{2\alpha}} \tag{11.92}$$

$$\langle 1|\boldsymbol{\mu}|2\rangle = \nabla\mu_0 \int_{-\infty}^{\infty} \Psi_1^* \frac{\eta}{\sqrt{\alpha}} \Psi_2\, d\eta = \frac{\nabla\mu_0}{\sqrt{\alpha}} \tag{11.93}$$

$$\langle 2|\boldsymbol{\mu}|3\rangle = \nabla\mu_0 \int_{-\infty}^{\infty} \Psi_2^* \frac{\eta}{\sqrt{\alpha}} \Psi_3\, d\eta = \frac{3\nabla\mu_0}{\sqrt{6\alpha}} \tag{11.94}$$

$$\langle 3|\boldsymbol{\mu}|4\rangle = \nabla\mu_0 \int_{-\infty}^{\infty} \Psi_3^* \frac{\eta}{\sqrt{\alpha}} \Psi_4\, d\eta = \frac{2\nabla\mu_0}{\sqrt{2\alpha}} \tag{11.95}$$

where $\nabla\mu_0 = (d\mu|dq)_0$, $\alpha = (4\pi^2\mu\nu_0/h)^{1/2}$, and use is made of the definite integral

$$\int_0^{\infty} x^{2\eta} e^{-ax^2}\, dx = \frac{1 \cdot 3 \cdot 5 \cdots (2n-1)}{2^{n+1} a^n} \left(\frac{\pi}{a}\right)^{1/2} \tag{11.96}$$

The intensity ratios are

$$R(1 \leftarrow 0):(2 \leftarrow 1) = \tfrac{1}{2} \qquad R(2 \leftarrow 1):(3 \leftarrow 2) = \tfrac{2}{3} \qquad R(3 \leftarrow 2):(4 \leftarrow 3) = \tfrac{3}{4}$$

As v increases the ratio approaches unity and for transitions between adjacent vibrational levels it will always lie between $\frac{1}{2}$ and 1. We see that the relative intensities of fundamental vibrational transitions are not very sensitive to the vibrational quantum numbers.

The effect of the dipole matrix elements on the relative intensities of rotational transitions of the same molecule is easily seen by use of Eq. (11.89). Hence

$$R\left(\frac{J+1 \leftarrow J}{J+2 \leftarrow J+1}\right) = \frac{\langle J|\boldsymbol{\mu}|J+1\rangle^2}{\langle J+1|\boldsymbol{\mu}|J+2\rangle^2} = \left(\frac{(2J+3)(J+1)}{(2J+1)(J+2)}\right) \tag{11.97}$$

For example,

$$R(1 \leftarrow 0):(2 \leftarrow 1) = \frac{3}{2} \qquad R(4 \leftarrow 3):(5 \leftarrow 4) = \frac{36}{35}$$

$$R(2 \leftarrow 1):(3 \leftarrow 2) = \frac{10}{9} \qquad R(5 \leftarrow 4):(6 \leftarrow 5) = \frac{55}{54}$$

$$R(3 \leftarrow 2):(4 \leftarrow 3) = \frac{21}{20}$$

As $J \to \infty$ the ratio becomes unity, hence the dipole matrix element has only a small effect on relative intensities.

If one attempts to compare intensities between like quantum transitions of different molecules then the dipole factors $\nabla\mu_0$ and μ_0 must be considered. If the frequencies of two transitions do not differ by more than an order of magnitude then the ratios $\nabla\mu_0/\nabla\mu_0'$ and μ_0/μ_0' will primarily determine the relative intensities for vibrational transitions and rotational transitions, respectively. The latter ratio is generally available or easily estimated but the former are generally obtained only from intensity measurements.

The remaining factor which affects transition intensities is the distribution of molecules in the various energy states. The distribution is considered to follow Boltzmann statistics

$$N_j = N_0 \frac{g_j}{g_0} e^{-(E_j - E_0)/kT} \tag{11.98}$$

The problem will first be approached by use of an example. For a simple diatomic molecule such as HCl the force constant is of the order of 10^2 N m^{-1} and $\mu \approx 10^{-27}$ kg. Therefore

$$\nu_0 = \frac{1}{2\pi}\sqrt{\frac{k}{\mu}} \approx \frac{1}{2\pi}\sqrt{\frac{10^2 \text{N m}^{-1}}{10^{-27}\text{ kg}}} \approx 5 \times 10^{13} \text{ Hz} \tag{11.99}$$

or

$$\bar{\nu}_0 = \frac{\nu_0}{c} = 1700 \text{ cm}^{-1} \tag{11.100}$$

In this case, at room temperature, the Boltzmann distribution for vibrational states gives

$$\frac{N_1}{N_0} = e^{-hc\bar{\nu}_0/kT} \approx 3 \times 10^{-4} \tag{11.101}$$

Comparison of the dipole matrix element and population ratios shows that at room temperature the population ratio is very small and in fact the intensity of the transition $v = 2 \leftarrow v = 1$ would be less than that of the transition $v = 1 \leftarrow v = 0$ by three orders of magnitude. For an ideal harmonic oscillator both transitions have the same frequency, and this difference would be of little consequence, but the discussion has served to point out that a very large fraction of the molecules will be in their lowest vibrational state at room temperature.

The population factor N_σ occurring in the expression for the absorption coefficient is that of an energy level denoted by a complete set of electronic, vibrational, and rotational quantum numbers, and must be considered in a more general manner than that of the previous example. In particular we wish to relate the population N_σ to the total population $N = \sum_{\sigma=1}^{\infty} N_\sigma$, and hence to the concentrations. The population of a state N_σ can be related to the total population as a fraction

$$f_\sigma = \frac{N_\sigma}{N} \tag{11.102}$$

This fraction is the product of the individual fractions in the electronic, vibrational, and rotational states under consideration

$$f_\sigma = f_{\text{elec}} \times f_{\text{vib}} \times f_{\text{rot}} \tag{11.103}$$

For most systems, even up to moderately high temperatures (1000 K) the fraction of molecules in the lowest electronic state is, for all practical purposes, unity. For example, the ratio of population in the ground and first excited state of a molecule having an electronic transition at 200 nm would be, at 1000 K,

$$f_{\text{elec}} = \frac{N_1}{N_0} = e^{-ch/\lambda kT} \approx e^{-70} \tag{11.104}$$

When a molecule has low vibrational frequency or is at an elevated temperature then there can be appreciable populations for the higher vibrational states. The fraction of molecules in a given vibrational state f_v is

$$f_v = \frac{N_v}{N} = \frac{N_0 e^{-(v+1/2)h\nu_0/kT}}{\sum_{j=0}^{\infty} N_0 e^{-(j+1/2)h\nu_0/kT}} \tag{11.105}$$

The denominator can be written

$$\sum_{j=0}^{\infty} e^{i(j+1/2)h\nu_0/kT} = e^{-h\nu/2kT} \sum_{j=0}^{\infty} e^{-jh\nu_0/kT} = \frac{e^{-h\nu_0/2kT}}{1 - e^{-h\nu_0/kT}} \tag{11.106}$$

therefore

$$f_v = e^{-vh\nu_0/kT}(1 - e^{-h\nu_0/kT}) \tag{11.107}$$

For example, for the KI molecule at 1000 K and having a vibrational frequency of $\nu_0 = 200\ \text{cm}^{-1}$ the fraction in the first excited vibrational state will be

$$f_1 = e^{-200hc/1000k}(1 - e^{-200hc/1000k}) \approx 0.2 \tag{11.108}$$

or the $v = 1$ state contains 20 percent of the molecules.

For rotational states the fraction in a particular level is given by

$$f_J = \frac{N_J}{N} = \frac{N_0(2J+1)e^{-hBJ(J+1)/kT}}{\sum_{J=0}^{\infty} (2J+1)e^{-hBJ(J+1)/kT}} \tag{11.109}$$

The factor $(2J + 1)$ occurs because of the degeneracy of each state with respect to the quantum number M. For rotational transitions, except in the very short (millimeter) wavelength region the factor $hBJ(J+1)/kT$ is much less than one for $J < 50$. For example, for KI with $B = 1,\ 825$ MHz, at 300 K and $J = 10$, $hBJ(J+1)/kT \approx 0.03$. For $na \ll 1$

$$\sum_{n=0}^{\infty} e^{-na} = \int_0^{\infty} e^{-ax}\,dx \tag{11.110}$$

Since $hBJ(J+1)kT \ll 1$ it follows that

$$\sum_{J=0}^{\infty} (2J+1)e^{-hBJ(J+1)/kT} = \int_0^{\infty} (2J+1)e^{-hBJ(J+1)/kT}\, dJ - \frac{kT}{hB} \tag{11.111}$$

If $hBJ(J+1)/kT$ is not less than unity one can use a series expansion

$$\sum (2J+1)e^{-hBJ(J+1)/kT} = \left(\frac{kT}{hB}\right) + \frac{1}{3} + \frac{1}{15}\left(\frac{hB}{kT}\right) + \frac{4}{315}\left(\frac{hB}{kT}\right)^2 + \cdots \tag{11.112}$$

and retain as many terms as are needed. For the case of $hB/kT \ll 1$ the rotational fraction becomes

$$f_J = \frac{(2J+1)hBe^{-hBJ(J+1)/kT}}{kT} \tag{11.113}$$

Since most rotational spectra are observed at room temperature or below and the frequencies are generally less than 100 MHz we have the experimental condition, $hB/kT \ll 1$, so for $J < 10$ the exponential, $\exp[hBJ(J+1)/kT]$, is approximately unity. Hence

$$f_J \approx (2J+1)\frac{hB}{kT} \tag{11.114}$$

Note that this factor will continuously increase with J so it is not valid for large values of J.

The total fraction of diatomic molecules in a state ψ_σ, having a low J value, then becomes

$$f = f_{\text{elec}} \times f_v \times f_J = (2J+1)\frac{hB}{kT}e^{-vh\nu_0/kT}(1 - e^{-h\nu_0/kT}) \tag{11.115}$$

The discussion so far has only been relative to diatomic molecules so the above fraction strictly applies only to that case.

11.9 ABSORPTION COEFFICIENTS FOR DIATOMIC VIBROTORS

The origin and nature of the factors which contribute to the absorption coefficients having now been discussed, it is instructive to summarize these contributions and examine the total presentation to see if suitable approximations may be used to simplify calculations. The absorption coefficient for the general case where all molecules are in the ground electronic state is given by

$$\alpha(\nu_{\sigma\sigma'}) = \left\{\frac{8\pi^2\nu_{\sigma\sigma'}}{3hc}\left[\frac{e^{-vh\nu_0/kT}(1-e^{-h\nu_0/kT})(2J+1)Ne^{-hBJ(J+1)/kT}}{kT/hB + \frac{1}{3} + \frac{1}{15}hB/kT + \frac{4}{315}(hB/kT)^2 + \cdots}\right]\right.$$
$$\times\left[\langle\sigma|\boldsymbol{\mu}|\sigma'\rangle^2(1-e^{-h\nu_{\sigma\sigma'}/kT})\left(\frac{\nu}{\nu_{\sigma\sigma'}}\right)\right]$$
$$\left.\times\left[\frac{\Delta\nu}{(\nu-\nu_{\sigma\sigma'})^2-(\Delta\nu)^2} + \frac{\Delta\nu}{(\nu+\nu_{\sigma\sigma'})^2+(\Delta\nu)^2}\right]\right\} \tag{11.116}$$

If one considers room temperature transitions between rotational states with no simultaneous change in vibrational state then $\nu_{\sigma\sigma'} = \nu_{J,J+1}$, and it is of the order of magnitude of 10–500 GHz. For this case $h\nu/kT$ lies between 0.0002 and 0.08 and the condition $h\nu/kT \ll 1$ is satisfied. Using the relationships for exponentials and series presented in Sect. 11.9 the following simplifications will occur

$$\frac{kT}{hB} + \frac{1}{3} + \frac{1}{15}\left(\frac{hB}{kT}\right) + \frac{4}{315}\left(\frac{hB}{kT}\right)^2 + \cdots \rightarrow \frac{kT}{hB} \tag{11.117}$$

$$1 - e^{-h\nu_{J,J+1}/kT} \rightarrow \frac{h\nu_{J,J+1}}{kT} \tag{11.118}$$

If the pressure is kept sufficiently low so that the line is not appreciably widened by collision-broadening then $\Delta\nu \ll \nu_{J,J+1}$. The condition $\Delta\nu \ll \nu_{\sigma\sigma'}$ is nearly always valid for the case of infrared and optical spectra. When this condition is true it follows that $\nu \approx \nu_{\sigma\sigma'}$, the second part of the shape function can be neglected, and the absorption coefficient becomes

$$\alpha(\nu_{J,J+1}) \approx \frac{8\pi^2\nu_{J,J+1}^2 Nf}{3ckT}\langle J|\boldsymbol{\mu}|J+1\rangle^2 \frac{\Delta\nu}{(\nu - \nu_{J,J+1})^2 + (\Delta\nu)^2} \tag{11.119}$$

where

$$f = \frac{(2J+1)hB}{kT}(e^{-vh\nu_0/kT})(1 - e^{-h\nu_0/kT})(e^{-hBJ(J+1)/kT}) \tag{11.120}$$

Since this coefficient is a function of frequency the absolute or integrated line intensity becomes

$$\begin{aligned} A &= \int_0^\infty \alpha(\nu_{\sigma\sigma'})\,d\nu \\ &= \frac{8\pi^2\nu_{J,J+1}^2 Nf}{3ckT}\langle J|\boldsymbol{\mu}|J+1\rangle^2 \left[\int_{-\infty}^{\infty} \frac{\Delta\nu}{(\nu - \nu_{J,J+1})^2 + (\Delta\nu)^2}\,d\nu\right] \end{aligned} \tag{11.121}$$

The change in integration limits occurs because strictly speaking there are two terms in the shape function, and integration of their sum from $0 \rightarrow \infty$ is equivalent to the integration of either from $-\infty \rightarrow \infty$. Since

$$\int_{-\infty}^{\infty} \frac{a\,dx}{x^2 + a^2} = \pi \tag{11.122}$$

it follows that

$$A = \frac{8\pi^3 Nf}{3ckT}\langle J|\boldsymbol{\mu}|J+1\rangle^2 \nu_{J,J+1}^2 \tag{11.123}$$

It is of interest to note that the absolute intensity increases as the square of the frequency, hence it is advantageous to work with high-frequency transitions when the population factors are low or the path length has to be short. It is also to be noted that the absolute intensity is independent of $\Delta\nu$, hence if a line is split by some perturbation such as the Stark effect, the sum of the absolute intensities of the multiple lines is equal to the absolute intensity of the unsplit transition.

REFERENCES

A. Specific

1. Born, M., and R. Oppenheimer, *Ann. Physik*, **84**, 457 (1927).
2. Perkeris, C. L., *Phys. Rev.*, **45**, 98 (1934).
3. Herzberg, G., *Spectra of Diatomic Molecules*, Van Nostrand, New York, NY, 1950, p. 97.

B. General

Barrow, G. M., *Introduction to Molecular Spectroscopy*, McGraw-Hill, New York, NY, 1962.

Gordy, W., and R. L. Cooke, *Microwave Molecular Spectra*, 2d ed., Wiley-Interscience, New York, NY, 1985.

Harmony, M. D., *Introduction to Molecular Energies and Spectra*, Holt, Rinehart, and Winston, New York, NY, 1972.

Herzberg, G., *Spectra of Diatomic Molecules*, Van Nostrand, Princeton, NJ, 1950.

Hollas, M. J., *High Resolution Spectroscopy*, Butterworths, London, GB, 1982.

King, G. W., *Spectroscopy and Molecular Structure*, Holt, Rinehart, and Winston, New York, NY, 1964.

Laidlaw, W. G., *Introduction to Quantum Concepts in Spectroscopy*, McGraw-Hill, New York, NY, 1970.

Townes, C. H., and A. Schawlow, *Microwave Spectroscopy*, McGraw-Hill, New York, NY, 1955.

PROBLEMS

11.1. Given $\omega_e = 2885.9\ \text{cm}^{-1}$ and assuming a Boltzmann distribution, calculate the relative populations of the $n = 0$ and $n = 1$ states of H^{35} at 100 K and at 300 K.

11.2. A diatomic molecule has a fundamental vibrational frequency, $\omega_e = 2000\ \text{cm}^{-1}$ and a reciprocal moment $B_e = 200$ MHz. Calculate the ratios for:

(*a*) The relative intensities of the $J+1 \leftarrow J$, $v = 0$ transitions for $J = 5$, 10, 15, 20, 30, 40 at 200 K and 400 K.

(*b*) The relative intensities of the $v = 1, J = 1 \leftarrow v = 0, J = 1$, $v = 2, J = 1 \leftarrow v = 0, J = 1$ and the $v = 2, J = 1 \leftarrow v = 1, J = 1$ transitions at 200, 500 and 1000 K.

11.3. Repeat Prob. 11.1 for $\omega_e = 200\ \text{cm}^{-1}$ and $B_e = 20{,}000$ MHz.

11.4. Calculate the absorption coefficients for the $J+1 \leftarrow J$ rotational transitions of $^{127}I^{35}Cl$ for $J = 0, 1, 2, 3, 4$. Assuming a Boltzmann distribution of energies calculate the relative populations of the $J = 0$, 1, 2, 3, 4 levels. Compare the behavior of the two quantities just calculated and justify their similarities and differences.

11.5. Given $B_e = 3000.0 \pm 0.1$ MHz, $\alpha_e = 25 \pm 0.05$ MHz and $\omega_e = 2000 \pm 0.5\ \text{cm}^{-1}$, determine the error one would have in a value of χ_e and $\chi_e\omega_e$ derived from

$$\alpha_e = 6\sqrt{\frac{\chi_e B_e^3}{\omega_e}} - \frac{6B_e^2}{\omega_e}$$

11.6. For $^{127}I^{35}Cl$ the centrifugal distortion constant is $D_e = 0.001$ MHz. What resolution (in cm^{-1}) would be necessary to detect this small an effect in a rotational vibrational spectrum at $384\ \text{cm}^{-1}$.

11.7. Show that the spherical harmonics Y_2^0 and Y_2^2 are eigenfunctions of the rotational wave equation.

11.8. Find the linear combinations $Y_2^2 \pm Y_2^{-2}$ and show that they are solutions of the rotational wave equation.

11.9. Find, for the $J = 1 \leftarrow 0$ transition, the numerical values of the dipole matrix elements for a molecule with $\mu_0 = 3$ debye.

11.10. Calculate the fraction of its time a harmonic oscillator having $\omega_e = 2000\ \text{cm}^{-1}$ spends outside its classical limits in the $n = 0$ and $n = 2$ states.

11.11. Show that for the harmonic oscillator the transition $n = 0 \rightarrow 2$ is forbidden.

11.12. Calculate the thermal distribution between two energy levels if their energies differ (*a*) by $10{,}000\ \text{cm}^{-1}$; (*b*) by 2×10^{-14} ergs. The temperature in each case is first supposed to be 300 K, then 1700 K.

11.13. An absorption spectrum gives the following measured wave numbers (in cm^{-1}); 2886; 5668; 8347; 10923; 13397. Draw the energy level diagram with the electron volt as the unit of energy.

11.14. Find the energy levels for a one-dimensional harmonic oscillator of mass $m = 6.2 \times 10^{-24}$ g submitted to a force constant $K = 5.8105$ dyne/cm.

CHAPTER 12

DIATOMIC VIBROTORS, SPECTRA

12.1 NATURE OF DIATOMIC VIBROTOR SPECTRA

In the previous chapter it was found that the most general expression for the energy levels of a diatomic vibrotor was that developed using the Dunham potential function. The general nature of spectra can be examined, without any loss of generality, by limiting the number of terms in the energy expression to five, the primary contributions of vibrational and rotation plus one each for anharmonicity, centrifugal distortion, and rotation-vibrational interaction. The five-term expression

$$\frac{E_{vJ}}{h} = (v + \tfrac{1}{2})\omega_e + B_e J(J+1) - (v + \tfrac{1}{2})^2 \chi_e \omega_e - D_e J^2(J+1)^2$$
$$- \alpha_e(v + \tfrac{1}{2})J(J+1) + \cdots \tag{12.1}$$

developed by use of the Morse potential function will be used as a starting point to discuss the nature of diatomic vibrotor spectra, and we will examine the need for the inclusion of higher-order terms when discussing specific examples. The constants in Eq. (12.1) are functions of other molecular parameters

$$\omega_e = \frac{1}{2\pi}\sqrt{\frac{k}{\mu}} \tag{12.2}$$

$$B_e = \frac{h}{8\pi^2 I_e} \tag{12.3}$$

$$I_e = \mu r_e \tag{12.4}$$

$$D_e = \frac{4B_e^3}{\omega_e^2} \tag{12.5}$$

$$\alpha_e = 6\left[\frac{\chi_e B_e^3}{\omega_e}\right]^{1/2} - \frac{6B_e^2}{\omega_e} \tag{12.6}$$

or are related to the parameters of the potential function

$$\omega_e = \frac{a}{2\pi}\sqrt{\frac{2D}{\mu}} \tag{12.7}$$

$$\chi_e = \frac{h\omega_e}{4D} \tag{12.8}$$

$$\alpha_e = \frac{3h^2\omega_e}{4\mu r_e^2 D}\left[\frac{1}{ar_e} - \frac{1}{a^2 r_e^2}\right] \tag{12.9}$$

Knowing the selection rules $\Delta v = \pm 1$, $\Delta J = \pm 1$, and the relative transition intensities we are now in a position to examine the total nature of the diatomic vibrotor spectra in different regions of the electromagnetic spectrum.

For the case of a transition involving only a change in the rotational quantum number J, with the vibrational quantum number v remaining unchanged, the frequency of the resulting transition is found from Eq. (12.1) to be

$$\nu_{J+1\leftarrow J} = \frac{E_{J+1} - E_J}{h} = 2B_e(J+1) - 4D_e(J+1)^3 - 2\alpha_e(v+\tfrac{1}{2})(J+1) \tag{12.10}$$

The first term on the right-hand side is the rigid rotor energy. The magnitude of B_e will vary from 627,700 MHz for a light molecule like HF to 3422 MHz for a heavy molecule like ICl. Except for very light molecules, whose transitions fall in the far infrared, the majority of pure rotational transitions fall in the microwave frequency region and are observed using a microwave spectrometer. The second term is a measure of the centrifugal distortion of the molecule as it undergoes rotation. The magnitude of the centrifugal distortion, as represented by the constant D_e, will depend on two factors: (1) The larger the moment of inertia, hence the smaller the magnitude of B_e, the smaller the centrifugal distortion will be. Therefore, centrifugal distortion is more of a problem for light molecules. (2) The weaker the interatomic bond or the heavier the atoms in the molecule the lower the vibrational frequency, hence the molecule is more susceptible to centrifugal distortion. Since ω_e can vary from 384 cm^{-1} for a heavy molecule like ICl to 4138 cm^{-1} for a light one like HF when one considers the magnitudes of B_e just mentioned it is found that D_e can vary from 0.0012 MHz for ICl to 64.5 MHz for HF. While the magnitude of D_e is insufficient to produce a large change in the frequency of a particular transition for heavier molecules nevertheless the change is well within the frequency measuring ability and resolution associated with a microwave spectrometer. The last term measures the effective contribution to the moment of inertia of the molecule due to vibration. The rotational-vibrational interaction constant α_e is a complicated function of ω_e, r_e and the parameters of the molecular potential function. Since the order of magnitude of r_e does not change from one diatomic molecule to another α_e is approximately inversely proportional to ω_e and μ. Using the previously given values of ω_e, B_e and the dissociation energies of $D(\text{HF}) = 6.4$ eV and $D(\text{ICl}) = 2.15$ eV, one estimates α_e (ICl) = 16 MHz and α_e (HF) = 23,000 MHz. The magnitudes of these terms are appreciable from the viewpoint of resolution and ability to measure frequencies but do not shift the frequencies from those of the rigid rotor approximation by more than a few percent except for very light molecules. Using $^{35}Cl^{19}F$ with $B_e = 15{,}484$ MHz, $D_e = 0.03$ MHz, $\alpha_e = 130$ MHz as an example and calculating relative intensities from Eq. (10.170) the distribution and relative intensities of the rotational transition from $1 \leftarrow 0$ to $15 \leftarrow 14$ at 300 K are shown in Fig. 12.1.

There are two features to be noted with regard to the use of Eq. (12.8). First, any accurate calculation of bond lengths based on the experimental spectrum must take into account the zero-point vibrational energy or else the reported error must allow for neglect of the factor. Noting that Eq. (12.3) can be written

$$\nu_{J+1\leftarrow J} = 2(J+1)[B_e - (v+\tfrac{1}{2})\alpha_e] - 4D_e(J+1)^3 \tag{12.11}$$

$$\nu_{J+1\leftarrow J} = 2B_v(J+1) - 4D_e(J+1)^3 \tag{12.12}$$

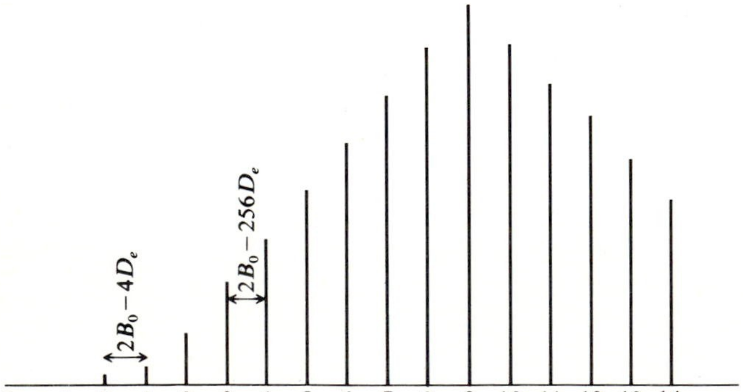

FIGURE 12.1
Rotational transitions of the $^{35}Cl^{19}F$ molecule in the ground vibrational state.

it is apparent that only B_v and D_e can be obtained from the observed frequencies. Molecules in the ground vibrational state have $B_v = B_0 = B_e - \frac{1}{2}\alpha_e$ while those in excited vibrational states give rise to a series of B_v values, one for each state. For rotational transitions occurring in the microwave region the frequencies of the transitions can be determined to 1 part in 10^7 to 10^8, or ± 0.01 MHz, at frequencies of the order of 25,000.00 MHz. Since J is an integer the error in B_0 is of the order of ± 0.01 MHz also. Considering an intermediate value of α_e, say 500 MHz, one finds that this contributes about 1 percent of the value of B_0. Neglect of this zero-point contribution, and the subsequent use of B_0 rather than B_e for determination of internuclear distances, will lead to an error in the determined distances of ± 0.0004 nm rather than the error of $\pm 10^{-7}$ nm, which could be obtained were a correct value of α_e to be combined with the observed B_0 value. This value of $\pm 10^{-7}$ nm is based on the assumption of no limitations due to uncertainties of atomic weights and fundamental constants, and is increased to $\pm 10^{-5}$ nm when these errors are included.

Since α_e is a function of B_e, ω_e, and χ_e it is impossible to determine its value from the measurement of rotational transitions for the ground vibrational state only. If one can observe the first vibrationally excited state transitions $\nu^1_{J+1\leftarrow J}$ and $\nu^1_{J'+1\leftarrow J'}$, then these, along with the corresponding ground state rotational transitions, give two pairs of equations

$$\nu^0_{J+1\leftarrow J} = 2B_0(J+1) - 4D_e(J+1)^3 \tag{12.13}$$

$$\nu^0_{J'+1\leftarrow J'} = 2B_0(J'+1) - 4D_e(J'+1)^3 \tag{12.14}$$

$$\nu^1_{J+1\leftarrow J} = 2B_1(J+1) - 4D_e(J+1)^3 \tag{12.15}$$

$$\nu^1_{J'+1\leftarrow J'} = 2B_1(J'+1) - 4D_e(J'+1)^3 \tag{12.16}$$

Thus from four experimental frequencies values of B_0, B_1, and D_e can be determined, and from

$$B_0 = B_e - \tfrac{1}{2}\alpha_e \tag{12.17}$$

$$B_1 = B_e - \tfrac{3}{2}\alpha_e \tag{12.18}$$

values of B_e and α_e are found. If ω_e and χ_e are known from vibrational spectra studies then a reasonable estimate of α_e can be obtained by using Eq. (12.7) and setting $B_e = B_0$.

The second factor which must be taken into account is centrifugal distortion. If we compare the energy levels of the diatomic rotor in the ground vibrational state as given by

$$\frac{E_J}{h} = J(J+1)B_0 - D_eJ^2(J+1)^2 \tag{12.19}$$

to those of a rigid rotor we observe that the levels of the real rotor tend to be decreased in separation from those of the rigid approximation. This effect is due to centrifugal distortion which, with increasing J, tends to increase the bond length, hence increase the effective moment of inertia and decrease the effective

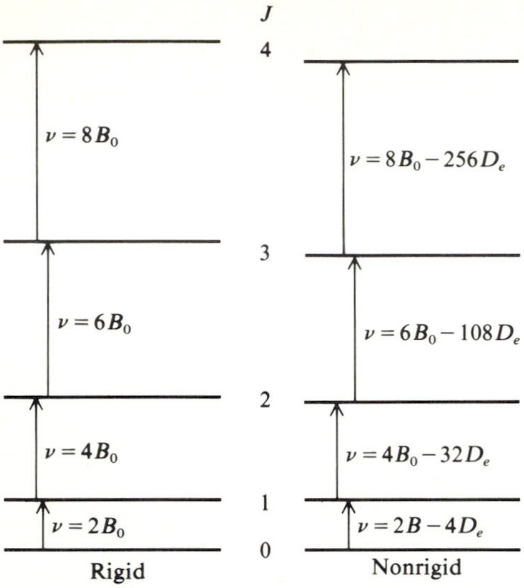

FIGURE 12.2
Effect of centrifugal distortion on the energy levels of a diatomic rotor.

rotational constant. This is illustrated by the energy level scheme shown in Fig. 12.2. The value of the centrifugal distortion constant D_e is obtained from the solution of Eqs. (12.11) and (12.12). Since D_e is related to B_e and ω_e the simultaneous determination of B_e and D_e allows the calculation of the fundamental vibrational frequency ω_e. Also, a knowledge of ω_e from vibrational spectra will allow one to estimate D_e values to aid in predicting spectra.

Using instrumentation which will make accurate measurements of relative intensities of microwave transitions enables one to make a direct determination of ω_e. Using the expression for the relative intensity (relative absorption coefficient) for the two transitions $\nu^1_{J+1\leftarrow J}$ and $\nu^0_{J+1\leftarrow J}$ given by

$$I_{\text{rel}} = \frac{f_1}{f_0} = \frac{N_1}{N_0} = e^{-h\nu_0/kT} \tag{12.20}$$

the value of ω_e can be obtained from the measurement of I_{rel}. This method can serve to determine vibrational frequencies lying in the experimentally difficult far infrared region (less than 50 cm^{-1}).

While it is in general a hypothetical situation we will consider the simple case of transitions which involve changes in vibrational quantum number only with no corresponding change in rotational quantum numbers. Such a transition would be between energy levels

$$\frac{E_v}{h} = (v + \tfrac{1}{2})\omega_e - (v + \tfrac{1}{2})^2\chi_e\omega_e \tag{12.21}$$

and would be given by

$$\nu_{v'\leftarrow v} = (v' - v)\omega_e - (v'^2 - v^2 + v' - v)\chi_e\omega_e \tag{12.22}$$

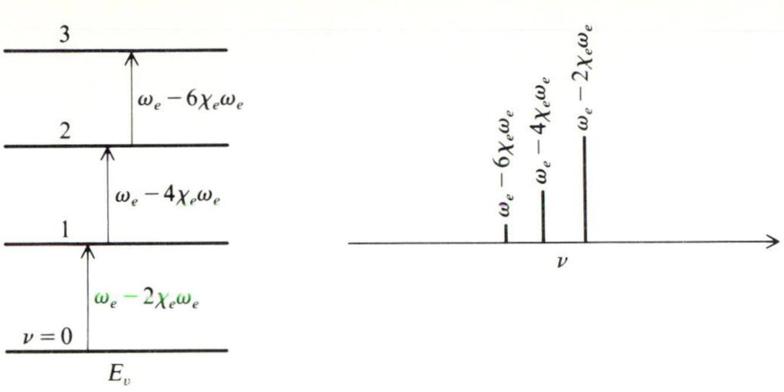

FIGURE 12.3
Effect of anharmonicity on the vibrational energy levels and spectrum of a diatomic molecule.

The product $\chi_e\omega_e$ is the anharmonicity constant. The introduction of anharmonicity and its effect on the potential functions and selection rules will be discussed in more detail in Chap. 13, and we will only briefly consider the problem at this point. For a diatomic molecule the effect of anharmonicity is illustrated in Fig. 12.3. Reference to Eq. (12.8) shows that the simultaneous determination of ω_e and χ_e can allow the dissociation energy of the molecule to be calculated, insofar as the Morse function is valid.

While transitions involving $\Delta J = \pm 1$ and $\Delta v = 0$ are observable in the microwave region, those with $\Delta v = \pm 1$ and $\Delta J = 0$, which would occur primarily in the infrared region, cannot be observed due to the fact that $\Delta J = 0$ is not an allowed selection rule for diatomic molecules. If we then consider the problem of transitions involving simultaneous changes in both v and J, while neglecting rotational-vibrational interactions (that is, $B_e = B_0$, $D_e = 0$, $\chi_e = 0$), the picture presented in Fig. 12.4 for the transition between the ground and the first excited vibrational states evolve. The energy of an individual level is given by

$$\frac{E_{Jv}}{h} = (v + \tfrac{1}{2})\omega_e + J(J+1)B_0 \tag{12.23}$$

and the transitions for $v = 0 \rightarrow v = 1$ by

$$\nu_R = \omega_e + 2B_0(J+1) \qquad \Delta J = 1 \tag{12.24}$$

$$\nu_P = \omega_e - 2B_0(J+1) \qquad \Delta J = -1 \tag{12.25}$$

Including typical values for the relative intensities of transitions, the spectrum for a molecule with $B \approx 25$ GHz is as shown in Fig. 12.4. The transitions lying above and below the fundamental vibrational frequency are referred to as the *R*-branch ($\Delta J = +1$) and *P*-branch ($\Delta J = -1$) transitions respectively.

The effect of centrifugal distortion on the spectrum shown in Fig. 12.4 is analogous to the effect produced in the case of pure rotational spectra, that is, the transitions tend to become more closely spaced. This effect is of little practical consideration, however, because even for a moderately light molecule such as

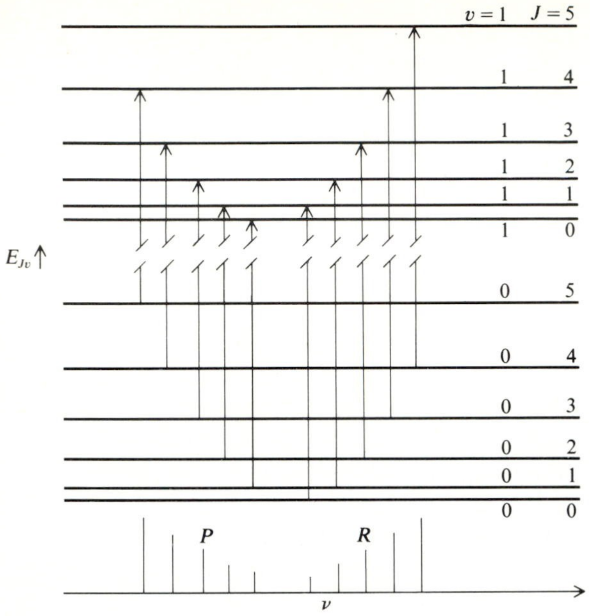

FIGURE 12.4
Energy levels and transitions for a diatomic vibrotor with no centrifugal distortion, vibrational-rotational interaction, or anharmonicity.

CO having $D_e = 0.2$ MHz the displacement of a $J = 10 \leftarrow J = 9$ transition from that of a rigid rotor would be 800 MHz, or $0.02\ \text{cm}^{-1}$, which would require an infrared spectrometer with very good resolution and accuracy.

Even though the effect of centrifugal distortion can, for the most part, be neglected when working in the infrared region, the interaction of vibration and rotation, which is manifested in the terms $B_v = B_e - (v + \frac{1}{2})\alpha_e$, cannot. For the transitions $v = 1 \leftarrow v = 0$, insofar as the R- and P-branch transitions are concerned, the effect of anharmonicity can generally be neglected since, to first-order, it will only affect the center of the entire spectrum and not the relative spacings. Neglecting centrifugal distortion and anharmonicity the energy levels of the diatomic vibrotors are given by

$$\frac{E_{JV}}{h} = (v + \tfrac{1}{2})\omega_e + J(J+1)[B_e - (v + \tfrac{1}{2})\alpha_e] \tag{12.26}$$

Considering the transition $v = 1 \leftarrow v = 0$ and $\Delta J = \pm 1$ the transition frequencies become, for the R branch,

$$\nu_R = \omega_e + 2B_1 + (2B_1 - B_0)J + (B_1 - B_0)J^2 \tag{12.27}$$

and, for the P branch,

$$\nu_P = \omega_e - (B_0 + B_1)J + (B_1 - B_0)J^2 \tag{12.28}$$

An illustration of such a spectrum exaggerated along the x axis, but with realistic relative intensities, is shown in Fig. 12.5.

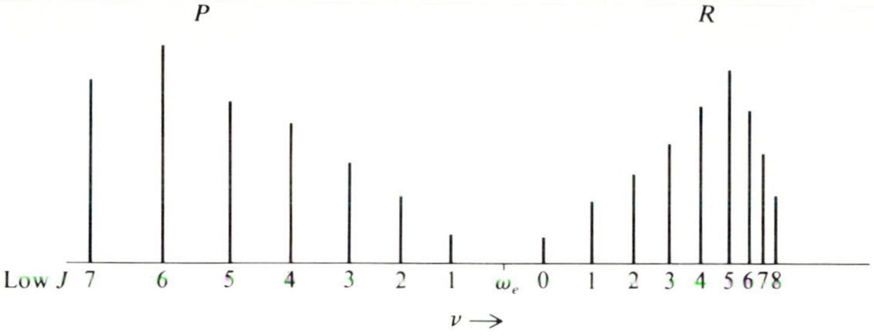

FIGURE 12.5
Spectrum of a diatomic vibrotor illustrating rotational-vibrational interaction.

Since the higher vibrational state has the greater moment of inertia $B_1 < B_0$. This means that the J^2 terms in Eqs. (12.25) and (12.26) will be negative. With increasing J values the spacings between the members of the P-branch increase and the spectrum extends further down in frequency. The R-branch transitions tend to converge at higher J values. At very large values of J the frequencies will actually start to move back to lower frequencies giving rise to a band head. Such behavior is frequently observed in the vibrational-rotational structure of electronic spectra.

The measurement of the P- and R-branch transition frequencies thus enables one to determine values of B_1 and B_0, hence B_e and α_e, and ultimately interatomic distances. The determination of several values of B_1 and B_0 from various pairs of transitions can, for very light molecules, sometimes allow one to observe effects of centrifugal distortion that otherwise would go undetected. Since it is not possible to measure frequencies in the infrared region as accurately as in the microwave region, ± 0.001 cm^{-1} (± 0.03 GHz) being typical, interatomic distances derived from values of B_e determined in the above manner are not as accurate as those determined from pure rotational spectra.

By careful consideration of the pattern of transitions shown in Fig. 12.5, or by the judicious selection of difference between various frequencies given by Eqs. (12.27) and (12.28), it is possible to readily determine B_1 or B_0 without the simultaneous solution of such a pair of equations. Taking the difference between the R- and P-branch transitions which start at the same J level one has

$$\nu_R(J+1 \leftarrow J) - \nu_P(J \rightarrow J-1) = 2B_1(2J+1) \tag{12.29}$$

while taking the difference between two transitions which end up at the same level gives

$$\nu_R(J+1 \leftarrow J) - \nu_P(J+2 \rightarrow J+1) = 2B_0(2J+3) \tag{12.30}$$

The last factor to be considered prior to the examination of the experimental results regarding several specific molecules is the effect of isotopic substitution on the observed spectra. The primary factor which will change upon changing isotopes in a diatomic molecule will be the reduced mass. From the definition of reduced mass $\mu = m_1 m_2/(m_1 + m_2)$ one can see that increasing the mass of

TABLE 12.1
Molecular parameters for isotopic pairs of diatomic molecules

Molecule	μ, amu	D, eV[b]	B_e, GHz	ω_e, cm^{-1}	r_e, nm	α_e, MHz	$\chi_e\omega_e$, cm^{-1}	D_e, MHz
$^{1}H^{35}Cl$	0.9799	4.43	317.51	2990	0.1275	9050	52	15.9
$^{2}H^{35}Cl$	1.9050	4.48	163.24	2090	0.1275	3352		
$^{2}H^{35}Cl$	1.9050	4.48	163.340	2990	0.1275	9050	52	15.9
$^{2}H^{37}Cl$	1.9100	.....	162.859	.....	0.1275	3352		
$^{127}I^{35}Cl$	27.4221	2.152	3.4223	384	0.2321	16.06	1.46	0.0012
$^{127}I^{37}Cl$	....	.....	3.2774	376	0.2321	15.05	.....	0.0011
$^{35}Cl^{19}F$	12.314	2.616	15.4837	793	0.1628	130.7	2.22	0.026
$^{37}Cl^{19}F$	....	.....	15.1892	778	0.1628	126.9	.....	0.025
$^{1}H^{19}F$	0.9573	6.40	628.17	4138	0.09171	23.100	90.1	
$^{2}H^{19}F$	1.8216	.....	330.21	2998	0.09170	8.790	45.1	

one atom will increase the reduced mass since the minimum value for any one atom is 1 amu. This increase in μ will consequently increase the moment of inertia I. Examination of Eqs. (12.2) through (12.9) shows that the end result of such a change is to decrease both B_e and ω_e, since the potential function, and hence the constant D, are not changed by isotopic substitution. Since μ enters into the expression for B_e as a first power and into the one for ω_e as a one-half power, a given change in μ will produce a larger change in B_e than in ω_e. In view of these statements, the increase of μ will result in a decrease in D_e, which again confirms the earlier statement that the effect of centrifugal distortion is larger in lighter molecules. Since an increase in μ, due to isotopic substitution, leads to a decrease in ω_e, the resulting value of χ_e will be decreased. α_e is a more complicated function of B_e and ω_e, but inspection of the defining equation indicates that it will be the difference between two terms both of which decrease with increased μ, hence α_e will decrease also. The conclusions reached above are substantiated by the data given in Table 12.1 [1].

With centrifugal stretching neglected the effect of isotopic substitution on rotational and rotational-vibrational spectra is shown pictorially for a diatomic molecule in Fig. 12.6. The primary feature which one observed from this figure is that each molecular species produces its own independent spectrum. The relative intensities of the two spectra will be equal to the relative abundance of the two isotopic species. The relative intensities of the various components of

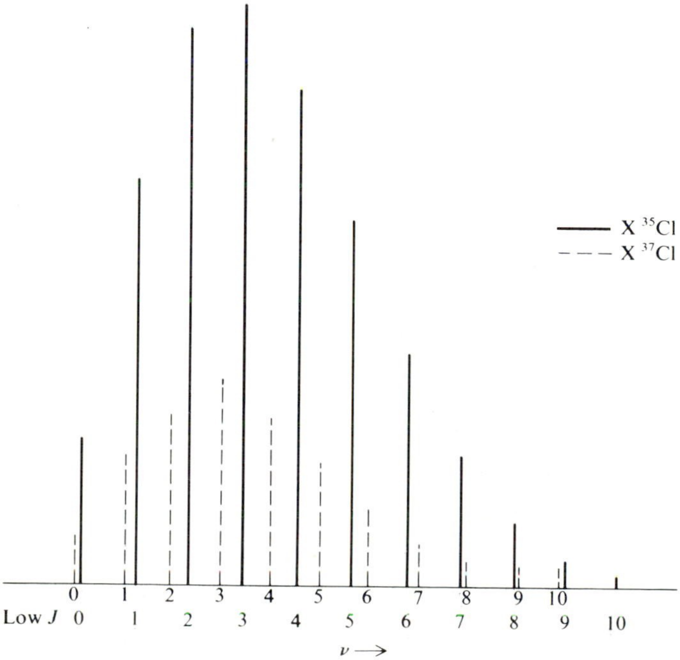

FIGURE 12.6
Effect of isotope substitution on the pure rotational spectrometer diatomic molecule.

each spectrum belonging to a given isotopic species is found in the manner previously described.

12.2 TYPICAL EXPERIMENTAL SPECTRA OF DIATOMIC MOLECULES

In this section we will consider several cases of rotational spectra in the microwave region and high-resolution vibrational-rotational spectra in the infrared. Experimental data will be presented and the methods of the previous sections used to analyze the data. For diatomic molecules the assignment of correct quantum numbers to the observed rotational transitions can be initiated by calculating a spectrum using normal covalent bond radii corrected for electronegatives. Due to the large differences between transition frequencies this method is generally unambiguous. In the examples involving rotational spectra which follow, some of the transitions considered are in reality split into hyperfine components by nuclear quadrupole coupling. This splitting is much less than the magnitude of the transition frequency and the frequencies given are the averages of the hyperfine components. Later we will consider this hyperfine interaction in detail but for the present it will suffice to say that the hyperfine interaction effects have been eliminated and the transition frequencies given are for the centers of the hyperfine patterns. For a diatomic molecule the average of the hyperfine components and not the separation of the components is the parameter which is dependent on the moment of inertia. When considering the following examples we will also show how to utilize auxiliary information to good advantage in some instances.

The high-temperature study of GeS by Hoeft [2] is a good example of the use of pure rotational spectroscopy to study a diatomic molecule. The observed transitions for two isotopic species are given in Table 12.2. Since data is available for several vibrational states the inclusion of a second rotational-vibrational interaction constant to convert Eq. (12.11) into the form

$$\nu^{v}_{J+1\leftarrow J} = 2[B_e - (v+\tfrac{1}{2})\alpha_e - (v+\tfrac{1}{2})^2\gamma_e](J+1) - 4D_e(J+1)3 \quad (12.31)$$

TABLE 12.2
Observed rotational transitions for GeS

Transition $J+1 \rightarrow J$	Vibrational state v	Observed frequency, MHz $^{74}Ge^{32}S$	$^{72}Ge^{32}S$
$1 \rightarrow 0$	0	11,163.72 ± 0.05	11,257.34 ± 0.05
	1	11,118.90 ± 0.05	11,211.88 ± 0.05
	2	11,073.96 ± 0.05	11,166.34 ± 0.05
	3	11,029.03 ± 0.05	11,120.90 ± 0.05
$3 \rightarrow 2$	0	33,490.95 ± 0.05	33,771.93 ± 0.05
	1	33,356.23 ± 0.05	33,635.44 ± 0.05
	2	33,221.51 ± 0.05	33,498.74 ± 0.05

TABLE 12.3
Rotational parameters for GeS

Molecule	B_0, MHz	B_e, MHz	α_e, MHz	D_e, kHz	r_e, nm
$^{74}Ge^{32}S$	5581.867 ± 0.009	5593.088 ± 0.014	22.443 ± 0.013	2.349	0.2012091 ± 0.0000059
$^{72}Ge^{32}S$	5628.692 ± 0.009	5640.064 ± 0.014	22.744 ± 0.013	2.388	0.2012085 ± 0.0000059

is warranted. Using the 11163.72, 1118.90, 11073.91, and 33490.95 MHz transitions for $^{74}Ge^{32}S$ the solution of four simultaneous equations of the form of Eq. (12.31) gives $B_e = 5593.037$ MHz, $\alpha_e = 22.354$ MHz, $\gamma_e = -0.029$ MHz, and $D_e = 2.188$ kHz.

Since there is more data than parameters it is best to use all possible combinations of frequencies and parameters and average the results. Doing this Hoeft obtained the results shown in Table 12.3. This work included studies of eight isotopic species, so the results in Table 12.3 reflect the use of more data than that given in Table 12.2 alone. The end result of this work gave an average value of $r_e = 0.2012088$ nm with an experimental-fitting error of ± 0.000002 nm. This illustrates the overall accuracy and precision with which internuclear distances of diatomic molecules can be obtained.

An example which shows how auxiliary data from vibrational spectra can be incorporated, and affords a look at the treatment of a large amount of experimental data, is the high-temperature microwave spectrum of AgCl which has been determined by Krisher and Norris [3]. The observed frequencies are listed in Table 12.4. The discussion of this molecule will involve more detailed points than the previous one and will inject a few concepts not previously mentioned. Since a number of excited vibrational states were observed in this

TABLE 12.4
Rotational transition frequencies for AgCl†, MHz

Transition	$^{107}Ag^{35}Cl$	$^{109}Ag^{35}Cl$	$^{107}Ag^{37}Cl$	$^{109}Ag^{37}Cl$
$J = 3 \rightarrow J = 2$				
$v = 0$	22,068.42	21,968.64	21,170.55	21,070.41
$v = 1$	21,961.54	....		
$v = 2$	....	21,755.81		
$J = 4 \rightarrow J = 3$				
$v = 0$	29,424.22	29,290.86	28,277.45	28,094.08
$v = 1$	29,281.61	29,149.44	‡	27,960.86
$v = 2$	29,139.00	29,008.65	‡	27,828.84
$v = 3$	28,997.87	28,867.36		
$v = 4$	28,856.21	28,726.62		
$v = 5$	28,714.58			

† Error in measurements is ±0.2 MHz. Frequencies are calculated centers of hyperfine patterns.

‡ Obscured by other more intense transitions.

work frequency expression given by Eq. (12.11) is insufficient to allow an adequate fit of the experimental data. In order to obtain a good fit of the experimental data it is necessary to include an additional rotational-vibrational interaction constant. Since the molecule is quite heavy and the centrifugal distortion is small the addition of additional terms involving distortion are unnecessary and the frequency of a transition is given by Eq. (12.31). The available data is sufficient to allow B_e, α_e, γ_e, and D_e to be calculated for the $^{107}Ag^{35}Cl$ and $^{109}Ag^{35}Cl$ species by using sets of four simultaneous equations. These values are given in Table 12.5.

Looking in more detail at the $^{107}Ag^{35}Cl$ species one can see the justification for the use of the extra constant γ_e in the frequency expression. For $v = 5$ the term $\gamma_e(v + \frac{1}{2})^2$ is equal to 0.6 MHz which is greater than six times the estimated error in B_e. The fact that the value of ω_e calculated from the determined value of B_e and D_e agrees so well with the value from infrared measurements indicates the validity of the value of D_e. The small value of this term is characteristic of heavy molecules.

A further look at the data shows that there are insufficient transitions having larger v values to allow the set of four molecular parameters to be calculated for the $^{107}Ag^{37}Cl$ and $^{109}Ag^{37}Cl$ species. For these two the value of γ_e can be determined by multiplying the value for the $^{107}Ag^{35}Cl$ species by ρ^3 where $\rho = (\mu/\mu')^{1/2}$ and μ is the reduced mass of the latter species. While the use of γ_e could have been omitted from the discussion with little loss of overall accuracy it has been included to point out the degree of precision accessible in the

TABLE 12.5
Molecular parameters for AgCl

	Isotopic species			
Parameter	^{107}**Ag**^{35}**Cl**	^{109}**Ag**^{35}**Cl**	^{107}**Ag**^{37}**Cl**	^{109}**Ag**^{37}**Cl**
B_e, MHz	3687.04 ± 0.09	3670.24 ± 0.09	3537.49 ± 0.03	3520.14 ± 0.05
α_e, MHz	17.855 ± 0.011	17.728 ± 0.011	16.766 ± 0.012	16.628 ± 0.012
γ_e, MHz	0.019 ± 0.004	0.019 ± 0.004	0.017 ± 0.004¶	0.017 ± 0.004¶
D_e, MHz	0.003 ± 0.003	[0.00007]§	0.002 ± 0.001	0.002 ± 0.001
ω_e, cm^{-1}	343.6	342.4	336.0	355.7¶
$\omega_e\chi_e$, cm^{-1}†	1.163	1.158¶	1.116¶	1.110¶
μ, amu	26.358016	26.477834	27.476495	27.606723
r_e, nm	0.228074	0.228080	0.228058	0.228082

† B. A. Brice, *Phys. Rev.*, **35**, 960 (1930).

‡ Calculated from B_e and D_e using $\omega_e^2 = 4B_e^3/D_e$.

§ Statistical error in B_e larger than effect of D_e value calculated.

¶ Calculated from isotopic ratios where $\rho = (\mu/\mu')^{1/2}$

$$\gamma'_e = \gamma_e\rho^3$$

$$\omega'_e = \omega_e\rho$$

$$\omega'_e\chi'_e = \omega_e\chi_e\rho^2$$

TABLE 12.6
Comparison of calculated and observed reciprocal moments of AgCl

	$^{109}Ag^{35}Cl$	$^{107}Ag^{37}Cl$	$^{109}Ag^{37}Cl$
B_e(expt.), MHz	3670.24	3537.49	3520.14
B_e(calc.), MHz	3670.35	3536.95	3520.27

determination of molecular parameters by microwave methods when using a suitable energy expression. The relations used to determine the ω_e and $\chi_e\omega_e$ values in Table 12.5 are obtained from Eqs. (12.2) through (12.9). Since there is a lack of good experimental data for the $^{107}Ag^{37}Cl$ species one would suspect that the r_e value determined from the corresponding B_e value would be the least accurate. The average of the other three values is $r_e = 0.228079 \pm 0.000004$ nm as compared to $r_e = 0.228058$ nm for this species.

This example affords the opportunity to show how the accuracy and internal consistency of the results can be tested. If one considers the equilibrium internuclear separation to be independent of the isotopic species then, since $B_e = h/8\pi^2\mu r_e^2$, the ratio of two rotational constants for two isotopic species i and j will be

$$\frac{B_e^i}{B_e^j} = \frac{\mu^i}{\mu^j} = \rho^2 \tag{12.32}$$

Hence, one can calculate the moments of all isotopic species based on one member. Such a tabulation is given in Table 12.6. One sees that the agreement for the $^{109}Ag^{35}Cl$ and $^{109}Ag^{37}Cl$ is very close to being within the experimental error but again, due to the lesser amount of data, there is a greater difference with the $^{107}Ag^{37}Cl$ species.

From the data available in Table 12.5 and by using Eqs. (12.2) through (12.9) it is possible to determine the values of a and D for the Morse potential curve of the molecule. Hence

$$D = \frac{h\omega_e^2}{4\omega_e\chi_e} \tag{12.33}$$

and

$$a = 2\pi\omega_e\sqrt{\frac{\mu}{2D}} \tag{12.34}$$

TABLE 12.7
Potential constants for AgCl

Species	a, m^{-1}	$D(J)$
$^{107}Ag^{35}Cl$	138.5 ± 0.1	303.7 ± 0.4
$^{109}Ag^{35}Cl$	135.4 ± 0.1	302.9 ± 0.4
$^{107}Ag^{37}Cl$	134.9 ± 0.1	302.5 ± 0.4
$^{109}Ag^{37}Cl$	134.8 ± 0.1	303.7 ± 0.4

The calculated values of D and a are given in Table 12.7. The consistency of these values shows that the ground state potential energy curves of the four species are almost identical and that the chemical and physical properties of all species are the same.

A good example of the analysis of vibrational-rotational spectra is afforded by the work of Hurlock, Alexander, Rao, and Dreska [4] on hydrogen iodide. The observed transitions for $^2H^{127}I$ and $^1H^{127}I$ are given in Tables 12.8 through

TABLE 12.8
Rotational-vibrational transitions for $^2H^{127}I$
($v = 1 \leftarrow v = 0$)

Transition	Observed	Calculated	Difference
	***R* branch, cm^{-1}**		
$R(0)$	1606.154	1606.149	+0.006
$R(1)$	1612.409	1612.410	−0.001
$R(2)$	1618.547	1618.545	+0.002
$R(3)$	1624.557	1624.554	+0.003
$R(4)$	1630.432	1630.436	−0.004
$R(5)$	1636.190	1636.189	+0.001
$R(6)$	1641.814	1641.812	+0.002
$R(7)$	....	1647.304	
$R(8)$	1652.659	1652.663	−0.004
$R(9)$	1657.884	1657.889	−0.005
$R(10)$	1662.973	1662.980	−0.007
$R(11)$	1667.935	1667.935	0.000
$R(12)$	1672.754	1672.753	+0.001
$R(13)$	1677.438	1677.433	+0.005
$R(14)$	1681.976	1681.973	+0.003
$R(15)$	1686.376	1686.373	+0.003
$R(16)$	1690.636	1690.631	+0.005
$R(17)$	1694.741	1694.746	−0.005
$R(18)$	1698.716	1698.718	−0.002
$R(19)$	1702.544	1702.544	
	***P* branch, cm^{-1}**		
$P(1)$	1593.262	1593.258	+0.004
$P(2)$	1596.630	1586.629	+0.001
$P(3)$	1579.879	1579.881	+0.003
$P(4)$	1573.010	1573.014	−0.004
$P(5)$	1566.033	1566.030	+0.003
$P(6)$	1558.923	1558.930	−0.007
$P(7)$	1551.720	1551.714	+0.006
$P(8)$	....	1544.386	
$P(9)$	....	1536.945	
$P(10)$	....	1529.394	
$P(11)$	1521.725	1521.733	−0.008
$P(12)$	1513.971	1513.965	+0.006

12.11 respectively. This example serves not only to illustrate the rotational-vibrational coupling but also to illustrate isotopic effects. By using the following equations developed from Eq. (12.1) with $v = 1 \leftarrow v = 0$

$$\nu_R = \omega_e - 2\chi_e\omega_e + (J+1)[(J+2)B_1 - JB_0] - 4D_e(J+1)^3 \tag{12.35}$$

$$\nu_P = \omega_e - 2\chi_e\omega_e + J[(J-1)B_1 - (J+1)B_0] + 4D_eJ^3 \tag{12.36}$$

$$\nu_R(J+1 \leftarrow J) - \nu_P(J \rightarrow J-1) = 2B_1(2J+1) - 4D_e(3J^2+3J+1) \tag{12.37}$$

$$\nu_R(J+1 \leftarrow J) - \nu_P(J+2 \rightarrow J+1) = 2B_0(2J+3) - 4D_e(J^2+1) \tag{12.38}$$

and any five transitions, one can evaluate the parameters in the equations. In this case there are sufficient transitions to allow for an overdetermination and cross-check of the parameters. The values for the rotational vibrational parameters, obtained by using the five transitions closest to the center for each isotopic species, are $B_1 = 3.1912\ \text{cm}^{-1}$, $B_0 = 3.2530\ \text{cm}^{-1}$, $D_e = -0.00037\ \text{cm}^{-1}$, $\alpha_3 = 0.0618\ \text{cm}^{-1}$, $B_e = 3.2839\ \text{cm}^{-1}$ for $^2H^{127}I$, and $B_1 = 6.2557\ \text{cm}^{-1}$, $B_0 = 6.4232\ \text{cm}^{-1}$, $D_e = -0.00062\ \text{cm}^{-1}$, $\alpha_e = 0.1675\ \text{cm}^1$, $B_e = 6.5070\ \text{cm}^{-1}$ for $^1H^{127}I$.

TABLE 12.9
Rotational-vibrational transitions for $^1H^{127}I$
($v = 1 \leftarrow v = 0$)

Transition	Observed	Calculated	Difference $\nu_0 - \nu_e$
	R branch, cm^{-1}		
$R(0)$	2242.087	2242.091	−0.004
$R(1)$	2254.257	2254.254	+0.003
$R(2)$	2266.071	2266.064	+0.007
$R(3)$	2277.510	2277.517	−0.007
$R(4)$	2288.616	2288.609	+0.007
$R(5)$	2299.330	2299.333	−0.003
$R(6)$	....	2309.686	+0.09
$R(7)$	....	2319.662	−0.04
$R(8)$	....	2329.257	
$R(9)$	....	2338.465	
	P branch, cm^{-1}		
$P(1)$	2216.723	2216.729	−0.006
$P(2)$	2203.541	2203.540	+0.001
$P(3)$	2190.025	2190.019	+0.006
$P(4)$	2176.168	2176.169	−0.001
$P(5)$	2161.990	2161.998	−0.008
$P(6)$	2147.516	2147.508	+0.008
$P(7)$	2132.707	2132.706	+0.001
$P(8)$	2117.597	2117.596	+0.001
$P(9)$	2102.181	2102.184	−0.003

TABLE 12.10
Rotational-vibrational transitions for $^2H^{127}I$
$(v = 2 \leftarrow v = 0)$

Transition	Observed	Calculated	Difference $\nu_0 - \nu_c$
	R **branch**		
$R(0)$	3165.620	3165.616	+0.004
$R(1)$	3171.628	3171.631	−0.003
$R(2)$	3177.395	3177.399	−0.004
$R(3)$	3182.913	3182.917	−0.004
$R(4)$	3188.187	3188.186	+0.001
$R(5)$	3193.202	3193.202	0.000
$R(6)$	3197.962	3197.967	−0.005
$R(7)$	3202.477	3202.477	0.000
$R(8)$	3206.738	3206.732	+0.006
$R(9)$	3210.732	3210.730	+0.002
$R(10)$	3214.471	3214.470	+0.001
$R(11)$	3217.953	3217.952	+0.001
$R(12)$	3221.178	3221.173	+0.005
$R(13)$	3224.128	3224.132	+0.004
$R(14)$	3226.826	3226.829	−0.003
$R(15)$	3229.260	3229.262	−0.002
$R(16)$	3231.430	3231.429	+0.001
	P **branch**		
$P(1)$	3152.851	3152.848	+0.003
$P(2)$	3146.090	3146.097	−0.007
$P(3)$	3139.099	3139.104	−0.005
$P(4)$	3131.870	3131.869	+0.001
$P(5)$	3124.404	3124.494	+0.010
$P(6)$	3116.686	3116.681	+0.005
$P(7)$	3108.729	3108.730	−0.001
$P(8)$	....	3100.543	
$P(9)$	3092.116	3092.120	−0.004
$P(10)$	3083.467	3083.464	+0.003
$P(11)$	3074.573	3074.575	−0.002

Since only the $v = 1 \leftarrow v = 0$ transition has been considered the values of ω_e and $\omega_e \chi_e \omega_e$ could not be determined. If the first overtone bands, as listed in Tables 12.10 and 12.11 for $^2H^{127}I$ and $^1H^{127}I$ respectively are included in the analysis, then one finds, using the values of B_1, B_0, α_e, and D_e from the analysis of the $v = 1 \leftarrow v = 0$ band and Eqs. (12.33) and (12.34) that $\omega_e = 1639.74\ \text{cm}^{-1}$, $\chi_e \omega_e = 19.984\ \text{cm}^{-1}$ for $^2H^{127}I$, and $\omega_e = 2309.51\ \text{cm}^{-1}$, $\chi_e \omega_e = 39.967\ \text{cm}^{-1}$ for $^1H^{127}I$.

The analysis just presented is the simplest method that can be used and does not take full advantage of the available data. Hurlock et al., performed a least-squares analysis of the data given in Tables 12.8 through 12.11 plus additional

TABLE 12.11
Rotational-vibrational transitions for $^{1}H^{127}I$
($v = 2 \leftarrow v = 0$)

Transition	Observed	Calculated	Difference $\nu_0 - \nu_c$
		R **branch**	
$R(1)$	4391.390	4391.391	−0.001
$R(2)$	4402.875	4402.864	+0.011
$R(3)$	4413.652	4413.640	+0.012
$R(4)$	4423.714	4423.715	−0.001
$R(5)$	4433.076	4433.083	−0.007
$R(6)$	4441.731	4441.739	−0.008
$R(7)$	4449.676	4449.679	−0.003
$R(8)$	4456.981	4456.897	−0.006
$R(9)$	4463.394	4463.389	+0.005
$R(10)$	4469.153	4469.151	+0.002
$R(11)$	4474.184	4474.177	+0.007
$R(12)$	4478.464	4478.462	+0.002
$R(13)$	4481.997	4482.003	−0.006
$R(14)$	4484.794	4484.793	+0.001
		P **branch**	
$P(1)$	4366.369	4366.373	−0.004
$P(2)$	4352.844	4352.839	+0.005
$P(3)$	4338.626	4338.628	−0.002
$P(4)$	4323.742	4323.744	−0.002
$P(5)$	4308.190	4308.195	−0.005
$P(6)$	4291.973	4291.983	−0.010
$P(7)$	4275.121	4275.115	+0.006
$P(8)$	4257.604	4257.595	+0.009
$P(9)$	4239.428	4239.429	−0.001
$P(10)$	4220.620	4220.622	−0.002
$P(11)$	4201.178	4201.179	−0.001

transition from the $v = 3 \leftarrow v = 0$ and $v = 4 \leftarrow v = 0$ bands of $^{1}H^{127}I$ and the $v = 3 \leftarrow v = 0$ band for $^{2}H^{127}I$. These results are shown in Table 12.12. It can be observed that this additional refinement produces some further, but not large, changes in the parameters. Due to the large amount of available data the least-squares refinement uses the Dunham energy expression

$$\frac{E_{JV}}{h} = \omega_e(v + \tfrac{1}{2}) - \chi_e\omega_e(v - \tfrac{1}{2})^2 - y_e\omega_e(v + \tfrac{1}{2})^3 - z_e\omega_e(v + \tfrac{1}{2})^4$$
$$+ [B_e - (v + \tfrac{1}{2})\alpha_e - (v + \tfrac{1}{2})^2\gamma_e]J(J + 1) - D_eJ^2(J + 1)^2 \quad (12.39)$$

which improves the fitting procedure. In addition to the observed frequencies,

TABLE 12.12
Spectroscopic parameters for hydrogen iodide, cm^{-1}

	$^1H^{127}I$	$^2H^{127}I$
$\nu_0(0 \rightarrow 1)$	2229.581 ± 0.006	1599.764 ± 0.004
B_e	6.5111 ± 0.0004	3.2893 ± 0.0002
α_e	0.16886 ± 0.00014	0.06082 ± 0.00006
γ_e	$(-9.5 \pm 0.2) \times 10^{-4}$	$(-1.76 \pm 0.14) \times 10^{-4}$
D_e	$(2.07 \pm 0.04) \times 10^{-4}$	$(5.31 \pm 0.14) \times 10^{-4}$
ω_e	2308.598 ± 0.018	1639.655 ± 0.018
$\chi_e\omega_e$	39.3610 ± 0.012	19.873 ± 0.012
$y_e\omega_e$	-0.0908 ± 0.018	-0.0459 ± 0.018

Tables 12.8 through 12.11 list the frequencies calculated using Eq. (12.39) and the deviation between observed and calculated frequencies.

REFERENCES

A. Specific

1. Gaydon, A. G., *Dissociation Energies and Spectra of Diatomic Molecules*, Chapman and Hall Ltd., London, GB, 1947. (Dissociation energies.)
   Herzberg, G., *Spectra of Diatomic Molecules*, Van Nostrand, Princeton, NJ, 1950. (Data for HCl and HF.)
   Townes, C. H., F. R. Merritt, and B. D. Wright, *Phys. Rev.*, **73**, 1334 (1948). (Data for ICl.)
   Gilbert, D. A., A. Roberts, and P. A. Griswold, *Phys. Rev.*, **76**, 1723L (1949); ibid, **77**, 742A (1950). (Data for FCl.)
   Cowan, M. J., and W. Gordy, *Bull. Am. Phys. Soc.*, **2**, 212 (1957). (Data for DCl.)
2. Hoeft, J., *Z. Naturforschg.*, **20A**, 826 (1965).
3. Krisher, L. C., and W. G. Norris, *J. Chem. Phys.*, **44**, 391 (1966).
4. Hurlock, S. C., R. M. Alexander, K. N. Rao, and N. Dreska, *J. Mol. Spect.*, **37**, 373 (1971).

B. General

See references at the end of Chap. 11.

PROBLEMS

12.1. The $J = 2 \leftarrow 1$ transition for $^{28}Si^{76}Se$ is at 23,292.18 MHz. Neglecting centrifugal distortion calculate the value of B_0 and r_0.

12.2. The vibrational spectrum of $H^{35}Cl$ has transitions at 2885.9, 5668.1, 8347.0 and 10,923.1 cm^{-1}. Calculate the best values for ω_e and $\chi_e\omega_e$.

12.3. A diatomic gas at high temperature shows a series of vibrational absorptions. By accident, some of the data are lost, but it is known that the absorption frequencies included 5600, 11,200, and 14,000 cm^{-1}. Determine the assignment of the spectrum and calculate ω_e and $\chi_e\omega_e$.

12.4. At a temperature above 900 K cesium chloride has sufficient vapor pressure to give a pure rotational spectra. The low vibrational frequency allows enough molecules to occupy excited vibrational levels at this temperature to permit rotational lines to be observed in the excited states. Below are given the $J = 6 \leftarrow 5$ rotational transition frequencies for two isotopic species in several vibrational states:

n	$^{133}Cs^{35}Cl$	n	$^{133}Cs^{37}Cl$
0	25,873.11 MHz	0	24,767.86 MHz
1	25,752.16	1	24,654.26
2	25,631.58	2	24,541.40
3	25,511.25		
4	25,390.36		
5	25,270.00		
6	25,150.1		
7	25,031.0		
8	24,911.2		

(*a*) Find B_0, B_e, and r_e for these two molecules. (Assume D_e to be negligible.)
(*b*) Calculate r_0, r_4, and r_8 for $^{133}Cs^{35}Cl$ and compare these values with r_e.

12.5. Consider a homonuclear diatomic molecule which has a force constant equal to that of Cl_2 ($\omega_e = 565\ \text{cm}^{-1}$). What will be the reduced mass of the molecule if its fundamental absorption frequency is equal to its average kinetic energy ($\times\, h^{-1}$) at 300 K?

12.6. For BrCl

Species	$J' \leftarrow J$	v	ν
$^{79}Br^{35}Cl$	$1 \leftarrow 0$	0	9080.73 MHz
$^{79}Br^{37}Cl$	$1 \leftarrow 0$	1	9034.14
$^{81}Br^{35}Cl$	$1 \leftarrow 0$	0	9018.40
$^{81}Br^{37}Cl$	$1 \leftarrow 0$	1	8972.41

(*a*) Calculate B_e and α_e for each species (in megahertz).
(*b*) If $\omega_e = 440\ \text{cm}^{-1}$ estimate D_e (in megahertz).
(*c*) Calculate r_e, r_0, and r_1 for $^{79}Br^{35}Cl$.
(*d*) For $^{79}Br^{35}Cl$ calculate $\chi_e\omega_e$ (in cm^{-1}).
(*e*) For $^{79}Br^{35}Cl$ calculate D and a and plot the Morse potential functions for $0 < r < 5A$.

12.7. The fundamental band of HCl, measured under high resolution [C.F. Meyers and A. A. Levin, *Physical Review*, **34**, 44 (1929)] shows each rotational line as a doublet cm^{-1}; 2599.00; 2625.74; 2651.97; 2677.73; 2703.06; 2727.75; 2752.03; 2775.79; 2798.78; 2821.49; 2843.56; 2865.09; 2906.25; 2925.78; 2944.81; 2963.24; 2980.90; 2997.78; 3014.29; 3029.96; 3044.88; 3059.07; 3072.76; 3085.62. For the "weak" lines the wave numbers are: 2597.43; 2624.03; 2650.36; 2675.90; 2701.29; 2726.01; 2750.31; 2773.77; 2796.88; 2819.51; 2841.59; 2862.99; 2904.16; 2923.69; 2942.71; 2961.08; 2978.68; 2995.66; 3012.16; 3027.69; 3042.62; 3056.84; 3070.51; 3083.28. Calculate the internuclear distance for the two isotopic species.

The harmonic band, as measured by the same authors, gives the following wave numbers: Strong components: 5468.55; 5496.97; 5525.04; 5551.68; 5577.25; 5602.05; 5624.81; 5647.03; 5687.81; 5706.21; 5723.29; 5739.29; 5753.88; 5767.50; 5779.54; 5790.54; 5799.94. Weak components: 5464.67; 5493.12; 5521.23; 5547.74; 5573.40; 5597.98; 5620.92; 5643.10; 5683.91; 5702.01; 5719.42; 5735.26; 5749.69; 5763.28; 5775.40; 5786.28; 5796.04. With the help of these values and of the values obtained for the fundamental frequency, calculate the ratio of the ω_es for the two isotopes, and compare it to the calculated value of this ratio.

12.8. The following spectral data are available for $^{133}Cs^{79}Br$.

Rotational transition	Vibrational state	Observed rotational frequency
$10 \leftarrow 9$	0	21588 MHz
$10 \leftarrow 9$	1	21514
$10 \leftarrow 9$	2	21441
$10 \leftarrow 9$	3	21366
$11 \leftarrow 10$	0	23747
$11 \leftarrow 10$	1	23666
$11 \leftarrow 10$	2	23583
$12 \leftarrow 11$	1	25816
$12 \leftarrow 11$	2	25649
$12 \leftarrow 11$	3	25550

(a) For CsBr calculate; B_1, B_0, B_e, D_e, D_e, α_e, ω_e, χ_e, D, and a.
(b) Find r_0 and r_e.

12.9. The vibrational frequencies of H_2, HD, and H_2 are respectively 4395.2, 3817.09, and 3118.4 cm^{-1}. Explain this difference quantitatively. Predict the values of $H^1—H^3$ and H^3_2.

12.10. The rotational constants of O_2 and $B_e = 1.44566$ cm^{-1} and $\alpha_e = 0.01579$. Calculate r_e and the mean value of r in each of the first four excited vibrational levels.

12.11. The vibrational constants of O_2 and $\omega_e = 1580.361$ cm^{-1} and $\chi_e\omega_e = 12.073$ cm^{-1}. Calculate the force constant for O_2.

12.12. The moment of inertia of $^{16}O^2H$ in the $^2\Sigma^+$ electronic state has been determined from a study of the electronic spectrum to be 9.194 ± 0.001 cm^{-1}. If it were possible to observe the pure rotational spectrum of this species and measure the $J = 1 \leftarrow 0$ transition to an accuracy of ± 0.01 MHz, what would be the relative precision of the r_0 values obtained by the two experiments, considering the species as a rigid rotor?

12.13. NaBr in the gas phase at 800 K has $\omega_e = 315$ cm^{-1} and $\chi_e\omega_e = 1.15$ cm^{-1}. Calculate the frequencies and relative intensities of the $n = 1 \leftarrow 0$, $2 \leftarrow 0$, $3 \leftarrow 0$, $2 \leftarrow 1$, and $3 \leftarrow 2$ transition at 800 K.

12.14. The observed vibrational frequencies of H_2 are at 4162, 8085, and 11779 cm^{-1}. Calculate the best value of ω_0, ω_e and $\chi_e\omega_e$ based on this data.

12.15. The observed rotational vibrational frequencies for $^{12}C^{16}O$ are at 2238.89; 2236.06; 2233.34; 2230.49; 2227.55; 2224.63; 2221.56; 2218.67; 2215.66; 2212.46; 2209.31; 2206.19; 2202.96; 2199.77; 2196.53; 2193.19; 2189.84; 2186.47; 2183.14; 2179.57; 2176.12; 2172.63; 2169.05; 2165.44; 2161.83; 2158.13; 2154.44; 2150.83; 2147.05; 2139.32; 2135.48; 2131.49; 2127.61; 2123.62; 2119.64; 2115.56; 2111.48; 2107.33; 2103.12; 2099.01; 2094.69; 2090.56; 2086.27; 2081.95; 2077.57; 2073.19; 2068.69; 2064.34; 2059.79; 2055.31; 2050.72; 2046.14 and for $^{13}C^{16}O$ are at 2159.43; 2156.50; 2153.43; 2144.03; 2140.83; 2137.49; 2134.30; 2124.20; 2120.92; 2117.46; 2113.98; 2110.41; 2092.35; 2088.56; 2084.85; 2081.12; 2069.52; 2065.73; 2061.81; 2057.81; 2053.79; 2049.75; 2045.61. (Note there are some missing transitions in the latter series.) For each molecule calculate $\nu(1 \leftarrow 0)$, ω_e, $\chi_e\omega_e$, B_0, B_1, B_e, α_e, I_e, and r_e. If the spectral data has an accuracy of ± 0.01 cm^{-1} estimate the accuracy of the calculated parameters.

CHAPTER 13

DIATOMIC VIBROTORS—ANHARMONICITY AND RAMAN EFFECT

13.1 ANHARMONICITY OF DIATOMIC MOLECULES

It was previously shown that by use of the Morse potential function a component frequency of a well-resolved rotational-vibrational band of a diatomic vibrotor can be expressed as a function of the five parameters ω_e, χ_e, B_e, α_e, and D_e. In principle it should be possible to determine these parameters from five experimental frequencies. Furthermore, it was shown that frequently the magnitude of D_e is sufficiently small that it is difficult to obtain accurate values of it from infrared spectra unless the observed frequencies are measured to within less than $\pm 0.1\ \text{cm}^{-1}$. If the terms in D_e are ignored the number of determinable parameters is reduced to four. Analysis of the analytical expressions for the relationships which give transitions provide values of B_0 and B_1, and therefore B_e and α_e, from measured differences between certain members of the band. Ignoring centrifugal distortion, the relationships for the transitions in the fundamental bands are

$$\nu_R = \omega_e - 2\chi_e\omega_e + 2B_1 + (3B_1 - B_0)J + (B_1 - B_0)J^2 \qquad (13.1)$$

$$\nu_P = \omega_e - 2\chi_e\omega_e - (B_1 + B_0)J + (B_1 - B_0)J^2 \qquad (13.2)$$

Since ω_e and $\chi_e\omega_e$ have the same coefficients in both equations it is impossible to determine more than their difference $\omega_e - 2\chi_e\omega_e$, regardless of which combination of frequencies one might use. Therefore ω_e cannot be determined directly from the observed spectra. By using the relationship

$$\alpha_e = 6\left(\frac{\chi_e B_e^3}{\omega_e}\right)^{1/2} - \frac{6B_e^2}{\omega_e} \tag{13.3}$$

χ_e can be determined indirectly from a knowledge of B_e, α_e, and ω_e. However, such a determination will generally result in a much greater error than if it were possible to obtain χ_e as a simple difference in a pair of frequencies. For molecules more complex than the diatomic case, it is considerably more difficult to obtain spectra of sufficient resolution to determine the rotational and vibrational parameters necessary to completely describe the system. This is due to the multiplicity of neighboring bands resulting from the several modes of vibration and the increase in the number of rotation-vibration interaction constants which must be used to describe the system. However, the use of high-resolution Fourier transform spectrometers are rapidly changing this. If overtone bands, (that is, $v = 0 \rightarrow v' = 2, 3, 4, \ldots \leftarrow v'' = 0$) can be observed then both ω_e and χ_e can be determined independently.

Examination of a typical potential curve such as Fig. 6.1, shows that its general shape is not parabolic and even in the region near $r = r_e$ the departure from parabolic is noticeable. The simplest representation of the departure from a parabolic curve can be made by the addition of a single cubic term to the harmonic potential. For the discussion of real systems it is necessary to use both x^3 and x^4 terms in the potential function to obtain a better fit of experimental data. However, to keep the details of the discussion simple, the following analysis will be initially concerned only with the x^3 term, and then the contribution of the x^4 term will be considered. The Hamiltonian for an oscillator perturbed in this manner will be

$$\mathcal{H} = \frac{\mathbf{P}_x^2}{2\mu} + \tfrac{1}{2}k\mathbf{x}^2 + k'\mathbf{x}^3 \tag{13.4}$$

In terms of the displacement from equilibrium, that is, $x = r - r_e$, the addition of a cubic term will displace both sides of the curve in the same direction rather than symmetrically. Since the cubic term is small compared to the quadratic term it can be considered as a perturbation and the Hamiltonian written as $\mathcal{H} = \mathcal{H}^0 + \mathcal{H}^1$, where $\mathcal{H}^1 = k'\mathbf{x}^3$.

The energy levels of this system are developed using the matrix method and perturbation theory. The matrix elements for the Hamiltonian of the harmonic oscillator are

$$H_{vv}^0 = \frac{1}{2\mu} P_{vv}^2 + \tfrac{1}{2}kx_{vv}^2 = (v + \tfrac{1}{2})h\nu_0 \tag{13.5}$$

and the matrix elements for the coordinate x are

$$x_{vw} = a\left(\frac{v+1}{2}\right)^{1/2} e^{-i(\gamma_v-\gamma_w)} e^{i(E_v-E_w)t/\hbar} \qquad (w = v+1) \tag{13.6}$$

$$x_{vw} = a\left(\frac{v}{2}\right)^{1/2} e^{-i(\gamma_v-\gamma_w)} e^{i(E_v-E_w)t/\hbar} \qquad (w = v-1) \tag{13.7}$$

$$x_{vw} = 0 \qquad (w \neq v \pm 1) \tag{13.8}$$

where

$$a = \left(\frac{\hbar^2}{k\mu}\right)^{1/4} \tag{13.9}$$

The diagonal matrix elements for the Hamiltonian of the anharmonic oscillator are those of $\underline{\mathscr{H}} = \underline{\mathscr{H}}^0 + \underline{\mathscr{H}}^1 = \underline{\mathscr{H}}^0 + k'\underline{\mathbf{x}}^3$ which to first-order are

$$E_v = H^0_{vv} + k'x^3_{vv} \tag{13.10}$$

The diagonal elements of $\underline{\mathbf{x}}^3$ are evaluated from those of $\underline{\mathbf{x}}$ by matrix multiplication and are all zero, indicating that the perturbation due to a cubic potential term does not contribute a first-order term to the energy. The second-order correction will be

$$E^{(2)}_v = H^{(2)}_{vv} + \sum_i{}' \frac{H^{(1)}_{vi}H^{(1)}_{iv}}{E^0_{vv} - E^0_{ii}} \tag{13.11}$$

Since there is only a single perturbing term in the potential there are no $H^{(2)}_{vv}$ elements. A complete evaluation of the matrix elements of $\underline{\mathbf{x}}^3$ shows that $x^3_{vw} = 0$ for $w \neq v \pm 1$, $v \pm 3$ and the nonzero terms are

$$x^3_{v,\,v+1} = \left\{a^3\left[\frac{(v+1)(2v+3)^2}{8}\right]^{1/2} + a^3\left[\frac{v^2(v+1)}{8}\right]^{1/2}\right\}\sigma_{v,\,v+1}\varepsilon_{v,\,v+1} \tag{13.12}$$

$$x^3_{v,\,v-1} = \left\{a^3\left[\frac{v(v+1)^2}{8}\right]^{1/2} + a^3\left[\frac{v(2v-1)}{8}\right]^{1/2}\right\}\sigma_{v,\,v-1}\varepsilon_{v,\,v-1} \tag{13.13}$$

$$x^3_{v,\,v+3} = \left\{a^3\left[\frac{(v+1)(v+2)(v+3)}{8}\right]^{1/2}\right\}\sigma_{v,v+3}\varepsilon_{v,\,v+3} \tag{13.14}$$

$$x^3_{v,\,v-3} = \left\{a^3\left[\frac{v(v-1)(v-2)}{8}\right]^{1/2}\right\}\sigma_{v,\,v-3}\varepsilon_{v,\,v-3} \tag{13.15}$$

where σ_{vw} and ε_{vw} are the phase on time factors discussed in Chap. 3.

The elements $H^{(1)}_{vw}$ are Hermitian, $H^{(1)}_{iv} = H^{(1)*}_{vi}$, so the second-order perturbation term becomes

$$E^{(2)}_v = \sum_i \frac{[H^{(1)}_{vi}]^2}{E^0_{vv} - E^0_{ii}} = (k')^2 \sum_i \frac{[x^3_{vi}]^2}{E^0_{vv} - E^0_{ii}} \tag{13.16}$$

Since the products $\sigma_{vi}\sigma^*_{iv}$ and $\varepsilon_{vi}\varepsilon^*_{iv}$ are both unity, the introduction of Eqs. (13.12) to (13.15) gives

$$E_v^{(2)} = -\frac{30a^6(k')^2}{8h\nu_0}[v^2 + v + \tfrac{11}{30}] \qquad (13.17)$$

If the perturbation includes an $\underline{\mathbf{x}}^4$ term it is found using the matrix relationship

$$x_{vv}^4 = \sum_i x_{vi}^2 x_{iv}^2 \qquad (13.18)$$

and the matrix elements x_{nm}^2 given in Chap. 3, that

$$x_{vv}^4 = \tfrac{3}{4}a^4[2v^2 + 2v + 1] \qquad (13.19)$$

Thus there is a first-order contribution to the energy from an x^4 potential term. Inclusion of this term in the Hamiltonian

$$\underline{\mathcal{H}} = \underline{\mathcal{H}}^0 + k'\underline{\mathbf{x}}^3 + k''\underline{\mathbf{x}}^4 \qquad (13.20)$$

then gives for the energy, when terms second-order in x^4 are neglected,

$$E_v = E_v^0 + E_v^{(1)} + E_v^{(2)} \qquad (13.21)$$

$$E_v = h\nu_0\left(v + \frac{1}{2}\right) + \frac{3a^4k''}{2}(v^2 + v + \tfrac{1}{2}) - \frac{30a^6(k')^2}{8h\nu_0}\left(v^2 + v + \frac{11}{30}\right) \qquad (13.22)$$

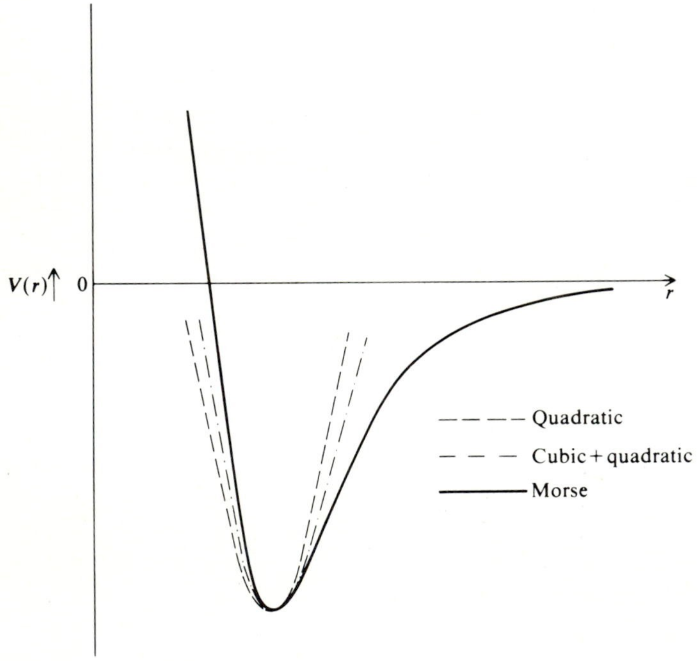

FIGURE 13.1
Diatomic oscillator potentials.

Multiplication and division of the perturbation terms by $h\nu_0$ gives, since $11/30 \approx \frac{1}{2}$,

$$\frac{E_v}{h} = \omega_e(v + \tfrac{1}{2}) - \chi'_e\omega_e(v^2 + v + \tfrac{1}{2}) \tag{13.23}$$

where

$$\chi'_e = \frac{3a^4k''}{2h\nu_0} - \frac{30a^6(k')^2}{8h^2\nu_0^2} \tag{13.24}$$

This expression is similar to that obtained by the use of the Morse potential function and the Pekeris method. It is not expected to be identical in view of the difference between the potential functions used. This point is illustrated in Fig. 13.1, where the quadratic, cubic, and Morse potential curves are compared. It is observed that the addition of the cubic term to the quadratic curve gives a curve approaching, but not coincidental with, the Morse curve.

Since the two approaches give such similar results the reader may wonder why the latter development was included. This was done to provide another illustration of the use of the matrix method and to point out the fact that there is generally no unique approach to the solution of a quantum mechanical problem.

13.2 ANHARMONIC OSCILLATOR SELECTION RULES AND WAVEFUNCTIONS

In order to examine the practical aspects of anharmonic oscillator behavior it is necessary to develop the selection rules for transitions between the energy states. Perturbation theory allows one to express the wavefunctions for the states of an anharmonic oscillator as a linear combination of those of the harmonic oscillator. In other words, the effect of the perturbation is to produce a mixing of the unperturbed states to form a new set of states for the perturbed system. Since the total contribution of each unperturbed wavefunction is identical, the mixing is of a conservative nature. For any real molecule this mixing will involve contributions from several states, but to simplify the discussion we will consider a hypothetical two-state example where a small fraction c^2 of the ψ_1^0 state of an oscillator becomes mixed with the ground state and vice versa. The two states may then be represented by

$$\psi_0 = (1 - c^2)^{1/2}\psi_0^0 + c\psi_1^0 \tag{13.25}$$

$$\psi_1 = c\psi_0^0 + (1 - c^2)^{1/2}\psi_1^0 \tag{13.26}$$

where the superscript ^{0} denotes the unperturbed wavefunction.

Using this simple pair of perturbed functions, and for simplicity considering the next state to be $\psi_2 = \psi_2^0$, the effect of the mixing on the selection rules can be demonstrated. For the harmonic oscillator, $\Delta v = \pm 1$ was found to be the valid selection rule. The dipole matrix element connecting the ground state with the next highest nonadjacent state of this perturbed oscillator will be

$$\langle 0|\mu|2\rangle = \int_{-\infty}^{\infty} [(1-c^2)^{1/2}\psi_0^{0*} + c\psi_1^{0*}]\left\{\mu_0 + \left(\frac{d\mu}{d\eta}\right)_0 \frac{\eta}{\alpha^{1/2}}\right\}\psi_2^0\, d\eta \tag{13.27}$$

Considering that

$$\int_{-\infty}^{\infty} \psi_v^* \mu_0 \psi_{v'}\, d\eta = 0 \qquad v \neq v' \tag{13.28}$$

the nonzero part of the dipole element becomes

$$\frac{1}{\alpha^{1/2}}\left(\frac{d\mu}{d\eta}\right)_0 \left\{(1-c^2)^{1/2}\int_{-\infty}^{\infty} \psi_0^{0*}\eta\psi_2^0\, d\eta + c\int_{-\infty}^{\infty} \psi_1^{0*}\eta\psi_2^0\, d\eta\right\} \tag{13.29}$$

Symmetry considerations employing the odd-even character of the integrands show the first integral to be zero and the second to be nonzero. Since c is nonzero, the element $\langle 0|\mu|2\rangle \neq 0$ or the selection rule has been relaxed to allow $\Delta v = \pm 2$. If the higher states are considered to be mixed also, then it is easily established that still-higher-order transitions are allowed. The $\Delta v = 1$ transition is commonly referred to as the *fundamental* and the higher-order ones as *harmonics* or *overtones.*

Although an elementary example was used to illustrate the relaxation of the normal selection rules it is possible to determine, in general, the coefficients of the terms making up the wavefunction of a particular state of the anharmonic oscillator. The unnormalized wavefunctions of the kth state of a perturbed system can be written

$$\psi_v = \psi_v^0 - \sum_w{}' \frac{H_{vw}^{(1)}}{E_v^0 - E_w^0}\psi^0 \tag{13.30}$$

where

$$H_{vw}^{(1)} = \int_{-\infty}^{\infty} \psi_v^{0*}\mathbf{H}^{(1)}\psi_\omega^0\, d\eta = \langle v|\mathbf{H}^{(1)}|\omega\rangle = k'\langle v|x^3|\omega\rangle \tag{13.31}$$

Again ignoring the second-order contribution from the $\underline{\mathbf{x}}^4$ term and using the matrix elements of $\underline{\mathbf{x}}^3$, the perturbed wavefunctions can be formulated. For example for $v = 4$

$$\psi_4 = \psi_4^0 - \frac{H_{34}^{(1)}}{E_3^0 - E_4^0}\psi_3^0 - \frac{H_{54}^{(1)}}{E_5^0 - E_4^0}\psi_5^0 - \frac{H_{14}^{(1)}}{E_1^0 - E_4^0}\psi_1^0 - \frac{H_{74}^{(1)}}{E_1^0 - E_7^0}\psi_7^0 \tag{13.32}$$

Likewise for $v = 0$

$$\psi_0 = \psi_0^0 - \frac{H_{10}^{(1)}}{E_1^0 - E_0^0}\psi_1^0 - \frac{H_{30}^{(1)}}{E_3^0 - E_0^0}\psi_3^0 \tag{13.33}$$

Examination of the dipole matrix element $\langle 0|\mu|4\rangle$ shows that it will contain a nonzero term having as a factor $\langle 3|\mu|4\rangle$. Hence it will be nonzero and the transition $\nu_{4\leftarrow 0}$ will be allowed.

The effect of anharmonicity is to mix the unperturbed wavefunctions in such a manner as to yield a new set of functions which are linear combinations

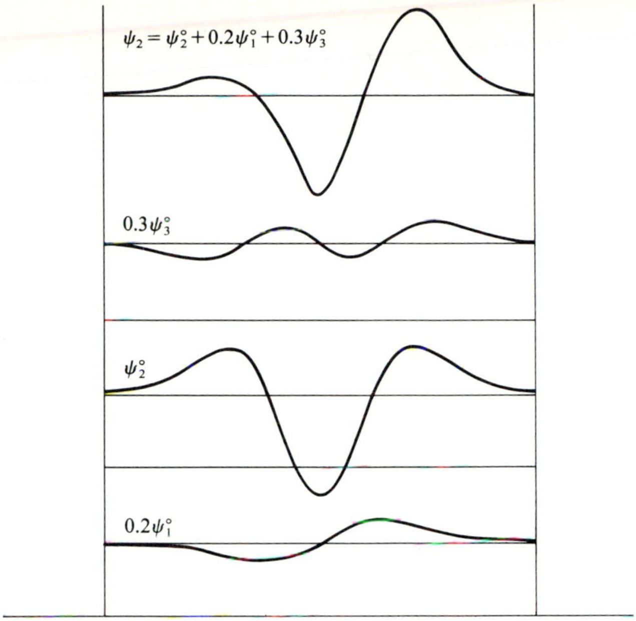

FIGURE 13.2
An example of a real oscillator wavefunction and its synthesis from ground state wavefunctions.

of the unperturbed ones. An example of such a perturbed function is illustrated in Fig. 13.2, in which the effect has been somewhat exaggerated to better illustrate the point.

13.3 SPECTRA OF ANHARMONIC OSCILLATORS

The vibrational energy levels of the anharmonic oscillator can be simply expressed by

$$\frac{E_v}{h} = \omega_e(v + \tfrac{1}{2}) - \chi_e\omega_e(v + \tfrac{1}{2})^2 \tag{13.34}$$

or more accurately by the relationship derived by use of higher-order perturbation theory

$$\frac{E_v}{h} = \omega_e(v + \tfrac{1}{2}) - \chi_e\omega_e(v + \tfrac{1}{2})^2 + y_e\omega_e(v + \tfrac{1}{2})^3 + z_e\omega_e(v + \tfrac{1}{2})^4 + \cdots \tag{13.35}$$

The energy levels are not evenly spaced as was the case for the harmonic oscillator, but are as illustrated in Fig. 13.3.

By using the infrared spectrum of HCl as an example we can develop an appropriate method for treatment of data to obtain the anharmonicity constants. The center frequencies of the infrared spectrum of $^1H^{35}Cl$ are given in Table 13.1.

$$E_5/h = \tfrac{11}{2}\omega_e - \tfrac{121}{4}\chi_e\omega_e + \tfrac{1331}{8}y_e\omega_e$$

$n = 5$

$$\nu_{0\to5} = 5\omega_e - 30\chi_e\omega_e + \tfrac{665}{4}y_e\omega_e$$

$$\nu_{4\to5} = \omega_e - 10\chi_e\omega_e + \tfrac{301}{4}y_e\omega_e$$

$$E_{4/h} = \tfrac{9}{2}\omega_e - \tfrac{81}{4}\chi_e\omega_e + \tfrac{729}{8}y_e\omega_e$$

$$\nu_{0\to4} = 4\omega_e - 20\chi_e\omega_e + 91\,y_e\omega_e$$

$$\nu_{3\to4} = \omega_e - 8\chi_e\omega_e + \tfrac{193}{4}y_e\omega_e$$

$$E_3/h = \tfrac{7}{2}\omega_e - \tfrac{49}{4}\chi_e\omega_e + \tfrac{343}{8}y_e\omega_e$$

$$\nu_{0\to3} = 3\omega_e - 12\chi_e\omega_e + \tfrac{171}{4}y_e\omega_e$$

$$\nu_{2\to3} = \omega_e - 6\chi_e\omega_e + \tfrac{109}{4}y_e\omega_e$$

$$E_2/h = \tfrac{5}{2}\omega_e - \tfrac{25}{4}\chi_e\omega_e + \tfrac{125}{8}y_e\omega_e$$

$$\nu_{0\to2} = 2\omega_e - 6\chi_e\omega_e + \tfrac{31}{2}y_e\omega_e$$

$$\nu_{1\to2} = \omega_e - 4\chi_e\omega_e + \tfrac{49}{4}y_e\omega_e$$

$$E_1/h = \tfrac{3}{2}\omega_e - \tfrac{9}{4}\chi_e\omega_e + \tfrac{27}{8}y_e\omega_e$$

$$\nu_{0\to1} = \omega_e - 2\chi_e\omega_e + \tfrac{13}{4}y_e\omega_e$$

$$E_0/h = \tfrac{1}{2}\omega_e - \tfrac{1}{4}\chi_e\omega_e + \tfrac{1}{8}y_e\omega_e$$

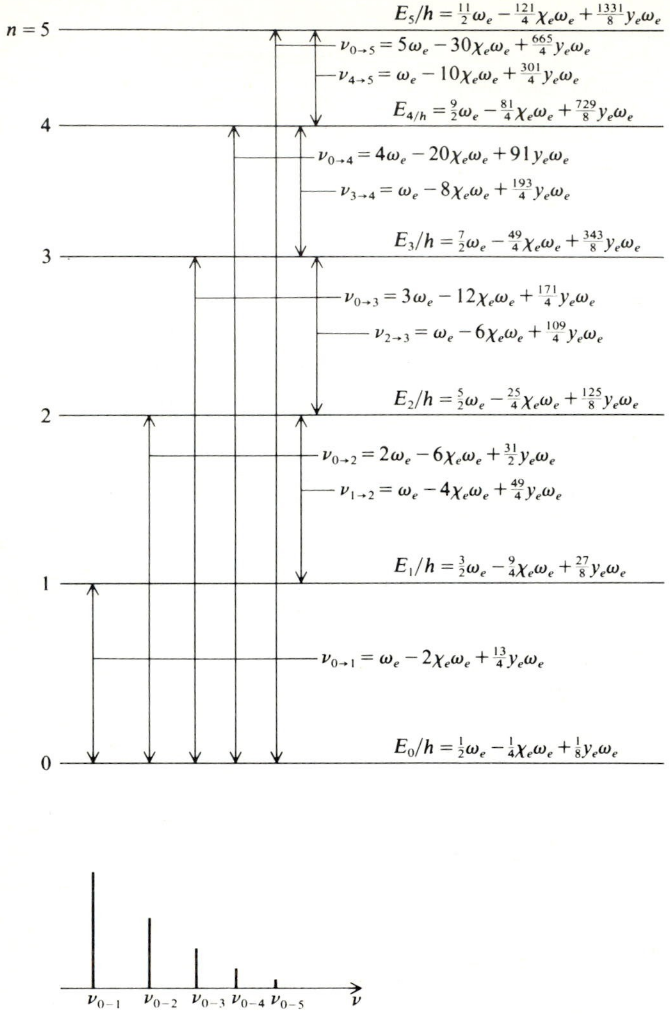

FIGURE 13.3
Energy levels and spectrum
of an anharmonic oscillator.

The algebraic approach would be to solve a set of simultaneous equations of the form

$$\nu_i = a_i\omega_e - b_i\chi_e\omega_e + c_i y_e\omega_e \tag{13.36}$$

to give the desired constants. However, frequency differences can often be used to advantage to avoid this lengthy mathematical procedure. Looking first at the energy of the ground state $v = 0$, one has

$$\frac{E_0}{h} = \frac{1}{2}\omega_e - \frac{1}{4}\chi_e\omega_e + \frac{1}{8}y_e\omega_e \tag{13.37}$$

TABLE 13.1
Infrared spectrum of HCl

Low v	Designation	Frequency, cm^{-1}	Δ	Δ^2
			2885.9	
0	Fundamental	2885.9		−103.7
			2782.2	
1	First overtone	5668.1		−103.3
			2678.9	
2	Second overtone	8347.0		−102.8
			2576.1	
3	Third overtone	10923.1		−102.6
			2473.5	
4	Fourth overtone	13396.6		

Using this as an arbitrary zero of energy reference the energies of the higher levels can be expressed as

$$\frac{E_v^0}{h} = \frac{E_v}{h} - \frac{E_0}{h} = v\omega_e - (v^2 + v)\chi_e\omega_e + \left(v^3 + \frac{3}{2}v^2 + \frac{3}{4}v\right)y_e\omega_e \tag{13.38}$$

or

$$\frac{E_v^0}{h} = \nu_{v\leftarrow 0} = v\omega_0 - v^2\chi_0\omega_0 + v^3 y_0\omega_0 \tag{13.39}$$

where

$$\omega_0 = \omega_e - \chi_e\omega_e + \tfrac{3}{4}y_e\omega_e \tag{13.40}$$

$$\chi_0\omega_0 = \chi_e\omega_e + \tfrac{3}{2}y_e\omega_e \tag{13.41}$$

$$y_0\omega_0 = y_e\omega_e \tag{13.42}$$

Since frequency and energy differences are independent of the choice of zero we find that

$$E_{v+1} - E_v = E_{v+1}^0 - E_v^0 \tag{13.43}$$

In general $y_e\omega_e \ll \chi_e\omega_e$ and in practice is often neglected. Hence if the cubic term in Eq. (13.38) is dropped, we find that the differences between successive frequency pairs Δ_{ij}, and the differences between successive energy levels, are the same. Therefore

$$\Delta_{v,\,v+1} = \nu_{v+1\leftarrow 0} - \nu_{v\leftarrow 0} = \frac{E_{v+1}^0}{h} - \frac{E_v^0}{h} \tag{13.44}$$

or

$$\Delta_{v,\,v+1} = \omega_0 - (2v+1)\chi_0\omega_0 = \omega_e - (2v+2)\chi_e\omega_e \tag{13.45}$$

Looking at the difference $E_1/h - E_0/h$, it is seen to be equal to

$$\frac{E_1^0}{h} = \nu_{1\leftarrow 0} = \omega_0 - \chi_0\omega_0 = \omega_e - 2\chi_e\omega_e \tag{13.46}$$

Reference to Fig. 13.3 will serve to confirm these relationships and the following ones. The second difference Δ^2 is defined as

$$\Delta^2_{v,\, v+2} = \Delta_{v+1,\, v+2} - \Delta_{v,\, v+1} \tag{13.47}$$

which, neglecting the cubic terms, becomes

$$\Delta^2_{v,\, v+2} = 2\chi_0\omega_0 = 2\chi_e\omega_e \tag{13.48}$$

Referring to the data in Table 13.1 we can now use the average value of Δ^2, 103.1 cm^{-1}, to obtain $\chi_e\omega_e = 51.55$ cm^{-1}. Using this value for $\chi_e\omega_e$ and Eq. (13.40), we get $\omega_e = 2989.1$ cm^{-1} and $\omega_0 = 2937.5$ cm^{-1}.

Since the a_i, b_i, and c_i are all functions of the vibrational quantum number v and, in general, four or five overtone transitions are observable, another method for analysis consists of using Eq. (13.36) as the basis of a least-squares fit of ν vs. v to obtain a best fit of the data.

The introduction of anharmonicity has resulted in the frequencies of the $\nu_{1\leftarrow 0}$ and $\nu_{2\leftarrow 1}$ transitions being displaced by $2\chi_e\omega_e$, which for diatomic molecules is of the order of 1–100 cm^{-1}. Since this is a substantial displacement compared to the resolution possible on modern instruments, one might expect both transitions to be observable. Referring to the expression for the absorption coefficient, Eq. (11.90), and using a vibrational frequency $\nu = 1000$ cm^{-1}, it is found that for $\exp - (h\nu/kT) \ll 1$ the absorption coefficient can be written

$$\alpha(\nu_{vv'}) = KN_v\nu_{vv'}\langle v|\mu|v'\rangle^2 S(\nu_{vv'}, \nu) \tag{13.49}$$

Previously it was shown that the ratio of the dipole moment matrix elements for this pair of transitions was 1 : 2, with $\langle 0|\mu|1\rangle^2$ being the smaller, hence this would indicate that the $\nu_{2\leftarrow 1}$ transition might be more intense since the frequencies are nearly the same. However, the population of the lower state has not been considered. Using the Boltzmann relationship and $\nu = 1000$ cm^{-1} the ratio $N_1/N_0 = 0.0089$, so the absorption coefficient for the $\nu_{2\leftarrow 1}$ transition would be about 2 percent of that of the $\nu_{1\leftarrow 0}$ transition. For sufficiently small values of ω_e or, at higher temperatures, the populations of the $v = 1$ level can be appreciable, and the $\nu_{2\leftarrow 1}$ transition or the "hot band," as it is referred to, can be observed. The relative intensities of the overtones and the fundamental are not affected by the population differences because they both originate from the ground state. The difference in the intensities for these transitions is primarily due to the differences in the magnitudes of the dipole moment matrix elements, although the increased frequencies tend to enhance the higher overtone intensities.

The variation of the ratio of the population of an excited state to that of the ground state can be easily depicted in a graphical manner as is done in Fig. 13.4 for several common temperatures. From the behavior shown on this chart we can see that at room temperature the population of the excited state will

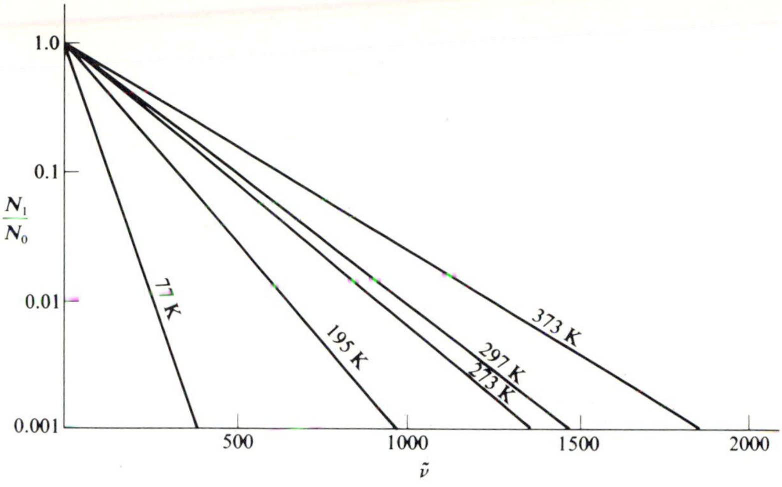

FIGURE 13.4
Relative population of an excited and the ground state vs. vibrational frequency for several common temperatures.

exceed 10 percent of that of the ground state only for transitions with frequencies less than about 500 cm^{-1}, so it would only be in this region that we would expect to observe a $\nu = 2 \leftarrow 1$ transition. Note that the effect of warming the sample is to increase the population of the upper state, hence at high temperatures one would expect to observe transitions originating at the $v = 1$ level, that is, $v = 2 \leftarrow 1$.

When analyzing overtone bands exhibiting sufficient resolution to resolve the rotational fine structure, the procedure is analogous to that previously discussed for the HI molecule. Once the position of the missing overtone Q-branches have been established they can be used along with that for the fundamental to evaluate the anharmonicity constants as was illustrated for HCl.

The effects of centrifugal distortion on the anharmonic oscillator are analogous to those for the harmonic oscillator. Except for the case of an extremely light molecule with a large value of D_e it requires exceptional instrument resolution to determine very accurate values of D_e from the rotational fine structure of vibrational transitions.

13.4 RAMAN EFFECT: CLASSICAL THEORY

Although the Raman effect has been known over a half-century, its employment as a spectroscopic method was quite limited until the development of the laser which provided a source of highly monochromatic radiation which resulted in greater resolution and opened the field to the study of colored compounds. The quantum mechanical description of the Raman effect will be considered in Sec. 13.6, but we will first describe the phenomenon from a classical viewpoint.

Unlike pure rotational or rotational-vibrational spectroscopy the Raman effect does not involve direct absorption of radiation. Instead it involves the scattering of incident radiation which has been modified by some internal change in the system. Internal changes can serve to either increase or decrease the energy of the scattered radiation. In this case one observes, in addition to the scattering of radiation at the incident frequency, ν_I (Rayleigh scattering), the scattering of radiation at frequencies both above and below the incident value. If the internal changes in the system are quantized then the observed scattered radiation will consist of a set of discrete spectral lines. This phenomenon, along with the conventional naming of the scattered radiation, is illustrated in Fig. 13.5. In practice the intensities of the anti-Stokes lines ($\nu > \nu_I$) are generally less than those of the Stokes lines ($\nu < \nu_I$).

The phenomenon of scattering is easily understood by considering the electrical nature of matter. Atoms and molecules consist of collections of oppositely charged particles whose relative positions can be altered by the application of external electric fields. This alteration leads to an electric dipole moment being induced into the system. The ease with which a molecule or atom may be distorted by an electric field is measured by the electric polarizability. For atoms where the symmetry is spherical, the polarizability will be the same in all directions (isotropic) and can be expressed by a single scalar quantity. For molecules with lower-than-spherical symmetry, the polarizability will not be the same along all directions (anisotropic), and is described by a tensor, analogous to the moment of inertia tensor. Using the symbol α_{ij} to denote the components of the polarizability tensor $\underline{\boldsymbol{\alpha}}$, the induced electric dipole moment $\boldsymbol{\mu}_I$, due to an external field $\mathbf{E}$, is defined by

$$\boldsymbol{\mu}_I = \underline{\boldsymbol{\alpha}} \cdot \mathbf{E} \tag{13.50}$$

or the components of $\boldsymbol{\mu}_I$ are

$$\begin{pmatrix} \mu_{Ix} \\ \mu_{Iy} \\ \mu_{Iz} \end{pmatrix} = \begin{pmatrix} \alpha_{xx} & \alpha_{xy} & \alpha_{xz} \\ \alpha_{yx} & \alpha_{yy} & \alpha_{yz} \\ \alpha_{zx} & \alpha_{zy} & \alpha_{zz} \end{pmatrix} \begin{pmatrix} E_x \\ E_y \\ E_z \end{pmatrix} \tag{13.51}$$

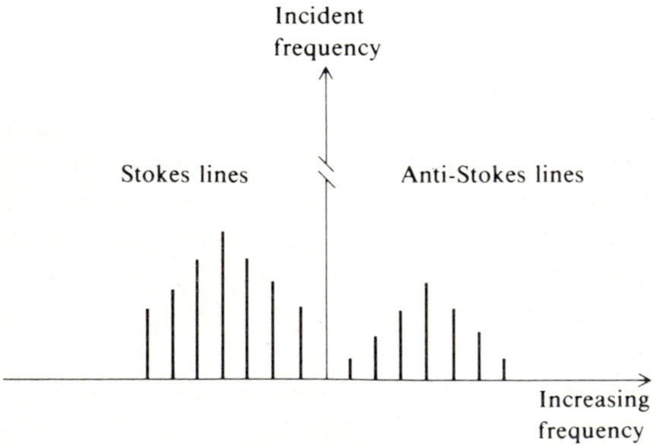

FIGURE 13.5
Schematic presentation of the Raman effect.

The polarizability tensor will be symmetric, and furthermore there will be a unique axis system where the polarizability tensor will be diagonal, that is, the components of the induced moments will be parallel to the external field if the field lies along any one of these axes. In this case, the polarizability tensor is

$$\underline{\boldsymbol{\alpha}} = \begin{pmatrix} \alpha_{x'x'} & 0 & 0 \\ 0 & \alpha_{y'y'} & 0 \\ 0 & 0 & \alpha_{z'z'} \end{pmatrix} \tag{13.52}$$

If one plots $(|\underline{\boldsymbol{\alpha}}|)^{-1/2}$ relative to the origin of a coordinate system the locus of such points defines an ellipsoid, whose axes are the principal axis system of the polarizability tensor and whose dimensions along these axes are inversely proportional to $(\alpha_{FF})^{-1/2}$. The diagrammatic use of this ellipsoid is useful for examination of the way $\underline{\boldsymbol{\alpha}}$ will change with rotation and vibration of the molecule. Figure 13.6 shows the cross sections of the polarizability ellipsoid for a diatomic molecule. Note that due to symmetry, $\alpha_{xx} = \alpha_{yy}$ when the z axis is taken as the unique direction.

The polarizability α, which is one of the factors in the Debye equation

$$\left(\frac{\varepsilon - 1}{\varepsilon + 2}\right)\frac{M}{P} = \frac{4\pi N_0}{3}\left(\alpha + \frac{\mu^2}{kT}\right) \tag{13.53}$$

is related to the polarizability tensor by

$$\langle \alpha \rangle = \tfrac{1}{3}(\alpha_{x'x'} + \alpha_{y'y'} + \alpha_{z'z'}) \tag{13.54}$$

We next examine the interaction of an external radiation field given by

$$E = E_0 \sin 2\pi\nu_I t \tag{13.55}$$

and an isotropic molecule. In this case, the induced electric dipole moment will be

$$\mu_I = \alpha E = \alpha E_0 \sin 2\pi\nu_I t \tag{13.56}$$

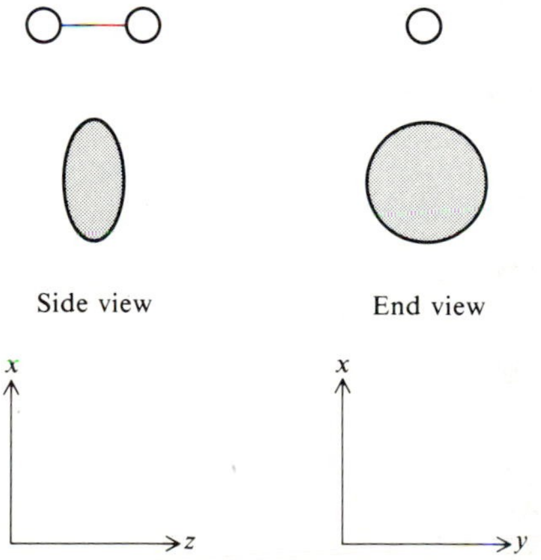

FIGURE 13.6
Polarizability ellipsoid cross sections for a diatomic molecule.

An oscillating dipole will, according to the laws of classical electrodynamics, emit radiation corresponding to the frequency of oscillation. The molecules will thus emit radiation proportional to $\sin 2\pi\nu_I t$. This is the unshifted scattering of the incident radiation.

If the system is undergoing some type of periodic internal motion, such as vibration or rotation, this may cause a periodic change in the polarizability, in which case the isotropic polarizability may be written as

$$\alpha' = \alpha + A \sin 2\pi\nu_i t \tag{13.57}$$

where ν_i is the frequency of the internal motion. The induced moment is then given by

$$\mu_I = \alpha' E = E_0(\alpha + A \sin 2\pi\nu_i t) \sin 2\pi\nu_I t \tag{13.58}$$

or

$$\mu_I = \alpha E_0 \sin 2\pi\nu_I t + \tfrac{1}{2}AE_0 \cos 2\pi(\nu_I + \nu_i)t + \tfrac{1}{2}AE_0 \cos 2\pi(\nu_I - \nu_i)t \tag{13.59}$$

and the emitted radiation will contain components at $\nu \pm \nu_i$ as well as at ν.

Examination of Eq. (13.59) shows that the prerequisite to the observation of the Raman effect is not only the occurrence of a periodic internal motion but also the association with this motion of a change in the polarizability of the molecule. This change can be either in the magnitude of the polarizability tensor components or in the direction of the principal axis system, since the latter will also lead to the presence of a changing dipole. The rotation of a molecule that does not have spherical symmetry will produce a change in the spatial orientation of the principal axis system. Hence, rotation of most molecules will provide a type of motion which gives rise to the Raman effect. The stretching of a bond as a result of the vibration of a molecule can also result in a change in a polarizability tensor component, hence this constitutes a second source for the Raman effect.

In order to correlate the vibration and rotation of molecules with the observed Raman spectrum it is necessary to look at the nature of polarizability changes and the appropriate quantum mechanical selection rules.

13.5 POLARIZABILITY

For a diatomic molecule we find that the polarizability tensor, in the principal axis frame, has two components since $\alpha_{x'x'} = \alpha_{y'y'} \neq \alpha_{z'z'}$. The polarizability ellipsoid is illustrated in Fig 13.6. Next consider the effect of rotation about the y inertial axis on the magnitude of the polarizability components relative to a laboratory frame of reference. The variation with rotation is illustrated in Fig. 13.7. Not only does the polarizability change with rotation, but its variation is at a frequency twice that of the rotation. This is true regardless of whether the molecule is homonuclear or heteronuclear since for either case the polarizability ellipsoid had $D_{\infty h}$ symmetry. For an asymmetric polyatomic molecule the only change from the diatomic case would be $\alpha_{x'x'} \neq \alpha_{y'y'}$, and for rotation about any one of the three principal inertial axes there would be a similar variation in polarizability.

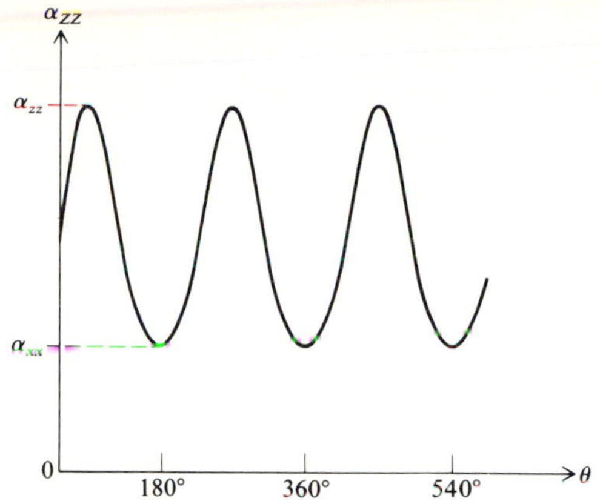

FIGURE 13.7
Variation of polarizability relative to a laboratory reference frame with rotation of a diatomic molecule.

The situation with regard to the effect of molecular vibration on the polarizability is not as simple as that for rotation but nevertheless is easily analyzed by a classical approach. When the bond in a diatomic molecule is extended by vibrational motion the positive nuclei move further apart and exert less attraction for the bonding electrons, thereby making them more easily displaced or increasing their polarizability both along and perpendicular to the bond direction. Compression of the bond produces just the opposite effect. Figure 13.8 shows qualitatively the effect on the polarizability tensor components and Fig. 13.9 shows the corresponding effect on the polarizability ellipsoid.

Since the change in bond length with vibration is small relative to the bond length only the effect of vibration in the vicinity of $r = r_e$ is needed to consider the polarizability change along a particular axis. The components of the polarizability tensor α_{gg} can then be given by a terminated Taylor series in $\xi = r - r_e$ about $r = r_e$

$$\alpha_{gg} = \alpha^0_{gg} + \left(\frac{\partial \alpha_{gg}}{\partial \xi}\right)_{\xi=0} \xi \tag{13.60}$$

for the harmonic oscillator

$$\xi = \xi_{\max} \sin 2\pi\nu_i t \tag{13.61}$$

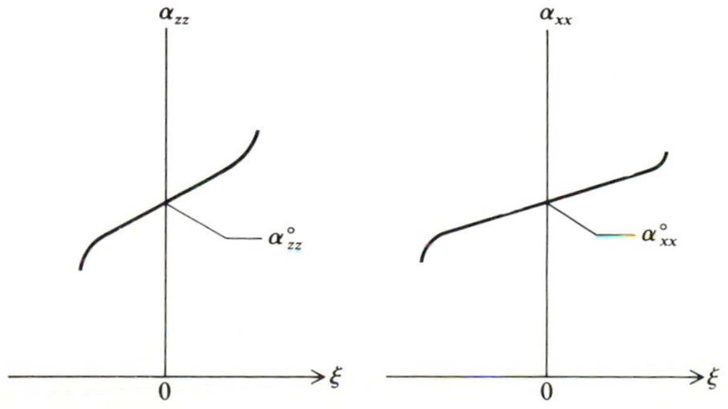

FIGURE 13.8
Changes in polarizability with vibration for a diatomic molecule.

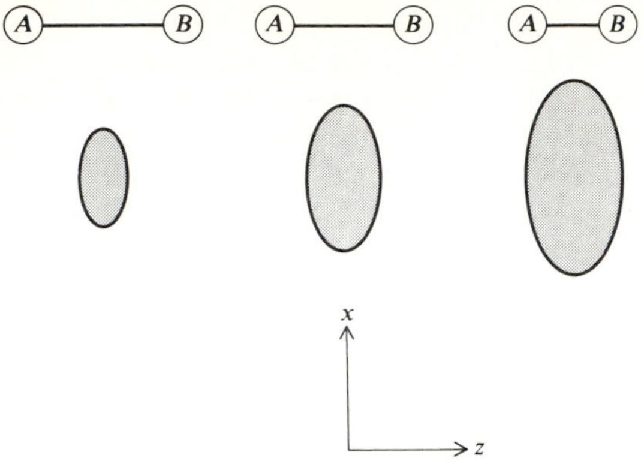

FIGURE 13.9
Changes in the polarizability ellipsoid with vibration for a diatomic molecule.

or

$$\alpha_{gg} = \alpha_{gg}^{0} + \left(\frac{\partial \alpha_{gg}}{\partial \xi}\right)_{\xi=0} \xi_{\max} \sin 2\pi\nu_i t \tag{13.62}$$

Comparing this to Eq. (13.57) one observes that

$$A = \xi_{\max}\left(\frac{\partial \alpha_{gg}}{\partial \xi}\right)_{\xi=0} \tag{13.63}$$

and the necessary condition for the occurrence of a vibrational Raman effect is that there be a nonzero slope for the polarizability-vs.-displacement curve at the equilibrium value of r.

13.6 RAMAN EFFECT: QUANTUM THEORY

On first examination the Raman effect and the phenomenon of nonresonant fluorescence appear to be similar in nature. Nonresonant fluorescence occurs when a molecule in a low-lying quantized energy state absorbs intense monochromatic radiation, is excited to a higher quantized state, and then decays back to an alternate lower state by emission of radiation having a frequency different from the exciting radiation. The Raman effect involves the interaction of a molecule with intense monochromatic radiation ν_I, having an energy content different from that necessary to excite the molecule to a quantized upper state. This interaction, referred to as inelastic scattering, results in the creation of an induced dipole given by

$$\boldsymbol{\mu}_I = \underline{\boldsymbol{\alpha}}\boldsymbol{\varepsilon} \tag{13.64}$$

In this condition the molecule is said to exist in a virtual state as opposed to a doubly quantized energy state. The existence of a molecule in a virtual state is essentially instantaneous with the molecule returning to some lower level with the concurrent emission of radiation of frequency ν. Since the level of the virtual state is dependent on the energy of the incident radiation and it has an infinitesimal lifetime, the radiation is effectively scattered rather than absorbed and reemitted.

If the final state of the molecule differs from the initial state than the scattered radiation will be shifted in frequency from the incident radiation, and the scattering is inelastic. The most commonly observed cases of the Raman effect involve molecules whose initial and final states involved different rotational or vibrational energy states. In both cases the final state of the molecule, and hence the frequency of the scattered radiation, is determined by quantum mechanical selection rules. This phenomenon is illustrated in Fig. 13.10*a* for the rotational Raman effect and in Fig. 13.10*b* for the vibrational Raman effect. If $\nu_I > \nu$ the scattered radiation is called Stokes scattering, and if $\nu_I < \nu$ it is referred to as anti-Stokes scattering. The net result of this process is to produce scattered radiation which has been shifted in frequency from the initial radiation by an amount equivalent to the difference between two quantized energy levels. If we add to this model of the Raman effect the necessary selection rules, then the basic tools for the analysis of spectra are at hand.

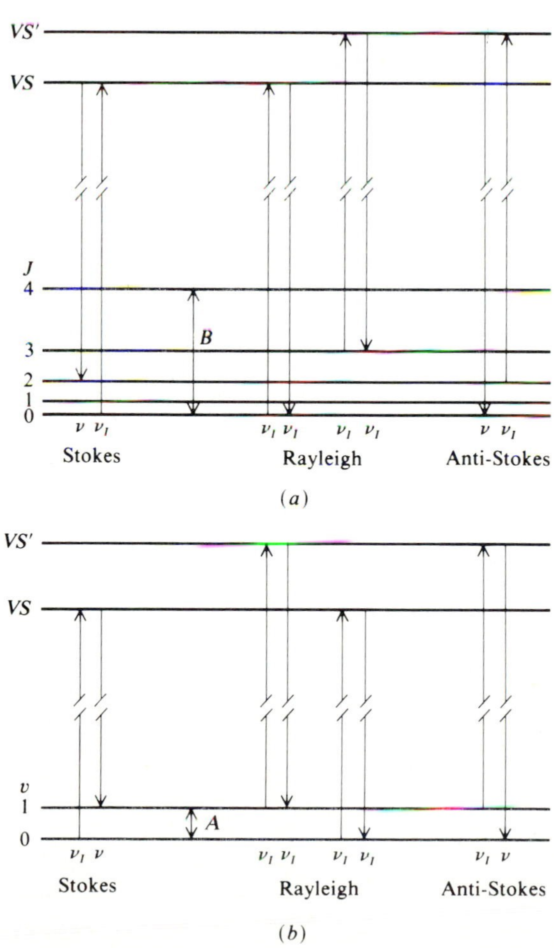

FIGURE 13.10
Raman scattering. (*a*) Rotational; (*b*) Vibrational.

Since the rotation of a diatomic molecule always leads to a change in the molecular polarizability relative to some space-fixed direction, a rotational Raman effect will be observed. Both the classical and quantum models have shown that there can exist a rotational Raman effect, but it remains to develop the necessary selection rules for a quantized molecular rotor. The general derivation of the rotational Raman selection rules has been formulated by Placzek [1], but we will consider a less complex approach based on the use of the dipole moment matrix elements. The selection rules, as for the case of the rigid rotor with a permanent dipole moment, can be determined from the nature of the dipole matrix elements. However, for the rotational Raman effect it will be the induced, rather than the permanent dipole moment that is effective in inducing transitions. The transition intensities and the selection rules are obtained from the matrix elements

$$\langle J, M | \boldsymbol{\mu}_I | J', M' \rangle = \int \psi^*_{JM} \boldsymbol{\mu}_I \psi_{J'M'} \, dv \tag{13.65}$$

where ψ_{JM} and $\psi_{J'M'}$ are the rotational wavefunctions of the initial and final states involved in the transition and $\boldsymbol{\mu}_I$ is the induced dipole moment. In general

$$\boldsymbol{\mu}_I = \mu_{Ix}\hat{\mathbf{i}} + \mu_{Iy}\hat{\mathbf{j}} + \mu_{Iz}\hat{\mathbf{k}} \tag{13.66}$$

so $\langle J | \mu_I | J' \rangle$ will be the sum of three terms, any one of which having a non-zero value will result in the transition being allowed. To simplify the discussion, consideration will be limited to one component, the induced moment in the space-fixed Z direction as given by

$$\mu_{IZ} = \alpha_{ZZ}E_Z + \alpha_{ZY}E_Y + \alpha_{ZX}E_X \tag{13.67}$$

At any instant in time, a molecule will be at some arbitrary position relative to the components of the electric field of the radiation, and will have three components of induced moment relative to a molecular-fixed axis system. The molecule-fixed induced moments μ_{Ig} will be related to the space-fixed component μ_{IZ} by the direction cosines Φ_{Fg}

$$\mu_{IZ} = \mu_{Ix}\Phi_{Zx} + \mu_{Iy}\Phi_{Zy} + \mu_{Iz}\Phi_{Zz} \tag{13.68}$$

The radiation field $\mathbf{E}$ will also have components along the molecular axes

$$E_g = E_X\Phi_{Xg} + E_Y\Phi_{Yg} + E_Z\Phi_{Zg} \tag{13.69}$$

For a diatomic molecule, the molecular-xyz-axis system will be the principal polarizability axis system, hence

$$\mu_{Ig} = \alpha_{gg}E_g \qquad (g = x, y, z) \tag{13.70}$$

Introducing Eq. (13.70) into Eq. (13.68) gives

$$\mu_{IZ} = \alpha_{xx}E_x\Phi_{Zx} + \alpha_{yy}E_y\Phi_{Zy} + \alpha_{zz}E_z\Phi_{Zz} \tag{13.71}$$

To limit the scope of the discussion, we will restrict the radiation field to one direction, $E_Z \neq 0$, and substitute Eq. (13.69) into Eq. (13.71), giving for the Z component of the induced dipole moment

$$\mu_{IZ} = [\alpha_{xx}\Phi^2_{Zx} + \alpha_{yy}\Phi^2_{Zy} + \alpha_{zz}\Phi^2_{Zz}]E_Z \tag{13.72}$$

For a linear molecule where $\alpha_{xx} = \alpha_{yy}$

$$\mu_{IZ} = [\alpha_{xx}(\Phi_{Zx}^2 + \Phi_{Zy}^2) + \alpha_{zz}\Phi_{Zz}^2]E_Z \tag{13.73}$$

Since the direction cosines are related by

$$\sum_{g=x,y,z} \Phi_{Zg}^2 = 1 \tag{13.74}$$

it follows that

$$\mu_{Iz} = [\alpha_{xx} + (\alpha_{zz} - \alpha_{xx})\Phi_{Zz}^2]E_z \tag{13.75}$$

The matrix element for the transition is then

$$\langle J, M|\boldsymbol{\mu}_{Ig}|J', M'\rangle = \langle J, M|J', M'\rangle\alpha_{xx}E_Z + \langle J, M|\boldsymbol{\Phi}_{Zz}^2|J', M'\rangle(\alpha_{zz} - \alpha_{xx})E_z \tag{13.76}$$

The first term in Eq. (13.76) contains the orthonormal integral for the rigid rotor wavefunctions.

$$\langle J, M|J', M'\rangle = \delta_{JJ'}\delta_{MM'} \tag{13.77}$$

This term gives rise to the undisplaced Rayleigh line. To evaluate the second term $\langle J, N|\boldsymbol{\Phi}_{Zz}^2|J', M'\rangle$ the matrix relation

$$\langle J, M|\boldsymbol{\Phi}_{Fg}^2|J', M'\rangle = \sum_{J''M''} \langle J, M|\boldsymbol{\Phi}_{Fg}|J'', M''\rangle\langle J'', M''|\boldsymbol{\Phi}_{Fg}|J', M'\rangle \tag{13.78}$$

and the direction cosine matrix elements are employed. For the Φ_{Zz} direction cosine matrix elements, the only nonzero terms are for $M = M'$, so Eq. (13.78) becomes

$$\langle J, M|\boldsymbol{\Phi}_{Zz}^2|J', M\rangle = \sum_{J''} \langle J, M|\boldsymbol{\Phi}_{Zz}|J'', M\rangle\langle J'', M|\boldsymbol{\Phi}_{Zz}|J', M\rangle \tag{13.79}$$

Since $\langle J, M|\boldsymbol{\Phi}_{Zz}|J', M\rangle = 0$ for $J \neq J'$, $J' \pm 1$, the only terms which need to be considered are those with $J' = J$, $J \pm 1$, $J \pm 2$. For $J' = J \pm 1$ each member of the summation will contain a term of the type $\langle J, M|\boldsymbol{\Phi}_{Zz}|J, M\rangle$ which, since $K = 0$ for the diatomic molecule, will be zero. Therefore, transitions for $J \rightarrow J \pm 1$ are forbidden. One of the remaining nonzero terms is

$$\langle J, M|\boldsymbol{\Phi}_{Zz}^2|J + 2, M\rangle = \langle J, M|\boldsymbol{\Phi}_{Zz}|J + 1, M\rangle\langle J + 1, M|\boldsymbol{\Phi}_{Zz}|J + 2, M\rangle \tag{13.80}$$

or

$$\langle J, M|\boldsymbol{\Phi}_{Zz}|J + 2, M\rangle = \left[\frac{(J + 1)^2 - M^2}{(2J + 1)(2J + 3)}\right]^{1/2}\left[\frac{(J + 2)^2 - M^2}{(2J + 3)(2J + 5)}\right]^{1/2} \tag{13.81}$$

and the other nonzero term is $\langle J, M|\boldsymbol{\Phi}_{Zz}^2|J - 2, M\rangle$. Therefore the selection rules are $\Delta J = \pm 2$. Note that the selection rules differ from those of pure rotational spectroscopy where $\Delta J = \pm 1$. In order to determine the complete coefficient arising from the matrix element it is necessary to sum over all possible M values, hence the transition intensity is proportional to

$$\sum_{M=-J}^{J} \langle J, M|\boldsymbol{\Phi}_{Zz}^2|J \pm 2, M\rangle(\alpha_{zz} - \alpha_{xx}) \tag{13.82}$$

For the discussion of Raman spectra of homonuclear diatomic molecules, an additional factor, the symmetry of the total wavefunction, must be considered. We will restrict the discussion to homonuclear molecules in Σ electronic states. For Σ molecules the electronic wavefunction ψ_e either remains unchanged, Σ^+, or changes sign, Σ^-, with a reflection through any plane parallel to the internuclear axis. It may also either remain the same, Σ_g, or change signs, Σ_u, with an inversion through the center of symmetry. The effect of an inversion of the entire molecule (equivalent to a C_2 rotation followed by a σ_v reflection) will be to leave a Σ^+ electronic wavefunction unchanged but change the sign of the Σ^- wavefunction. The vibrational wave function ψ_v will be unaffected by an inversion since the vibration is a function only of internuclear separation. The rotational wavefunction ψ_r may remain unchanged (positive) or change sign (negative) on inversion. For J = even (odd), ψ_r is + (−). Not considering nuclear spin wavefunctions the molecular wavefunction is $\psi_m = \psi_e\psi_v\psi_r$. This wavefunction may remain unchanged or change sign with respect to an inversion. The rotational levels of the homonuclear diatomic molecules are classified by the behavior of ψ_m with respect to parity. If ψ_m remains unchanged upon inversion, then ψ_m is positive (+). For this case the rotational levels of a Σ^- state are negative (−) or positive (+) depending on whether J is even or odd. The total wavefunction will also be affected by an exchange of the nuclei. Considering an exchange of nuclei as a total molecular inversion followed by an electronic inversion, it is found that the first inversion leaves ψ_m unchanged for positive rotational levels and causes ψ_m to change sign for negative rotational levels. The second inversion causes ψ to remain unchanged for Σ_g states and vice versa. Therefore, the positive rotational states are symmetric for Σ_g wavefunctions, and the negative rotational states are symmetric for Σ_u wavefunctions. Thus the symmetry of rotational states will alternate. The symmetries for the different possible cases are shown in Fig. 13.11.

For dipole transitions in absorption spectra it was determined that the selection rules were found by examining a dipole integral of the form $\int \psi_m \boldsymbol{\mu} \psi_l \, dV$.

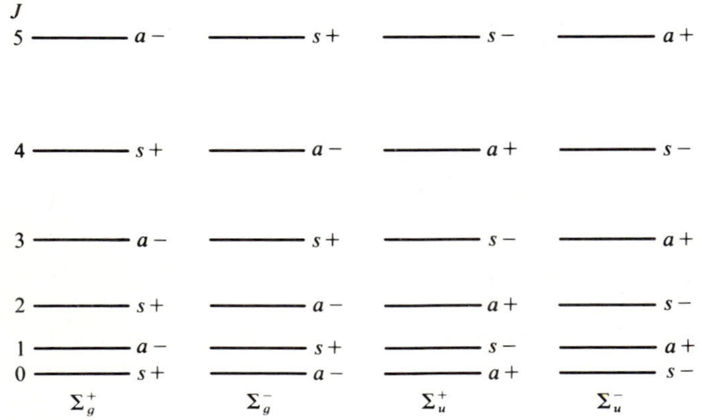

FIGURE 13.11
Symmetry of rotational levels for homonuclear diatomic molecules.

For a pure rotational transition there is no change in the electronic or vibrational wavefunctions, so ψ_m and ψ_l are two rotational wavefunctions and the zero-nonzero character of the integral depends on the parity (positive-negative character) of these functions. Since μ has the parity of a Cartesian coordinate the integral will be zero unless ψ_m and ψ_l have opposite parities, hence the dipolar selection rules, $+ \nleftrightarrow +$, $- \nleftrightarrow -$, and $+ \leftrightarrow -$.

For a homonuclear molecule with two equivalent nuclei the wavefunctions ψ_m are further classified as symmetric, s, or antisymmetric, a, depending on whether ψ_m changes sign or remains unchanged with respect to an interchange of nuclei. This additional factor must be considered when evaluating selection rules. Integrals of the type $\int \psi_m^a \mu \psi_l^s \, dV$, will change sign with an interchange of nuclei. Since the value of the integral cannot depend on the manner in which the nuclei are labeled it must therefore be zero and we have the additional selection rules $a \leftrightarrow a$, $s \leftrightarrow s$, $s \nleftrightarrow a$. For rotational Raman transitions it was previously established that $\Delta J = \pm 2$. Reference to Fig. 13.11 then shows, since $s \leftrightarrow s$ and $a \leftrightarrow a$ transitions are allowed, that a Raman spectrum can be observed for homonuclear molecules. Thus Raman spectroscopy provides a technique for the determination of structural parameters of homonuclear diatomic molecules.

If the effects of nuclear spin are considered, interesting observed alterations of spectral intensities can be explained. For example, it is observed that (1) the rotational Raman transitions of $^1H_2$ originating from odd J levels are about three times the intensity of the neighboring transitions originating from an even J level, and (2) those of $^{16}O_2$ originating from even J levels are missing. The effects of nuclear spin are incorporated by inclusion of a nuclear spin wavefunction ψ_n in the total wavefunction, giving

$$\psi = \psi_e \psi_v \psi_r \psi_n \tag{13.83}$$

Nuclei with half-integer spins, $I = n + \frac{1}{2}$ (n = integer), are fermions and obey Fermi-Dirac statistics, the same as electrons. The generalized Pauli principle states that the total wavefunctions of a system must be antisymmetric with respect to the exchange of any pair of identical fermions. Since it has been shown earlier that both ψ_v and ψ_e (ground state) are symmetric with respect to nuclear exchange the symmetry of the total wavefunction will be determined by the product $\psi_r \psi_n$. For Σ_g^+ electronic states and even J, ψ_r will be symmetric with respect to nuclear exchange and, for odd , antisymmetric. Thus, for the function $\psi_r \psi_n$ to be antisymmetric, symmetric values of ψ_n must be associated with ψ_r having odd J and vice versa.

Consider a Σ_g^+ diatomic molecule with two identical nuclei having $I = \frac{1}{2}$. There are four possible spin wavefunctions, the symmetric functions, $\alpha\alpha$, $2^{-1/2}(\alpha\beta + \beta\alpha)$, $\beta\beta$, and the antisymmetric function, $2^{-1/2}(\alpha\beta - \beta\alpha)$. Normally interchange between symmetric and antisymmetric nuclear spin states is forbidden so there will be three times as many symmetric as antisymmetric states. Coupling the appropriate ψ_r with these spin states to give antisymmetric functions shows that rotational levels with odd J will have approximately three times the population of the even J states, and the transitions originating from the states will reflect these population differences in their intensities. Were all of the rotational levels equally populated, then the intensity ratios would be strictly 3:1; however, the Boltzmann distribution of rotational states leads to the factor of 3 being somewhat approximate. One instance where this ratio is not strictly observed is for light molecules at low temperatures where there are appreciable differences in Boltzmann factors between adjacent rotational states.

Nuclei having integer spin value I are bosons, and obey Bose-Einstein statistics. Diatomic molecules having such nuclei must have, in accordance with the Pauli principle, total wavefunctions which are symmetric with respect to nuclei exchange. For example, $^2H_2$ or $^{14}N_2$ having $I = 1$ with $M_I = 0, \pm 1$ will have a total of nine ψ_n functions, six symmetric and three antisymmetric. The symmetric spin states will be combined with the even J rotational states and vice versa, hence, transitions originating from the even J states will have approximately double the intensities of those originating from the odd states.

An interesting phenomenon occurs when a molecule is composed of two identical nuclei with $I = 0$. In this case, the nuclei are bosons, and for a Σ_g^+ electronic state such as C_2, the total wavefunctions must be symmetric. Since the statistical weight factors for two identical nuclei are given by

$$g_n^s = (2I + 1)(I + 1) \tag{13.84}$$

$$g_n^a = (2I + 1)I \tag{13.85}$$

it is noted that the antisymmetric nuclear states will have $g_n^a = 0$ and are missing. Thus only those states with both symmetrical ψ_r and ψ_n will be populated and only transitions originating from the even J levels will be observed. Looking at Fig. 13.11, it is observed that the same will be true for Σ_u^- states where transitions will originate only from the odd J levels for molecules in Σ_g^- and Σ_u^+ electronic states.

For $^{16}O_2$, although $I = 0$, the situation is somewhat different since the ground electronic state of $^{16}O_2$ is $^3\Sigma_g^-$ and is antisymmetric. Hence, for the total wavefunctions to be symmetric with respect to nuclear exchange, the product function $\psi_r\psi_n$ must be antisymmetric. Since only the symmetric nuclear spin state exists, this requires that only the odd J rotational states be populated, and the rotational Raman spectrum will contain only transitions originating from these levels. The effects of nuclear spin are summarized in Table 13.2.

The vibrational selection rules are established by determining the value of the matrix elements

$$\langle v | \boldsymbol{\mu}_I | v' \rangle = \int \psi_v^* \mu_I \psi_{v'} \, dv \tag{13.86}$$

where ψ_v and $\psi_{v'}$ are the vibrational wavefunctions. Since

$$\mu_{Ig} = \alpha_{gg} E_g + \alpha_{gg'} E_{g'} + \alpha_{gg''} E_{g''} \tag{13.87}$$

where $g, g', g'' = x, y, z$ taken in cyclic order to obtain the three components of μ_I, and

$$\langle v | \boldsymbol{\mu}_I | v' \rangle = \sum_{g=x,y,z} \langle v | \boldsymbol{\mu}_{Ig} | v' \rangle^2 \tag{13.88}$$

TABLE 13.2
Symmetries and statistical weights of the energy levels of homonuclear diatomic molecules

		Σ_g^+ Electronic state, $\psi_e(s)$									Σ_g^- Electronic state, $\psi_e(a)$								
		Fermions			Bosons						Fermions			Bosons					
		$I=\frac{1}{2}$			$I=1$			$I=0$			$I=\frac{1}{2}$			$I=1$			$I=0$		
J	ψ_r	ψ_n	g_n	ψ	ψ_n	g_n	ψ	ψ_n	g_n	ψ	ψ_n	g_n	ψ	ψ_n	g_n	ψ	ψ_n	g_n	ψ
5	a	s	3	a	a	3	s		0		a	1	a	s	6	s	s	1	s
4	s	a	1	a	s	6	s	s	1	s	s	3	a	a	3	s	a	0	
3	a	s	3	a	a	3	s		0		a	1	a	s	6	s	s	1	s
2	s	a	1	a	s	6	s	s	1	s	s	3	a	a	3	s	a	0	
1	a	s	3	a	a	3	s		0		a	1	a	s	6	s	s	1	s
0	s	a	1	a	s	6	s	s	1	s	s	3	s	a	3	s	a	0	

† For molecules like O_2 which, due to the presence of unpaired electrons have electronic angular momenta which can couple with the rotational angular momentum, J is the total angular momentum quantum number.

it follows that

$$\langle v | \boldsymbol{\mu}_{Ig} | v' \rangle = \langle v | \alpha_{gg} | v' \rangle E_g^+ \langle v | \alpha'_{gg} | v' \rangle E_{g'} + \langle v | \alpha_{gg''} | v' \rangle E_{g''} \tag{13.89}$$

As in the classical analysis, we find that the occurrence of transitions depends on the polarizability tensor components. A vibrational Raman transition will be allowed if any one of the matrix elements $\langle v | \alpha_{gg'} | v' \rangle$ is nonzero. Integrals of the form

$$\int \psi_v^* \alpha_{gg'} \psi_{v'} \, dV = \left(\frac{\partial \alpha_{gg'}}{\partial x} \right)_0 \int \psi_v^* x \psi_{v'} \, dx + \cdots \tag{13.90}$$

can be nonzero only if the integrand is totally symmetric. For the diatomic harmonic oscillator the product $\psi_v^* \psi_{v'}$ will be even if $v' = v$, $v \pm 2$, etc., and odd if $v' = v \pm 1$, $v \pm 3$, etc. This is easily verified since the symmetry of ψ_v is that of the Hermite polynomial $h_v(q)$, and the latter is even for v even and odd for v odd. Since x is an odd function this requires that $\psi_v^* \psi_{v'}$ be odd and the selection rule becomes $\Delta V = \pm 1$ when considering the other properties of the Hermite functions. If one considers anharmonicity then it is found, just as for pure vibrational spectra, that $\Delta v = \pm 2, \pm 3, \ldots$ are allowed.

13.7 THE NATURE OF RAMAN SPECTRA

For a diatomic molecule in the gas phase, both a rotational and a rotational-vibrational Raman effect may be observed. The rotational Raman effect for a rigid diatomic molecule will give rise to a set of transitions as shown in Fig. 13.12. Since the selection rule is $\Delta J = \pm 2$ the Stokes ($J + 2 \leftarrow J$, $\nu < \nu_I$) transitions will be given by

$$\nu^S = \nu_I - 2B(2J + 3) \tag{13.91}$$

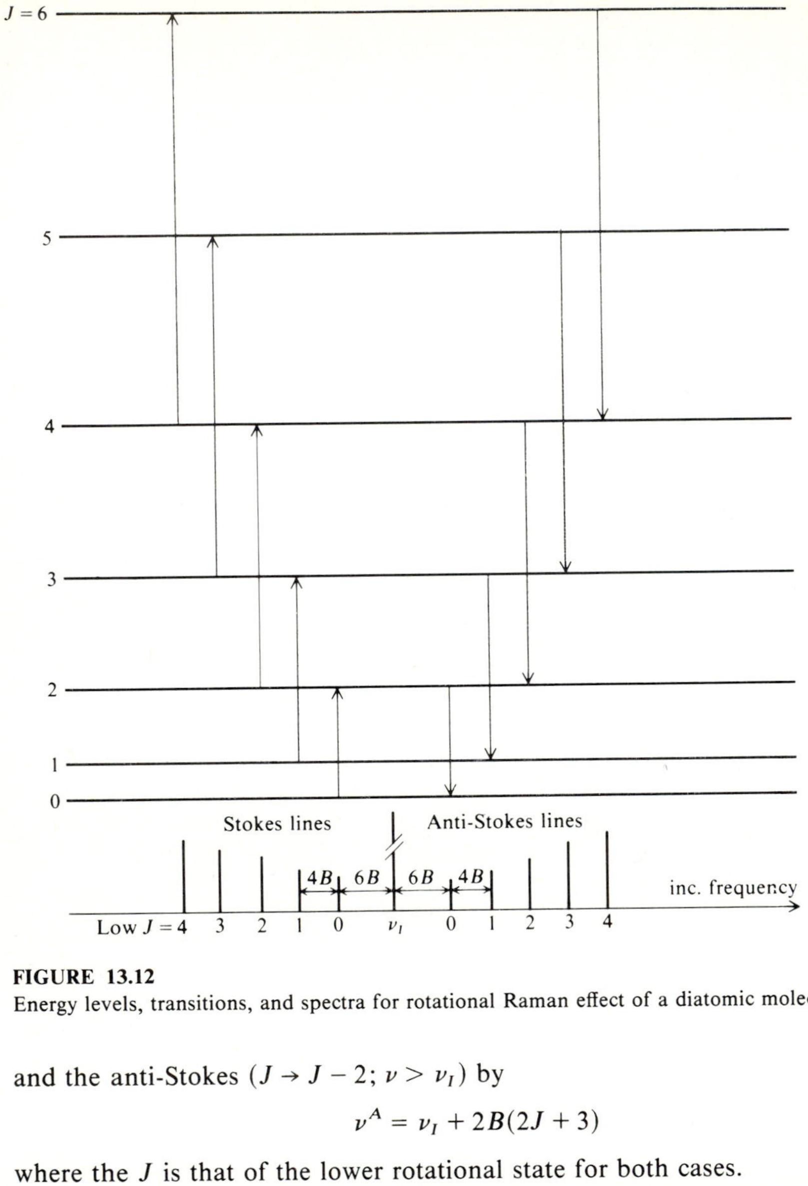

FIGURE 13.12
Energy levels, transitions, and spectra for rotational Raman effect of a diatomic molecule.

and the anti-Stokes ($J \rightarrow J-2$; $\nu > \nu_I$) by

$$\nu^A = \nu_I + 2B(2J+3) \tag{13.92}$$

where the J is that of the lower rotational state for both cases.

The exciting radiation for the observation of Raman spectra for work prior to the availability of lasers was one of the strong emission lines of atomic mercury located in the near ultraviolet region of the spectrum. The development of the laser has given Raman spectroscopy a substantial boost as a structural tool by providing a source of radiation with an exceptionally narrow bandwidth. The exciting radiation will be of the order of 400 nm ($\bar{\nu} = 2.5 \times 10^4\ \text{cm}^{-1}$) while for a typical diatomic molecule with $B_0 = 9000$ MHz (0.3 cm^{-1}) the rotational transitions will be separated by about 1.2 cm^{-1} and clustered about the exciting line.

It is necessary for the linewidth of the exciting radiation to be as narrow as possible and for the resolution of the spectrometer optics to be exceptionally good to observe pure rotational Raman spectra.

Molecules in general are nonrigid so when the effects of centrifugal distortion and vibrational-rotational interaction are included the energy levels are given by

$$\frac{E_{vJ}}{h} = [B_e - (v + \tfrac{1}{2})\alpha_e]J(J+1) - D_e J^2(J+1)^2$$

$$= B_v J(J+1) - D_e J^2(J+1)^2 \tag{13.93}$$

The frequencies of the Stokes lines then become

$$\nu^S = \nu_I - 2B_v(2J+3) + 4D_e(2J^3 + 9J^2 + 15J + 9) \tag{13.94}$$

while those of the anti-Stokes lines are

$$\nu^A = \nu_I + 2B_v(2J+3) - 4D_e(2J^3 + 9J^2 + 15J + 9) \tag{13.95}$$

Except for very light molecules, D_e is sufficiently small that the detection of displacements due to centrifugal distortion requires very good instrumentation. For example, for the relatively light $^1H^{35}Cl$ molecule with $D_e = 15.98$ MHz the shift in a $J = 4 \leftarrow 2$ transition would only be 0.2 cm^{-1}. Unless the vibrational frequency is very low the majority of the molecules will be in the ground vibrational state and the analysis of the Raman spectrum will only give B_0.

Since vibrational frequencies are of the order of 50–4000 cm^{-1}, the vibrational Raman effect will give rise to transitions separated from the exciting frequency by these amounts. In the gas phase, each vibrational transition will consist of a number of closely spaced components analogous to the rotational-vibrational absorption spectrum. For the vibrational Raman effect, the sets of rotational fine structure lines are referred to as the O branch ($\Delta J = -2$), Q branch ($\Delta J = 0$), and S branch ($\Delta J = 2$). For a $v = 1 \leftarrow 0$ vibrational transition, ignoring anharmonicity and centrifugal distortion, the energy levels involved in the transition will be

$$E_{1J} = \tfrac{3}{2}\omega_e + B_1 J(J+1) \tag{13.96}$$

$$E_{0J} = \tfrac{1}{2}\omega_e + B_0 J(J+1) \tag{13.97}$$

The anti-Stokes transitions will be given by

$$\nu_S^A = \nu_I + \omega_e - 6B_0 - (5B_0 - B_1)J - (B_0 - B_1)J^2 \tag{13.98}$$

for the S branch ($1, J \rightarrow 0, J+2$), by

$$\nu_Q^A = \nu_I + \omega_e + (B_0 - B_1)J + (B_0 - B_1)J^2 \tag{13.99}$$

for the Q branch ($1, J \rightarrow 0, J$), and by

$$\nu_O^A = \nu_I + \omega_e + 6B_1 + (5B_1 - B_0)J + (B_1 - B_0)J^2 \tag{13.100}$$

for the O branch ($1, J+2 \rightarrow 0, J$). In all cases J is that of the $v = 0$ level. Correspondingly the Stokes transitions will be given by

$$\nu_S^S = \nu_I - \omega_e - 6B_1 - (5B_1 - B_0)J - (B_1 - B_0)J^2 \tag{13.101}$$

for the S branch $(1, J+2 \leftarrow 0, J)$, by

$$\nu_Q^S = \nu_I - \omega_e - (B_1 - B_0)J - (B_1 - B_0)J^2 \qquad (13.102)$$

for the Q branch $(1, J \leftarrow 0, J)$, and by

$$\nu_0^S = \nu_I - \omega_e + 6B_0 + (5B_0 - B_1)J + (B_0 - B_1)J^2 \qquad (13.103)$$

for the O branch $(1, J \leftarrow 0, J+2)$. These transitions are shown schematically in Fig. 13.13. For most diatomic molecules the magnitude of ω_e is such that the $v = 1$ vibrational level is scarcely populated. Since the anti-Stokes transitions originate from molecules in the $v = 1$ state, these lines will generally be weak or completely absent.

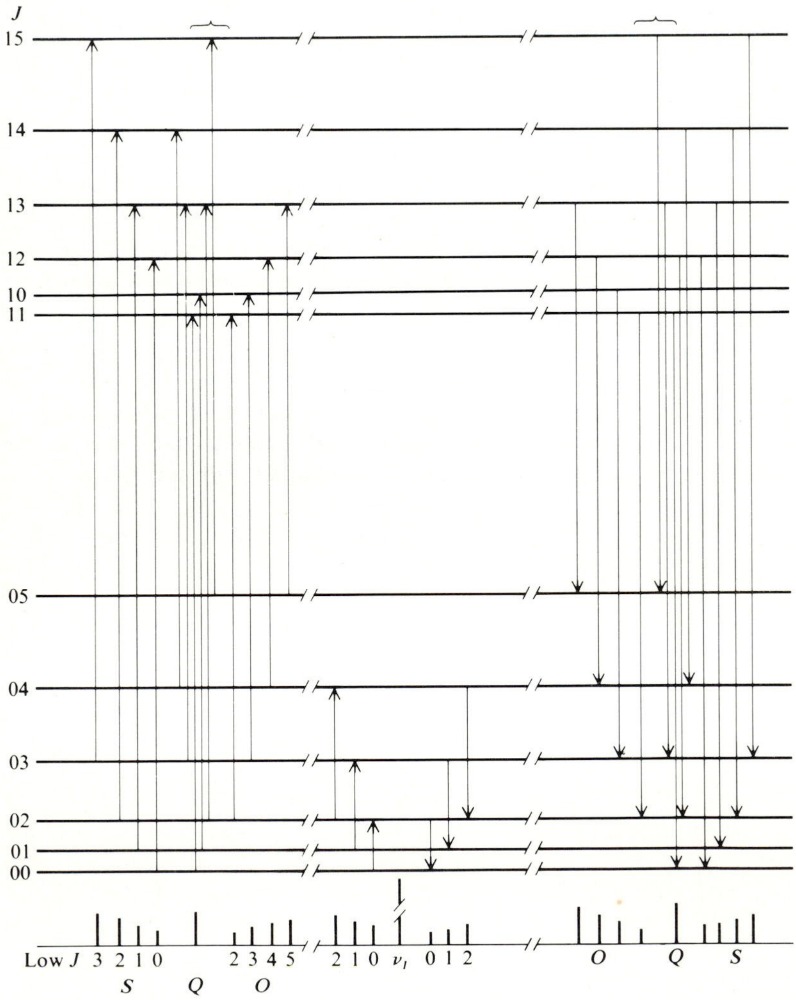

FIGURE 13.13
Energy levels, transitions, and spectra for vibrational-rotational Raman effect for a diatomic molecule.

In the condensed phases the rotational fine structure that is present in the gas is absent due to molecular collisions, and the Raman spectrum of a diatomic molecule is simplified to a set of broad transitions displaced from the incident frequency by some multiple of the vibrational frequency as modified by the anharmonicity. If the vibrational energy levels without rotational fine structure are given by

$$\frac{E_v}{h} = (v + \tfrac{1}{2})\omega_e - (v + \tfrac{1}{2})^2 \chi_e \omega_e \tag{13.104}$$

then the Raman transitions, allowing for $\Delta v \geqq 1$ transitions being possible, are given by

$$\nu = \nu_I \pm [v'\omega_e - v'(v' + 1)\chi_e \omega_e] \tag{13.105}$$

where v' is the vibrational quantum number of the upper state and the + sign refers to the anti-Stokes lines. Again, for the diatomic molecule, the anti-Stokes lines will be weak or absent.

13.8 EXAMPLES OF RAMAN SPECTRA

Although modern Raman spectroscopy employs laser sources, the results of an older investigation of fluorine by Andrychuk [2] provides an excellent example of the use of the rotational Raman effect to obtain structural information about a nonpolar molecule where conventional rotational spectroscopy cannot be employed. The spectrum was recorded photographically using a spectrometer with a high-intensity mercury arc. Due to the long exposure time necessary to record the Raman spectrum of the gaseous sample, the attendant scattering of the incident radiation obscured the lower-lying transitions for both the Stokes and anti-Stokes lines. The nuclear spin of F is $I = \frac{1}{2}$ so there is an alteration in intensities. Only the stronger set of lines have been used for measurements. The resulting data are summarized in Table 13.3 where $\Delta\nu$ is the displacement from the incident frequency.

Using Eqs. (13.94) and (13.95) the separations between pairs of lines with adjacent J values and $v = 0$ can be expressed as

$$\Delta\nu = 2B_0(2J + 3) - 4D_e(2J^3 + 9J^2 + 15J + 9) \tag{13.106}$$

or

$$\Delta\nu = (2B_0 - 3D_e)(2J + 3) - D_e(2J + 3)^3 \tag{13.107}$$

This equation can be put into a form convenient for analysis of the data for F_2 by dividing by $(2J + 3)$. One can now plot $\Delta\nu/(2J + 3)$ vs. $(2J + 3)^2$ and obtain B_0 and D_e from the intercept and slope as shown in Fig. 13.14. From this plot using a least-squares fit to get the best straight line one obtains $(2B_0 - 3D_e) = 1.766\ \text{cm}^{-1}$ and $D_e = 0.0000038\ \text{cm}^{-1}$. From these values, noting that D_e is insignificant, we get $B_0 = 0.883 \pm 0.001\ \text{cm}^{-1}$ and $r_0 = 0.1418 \pm 0.0002$ nm.

Contrasted to this investigation are the results of a laser Raman investigation of N_2 by Butcher, Willetts, and Jones [3]. The raw data as shown in Table 13.4, where the frequency displacement measurements are accurate to $\pm 0.005\ \text{cm}^{-1}$,

TABLE 13.3
Raman spectral data for gaseous F_2

	Stokes lines		Anti-Stokes lines	
J	$\Delta\nu$, cm^{-1}	I, rel	$\Delta\nu$, cm^{-1}	I, rel
5	−22.57	35		
6		16		
7	−29.88	31	28.89	12
8		9		6
9	−36.93	19	36.93	10
10		4		3
11	−44.00	15	44.10	12
12		3		5
13	−51.07	16	51.06	13
14		5		6
15	−58.16	11	58.17	11
16		5		3
17	−65.18	11	65.16	10
18		5		3
19	−72.09	10	72.13	7
20		3		3
21	−79.12	8	78.98	7
22		3		3
23	−86.27	7	86.36	5
24		2		
25	−93.35	5		
26				
27	−100.19			
28				
29	−106.70			

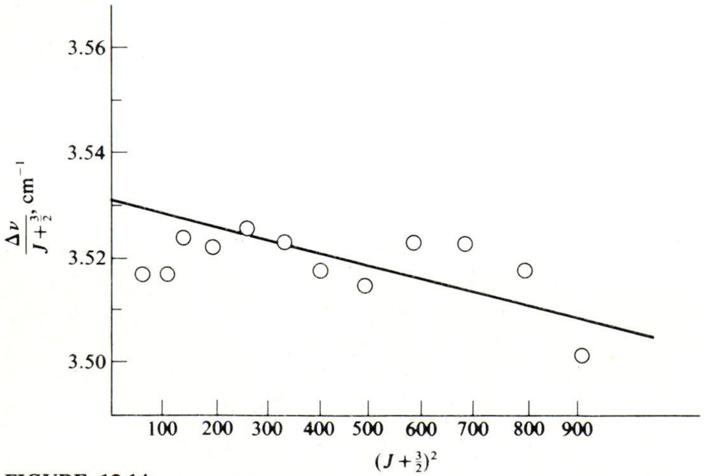

FIGURE 13.14
Plot of Raman data for F_2.

TABLE 13.4
Rotational Raman data for N_2

J	$\Delta\nu_{ob}$	$\Delta\nu_{calc}$	$\Delta\nu_o - \Delta\nu_c$
1	19.8936	19.8943	−0.0007
2	27.8499	27.8511	−0.0012
3	35.8077	35.8070	0.0007
4	43.7629	43.7617	0.0012
5	51.7193	51.7149	0.0044
6	59.6699	59.6665	0.0034
7	67.6145	67.6160	−0.0015
8	75.5619	75.5633	−0.0014
9	83.5044	83.5082	−0.0038
10	91.4490	91.4503	−0.0013
11	99.3901	99.3893	0.0008
12	107.3244	107.3251	−0.0007
13	115.2565	115.2573	−0.0008
14	123.1847	123.1857	−0.0010
15	131.1137	131.1100	0.0037
16	139.0299	139.0300	−0.0001

illustrates the magnitude of the improvement brought about by use of lasers. The data are analyzed in the same manner as was used for F_2. The plot of $\Delta\nu(J+\frac{3}{2})$ vs. $(J+\frac{3}{2})^2$ shown in Fig. 13.15 further illustrates the precision of the data. From these data one obtains $B_0 = 1.989506 \pm 0.000027\ \text{cm}^{-1}$ and $D_e = (5.48 \pm 0.06) \times 10^{-6}\ \text{cm}^{-1}$. This leads to an internuclear separation of $r_0 = 0.110010 \pm 0.000001$ nm.

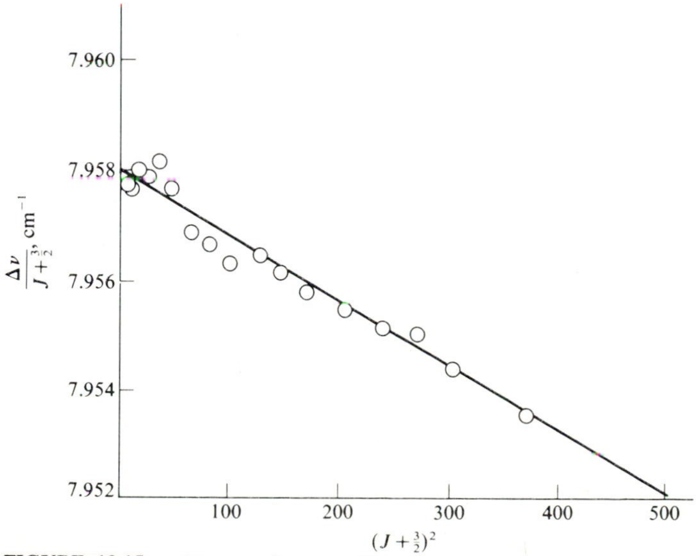

FIGURE 13.15
Plot of Raman data for N_2.

TABLE 13.5
Raman data for Cl_2

	$\Delta\nu$, cm^{-1}	
Species	$v = 1 \leftarrow 0$	$v = 2 \leftarrow 1$
$^{35}Cl_2$	548.4 ± 0.1	543.1 ± 0.2
$^{35}Cl^{37}Cl$	540.9 ± 0.1	535.7 ± 0.1
$^{37}Cl_2$	533.4 ± 0.1	528.2 ± 0.2

The vibrational Raman effect for diatomic molecules is rather simple, especially if the liquid state is used. In the solid state, additional complications due to lattice modes of vibration often arise. Typical of a vibrational Raman study of a nonpolar molecule is the investigation of liquid Cl_2 by Stammreich and Forneris [4]. The displacements, from the 5875.62 A line of helium, of the Stokes lines for the three isotopic species and for two vibrational states are given in Table 13.5. Note that due to the low value of the vibrational frequency a higher-order transition is observed because even at room temperature about 10 percent of the molecules will be in the $v = 1$ vibrational state. Allowing for anharmonicity these observed shifts will be given by

$$\Delta\nu = \omega_e - 2\chi_e\omega_e - 2v\chi_e\omega_e \tag{13.108}$$

For $^{35}Cl_2$ using the pair of equations

$$548.4 = \omega_e - 2\chi_e\omega_e \tag{13.109}$$

$$543.1 = \omega_e - 4\chi_e\omega_e \tag{13.110}$$

one calculates $\omega_e = 533.1$ cm^{-1} and $\chi_e\omega_e = 2.5$ cm^{-1}.

REFERENCES

A. Specific

1. Placzek, E., *Marx Handbuch der Radiologie*, VI, **2**, 205 (1934).
2. Andrychuk, D., *Canadian J. Phys.*, **29**, 151 (1951).
3. Butcher, R. J., D. V. Willetts, and W. J. Jones, *Proc. Roy. Soc.*, ser. A, **324**, 231 (1971).
4. Stammreich, H., and R. Forneris, *Spectrochim. Acta.* **17**, 775 (1961).

B. General

Demtroder, W., *Laser Spectroscopy*, Springer-Verlag, Berlin, 1981.
Herzberg, G., *Molecular Spectra and Molecular Structure*, vol. I: *Diatomic Molecules*, Van Nostrand Reinhold, New York, NY, 1950.
Hollas, J. M., *High Resolution Spectroscopy*, Butterworths, London, GB, 1982.
Koningstein, J. A., *Introduction to the Theory of the Raman Effect*, D. Reidel Publishing, Co., Dordrecht, NL, 1972.
Long, D. A., *Raman Spectroscopy*, McGraw-Hill, London, GB, 1977.

PROBLEMS

13.1. NaBr in the gas phase at 800 K has $\omega_e = 315\ \text{cm}^{-1}$ and $\chi_e\omega_e = 1.15\ \text{cm}^{-1}$. Calculate the frequencies and relative intensities of the $\Delta\nu = 1 \leftarrow 0$, $2 \leftarrow 0$, $3 \leftarrow 0$, $2 \leftarrow 1$, and $3 \leftarrow 2$ transitions at 800 K.

13.2. Microwave spectroscopic investigations have determined the rotational constants of $^{16}O^{1}H$, $^{69}Ga^{127}I$, and $^{32}S^{16}O$ to be 565, 738, 1706.86, and 24,584.35 MHz respectively. For irradiation of these gases with a 5145 A laser calculate the expected positions of the first six Stokes lines in the rotational Raman spectra (1) considering the molecules as rigid rotors in the ground vibrational state, and (2) considering the molecules as nonrigid rotors with a centrifugal distortion constant D_e, equal to 0.01 percent of the B_e value.

13.3. The rotational Raman spectrum of nitrous oxide is shown in the diagram below (page 384). Considering that a linear molecule has the same type of spectrum as a diatomic molecule, fit the spectrum and calculate the best value for the moment of inertia of N_2O.

13.4. The rotational constant of $^{16}O^{2}H$ in the $^{2}\Sigma^{+}$ electronic state has been determined from a study of the electronic spectrum to be $9.194 \pm 0.001\ \text{cm}^{-1}$. If it were possible to observe the pure rotational spectrum of this species and measure the $J = 1 \leftarrow 0$ transition to an accuracy of ± 0.01 MHz what would be the relative precision of the r_0 values obtained by the two experiments, considering the species as a rigid rotor?

13.5. The rotational Raman spectrum of air is shown in the diagram below (page 385). Fit the spectrum and determine the best internuclear distances for O_2.

13.6. Calculate the relative rotational populations of the lowest four rotational states of $^{1}H_2$ and $^{19}F_2$ at 20 and 300 K. Combine these results with the nuclear statistics and predict the relative intensities of the first four rotational transitions.

13.7. Riefer and Bernstein (*J. Raman Spect.*, **1**, 417 (1973)) measured the vibrational Raman spectrum of I_2 in several solvents as shown in the following table.

Raman data for I_2 in various solutions

Solvent	ν, cm^{-1} $0 \to 1$	ν, cm^{-1} $0 \to 2$	ν, cm^{-1} $0 \to 3$
CCl_4	211.4	421.4	630.3
$CHCl_3$	210.9	420.4	629.1
CS_2	207.0	412.7	
n-Hexane	211.2	421.3	630.2
Cyclohexane	211.4	421.6	630.2
Benzene	205.2	407.9	
p-Xylene	203.3	404.6	604.7

For I_2 in each solvent calculate where possible values for ω_e, $\chi_e\omega_e$ and $y_e\omega_e$. For I_2 in the gas phase $\omega_e = 214.534\ \text{cm}^{-1}$, $\chi_e\omega_e = 0.6070\ \text{cm}^{-1}$ and $y_e\omega_e = -0.0016\ \text{cm}^{-1}$. Suggest a reason for the differences between the gas and solution I_2 data.

13.8. Using the data in Table 13.1 determine ω_e, $\chi_e\omega_e$, and $y_e\omega_e$ by means of a nonlinear least-squares analysis.

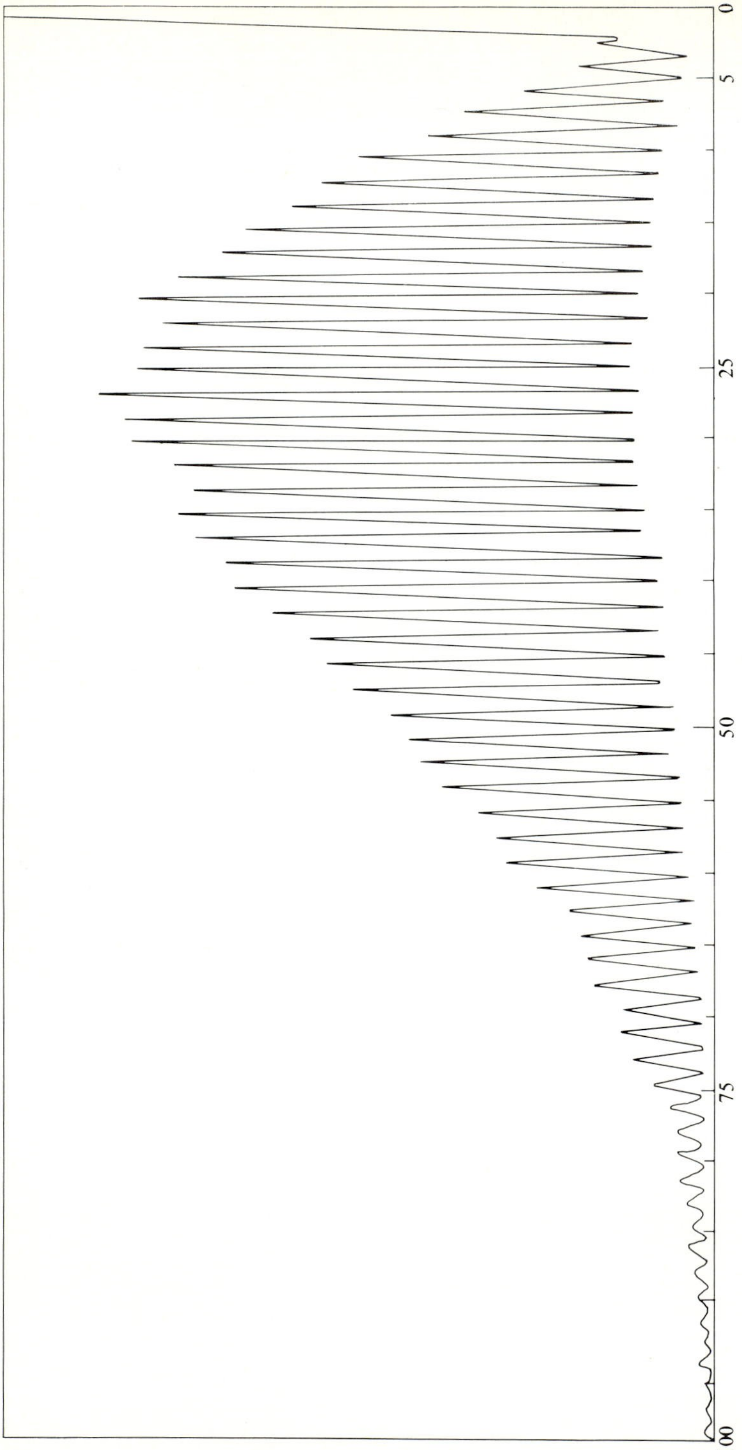

FIGURE P13.3

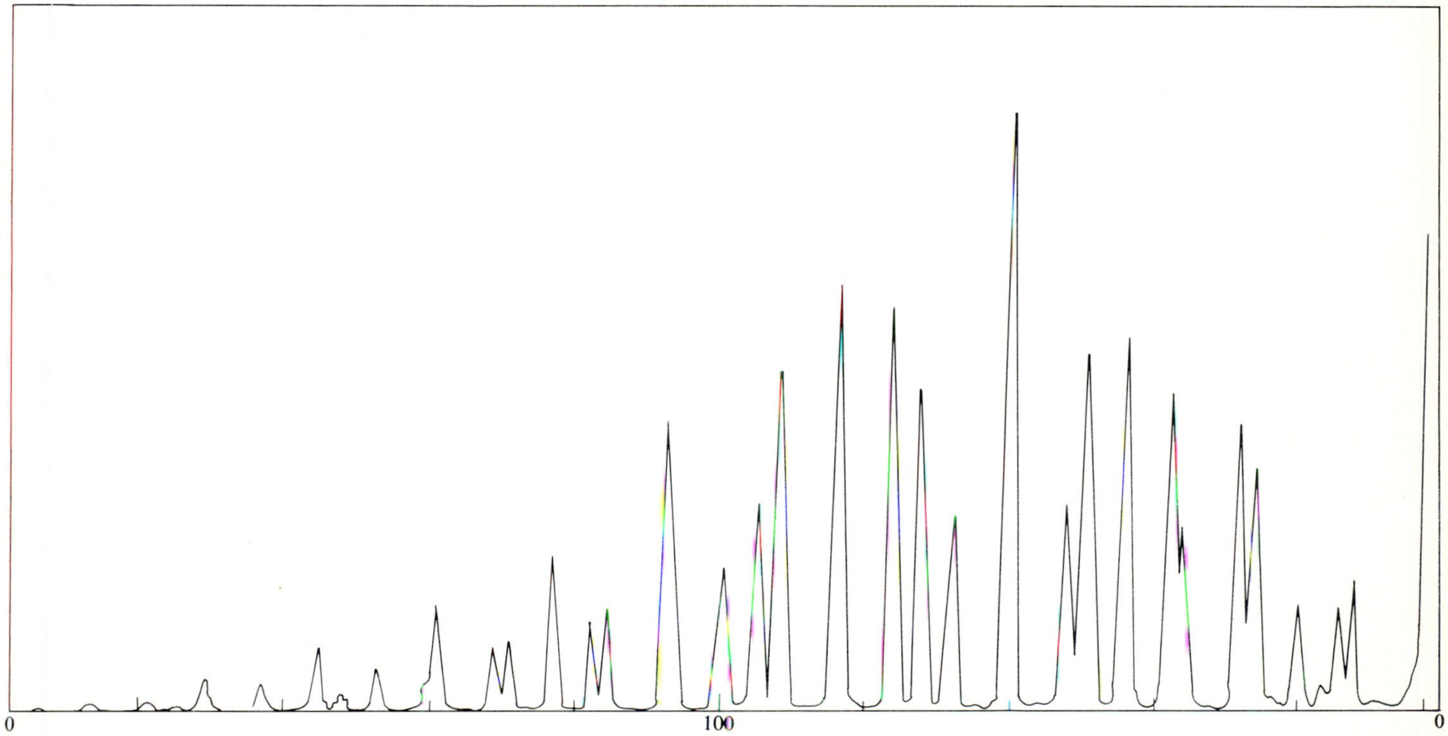

FIGURE P13.5

CHAPTER 14

DIATOMIC VIBROTORS— FIELD EFFECTS AND HYPERFINE INTERACTIONS

14.1 INTRODUCTION

This chapter will investigate changes in the nature of the rotational spectra of diatomic molecules which occur when the molecules are subjected to external fields or are perturbed by internal interactions other than rotational-vibrational interaction. Of particular interest will be the use of these effects to identify transitions and the determination of additional parameters which can be related to the structures of the molecules.

14.2 STARK EFFECT: ENERGY LEVELS AND TRANSITIONS[1]

When an electric dipole $\boldsymbol{\mu}$ is subjected to an electric field $\mathscr{E}$ there will occur an electrostatic interaction which is dependent on the magnitudes of the field and

[1] The discussion of the Stark effect in this chapter will be limited to $^1\Sigma$ electronic state molecules where the total angular moment is that due only to molecular rotation.

the dipole, and on the relative orientation of the two. The classical energy of such an interaction is given by

$$E = -\boldsymbol{\mu} \cdot \boldsymbol{\mathscr{E}} = -|\mu\mathscr{E}| \cos\theta \tag{14.1}$$

where θ is the angle between the dipole and the external field. The normal criterion for a molecule to exhibit a pure rotational spectrum is that it possesses a permanent dipole moment which can interact with an external electric field to produce changes in the energy of rotation of the molecule. It is also possible for this interaction to occur in a molecule without a permanent dipole moment if one is induced by an unsymmetrical vibration. For electric fields of the order of 1000 V/cm and a dipole moment of 2 debye, the energy of the interaction is about 1000 MHz.[2] Compared to the magnitude of rotational energies, 10–500 GHz, this is small and can be considered as a perturbation.

Before considering the quantum mechanical treatment of the Stark effect let us examine a semiclassical description. Although the first-order Stark effect is absent for diatomic and linear molecules, it will be considered at this point since it will be encountered in the discussion of the symmetric top and it provides an elementary qualitative introduction to the phenomenon. For a nonlinear molecule possessing cylindrical symmetry and an electric dipole directed along the symmetry axis, the free rotation can be described in terms of its total angular momentum. Since there is motion about the symmetry axis there will be a component of the angular momentum along the symmetry axis. This component will precess rapidly about the direction of the total angular momentum. The result of this precessional motion is to conserve the component of the electric dipole moment along the axis of total angular momentum. These relationships are illustrated in Fig. 14.1*a*. Classically the electric dipole component lying along the axis of total momentum is given by the product of the ratio of the momenta times the dipole moment $\mu_J = \mu(|K|/|J|)$. For the quantum mechanical rotor, the total angular momentum is $\hbar[(J+1)J]^{1/2}$ and the component about a fixed body z axis is $K\hbar$, so

$$\mu_J = \frac{\mu K}{[J(J+1)]^{1/2}} \tag{14.2}$$

When an external electric field $\mathscr{E}$ is applied at an angle θ to the direction of the total angular momentum **J**, the electric dipole component μ_J will interact with the field giving rise to an energy of interaction

$$E^{(1)} = -\boldsymbol{\mu}_J \cdot \boldsymbol{\mathscr{E}} = \mu_J \mathscr{E} \cos\theta \tag{14.3}$$

This interaction results in the precession of $\boldsymbol{\mu}_J$ about the direction of $\boldsymbol{\mathscr{E}}$. The classical projection of the total angular momentum on a space-fixed Z axis is

---

[2] 1 μE (debye) (V/cm) = 0.50344 h(MHz).

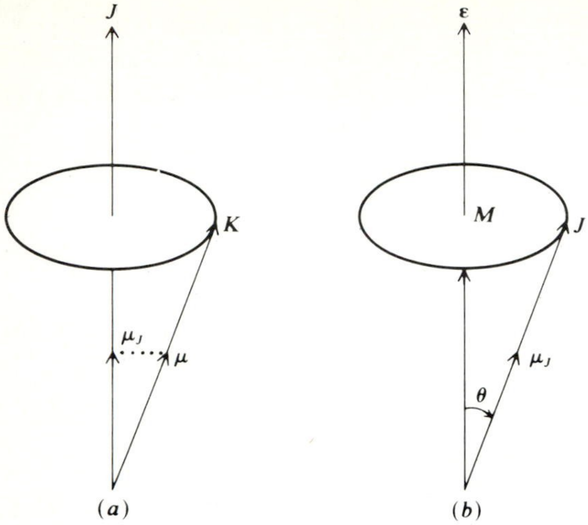

FIGURE 14.1
Vector relationships for first-order Stark effect. (*a*) Relationship of *J* and *K* with no field; (*b*) Relationship of *J* and *M* in the presence of an electric field.

related to the ratio $|M|/|J|$ or, using the quantum mechanical magnitude of **J**,

$$\cos\theta = \frac{M}{[J(J+1)]^{1/2}} \tag{14.4}$$

Combining Eqs. (14.2), (14.3), and (14.4) gives an expression the first-order Stark effect

$$E^{(1)} = -\frac{\mu \mathscr{E} K M}{J(J+1)} \tag{14.5}$$

For a diatomic molecule there is no component of angular momentum about the symmetry axis, hence $K = 0$ and $E^{(1)} = 0$.

For a diatomic molecule, the rotational angular momentum **J** is perpendicular to the direction of the dipole moment as shown in Fig. 14.2. Considering

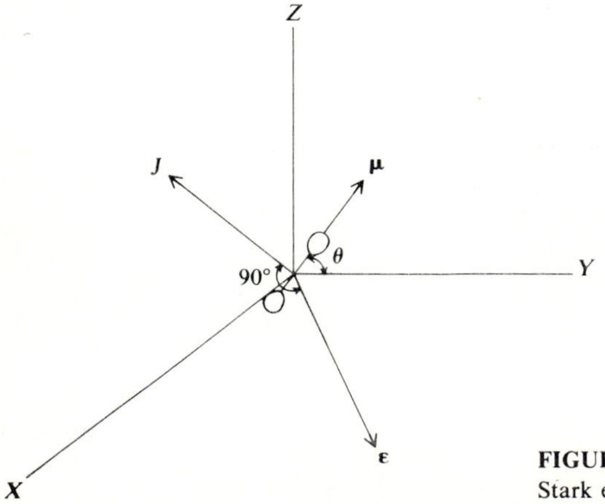

FIGURE 14.2
Stark effect relationships for a diatomic molecule.

an electric field oriented perpendicular to the angular momentum vector one notes that the interaction of the molecular dipole moment with the field will be dependent on its orientation. In this case the fractional difference between the time the dipole spends oriented away from and with the field will be proportional to the ratio of the electrostatic energy to the rotational energy [9]

$$F \propto \frac{\boldsymbol{\mu} \cdot \boldsymbol{\mathscr{E}}}{\frac{1}{2} I \omega^2} \tag{14.6}$$

The change in the energy of the molecule in the electric field will be the energy in the field times this fractional difference

$$\Delta E \propto F \boldsymbol{\mu} \cdot \boldsymbol{\mathscr{E}} \tag{14.7}$$

Using the quantum mechanical rotational energy

$$E_r = hBJ(J+1) \tag{14.8}$$

this becomes

$$\Delta E \propto \frac{(\boldsymbol{\mu} \cdot \boldsymbol{\mathscr{E}})^2}{hBJ(J+1)} \tag{14.9}$$

and we have a second-order Stark effect. Having taken this semiclassical qualitative look at the Stark effect we will now proceed to a rigorous quantum mechanical development of the effect.

The Hamiltonian for the rigid diatomic rotor in an external electric field is given by

$$\mathscr{H} = \mathscr{H}^0 + \mathscr{H}^{(1)} = \mathscr{H}^0 - \mu \mathscr{E} \cos \theta \tag{14.10}$$

For μ parallel to the molecule-fixed z axis and the electric field in the space-fixed Z direction $\cos \theta$ is the direction cosine, Φ_{Zz} and the matrix elements for the total energy of the system are

$$\begin{aligned} \langle J, K, M | \mathscr{H} | J', K', M' \rangle &= \langle J, K, M | \mathscr{H}^0 | J', K', M' \rangle \\ &\quad - \mu \mathscr{E} \langle J, K, M | \boldsymbol{\Phi}_{Zz} | J', K', M' \rangle \end{aligned} \tag{14.11}$$

The first term on the right is the unperturbed rigid rotor energy $E_J = hBJ(J+1)$. Since the direction cosine matrix elements are nonzero for $J' = J, J \pm 1$, one has a nondiagonal energy matrix, and perturbation methods are used to evaluate the energies of the system. Considering that the matrix elements of $\boldsymbol{\Phi}_{Zz}$ are diagonal in M, the energy of the perturbed rotor is

$$E_{JM} = E_j^0 + E_{JM}^{(1)} + E_{JM}^{(2)} = H_{JJ}^0 + H_{JJ}^{(1)} + \sum_i{}' \frac{H_{Ji}^{(1)} H_{iJ}^{(1)}}{E_{JJ}^0 - E_{ii}^0} J \tag{14.12}$$

where the H_{JJ}^0 terms are the energies of the unperturbed rotor states. The first-order correction to the energy is

$$H_{JJ}^{(1)} = -\mu \mathscr{E} \langle J, K, M | \boldsymbol{\Phi}_{Zz} | J, K, M \rangle \tag{14.13}$$

Referring to Table 3.2, this is evaluated to give

$$H_{JJ}^{(1)} = \frac{-\mu \mathscr{E} KM}{J(J+1)} \tag{14.14}$$

Note that this is the same result obtained classically in Eq. (14.5). For the linear molecule $K = 0$, since there is no component of angular momentum about the symmetry axis, and the first-order term vanishes.

The only nonzero terms in the second-order summation are those with $i = J$, $J \pm 1$, since the only nonzero direction matrix elements $(JKM|\Phi_{Fg}|J'K'M')$, are those with $J'' = J, J \pm 1$. Recalling that $H_{Ji} = H_{ij}^*$ and that the summation omits the $i = J$ term, the second-order term is written as

$$E_{JM}^{(2)} = \frac{(H_{J,J-1}^{(1)})^2}{E_J^0 - E_{J-1}^0} + \frac{(H_{J,J+1}^{(1)})^2}{E_J^0 - E_{J+1}^0}$$

$$= \frac{\mu^2 \mathscr{E}^2}{hB_v}\left[\frac{\langle J, K, M|\Phi_{Zz}|J-1, K, M\rangle^2}{2J} - \frac{\langle J, K, M|\Phi_{Zz}|J+1, K, M\rangle}{2(J+1)}\right] \tag{14.15}$$

Again employing Table 3.2 gives

$$E_{JM}^{(2)} = \frac{\mu^2 \mathscr{E}^2}{2hB_v}\left\{\frac{(J^2 - K^2)(J^2 - M^2)}{J^3(2J-1)(2J+1)} - \frac{[(J+1)^2 - K^2][(J+1)^2 - M^2]}{(J+1)^3(2J+1)(2J+3)}\right\} \tag{14.16}$$

for the diatomic rotor $K = 0$, and this expression reduces to

$$E_{JM}^{(2)} = \frac{\mu^2 \mathscr{E}^2}{2hB_v} \frac{J(J+1) - 3M^2}{J(J+1)(2J-1)(2J+3)} \tag{14.17}$$

Since $M = 0, \pm 1, \pm 2, \ldots, \pm J$ and it occurs only as a square in the energy expression, the effect of an external electric field is to split each energy level into a set of $J + 1$ levels. Were the first-order effect operative for the diatomic molecule, then the degeneracy of the levels would be completely removed giving $2J + 1$ separate levels. The field split energy levels for a diatomic molecule are shown in Fig. 14.3.

14.3 START EFFECT: SELECTION RULES AND INTENSITIES

The selection rules for both the J and M quantum numbers were developed during the derivation of the direction cosine matrix elements and are obtained by looking at the quantum numbers associated with the nonzero direction cosine elements. In Chap. 11 it was found that the determination of the dipole moment

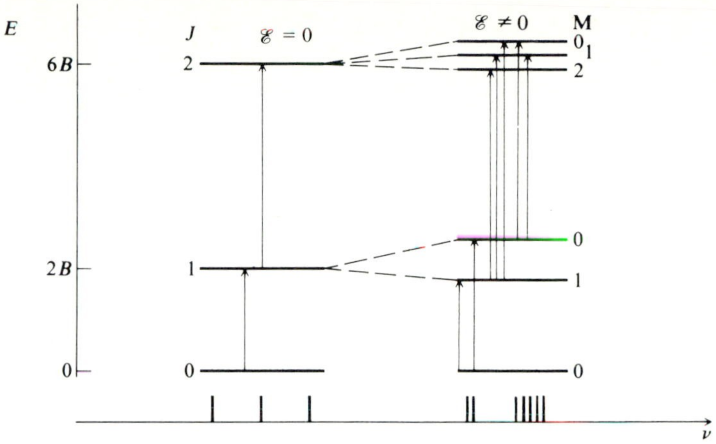

FIGURE 14.3
Stark effect for a diatomic molecule (not scaled) (only $\Delta M = 0$ transitions illustrated).

matrix element for a transition $J' \leftarrow J$ involved a sum over the states degenerate in M. The dipole moment matrix elements for transitions between states of different M values are just the individual members of those summations. They will be of the form

$$\langle J, M|\boldsymbol{\mu}|J', M'\rangle = \langle J, M|\mu_0 \cos\theta|J', M'\rangle = \mu_0\langle J, M|\boldsymbol{\Phi}_{Zz}|J', M'\rangle \quad (14.18)$$

and are directly related to the direction cosine matrix elements. The occurrence of nonzero matrix elements for $J' = J$, $J \pm 1$ and $M' = M$, $M \pm 1$ give as the selection rules $\Delta J = 0, \pm 1$ and $\Delta M = 0, \pm 1$. However, since $K = 0$ for a diatomic rotor $\Delta J = 0$ transitions are not allowed. The nature of the spectrum for the $\Delta M = 0$ transitions, not allowing for differences in intensities of components, is illustrated in Fig. 14.3.

Using the selection rules it is possible to derive explicit expressions for particular transitions.

In general there are $3J + 2$ transitions originating from a particular J level, but all of these are not normally observed in a single experiment. The reduction in the number actually observed is due to the fact that the operation of the selection rules is subject to the relative orientation of the static electric field which gives rise to the Stark effect, and the electric vector of the radiation field which is interacting with the sample and giving rise to the absorption. Considering a molecule as a classical electrostatic dipole enables one to understand the reason for this phenomenon. In the presence of a static electric field, the dipole will precess about the field direction, the Z axis, with a component in the field direction. This component is proportional to the projection of **J** on the Z axis, given by M. For a large number of dipoles, they will be distributed among the possible M values, for a given J, in a manner proportional to $(1 - E_{JM}/kT)$. If the radiation field is applied with its electric vector parallel to the Z axis, it is not possible for it to exert a torque on the resultant dipoles, hence energy can be absorbed only by increasing the value of J without changing M. For this geometry $\Delta M = 0$ and the number of observed Stark components for a transition of $J + 1 \leftarrow J$ is equal to J. For this experimental arrangement the displacements of the Stark components from the zero field transition $2B_v(J + 1)$ are given by

$$\Delta\nu_{J+1\leftarrow J}^{\Delta M=0} = \frac{E_{J+1,M}^{(2)}}{h} - \frac{E_{J,M}^{(2)}}{h}$$

$$= \frac{\mu^2\mathscr{E}^2}{h^2B_v}\left[\frac{3M^2(8J + 16J + 5) - 4J(J + 1)^2(J + 2)}{J(J + 1)(J + 2)(2J - 1)(2J + 1)(2J + 3)(2J + 5)}\right] \quad (14.19)$$

If the Stark and radiation fields are perpendicular then a torque will always be exerted on the molecule by the radiation field and M will change in accordance with the selection rule $\Delta M = \pm 1$.

Both the number and relative intensity of the Stark components are useful criteria for identification of the rotational quantum number associated with a transition. The relative intensities can be found by consideration of the absorption coefficient for the diatomic rotor as derived in Chap. 11. Ignoring the shape factor and considering the case where $h\nu/kT < 1$

$$\alpha(\nu_{J+1,M'\leftarrow J,M}) = \frac{8\pi^2\nu_{J+1,M'\leftarrow J,M}N_{JM}}{3ckT\Delta\nu}\langle J, K, M|\boldsymbol{\mu}|J + 1, K, M'\rangle^2 \quad (14.20)$$

We can greatly simplify the calculation of relative intensities by first considering the magnitude of the quantities involved. Looking at two transitions $J + 1$, $M + 1 \leftarrow J$, M, and $J + 1$, $M + 2 \leftarrow J$, $M + 1$ the transition frequencies are found using Eq. (14.17) to be

$$\nu_{J+1,M+1\leftarrow J,M} = 2B_v(J + 1) + \frac{\mu^2\mathscr{E}^2}{B_vh^2}[M^2f(J) - f'(J)] \quad (14.21)$$

$$\nu_{J+1,M+2\leftarrow J,M+1} = 2B_v(J + 1) + \frac{\mu^2\mathscr{E}^2}{B_vh^2}[(M + 1)^2f(J) - f'(J)] \quad (14.22)$$

The second term in each of these expressions is small compared to the first, and the second terms differ only slightly so to a good approximation $\nu_{J+1,M+1\leftarrow J,M} \approx \nu_{J+1,M+2\leftarrow J,M+1}$. Furthermore the ratio of the populations of the lower states of these two transitions will be given by:

$$N_{J,M+1} = N_{J,M}e^{-(E_{J,M+1}-E_{J,M})/kT} \quad (14.23)$$

Using Eq. (14.17), the precise value of the ratio can be calculated; however, without any numerical evaluation it is observed that the ratio is approximately unity since $\mu^2\mathscr{E}^2/hB_v$ will be small and we will be taking the difference $[3(M + 1)^2 - 3M^2]/f''(J)$ where $f''(J)$ will be larger than $3(M + 1)^2 - 3M^2$. Using these two approximations, the ratio of the absorption coefficients of two transitions reduces to

$$R_{M+1,M} = \frac{\alpha(\nu_{J+1,M+1\leftarrow J,M})}{\alpha(\nu_{J+1,M+2\leftarrow J,M+1})} = \frac{\langle J, K, M|\boldsymbol{\mu}|J+1, K, M+1\rangle^2}{\langle J, K, M+1|\boldsymbol{\mu}|J+1, K, M+2\rangle^2} \tag{14.24}$$

or to a good approximation of the relative intensities of the Stark components are obtained directly from the ratios of the direction cosine matrix elements $\langle J, M|\Phi_{Zz}|J'M'\rangle$.

14.4 STARK EFFECT: HIGH FIELDS

When one observes Stark transitions using large static fields such that $\mu\mathscr{E} \approx h\nu$ then simple first- or second-order perturbation theory is insufficient to accurately describe the observed spectra. If a first-order Stark effect is exhibited then the addition of the second-order correction will generally suffice; however, for linear molecules, the first-order effect is absent and terms higher than second-order must be used to describe the system. The development of higher-order perturbation terms is a straightforward, but algebraically lengthy, extension of the procedure given in Sec. 3.5. The inclusion of higher-order terms in Eqs. (3.94), (3.95), and (3.97) will allow one to obtain additional relations like Eqs. (3.101) to (3.103). From these the higher-order terms can be derived. Since there is little in the way of new concepts to be gained by the algebraic manipulations leading to higher-order perturbation terms, it will suffice at this point to state the results and give important references [1, 2]. For a linear molecule, having all of the $H_{JJ}^{(1)}$ terms equal to zero, the third-order perturbation term vanishes and the fourth is the next term to contribute [2]

$$\begin{aligned} E_{JM}^{(4)} = \frac{\mu^4\mathscr{E}^4}{8h^3B^3}\Bigg\{ & \frac{[(J-1)^2-M^2][J^2-M^2]}{(2J-3)(2J+1)J^2(2J-1)^3} \\ & - \left[\frac{[(J+1)^2-M^2][(J+2)^2-M^2]}{(2J+1)(2J+5)(J+1)^2(2J+3)^3}\right] \\ & \times \left[\frac{[J^2-M^2][(J+1)^2-M^2]}{(2J-1)(2J+3)(2J+1)^2(J+1)^2J^2}\right] \\ & + \frac{[(J+1)^2-M^2]^2}{(2J+3)^2(2J+1)^2(J+1)^3} \\ & - \frac{[J^2-M^2]^2}{(2J+1)^2(2J-1)^2J^3}\Bigg\} \end{aligned} \tag{14.25}$$

A look at the magnitudes of the terms involved allows some insight into the degree of experimental refinement necessary for the analysis of a particular molecular system. Table 14.1 summarizes the magnitudes of the various terms involved. These terms are multiplied by factors which are functions of J and M. All of these factors are of a form such that they never exceed unity, hence the

TABLE 14.1
Magnitudes of Stark parameters
($\nu_0 = 25$ GHz)

	$\mu\mathscr{E}$, MHz	$\frac{\mu^2\mathscr{E}^2}{\nu_0}$, MHz	$\frac{\mu^4\mathscr{E}^4}{\nu_0^3}$, MHz
$\mathscr{E} = 500$, V/cm^{-1}			
$\mu = 0.1$ D	25	0.03	0.00000004
0.5 D	126	0.64	0.00002
1.0 D	253	2.56	0.0003
3.0 D	760	23.10	0.02
$\mathscr{E} = 1000$, V cm^{-1}			
$\mu = 0.1$ D	51	0.10	0.0000004
0.5 D	253	2.56	0.0003
1.0 D	507	10.28	0.004
3.0 D	1,520	92.42	0.34
$\mathscr{E} = 5000$, V cm^{-1}			
$\mu = 0.1$ D	253	2.56	0.0003
0.5 D	1,266	14.11	0.16
1.0 D	2,533	256.64	2.63
3.0 D	7,598	2309.18	213.29
$\mathscr{E} = 10{,}000$, V/cm^{-1}			
$\mu = 0.1$ D	507	10.28	0.004
0.5 D	2,533	256.64	2.63
1.0 D	5,065	1026.17	42.12
3.0 D	15,195	9235.52	3411.79

terms in Table 14.1 represent maximum changes in energy levels. It is observed then that at high fields and for large dipole moments the higher-order terms can become important. Their importance is minimized somewhat for $\Delta M = 0$ transitions because one is taking the difference between two J, M dependent factors. For example, comparing $E^{(2)}_{5,5}$, $\Delta\nu^{(2)M=1}_{6\leftarrow5}$ and $\Delta\nu^{(4)M=1}_{6\leftarrow5}$ for a linear molecule by substituting $J = 5$ and $M = 1$ into Eqs. (14.17), (14.19), and (14.25) respectively, one has

$$E^{(2)}_{5,1} \approx \left(\frac{1}{130}\right)\frac{\mu^2\mathscr{E}^2}{\nu_0}$$

$$\Delta\nu^{(2)M=1}_{5\leftarrow5} \approx \left(\frac{-1}{253}\right)\frac{\mu^2\mathscr{E}^2}{\nu_0}$$

$$\Delta\nu^{(4)M=1}_{6\leftarrow5} \approx \left(\frac{1}{4820}\right)\frac{\mu^4\mathscr{E}^4}{\nu_0^3}$$

One observes that the fourth-order term will lead to changes of 0.1 MHz or greater for $\mu > 0.5$ D and $\mathscr{E} > 5000$ V cm^{-1}. An alternate treatment, and one which has generally replaced this perturbation method, is to take advantage of modern high-speed computers to numerically solve the energy matrices. However, Eq. (14.25) does provide a rapid method to obtain orders of magnitude calculations. The last section will present an example of the Stark effect.

14.5 NUCLEAR HYPERFINE INTERACTIONS IN ISOLATED MOLECULES—THEORY

In Chap. 10 we developed the theory and discussion applications of the interaction of nuclear quadrupole moments and electric fields in rigid lattic systems. These concepts will now be extended to the interactions which occur in an isolated molecule undergoing quantized rotation. The Hamiltonian in this case has the same form as before, which in terms of a Cartesian axis system is given by

$$\mathscr{H}_Q = -\frac{1}{6}\underline{\mathbf{Q}} \cdot \underline{\nabla\mathbf{E}} = \frac{1}{6}\sum_i \sum_j Q_{ij} \nabla E_{ij} = \frac{1}{6}\sum_i \sum_j Q_{ij} V_{ij} \tag{14.26}$$

The components of the quadrupole moment tensor will be the same as those in a solid

$$Q_{ij} = \frac{eQ}{I(2I-1)}\left[\frac{3}{2}(\mathbf{I}_i\mathbf{I}_j + \mathbf{I}_j\mathbf{I}_i) - \delta_{ij}\mathbf{I}^2\right] \tag{14.27}$$

but now the charge distribution which contributes to $\underline{\nabla\mathbf{E}}$ will be that due to the molecular rotation rather than that due to a rigid lattice. The procedure for establishing the form of the $\underline{\nabla\mathbf{E}}$ components is analogous to that used to derive the $\underline{\mathbf{Q}}$ tensor elements in Sec. 10.2. As with $\underline{\mathbf{Q}}$, the $\underline{\nabla\mathbf{E}}$ elements can all be related to a single scalar quantity and an appropriate set of quantum numbers.

The basic defining relationship for the $\underline{\nabla\mathbf{E}}$ components is

$$V_{ij} = \int_{\text{elect}} \frac{\rho(xyz)}{r^5}[3x_ix_j - \delta_{ij}r^2]\,dV \tag{14.28}$$

where the x_i are electronic coordinates and $\rho(x, y, z)$ is the electronic charge distribution. Since the components of $\mathbf{r}$ commute, $\underline{\nabla\mathbf{E}}$ is a symmetric tensor. The Laplacian condition requires the trace to be zero, hence $\underline{\nabla\mathbf{E}}$ is a second-rank symmetric traceless tensor. It is to be remembered at this point that the coordinate system is collinear with the one which describes the components of the $\underline{\mathbf{Q}}$ tensor. The dependency of $\underline{\nabla\mathbf{E}}$ on the rotation of the molecule is expressed in the form of a set of matrix elements of the form $\langle J', M'|\nabla\mathbf{E}|J, M\rangle$ or for the individual components, $-\langle J', M'|V_{ij}|J, M\rangle$, where J and M are the rotational quantum numbers. It is assumed that the rotation is sufficiently fast, that the nucleus experiences an average electronic field, hence J is a good quantum number. Employing the theorem discussed in App. M gives

$$\langle J', M'|\mathbf{V}_{ij}|J, M\rangle = C\langle J', M'|\tfrac{3}{2}(\mathbf{J}_i\mathbf{J}_j + \mathbf{J}_j - \mathbf{J}_i) - \delta_{ij}\mathbf{J}^2|J, M\rangle \tag{14.29}$$

where C is a proportionality constant.

Taking the Z axis to be a space-fixed axis, one can define a scalar electric field gradient q_J as the expectation value for V_{ZZ} in the state $J = M$. In this state the component of J along Z is a maximum, thus

$$q_J = \frac{1}{e} \langle J, J | \mathbf{V}_{\mathbf{ZZ}} | J, J \rangle \tag{14.30}$$

Considering that the electronic distribution in the molecule is described by a total electronic wavefunction ψ then

$$eq_J = \int_{\text{elect}} \rho(xyz) \left(\frac{3z^2 - r^2}{r^5} \right) dV = \int_{\text{elect}} \psi^* \left(\frac{3 \cos^2 \theta - 1}{r^3} \right) \psi \, dv \tag{14.31}$$

where θ is the polar angle between the spaced-fixed Z axis and the radius vector r which locates an electron. Incorporating Eq. (14.29) gives

$$eq_J = \langle J, J | \mathbf{V}_{\mathbf{ZZ}} | J, J \rangle = C \langle J, J | \tfrac{3}{2}(\mathbf{J_Z J_Z} + \mathbf{J_Z J_Z}) - \mathbf{J}^2 | J, J \rangle \tag{14.32}$$

Using the angular momentum operators from Table 3.1 this reduces to

$$eq_J = CJ(2J - 1) \tag{14.33}$$

Therefore,

$$V_{ij} = \frac{eq_J}{J(2J - 1)} \left[\frac{3}{2} (\mathbf{J}_i \mathbf{J}_j + \mathbf{J}_j \mathbf{J}_i) - \delta_{ij} \mathbf{J}^2 \right] \tag{14.34}$$

Using Eqs. (10.50), (10.52), and (10.55) which define the irreducible *EFG* tensor components, and introducing the above expansion for V_{ij}, they can be written

$$(\nabla E)_2^0 = \frac{1}{2} \frac{eq_J}{J(2J - 1)} [3\mathbf{J}_z^2 - \mathbf{J}^2] \tag{14.35}$$

$$(\nabla E)_2^{\pm 1} = \mp \frac{3}{2\sqrt{6}} \frac{eq_J}{J(2J - 1)} [\mathbf{J}_z \mathbf{J}_\pm + \mathbf{J}_\pm \mathbf{J}_z] \tag{14.36}$$

$$(\nabla E)_2^{\pm 2} = \frac{3}{2\sqrt{6}} \frac{eq_J}{J(2J - 1)} [\mathbf{J}_\pm^2] \tag{14.37}$$

By using either summation

$$\mathcal{H}_Q = \frac{1}{6} \sum_i \sum_j Q_{ij} V_{ij} = \sum_{-m}^{m} Q_2^{\pm m} (\nabla E)_2^{\pm m} \tag{14.38}$$

the Hamiltonian may be written

$$\mathcal{H}_Q = \frac{1}{6} \frac{e^2 Q q_J}{I(2I - 1)J(2J - 1)} \sum_i \sum \left\{ \left[\frac{3}{2} (\mathbf{I}_i \mathbf{I}_j + \mathbf{I}_j \mathbf{I}_i) - \delta_{ij} \mathbf{I}^2 \right] \times \left[\frac{3}{2} (\mathbf{J}_i \mathbf{J}_j + \mathbf{J}_j \mathbf{J}_i) - \delta_{ij} \mathbf{J}^2 \right] \right\} \tag{14.39}$$

This relation can be transformed to an alternate form which is more convenient and embodies the vector properties of **I** and **J** (that is, $\mathbf{I} = I_X\hat{\mathbf{i}} + \mathbf{I}_y\hat{\mathbf{j}} + \mathbf{I}_Z\hat{\mathbf{k}}$; $\mathbf{J} = J_X\hat{\mathbf{i}} + J_Y\hat{\mathbf{j}} + J_Z\hat{\mathbf{k}}$). Since **I** and **J** commute, that is, the nuclear angular momentum and rotation angular momentum can both be simultaneously determined, it follows that

$$\sum_i\sum_j \mathbf{I}_i\mathbf{I}_j\mathbf{J}_i\mathbf{J}_j = \left(\sum_i \mathbf{I}_i\mathbf{J}_i\right)\left(\sum_j \mathbf{I}_j\mathbf{J}_j\right) = (\mathbf{I}\cdot\mathbf{J})^2 \tag{14.40}$$

$$\sum_i\sum_j \mathbf{I}_i\mathbf{I}_j\delta_{ij}\mathbf{J}^2 = \left(\sum_i \mathbf{I}_i^2\right)\mathbf{J}^2 = \mathbf{I}^2\mathbf{J}^2 \tag{14.41}$$

$$\sum_i\sum_j \delta_{ij}\mathbf{I}^2\mathbf{J}^2 = 3\mathbf{I}^3\mathbf{J}^2 \tag{14.42}$$

$$\sum_i\sum_j \mathbf{I}_j\mathbf{I}_i\mathbf{J}_i\mathbf{J}_j = \sum_i\sum_j \mathbf{I}_i\mathbf{I}_j\mathbf{J}_j\mathbf{J}_i = (\mathbf{I}\cdot\mathbf{J})^2 + \mathbf{I}\cdot\mathbf{J} \tag{14.43}$$

Using these relationships, the Hamiltonian becomes

$$\mathscr{H}_Q = \frac{1}{2}\frac{e^2Qq_J}{I(2I-1)J(2J-1)}\left[3(\mathbf{I}\cdot\mathbf{J})^2 + \frac{3}{2}(\mathbf{I}\cdot\mathbf{J}) - \mathbf{I}^2\mathbf{J}^2\right] \tag{14.44}$$

If a molecule has more than a single quadrupolar nucleus present then the total Hamiltonian is a sum over all nuclei

$$\mathscr{H}_Q = \frac{1}{2}\sum_k e^2Q_kq_J\left[\frac{3(\mathbf{I}^k\cdot\mathbf{J})^2 + \frac{3}{2}(\mathbf{I}^k\cdot\mathbf{J}) - (\mathbf{I}^k)^2\mathbf{J}^2}{I^k(2I^k-1)J(2J-1)}\right] \tag{14.45}$$

Note that the index k, which denotes a particular nucleus, is written as a superscript on **I** to prevent confusion with components of **I**.

To complete this section we will now use Eq. (14.44) for a single quadrupolar nucleus, and derive the Hamiltonian matrix elements. The total Hamiltonian for a rotating molecule can be written $H = H_R + H_Q$, where H_R is the rotational Hamiltonian. The magnitude of H_Q will be of the order of e^2Qq_J which, for the majority of elements of interest, lies below 500 MHz. The magnitude of H_R will generally be in the range of 3-500 GHz. So the quadrupole interaction can be considered as a perturbation. Accordingly the first-order perturbation energy matrix elements will be

$$\langle J, I|\mathscr{H}_Q|J, I\rangle = \langle J, I|e^2Qq_J\left[\frac{3(\mathbf{I}\cdot\mathbf{J})^2 + \frac{3}{2}(\mathbf{I}\cdot\mathbf{J}) - \mathbf{I}^2\mathbf{J}^2}{2I(2I-1)J(2J-1)}\right]|J, I\rangle \tag{14.46}$$

Since I and J are both good quantum numbers, the nuclear and angular momenta will vector-couple to give a resultant total angular momentum $\mathbf{F} = \mathbf{I} + \mathbf{J}$, where $F = J + I, J + I - 1, \ldots, |J - I|$. This is illustrated for $J = 2$, $I = \frac{3}{2}$ in Fig. 14.4.

Evaluating the scalar product $\mathbf{I}\cdot\mathbf{J}$ in terms of the quantum numbers J, I, and F allows Eq. (14.46) to be reduced to a simpler form where q_J will be the only term which is explicitly dependent on the nature of the molecule involved. Since **F** is the total angular momentum of the system, it will have eigenstates

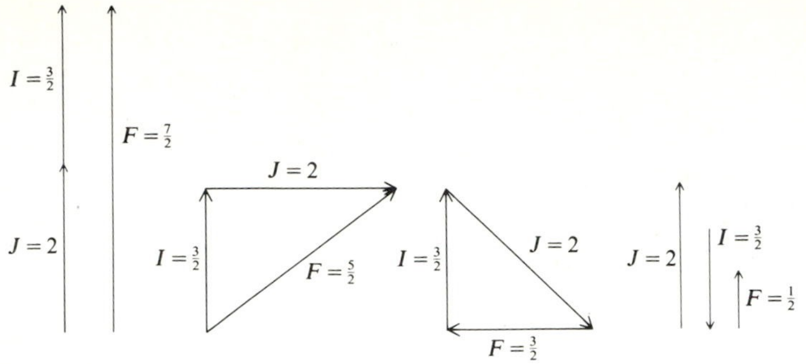

FIGURE 14.4
Vector coupling of **I** and **J** for $I = \frac{3}{2}$, $J = 2$.

given by

$$\mathbf{F}^2\psi_F = F(F+1)\hbar^2\psi_F \tag{14.47}$$

where ψ_F is the total angular momentum wavefunction of the system.

The eigenvalues of $\mathbf{I}\cdot\mathbf{J}$ are evaluated by considering that

$$(\mathbf{I}+\mathbf{J})\cdot(\mathbf{I}+\mathbf{J}) = (\mathbf{I}+\mathbf{J})^2 = \mathbf{F}^2 = \mathbf{I}^2 + 2\mathbf{I}\cdot\mathbf{J} + \mathbf{J}^2 \tag{14.48}$$

or

$$\mathbf{I}\cdot\mathbf{J} = \tfrac{1}{2}(\mathbf{F}^2 - \mathbf{I}^2 - \mathbf{J}^2) \tag{14.49}$$

and the eigenvalues will be $(\frac{1}{2})[F(F+1) - I(I+1) - J(J+1)]$. The matrix elements for H_Q are then

$$\langle J, I, F|\mathscr{H}_Q|J, I, F\rangle = e^2Qq_J\frac{\frac{3}{4}C(C+1) - I(I+1)J(J+1)}{2I(2I-1)J(2J-1)} \tag{14.50}$$

where $C = F(F+1) - I(I+1) - J(J+1)$.

The electric field gradient tensor component employed in Eq. (14.50) is actually an average value taken with respect to a space-fixed Z direction

$$q_J = \left(\frac{\partial^2 V}{\partial Z^2}\right)_{\text{av}} \tag{14.51}$$

When discussing the properties of an individual molecule, it is more convenient to use a component relative to some molecular axis system. The transformation of the EFG tensor components from a space-fixed axis system, $F = X, Y, Z$ to a molecule-fixed axis system by $q = x, y, z$ is given by

$$\left(\frac{\partial^2 V}{\partial F^2}\right)_{\text{av}} = (\Phi_{Fx}^2 q_{xx} + \Phi_{Fy}^2 q_{yy} + \Phi_{Fz}^2 q_{zz} + 2\Phi_{Fx}\Phi_{Fz}q_{xy} + 2\Phi_{Fx}\Phi_{Fz}q_{xz}$$
$$+ 2\Phi_{Fy}\Phi_{Fz}q_{yz})_{\text{av}} \tag{14.52}$$

where the Φ_{Fg}'s are the direction cosines. If the z axis of the molecular axis system is the direction of the electronic angular momentum, then x, y, z will be the EFG tensor principal axis system, and the last three terms in this equation vanish. Looking back at Eq. (14.31) it is noted that $q_J \equiv q_{ZZ}$ is the average of $q_{\rm op} = (3\cos^2\theta - 1)/r^3$ over the rotational and electronic wavefunctions. The components q_{gg}, however, are just the average values of $q_{\rm op}$ over the electronic wavefunctions so that the rotational averaging is given by the occurrence of the Φ_{Fg}'s. If θ is the angle between the Z axis and the z axis, and the x axis is taken to be collinear with the X axis then

$$q_J = q_{ZZ} = q_{zz}\cos^2\theta_{Zz} + q_{yy}\sin^2\theta_{Zz} \tag{14.53}$$

The use of Laplace's equation $q_{xx} + q_{yy} + q_{zz} = 0$, and the assumption of cylindrical symmetry of the charge distribution $q_{xx} = q_{yy}$, simplifies this expression further. It should be recognized, however, that this step does involve an approximation. The use of Laplace's equation assumes that the charge distribution giving rise to $\underline{\nabla \mathbf{E}}$ lies outside of the nucleus. This condition is not strictly met because of the finite probability of extranuclear electrons being found inside the nucleus. This probability is the largest, however, for electrons in s orbitals and, due to the spherical distribution of charge in such orbitals, they will not provide any contribution to the EFG tensor. The probability for the occurrence of p, d, and f electrons within the nucleus is sufficiently small it can be neglected. Thus $q_{xx} = q_{yy} = -\frac{1}{2}q_{zz}$ and

$$q_J = \left(\frac{\partial^2 V}{\partial Z^2}\right)_{\rm av} = \left(\frac{\partial^2 V}{\partial z^2}\right)\left(\frac{3\cos^2\theta - 1}{2}\right)_{\rm av} \tag{14.54}$$

The matrix elements for the quadrupolar interaction in rotating molecules will then be given by

$$(J, I, F|\mathscr{H}_Q|J, I, F) = e^2Qq_{zz}\left(\frac{3\cos^2\theta - 1}{2}\right)_{\rm av}\left[\frac{\frac{3}{4}C(C+1) - I(I+1)J(J+1)}{2I(2I-1)J(2J-1)}\right] \tag{14.55}$$

where

$$q_{zz} = \left(\frac{\partial^2 V}{\partial z^2}\right) = \int_{\rm elect} \psi_{\rm el}^*\left(\frac{3\cos^2\theta - 1}{r^3}\right)\psi_{\rm el}\,dv \tag{14.56}$$

$\psi_{\rm el}$ is the electronic wavefunction of the molecule, and θ is the angle between the molecular z axis and the vector locating an electron relative to the z axis.

$$\left(\frac{3\cos^2\theta - 1}{2}\right)_{\rm av} = \int_{\rm rot} \psi_r^*\left(\frac{3\cos^2\theta - 1}{2}\right)\psi_r\,dv \tag{14.57}$$

and ψ_r is the rotational wavefunction of the molecule. Thus, the background for the discussion of hyperfine interactions in various types of molecules has been formulated. The process used to evaluate $[(3\cos^2\theta - 1)/2]_{\rm av}$ will be discussed in the next section.

14.6 NUCLEAR QUADRUPOLE HYPERFINE INTERACTIONS IN DIATOMIC MOLECULES

Since Eqs. (14.54) to (14.57) were derived using the condition of cylindrical symmetry for the electronic field of the molecule we have only to evaluate the integral in Eq. (14.57) using the diatomic rotor rotational wavefunctions to obtain the orientation-dependent part of the Hamiltonian matrix elements, Eq. (14.55). The e^2Qq_{zz} term is a molecular property which is independent of orientation so the entire orientation-dependence is embodied in the $[(3\cos^2\theta - 1)/2]_{\text{av}}$ term.

The orientation-dependent term may be evaluated in two ways. The first method is to use the conventional average value technique of quantum mechanics,

$$\left(\frac{3\cos^2\theta - 1}{2}\right)_{\text{av}} = \frac{1}{2}\int_0^{\pi}\int_0^{2\pi} \psi_{JJ}^*(3\cos^2\theta - 1)\psi_{JJ}\sin\theta\, d\theta\, d\phi \qquad (14.58)$$

where ψ_{JJ} is the rotational wavefunction for the state $M = J$ and was found previously to be given by

$$\psi_{JJ} = \left[\frac{(2J+1)}{4\pi(2J)!}\right]^{1/2} P_J^J(\cos\theta)e^{iJ\phi} \qquad (14.59)$$

Substitution of Eq. (14.59) into Eq. (14.58) with subsequent integration gives

$$\left(\frac{3\cos^2\theta - 1}{2}\right)_{\text{av}} = \frac{-J}{2J+3} \qquad (14.60)$$

A second and simpler method is to employ the direction cosine matrix elements. The average value will be given by

$$\left(\frac{3\cos^2\theta - 1}{2}\right)_{\text{av}} = \frac{3\langle J, K, M|\boldsymbol{\Phi}_{Zz}|J', K', M'\rangle^2 - 1}{2} \qquad (14.61)$$

Keeping in mind that for the diatomic molecule, $K = 0$ and only $M = J$ elements are used,

$$\left(\frac{3\cos^2\theta - 1}{2}\right)_{\text{av}} = \frac{\left[3\sum_{J'}\langle J, 0, J|\boldsymbol{\Phi}_{Zz}|J', 0, J'\rangle\langle J', 0, J'|\Phi_{Zz}|J, 0, J\rangle - 1\right]}{2} = \frac{-J}{2J+3} \qquad (14.62)$$

For the diatomic rotor the quadrupolar energy term becomes

$$E_{JIF}^Q = \langle J, I, F|\mathscr{H}_Q|J, I, F\rangle = e^2Qq_{zz}f(I, J, F) \qquad (14.63)$$

where

$$f(I, J, F) = \frac{\frac{3}{4}C(C+1) - J(J+1)I(I+1)}{I(2I-1)(2J-1)(2J+3)} \qquad (14.64)$$

and $C = F(F+1) - I(I+1) - J(J+1)$.

The selection rules for J were previously found to be $\Delta J = 0, \pm 1$. Since I denotes the nuclear spin and the nucleus does not interact with the radiation field, $\Delta I = 0$. Therefore, since **J** and **I** couple to give **F**, the selection rules for the latter must be $\Delta F = 0, \pm 1$.

The effect of nuclear quadrupole interactions on the rotational spectrum of a diatomic molecule will be illustrated by using a hypothetical molecule MX, where the nuclear spins of M and X are 0 and $\frac{3}{2}$, respectively. For $I = \frac{3}{2}$ and $J \geqq 2$ there will be four values of F: $(J + \frac{3}{2})$, $(J + \frac{1}{2})$, $(J - \frac{1}{2})$, $(J - \frac{3}{2})$. For a given value of J there will be four different values of $f\ (J, I, F)$, corresponding to the four values of F. The net result is to split the rotational state into a set of four closely spaced levels. The spacing will be of the order of e^2Qq_{zz}, which varies from 0 to 1 GHz for the majority of quadrupolar nuclei. Except for a few nuclei with very large coupling constants, ^{127}I, ^{79}Br, and ^{81}Br, for example, most values will lie below 100 MHz. Since rotational energy levels generally exceed 1 GHz, this shows that the effect can indeed be treated as a perturbation. The nuclear quadrupole splitting of the pair of energy levels with rotational quantum numbers J and $J + 1$ is illustrated in Fig. 14.5.

Employing the selection rules $\Delta F = 0, \pm 1$ leads to the prediction of the nine transitions shown in Fig. 14.5. The net effect is to split the single $J + 1 \leftarrow J$ transition into a closely spaced set of lines called hyperfine components. The total intensity of the nine transitions will be equal to the intensity of a comparable transition of a molecule having the same dipole moment and no quadrupolar nucleus. The relative intensities of the hyperfine components are of interest because, coupled with the splittings, they can be an aid in the assignment of the correct quantum numbers of a transition.

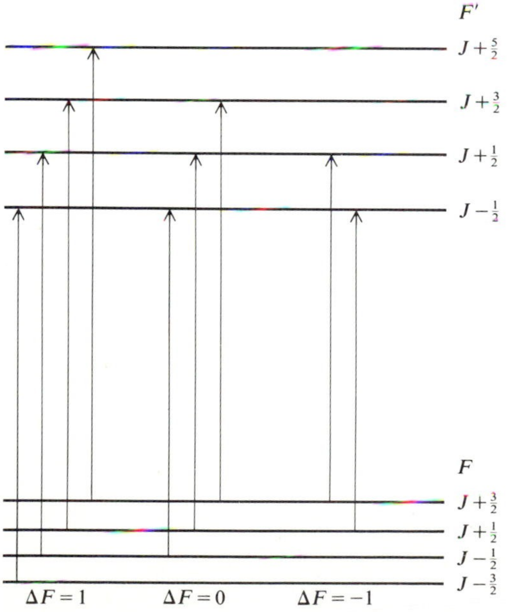

FIGURE 14.5
Energy level scheme and transitions for a diatomic molecule with a single quadrupole nucleus with $I = \frac{3}{2}$.

TABLE 14.2
Relative intensities of hyperfine components

ΔJ	ΔF	I_{rel}
+1	+1	$\frac{A(F)\cdot A(F-1)}{F}$
+1	−1	$\frac{B(F)\cdot B(F-1)}{F}$
+1	0	$\frac{(2F+1)[A(F)\cdot B(F)]}{F(F+1)}$
0	+1	$\frac{A(F)\cdot B(F-1)}{F}$
0	−1	$\frac{A(F)\cdot B(F-1)}{F}$
0	0	$\frac{(2F+1)[D(F)]^2}{F(F+1)}$
−1	+1	$\frac{B(F)\cdot B(F-1)}{F}$
−1	−1	$\frac{A(F)\cdot A(F-1)}{F}$
−1	0	$\frac{(2F+1)[A(F)\cdot B(F)]}{F(F+1)}$

$A(F) = (F+J)(F+J+1) - I(I+1)$
$B(F) = (I+1)I - (F-J)(F-J+1)$
$D(F) = F(F+1) + J(J+1) - I(I+1)$
J, F = quantum numbers of lower state

TABLE 14.3
Selected examples of relative intensities of hyperfine components

			I_{rel}				
			$J+1 \leftarrow J$			$J \leftarrow J$	
I	J	F	$F+1 \leftarrow F$	$F \leftarrow F$	$F-1 \leftarrow F$	$F \leftarrow F+1$ $F+1 \leftarrow F$	$F \leftarrow F$
1	1	2	46.6	8.33	0.56	....	41.7
		1	25.0	8.33	.....	13.9	8.3
		0	11.1	....	.....	11.1	0
1	5	6	38.5	0.93	0.006	....	38.3
		5	32.4	0.93	.....	1.09	31.1
		4	27.3	....	.....	1.09	26.2
1	10	11	36.2	0.28	0.001	....	36.2
		10	33.1	0.28	.....	0.3	32.7
		9	30.2	....	.....	0.3	29.9

A convenient summary of the relative intensities of the hyperfine components is given in Table 14.2. A tabulation of these values is found in several monographs on atomic spectra, such as that by Condon and Shortley [3]. Looking at such a listing of values or at the general formulations in Table 14.2, a fortunate simplification emerges. The bulk of the intensity is concentrated in only a part of the components and as J increases this number decreases. This is illustrated by the data for the spin $I = 1$ case shown in Table 14.3. Thus, for example, a $J = 6 \leftarrow 5$ transition would have three strong and three very weak components.

14.7 MAGNETIC INTERACTIONS IN PARAMAGNETIC MOLECULES

Although the majority of diatomic molecules have $^1\Sigma$ ground states and exhibit only weak magnetic interactions, there are a few stable molecules such as O_2 and NO and free radical species like OH, ClO, and CH which exhibit strong couplings between the angular momenta due to rotation and that due to the orbital motion of the unpaired electron. The recent emergence [4] of rotational spectroscopy as a technique for investigating these species warrants a brief summary of this topic.

There are a variety of ways in which the electronic, spin, and rotational angular momenta in a molecule can couple. In fact, it is seldom that the coupling is exactly the same in any two molecules. This problem is handled by the use of several ideal types of couplings known as Hund's cases. A specific molecule is then considered in terms of the ideal case closest to its behavior or as an intermediate case. A complete discussion of all Hund's cases has been presented by Herzberg [5] and Mizushima [6], and is beyond the scope of this chapter. The discussion of magnetic hyperfine interactions presented will be restricted to molecules which can be represented by Hund's case(a) or Hund's case(b).

Unless all nuclei in a molecule have zero spin, there will be a simultaneous interaction of the nuclear spin angular momentum, the electronic orbital and spin angular momenta, and the rotational angular momentum. The initial discussion will be limited to the assumed zero nuclear spin case. Hund's case(a) involves a system where both the total electronic orbital angular momentum **L**, the total electron spin angular momentum **S**, precess rapidly about the internuclear axis, producing axial components Λ and Σ which couple to give a total electronic angular momentum $\Omega = |\Lambda + \Sigma|$, which is directed along the axis. This total electronic angular momentum then couples with the rotational angular momentum **R** which will be perpendicular to the molecular axis, to give a total angular momentum **J**, as illustrated in Fig. 14.6. If the molecule has an even number of electrons, Ω will be an integer and the quantum number describing the total momentum, **J**, will be an integer. If there are an odd number of electrons, then Ω and the rotational quantum number will be half-integer values.[3]

---

[3] Note the change in the use of **J** to denote the total angular momentum while the rotational angular momentum is represented by **R**.

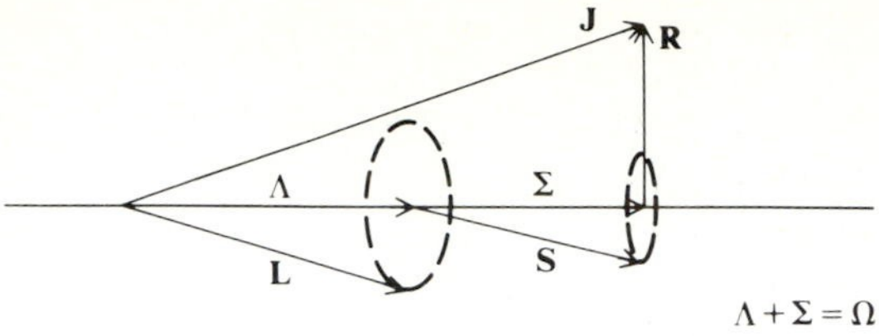

FIGURE 14.6
Vector diagram of Hund's case(a).

Hund's case(b) represents the situation where the electron spin momentum **S** is uncoupled from the symmetry axis. In this case, the axially coupled electron angular momentum Λ, and the rotational angular momentum **R**, couple to form an intermediate resultant **N**, which in turn couples with **S** to form the total angular momentum **J**, as shown in Fig. 14.7.

As was mentioned earlier, for any real molecule complete conformity to these or any of the idealized coupling cases is highly unlikely. In many cases, however, the deviations are small and these simple coupling schemes form an adequate basis for discussion. It is important to recognize, nevertheless, that many complex coupling situations exist and that they can be evaluated. Any change in the quantum number Ω will be an electronic transition lying far above the range of normal rotational transitions. Pure rotational transitions will then have $\Delta\Omega = 0$ and the calculation of any rotational frequency will involve the difference $E_{J',\Omega} - E_{J,\Omega}$.

The classical energy of the diatomic molecule will be

$$E_r = \frac{1}{2}\left(\frac{P_x^2}{I_x} + \frac{P_y^2}{I_y} + \frac{P_z^2}{I_z}\right) \tag{14.65}$$

where $I_x = I_y = I_B$ and I_z is the pseudo moment due to the electrons. Rewriting the energy as

$$E = \frac{1}{2}\left(\frac{P_x^2}{I_B} + \frac{P_y^2}{I_B} + \frac{P_z^2}{I_B}\right) + \frac{1}{2}\left(\frac{P_z^2}{I_z} - \frac{P_z^2}{I_B}\right) \tag{14.66}$$

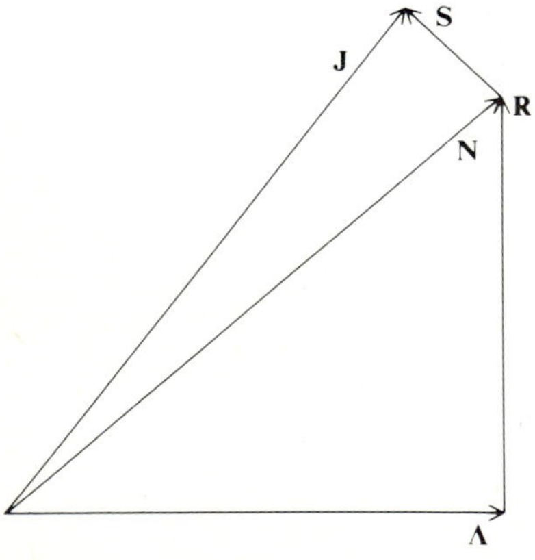

FIGURE 14.7
Vector diagram of Hund's case(b).

incorporating the angular momentum elements from Table 3.1, and recognizing that for Hund's case(a), the quantum numbers K and Ω describe analogous momenta about the z axis, the matrix elements of the Hamiltonian of the rigid rotor are

$$\langle J, \Omega, M|\mathscr{H}_r|J', \Omega', M'\rangle = \frac{1}{2I_B}\langle J, \Omega, M|\mathbf{P}^2|J', \Omega', M'\rangle + \left(\frac{1}{I_B} - \frac{1}{I_z}\right)\langle J, \Omega, M|\mathbf{P}_z^2|J', \Omega', M'\rangle \quad (14.67)$$

The energy levels are given by

$$E_{J,\Omega} = hB_v[J(J+1) - \Omega^2] + hA'\Omega^2 \quad (14.68)$$

and the transition frequencies by

$$\nu_{J+1\leftarrow J} = 2B_v(J+1) \quad (14.69)$$

On the surface it would appear that the presence of an unpaired electron has no effect on the general appearance of the rotational spectrum of a diatomic molecule. If all paramagnetic diatomic molecules obeyed Hund's case(a) this would be true. The deviation of real molecular behavior from this ideal model leads to several experimentally observed phenomena of interest.

Molecules in Σ states having $\Lambda = 0$ and $S \neq 0$ comprise one special group. In these systems, there is no field due to the orbital motion and the spin is magnetically coupled to $\mathbf{N}$ by the field produced by molecular rotation. This is Hund's case(b) with $\mathbf{\Lambda} = 0$ so that $\mathbf{R} \equiv \mathbf{N}$. In this case the total angular momentum will have allowed values of

$$J = N + S, N + S - 1, N + S - 2, \ldots, |N - S| \quad (14.70)$$

Consideration of a $^2\Sigma$ molecule will further illustrate this case. For a $^2\Sigma$ molecule having $S = \frac{1}{2}$ the magnetic moment due to the spin is coupled to that produced by molecular rotation. This rotation coupling adds a contribution proportional to $\mathbf{S} \cdot \mathbf{N}$ to the Hamiltonian of the molecule if it is assumed that the magnetic field produced by the molecular rotation is proportional to the angular momentum, $\mathbf{N}$. Referring to Fig. 14.7 and using the law of cosines shows that the interaction term $\mathbf{S} \cdot \mathbf{N}$ can be written as

$$\mathbf{S} \cdot \mathbf{N} = \tfrac{1}{2}(\mathbf{J}^2 - \mathbf{N}^2 - \mathbf{S}^2) \quad (14.71)$$

The rigid rotor Hamiltonian as a function of the momentum vectors will be

$$\mathscr{H} = \frac{\mathbf{N}^2}{2I_B} + \frac{\gamma'}{2}(\mathbf{J}^2 - \mathbf{N}^2 - \mathbf{S}^2) \quad (14.72)$$

The Hamiltonian is diagonal in N, S, and J, so incorporation of the appropriate elements from Table 3.1 and expressing the spin-rotation coupling constant γ_{sr} in units of s gives

$$E_{J,N,S} = hBN(N+1) + \frac{h\gamma_{sr}}{2}[J(J+1) - N(N+1) - S(S+1)] \quad (14.73)$$

where $N = 0, 1, 2, \ldots, S = \pm\frac{1}{2}$, and $J = N \pm \frac{1}{2}$. The selection rule for N will be that for the rigid diatomic rotor $\Delta N = \pm 1$, and that for J follows as $\Delta J = 0, \pm 1$. Figure 14.8 shows that the spin-rotation interaction leads to a splitting of each energy level having $N > 0$ and a resulting splitting of the transitions. From the measurement of the frequencies of the rotational transitions the spin-rotation interaction constant γ_{sr} can be determined.

For molecules in Π or Δ states, where $\Lambda \neq 0$ deviations from Hund's cases, are common. For diatomic hydrides the coupling is generally intermediate between case(a) and case(b). For heavier molecules there is frequently deviation from Hund's case(a) behavior due to an interaction of the molecular and electron momenta. This interaction can lead to the uncoupling of both **L** and **S** from the internuclear axis. The spin uncoupling leads to the necessity of replacing the rotational constant B with an effective value B_{eff}, while the uncoupling of **L** gives rise to the phenomenon known as Λ-doubling. The behavior of a $^2\Pi$ molecule will be used to illustrate these points.

For a $^2\Pi$ molecule having $\Lambda = 1$ and $\Sigma = \frac{1}{2}$ there will be two values of Ω, $\frac{1}{2}$, and $\frac{3}{2}$, leading to two electronic states, $^2\Pi_{1/2}$ and $^2\Pi_{3/2}$. The rotational angular momentum can have values of $0 = 0, 1, 2, 3, \ldots$, and since $\mathbf{J} = \mathbf{R} + \mathbf{\Omega}$, the total rotational quantum numbers J will be, as shown in Fig. 14.9, half-integer values in both states.

The effect of spin uncoupling is to produce a mixing of the two electronic states. When the uncoupling is small, $2JB \ll |\Lambda A|$, where A is the spin-orbit coupling constant, the effect is best handled by replacement of B with

$$B_{\text{eff}} = B\left(1 \pm \frac{B}{\Lambda A} + \cdots\right) \tag{14.74}$$

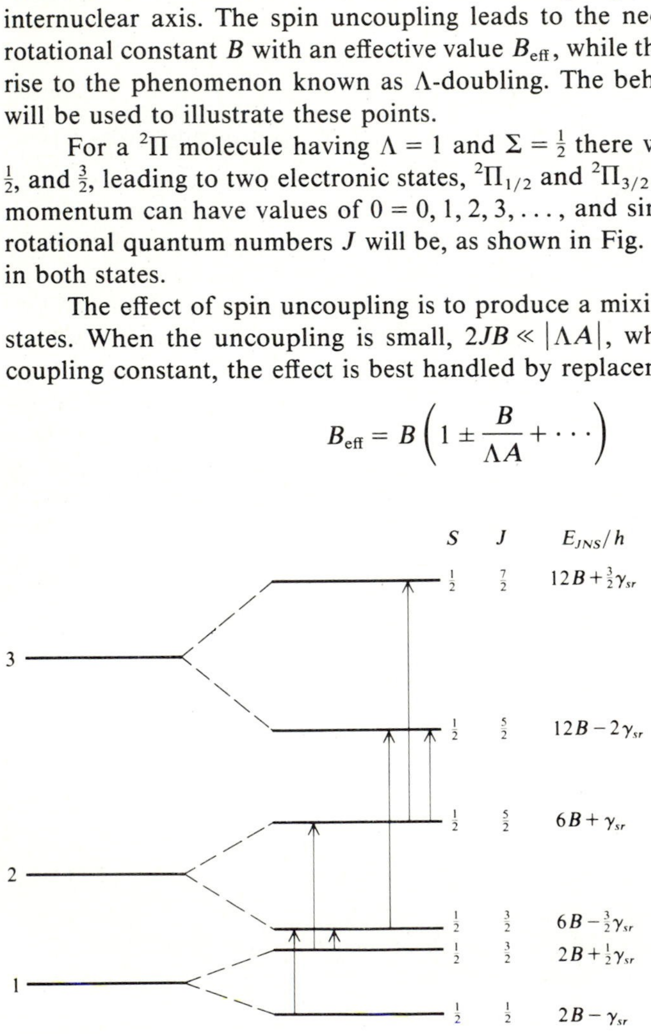

FIGURE 14.8
Effect of spin-rotation interaction in a $^2\Sigma$ molecule.

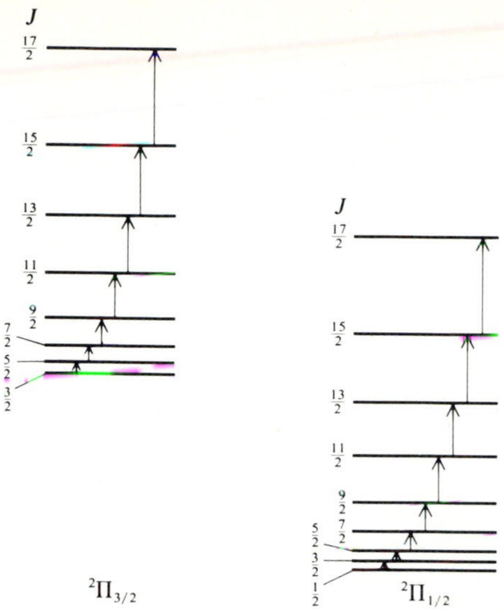

FIGURE 14.9
Energy levels and transitions for a molecule in a $^2\Pi$ electronic state.

Since $\boldsymbol{\Omega}$ results from the vector addition of $\boldsymbol{\Lambda}$ and $\boldsymbol{\Sigma}$, the scalar Ω^2 will be equal to $(\Lambda + \Sigma)^2$ or the energy given by Eq. (14.68) will be degenerate in Λ. The coupling of **L**, which is due to the magnetic field produced by the molecular rotation, gives rise to a splitting of the doubly degenerate orbital states. The lifting of the Λ degeneracy by the uncoupling is the result of the admixing of the Π ground states with nearby Σ states of the molecule and is reflected in the fact that the Λ-doubling constants depend on the energy separations of these states, $h\nu_e$. The derivations for the necessary relationships for the various coupling cases are beyond the intentions of this presentation, but have been published in several sources [5–9]. For $^1\Pi$ states where $\Sigma = 0$ the splitting is given by

$$\Delta E_\Lambda = hq_\Lambda J(J+1) \tag{14.75}$$

where $q_\Lambda = 4B^2/\nu_e$. For the $^2\Pi_{1/2}$ and $^2\Pi_{3/2}$ states of molecules with behavior approximating Hund's case(a) the first-order separations are [10]

$$\Delta E_\Lambda = hq_\Lambda^a(J + \tfrac{1}{2}) \tag{14.76}$$

where $q_\Lambda^a = 4AB/\nu_e$ and

$$\Delta E_\Lambda = hq_\Lambda^b(J - \tfrac{1}{2})(J + \tfrac{1}{2})(J + \tfrac{3}{2}) \tag{14.77}$$

where $q_\Lambda^b = 8B^3/A\nu_e$.

In the discussions of diatomic molecules possessing electronic angular momentum, the model used to this point is the rigid rotor. As with diamagnetic molecules there remains the effects of centrifugal distortion and rotation-vibration interactions, which must be included in any accurate energy level expression. If this is done, Eq. (14.68) becomes

$$E_{J,\Omega} = hB_v[J(J+1) - \Omega^2] + hA'\Omega^2 - hD_eJ^2(J+1)^2 \tag{14.78}$$

For molecules in $^2\Sigma$ states, Eq. (14.73) will be modified to

$$E_{J,N,S} = hB_v N(N+1) - hD_e N^2(N+1)^2 + \frac{h\gamma_{sr}}{2}[J(J+1) - N(N+1) - S(S+1)] \quad (14.79)$$

Additional terms will also be incorporated into the energy level expressions for the $^2\Pi$ states which, when the effects of spin uncoupling and Λ-doubling are included, become [10]

$$(^2\Pi_{1/2}) \qquad E_J^{\pm} = hB_v\left(1 - \frac{B_v}{A}\right) J(J+1) \pm \tfrac{1}{2}hq^a(J+\tfrac{1}{2}) - hD_e[(J-\tfrac{1}{2})(J+\tfrac{1}{2})^2(J+\tfrac{3}{2}) + 1] \quad (14.80)$$

$$(^2\Pi_{3/2}) \qquad E_J^{\pm} = hB_v\left(1 - \frac{B_v}{A}\right) J(J+1) \pm \tfrac{1}{2}hq^b(J-\tfrac{1}{2})(J+\tfrac{1}{2})(J+\tfrac{3}{2}) - hD_e[(J-\tfrac{1}{2})(J+\tfrac{1}{2})^2(J+\tfrac{3}{2}) + 1] \quad (14.81)$$

For molecules containing nuclei with $I \neq 0$ in addition to the coupling of electronic angular momentum to the rotational angular momentum it can simultaneously couple with nuclear angular momenta, via the associated magnetic moments to give rise to magnetic hyperfine interactions. For diatomic molecules, other than $^1\Sigma$ type, this interaction can be of a magnitude comparable to nuclear electric quadrupole hyperfine interactions. For $^1\Sigma$ molecules where there is no resulting electronic angular momenta, there is only a weak coupling between nuclear moments or between nuclear moments and rotational magnetic moments. In either case, the resulting hyperfine interactions are small and generally neglected. For Π or Δ states, the observable hyperfine interactions involve either nuclear spin–electron orbital or electron spin–nuclear spin interactions. An oversimplified model of the former would be the interaction of the magnetic field due to the orbital motion and parallel to **L**, the orbital angular momentum, with the nuclear magnetic moment producing an energy term $\mathscr{H}_{IL} = a\mathbf{I} \cdot \mathbf{L}$. The interaction of the nuclear and electron spins give rise to two energy contributions, a classical dipole-dipole interaction of the form $\mathscr{H}_{IS} = b\mathbf{I} \cdot \mathbf{S}$ and the Fermi contact term of the form $H'_{IS} = a_0\,(\mathbf{I} \cdot \hat{\mathbf{k}})(\mathbf{S} \cdot \hat{\mathbf{k}})$ for isotropic coupling. This term is of special significance for s electrons since the spatial region of their wavefunctions overlaps the nuclear volume. The electron-nuclear interaction Hamiltonian for the magnetic hyperfine interaction of a single electron in a diatomic molecule is [5]

$$\mathscr{H}_{Ie} = \mathscr{H}_{IL} + \mathscr{H}_{IS} + \mathscr{H}'_{IS} = a\mathbf{I} \cdot \mathbf{L} + b\mathbf{I} \cdot \mathbf{S} + a_0(\mathbf{I} \cdot \hat{\mathbf{k}})(\mathbf{S} \cdot \hat{\mathbf{k}}) \quad (14.82)$$

where [11]

$$a = \frac{2\beta\beta_N}{I}\left(\frac{1}{r^3}\right) \quad (14.83)$$

$$b = \frac{2\beta\beta_N}{I}\left[\frac{8\pi}{3}\psi^2(0) - \frac{3\cos^2\theta - 1}{2r^3}\right]_{\text{av}} \quad (14.84)$$

$$a_0 = \frac{3\beta\beta_N}{I}\left(\frac{3\cos^2\theta - 1}{r^3}\right)_{\text{av}} \quad (14.85)$$

and $\hat{\mathbf{k}}$ = unit vector along molecular axis
β = Bohr magneton
β_N = nuclear magneton
$\mathbf{S}$ = electron spin
$\mathbf{r}$ = distance between the nucleus and the electron
$\psi^2(0)$ = electron density at the nucleus
θ = angle between the molecular axis and r
L = total orbital angular momentum

For Π and Δ diatomic molecules $\mathbf{L}$ has an average nonzero value $\Lambda\hat{\mathbf{k}}$ where Λ is the component along the internuclear direction and the first term in the Hamiltonian becomes

$$\mathscr{H}_{IL} = \frac{2\beta\beta_N\Lambda}{I}\left(\frac{1}{r^3}\right)_{\text{av}} \mathbf{I}\cdot\hat{\mathbf{k}} = a\Lambda\mathbf{I}\cdot\hat{\mathbf{k}} \tag{14.86}$$

There are several coupling situations possible. Frosch and Foley [12] have considered these individually and in detail. We will examine one case which will be applicable to the NO molecule. This is Hund's case(a_β) and involves the same basic momentum coupling as in Hund's case(a), but in addition has the nuclear spin **I** coupled to **J** giving a new total angular momentum **F**. This is illustrated in Fig. 14.10.

If any effects due to Λ-doubling are ignored, the energy for the (a_β) case may be written, using Eqs. (14.82) through (14.85), as

$$\mathscr{H}_{Ie} = [a\Lambda + (b + c)\Sigma]\mathbf{I}\cdot\hat{\mathbf{k}} \tag{14.87}$$

This relationship recognizes that $\mathbf{S}\cdot\hat{\mathbf{k}} = \Sigma$ and $\mathbf{I}\cdot\mathbf{S} = (\mathbf{I}\cdot\hat{\mathbf{k}})(\mathbf{S}\cdot\hat{\mathbf{k}})$ when **S** is coupled to the internuclear axis. This is due to the fact that the precession results in the effective values of the S_x and S_y components being zero. The matrix elements of $\mathscr{H}_{Ie}$ are obtained from those of $\mathbf{I}\cdot\hat{\mathbf{k}}$ which can be reduced by a vector transformation to

$$\mathbf{I}\cdot\hat{\mathbf{k}} = \frac{(\mathbf{I}\cdot\mathbf{J})(\mathbf{J}\cdot\hat{\mathbf{k}})}{J(J+1)} \tag{14.88}$$

Since $\mathbf{J}\cdot\hat{\mathbf{k}} = \Omega$ the matrix elements for the energy become

$$\langle I, J, F|\mathscr{H}_{Ie}|I, J, F\rangle = \frac{[a\Lambda + (b + a_0)\Sigma]\Omega}{J(J+1)}\langle I, J, F|\mathbf{I}\cdot\mathbf{J}|I, J, F\rangle \tag{14.89}$$

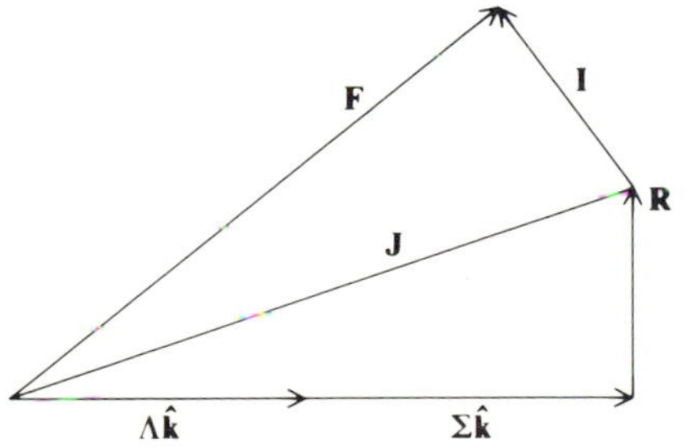

FIGURE 14.10
Vector diagram for Hund's case(a_β).

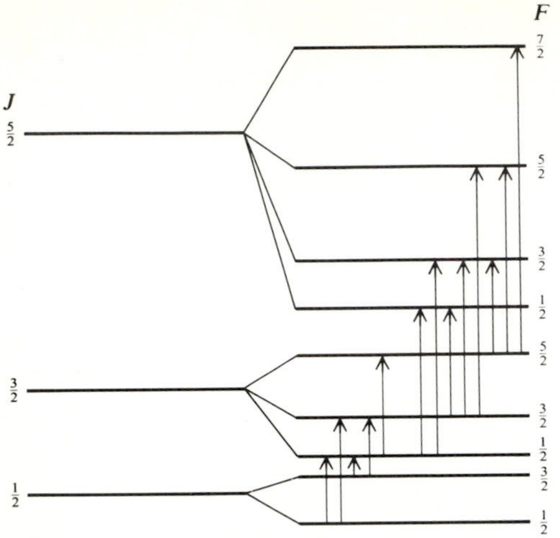

FIGURE 14.11
Energy level scheme for a $^2\Pi_{1/2}$ state diatomic molecule with magnetic hyperfine interactions (neglecting possible Λ-doubling).

and it was determined earlier that

$$\langle I, J, F|\mathbf{I}\cdot\mathbf{J}|I, J, F\rangle = \tfrac{1}{2}[F(F+1) - I(I+1) - J(J+1)] \tag{14.90}$$

The selection rules for I, J, and F are the same as derived for nuclear quadrupole transitions $\Delta J = 0, \pm 1$, $\Delta I = 0$, $\Delta F = 0, \pm 1$.

In general the factor $[a\Lambda + (b + a_0)\Sigma]$ for paramagnetic molecules will be of the same order of magnitude as nuclear electric quadrupole interactions. A typical energy level scheme for a diatomic molecule with $\Lambda = 1$, $\Sigma = \frac{1}{2}$ and having a single nucleus with $\mathbf{I} = 1$ is shown in Fig. 14.11.

14.8 ZEEMAN EFFECT: DIAMAGNETIC DIATOMIC MOLECULES

The Zeeman effect, or the interaction of molecular rotation with an external magnetic field, is of importance for the determination of several molecular parameters. The magnitude of the Zeeman effect is much smaller for diamagnetic molecules than for paramagnetic ones, but the former constitutes a more extensive class. This would be expected since the rotational magnetic moments of the former are much smaller. Within the past few years, there has been considerable activity in this area of molecular spectroscopy. The basic theory has been extended to the point where there are well-developed relationships between the experimentally observed transition frequencies and the components of the molecular g tensor, the magnetic susceptibility tensor, and the nuclear magnetic shielding tensor [13]. This discussion will not consider such detail with regard to the Zeeman effect, and will examine only the most elementary concepts.

The molecular rotation of a $^1\Sigma$ molecule gives rise to a rotational magnetic moment due to the coupling of the ground and excited electronic states. An elementary approach is to express the molecular rotation magnetic moment μ_J

as a product of a molecular g factor g_J, which is analogous to the nuclear g factor, and the rotational angular momentum

$$\boldsymbol{\mu}_J = g_J \beta_N \mathbf{J} \tag{14.91}$$

For a $^1\Sigma$ diatomic molecule with both nuclei having $I = 0$, this will be the only contribution to the magnetic moment of the system and the Zeeman energy will be given by

$$E_H = -\boldsymbol{\mu}_J \cdot \mathbf{H} = -g_J \beta_N \mathbf{J} \cdot \mathbf{H} \tag{14.92}$$

But $\mathbf{J} \cdot \mathbf{H} = JH \cos\theta$, where H is taken along the space-fixed Z direction and $J\cos\theta = J_Z$, so the total energy becomes $E_{JM} = E_J^0 + E_{HM} = E_J^0 - g_J\beta_N MH$. If there are no magnetic nuclei in the molecular rotation, then the energy level scheme is shown in Fig. 14.12. The selection rules are determined from the nonzero matrix elements of the type $\langle JM|\boldsymbol{\mu}_J|J'M'\rangle = \mu_J\langle JM|\Phi_{Zz}|J'M'\rangle$ to be $\Delta M = 0, \pm 1$. The $\Delta M = 0$ transitions, shown by $\nu_{\pi i}$ in Fig. 14.12, occur when H is parallel to the radiation field electric vector while the $\Delta M = \pm 1$ transition, denoted by $\nu_{\sigma i}$, occur when H is perpendicular to the electric vector. If g_J does not change significantly between two adjacent rotational states the frequencies of the Zeeman transitions are given by

$$\nu = \nu^0_{J+1\leftarrow J} - [g_{J+1}\Delta M + (g_J - g_{J+1})M]\frac{\beta_N H}{h} = \nu^0_{J+1\leftarrow J} - \frac{g_J\beta_N H}{h}\Delta M \tag{14.93}$$

For H in tesla and ν in megahertz this becomes

$$\nu = \nu^0_{J+1\leftarrow J} - (762 \times 10^{-2}) g_J H \Delta M \tag{14.94}$$

The proceeding development, although giving the form of the observed spectra, has done so by the arbitrary introduction of the molecular g factor. A more detailed study of the magnetic interactions in the diatomic molecule can serve to show the significance of the molecular g factor. The total magnetic

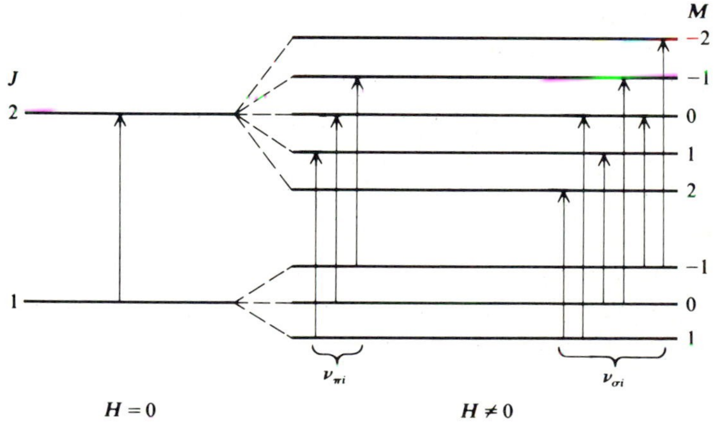

FIGURE 14.12
Typical Zeeman energy levels for a $^1\Sigma$ diatomic molecule with $I_1 = I_2 = 0$.

interaction Hamiltonian includes terms to account for nuclear spin interactions as well as the interaction of an external field. For a diatomic molecule with one nonzero nuclear spin and considering only a single excited electronic state it is

$$\mathscr{H}^{(1)} = -\frac{1}{I_x}\mathbf{J}\cdot\mathbf{L} - g_I\beta_N\frac{Ze}{crI_x}\mathbf{I}\cdot\mathbf{J} + a\mathbf{I}\cdot\mathbf{L} + \beta\mathbf{H}\cdot\mathbf{L} - \boldsymbol{\mu}_I\cdot\mathbf{H} \tag{14.95}$$

where $\beta\mathbf{H}\cdot\mathbf{L}$ is the interaction between the external field and the rotational magnetic moment and $\boldsymbol{\mu}_I\cdot\mathbf{H}$ is the interaction between the external field and the spin of the $I \neq 0$ nucleus. The total perturbation energy is given by

$$E^{(1)} = \frac{\langle 0|\mathscr{H}^{(1)}|n\rangle^2}{(E_0^0 - E_n^0)} \tag{14.96}$$

where $\langle 0|\mathscr{H}^{(1)}|n\rangle$ is the matrix element of $\mathscr{H}^{(1)}$ between the ground and excited electronic state. The terms involving $\mathbf{I}\cdot\mathbf{J}$ and $\boldsymbol{\mu}_I\cdot\mathbf{H}$ are both independent of the electronic angular momentum and can be considered separately from the other three terms. Hence

$$E^{(1)} = \frac{\langle 0|-(1/I_x)\mathbf{J}\cdot\mathbf{L} + a\mathbf{I}\cdot\mathbf{L} + \beta\mathbf{H}\cdot\mathbf{L}|n\rangle^2}{(E_0^0 - E_n^0)} - g_I\beta_N\frac{Ze}{CrI_x}\mathbf{I}\cdot\mathbf{J} - \boldsymbol{\mu}_I\cdot\mathbf{H} \tag{14.97}$$

Expanding this expression gives

$$E^{(1)} = \frac{\langle 0|-(1/I_x)\mathbf{J}\cdot\mathbf{L} + a\mathbf{I}\cdot\mathbf{L} + \beta\mathbf{H}\cdot\mathbf{L}|n\rangle\langle n|-(1/I_x)\mathbf{J}\cdot\mathbf{L} + a\mathbf{I}\cdot\mathbf{L} + \beta\mathbf{H}\cdot\mathbf{L}|0\rangle}{(E_0^0 - E_n^0)} - g_I\beta_N\frac{Ze}{CrI_x}\mathbf{I}\cdot\mathbf{J} - \boldsymbol{\mu}_J\cdot\mathbf{H} \tag{14.98}$$

or

$$\begin{aligned}
E^{(1)} &= \frac{1}{I_x^2}\frac{\langle 0|\mathbf{J}\cdot\mathbf{L}|n\rangle\langle n|\mathbf{J}\cdot\mathbf{L}|0\rangle}{(E_0^0 - E_n^0)} + a^2\frac{\langle 0|\mathbf{I}\cdot\mathbf{L}|n\rangle\langle n|\mathbf{I}\cdot\mathbf{L}|0\rangle}{(E_0^0 - E_n^0)} \\
&\quad + \beta^2\frac{\langle 0|\mathbf{H}\cdot\mathbf{L}|n\rangle\langle n|\mathbf{H}\cdot\mathbf{L}|0\rangle}{(E_0^0 - E_n^0)} \\
&\quad - \frac{a}{I_x}\frac{\langle 0|\mathbf{J}\cdot\mathbf{L}|n\rangle\langle n|\mathbf{I}\cdot\mathbf{L}|0\rangle + \langle 0|\mathbf{I}\cdot\mathbf{L}|n\rangle\langle n|\mathbf{J}\cdot\mathbf{L}|0\rangle}{(E_0^0 - E_n^0)} \\
&\quad - \frac{\beta}{I_x}\frac{\langle 0|\mathbf{J}\cdot\mathbf{L}|n\rangle\langle n|\mathbf{H}\cdot\mathbf{L}|0\rangle + \langle 0|\mathbf{H}\cdot\mathbf{L}|n\rangle\langle n|\mathbf{J}\cdot\mathbf{L}|0\rangle}{(E_0^0 - E_n^0)} \\
&\quad + a\beta\frac{\langle 0|\mathbf{I}\cdot\mathbf{L}|n\rangle\langle n|\mathbf{H}\cdot\mathbf{L}|0\rangle\langle 0|\mathbf{H}\cdot\mathbf{L}|n\rangle\langle n|\mathbf{I}\cdot\mathbf{L}|0\rangle}{(E_0^0 - E_n^0)} \\
&\quad - g_I\beta_N\frac{Ze}{CrI_x}\mathbf{I}\cdot\mathbf{J} - \boldsymbol{\mu}_I\cdot\mathbf{H}
\end{aligned} \tag{14.99}$$

The first term gives the contribution of the electrons to the overall rotation, the second is a pseudoquadrupole term, and the fourth and seventh are the magnetic hyperfine interaction terms, all of which were discussed in Sec. 14.9. The third, sixth, and eighth terms are independent of molecular rotation and are of no further interest. The third term represents the interaction between the external magnetic field and a field-induced magnetic dipole in the orbital electrons. The eighth term is the basic NMR interaction. The sixth term describes the interaction between the nuclear magnetic moment and a magnetic field produced by angular momentum induced in electrons by the field. This is the chemical shift effect observed in NMR. The fifth term describes the interaction of the magnetic moment produced by the molecular rotation and an external magnetic field, the Zeeman effect.

The shift from the zero field energy level due to the Zeeman effect is given by

$$E_H = -\frac{\beta}{I_x}\frac{\langle 0|\mathbf{J}\cdot\mathbf{L}|n\rangle\langle n|\mathbf{H}\cdot\mathbf{L}|0\rangle + \langle 0|\mathbf{H}\cdot\mathbf{L}|n\rangle\langle n|\mathbf{J}\cdot\mathbf{L}|0\rangle}{(E_0^0 - E_n^0)} \tag{14.100}$$

For the diatomic molecule, where $L_x = L_y$, $L_z = 0$ this expression reduces to

$$E_H = -\frac{2\beta_N}{I_x}\frac{\langle 0|\mathbf{L}_x|n\rangle^2}{E_0^0 - E_n^0}\mathbf{J}\cdot\mathbf{H} \tag{14.101}$$

Comparison of this expression to that given by Eq. (14.92) shows that the molecular g factor is related to the electronic structure of the molecule via

$$g_J = \frac{2}{I_x}\frac{\langle 0|\mathbf{L}_x|n\rangle^2}{E_0^0 - E_n^0} \tag{14.102}$$

Thus either approach leads to the conclusion that there is a molecular property which is structure-dependent and can be experimentally determined from a Zeeman experiment.

14.9 SELECTED EXAMPLES

The chapter to this point has reviewed the effect of several perturbations on the rotational spectra of diatomic molecules. In this section several examples which illustrate the experimental aspects of these effects are presented. Where possible the examples consider a single perturbing effect, but it should be realized that there are many situations, particularly paramagnetic molecules and molecules having nuclear quadrupole interactions, where two or more perturbing effects are simultaneously operative.

Carbon Monoxide—Stark Effect

The Stark effect for the $J = 1 \leftarrow 0$ transition of $^{12}C^{16}O$ at 115,271.20 MHz has been measured at 195 K by Burrus [14]. The spectrometer was calibrated using the HCN transition at 88.6 GHz so the absolute error in the experimentally determined dipole moment reflects the uncertainty in the dipole moment of HCN 3.00 ± 0.02 debye. The resulting frequency shift, $\Delta\nu$, vs. $\mathscr{E}^2$ is shown in Table 14.4. Using the first three transitions for $^{12}C^{16}O$: $J = 1 \leftarrow 0$, $\nu = 115{,}271.20$ MHz, $J = 2 \leftarrow 1$, $\nu = 230{,}537.97$ MHz, and $J = 3 \leftarrow 2$, $\nu = 345{,}795.90$ MHz [15] with Eqs. (12.11) through (12.13) gives $B_e = 57{,}898.57$ MHz, $D_e = 183.8$ kHz, and $\alpha_e = 525.2$ MHz. Equation (14.20) provides for a $J = 1 \leftarrow 0$ transition a frequency shift of

$$\Delta\nu = \frac{4\mu^2\varepsilon^2}{15B_e h^2} \tag{14.103}$$

Substitution of $\Delta\nu$ and ε^2 values from Table 14.4 could lead to the calculation of nine values for μ which could be averaged to get the best value. A more generally used technique, however, is that of plotting $\Delta\nu$ vs. $\mathscr{E}^2$, obtaining the best linear least-squares fit, and calculating μ from the resulting slope. Using the latter method, as shown in Fig. 14.13, one obtains $\mu = 0.112 \pm 0.005$ D.

Deuterium Chloride—Nuclear Quadrupole Coupling

The hyperfine transitions for DCl as observed by Cowan and Gordy [16] are shown in Table 14.5. The extraction of the unsplit frequency, which is needed to determine the internuclear distances, and the quadrupole coupling constants require the use of Eqs. (14.63) and (14.64). From the latter, the appropriate values of $f(I, J, F)$ are found to be as shown in Fig. 14.14. Three equations of the form $\nu_i = \nu_0 + a_i e^2Qq$ can be solved to obtain ν_0 and e^2Qq_{zz}. For example, the data for $D^{35}Cl$ yields $\nu_0 = 323{,}295.77 \pm 0.13$ MHz and $e^2Qq_{zz} = 67.3 \pm 0.7$ MHz. For very light molecules where the rotational constant B_e is very large it is often difficult or impossible to observe a sufficient number of rotational transitions to independently determine B_e, D_e, and α_e. Also, since the molecules are light, D_e

TABLE 14.4
Stark effect for carbon monoxide

$\mathscr{E}^2$, V^2 cm^2, ±0.5%	$\Delta\nu$, MHz
0	0
1.623×10^6	
6.853	0.100
16.008	0.245
27.750	0.435
42.708	0.610
60.092	0.875
81.040	1.175
106.83	1.560
132.32	1.945
163.92	2.375

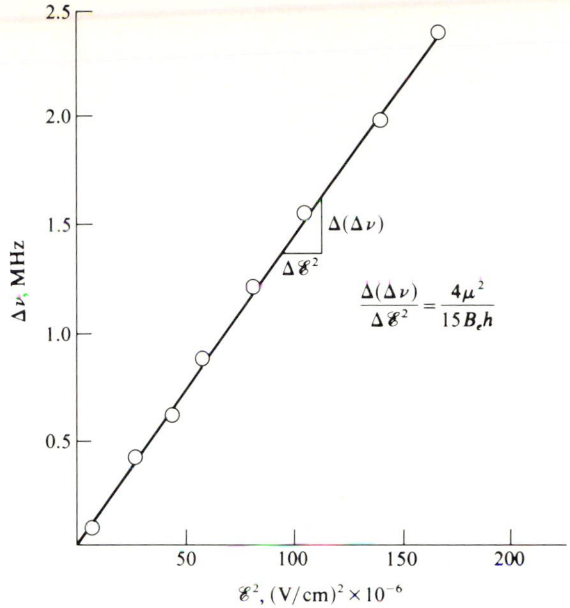

FIGURE 14.13
Stark effect for CO.

is generally of appreciable magnitude. In situations of this type, particularly for diatomic molecules, there is often sufficient information available from vibrational-rotational spectra in the infrared region to complement the microwave data. For $D^{35}Cl$ the infrared data gives $D_e = 4.11$ MHz and $\alpha_e = 3667$ MHz [17]. Use of Eqs. (12.8) and (12.11) yields $B_0 = 161{,}656.10 \pm 0.07$ MHz and $B_e = 163{,}340 \pm 1$ MHz. Since the accuracy of the infrared data is of the order ± 0.02 cm^{-1}, the errors associated with D_e and α_e are at least an order of magnitude greater than those obtained for CO in the previous section, hence the values of B_0 and B_e are not as accurate as those based on microwave measurements alone. Often, as is the case for DCl, the relative error in the B_e and B_0 values are less than those of the atomic masses and the physical constants involved (for example, $h = (6.62559 \pm 0.00016) \times 10^{-34}$ J-s). This leads to comparable relative errors for

TABLE 14.5
Microwave spectrum of DCl

Molecule	Transition $J_F \to J'_F$	Relative intensity	Observed frequency, MHz
$D^{35}Cl$	$\nu_1 = 0_{3/2} \to 1_{1/2}$	33	$323{,}312.52 \pm 0.13$
	$\nu_2 = 0_{3/2} \to 1_{5/2}$	100	$323{,}299.17 \pm 0.13$
	$\nu_3 = 0_{3/2} \to 1_{1/2}$	67	$323{,}282.28 \pm 0.13$
$D^{37}Cl$	$\nu_1 = 0_{3/2} - 1_{1/2}$	33	$322{,}362.94 \pm 0.13$
	$\nu_2 = 0_{3/2} - 1_{5/2}$	100	$322{,}352.33 \pm 0.13$
	$\nu_3 = 0_{3/2} - 1_{3/2}$	67	$322{,}339.09 \pm 0.13$

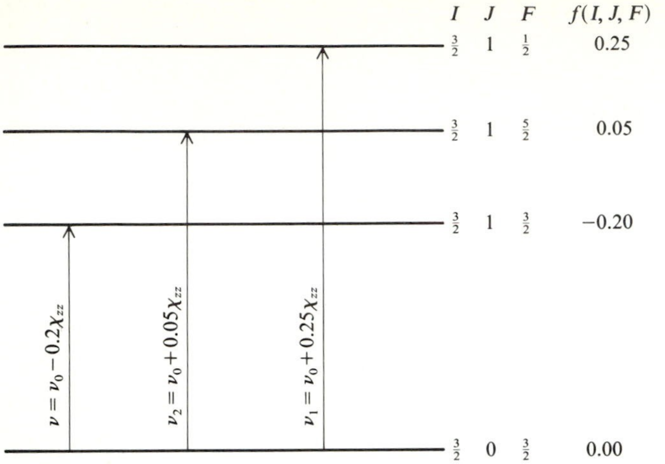

FIGURE 14.14
Energy level-transition diagram for the $J = 1 \leftarrow 0$ transition at DCl.

r_e and r_0. For example, for $D^{35}Cl$ $r_0 = (h/8\pi^2\mu B_0)^{1/2} = 1.28125 \pm 0.00001$ cm and $r_e = (h/8\pi^2\mu B_e)^{1/2} = 1.27462 \pm 0.00001$ cm.

Deuterium Iodide—Zeeman Effect

Molecules which are suitable as examples of a single perturbing effect are relatively uncommon. To illustrate the Zeeman effect of a $^1\Sigma$ diatomic molecule we will examine the case of $^2H^{127}I$ which was studied by Burrus [18]. This molecule has a nucleus with $I = \frac{5}{2}$, so there will be nuclear quadrupole hyperfine structure present. In this case the nuclear spin momentum and the molecular rotational angular momentum are coupled to give a resultant angular momentum **F**. In the presence of an external magnetic field **F** will precess about the field and have a projection M_F on the field direction; thus the quantum number that specifies the Zeeman energies will be M_F rather than M_J. Both **I** and **J** will precess about **F** to produce an effective moment $\boldsymbol{\mu}_F$ which will then interact with the external field, as shown in Fig. 14.15. The Hamiltonian is

$$\mathscr{H}_H = -\boldsymbol{\mu}_F \cdot \mathbf{H} \tag{14.104}$$

The moment associated with **F** will be of a comparable magnitude to those associated with **I** and **J**. Defining another g factor, g_F, allows the Zeeman energy to be expressed as

$$E_H = -g_F \beta_N H M_F \tag{14.105}$$

and the frequency of a Zeeman component becomes

$$\nu = \nu_0 + \frac{\beta_N H}{h}[(g_F - g_{F'})M_F - g_{F'}\Delta M_F] \tag{14.106}$$

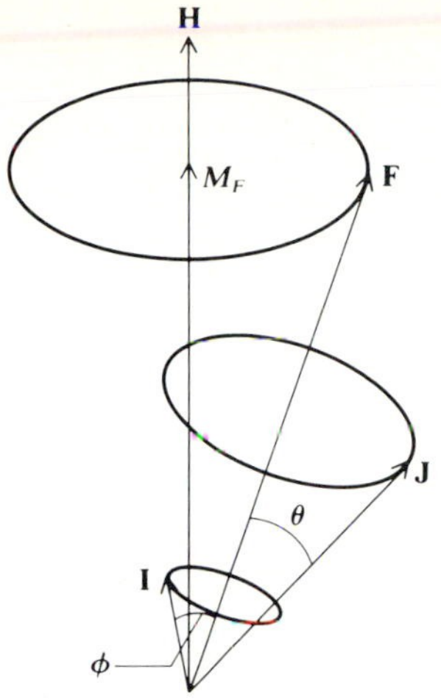

FIGURE 14.15
Vector relations for coupling of **I**, **J**, and **F** in a magnetic field.

where $\nu_0 = \nu_{J+1, F' \leftarrow J, F}$. The analysis of the Zeeman splittings of a transition can yield a measurement of g_F and $g_{F'}$. We can go one step further and evaluate g_F in terms of g_J and g_I. Since **J** and **I** precess about **F**, the value of $\boldsymbol{\mu}_F$ will be the sum of the projections of $\boldsymbol{\mu}_I$ and $\boldsymbol{\mu}_J$ on **F**

$$\boldsymbol{\mu}_F = [(\boldsymbol{\mu}_J \cdot \mathbf{F}) + (\boldsymbol{\mu}_I \cdot \mathbf{F})]\mathbf{F} \tag{14.107}$$

where **F** is a unit vector, $\mathbf{F} = \mathbf{F}/F(F+1)$. Therefore,

$$\boldsymbol{\mu}_F = \frac{g_J \beta_N \mathbf{J} \cdot \mathbf{F}}{[F(F+1)]^{1/2}} + \frac{g_I \beta_N \mathbf{I} \cdot \mathbf{F}}{[F(F+1)]^{1/2}} \tag{14.108}$$

However, $\boldsymbol{\mu}_F = [g_F \beta_N F(F+1)]$ so

$$g_F = \frac{g_J}{F(F+1)} \mathbf{J} \cdot \mathbf{F} + \frac{g_I}{F(F+1)} \mathbf{I} \cdot \mathbf{F} \tag{14.109}$$

In terms of the angles shown in Fig. 14.15

$$g_F = g_J \left(\frac{J(J+1)}{F(F+1)} \right)^{1/2} \cos\phi + g_I \left(\frac{I(I+1)}{F(F+1)} \right)^{1/2} \cos\theta = g_J \alpha_J(I, F) + g_I \alpha_I(J, F) \tag{14.110}$$

Applying the law of cosines to the vectors in Fig. 14.15 gives

$$|\mathbf{I}|^2 = |\mathbf{F}|^2 + |\mathbf{J}|^2 - 2|\mathbf{F}|\,|\mathbf{J}| \cos\phi \tag{14.111}$$

or

$$I(I+1) = F(F+1) + J(J+1) - 2[J(J+1)F(F+1)]^{1/2} \cos\phi \tag{14.112}$$

and

$$J(J+1) = F(F+1) + I(I+1) - 2[I(I+1)F(F+1)]^{1/2} \cos\theta \tag{14.113}$$

Therefore

$$\alpha_I(J, F) = \frac{F(F+1) + I(I+1) - J(J+1)}{2F(F+1)} \tag{14.114}$$

$$\alpha_J(I, F) = \frac{F(F+1) + J(J+1) - I(I+1)}{2F(F+1)} \tag{14.115}$$

Frequently the observed hyperfine splittings of a rotational transition cannot be satisfactorily explained by first-order nuclear quadrupole interactions as given by Eq. (14.63), and second-order perturbation theory is needed. The second-order nuclear quadrupole interaction energy will be

$$E_Q^{(2)} = \sum_{j'}' \frac{|\langle J, I, F|\mathscr{H}_Q|J', I, F\rangle|^2}{(E_j^0 - E_{j'}^0)} \tag{14.116}$$

This second-order term arises because the coupling of **I** to **J** perturbs the rotational wavefunction and J is no longer a good quantum number. The total angular momentum quantum number F, and the nuclear spin quantum number I, are still good numbers, however, so the matrix elements are still diagonal in I and F. The matrix elements necessary to evaluate the second-order effect have been developed by Racah [19] and the second-order energy evaluated by Townes [20]

$$E_Q^{(2)} = \frac{|\langle J, I, F|\mathscr{H}_Q|J+2, I, F\rangle|^2}{(E_{J+2}^0 - E_J^0)} + \frac{|\langle J, I, F|\mathscr{H}_Q|J-2, I, F\rangle|^2}{(E_J^0 - E_{J-2}^0)} \tag{14.117}$$

where

$$\begin{aligned}
&\langle J, I, F|\mathscr{H}_Q|J+2, I, F\rangle \\
&\quad = (3e^2Qq_{zz})^2 \frac{(F+I+J+3)(F+I+J+2)(J+I-F+2)}{\{16I(2I-1)(2J+3)\}^2(2J+1)(2J+3)} \\
&\qquad \times [(J+I-F+1)(J+F-I+2)(J+F-I+1) \\
&\qquad\qquad \times (I+F-J-1)(I+F-J)] \\
&\quad = (3e^2Qq_{zz})^2 f'(F, I, J)
\end{aligned} \tag{14.118}$$

For the $J = 1 \leftarrow 0$ transition the second part of Eq. (14.117) will vanish. To the extent that centrifugal distortion and rotational-vibrational interactions may be neglected the second-order energy is given by

$$E_Q^{(2)} = \frac{9(e^2Qq_{zz})^2 f'(F, I, J)}{2B_0(2J+3)} \tag{14.119}$$

The splittings of the energy levels due to quadrupolar interactions are given by

$$E_Q = E_Q^{(1)} + E_Q^{(2)} = -e^2Qq_{zz}Y(J, I, F) + \frac{9(e^2Qq_{zz})^2 f'(F, I, J)}{2B_0(2J+3)} \tag{14.120}$$

and the transition frequencies will be

$$\nu_{J+1,F' \leftarrow J,F} = \nu_J - (e^2Qq_{zz})[Y(J+1, I, F') - Y(J, I, F)] + \frac{9(e^2Qq_{zz})^2}{2B_0}\left[\frac{f'(J+1, I, F')}{(2J+5)} - \frac{f'(J, I, F)}{(2J+3)}\right] \tag{14.121}$$

The observations of two hyperfine splittings will, in principle, allow the value of e^2Qq_{zz} to be determined.

In addition to the second-order nuclear electric quadrupole hyperfine interaction in DI, a small zero field magnetic interaction between the nuclear and rotational moments has also been observed [21]. This interaction can be considered to be due to the small magnetic field associated with the nuclear spin interacting with the rotational moment. The effect on the rotational energy level will be given by

$$E_I = \tfrac{1}{2}C_I[F(F+1) - J(J+1) - I(I+1)] = \tfrac{1}{2}C_I C(J, I, F) \tag{14.122}$$

so that total zero-field energy for the (J, F, I) state will be

$$E(J, I, F) = E_0(J) + E_Q(J, I, F) + E_I(J, I, F) \tag{14.123}$$

In addition to the terms given in Eq. (14.123) there will be a fourth term

$$\nu_I = \tfrac{1}{2}C_I[C(J+1, I, F') - C(J, I, F)] \tag{14.124}$$

There are now a sufficient number of analytical expressions to evaluate the experimental data on $^2H^{127}I$. The experimentally observed frequencies for the

TABLE 14.6
Observed frequencies for the $J = 1 \leftarrow 0$ transition of $^2H^{127}I$

		$H \neq 0$			
		π components		**σ components**	
Transition $F \rightarrow F'$	$H = 0$ ν, **MHz**	$\lvert M_F \rvert$ $\Delta M_F = 0$	$\lvert \Delta\nu \rvert$, **MHz**	$\lvert M_F \rvert \rightarrow \lvert M'_F \rvert$ $\Delta M_F = \pm 1$	$\lvert \Delta\nu \rvert$, **MHz**
$\frac{5}{2} \rightarrow \frac{3}{2}$	195,322.73 ± 0.2	$\frac{3}{2}$	4.69	$\frac{1}{2} \rightarrow \frac{3}{2}$	13.26
		$\frac{1}{2}$	1.56	$\frac{1}{2} \rightarrow \frac{1}{2}$	10.11
				$\frac{3}{2} \rightarrow \frac{1}{2}$	6.96
				$\frac{5}{2} \rightarrow \frac{3}{2}$	3.83
$\frac{5}{2} \rightarrow \frac{7}{2}$	195,159.65 ± 0.2	$\frac{5}{2}$	5.54	$\frac{5}{2} \rightarrow \frac{3}{2}$	
		$\frac{3}{2}$	3.33	$\frac{3}{2} \rightarrow \frac{1}{2}$	9.71
		$\frac{1}{2}$	1.10	$\frac{1}{2} \rightarrow \frac{1}{2}$	7.46
				$\frac{1}{2} \rightarrow \frac{3}{2}$	5.19
				$\frac{3}{2} \rightarrow \frac{5}{2}$	2.98
				$\frac{5}{2} \rightarrow \frac{7}{2}$	
$\frac{5}{2} \rightarrow \frac{5}{2}$	194,776.15 ± 0.2	$\frac{5}{2}$	2.24	$\frac{5}{2} \rightarrow \frac{3}{2}$	9.91
		$\frac{3}{2}$	1.33	$\frac{3}{2} \rightarrow \frac{1}{2}$	9.96
		$\frac{1}{2}$	0.44	$\frac{1}{2} \rightarrow \frac{1}{2}$	8.11
				$\frac{1}{2} \rightarrow \frac{3}{2}$	7.21
				$\frac{3}{2} \rightarrow \frac{5}{2}$	6.31

TABLE 14.7
Calculated zero-field frequencies for $J = 1 \leftarrow 0$ transition of $^2H^{127}I$

Transition $F \rightarrow F'$	$\Delta\nu_Q^{(1)}$, MHz	$\Delta\nu_Q^{(2)}$, MHz	$\Delta\nu_I$, MHz	ν_{calc}, MHz	ν_{exp}, MHz
$\frac{5}{2} \rightarrow \frac{3}{2}$	255.26	0.18	−0.64	195,322.73	195,322.73
$\frac{5}{2} \rightarrow \frac{7}{2}$	91.16	6.10	0.46	195,159.65	195,159.65
$\frac{5}{2} \rightarrow \frac{5}{2}$	−291.72	0.13	−0.18	194,776.16	194,776.15

$J = 1 \leftarrow 0$ transitions for both $H = 0$ and $H = 10{,}080 \pm 10$ gauss are given in Table 14.6. The Zeeman data is given as displacements from the $\nu_{1,F' \leftarrow 0,F}$ transition. Using the data for $H = 0$ and Eqs. (14.121) and (14.124), the nuclear quadrupole interaction terms are calculated to be as shown in Table 14.7. The fit shown is obtained with parameters of $e^2Qq_{zz} = 1823.27 \pm 0.5$ MHz, $C_I = 0.18 \pm 0.05$ MHz, and $\nu_0 = 195{,}067.93 \pm 0.15$ MHz. By using Eq. (14.106), g_F can be evaluated from the Zeeman splittings. The procedure is easily followed by referring to Fig. 14.16, which shows that the shift in a spectral line for a $\Delta M = \pm 1$ transition is the same magnitude but of opposite sign for the members of the pairs of transitions $F \rightarrow F \pm 1$ and $-F \rightarrow -(F \pm 1)$. Using Eqs. (14.114) and (14.115) the α coefficients for the $1_{7/2} \leftarrow 0_{5/2}$ transition are calculated to be $\alpha_{J=1}(\frac{5}{2}, \frac{7}{2}) = \frac{2}{7}$, $\alpha_{J=1}(\frac{5}{2}, \frac{5}{2}) = 0$, $\alpha_1(1, \frac{7}{2}) = \frac{45}{63}$, and $\alpha_I(0, \frac{5}{2}) = 1$. Using the π component $M_F = \frac{1}{2} \rightarrow \frac{1}{2}$ having a frequency $\nu = \nu_0 + 1.10$ MHz gives

$$1.10 \text{ MHz} = \nu - \nu_0 = [-\tfrac{1}{2}(\tfrac{2}{7}g_1 + \tfrac{45}{63}g_I) + \tfrac{1}{2}g_I](7.62 \times 10^{-4})H \quad (14.125)$$

Using $g_I = +1.113$ and $H = 10{,}080$ gauss this gives $g_1 = +0.1056$. The σ component $M_F = -\frac{3}{2} \rightarrow \frac{5}{2}$ has a frequency $\nu = \nu_0 + 2.98$ MHz so

$$2.98 \text{ MHz} = [\tfrac{5}{2}(\tfrac{2}{7}g_1 + \tfrac{45}{63}g_I) - \tfrac{3}{2}g_I](7.62 \times 10^{-4})H \quad (14.126)$$

or $g_1 = 0.0952$. The use of all possible transitions for the evaluations of g_1 gives as an average value $g_1 = 0.096 \pm 0.010$.

Nitric Oxide—A paramagnetic species

The nitric oxide molecule has probably been studied more by rotational spectroscopy than any other species except for NH_3 and OCS. It is actually a complex molecule from a spectroscopic point of view, not being exactly represented by any single one of Hund's cases and having a mix of electronic wavefunctions due to the proximity of the ground, $^2\Pi_{1/2}$, and first excited, $^2\Pi_{3/2}$, electronic states. We will consider only the zero field spectrum of the ground state. Numerous papers dealing with various aspects of the NO spectrum exist [12, 22–31]. The student who has further interests in these topics can review these references for a more complete discussion.

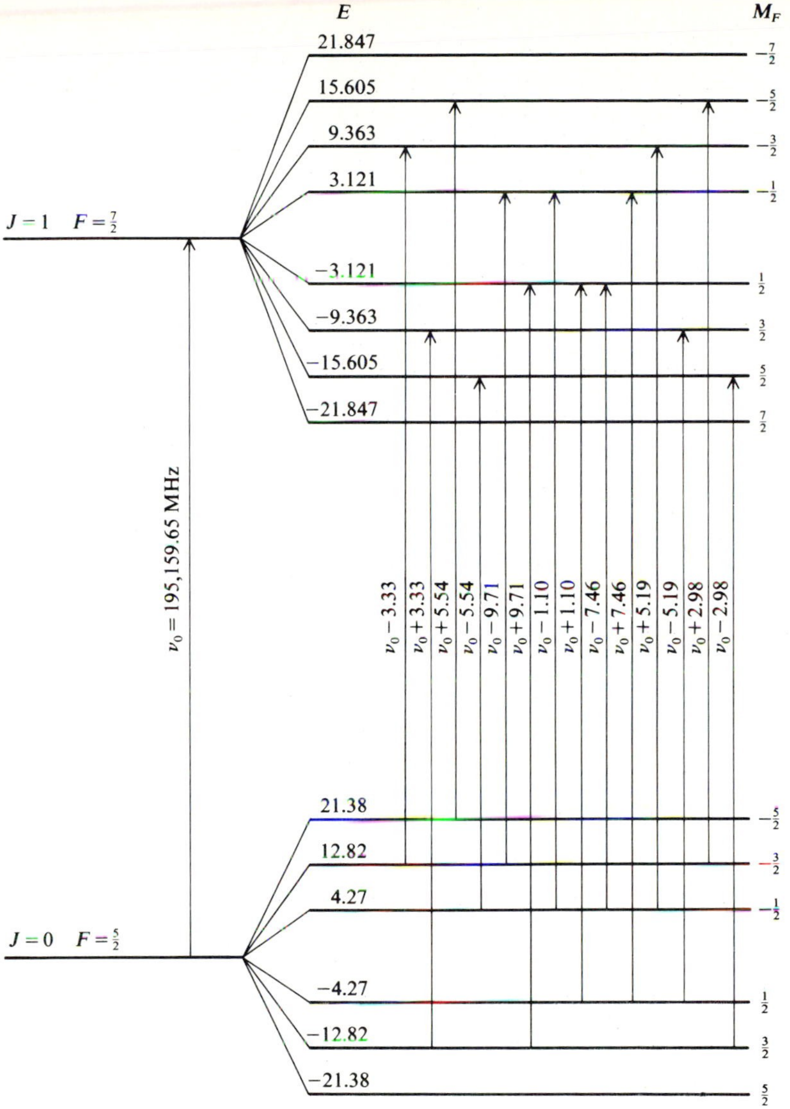

FIGURE 14.16
Zeeman energy levels and transitions for $1_{7/2} \leftarrow 0_{5/2}$ transition of $^2H^{127}I$.

To a first approximation the nitrogen nuclear spin perturbation to the rotational energy levels of $^{14}N^{16}O$ in the $^2\Pi_{1/2}$ state will be given by

$$E(J, I, F) = \frac{[a\Lambda + (b+c)\Sigma]\Omega}{2J(J+1)}[F(F+1) - I(I+1) - J(J+1)] \quad (14.127)$$

where $E(J, I, F)$ is in hertz. If this were the only consideration then the energy level scheme and spectrum would be as shown in Fig. 14.11. However, there is an additional interaction between the rotational and electronic motions of the molecule which leads to Λ-doubling. The removal of the degeneracy in Λ results in a splitting of the energy levels but the exact formulation depends on the type of coupling (Hund's case) and the type of electronic state. For the two lowest electronic levels of NO, assuming Hund's case(a) is valid, the splittings are given by

$$^2\Pi_{1/2} \qquad \Delta E_{\Lambda,J} = \rho(J + \tfrac{1}{2}) \tag{14.128}$$

$$^2\Pi_{3/2} \qquad \Delta E_{\Lambda,J} = q(J^2 - \tfrac{1}{4})(J + \tfrac{3}{2}) \tag{14.129}$$

where

$$\rho = \frac{4AB_n}{\Delta E_e} \qquad q = \frac{8B_n^3}{A\Delta E_e} \tag{14.130}$$

A is a measure of the magnitude of the coupling between S and Λ, and ΔE_c is the difference in energy between the two lowest electronic states, 124 cm^{-1} for NO. Letting

$$(a\Lambda + (b + c)\Sigma)\Omega = K_\Omega \tag{14.131}$$

$$F(F + 1) - I(I + 1) - J(J + 1) = C(J, I, F) \tag{14.132}$$

and allowing for centrifugal distortion, the energy levels are given by

$$E_{JF} = B_n J(J + 1) - D_e J^2(J + 1)^2 \pm \tfrac{1}{2}\Delta E_{\Lambda,J} + \frac{K_\Omega C(J, I, F)}{2J(J + 1)} \tag{14.133}$$

For the $^2\Pi_{1/2}$ state with $\Lambda = 1$, $\Sigma = \frac{1}{2}$, and $\Omega = \frac{1}{2}$, the energy level diagrams and allowed transitions are shown in Fig. 14.17.

The qualitative form of the rotational spectrum of the $^2\Pi_{1/2}$ state of NO as given by Eq. (14.133) and Fig. 14.17 has been shown by Gallagher, Bedard, and Johnson [31] to be a moderately good representation of the true energy level scheme. Using a more general relationship for the electronic-rotational coupling, they represented the energy levels by

$$E_{JF} = B_n[(J + \tfrac{1}{2})^2 - \Lambda^2] \pm \tfrac{1}{2}[4B_n^2(J + \tfrac{1}{2})^2 + A^2\Lambda^2 - 4AB_n\Lambda^2]^{1/2} - D_e J^2(J + 1)^2 + \frac{K_\Omega}{2J(J + 1)} C(J, I, F) \pm \tfrac{1}{2}\Delta E_{\Lambda,J} \tag{14.134}$$

For strong coupling of Λ and Σ where $A \gg B_n$, this expression reduces to

$$E_{J,F} = B_n\left(1 \pm \frac{B_n}{A\Lambda^2}\right) J(J + 1) - D_e J^2(J + 1)^2 + \frac{K_\Omega}{2J(J + 1)} C(J, I, F) \pm \Delta E_{\Lambda,J} \tag{14.135}$$

For the $^2\Pi_{1/2}$ state, this becomes

$$E_{J,F} = B_{n,\text{eff}} J(J + 1) - D_e J^2(J + 1)^2 + \frac{K_\Omega}{2J(J + 1)} C(J, I, F) \pm \tfrac{1}{2}\Delta E_{\Lambda,J} \tag{14.136}$$

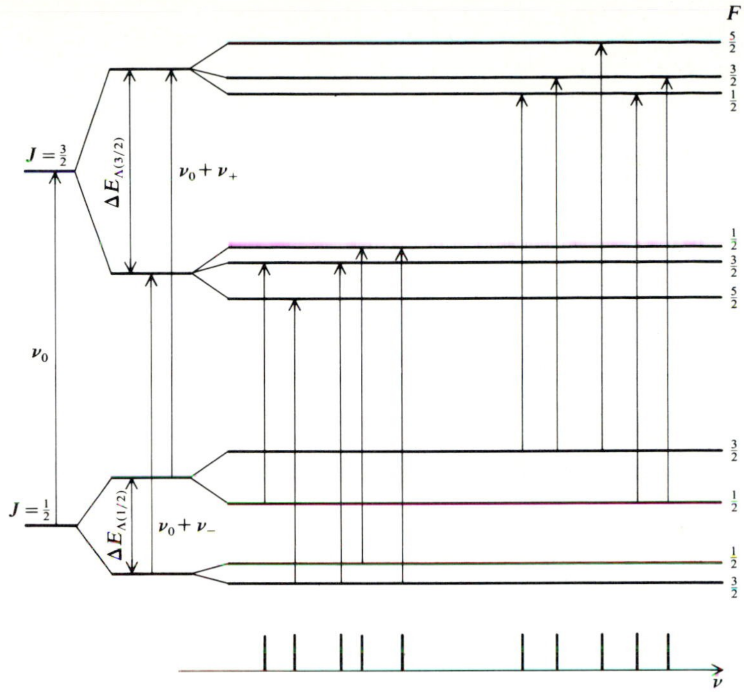

FIGURE 14.17
Schematic energy level diagram and spectrum for the $^2\Pi_{1/2}$ state of NO.

where

$$B_{n,\,\text{eff}} = B_n\left(1 - \frac{B_n}{A}\right) \tag{14.137}$$

Equation (14.136) is analogous to Eq. (14.135) with $B_{n,\,\text{eff}}$ replacing B_n so the schematic diagrams of Fig. 14.17 remain valid. The observed transitions are shown in Table 14.8. The frequencies of the observed transitions of the ground electronic state will be given, to the extent that Hund's case(a) is valid, by

$$\nu_{J+1,F'\to J,F} = 2B_{0,\,\text{eff}}(J+1) - 4D_e(J+1)^3 + K_\Omega\left[\frac{C(J+1,I,F')}{2(J+1)(J+2)} - \frac{C(J,I,F)}{2J(J+1)}\right]$$
$$\pm (\Delta E_{\Lambda,J+1} - \Delta E_{\Lambda,J}) \tag{14.138}$$

Gallagher, Bedard, and Johnson modified the magnetic interaction terms to allow for a slight deviation from Hund's case(a) and used

$$E_{J,F,I} = \frac{K_\Omega}{2J(J+1)}C(J,I,F) \pm d(J+\tfrac{1}{2})\frac{C(J,I,F)}{2J(J+1)} \tag{14.139}$$

where

$$d = \frac{g_I\beta\beta_N}{2}\left(\frac{3\sin^2\theta}{r^3}\right)_{\text{av}} \tag{14.140}$$

TABLE 14.8
Observed rotational transition for the $J = \frac{3}{2} \leftarrow \frac{1}{2}$ transition of the $^2\Pi_{1/2}$ state of NO

Λ doublet component	$F \rightarrow F'$	ν, MHz
Lower	$\frac{3}{2} \rightarrow \frac{5}{2}$	150,176.30 ± 0.25
	$\frac{1}{2} \rightarrow \frac{3}{2}$	150,195.52
	$\frac{3}{2} \rightarrow \frac{3}{2}$	150,218.57
	$\frac{1}{2} \rightarrow \frac{1}{2}$	150,225.47
	$\frac{3}{2} \rightarrow \frac{1}{2}$	150,245.38
Upper	$\frac{3}{2} \rightarrow \frac{1}{2}$	150,375.02
	$\frac{3}{2} \rightarrow \frac{3}{2}$	150,438.92
	$\frac{3}{2} \rightarrow \frac{5}{2}$	150,546.25
	$\frac{1}{2} \rightarrow \frac{1}{2}$	150,580.38
	$\frac{1}{2} \rightarrow \frac{3}{2}$	150,644.11

The coordinates θ and r are defined following Eq. (14.85) and the (+) signs applies to the upper Λ doublet. Combining Eqs. (14.135) and (14.139), letting $J = \frac{1}{2}$ and $\Omega = \frac{1}{2}$, the rotational transitions for the $^2\Pi_{1/2}$ state will be given by

$$\nu_{3/2,F'\leftarrow 1/2,F} = 3B_{0,\text{eff}} - 13.5D_e + \nu_{\pm} - \frac{8}{15}K_{1/2}\left[C\left(\frac{3}{2}, 1, F'\right) - C\left(\frac{1}{2}, 1, F\right)\right.$$
$$\left. + \frac{4d}{30}C\left(\frac{3}{2}, 1, F'\right) \mp \frac{10d}{30}C\left(\frac{1}{4}, 1, F\right)\right] \tag{14.141}$$

where

$$\nu_{\pm} = \frac{1}{2h}\left(\pm\Delta E_{\Lambda,3/2} \mp \Delta E_{\Lambda,1/2}\right) \tag{14.142}$$

and the upper signs apply to the upper Λ doublet. Equation (14.141) contains a total of six terms, $B_{0,\text{eff}}$, D_e, ν_+, ν_-, $K_{1/2}$, and d. Using the ten observed transitions they were evaluated to be

$$B_{0,\text{eff}} = 50{,}124.17 \text{ MHz}$$
$$\nu_+ = 150{,}549.90 \text{ MHz}$$
$$\nu_- = 150{,}194.76 \text{ MHz}$$
$$K_{1/2} = 46.3 \text{ MHz}$$
$$d = 112.6 \text{ MHz}$$
$$D_e = 0.53 \times 10^{-6} \text{ cm}^{-1} = 0.016 \text{ MHz}$$

where the value of D_e was taken from an infrared study [32]. Using $A = 124.2 \text{ cm}^{-1}$ and $\alpha_e = 0.0178 \text{ cm}^{-1}$ from the infrared work, additional parameters may be calculated:

$$B_0 = 50{,}817.73 \text{ MHz}$$

$$I_0 = 16.5089 \times 10^{-40} \text{ g cm}^2$$

$$r_0 = 1.1539 \text{ A}$$

$$B_e = 51{,}084.5 \text{ MHz}$$

$$I_e = 16.4226 \times 10^{-40} \text{ g cm}^2$$

$$r_e = 1.1509 \text{ A}$$

$$\nu_0 = 150{,}372.20 \text{ MHz}$$

$$\rho = 355.2 \text{ MHz}$$

Using the parameters obtained the eight hyperfine line separations as calculated show an average deviation of ±0.77 MHz. This indicates that while the formulation used is good it is not exact since this deviation is larger than the experimental errors in the line separation measurements, ±0.2 MHz.

These examples have served not only to demonstrate the utility of the concepts and analytical relationships developed in the chapter but also to emphasize the complexities which are encountered even in the rotational spectra of simple molecules. They also serve to show that each molecule must be considered individually and general theories must often be modified to accommodate specific cases.

It should also be recognized that the perturbations discussed are not all inclusive and that there are several other major phenomenon, such as higher-order Stark effects, magnetic interactions in diamagnetic molecules, and the Zeeman effect in paramagnetic molecules, which have been omitted.

REFERENCES

A. Specific

1. Hughes, H. K., *Phys. Rev.*, **76**, 1675 (1949).
2. Scharpen, L. H., J. S. Poyenter, and V. W. Laurie, *J. Chem. Phys.*, **46**, 2431 (1967).
3. Condon, E. U., and G. H. Shortley, *Theory of Atomic Spectra*, Cambridge University Press, Cambridge, England, 1953, p. 241.
4. Hirota, E., *High Resolution Spectroscopy of Transient Molecules*, Springer-Verlag, Berlin, 1985.
5. Herzberg, G., *Molecular Structure and Molecular Spectra I: Diatomic Molecules*, D. Van Nostrand Co., New York, NY, 1950.
6. Mizushima, M., *The Theory of Rotating Diatomic Molecules*, Wiley-Interscience, New York, NY, 1975.
7. Hollas, J. M., *High Resolution Spectroscopy*, Butterworths, London, GB (1982), ch. 6.
8. King, G. B., *Spectroscopy and Molecular Structure*, Holt, Rinehart & Winston, New York, NY, 1964, ch. 6.
9. Townes, C. H., and A. L. Schawlow, *Microwave Spectroscopy*, McGraw-Hill, New York, NY 1955.
10. Van Vleck, J. H., *Phys. Rev.*, **3**, 467 (1929).
11. ———, *Rev. Mod. Phys.*, **23**, 213 (1951).
12. Frosch, R. A., and H. M. Foley, *Phys. Rev.*, **88**, 1337 (1952).
13. Sutter, S. H., and W. H. Flygare, "The Molecular Zeeman Effect," in *Topics in Current Chemistry*, 63: *Bonding and Structure*, Springer-Verlag, Berlin, 1976, pp. 86–196.

14. Burrus, C. A., *J. Chem. Phys.*, **28**, 427 (1958).
15. Wacher, P. F., M. Mizushima, J. D. Peterson, and J. R. Ballard, *Microwave Spectral Tables, NBS Monograph* 70, vol. I, U.S. Govt. Printing Office, Washington, DC, 1964.
16. Cowan, M., and W. Gordy, *Phys. Rev.*, **111**, 209 (1958).
17. Pickworth, J., and H. W. Thompson, *Proc. Roy. Soc.* (*London*), **A218**, 37 (1953).
18. Burrus, C. A., *J. Chem. Phys.*, **30**, 976 (1959).
19. Racah, G., *Phys. Rev.*, **12**, 438 (1942).
20. Bardeen, J., and C. H. Townes, *Phys. Rev.*, **73**, 647, 1204 (1948).
21. Gordy, W., and C. A. Burrus, *Phys. Res.* **92**, 1437 (1953).
22. Mizushima, M., *Phys. Rev.*, **94**, 569 (1954).
23. Chin, C., and M. Mizushima, *Phys. Rev.*, **100**, 1726 (1955).
24. Mizushima, M., *Phys. Rev.*, **109**, 1557 (1958).
25. Burrus, C. A., and W. Gordy, *Phys. Rev.*, **92**, 1437 (1953).
26. Favero, P. G., A. M. Mirri, and W. Gordy, *Phys. Rev.*, **114**, 1534 (1959).
27. Beringer, R. A., and J. G. Castle, Jr., *Phys. Rev.*, **78**, 581 (1950).
28. Gallagher, J. J., and C. M. Johnson, *Phys. Rev.*, **103**, 1727 (1956).
29. Burrus, C. A., and J. D. Graybeal, *Phys. Rev.*, **109**, 1553 (1958).
30. Mizushima, M., J. J. Cox, and W. Gordy, *Phys. Rev.*, **98**, 1034 (1955).
31. Gallagher, J. S., F. D. Bedard, and C. M. Johnson, *Phys. Rev.*, **93**, 729 (1954).
32. Gillette, R. H., and E. H. Eyster, *Phys. Rev.*, **56**, 1113 (1939).

B. General

See Specific References 5-9.

Gordy, W., and R. L. Cook, *Microwave Molecular Spectra*, 2d ed., Wiley-Interscience, New York, NY, 1984.

Wollrab, J. E., *Rotational Spectra and Molecular Structure*, Academic Press, New York, NY, 1967.

PROBLEMS

14.1. For a diatomic molecule having a dipole moment, $\mu = 2.0$ D make plots of (a) $\Delta\nu$ vs. ε and (b) $\Delta\nu$ vs. ε^2 for the components of the $J = 3 \leftarrow 2$ transition. Allow ε to vary from 0 to 5000 V cm^{-1}.

14.2. For the diatomic molecule $^{35}Cl^{19}F$ construct line spectra diagrams showing the positions and relative intensities of hyperfine components of the $J = 1 \leftarrow 0, J = 3 \leftarrow 2$, $J = 6 \leftarrow 5$, $J = 10 \leftarrow 9$, and $J = 15 \leftarrow 14$ transitions. Note the general behavior with increasing J.

14.3. What would be the difference between the values of the dipole moment calculated for $^{12}C^{16}O$ using only the $J = 1 \leftarrow 0$ transition, $\nu = 115{,}271.20$ MHz to obtain B_0 rather than using $B_e = 57{,}898.57$ MHz.

14.4. Gordy and Cox [*Phys. Rev.*, **101**, 1298 (1956)] found the Zeeman σ-components of the $J = 1 \leftarrow 0$ transition for $^{12}C^{16}O$ ($\nu_0 = 115{,}271.20$ MHz) to have a separation of 4.08 MHz at 10,000 gauss. (*a*) Calculate the molecular g factor for the CO. (b) Calculate the frequencies of the two σ components.

14.5. Burrus [*J. Chem. Phys.*, **30**, 976 (1959)] has observed the hyperfine components of the $J = 1 \leftarrow 0$ transition of $^2H^{79}Br$ to be

$F \rightarrow F'$	ν
$\frac{3}{2} \rightarrow \frac{1}{2}$	254,571.69
$\frac{3}{2} \rightarrow \frac{3}{2}$	254,810.69
$\frac{3}{2} \rightarrow \frac{5}{2}$	254,678.37

Assuming first-order nuclear quadrupole interaction only determine (a) e^2Qq_{zz} for ^{79}Br($I = \frac{3}{2}$) and (b) the location of the hypothetical unsplit resonance frequency ν_0.

14.6. Using the data of Prob. 14.5 evaluate e^2Qq_{zz} and ν_0 considering second-order nuclei quadrupole interactions.

14.7. For the $F = \frac{3}{2} \rightarrow \frac{5}{2}$ transition of D^{79}Br given in Prob. 14.5 the following splittings were observed in a magnetic field of 10,080 gauss.

Π components		**σ components**	
$\lvert M_F \rvert$ $\Delta M_F = 0$	$\Delta\nu$	$\lvert M_F \rvert \rightarrow \lvert M'_F \rvert$ $\Delta M_F = \pm 1$	$\Delta\nu$
$\frac{3}{3}$	6.07 MHz	$\frac{3}{2} \rightarrow \frac{1}{2}$	13.55 MHz
$\frac{1}{2}$	2.04	$\frac{1}{2} \rightarrow \frac{1}{2}$	9.55
		$\frac{1}{2} \rightarrow \frac{3}{2}$	5.46
		$\frac{3}{2} \rightarrow \frac{5}{2}$	1.45

Taking g_I (^{79}Br) = +1.396 calculate the best value of g_J for DBr using these data.

14.8. The Zeeman spectrum of a linear molecule is analogous to that of a diatomic molecule. Cox and Gordy [*Phys. Rev.*, **101**, 1289 (1956)] found the separations between the Zeeman doublets $2\Delta\nu$ for several $^{16}O^{12}C^{32}S$ transitions to be:

$J' \leftarrow J$	ΔM_F	H, **gauss**	$2\Delta\nu$, **MHz**
8 ← 7	±1	9812	0.442
10 ← 9	±1	9812	0.442
12 ← 11	±1	9812	0.442
14 ← 13	±1	9812	0.430

Determine the value of g_J for these transitions.

CHAPTER 15

ELECTRONIC SPECTRA OF DIATOMIC MOLECULES

15.1 INTRODUCTION

In Chap. 6 it was noted that in many situations the Born-Oppenheimer separation of electronic, vibrational, and rotational energies of molecules was suitably valid to independently develop theories and discuss spectra resulting from transitions involving these energy levels. When the molecular systems are simple and the resolution of available spectroscopic instrumentation remains low, this approach is valid. However, the sophistication of current instruments is such that the effects on rotational and vibrational spectra due to interactions of these different motions can be observed. The current state of the art in optical (visible, ultraviolet) spectroscopy, especially with the advent of lasers as spectrophotometer sources, is such that spectral line structure due to the interaction of both vibrational and rotational motions with electronic motion is readily observed. Although electronic spectra are much more complex and difficult to analyze than diatomic vibrator spectra, the problem is not insurmountable. It is possible to determine many of the same molecular parameters from electronic spectra as could be obtained from rotational and vibrational spectra. In practice, however, the accuracy of the parameters is generally lower.

15.2 ENERGY LEVELS OF DIATOMIC MOLECULES

By combining the expression for the vibrational and rotational energies of diatomic molecules [Eqs. (11.29) or (11.36)] with that for the electronic energy [Eq. (6.26)], it is possible to have an analytical expression which will characterize the energy levels of the molecule. In recent years the resolution and accuracy of measurement of spectra in the visible region, by means of laser sources, has begun to approach that in the microwave region. However, except for a limited number of very high resolution studies, there is little need, for many molecules, to consider the perturbations arising from centrifugal distortion and rotational hyperfine interactions. For example, the resolution of ^{14}N hyperfine splitting will require resolution of the order of 0.0001 cm^{-1}. However, the effects of anharmonicity cannot be ignored.

The consideration of electronic spectra will proceed by looking at the coarse features of the energy levels to establish their basic relationships for several cases, relating these levels to the potential energy curves, developing the selection rules for transitions, and finally analyzing the details of the spectra of selected systems.

In very general terms, the total of the molecular energy due to electronic structure, vibration, and rotation can be written as

$$E = E_e + E_v + E_r \tag{15.1}$$

Note that in this treatment there is no inclusion of translational energy. Simple "particle-in-a-box" calculation shows that the energy differences between translational states are several orders of magnitude less than the differences between rotational energy states. Also, for most systems, the Hamiltonian can be separated into separate translation and internal energy terms. The manifold of energy levels will then appear as shown in Fig. 15.1. For purposes of qualitatively illustrating the behavior relative to simultaneous changes in both electronic and vibrational energies, diagrams such as Fig. 15.2, which relate the vibrational energy levels to the electronic potential curve, are useful. The conventional terminology for representing the energies illustrated by such curves is $T \equiv$ electronic energy term, $G_v \equiv$ vibrational energy term, and $F_J \equiv$ rotational energy term, all expressed in comparable units, generally cm^{-1}. The lower of a pair of levels is designated with a double prime, for example, T'' and the upper with a single prime. The allowed transitions are determined by a combination of specific selection rules for electronic, vibrational, and rotational transitions. These will be considered in detail following a qualitative look at spectra based on simple principles.

In more explicit terms Eq. (15.1) may be written for the ground electronic state as

$$E'' = T''_e + G''_{v''} + F''_{v''}(J'') \tag{15.2}$$

and for an excited state as

$$E' = T'_e + G'_{v'} + F'_{v'}(J') \tag{15.3}$$

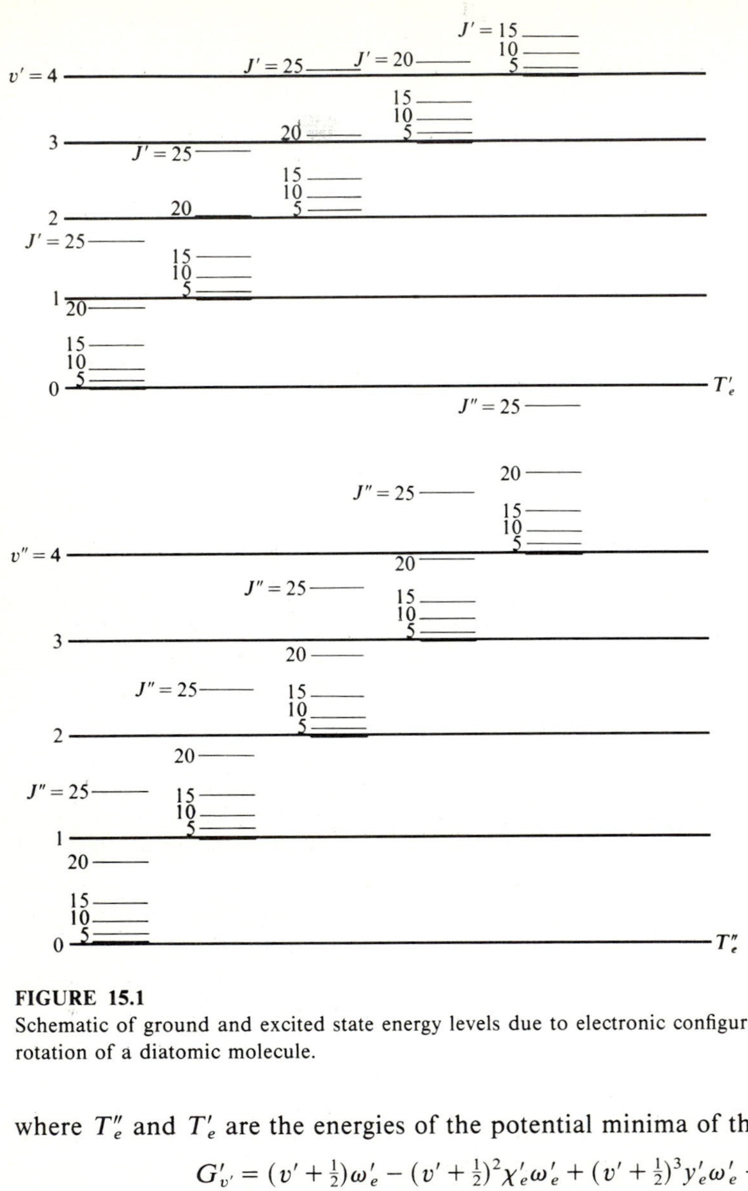

FIGURE 15.1
Schematic of ground and excited state energy levels due to electronic configuration, vibration, and rotation of a diatomic molecule.

where T''_e and T'_e are the energies of the potential minima of the two states and

$$G'_{v'} = (v' + \tfrac{1}{2})\omega'_e - (v' + \tfrac{1}{2})^2\chi'_e\omega'_e + (v' + \tfrac{1}{2})^3 y'_e\omega'_e - \cdots \tag{15.4}$$

$$F'_{v'}(J') = B'_{v'}J'(J' + 1) - D'_{ev'}J'^2(J' + 1)^2 - \cdots \tag{15.5}$$

If the difference terms $G'_v - G''_v$ and $F'_{v'}(J') - F''_{v''}(J'')$ are very small compared to $T' - T''$, the instrument resolution is sufficiently low or the molecules are in a condensed phase where rotational and vibrational motions are quenched or perturbed by collisions then, to a first approximation and assuming the transition is allowed, the electronic transition will be given

$$\nu_e = T'_e - T''_e \tag{15.6}$$

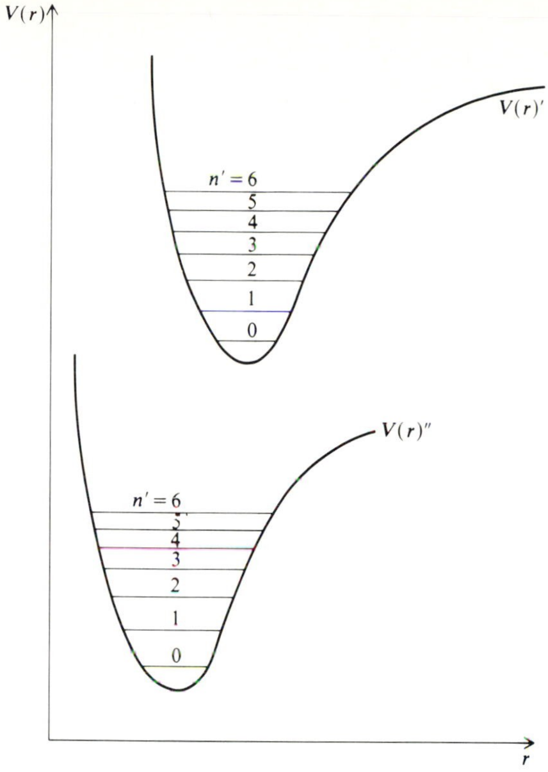

FIGURE 15.2
Potential energy curves with superimposed vibrational energy levels for the ground and an excited state of a diatomic molecule.

In practice such transitions will generally be very broad due to the unresolved rotational and vibrational structure.

The allowed electronic transitions for diatomic molecules will first be considered independent of vibrational and rotational effects and the results then combined with the selection rules already developed for vibration and rotation.

15.3 SYMMETRY OF ELECTRONIC, VIBRONIC, AND ROVIBRONIC STATES

In Chap. 6 the methods used to determine the symmetries of electronic states were reviewed in some detail. Those procedures can be summarized as

$$\Gamma(\psi_e) = \prod_i \Gamma(\psi_i^{MO}) \tag{15.7}$$

where $\Gamma(\psi_e)$ is the symmetry species of the electronic state characterized by the spatial electronic wavefunction ψ_e and $\Gamma(\psi_i^{MO})$ are the symmetry species for all of the occupied molecular orbitals of the molecule. When applying this relationship keep in mind that the symmetry species for any completely filled molecular shell will be the totally symmetric species of the molecular point group and, since the electrons in the filled molecular shell are spin-paired, $S = 0$ for these shells.

For example, the ground state of the homonuclear N_2 molecule with the electronic structure $KKLL(\pi_u 2p)^4(\sigma_g 2p)^2$ will belong to

$$\Gamma(\psi_e'') = \Sigma_g^+ \times \Sigma_g^+ \times \Sigma_g^+ \times \Sigma_g^+ \times \Sigma_g^+ \times \Sigma_g^+ = \Sigma_g^+ \tag{15.8}$$

since $S = 0$ the ground state will be a $^1\Sigma_g^+$. For an excited configuration given by $\ldots(\pi_u 2p)^4(\sigma_g 2p)^1(\pi_g^* 2p)^1$, where the $KKLL$ has been omitted in the interest of simplicity, the direct product shows that the state will belong to

$$\Gamma(\psi_e') = \Sigma_g^+ \times \Sigma_g^+ \times \Pi_g = \Pi_g \tag{15.9}$$

Since there are two partially filled molecular orbitals, we can have $S = 0$ or $S = 1$ and the corresponding states $^1\Pi_g$ or $^3\Pi_g$. Another possible excited configuration would be $\ldots(\pi_u 2p)^3(\sigma_g 2p)^2(\pi_g^* 2p)$ which gives rise to three states of symmetry species given by (see App. N)

$$\Gamma(\psi_e') = \pi_u \times \Sigma_g^+ \times \Pi_g = \Sigma_u^+ + \Sigma_u^- + \Delta_u \tag{15.10}$$

Again considering that $S = 0$ or $S = 1$ there will be six states $^1\Sigma_u^+$, $^3\Sigma_u^+$, $^1\Sigma_u^-$, $^3\Sigma_u^-$, $^1\Delta_u$, and $^3\Delta_u$.

The example just presented, N_2, is a case where unpaired electrons occur in two distinct and nondegenerate molecular orbitals. In some molecules, O_2 for example, unpaired electrons populate members of a degenerate set of molecular orbitals and the occurrence of some orbital and spin combinations are forbidden by the Pauli principle. The ground state configuration of O_2 is $\ldots(\sigma_g 2p)^2(\pi_u 2p)^4(\pi_g^* 2p)^2$ and (see App. N)

$$\Gamma(\psi_e'') = \Sigma_g^+ \times \Sigma_g^+ \times (\Pi_g \times \Pi_g) = \Pi_g \times \Pi_g = \Sigma_g^+ + \Sigma_g^- + \Delta_g \tag{15.11}$$

The direct product of two identical symmetry species will contain an antisymmetric and a symmetric part [1]. The spin wavefunction due to two electrons can be symmetric, $\psi_s^s = \alpha\alpha$, $\beta\beta$ [or $(1/\sqrt{2})(\alpha\beta + \beta\alpha)$] or antisymmetric, $\psi_s^a = (1/\sqrt{2})(\alpha\beta - \beta\alpha)$. The Pauli principle requires the total wavefunction to be antisymmetric, hence ψ_s^s can only combine with Σ_g^- and ψ_s^a can combine with either Σ_g^+ or Δ_g. The allowed symmetry states are then $^3\Sigma_g^-$, $^1\Sigma_g^+$, and $^1\Delta_g$. Hunds' rule [2] states that for a ground electronic state configuration the term of highest multiplicity lies lowest and predicts that the ground state of O_2 will contain unpaired electrons, a fact substantiated by experiment.

The characterization of the states of a heteronuclear molecule, particularly those with two atoms of comparable size, such as CO or NO, is similar to that just discussed for homonuclear molecules with the exception that the molecule will no longer possess a center of symmetry with respect to which a gerade-ungerade character for the wavefunctions can be designated. For example, the NO molecule will have a ground state configuration of $\ldots(\sigma 2p)^2(\pi 2p)^4(\pi^* 2p)^1$ and

$$\Gamma(\psi_e'') = \Sigma^+ \times \Sigma^+ \times \pi = \Pi \tag{15.12}$$

The single unpaired electron results in a $^2\Pi$ ground state. Excitation of this ground state could lead to the $\ldots(\sigma 2p)(\pi 2p)^4(\pi^* 2p)^2$ configuration which has

$$\Gamma(\psi_e') = \Sigma^+ \times (\pi \times \pi) = \Sigma^+ + \Sigma^- + \Delta \tag{15.13}$$

The designation of allowed multiplicities is considered in two steps. The two electrons in the $\pi^2 2p$ molecular orbital can have either symmetric ($S = 1$) or antisymmetric ($S = 0$) spin functions which can couple with the appropriate electron orbital wavefunction to satisfy the Pauli principle. The resulting allowed terms will be $^3\Sigma^-$, $^1\Sigma^+$, and $^1\Delta$. The lone electron in the $\sigma 2p$ molecular orbital can couple either in a parallel or antiparallel manner with those in the $\pi^* 2p$ orbital giving a total of four possible terms $^2\Sigma^-$, $^4\Sigma^-$, $^2\Sigma^+$, and $^2\Delta$.

When the heteronuclear molecule is of the type HX the resulting molecular orbitals no longer exhibit the forms depicted in Fig. 6.14, but they can still be characterized in terms of a linear combination of the hydrogen $1s$ atomic orbital and either an ns, np_z, or sp hybrid atomic orbital of comparable energy on X,

$$\psi^{MO} = a_1\psi_{1s}(H) + a_2\psi_{np_z}(X) \tag{15.14}$$

where a_1/a_2 will not be close to unity as was the case for homonuclear molecular orbitals. For such a molecule, ψ^{MO} will be occupied with a pair of electrons with $S = 0$, so the molecular term will be $^1\Sigma^+$. If we consider the molecule to be composed of an X atom having seven valence electrons, then one example of an excited state of the HX molecule will be that which occurs when an electron is promoted from one of the filled π atomic orbitals into the σ^* antibonding orbital. This results in two possible states $^1\Pi$ and $^3\Pi$.

The symmetry of the electronic wavefunctions for homonuclear diatomic molecules is characterized by two operations: (1) The inversion of the electron coordinates through the center of symmetry giving the g or u designation; (2) The reflection of all particles through a plane containing the two nuclei giving the + or − designation. For heteronuclear molecules, the inversion of electronic coordinates will lead to a different wavefunction rather than one just changed in sign. The + or − designation in this case still refers to the behavior of wavefunctions relative to reflection in the plane containing the nuclei.

The symmetry of the total molecular wavefunction not only depends on the electronic structure but also on the symmetry of the vibrational and rotational wavefunctions. An examination of the diatomic rotational wavefunctions

$$\psi_J = N_J P_J^{|M|}(\cos\theta)e^{iM\phi} \tag{15.15}$$

indicates that an inversion of coordinates, $\theta \to \pi - \theta$ and $\phi \to \phi + \pi$, results in a change of sign for ψ_J if J is odd and no change of sign if J is even. A − or + notation is used to designate the wavefunctions for these two cases. The term odd (−) or even (+) parity is used to denote this property. The symmetry of the total wavefunction

$$\psi_T = \psi_e\psi_v\psi_r \tag{15.16}$$

will be determined by that of the product function $\psi_e\psi_v\psi_r$. For a heteronuclear

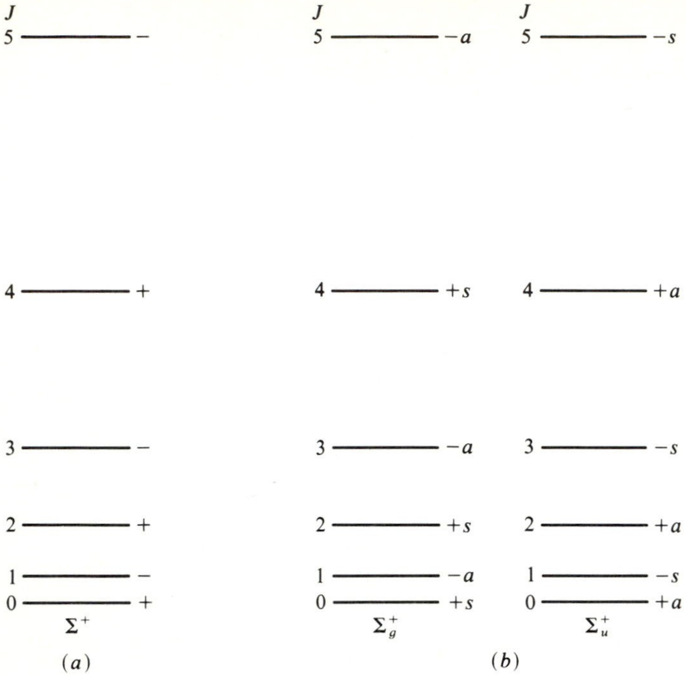

FIGURE 15.3
Symmetry properties of rotational states for (*a*) heteronuclear and (*b*) homonuclear diatomic molecules in $^1\Sigma^+$ ground electronic states.

diatomic molecule in the ground electronic and ground vibrational states the symmetry of the accessible rotational states is illustrated in Fig. 15.3.

For homonuclear molecules an additional symmetry consideration arising due to the presence of two identical nuclei must be considered. Such molecular states are denoted as symmetric (s) or antisymmetric (a) depending if the total wavefunction remains the same or changes sign upon interchanging the two identical nuclei. Depending on the symmetry of the electronic part of the total wavefunction it is possible to have two sets of combinations among the $\pm$ and the *a-s* characteristics of the rotational wavefunctions. These symmetry relationships are shown in Fig. 15.3 also.

We next need to examine in more detail the full symmetry of the electronic wavefunctions and the possible combinations with the rotational wavefunctions to give total wavefunctions. If the electronic state of a heteronuclear molecule is Σ^+, the total wavefunction will be + for a positive rotational function (even J) and will be − for a negative rotational function (odd J). For a Σ^- state just the reverse is true. For homonuclear molecules the additional symmetry, denoted by g or u, must be included and the final states are characterized further by an s or a notation. The various allowed combinations are best illustrated by use of a Herzberg type diagram [3] shown in Fig. 15.4.

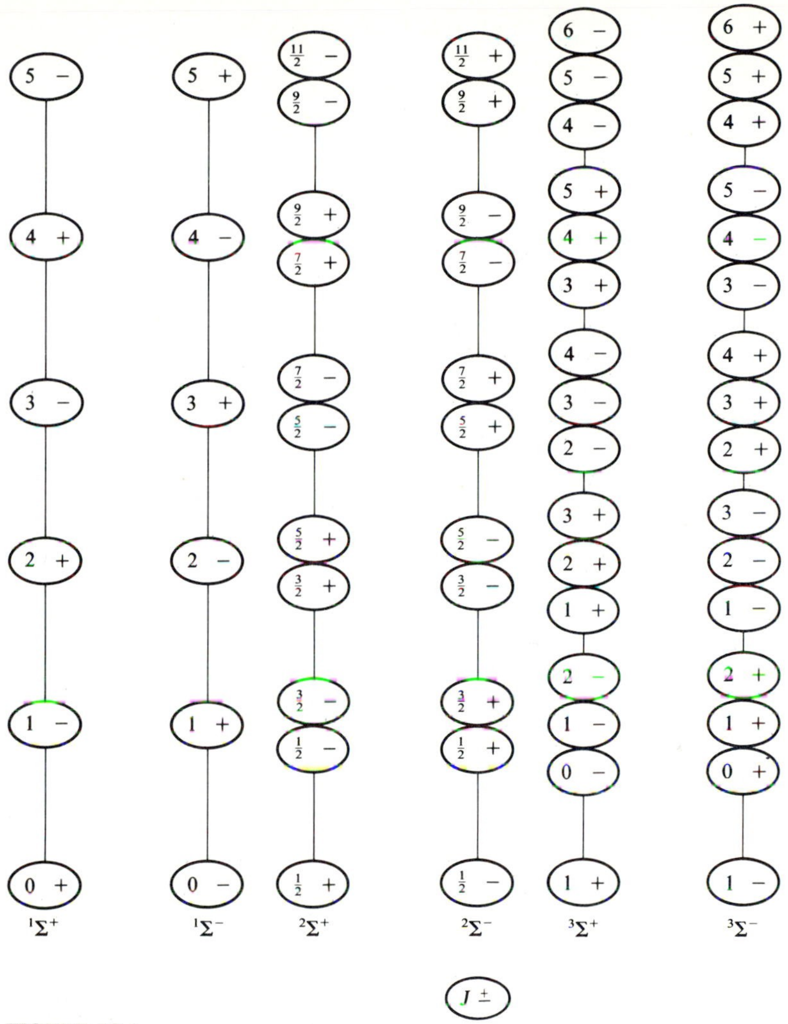

FIGURE 15.4
Symmetry properties of the rotational levels of diatomic molecules in Σ electronic states.

For states of higher total angular momentum, Π, Δ, Φ, etc., the twofold degeneracy is removed by an increasing amount of rotation, the phenomenon of Λ-doubling. For these states the members of the pair have the same value of J but one is + and the other −.

15.4 ELECTRIC DIPOLE SELECTION RULES FOR ELECTRONIC SPECTRA

Having established the symmetry of the states of diatomic molecules we next examine the selection rules which determine which states will be connected via a spectroscopic transition. As indicated in Chap. 7 the basic criteria for the occurrence of a transition is the existence of a nonzero dipole integral

$$\int \psi_{T'}^{*} \mu \psi_{T''}\, dV$$

In Secs. 11.7 and 13.6 the selection rules for pure rotational spectra and rotational Raman spectra of diatomic molecules were found to be $\Delta J = \pm 1$ and $\Delta J = \pm 2$ respectively. In reference to the symmetries of the rotational states these rules translate to "Only unlike levels combine" for pure rotational spectra

$$+ \leftrightarrow - \qquad - \leftrightarrow + \qquad - \not\leftrightarrow - \qquad + \not\leftrightarrow +$$

and "Only like levels combine" for rotational Raman spectra

$$+ \not\leftrightarrow - \qquad - \not\leftrightarrow + \qquad - \leftrightarrow - \qquad + \leftrightarrow +$$

For the dipole integral to be nonzero the integrand must be even, or the direct product $\Gamma(\psi''_T) \times \Gamma(\mu) \times \Gamma(\psi'_T)$ must either be equal to or contain the totally symmetric representation of the molecular group. Since μ is a sum of three Cartesian components (μ_g, $g = x, y, z$), each of which is a product of a charge and a coordinate, the direct product $\Gamma(\psi''_T) \times \Gamma(\psi'_T)$ will have to contain one or a combination of $\Gamma(x)$, $\Gamma(y)$, or $\Gamma(z)$ in order to have a nonzero integral. Having already established the selection rules for J we now examine the behavior of the electronic states. This necessitates the determination of the representation for the direct product $\Gamma(\psi''_e) \times \Gamma(\psi'_e)$.

Due to the possibility of different types of coupling between the orbital and spin angular moments in molecules, i.e., Hund's cases, the discussion in this section will concentrate on the development of the general electronic selection rules, and applications to the various Hund's cases will be considered in the examples.

Referring to the $C_{\infty v}$ character table (App. J), we determine that z belongs to the Σ^+ representation and (x, y) belongs to the Π representation. Therefore, for a transition between the two states ψ''_e and ψ'_e to be observed, the direct product $\Gamma(\psi''_e) \times \Gamma(\psi'_e)$ must contain either the Σ^+ or the Π representation. Looking at the table of direct products in App. N this criterion indicates that $\Sigma^+ \leftrightarrow \Sigma^+$, $\Sigma^+ \leftrightarrow \Pi$, $\Sigma^- \leftrightarrow \Sigma^-$, $\Sigma^- \leftrightarrow \Pi$, $\Pi \leftrightarrow \Pi$, $\Pi \leftrightarrow \Delta$ transitions should be allowed while $\Sigma_\pm \leftrightarrow \Delta$ and $\Pi \leftrightarrow \Phi$ transitions should be forbidden. For homonuclear molecules belonging to the $D_{\infty h}$ point group, $\Gamma(\psi''_e) \times \Gamma(\psi'_e)$ must contain either Σ_u^+ or Π_u. For this case the transitions $\Sigma_u^+ \leftrightarrow \Sigma_g^+$, Σ_g^-, $\Sigma_u^- \leftrightarrow \Sigma_g^-$, $\Pi_g \leftrightarrow \Pi_u$, $\Sigma_u^\pm \leftrightarrow \Pi_g$, $\Sigma_g^\pm \leftrightarrow \Pi_u$, $\Pi_g \leftrightarrow \Delta_u$, and $\Pi_u \leftrightarrow \Delta_g$ are allowed while $\Sigma_g^\pm \not\leftrightarrow \Sigma_g^\mp$, $\Sigma_u^\pm \not\leftrightarrow \Sigma_u^\pm$, $\Sigma_g^\pm \not\leftrightarrow \Sigma_u^\mp$, $\Pi_g \not\leftrightarrow \Pi_g$, $\Pi_u \not\leftrightarrow \Pi_u$ are forbidden.

An examination of the previous results permits the establishment of a set of three primary selection rules which are applicable to systems where Λ is a good quantum number.

1. $\Delta\Lambda = 0, \pm 1$
2. $g \not\leftrightarrow g$, $u \not\leftrightarrow u$, $g \leftrightarrow u$ (for homonuclear molecules)
3. $+ \not\leftrightarrow -$, $+ \leftrightarrow +$, $- \leftrightarrow -$ (for $\Sigma \leftrightarrow \Sigma$ transitions)

The remaining rules to be considered are those relating to the quantum numbers S, Σ, and Ω.

As in the case of atoms, so long as there are no highly charged nuclei and no spin-orbital momentum coupling exists there can be no change in multiplicity of the states involved in an electronic transition [4]. This criterion then leads to three additional selection rules

4. $\Delta S = 0$

5. $\Delta\Sigma = 0$

6. $\Delta\Omega = 0, \pm 1$

If we consider the BH molecule whose first six energy levels are given in Fig. 6.11, it is found that the ${}^1\Sigma^+ \leftrightarrow {}^1\Pi$ and ${}^1\Sigma^+ \leftrightarrow {}^1\Sigma^+$ transitions are allowed while the ${}^1\Sigma^+ \leftrightarrow {}^3\Pi$ and ${}^1\Sigma^+ \leftrightarrow {}^3\Sigma^-$ transitions are forbidden.

As in the case of atoms, the occurrence in a molecule of nuclei with high charge can lead to a breakdown of the $\Delta S = 0$ selection rule [5]. This breakdown is due to spin-orbital interaction and leads to a mixing of states of different multiplicity. A classical example of the occurrence of a $\Delta S = 1$ transition is the ${}^1\Sigma^+ \rightarrow {}^3\Pi$ transition for CO which occurs at 48,687 cm^{-1}. When transitions of this type occur they are at the expense, intensitywise, of normal $\Delta S = 0$ transitions, and the phenomenon is referred to as "intensity stealing". Figure 15.5, constructed from data in Huber and Herzberg [6], shows the singlet and triplet manifolds for CO and will be used to discuss this effect.

The spin-orbital coupling can lead to the 48,687 cm^{-1} transition "stealing" intensity either by the ${}^1\Sigma^+$ level mixing with one or more of the triplet manifold or the ${}^3\Pi$ level mixing with one or more of the singlet levels. This mixing is somewhat analogous to that occurring in the anharmonic oscillator where the normal $\Delta v = \pm 1$ selection rule is broken. Barring restrictions due to symmetry, the $X\,{}^1\Sigma^+$ state could mix with any of the triplet states; however, mixing most frequently occurs when the energy differences are minimal. Since both the $X\,{}^1\Sigma^+ \rightarrow A\,{}^1\Pi$ and the $a\,{}^3\Pi \rightarrow a\,{}^3\Sigma^+$ transitions are allowed, the $X\,'\Sigma^+ \rightarrow a\,{}^3\Pi$ transition can "steal" intensity from either of these. Generally the intensity of the "forbidden" transition comes from only one allowed transition, this being determined by symmetry considerations. A singlet-triplet transition will be allowed if the direct product $\Gamma(\psi_s) \times \Gamma(\psi_t)$ contains or equals a representation associated with x, y, or z. The same criterion holds for the connecting of singlet and triplet levels. Since the triplet level wavefunction is a product of orbital and spin functions and the spin components belong to the same representations as the rotations R_x, R_y, and R_z, it follows that there will be mixing of states if either the triplet member of the transition is connected to a singlet of comparable energy by a rotational symmetry species or the singlet member is connected to a nearby triplet in a similar manner. This means that for couplings to occur either (1) the direct products $\Gamma(X\,{}^1\Sigma^+) \times \Gamma(R_g)$ must contain the symmetry species of a connected triplet state $\Gamma(\psi^T)$, or (2) the direct product $\Gamma(\psi^s) \times \Gamma(R_g)$ must contain the representation $\Gamma(a\,{}^3\Pi)$ of the upper level of the singlet-triplet transition. Looking

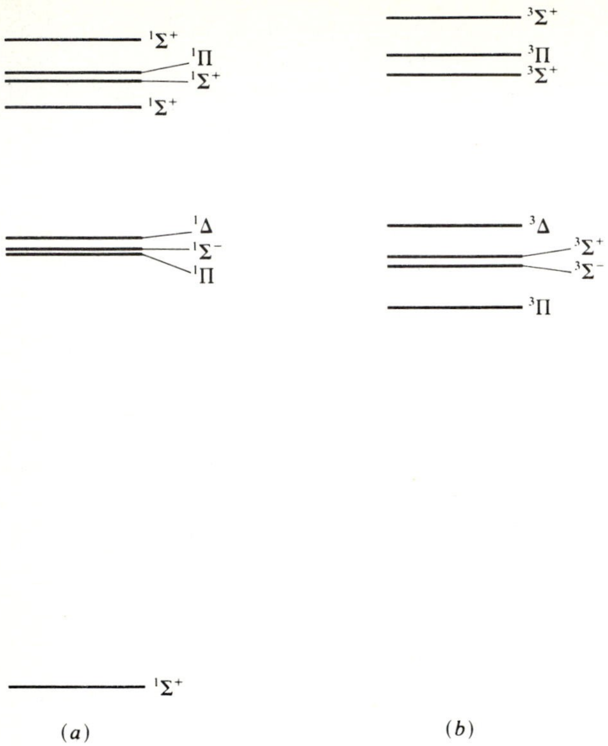

FIGURE 15.5
Energy level diagram for CO showing (*a*) the singlet manifold and (*b*) the triplet manifold.

at possible combinations of the first type we have

$$\Sigma^+ \times \Gamma(R_z) = \Sigma^+ \times \Sigma^- = \Sigma^+ \tag{15.17}$$

$$\Sigma^+ \times \Gamma(R_x, R_y) = \Sigma + \ \times \Pi = \Pi \tag{15.18}$$

Therefore, the $^1\Sigma^+$ state could couple with $c^3\Pi$ state and "steal" intensity from the $a^3\Pi \rightarrow c^3\Pi$ transition. For combinations of the second type we have

$$\Pi \times \Gamma(R_z) = \Pi \times \Sigma^- = \Pi \tag{15.19}$$

$$\Pi \times \Gamma(R_x, R_y) = \Pi \times \Pi = \Sigma^+ + \Sigma^- + \Delta \tag{15.20}$$

$$\Delta \times \Gamma(R_z) = \Delta \times \Sigma^- = \Delta \tag{15.21}$$

$$\Delta \times \Gamma(R_x, R_y) = \Delta \times \Pi = \Pi + \Phi \tag{15.22}$$

$$\Sigma^+ \times \Gamma(R_z) = \Sigma^+ \times \Sigma^- = \Sigma^- \tag{15.23}$$

$$\Sigma^+ \times \Gamma(R_x, R_y) = \Sigma^+ \times \Pi = \Pi \tag{15.24}$$

Equations (15.19), (15.22) and (15.24) show that the lowest triplet state could couple with any of the $^1\Sigma$, $^1\Pi$, or $^1\Delta$ states. Energy considerations indicate that the most probable coupling will be between either the $a^3\Pi$ and $A^1\Pi$, or the $a^3\Pi$ and $D^1\Delta$ states with the intensity being "stolen" from either the $X^1\Sigma^+ \rightarrow A^1\Pi$ or the $X^1\Sigma^+ \rightarrow A^1\Pi$ or the $X^1\Sigma^+ \rightarrow D^1\Delta$ transition. Since the latter transition has $\Delta\Lambda = 2$ it is a "forbidden" transition and the attendant loss of intensity must come from the former transition.

Another example [7] would be the $X^1\Sigma_g^+ \rightarrow a^3\Sigma_u^+$ transition in P_2 which has been observed at 18,687 cm^{-1}. The ordering of the lower states of P_2 is $X^1\Sigma_g^+ < a^3\Sigma_u^+ < b^3\Pi_g < A^1\Pi_g < C^1\Sigma_u^+ < c^3\Pi_u < B^1\Pi_u$. Since this molecule belongs to the $D_{\infty h}$ point group $\Gamma(R_z) = \Sigma_g^-$, $\Gamma(R_x, R_y) = \Pi_g$, and

$$\Sigma_g^+ \times \Sigma_{g-} = \Sigma_g^- \tag{15.25}$$

$$\Sigma_g^+ \times \Pi_g = \Pi_g \tag{15.26}$$

Thus the $^1\Sigma_g^+$ singlet state may combine with either a Σ_g^- or a Π_g triplet state and permit the $X^1\Sigma_g^+ \leftrightarrow a^3\Sigma_u^+$ to "steal" intensity from either an $a^3\Sigma_u^+ \leftrightarrow {}^3\Sigma_g^-$ or an $a^3\Sigma_u^+ \leftrightarrow {}^3\Pi_g$ transition. Since the former transition is parity-forbidden (rule 3), the latter transition will be the one to lose intensity. The second approach shows that the possible coupling of the $a^3\Sigma_u^+$ state to singlet states can be found by considering the following direct products:

$$\Pi_g \times \Sigma_g^- = \Pi_g \tag{15.27}$$

$$\Pi_g \times \Pi_g = \Sigma_g^+ + \Sigma_g^- + \Delta_g \tag{15.28}$$

$$\Sigma_u^+ \times \Sigma_g^- = \Sigma_u^- \tag{15.29}$$

$$\Sigma_u^+ \times \Pi_g = \Pi_u \tag{15.30}$$

$$\Pi_u \times \Sigma_g^- = \Pi_u \tag{15.31}$$

$$\Pi_u \times \Pi_g = \Sigma_u^+ + \Sigma_u^- + \Delta_u \tag{15.32}$$

The only product containing the Σ_u^+ species corresponding to the lowest triplet state is Eq. (15.32), hence the $X^1\Sigma_g^+ \leftrightarrow a^3\Sigma_u^+$ transition can also "steal" intensity from an $X^1\Sigma_g^+ \leftrightarrow {}^1\Pi_u$ transition. Thus, the normally forbidden $X^1\Sigma_g^+ \leftrightarrow a^3\Sigma_u^+$ transition can be observed due to the spin-orbital interaction between either the $X^1\Sigma_g^+$ and the $b^3\Pi_g$ states or the $a^3\Sigma_u^+$ and the $B^1\Pi_u$ states.

15.5 COUPLING OF ELECTRONIC AND ROTATIONAL ANGULAR MOMENTUM

The total energy or angular momentum of a diatomic molecule is a combination of the separate momenta due to electronic orbital motion, electron spin, and molecular rotation. In some cases, experimental spectra are of such accuracy and resolution that nuclear spin must also be considered, but our discussion will be limited to considering only the first three. The most elementary case is that where all electrons in the molecule are paired, $S = 0$, and the electronic orbital momentum is zero, $\Lambda = 0$, so one has a $^1\Sigma$ state and the total angular momentum is just that due to molecular rotation. The discussion of paramagnetic molecules in Chap. 14 introduced this type of coupling and in this section those primary features will be extended to cover four common Hund's cases.

Hund's case(a) applies to a molecule where the electronic angular momentum and the electron spin momentum are individually strongly coupled to the internuclear electrostatic field and have well-defined internuclear projections specified by nonzero values of $\mathbf{\Lambda}$, $\mathbf{\Sigma}$, and $\mathbf{\Omega}$. The total electronic momentum $\mathbf{\Omega}$ is then coupled to the molecular rotational angular momentum **R** to give a resultant total momentum **J**. Note that in this discussion R replaces the J used in Chaps. 11 and 12, and J denotes the total angular momentum.[1] The nature of this coupling is shown by the vector diagrams in Fig. 15.6*a* and two typical energy level schemes which illustrate the spin-orbital splitting of the level into the multiplets having different Ω values are shown in Figs. 15.6*b* and *c*. Note that Hund's case(a) is not applicable to Σ states. The rotational energy for case(a) in the absence of centrifugal distortion will be given by

$$F_v(J) = B_v[J(J+1) - \Omega^2] \tag{15.33}$$

where J can have values of $J = \Omega, \Omega + 1, \Omega + 2, \ldots$ but cannot be less than Ω.

When $\Lambda = 0$ (Σ states), the absence of the magnetic field normally present when $\Lambda \neq 0$ prevents the spin momentum from coupling to the internuclear axis and giving a well-defined quantum number Σ. For some light molecules having $\Lambda \neq 0$, the magnetic field is such that S is only weakly or not coupled to the internuclear axis. For such systems, which comprise Hund's case(b), the spin momentum **S** couples either directly with the molecular rotational angular momentum **R** or with a momentum **N** resulting from the coupling of $\mathbf{\Lambda}$ and **R** to give the total momentum **J**. These coupling schemes are shown in Fig. 15.7*a* for the former and in Fig 15.7*b* for the latter. Examples of two energy level schemes are shown in Figs. 15.7*c* and *d*. **N**, which is the total angular momentum except for spin can have values of $N = \Lambda, \Lambda + 1, \Lambda + 2, \ldots$. If $\Lambda = 0$, then $\mathbf{N} \equiv \mathbf{R}$ and $N = 0, 1, 2, 3, \ldots$. The total angular momentum **J** will have values of $J = (N + S), (N + S - 1), (N + S - 2), \ldots |N - S|$. The rotational energy term values for a $^2\Sigma$ state have been found to be [8]

$$F_v(N) = B_v N(N+1) + \tfrac{1}{2}aN \tag{15.34}$$

$$F_v(N) = B_v N(N+1) - \tfrac{1}{2}aN \tag{15.35}$$

where a is a splitting constant which is generally very small compared to B_v. More complex expressions for other states such as the $^3\Sigma$ are available in the literature [9, 10, 11].

---

[1] There is a variation among the symbols used in the literature to denote the momentum vectors for the various coupling schemes. A survey of 10 sources shows (**O**, **N**, **J**), (**N**, **K**, **J**) and (**R**, **N**, **J**) all to be used to denote (molecular, intermediate and total angular momenta).

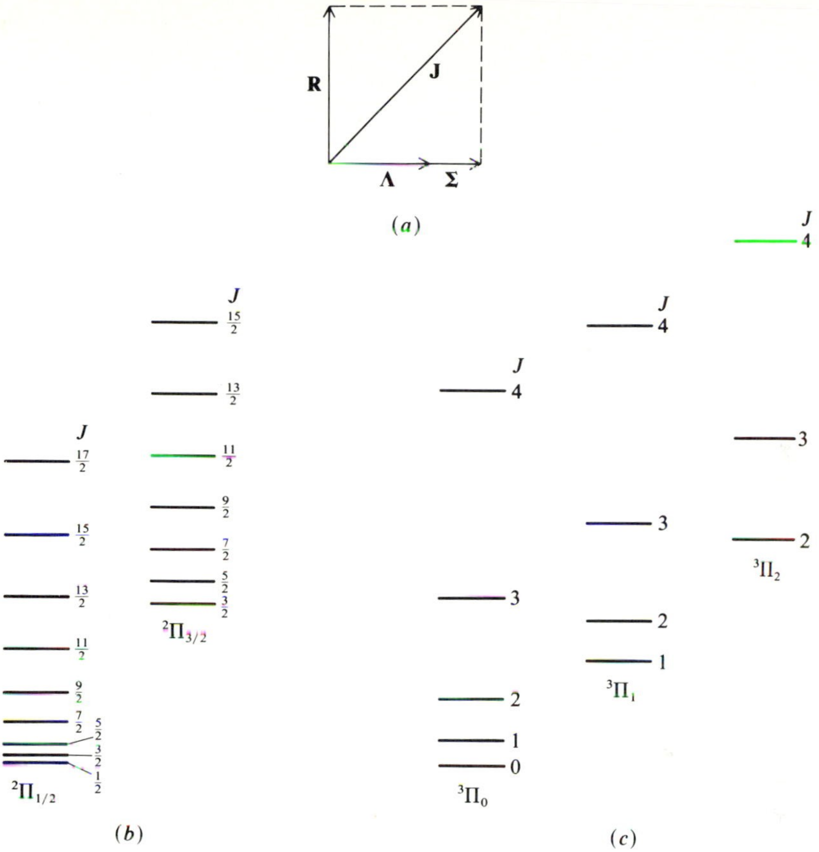

FIGURE 15.6
Hund's case(a) illustrating (*a*) the vector coupling, (*b*) the energy levels of a $^2\Pi$ state, and (*c*) the energy levels of a $^3\Pi$ state.

For some heavy molecules, the interaction between the electronic orbital momentum $\mathbf{\Lambda}$ and the spin momentum $\mathbf{S}$ is stronger than the interaction of the separate momenta with the electrostatic field along the internuclear axis, and neither Λ nor Σ are good quantum numbers. This is Hund's case(c) and is illustrated in Fig. 15.8. For this case only S, Ω, and J are good quantum numbers. Since the same momenta as in case(a) are ultimately coupled, the rotational energy term will be given by Eq. (15.33).

Hund's case(d) occurs when the coupling between the orbital angular momentum $\mathbf{L}$ and the molecular momentum $\mathbf{R}$ is much stronger than that between $\mathbf{L}$ and the electrostatic field. In this case, $\mathbf{L}$ and $\mathbf{R}$ first couple to form a resultant angular momentum $\mathbf{N}$, which will have values $N = (R + L)$, $(R + L - 1)$, $(R + L - 2), \ldots |R - L|$, and then $\mathbf{N}$ couples with $\mathbf{S}$ to form the total angular momentum $\mathbf{J}$. These couplings are illustrated in Fig. 15.9*a*. To a first approximation the

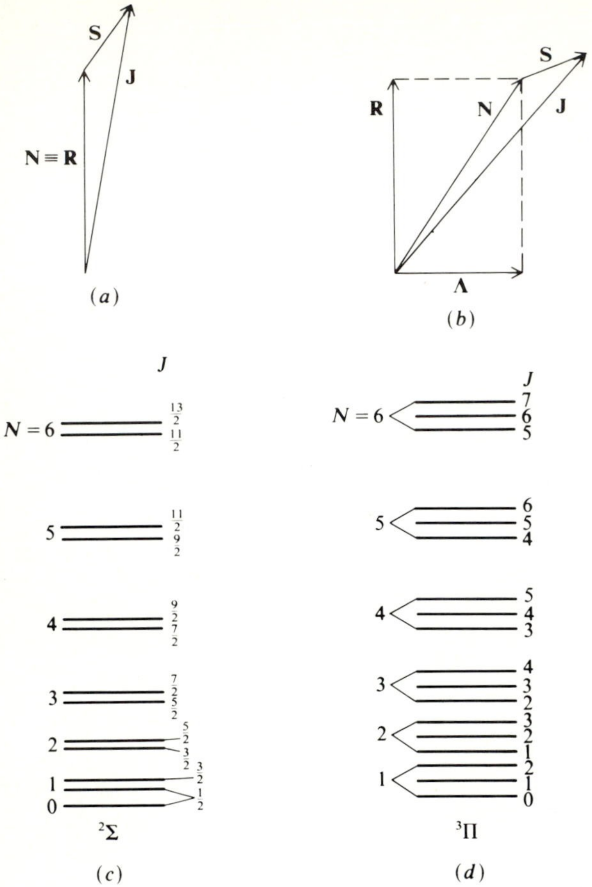

FIGURE 15.7
Hund's case(b) showing (*a*) the coupling with $\Lambda = 0$, (*b*) the coupling with $\Lambda \neq 0$, and the energy level schemes for (*c*) a $^2\Sigma$ state and (*d*) a $^3\Sigma$ state.

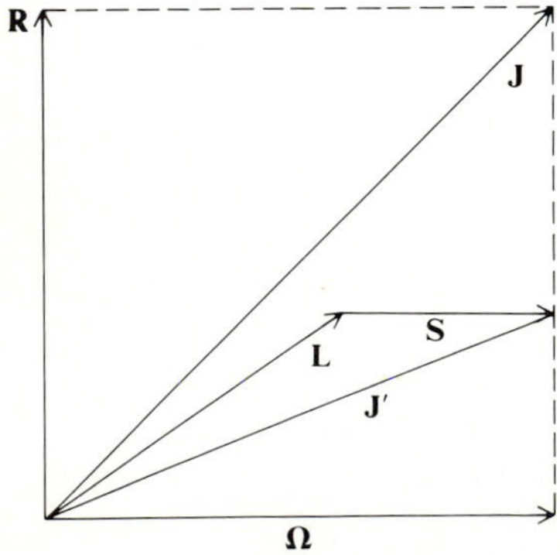

FIGURE 15.8
Hund's case(c) showing the vector coupling.

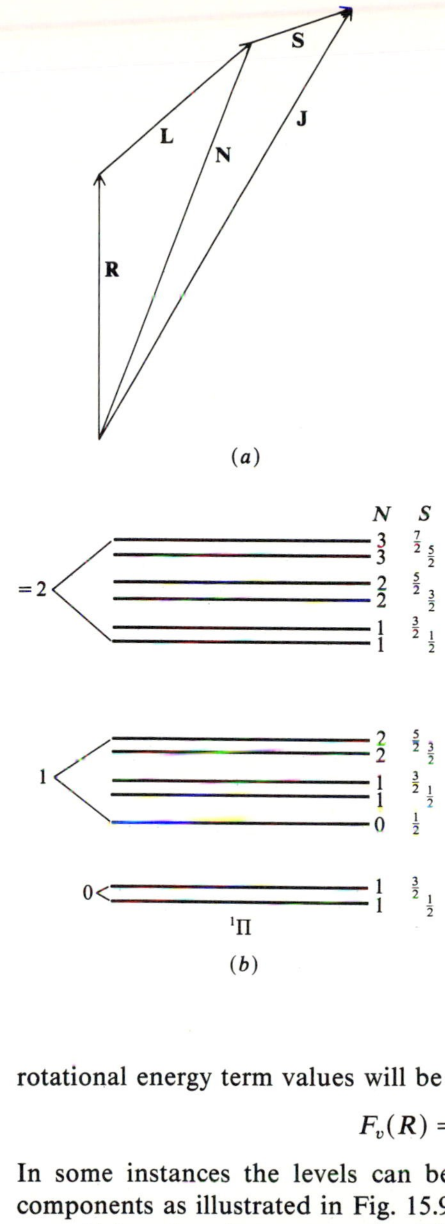

FIGURE 15.9
Hund's case(d) showing (*a*) the vector coupling and (*b*) a typical energy level diagram.

rotational energy term values will be analogous to that of a $^1\Sigma$ state

$$F_v(R) = B_v R(R+1) \tag{15.36}$$

In some instances the levels can be split further into $2L + 1$ closely spaced components as illustrated in Fig. 15.9b.

Although we have summarized the characteristics of four distinctive coupling cases it is rare that the behavior of an individual molecule conforms directly to any one. More often than not the behavior of a given molecule will be intermediate to two cases and the coupling and energy level schemes must be treated on an individual basis. The treatment of such systems is beyond the scope of this book and the student is referred to the monographs by Herzberg [9] and Mizushima [11] for a more extensive treatment.

15.6 ANALYSIS OF DIATOMIC VIBRONIC SPECTRA

Although high-resolution electronic spectra of gaseous diatomic molecules exhibits well-resolved rotational structure, it will be advantageous to first look at the hypothetical case where rotational structure is ignored. This will allow us to examine the gross features of electronic spectra and establish the first relationships necessary for spectral analysis. Another advantage of this approach is that the results obtained will closely relate to the features of electronic spectra in condensed phases where the rotation is quenched and the rotational structure is unresolved and serves only to broaden the observed transitions.

If it is assumed that the transition from an electronic state T'' to another T' is allowed, and if the effects of rotation are temporarily ignored, then the observed spectral transitions (in cm^{-1}) for the molecule will be given by the relationship

$$\nu = \frac{E' - E''}{ch} = T'_e - T''_e + G'_{v'} - G''_{v''} \tag{15.37}$$

where T'_e and T''_e are the electronic potential minima for the two states and

$$G_v = (v + \tfrac{1}{2})\omega_e - (v + \tfrac{1}{2})^2 \chi_e \omega_e - \cdots \tag{15.38}$$

Due to the different characteristics of the potential curves for the various electronic states of a molecule the spacing of the vibrational levels will change when going from one electronic state to another and hence ω_e will change. In general then $\omega'_e \neq \omega''_e$ and $\chi'_e\omega'_e \neq \chi''_e\omega''_e$. Substituting $\nu_e = T'_e - T''_e$, the frequencies of the transitions are given by

$$\nu = \{\nu_e + [(v' + \tfrac{1}{2})\omega'_e - (v' + \tfrac{1}{2})^2\chi'_e\omega'_e] - [(v'' + \tfrac{1}{2})\omega''_e - (v'' - \tfrac{1}{2})^2\chi''_e\omega''_e]\} \tag{15.39}$$

Letting ν_{00} be the transition between states having $v' = v'' = 0$

$$\nu_{00} = \nu_e + \left(\frac{\omega'_e}{2} - \frac{\chi'_e\omega'_e}{4}\right) - \frac{\omega''_e}{2} - \frac{\chi''_e\omega''_e}{4} \tag{15.40}$$

Equation (15.39) reduces to

$$\nu = \nu_{00} + (\omega'_0 v' - \chi'_0\omega'_0 v'^2) - (\omega''_0 v'' - \chi''_0\omega''_0 v''^2) \tag{15.41}$$

where $\omega_0 = \omega_e - \chi_e\omega_e$ and $\chi_0\omega_0 = \chi_e\omega_e$. For the analysis of spectra having frequencies measured to a higher accuracy, it is necessary to include higher-order anharmonicity terms as shown in Eq. (15.4).

The general nature of these vibronic transitions can be ascertained from Eq. (15.41) without including higher-order anharmonicity terms. For the transitions from the $v'' = 0$ to the v' levels expressions for the frequencies are listed, along with first and second differences, in Table 15.1. A typical order-of-magnitude calculation gives the results tabulated in Table 15.2, examination of which shows that the spectrum will consist of several series of transitions. The vertical columns list series consisting of transitions with a common lower vibrational quantum number v'', and having first and second differences related to ω'_e and $\chi'_e\omega'_e$. The

TABLE 15.1
Expressions for the vibronic transition frequencies and frequency differences of a diatomic rotor assuming no rotational interactions

v'	v''	**Transition frequency**	Δ_1	Δ_2
0	0	$\nu_e + \frac{1}{2}(\omega'_e - \omega''_e) - \frac{1}{4}(\chi'_e\omega'_e - \chi''_e\omega''_e)$		
1	0	$\nu_e + \frac{1}{2}(3\omega'_e - \omega''_e) - \frac{1}{4}(9\chi'_e\omega'_e - \chi''_e\omega''_e)$	$\omega'_e - 2\chi'_e\omega'_e$	
2	0	$\nu_e + \frac{1}{2}(5\omega'_e - \omega''_e) - \frac{1}{4}(25\chi'_e\omega'_e - \chi''_e\omega''_e)$	$\omega'_e - 4\chi'_e\omega'_e$	$2\chi'_e\omega'_e$
3	0	$\nu_e + \frac{1}{2}(7\omega'_e - \omega''_e) - \frac{1}{4}(49\chi'_e\omega'_e - \chi'_e\omega''_e)$	$\omega'_e - 6\chi'_e\omega'_e$	$2\chi'_e\omega'_e$
4	0	$\nu_e + \frac{1}{2}(9\omega'_e - \omega''_e) - \frac{1}{4}(81\chi'_e\omega'_e - \chi'_e\omega''_e)$	$\omega'_e - 8\chi'_e\omega'_e$	

TABLE 15.2
Hypothetical vibronic transitions and differences for a diatomic molecule having $\nu_e = 30{,}000\ \text{cm}^{-1}$, $\omega'_e = 2200\ \text{cm}^{-1}$, $\omega''_e = 2000\ \text{cm}^{-1}$, $\chi'_e\omega'_e = 20\ \text{cm}^{-1}$, $\chi''_e\omega''_e = 30\ \text{cm}^{-1}$, and rotational interactions ignored

Transition frequencies, cm^{-1}

	v'					
v''	0	1	2	3	4	5
0	30,097	28,037	25,917	23,737	21,497	19,197
1	32,337	30,277	28,157	25,977	23,737	21,437
2	34,617	32,557	30,437	28,257	26,017	23,717
3	36,937	34,877	32,757	30,577	28,337	26,037
4	39,297	37,237	35,117	32,937	30,697	28,397
5	41,697	39,637	37,517	55,337	33,097	30,797

First frequency differences, Δ_1, cm^{-1}

	v''					
$v'_2 - v'_1$	0	1	2	3	4	5
1-0	2240	2240	2240	2240	2240	2240
2-1	2280	2280	2280	2280	2280	2280
3-2	2320	2320	2320	2320	2320	2320
4-3	2360	2360	2360	2360	2360	2360
5-4	2400	2400	2400	2400	2400	2400

Secondary frequency differences, Δ_2, cm^{-1}

	v''					
$(v'_2 - v'_1)_2 - (v'_2 - v'_1)_1$	0	1	2	3	4	5
(2-1) − (1-0)	40	40	40	40	40	40
(3-2) − (2-1)	40	40	40	40	40	40
(4-3) − (3-2)	40	40	40	40	40	40
(5-2) − (4-3)	40	40	40	40	40	40

horizontal rows list series consisting of transitions with a common upper vibrational quantum number, and having first and second differences related to ω_e'' and $\chi_e''\omega_e''$. Figure 15.10 illustrates the relationship between the energy levels and transition frequencies for this hypothetical case.

There are two relationships among the observed series of vibronic transitions which are used to characterize observed spectra. The first of these, illustrated by Table 15.2 and on the right side of Fig. 15.10, is a progression. A progression of vibronic transitions consists of a series of lines all of which originate from or terminate at some common vibrational state and have increasing or decreasing values of $\Delta\nu$. In contrast, a sequence of vibronic transitions is a series of lines, of fixed $\Delta v = 0, \pm1, \pm2, \ldots$, where the lower vibrational state increases by 1 for each successive transition.

While this cursory treatment illustrates the general nature of the form of vibronic transitions it must be emphasized that for real molecules the spectra are more complex. In particular, the occurrence of rotational interactions introduces

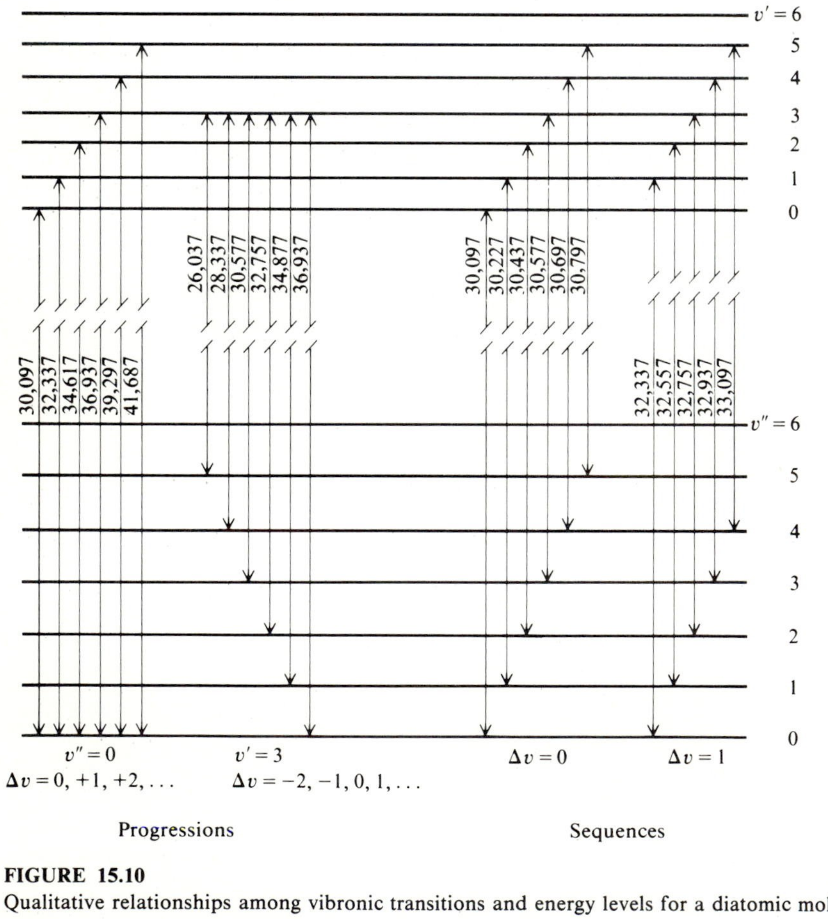

FIGURE 15.10
Qualitative relationships among vibronic transitions and energy levels for a diatomic molecule.

an additional manifold of energy states which results in each of the vibronic transitions, being the lead member of an extended series or band of rovibronic transitions. Also, in real molecules there is generally a need to include higher-order terms to fully account for anharmonicity, and the first and second differences are not as regular as indicated by the hypothetical case.

Although the observation of molecular transitions involving rotation and vibration generally involves the absorption of radiation by the sample, the observation of electronic spectra more frequently uses emission techniques. This is due to the fact that the majority of the electronic ground state molecules in a sample are also in the ground vibrational state, unless ω_e'' is low, and only $v'' = 0 \rightarrow v'$ transitions can be observed. Excitation of molecules by means of an electrical discharge or in a plasma can result in the production of a large number of species in both electronic and vibrational excited states. The decay of these species to lower states can provide a very rich emission spectrum.

Some elementary aspects of the potential functions and the harmonic oscillator wave functions of a diatomic molecule serve to describe several additional features of vibronic spectra, including information on the relative intensities of

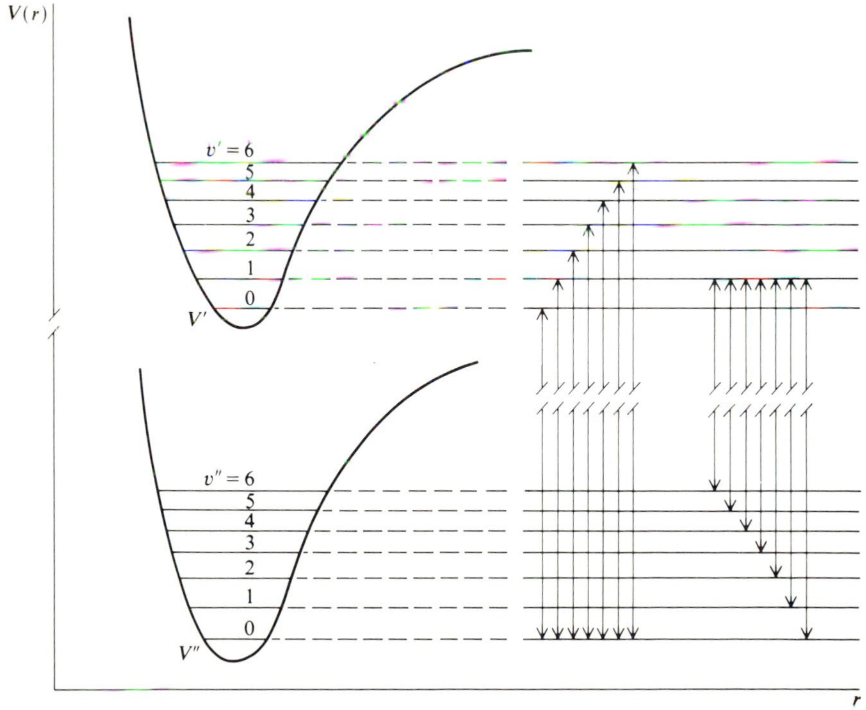

FIGURE 15.11
Illustration of the relationship between potential curves of a diatomic molecule in the ground and in an excited electronic state when the e is a small difference in equilibrium bond lengths. (The left-hand series of transitions is for a $v'' = 0 \rightarrow v'$ progression while the right-hand series is for a $v'' \rightarrow v' = 1$ progression.)

the transitions. The observation of a normal progression of vibronic bands occurs when the potential curves are related in the manner illustrated in Fig. 15.11. In this case the equilibrium displacement, r'_e, in the upper state does not differ greatly from that of the ground state, r''_e, and there is a near vertical relationship of the curves and several progressions are realizable.

This vertical relationship illustrated in Fig. 15.11 is a manifestation of the Franck-Condon principle [12], which states:

> The time required for an electronic transition to occur is so short relative to that required for a cycle of vibration that effectively the nuclear configuration of the molecule remains fixed during the electronic transition and readjusts to the new potential following the emission or absorption of radiation. Thus, the absorption of radiation by the ground state molecule, or conversely, the emission by an excited state molecule, results in the molecule being transformed to a different energy state, represented by a different potential curve, but with no change in internuclear separation.

If the upper electronic state potential curve is appreciably shifted to the right, relative to the ground state curve, then excitation from the ground state $v'' = 0$ results in the system ending up on the upper curve at a point above the asymptote. In this region there is no longer a bound state with quantized vibrational levels and the observed spectrum degrades into a continuum as opposed

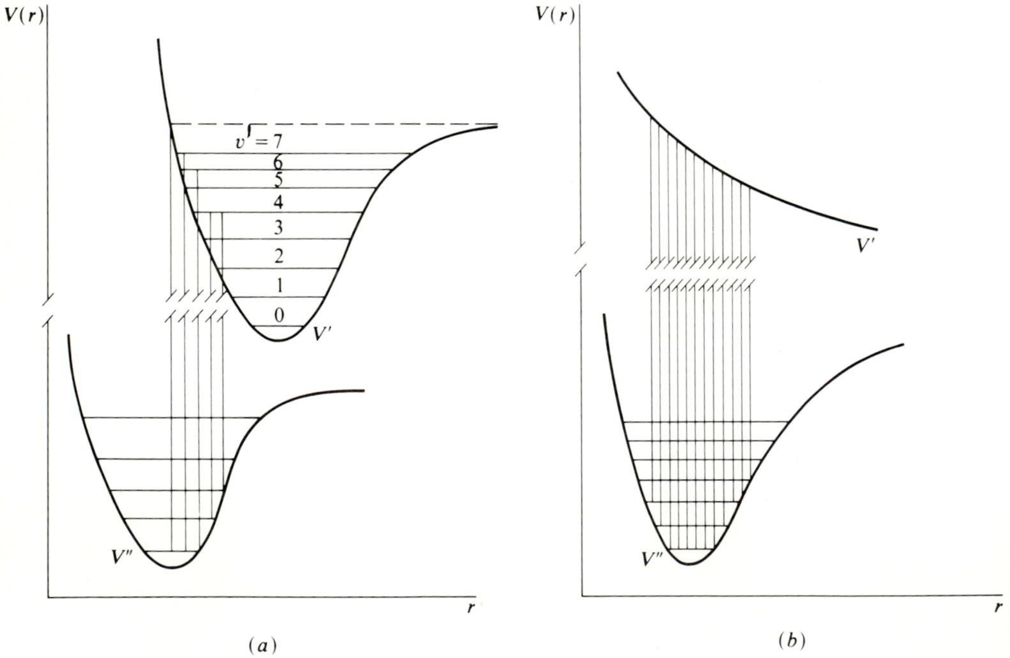

FIGURE 15.12
Illustrations of the occurrence of an absorption continuum due to (*a*) the shift of the upper state potential function relative to the ground state and (*b*) the upper state being an antibonding state.

to a discrete line spectrum. If the upper state is an antibonding state then the potential curve has no minimum and excitation leads to an absorption continuum and dissociation of the molecule. These two phenomena are illustrated in Fig. 15.12.

Using the Franck-Condon principle, pairs of potential curves on which are superimposed the harmonic oscillator wavefunctions can be used to qualitatively illustrate the relative intensities of the vibronic transitions. Figure 15.13 illustrates

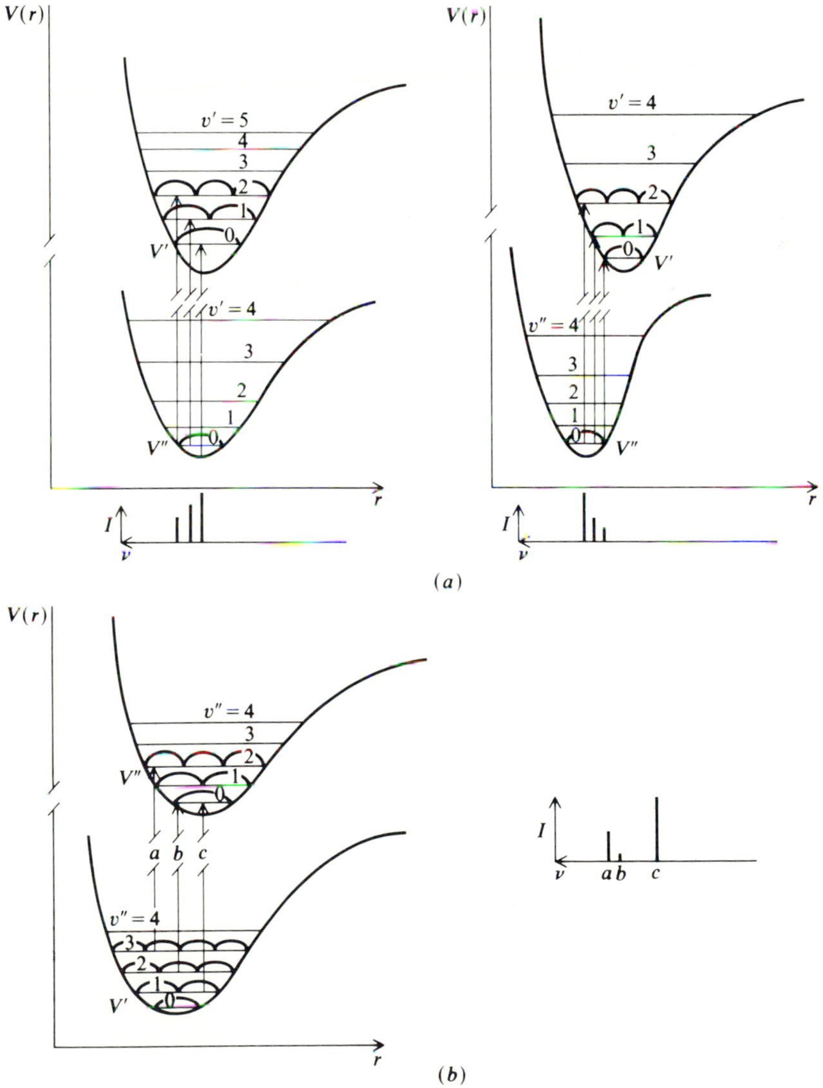

FIGURE 15.13
Illustration of the application of the Franck-Condon principle to electronic transitions originating at (*a*) $v'' = 0$ levels, (*b*) $v'' \neq 0$ levels. (Qualitative sketches of ψ_v^2 shown.)

three separate cases where the absorption originates from the $v'' = 0$ ground state level. The vibrational ground state wavefunction has its maximum at $r = r_e$ so the majority of the transitions can be considered to start at that point. There will be a distribution of oscillators along the horizontal axis direction which will result in other vertical transitions. The nature of the wavefunction for the upper state will also influence the intensity. For example, a transition which would terminate on a node would have a low intensity. The Franck-Condon principle is equally useful for consideration of transitions originating from levels other than those with $v'' = 0$. This is also illustrated in Fig. 15.13.

The Franck-Condon principle which has so far been viewed in a qualitative manner can be viewed in a more quantitative quantum mechanical manner. The intensity of a vibronic transition will be proportional to the integral

$$I = \int \psi_T'^{*} \boldsymbol{\mu}_T'' \, dV \tag{15.42}$$

where ψ_T' and ψ_T'' are the vibronic wavefunctions for the two states involved in the transition. Invoking the Born-Oppenheimer approximation ψ_T can be written as a product function $\psi_e \psi_v$. Writing the moment $\boldsymbol{\mu} = \boldsymbol{\mu}_e + \boldsymbol{\mu}_n$, where μ_e and μ_e are contributions due to electrons and nuclei respectively the integral becomes

$$I = \int\int \psi_e'^{*} \psi_{v'}'^{*} (\boldsymbol{\mu}_e + \boldsymbol{\mu}_n) \psi_e'' \psi_{v''}'' \, dV_e \, dV_v \tag{15.43}$$

Considering that $\boldsymbol{\mu}_e$ and $\boldsymbol{\mu}_n$ are separate functions of electronic and nuclear coordinates, the integral can be written

$$I = \int \psi_e'^{*} \psi_e'' \, dV_e \int \psi_{v'}'^{*} \boldsymbol{\mu}_n \psi_{v''}'' \, dV_v + \int \psi_{v'}'^{*} \psi_{v''}''^{*} \, dV_v \int \psi_e'^{*} \boldsymbol{\mu}_e \psi_e'' \, dV_e \tag{15.44}$$

Since the electronic states are orthogonal

$$\int \psi_e'^{*} \psi_e'' \, dV_e = 0 \tag{15.45}$$

the integral reduces to

$$I = I_e \int \psi_{v'}'^{*} \psi_{v''}'' \, dV_v \tag{15.46}$$

where I_e is the electronic dipole integral discussed in Sec. 15.4. The integral $\int \psi_{v'}'^{*} \psi_{v''}'' \, dV_v$ is a vibrational overlap integral which measures the degree of overlap of the upper and lower vibrational wavefunctions, and its square is known as the Franck-Condon factor.

Another phenomenon which can lead to an abnormal type of electronic spectrum is predissociation. This occurs when there is an interaction of two potential energy curves in a manner such that upon excitation the molecule moves from the stable region of one curve to the unstable region of the other by virtue of an interstate crossing or radiationless transition. This is illustrated in Fig. 15.14 for (a) the case of a low-lying antibonding state approaching the bonding ground state and (b) the case of an overlapping of two excited states. In either case the

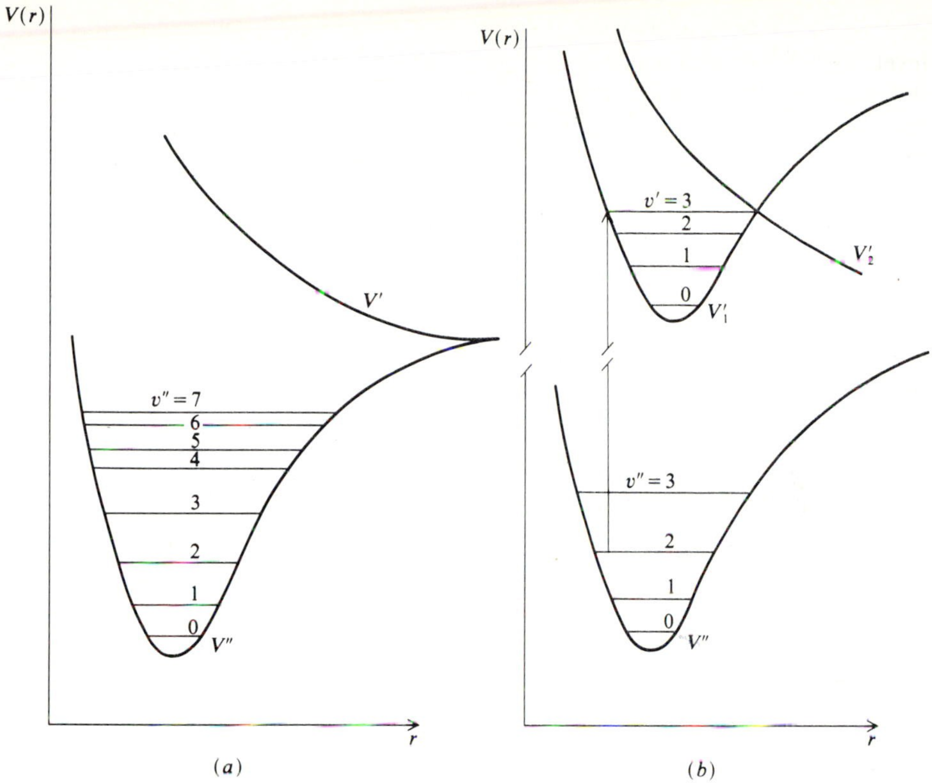

FIGURE 15.14
Illustration of predissociation from (*a*) the ground electronic state, and (*b*) an excited electronic state.

wavefunctions of two states share the same spatial region, so when a molecule is in a configuration such as to satisfy either function there will be a finite probability for it to behave according to either potential function. In the former case only vibrational excitation is necessary to cause predissociation while in the latter, vibronic excitation is needed.

Having discussed vibronic spectra in an idealized manner we now need to extend these concepts to real systems. This is done by allowing for anharmonicity. Continuing to ignore rotational effects and combining Eqs. (15.38) and (15.41) gives

$$\nu = \nu_{00} + (v'\omega_0' - v''\omega_0'') - (v'^2\chi_0'\omega_0' - v''^2\chi_0''\omega_0'') + (v'^3y_0'\omega_0' - v''^3y_0''\omega_0'') + \cdots \tag{15.47}$$

where

$$\omega_0 = \omega_e - \chi_e\omega_e + \tfrac{3}{4}y_e\omega_e \tag{15.48}$$

$$\chi_0\omega_0 = \chi_e\omega_e - \tfrac{3}{2}y_e\omega_e \tag{15.49}$$

$$y_0\omega_0 = y_e\omega_e \tag{15.50}$$

15.7 DIATOMIC VIBRONIC SPECTRA—EXAMPLE

A typical set of vibronic frequencies and frequency differences are those for the ^{7}LiH molecule tabulated in Table 15.3 [13]. Inspection of the data in this table reveals that the basic pattern is like that in Table 15.2 but the second differences are not constant.

Two useful features of the data shown in Table 15.3 are the consistencies of the differences between adjacent members in different columns as one goes across the table, or the differences between adjacent members in different rows as one goes down the table. This feature of the spectrum can serve as a guide to having made a correct vibrational assignment but by itself is not an absolute criterion. A relatively straightforward way to evaluate the vibrational constants, ω_0, $\chi_0\omega_0$, and $y_0\omega_0$, for systems where the data are not of very high precision is to employ these and successive differences with the limitation that the third differences are constant. Table 15.4 which was constructed using Eq. (15.48) shows that if terms involving powers of $(v+\frac{1}{2})^n$ with $n \geqq 3$ are neglected, this is indeed true. If the accuracy of the measured frequencies is of the same order of magnitude as the average deviation of the observed Δ_3 terms, this is a valid method. Use of Table 15.4 and the selected difference data in Table 15.5 enables one to calculate $y_0'\omega_0'$ from the average values of the Δ_3. The Δ_2 values can then be modified by this calculated parameter and an average value of $x_0'\omega_0'$ determined.

TABLE 15.3
Vibronic frequencies and frequency differences (Δ_1) for the $X^1\Sigma^X - A^1\Sigma^+$ band of the ^{7}LiH molecule
Frequencies are for the band origins and are given in cm^{-1}

	v''						
v'	0	Δ_1	1	Δ_1	2	Δ_1	3
0					23,268.52		
Δ_1					280.96		
1			24,864.32	−1314.84	23,549.48		
Δ_1			312.97		312.96		
2	26,537.02	−1359.73	25,177.29	−1314.85	23,862.44	−1270.89	22,591.55
Δ_1	335.76		335.73		335.71		335.69
3	26,872.78	−1359.76	25,513.02	−1314.87	24,198.15	−1270.91	22,927.24
Δ_1	352.80		352.78		352.80		
4	27,225.58	−1359.78	25,865.80	−1314.85	24,550.95		
Δ_1	365.83		365.87				
5	27,591.41	−1359.74	26,231.67				
Δ_1	375.64		375.56				
6	27,967.05	−1359.82	26,607.23				
Δ_1	382.68						
7	28,349.73						
Δ_1	387.55						
8	28,737.28						

TABLE 15.4
Expressions for the analysis of a $v'' = 0$ progression

Transition $v'' \to v'$	Frequency	Δ_1	Δ_2	Δ_3
$0 \to 0$	$\nu = \nu_{00}$			
		$\omega_0' - \chi_0'\omega_0' + y_0'\omega_0'$		
$0 \to 1$	$\nu = \nu_{00} + \omega_0' - x_0'\omega_0' + y_0'\omega_0'$		$-2\chi_0'\omega_0' + 6y_0'\omega_0'$	
		$\omega_0' - 3\chi_0'\omega_0' + 7y_0'\omega_0'$		$6y_0'\omega_0'$
$0 \to 2$	$\nu = \nu_{00} + 2\omega_0' - 4\chi_0'\omega_0' + 8y_0'\omega_0'$		$-2\chi_0'\omega_0' + 12y_0'\omega_0'$	
		$\omega_0' - 5\chi_0'\omega_0' + 19y_0'\omega_0'$		$6y_0'\omega_0'$
$0 \to 3$	$\nu = \nu_{00} + 3\omega_0' - 9\chi_0'\omega_0' + 27y_y'\omega_0'$		$-2\chi_0'\omega_0' + 18y_0'\omega_0'$	
		$\omega_0' - 7x_0'\omega_0' + 37y_0'\omega_0'$		$6y_0'\omega_0'$
$0 \to 4$	$\nu = \nu_{00} + 4\omega_0' - 16\chi_0'\omega_0' + 64y_0'\omega_0'$		$-2\chi_0'\omega_0' + 24y_0'\omega_0'$	
		$\omega_0' - 9x_0'\omega_0' + 61y_0'\omega_0'$		$6y_0'\omega_0'$
$0 \to 5$	$\nu = \nu_{00} + 5\omega_0' - 25\chi_0'\omega_0' + 125y_0'\omega_0'$		$-2\chi_0'\omega_0' + 30y_0'\omega_0'$	

The process is then repeated using Δ_1 to obtain ω_0'. It is to be noted that the experimental error in the data as reflected by the random behavior of the Δ_3 values indicates an uncertainty in the value of $y_e'\omega_e'$ of the order of $\pm 0.1\ \text{cm}^{-1}$.

Using the data in the two left blocks of Table 15.5, Eq. (15.50), Table 15.5, and the average value of Δ_3 ($3.04 \pm 0.6\ \text{cm}^{-1}$) gives $y_0'\omega_0' = y_e'\omega_e' = 0.51 \pm 0.1\ \text{cm}^{-1}$. Note that for this molecule all Δ_3 are negative, hence a negative value for $y_0'\omega_e'$. Using the expressions in the Δ_2 column of Table 15.4 and the left-hand Δ_2 values from Table 15.5 with $y_0'\omega_0' = 0.51\ \text{cm}^{-1}$ one obtains an average value $x_0'\omega_0' = 12.8 \pm 0.3\ \text{cm}^{-1}$ which, when incorporated into Eq. (15.49), gives $x_e'\omega_e' = -13.6 \pm 0.4\ \text{cm}^{-1}$. These two values are then used with the expressions in the Δ_1 column of Table 15.4 and the differences in the left-hand Δ_1 column of Table

TABLE 15.5
Selected frequency differences, cm^{-1}, for the $X^1\Sigma^+ - A^1\Sigma$ vibronic band of LiH

v'	v''	Δ_1	Δ_2	Δ_3	v''	v'	Δ_1	Δ_2	Δ_3
2	0				0	2			
		335.76					−1359.73		
3	0		17.04		1	2		44.88	
		352.80		−4.01			−1314.85		−0.92
4	0		13.03		2	2		43.96	
		365.83		−3.22			−1270.89		
5	0		9.81		2	3			
		375.64		−2.77					
6	0		7.04						
		382.68		−2.17					
7	0		4.87						
		387.55							
8	0								

15.3 to obtain an average of $\omega_0' = 281.3 \pm 10\ \text{cm}^{-1}$. Using Eq. (15.48) then gives $\omega_e' = 268.1 \pm 10\ \text{cm}^{-1}$. These figures reflect the accuracy obtained using optical spectrographs with film recording. Modern laser systems are readily capable of giving data with accuracies with two orders of magnitude better. Subjecting the data in the right-hand block of Table 15.5 to a similar treatment with sign reversals in the equations for the Δ_n given in Table 15.4 and assuming uncertainties comparable to those for the v' state allows one to determine $y_0''\omega_0'' = 0.15 \pm 0.1\ \text{cm}^{-1}$, $x_0''\omega_0'' = 22.9 \pm 0.3\ \text{cm}^{-1}$, and $\omega_0'' = 1382.5 \pm 10\ \text{cm}^{-1}$.

When a large number of bands can be observed, the components identified, and the frequencies of the band origins well determined, a least-squares analysis may be used. Equation (15.42) can be employed to define general expressions for frequency differences of progressions. For a $v'' = 0$ progression

$$\Delta G = [\nu_e + G_{v'2}' - G_0''] - [\nu_e + G_{v'1}' - G_0''] = G_{v'2}' - G_{v'1}' \qquad (15.51)$$

where v_1' and v_2' are successive vibrational quantum numbers ($v_2' > v_1'$) in the progression. This relationship may be written

$$\Delta G = [(v_2' + \tfrac{1}{2})\omega_e' - (v_1' + \tfrac{1}{2})\omega_e'][(v_2' + \tfrac{1}{2})^2\chi_e'\omega_e' - (v_1' + \tfrac{1}{2})^2\chi_e'\omega_e'] + [(v_2' + \tfrac{1}{2})^3 y_e'\omega_e' - (v_1' + \tfrac{1}{2})^3 y_e'\omega_e'] + \cdots \qquad (15.52)$$

For successive members of the progression v_2' will always be $v_1' + 1$, so

$$\Delta G = \omega_e' - 2\chi_e'\omega_e'(v_1' + 1) + y_e'\omega_e'(3v_1^3 + 6v_1 + \tfrac{13}{4}) \qquad (15.53)$$

This equation may be used to evaluate ω_e', χ_e', and y_e' from the Δ_1 values by either using a limited set of simultaneous equations or by means of a least-squares fit of a set of Δ_1 values. For the former method using the data in Table 15.5 the following relationships emerge:

$$v_1' = 2 \qquad 335.76 = \omega_e' - 6\chi_e'\omega_e' + 27.25 y_e'\omega_e'$$

$$v_1' = 3 \qquad 352.80 = \omega_e' - 8\chi_e'\omega_e' + 48.25 y_e'\omega_e'$$

$$v_1' = 4 \qquad 365.83 = \omega_e' - 10\chi_e'\omega_e' + 75.25 y_e'\omega_e'$$

Solution of these equations yields $\omega_e' = 260.8\ \text{cm}^{-1}$, $\chi_e'\omega_e' = -15.53\ \text{cm}^{-1}$, and $y_e'\omega_e' = -0.668\ \text{cm}^{-1}$. These values differ from those obtained by using an average value of Δ_3. Both values of ω_e' differ from the literature value [13] of 236.225 cm^{-1} which was obtained using a complete vibrational-rotational analysis. This points out the limitations of these approximate methods.

The band origins ν_{00} must be found by analysis of the rotational fine structure of the bands as will be described in the Sec. 15.9. If it is not possible to analyze this fine structure then approximate values of the vibrational constants may be found by using the frequencies of the observed band heads, the point from which the transitions tend to spread out in the long wavelength direction.

15.8 DIATOMIC VIBRONIC SPECTRA—ISOTOPE EFFECT

Frequently it is very difficult to make an absolute assignment of the v' and v'' values of a given vibronic transition. This problem can frequently be overcome if the same band can be observed for two different isotopic species. Since the potential function for a diatomic molecule depends almost entirely on the electronic and nuclear bonding interactions and not on the mass, the relative ω values will depend only on the relative masses. For a parent molecule

$$\omega_e = \frac{1}{2\pi}\sqrt{\frac{k}{\mu}} \tag{15.54}$$

while for another isotopic species

$$\omega_{ei} = \frac{1}{2\pi}\sqrt{\frac{k}{\mu_i}} \tag{15.55}$$

hence

$$\frac{\omega_{ei}}{\omega_e} = \left(\frac{\mu}{\mu_i}\right)^{1/2} = \rho \tag{15.56}$$

Recalling, insofar as the Morse potential function is valid, that

$$\chi_e = \frac{h\omega_e}{4D} \tag{15.57}$$

where D is independent of the mass, it follows that

$$\frac{\chi_{ei}}{\chi_e} = \frac{\omega_{ei}}{\omega_e} = \rho \tag{15.58}$$

The frequencies of the vibronic transitions for the parent and the other isotopic species will be given to the first anharmonicity term by

$$\nu = \nu_e + \omega_e'(v' + \tfrac{1}{2}) - \omega_e''(v'' + \tfrac{1}{2}) - \chi_e'\omega_e'(v' + \tfrac{1}{2})^2 + \chi_e''\omega_e''(v'' + \tfrac{1}{2})^2 \tag{15.59}$$

and

$$\nu_i = \nu_e + \rho\omega_e'(v' + \tfrac{1}{2}) - \rho\omega_e''(v'' + \tfrac{1}{2}) - \rho^2\chi_e'\omega_e'(v' + \tfrac{1}{2})^2 + \rho^2\chi_e''\omega_e''(v'' + \tfrac{1}{2})^2 \tag{15.60}$$

Since $\rho \approx 1$ and $\chi_e\omega_e \ll \omega_e$ as a first approximation it is safe to ignore the last two terms in both equations. Dividing this modified Eq. (15.60) by ρ, rearranging and combining with the modified Eq. (15.59) gives

$$\frac{\nu_i - \nu_e}{\nu - \nu_e} = \rho \tag{15.61}$$

We can observe from this equation, since $\rho < 1$ when ν_1 corresponds to the heavier isotope, that the entire band system of the heavier isotope will be contracted relative to the lighter one. If one is using band head frequencies rather

TABLE 15.6
Vibronic band origins, cm^{-1}, for $^1\Sigma_g^-$ $^1\Pi_u$ transitions of Li_2 molecules

v''	v'	$\nu(^7Li_2)$	$\nu_i(^6Li^7Li)$	$(\nu_i - \nu_e)/(\nu - \nu_e)$
0	0	20,397.79	20,395.24	1.0613
0	1	20,633.32	20,670.39	1.0316
0	2	20,919.78	20,939.54	1.0411
0	3	21,171.42	21,198.91	1.0376
1	2	20,571.73	20,576.40	1.0353
1	3	20,825.08	20,839.65	1.0378
1	4	21,069.55	21,092.51	1.0364
		$\nu_i = 20{,}439.4$ cm^{-1}	$[\mu(^7Li_2)/\mu(^7Li^6Li)]^{1/2} = 1.0408$	

than band origin frequencies for ν and ν_i then the presence of a rotational isotope effect can complicate the analysis and make the assignment more difficult and less certain.

If one examines the ratios based on the band origins for the Li_2 molecules given in Table 15.6 [14], it is found that they are very close to the isotopic ratio $\rho = 1.0408$. The validity of this technique is further confirmed by taking as an example the $v'' = 0 \rightarrow v' = 1$ band origins and assigning them to the closest experimentally observed lines, which have been assigned as $\nu(v'' = 6 \rightarrow v' = 10) =$ 20722.4 cm^{-1} and $\nu(v'' = 5 \rightarrow v' = 8) = 20{,}638.98$ cm^{-1}. If the assignment is $\nu(v'' = 0 \rightarrow v' = 1) = 20{,}663.32$ cm^{-1} and $\nu_i(v'' = 0 \rightarrow v' = 1) = 20{,}722.40$ cm^{-1}, $\rho = 1.26$, and if it is $\nu(v'' = 0 \rightarrow v' = 1) = 20{,}638.98$ cm^{-1} and $\nu_i(v'' = 0 \rightarrow v' = 1) =$ 20,670.39 cm^{-1}, $\rho = 9.4$. One then observes that both the consistency of the ratios given in Table 15.6 and their proximity to the isotopic ratio helps to confirm the assignment.

15.9 ANALYSIS OF DIATOMIC ROVIBRONIC SPECTRA— GENERAL TREATMENT

The nature of the rotational five structures of an electronic transition is highly dependent on the states involved in the transition. Although the general nature of the energy level scheme is as illustrated in Fig. 15.1, the details will vary depending on the values of L and S on the type of momenta coupling found for a particular molecule. The qualitative aspects of an idealized spectrum for a $^1\Sigma \leftrightarrow {}^1\Sigma$ type transition is shown in Fig. 15.15. This diagram illustrates a limited set of the series of fine structures associated with two vibronic transitions, a $v'' = 0 \rightarrow v' = 0$ (00) band and a $v'' = 0 \rightarrow v' = 1$ (01) band. As with rotational-vibrational spectra there will be three branches, $R(J' - J'' = 1)$, $Q(J' - J'' = 0)$, and $P(J' - J'' = -1)$ associated with each transition. Depending on the spacing

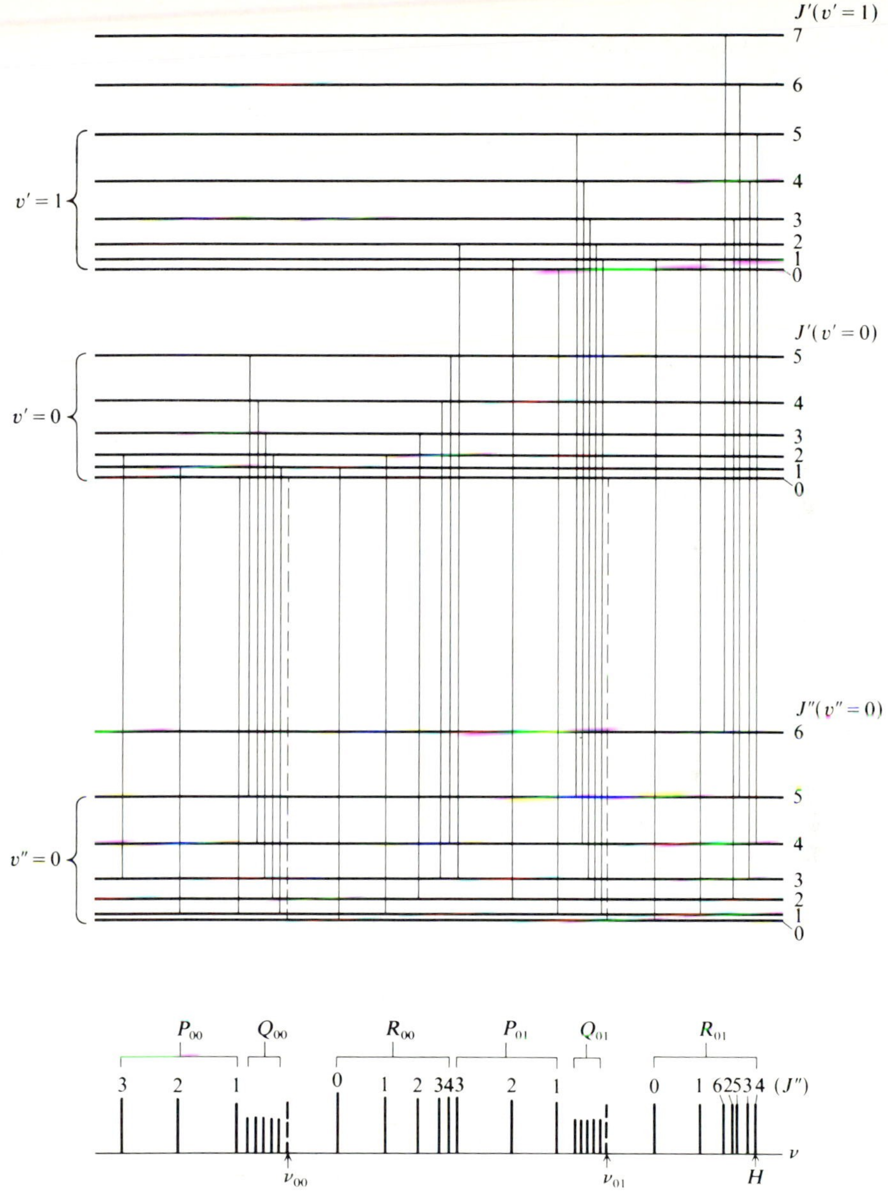

FIGURE 15.15
Hypothetical band spectrum for a diatomic molecule and its relationship to the molecular energy levels.

of the vibrational energy levels the bands may possibly overlap. Each band will be characterized by a band origin denoted by $\nu_{v''v'}$ on the frequency axis and by a band head denoted for the ν_{01} transition by point H. The band head, as discussed in Chap. 12, arises from the J^2 dependency of $(B'_{v'} - B''_{v''})$ becoming dominant for large J values. In practice the value of J at which a band head occurs is much larger than the $J'' = 4$ value which was used for illustrative purposes in the figure. Another point to note relates to the transitions at the band origin. For a $^1\Sigma \leftrightarrow {}^1\Sigma$ electronic transition the selection rule for changes in rotational energies is $\Delta J = \pm 1$ so the Q branches are missing. For transitions from $^1\Sigma$ states to electronic states with $\Lambda \neq 0$, and hence a low J value of at least 1, the selection rule $\Delta J = 0$ is applicable, but there will be no $J = 0$ level for the upper state, and the transitions for the P, Q and R branches will have J'' values of 2, 1, and 0 respectively.

Examination of Fig. 15.15 shows that the frequencies of the band origins are independent of J and are just the vibronic transition frequencies discussed in Sec. 15.6. We can now develop an analytical expression for the analysis of the fine structure by combining Eqs. (15.2) and (15.3) to give

$$\nu = (T'_e - T''_e) + (G'_{v'} - G''_{v''}) + (F'_{v'}(J') - F''_{v''}(J'')) \tag{15.62}$$

or

$$\nu = \nu_0 + (F'_{v'}(J') - F''_{v''}(J'')) \tag{15.63}$$

where ν_0 is the frequency corresponding to the band origin. Incorporation of Eqs. (15.5) leads to

$$\nu = \nu_0 + [B'_{v'}J'(J'+1) - B''_{v''}J''(J''+1)] + [D'_{ev'}J'^2(J'+1)^2 - D''_{ev''}J''^2(J''+1)^2] \tag{15.64}$$

The errors in the experimental values of rotational constants determined by emission spectroscopy are of the order of 1 percent and those using laser spectroscopy 2 orders of magnitude less. Except for very light molecules, D_e values tend to be less than 0.001 percent of B values. Therefore, even though centrifugal distortion is of importance in pure rotational spectroscopy, for the analysis of many electronic spectra of heavy molecules the terms involving D_e are ignored, unless very high resolution spectra are available.

Equation (15.64) may now be rewritten for each type of transition $R(J' = J'' + 1)$, $Q(J' = J'')$, and $P(J' = J'' - 1)$:

$$\nu_R = \nu_0 + 2B'_{v'} + (2B'_{v'} - B''_{v''})J'' + (B'_{v'} - B''_{v''})J''^2 \tag{15.65}$$

$$\nu_Q = \nu_0 + (B'_{v'} - B''_{v''})J''(J''+1) \tag{15.66}$$

$$\nu_P = \nu_0 - (B'_{v'} + B''_{v''})J'' + (B'_{v'} - B''_{v''})J''^2 \tag{15.67}$$

One recognizes that, with the exception of ν_0 now containing an electronic component, those equations are analogous to those developed for the analysis of rotational-vibrational spectra in Chap. 13. There will be primary difference, however. For the case of the diatomic vibrotor the reciprocal moments incorporated into the equations were for different vibrational states of the same electronic state. For a given electronic state the value of B_v will decrease with increasing v as the effective internuclear separation increases. Therefore, the general appearance of all diatomic rotational-vibrational spectra will be the same. For rovibronic spectra, however, $B'_{v'}$ and $B''_{v''}$ characterize molecules in different electronic states, and there is no consistency in the order of their magnitudes. For example, the following values (cm^{-1}) of B_e for $^{12}C^{16}O$ in the listed states (increasing order of energy) have been observed [14]. $^1\Sigma^+$ (1.93), $^3\Pi$ (1.69) $^3\Sigma^+$ (1.34), $^3\Sigma^-$ (1.28), $^1\Pi$ (1.61). Contrasted to this are the values for $^{14}N^{16}O$: $^2\Pi_{1/2}$ (1.67), $^2\Pi_{3/2}$ (1.72), $^2\Sigma^+$ (1.99). These differences in B_e values among electronic states are to be contrasted with the difference between electronic ground state values of B_0 and B_1 of the order of 0.02 cm^{-1}. Only the R-branch transitions can exhibit a band head phenomenon in a rotational-vibrational spectrum and then generally only at a very high J value ($J \approx 60$–100). Either R-branch or P-branch rovibronic transitions can exhibit this feature depending on the relative magnitudes of $B'_{v'}$ and $B''_{v''}$.

The analysis of rovibronic spectra is aided by the use of Fortrat diagrams based on a general equation which is derived from Eqs. (15.65) and (15.67). If Eq. (15.65) is rearranged to include the term $(B'_{v'} + B''_{v''})$ as a factor, and compared to Eq. (15.67), it is observed that the single equation

$$\nu = \nu_0 + (B'_{v'} + B''_{v''})m + (B'_{v'} - B''_{v''})m^2 \qquad (15.68)$$

with $m = J'' + 1$ for R-branch and $m = -J''$ for P-branch transitions results. A typical Fortrat diagram is illustrated in Fig. 15.16. For those spectra where Q-branch transitions exist, Eq. (15.66) with $J'' = m$ may be used to include this branch on the Fortrat diagram.

Another useful aspect of Eq. (15.68) is its use in rapidly providing approximate values for the rotational constants from a minimum of measurements. If this equation is viewed as a function which is continuous in the variable m, then it will exhibit an extremum at the value of m_H corresponding to the band head

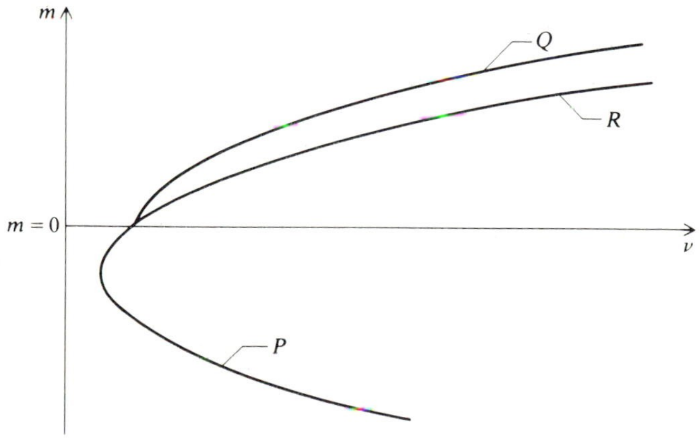

FIGURE 15.16
Fortrat diagrams for a hypothetical transition.

frequency ν_H. Thus,

$$\frac{d\bar{\nu}}{dm} = (B'_{v'} + B''_{v''}) + 2(B'_{v'} - B''_{v''})m = 0 \tag{15.69}$$

or

$$m_H = -\frac{(B'_{v'} + B'_{v''})}{2(B'_{v'} - B''_{v''})} \tag{15.70}$$

Substitutions of this expression into Eq. (15.68) will now give

$$\nu_H - \nu_0 = -\frac{(B'_{v'} + B''_{v''})^2}{4(B'_{v'} - B''_{v''})} \tag{15.71}$$

In general the solution of Eq. (15.70) will not give an integer value of m_H, but it will be sufficiently close to be useful.

The evaluation of rotational constants can also proceed via a difference analysis as was done for rotational-vibrational spectra. Equations (15.65) to (15.67) may also be written

$$\nu_R = \nu_0 + F'_{v'}(J'' + 1) - F''_{v''}(J'') \tag{15.72}$$

$$\nu_Q = \nu_0 + F'_{v'}(J'') - F''_{v''}(J'') \tag{15.73}$$

$$\nu_P = \nu_0 + F'_{v'}(J'' - 1) - F''_{v''}(J'') \tag{15.74}$$

If we define a difference function $\Delta_2 F''(J'')$ as

$$\Delta_2 F'(J'') = F'_{v'}(J'' + 1) - F'_{v'}(J'' - 1) \tag{15.75}$$

and incorporate Eqs. (15.72), (15.74), (15.65) and (15.67) into this expression there emerges the relationship

$$\Delta_2 F'(J'') = \nu_R(J'') - \nu_P(J'') = 2B'_{v'}(2J'' + 1) \tag{15.76}$$

where $\Delta_2 F'(J'')$ is the difference between R- and P-branch transitions originating from a common level J''. A comparable treatment of a second difference term

$$\Delta_2 F''(J'') = F''_{v''}(J'' + 1) - F''_{v''}(J'' - 1) \tag{15.77}$$

leads to

$$\Delta_2 F''(J'') = \nu_R(J'' - 1) - \nu_P(J'' + 1) = 2B''_{v''}(2J'' + 1) \tag{15.78}$$

where $\Delta_2 F''(J'')$ is the difference between R- and P-branch transitions terminating at a common level.

For some transitions the Q branch will be forbidden or weak but for those where they exist some additional relationships which are readily obtained using Eqs. (15.72) to (15.74) can be useful for the analysis of spectra. In the absence of any multiplicities of energy levels due to values of $S \geqq \frac{1}{2}$ or Λ-type doubling a series of first differences are defined as

$$\Delta_1 F''(J'') = F''_{v''}(J'' + 1) - F''_{v''}(J'') = \nu_R(J'') - \nu_Q(J'' + 1) = \nu_Q(J'') - \nu_P(J'' + 1) \tag{15.79}$$

$$\Delta_1 F'(J'') = F'_{v'}(J''+1) - F'_{v'}(J') = \nu_R(J'') - \nu_Q(J'') = \nu_Q(J''+1) - \nu_P(J''+1) \tag{15.80}$$

In many instances it is not possible to ignore multiplicities or Λ-type doubling and these combination relations must be modified. This will be considered with respect to individual cases in Sec. 15.10.

Although the difference analysis just presented is useful for the assignment and analysis of rovibronic spectra, frequently it alone or in combination with a Fortrat analysis [Eqs. (15.70) and (15.71)] is insufficient to be certain of an assignment. One problem which arises is the absence of an observed transition at the band origin either due to the $\Delta J = 0$ selection rule being operative for Σ-Σ transitions or the absence or $J' = 0$ rotational levels for transitions involving states having $\Lambda \neq 0$. Not only are the positions of band origins needed for correct assignments but also they are less affected by perturbations and are best suited for obtaining accurate molecular constants.

Combining Eqs. (15.65) and (15.67) for the R-branch transitions originating at $J'' - 1$ and the P-branch transition originating at J'' gives the expression

$$\nu_R(J''-1) + \nu P(J'') = 2\nu_0 + 2(B'_{v'} - B''_{v''})J^2 \tag{15.81}$$

If the left-hand frequency sum is plotted vs. J^2 and the resulting points fit using a linear least-squares analysis, the slope gives $2(B'_{v'} - B''_{v''})$ and the intercept is $2\nu_0$. If a transition has a resolved Q branch then Eq. (15.66) may be used to construct a plot of ν_Q vs. $J''(J''+1)$ which will have a slope of $(B'_{v'} - B''_{v''})$ and an intercept of ν_0.

15.10 ANALYSIS OF DIATOMIC ROVIBRONIC SPECTRA—PARTICULAR CASES

In the last section the analysis of an idealized $^1\Sigma \leftrightarrow {}^1\Sigma$ rovibronic transition was discussed in some detail. The methods presented are applicable to all types of diatomic molecules; however, the detailed energy level diagrams and the specifics of the assignments will depend on the type of angular momentum coupling which exists for any individual molecule. Prior to a discussion of selected experimental examples we will look at the specific energy level diagrams and selection rules relating to some common cases. The nature of the energy level structure of diatomic molecules was reviewed in Chap. 6, and the electronic selection rules and various coupling cases were presented earlier in this chapter, so this section will collect together the important features of these topics as they relate to the individual cases.

The normal selection rules which are applicable to diatomic rovibronic transitions are summarized in Table 15.7. In order to apply these selection rules it is necessary to ascertain the symmetry (parity) of each the total wavefunctions involved in a transition. The parities of the rotational wavefunctions for states

TABLE 15.7
Diatomic rovibronic selection rules

	Heteronuclear	Homonuclear
Molecular rotation	$\Delta R = \pm 1$; $+ \leftrightarrow -$	$\Delta R = \pm 1$; $+ \leftrightarrow -$
Vibration	$\Delta v = \pm$ integer	$\Delta v = \pm$ integer
Orbital electronic	$\Delta\Lambda = 0, \pm 1$; $+ \leftrightarrow +$, $- \leftrightarrow -$	$\Delta\Lambda = 0, \pm 1$; $+ \leftrightarrow +$, $- \leftrightarrow -$; $g \leftrightarrow u$
Electron spin	$\Delta S = 0$	$\Delta S = 0$
Total angular momenta	$\Delta J = 0, 1$ $+ \leftrightarrow -$	$\Delta J = 0, \pm 1$; $+ \leftrightarrow -$
Common electronic transitions	$\Sigma^+ \leftrightarrow \Sigma^+$, $\Sigma^- \leftrightarrow \Sigma^-$ $\Pi \leftrightarrow \Pi$, $\Delta \leftrightarrow \Delta$ $\Sigma^+ \leftrightarrow \Pi$, $\Sigma^- \leftrightarrow \Pi$ $\Pi \leftrightarrow \Delta$	$\Sigma_g^+ \leftrightarrow \Sigma_u^+$, $\Sigma_g^- \leftrightarrow \Sigma_u^-$ $\Pi_g \leftrightarrow \Pi_u$, $\Delta_g \leftrightarrow \Delta_u$ $\Sigma_g^+ \leftrightarrow \Pi_u$, $\Sigma_u^+ \leftrightarrow \Pi_g$ $\Sigma_g^- \leftrightarrow \Pi_u$, $\Sigma_g^- \leftrightarrow \Pi_u$ $\Pi_g \leftrightarrow \Delta_u$, $\Pi_u \leftrightarrow \Delta_g$

having integer ($S = 0$) and half-integer ($S \neq 0$) values are given by $+$ for $(-1)^J > 0$, $-$ for $(-1)^J < 0$, $+$ for $(-1)^{J-1/2} > 0$ and $-$ for $(-1)^{J-1/2} < 0$. The symmetry of the total wavefunction for any particular state of a heteronuclear molecule will be determined by use of the products

$$(+)_r \times (+)_e = (-)_r \times (-)_e = (+)_t \tag{15.82}$$

$$(+)_r \times (-)_e = (-)_r \times (+)_e = (-)_t \tag{15.83}$$

where the subscripts denote rotational, electronic, and total wavefunctions. For homonuclear molecules they become

$$(+)_r \times (g)_e = (-)_r \times (u)_e = (S)_t \tag{15.84}$$

$$(+)_r \times (u)_e = (-)_r \times (g)_e = (a)_t \tag{15.85}$$

Employing a diagram based on those derived by Herzberg [9], the symmetries of the first few energy levels for several common electronic states in addition to those shown in Fig. 15.4 are illustrated in Figs. 15.17 and 15.18. In these figures energy increases upward but the separations are constant and not proportional to the real splittings in any particular molecule. For singlet states of molecules having $\Lambda \geqq 1$ and exhibiting Λ-type doubling, the doublet levels are denoted by adjacent pairs of ovals. For Σ states of multiplicity greater than 1, adjacent ovals denote the multiplet levels for a given N value. The extent of adjacent ovals is increased for cases where both Λ-type doubling and spin multiplicities occur. Each oval in the figures contains the total J value for the level and the symmetry (parity) of the state. For the figure relating to homonuclear molecules the additional *a-s* symmetry property is included. We previously discussed the phenomenon of Λ-type doubling and pointed out that the net effect of it was to remove the Λ degeneracy in the energy levels having $\Lambda \geqq 1$ and thus produce two closely spaced electronic levels denoted by ψ_e^+ and ψ_e^-. The ordering of these states will vary depending on the individual molecule, but in general

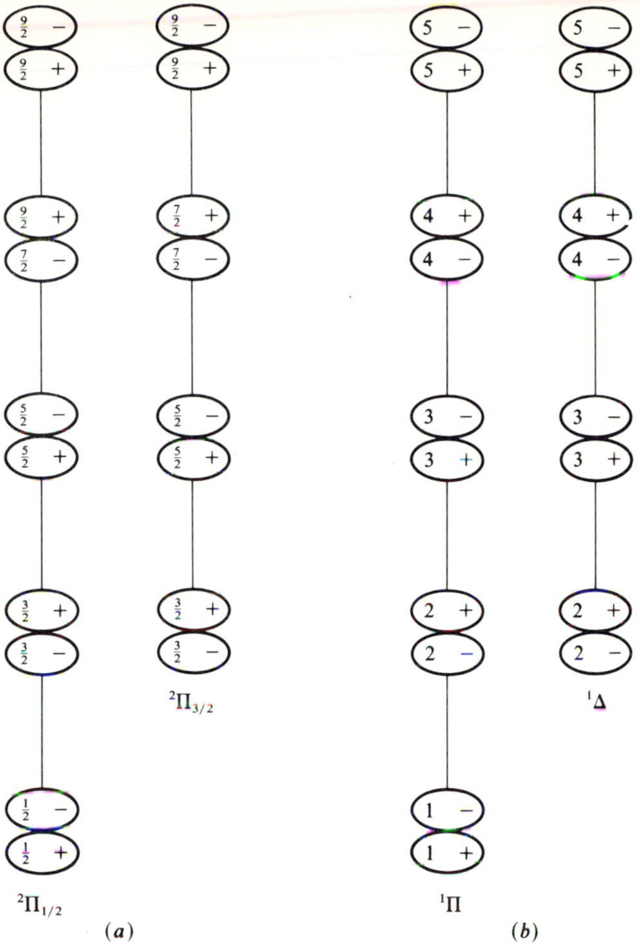

FIGURE 15.17
Herzberg-type state symmetry diagram for heteronuclear molecules. (*a*) Σ States-Hund's case(a) or case(b); (*b*) $^1\Pi$ and $^1\Delta$ states (effect of Λ-doubling shown).

will exhibit the same ordering as the levels associated with Σ states. There will also be cases where the $\Delta J = 0$ rule is valid. The selection rules for the total angular momentum relative to these latter cases are shown in the middle of Table 15.7 where ± denotes the parity of the total angular momentum state. At the bottom of the table are listed some of the more commonly observed electronic transitions. The concepts embodied in this table will be elaborated on to a further extent by the presentation of the energy level diagrams for particular types of transitions. In the presentations which follow the upper level is denoted by the first symbol.

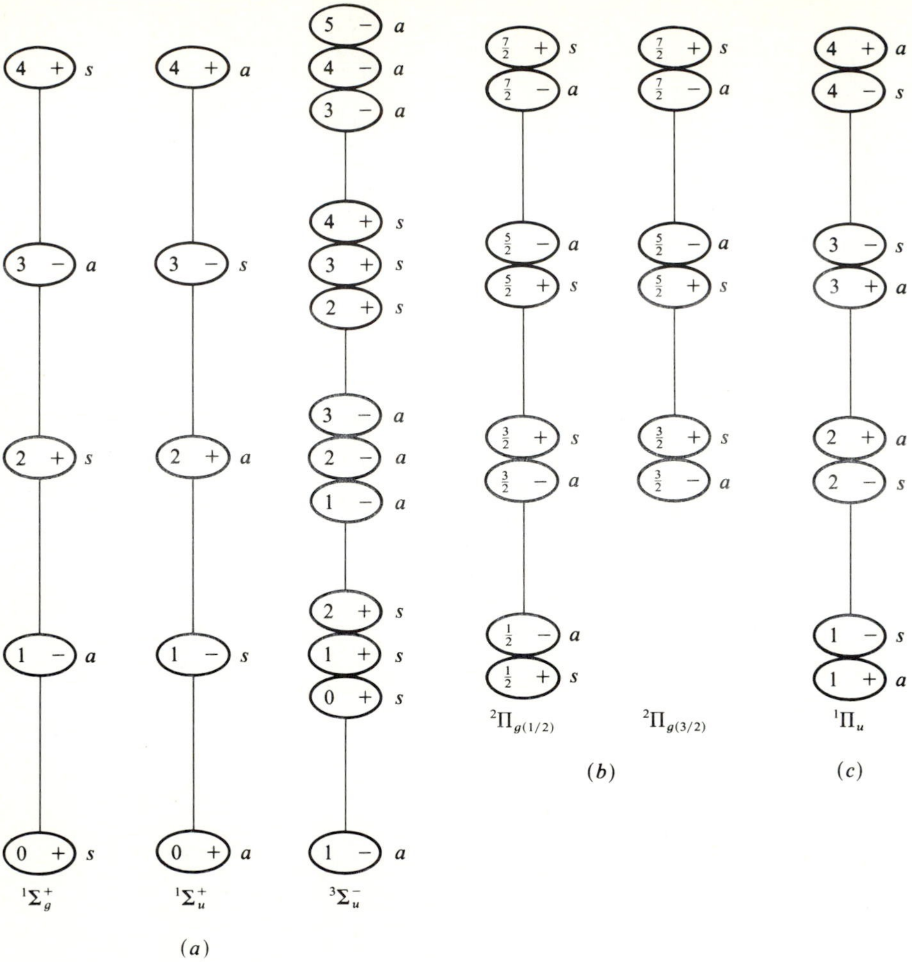

FIGURE 15.18
Herzberg-type state symmetry diagram for homonuclear molecules. (*a*) Σ States-Hund's case(a) or case(b); (*b*) $^2\Pi$ State-Hund's case(a); (*c*) Singlet states—$\Lambda \geqq 1$.

$^1\Sigma^+ \leftrightarrow {}^1\Sigma^+$ Transitions—Heteronuclear Diatomic Molecule

The rotational fine structure of the energy levels and the transition for a heteronuclear diatomic molecule undergoing an emission or absorption process between a $^1\Sigma^+$ ground electronic state with $v = v''$ and a $^1\Sigma^+$ excited state with $v = v'$ is shown in Fig. 15.19. The allowed rotational fine structure components of the transition are indicated as vertical lines. Below the level diagram is a qualitative line spectrum, part (*b*). Figure 15.19(*c*) is a Herzberg-type diagram [9] which depicts the successive rotational states for each electronic-vibrational level as a

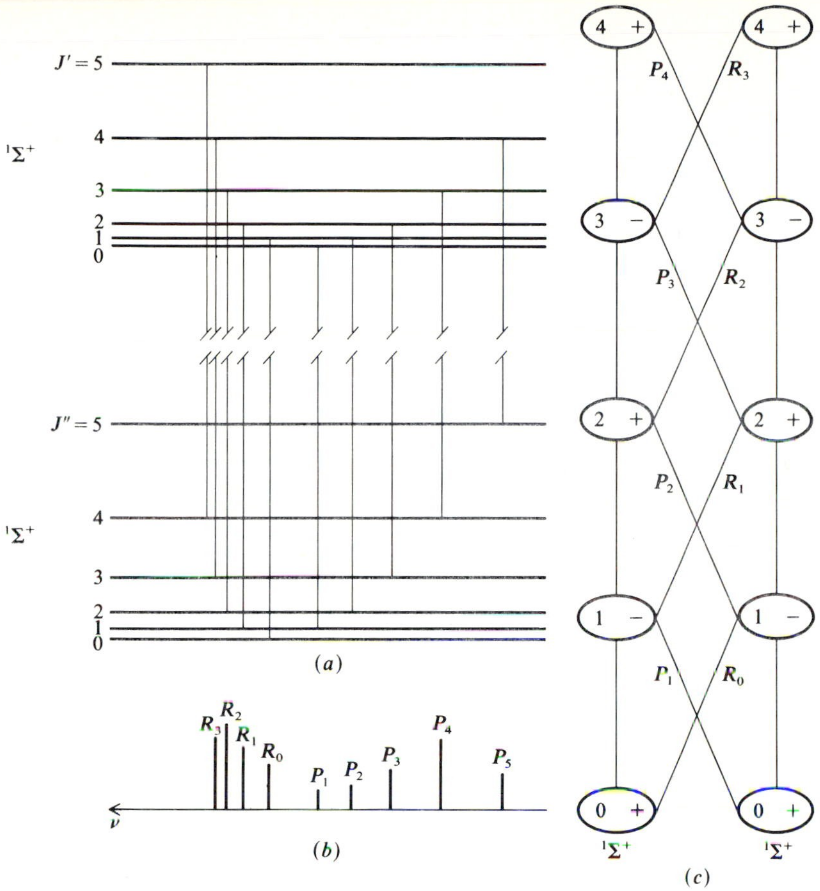

FIGURE 15.19
$^1\Sigma^+ \leftrightarrow {}^1\Sigma^+$ rovibronic transition for a heteronuclear molecule. (*a*) Energy-level diagram; (*b*) Line spectrum; (*c*) Herzberg-type diagram showing allowed transitions.

linear array of ovals containing notations relative to the symmetries of the states. For this type of a transition where $\Lambda = 0$ for both states there is no distinction to be made between Hund's case(a) and Hund's case(b) insofar as the spectral fine structure is concerned. The total angular momentum will be due only to molecular rotation, so the $\Delta J \pm 1$ selection rule is applicable and only *R*- and *P*-branch transitions will be observed. In the Herzberg-type diagram the *R*-branch transitions will be represented by connecting lines of positive slope. The lines with negative slopes denote the *P*-branch transitions. This same energy level scheme will be applicable to a $^1\Sigma^- \leftrightarrow {}^1\Sigma^-$ transition if there is a reversal of the parity of the levels.

$^1\Sigma_g^+ \leftrightarrow {}^1\Sigma_u^+$ Transitions—Homonuclear Diatomic Molecule

With homonuclear diatomic molecular transitions the behavior is similar to that for heteronuclear molecules with added constraints due to the additional symmetry selection rules and the intensity variation due to nuclear spin statistics. Figure 15.20, which follows the same format as Fig. 15.19, illustrates the behavior of a homonuclear molecule. In this case, assuming that the molecules are bosons, the dashed lines represent the transitions of lower or missing intensities due to the spin statistics. Examination of Fig. 15.20 shows that for the $I = 0$ case where alternate transitions are absent the P-branch transitions do not appear to form

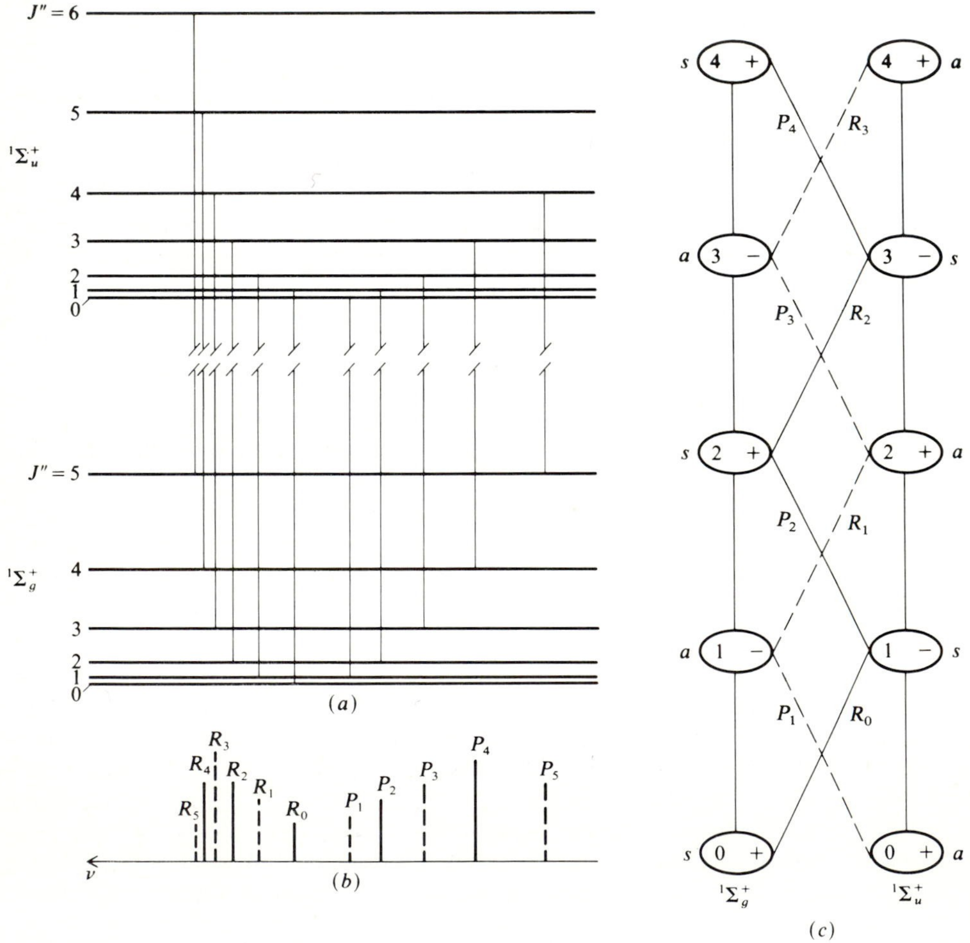

FIGURE 15.20
$^1\Sigma_g^+ \leftrightarrow {}^1\Sigma_u^+$ rovibronic transitions for a homonuclear molecule. (a) Energy-level diagram; (b) Line spectrum; (c) Herzberg-type diagram showing allowed transitions.

a continuation of the R-branch series due to the absence of both the $J'' = 0$ to $J' = 0$ and the $P(1)$ transitions.

$^2\Sigma^+ \leftrightarrow {}^2\Sigma^+$ Transitions—Heteronuclear Diatomic Molecule

The majority of molecules exhibiting transitions of this type belong to Hund's case(b). For the general situation, Ω is not defined, and **R** couples with **Λ** to give the intermediate momentum **N**, which in turn couples with **S** to give **J**. For this case the good quantum numbers will be S, N, and J where $J = N + S, N + S - 1, \ldots, |N - S|$ and $N = \Lambda, \Lambda + 1, \Lambda + 2, \ldots$. For $^2\Sigma$ molecules, where $\Lambda = 0$ and $S = \pm\frac{1}{2}$, the allowed values of N are just those for the molecular rotational angular momenta R. The rotational terms in Eqs. (15.2) and (15.3) are thus the ones given by Eqs. (15.34) and (15.35). The general selection rules for case(b) are $\Delta S = 0$, $\Delta N = 0, \pm 1$, and $\Delta J = 0, \pm 1$; however, for Σ states the $\Delta N = 0$ transitions are forbidden. The net result, as shown in Fig. 15.21, is a splitting of each level specified by a value of N into a pair of closely spaced levels and a resulting tripling of all but the two transitions associated with the $N = 0$ states. In this figure the transitions are designated by N'', $R(N'')$, or $P(N'')$, and subscripts 1 or 2 are added to distinguish between J'' values of $N'' + \frac{1}{2}$ and $N'' - \frac{1}{2}$. The $Q(N'')$ transitions are subscripted 12 to denote a transition between $N' + \frac{1}{2}$ and $N'' - \frac{1}{2}$ levels and 21 to denote one between $N' - \frac{1}{2}$ and $N'' + \frac{1}{2}$ levels.

$^1\Pi \leftrightarrow {}^1\Sigma^+$ Transitions—Heteronuclear Diatomic Molecule

This is one of the simplest transitions in which the rotational levels of one state are split by Λ-type doubling. Since $\Lambda'' = 0$ there is no distinction between the spectra of case(a) and case(b) molecules. For these transitions, $\Delta\Lambda$ will always be nonzero so the $\Delta N \neq 0$ rule is no longer applicable and the final selection rules for J will be $\Delta J = 0, \pm 1$. As shown in Fig. 15.22, the Π state rotational levels will be split by Λ-type doubling with the splitting given by

$$\Delta E' = hq_\Lambda J'(J' + 1) \tag{15.86}$$

Although our first inclination would be to surmise that the Λ-type doubling would cause a doubling of the observed transitions, the occurrence of the symmetry selection rule, $+ \leftrightarrow -$, prevents this from happening. This is a case where the combination relationships given by Eqs. (15.79) and (15.80) must be modified to allow for the effect of Λ-type doubling. Due to the alternation of parity of the members of the doublets, alternate Q-branch transitions will terminate at alternating $+$ or $-$ levels. In this case the combinations relationships become

$$\nu_R(J'') - \nu_Q(J'') = \nu_Q(J'' + 1) - \nu_P(J'' + 1) + \varepsilon \tag{15.87}$$

and

$$\nu_R(J'') - \nu_Q(J'' + 1) = \nu_Q(J'') - \nu_P(J'' + 1) + \varepsilon \tag{15.88}$$

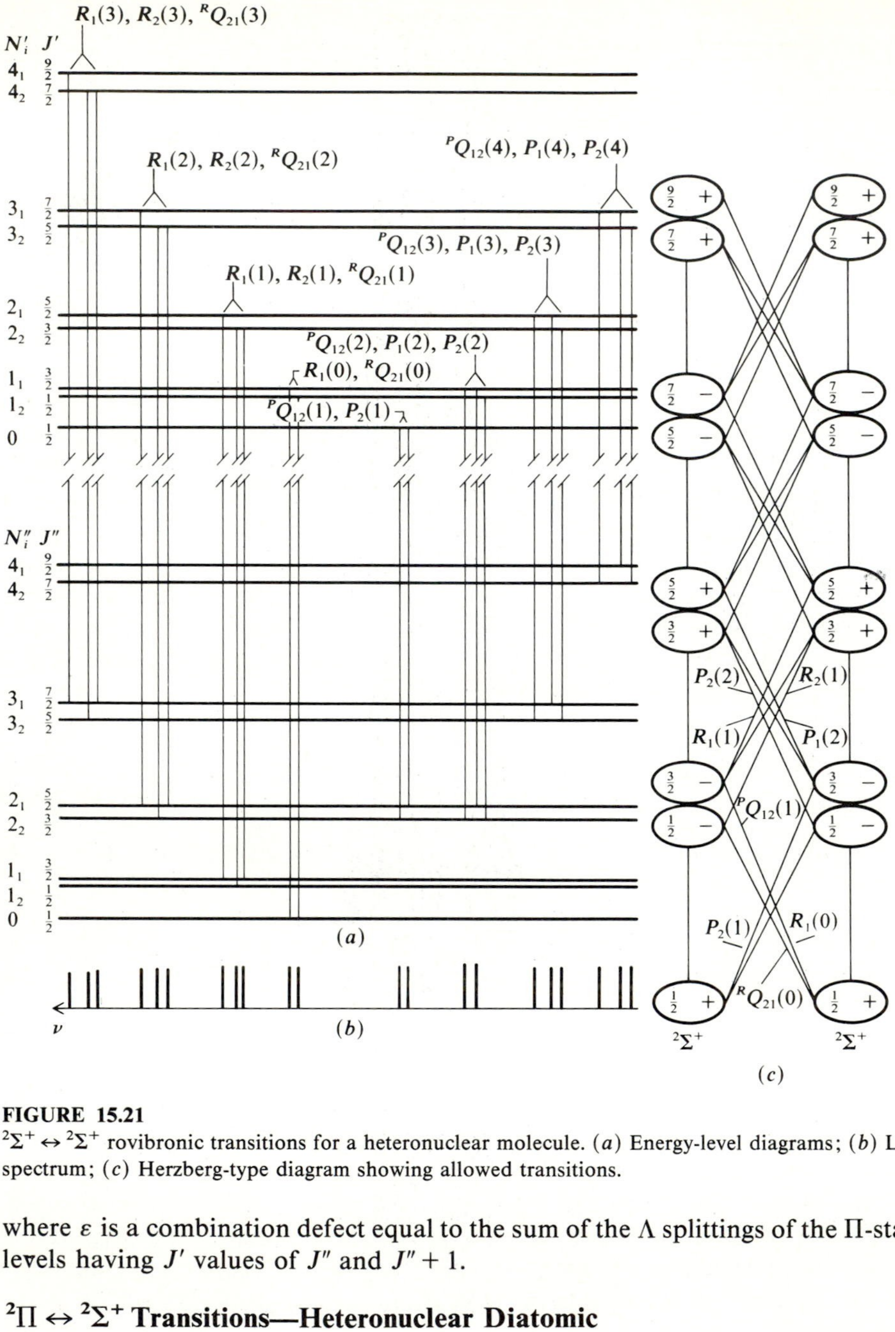

FIGURE 15.21
$^2\Sigma^+ \leftrightarrow {}^2\Sigma^+$ rovibronic transitions for a heteronuclear molecule. (*a*) Energy-level diagrams; (*b*) Line spectrum; (*c*) Herzberg-type diagram showing allowed transitions.

where ε is a combination defect equal to the sum of the Λ splittings of the Π-state levels having J' values of J'' and $J'' + 1$.

$^2\Pi \leftrightarrow {}^2\Sigma^+$ Transitions—Heteronuclear Diatomic Molecule

The last particular case to be considered is one for which the appearance of spectra will be highly dependent on which of Hund's cases is applicable to the

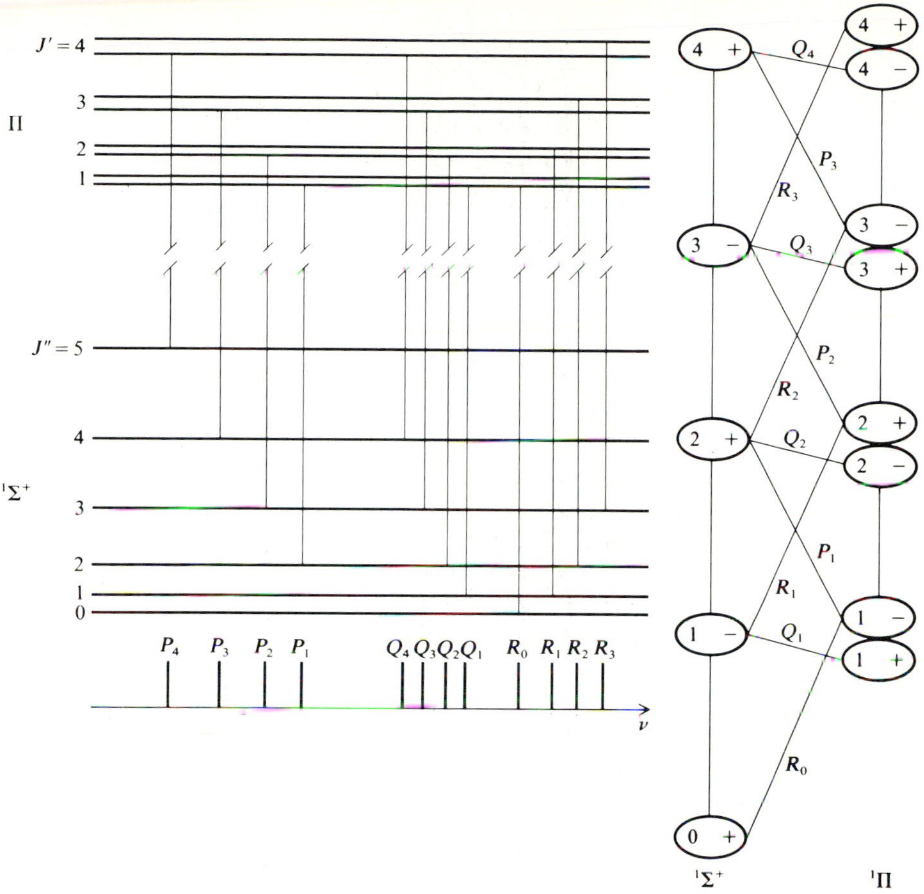

FIGURE 15.22
$^1\Pi \leftarrow {}^1\Sigma^+$ rovibronic transition for a heteronuclear molecule. (*a*) Energy-level diagram; (*b*) Line spectrum; (*c*) Herzberg-type diagram showing allowed transitions.

Π state, the $^2\Sigma$ states always belonging to case(b). The ideal situation for the $^2\Pi$ state belonging to Hund's case(b) is shown in Fig. 15.23 where the energy-level splitting due to both spin multiplicity and Λ-type doubling are delineated. Two features common to previous cases also show up in this case. For the $^2\Sigma^+$ state the parities of the doublet levels due to $S = \pm\frac{1}{2}$ are both the same for a given value of the molecular angular momentum *N*, but show an alternation with alternate values of *N*. For the $^2\Pi$ state there will be a splitting of each level having a total angular momentum denoted by J' into a Λ-type doublet of opposite parities in addition to the splitting of each level characterized by *N* by the spin multiplicity.

When the spectra of individual molecules are examined, it is generally found that they do not conform exactly to the scheme just presented. They can in fact range all the way from this situation to that where the $^2\Pi$ level belongs to Hund's

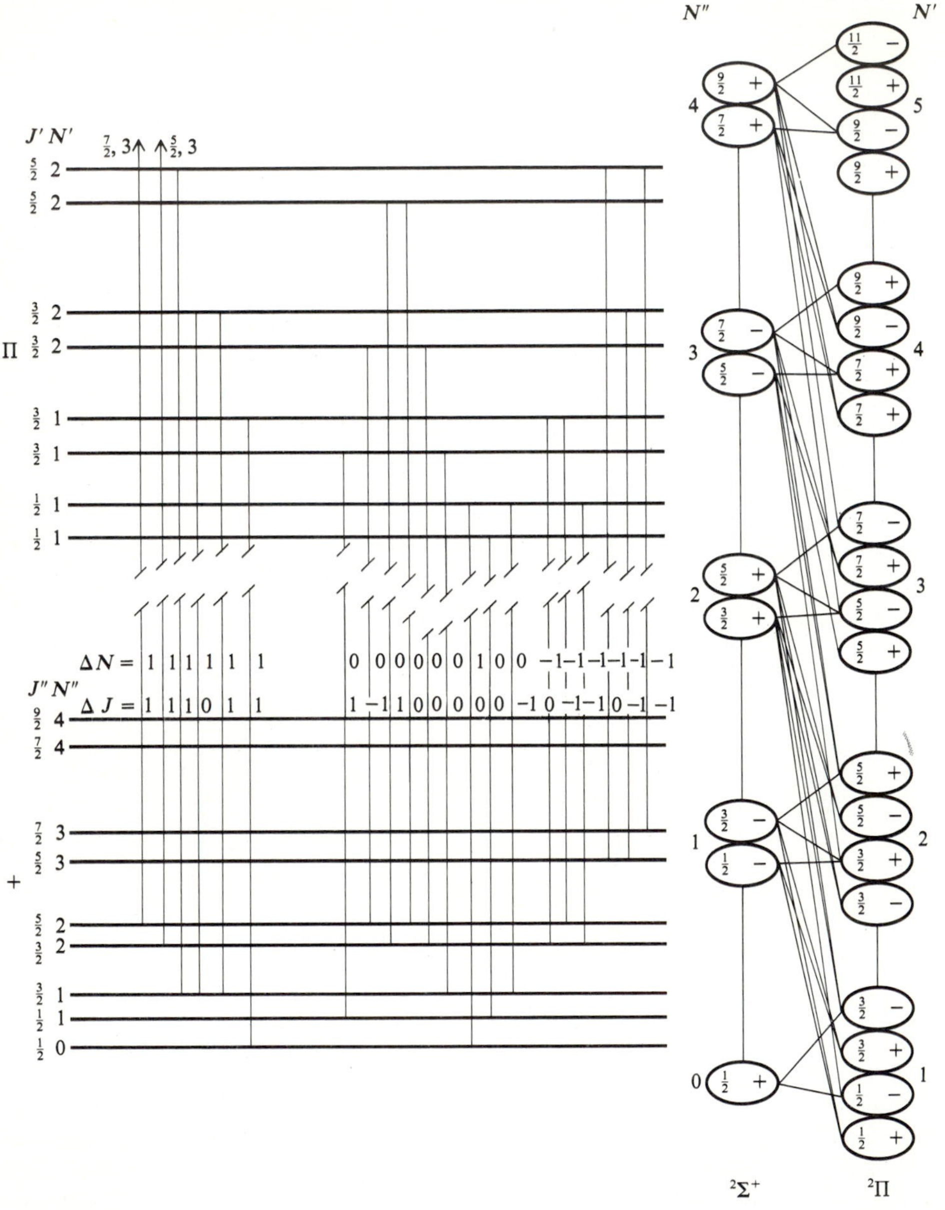

FIGURE 15.23
$^2\Pi \leftrightarrow {}^2\Sigma^+$ rovibronic transition for a heteronuclear molecule. (*a*) Energy-level diagram; (*b*) Line spectrum; (*c*) Herzberg-type diagram showing allowed transitions.

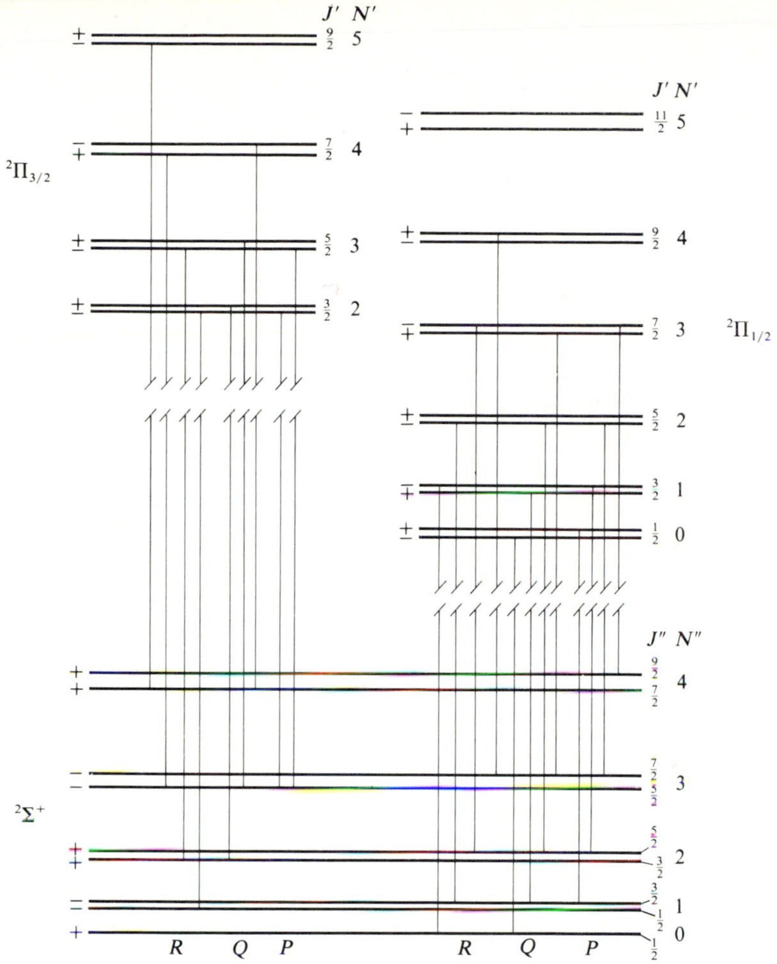

FIGURE 15.24
$^2\Pi_{1/2} \leftrightarrow {}^2\Sigma^+$ and $^2\Pi_{3/2} \leftrightarrow {}^2\,{}^2\Sigma^+$ rovibronic transitions, energy-level diagram.

case(a) and the behavior is as shown in Fig. 15.24. One point to note upon comparing Figs. 15.23 and 15.24 is the reversal of parity of the Λ-type doublets of the $^2\Pi_{1/2}$ levels. Also in many cases either the Λ-type doublet separations or the multiplet splittings are so small that they are experimentally unresolvable and the observed spectrum of a molecule exhibits simplification relative to those predicted by these diagrams. If both effects are unresolvable, the spectrum reduces to the three-branch spectrum of the $^1\Pi \leftrightarrow {}^1\Sigma$ type. For unresolved Λ-type doubling, the triplets such as P_1, P_2, $^PQ_{12}$, or R_1, R_2, $^RQ_2$, reduce to multiplet doublets.

It is apparent by this stage of the presentation that there is a near endless variety of combinations of upper and lower states, either of which may be specific Hund's cases or various combinations of cases. The object of this chapter was only to give an introductory background so that the interested student can proceed to more advanced works [9, 11].

15.11 DIATOMIC ROVIBRONIC SPECTRA—EXAMPLE

In this section we will use literature data for molecules which are close to the ideal ones reviewed in Sec. 15.10. In some cases there is not a precise correspondence but they are sufficiently close to allow us to carry out a reasonable application of the methods presented in the last section. A brief discussion of the variations of the examples from ideal behavior and the additional factors considered to get the assignment will also be added.

The analysis of a limited set of data is afforded by that of Ringstrom [15] for the $v' - v'' = 3\text{-}1$ component of the $^1\Sigma^+ \leftrightarrow {}^1\Sigma^+$ band of CuH in the region of 36,490-44,050 cm^{-1}. As shown in Table 15.8, only *P*- and *R*-branch transitions are observed. Examination of this data indicates no presence of a band head in the range of the data. The data, when analyzed by using the differences enumerated by Eqs. (15.77) and (15.78), produce the results shown in Table 15.8. The nonideal behavior of CuH is readily demonstrated by the changing of values of B_3' and B_1'' as J'' increases. The best value of the rotational constants are found by plotting B_v vs. $(J'' + 1)$ and extrapolating to $J'' = 0$. Values of $B_1'' = 7.78\ \text{cm}^{-1}$ and $B_3' =$

TABLE 15.8
Rovibronic spectral frequencies and difference analysis for the $^1\Sigma^+ \leftrightarrow {}^1\Sigma^+$, $v' - v'' = 3\text{-}1$ transition of CuH

J''	$R(J'') = \nu_R(J'')$, cm^{-1}	$P(J'') = \nu_P(J'')$, cm^{-1}	Δ_2F'',† cm^{-1}	$\Delta F''$,‡ cm^{-1}	B_3', cm^{-1}	B_1'', cm^{-1}
2	38,503.28					
3	38,483.16	38,439.10	44.06	105.62	3.1471	7.5443
4	38,454.08	38,397.66	56.42	135.69	3.1344	7.5383
5	38,416.37	38,347.47	68.90	165.71	3.1318	7.5323
6	38,369.67	38,288.37	81.30	195.61	3.1269	7.5235
7	38,314.28	38,220.76	93.52	225.32	3.1173	7.5107
8	38,249.90	38,144.35	105.55	254.80	3.1044	7.4941
9	38,176.83	38,059.48	117.35	284.04	2.0082	7.4747
10	38,095.07	37,965.86	129.21	312.91	3.0764	7.4502
11	38,004.47	37,863.89	140.58	......	3.0561	
12	37,905.31	....	......	370.04	......	7.4008
13	37,797.56	37,634.43	163.13	398.24	3.0209	7.3749
14	37,681.23	27,507.07	174.16	426.18	3.0028	7.3479
15	37,556.31	37,371.38	184.93	453.80	2.9827	7.3194
16	37,422.95	37,227.43	195.52	480.79	2.9624	7.2847
17		37,075.52				
18		36,915.44				

† $\Delta_2F'' = R(J'') - P(J'')$.
‡ $\Delta_2F'' = R(J'' - 1) - P(J''+1)$ (difference is listed following J'' value).

3.14 cm^{-1} are found using this technique. This value of B_3' compares favorably with that obtained by Ringstrom [16], $B_3' = 3.156\ cm^{-1}$, where interactions with nearby perturbing levels were considered. The value of B_1'' is close to that of $B_1'' = 7.81\ cm^{-1}$ determined by Heimer and Heimer [17]. If one contrasts the values of B_3' and B_1'' to those of comparable vibrational states belonging to the same electronic state it is observed that there is a much greater difference. Although the spectrum just examined was of insufficient resolution to have separately recorded separate transitions for the $^{63}Cu^1H$ and $^{65}Cu^1H$ species, we can nevertheless make a rough calculation of the internuclear distance. Using the relationship B_v (MHz) $\times I_v$ (amu A^2) = 505376 and $I_v = \mu r_v^2$ gives values of $r_1'' = 0.148$ nm and $r_3' = 0.233$ nm. This indicates that in the higher $^1\Sigma^+$ states the bonding is considerably weaker.

REFERENCES

Specific

1. Harris, D. C., *Symmetry and Spectroscopy*, Oxford University Press, New York, NY, 1978.
2. Hund, F., *Z. Physik*, **63**, 719 (1930).
3. Herzberg, G., *Molecular Spectra and Molecular Structure: I. Spectra of Diatomic Molecules*, 2d ed., Van Nostrand Reinhold, New York, NH, 1950, p. 238.
4. Ibid., pp. 219–226.
5. Ibid, p. 242.
6. Huber, K. P., and G. Herzberg, *Molecular Spectra and Molecular Structure: IV. Constants of Diatomic Molecules*, Van Nostrand Reinhold, New York, NY, 1979, pp. 158–170.
7. Ibid., pp. 518–521.
8. Mulliken, R. S., *Rev. Mod. Ph.*, **2**, 60 (1930).
9. Herzberg, G., loc. cit., chap. V.
10. King, G. W., *Spectroscopy and Molecular Structure*, Holt, Reinhart & Winston, New York, NY (1964), chap. 6.
11. Mizushima, M., *The Theory of Rotating Diatomic Molecules*, John Wiley and Sons, New York, NY, 1975, chap. 2.
12. Condon, E. U., *Phys. Rev.*, **32**, 858 (1928).
13. Crawford, F. H., and T. Torgensen, Jr., *Phys. Rev.*, **47**, 932 (1935).
14. Loomis, F. W., and R. E. Nusbaum, *Phys. Rev.*, **38**, 1447 (1931).
15. Huber, K. P., loc cit., pp. 158–170, 466–481.
16. Ringstrom, U., *Can. J. Phys.*, **46**, 2291 (1968).
17. Heimer, V. A., and T. Heimer, *Z. for Physik*, **84**, 232 (1933).

General

See specific references 3, 6, 10, and 11.
Dunford, H. B., *Elements of Diatomic Molecular Spectra*, Addison-Wesley, Reading, MS, 1968.
Hollas, J. M., *High Resolution Spectroscopy*, Butterworths, London, G.B., 1982.

PROBLEMS

15.1. Construct the energy-level diagram and the Herzberg-type transition diagrams for the following transitions:
$^1\Sigma^- \leftrightarrow {}^1\Sigma^-$ heteronuclear molecule
$^1\Sigma^- \leftrightarrow {}^1\Sigma^-$ homonuclear molecule

$^2\Sigma^{+u} \leftrightarrow {}^2\Sigma_u^{+g}$ homonuclear molecule
$^1\pi_g^g \leftrightarrow {}^1\pi_u$

15.2. In the $^3\Sigma$-$^3\Sigma$ ultraviolet band systems for both O_2 and S_2 the even-numbered rotational transitions are missing. Rationalize this behavior by means of an appropriate energy-level diagram.

15.3. The following vibronic band origins (cm^{-1}) have been observed for the $X^1\Sigma^+ = A^1\pi$ transition of BeO. Calculate the best values for ω_e and $\chi_e\omega_e$ for both electronic states. Find the differences between the $v = 0$ levels of the electronic states.

	v''		
v'	0	1	2
2	12,569.95		
3	13,648.43	12,184.83	
4	14,710.85	13,246.85	
5	15,757.50	14,294.40	12854.15
6	16,788.95	15,325.05	13885.32
7		16,341.15	14,900.88
8		17,342.25	15,901.75
9			16,888.15

15.4. The vibronic band origins (in cm^{-1}) for the $X^1\Sigma_g^+ - A^1\Sigma_u^+$ transitions of $^7Li_2$ are listed below. Determine the best values for ω_e and $\chi_e\omega_e$ for both electronic states and find the difference between the $\nu = 0$ levels of the electronic states.

	v''					
v'	0	1	2	3	4	5
0	14,020	13,663	13,302	12,934	12,562	12,148
1	14,279	13,922	13,560	13,193	12,820	12,443
2	14,541	14,184	13,822	13,455	13,082	12,705
3	14,805	14,449	14,087	13,720	13,347	12,970
4	15,074	14,717	14,355	13,988	13,616	12,238
5	15,345	14,989	14,627	14,260	13,887	13,509

15.5. The electronic spectra band origin ν_∞ data for diatomic PN is given below. Using this data determine the values of ω_e and $\chi_e\omega_e$ for the upper and lower electronic states, and the energy difference T_e between the electronic states.

	v''				
v'	0	1	2	3	4
0	39,698.8	38,376.5	37,068.7		
1	40,786.2	39,467.2	38,155.5	36,861.3	
2	41,859.1	40,536.2	.....	37,932.9	36,652.5
3		41,597.4	40,288.3		37,712.5
4			41,331.2		38,756.4

15.6. Below are given the (1 − 1) rovibronic components (in cm^{-1}) of the $X^2\Sigma^+ - A^2\Sigma^+$ transitions of CN. (*a*) Construct a Fortrat diagram using this data. (*b*) Determine the values for B_1' and B_1''.

25,744.73, 25,745.34, 25,746.08, 25,746.00, 25,748.02, 25,749.19, 25,750.52, 25,751.98, 25,753.47, 25,755.89, 25,757.26, 25,759.29, 25,761.47, 25,763.75, 25,766.16, 25,768.72, 25,771.35, 25,774.23, 25,777.19, 25,780.32, 25,783.53, 25,786.90, 25,790.41, 25,794.03, 25,801.81, 25,805.50, 25,810.01, 25,814.23, 25,818.77, 25,828.88, 25,828.06, 25,833.02, 25,837.97, 25,843.13, 25,848.40, 25,853.77, 25,859.38, 25,864.00, 25,870.60, 25,876.60, 25,882.73, 25,888.91, 25,895.22, 25,902.63, 25,908.26, 25,914.95, 25,921.86, 25,928.83.

CHAPTER 16

VIBRATION OF POLYATOMIC MOLECULES

16.1 INTRODUCTION

The discussion of the vibrational spectroscopy of diatomic molecules established the fact that the force constant of a molecule could be calculated from the vibrational frequencies. When we examine a polyatomic molecule it is found that the internal motion can be characterized by a set of force constants which, in a general harmonic oscillator formulation, contains one constant for each pairwise interaction in the molecule. This chapter examines the relationship of these force constants to the multiplicity of vibrational transitions which are observed. Although the vibrational transitions of polyatomic molecules, when observed using adequate spectral resolution, exhibit rotational fine structure, no detailed discussion of this subject is included.

When one attempts to adapt the concepts developed for diatomic molecules to polyatomic molecules it is discovered that it is not a simple extension. In the discussion of the nonrigid diatomic molecule it was found that it required three coordinates to specify the location of the center of mass of the system and two to specify the orientation of the internuclear axis relative to a coordinate system located at the center of mass. Since the total number of coordinates needed to specify the locations of N independent particles is $3N$, this left one coordinate available to specify the position of one atom relative to the other along the bond direction.

For a molecule containing three or more atoms, although $3N$ coordinates are necessary to specify the positions of all atoms, the number necessary to specify the internal motion depends on the shape of the molecule. For any polyatomic molecule the specification of the center of mass requires three coordinates leaving $3N - 3$ available to describe the rotational and vibrational motions. For a linear molecule, the restraint that all mass points must lie on a line allows the rotational motion to be specified by two coordinates while for a nonlinear molecule, three coordinates are required to define the rotation. This argument leads to the conclusion that the vibrational motion of a linear molecule is specified by $3N - 5$ coordinates while that of a nonlinear molecule requires $3N - 6$. The number of coordinates necessary to specify the vibrational motion of a molecule are referred to as the vibrational degrees of freedom.

A general set of force constants for pairwise interactions, although a basis for a satisfactory mathematical description of the vibrational problem, is not easily incorporated into a simple physical interpretation. This chapter discusses the problem of selecting coordinates which will allow the use of a set of force constants which directly relate to the stretching and bending of chemical bonds and the prediction of the frequencies of radiation absorbed due to the vibrations in polyatomic molecules. The problem is approached by first introducing the concept of normal modes of vibration and showing how they may be determined for a simple system. We then proceed to discuss a more generalized method based on the use of group theory and matrix methods. No attempt is made to explore all of the sophisticated methods that have been developed for analysis of vibrations, but rather to discuss and present examples of the basic concepts and methods which the reader can use as a background for further study.

16.2 CLASSICAL VIBRATION OF A MANY-BODY SYSTEM

A linear three-mass system constrained to move in one dimension is used to demonstrate that each vibrational frequency can be related to a complex coupled motion of the individual atoms within the molecule. These coupled motions, of which there is one for each vibrational degree of freedom, are called normal modes of vibration. These normal modes are further characterized by the fact that such motions leave the center of mass of the molecule unchanged and that the motions of the individual atoms are in phase, that is, for a given normal mode they all reach the turning points of their motion simultaneously.

This linear system is shown in Fig. 16.1. The displacements of the atoms from their equilibrium positions are denoted by the ξ_i. It is assumed that the ξ_i are all small compared to the r_{ij} so that the system obeys Hooke's law.

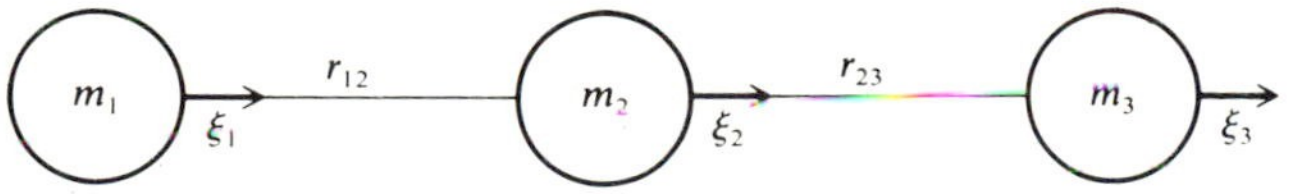

FIGURE 16.1
Hypothetical one-dimensional linear molecule.

For this hypothetical one-dimensional molecule the form of the vibrations will be related to the nature of the bonds. This relationship is characterized by a force constant associated with each bond. It is expedient at this point to introduce two new coordinates, $(\xi_1 - \xi_2)$ and $(\xi_2 - \xi_3)$, which specify the change in the individual bond lengths r_{12} and r_{23} with vibration. The potential energy V, to terms quadratic in the displacement, is given by

$$V = \tfrac{1}{2}k_{12}(\xi_1 - \xi_2)^2 + \tfrac{1}{2}k_{23}(\xi_2 - \xi_3)^2 \tag{16.1}$$

In order to determine the resulting motion we evaluate the force acting on each atom i, $F_i = -dV/d\xi_i$. Equation (16.1) then leads to

$$F_1 = -\frac{\partial V}{\partial \xi_1} = k_{12}(\xi_1 - \xi_2) \tag{16.2}$$

$$F_2 = -\frac{\partial V}{\partial \xi_2} = k_{12}(\xi_1 - \xi_2) - k_{23}(\xi_2 - \xi_3) \tag{16.3}$$

$$F_3 = -\frac{\partial V}{\partial \xi_3} = k_{23}(\xi_2 - \xi_3) \tag{16.4}$$

Since Hooke's law conditions were assumed, the motion will be harmonic.

A set of equations of motion for the system can be developed using either Newtonian, Lagrangian, or Hamiltonian mechanics. In order to keep the initial discussion simple, the first of these will be employed. The application of Newton's second law, $F = ma$, to this system can be simplified by introducing mass weighted coordinates, defined by $\zeta_i = \xi_i/\sqrt{\mu_i}$, where $\mu_i = 1/m_i$. Newton's second law is then expressed as

$$F_i = m_i a_i = m_i \ddot{\xi}_i = \mu_i^{-1/2}\ddot{\zeta}_i \tag{16.5}$$

Introducing the force relationships, Eqs. (16.2) to (16.4), one obtains the equations of motion

$$\mu_1^{-1/2}\ddot{\zeta}_1 = -k_{12}(\mu_1^{1/2}\zeta_1 - \mu_2^{1/2}\zeta_2) \tag{16.6}$$

$$\mu_2^{-1/2}\ddot{\zeta}_2 = k_{12}(\mu_1^{1/2}\zeta_1 - \mu_2^{1/2}\zeta_2) - k_{23}(\mu_2^{1/2}\zeta_2 - \mu_3^{1/2}\zeta_3) \tag{16.7}$$

$$\mu_3^{-1/2}\ddot{\zeta}_3 = k_{23}(\mu_2^{1/2}\zeta_2 - \mu_3^{1/2}\zeta_3) \tag{16.8}$$

This set of second-order differential equations has general solutions of the form

$$\zeta_{ik} = A_{ik}\cos(2\pi\nu_k t - \delta_k) \tag{16.9}$$

where A_{ik} will be the maximum displacement of m_i for the kth normal mode. If we can evaluate the A_{ik} and ν_k, we will have a description of the internal motion of the system. This is done by twice differentiating Eq. (16.9) to give $\ddot{\zeta}_{ik} = -4\pi^2\nu_k^2\zeta_{ik} = -\lambda_k\zeta_{ik}$, followed by substitution into Eqs. (16.6) to (16.8)

$$(k_{12}\mu_1 - \lambda_k)A_{1k} + k_{12}\mu_1^{1/2}\mu_2^{1/2}A_{2k} = 0 \tag{16.10}$$

$$-k_{12}\mu_1^{1/2}\mu_2^{1/2}A_{1k} + (k_{12}\mu_2 + k_{23}\mu_2 - \lambda_k)A_{2k} - k_{23}\mu_2^{1/2}\mu_3^{1/2}A_{3k} = 0 \tag{16.11}$$

$$-k_{23}\mu_2^{1/2}A_{2k} + (k_{23}\mu_3 - \lambda_k)A_{3k} = 0 \tag{16.12}$$

This set of secular equations will have a nontrivial solution only if the determinate vanishes

$$\begin{vmatrix} (k_{12}\mu_1 - \lambda_k) & -k_{12}\mu_1^{1/2}\mu_2^{1/2} & 0 \\ -k_{12}\mu_1^{1/2}\mu_2^{1/2} & (k_{12}\mu_2 + k_{23}\mu_2 - \lambda_k) & -k_{23}\mu_2^{1/2}\mu_3^{1/2} \\ 0 & -k_{23}\mu_2^{1/2}\mu_3^{1/2} & (k_{23}\mu_3 - \lambda_k) \end{vmatrix} = 0 \qquad (16.13)$$

There are three roots for this determinant, designated by λ_k, each corresponding to a different mode of motion. The frequency of the kth normal mode is $\nu_k = \lambda_k^{1/2}/2\pi$.

The secular determinant gives a cubic equation in λ_k, which in general will be nonfactorable. We can, however, without losing sight of the basic concepts, simplify the present system so that we can quickly obtain the end results and examine the nature of the motion. This is done by considering our hypothetical molecule to have $D_{\infty h}$ symmetry, in which case $\mu_1 = \mu_3$ and $k_{12} = k_{23} = k$. For this case the determinant becomes

$$(\mu_1 k - \lambda_k)(2\mu_2 k - \lambda_k)(\mu_3 k - \lambda_k) - 2\mu_1^{1/2}\mu_2^{1/2}k^2(\mu_1 k - \lambda_k) = 0 \qquad (16.14)$$

having roots

$$\lambda_1 = \mu_1 k \qquad \lambda_2 = 0 \qquad \lambda_3 = (2\mu_2 + \mu_1)k \qquad (16.15)$$

The relative amplitudes of the motion of the individual masses can now be found for each normal mode by successive insertion of the roots into the secular equations. The A_{ik}, which denotes the amplitude of the motion of m_i undergoing a normal vibration designated by ν_k and associated with the root λ_k, can be found by substitution of the λ_k into the secular equations, for example, if $\lambda_1 = \mu_1 k$

$$(k\mu_1 - k\mu_1)A_{11} - k\mu_1^{1/2}\mu_2^{1/2}A_{21} = 0 \qquad (16.16)$$

$$-k\mu_1^{1/2}\mu_2^{1/2}A_{11} - (2k\mu_2 - k\mu_1)A_{21} - k\mu_1^{1/2}\mu_2^{1/2}A_{31} = 0 \qquad (16.17)$$

$$-k\mu_1^{1/2}\mu_2^{1/2}A_{21} + (k\mu_1 - k\mu_1)A_{31} = 0 \qquad (16.18)$$

To satisfy Eq. (16.16) A_{21} must be equal to zero. Equation (16.17) then gives $A_{11} = -A_{31}$. Since it is never possible to obtain more than a set of ratios for the variables from the secular equations, we need one additional condition to calculate the values of the A_{ik}. This condition involves normalizing the A_{ik} terms to unity

$$\sum_{i=1}^{3} A_{ik}^2 = 1 \qquad k = 1, 2, 3 \qquad (16.19)$$

Using this additional relationship $A_{11} = -A_{31} = 1/\sqrt{2}$ and $A_{21} = 0$. Following the same procedure, but using λ_2 and λ_3, the normalized amplitudes of the other two modes become

$$A_{12} = A_{32} = \left[\frac{\mu_2}{\mu_1 + 2\mu_2}\right]^{1/2} \qquad (16.20)$$

$$A_{22} = \left[\frac{\mu_1}{\mu_1 + 2\mu_2}\right]^{1/2} \qquad (16.21)$$

and

$$A_{13} = A_{33} = \left[\frac{\mu_1}{2(\mu_1 + 2\mu_2)}\right]^{1/2} \tag{16.22}$$

$$A_{23} = -2\left[\frac{\mu_2}{2(\mu_1 + 2\mu_2)}\right]^{1/2} \tag{16.23}$$

The amplitudes of the atom motions for these three modes are qualitatively illustrated in Fig. 16.2.

Examinations of these motions shows that the second is a translational mode since each of the atoms moves in the same direction by an equal amount, hence moving the center of mass in space. Considering that the three masses are constrained to move in one dimension, only three Cartesian coordinates are needed to specify the total motion. The three Cartesian coordinates give rise to three normal modes with that one corresponding to λ_2 describing the translation of the system as a whole. The other two modes are true vibrations which leave the center of mass of the system undisplaced.

Having defined the nature of vibrational motion in a relatively simple polyatomic molecule let us next generalize these findings to a more complicated system. In order to facilitate the development of the necessary relationships the Cartesian coordinates of the N atoms in a molecule, $x_1, y_1, z_1, x_2, y_2, z_2, \ldots, x_N, y_n, z_N$, will be represented by $\xi_1, \xi_2, \xi_3, \xi_4, \ldots, \xi_{3N}$. Since the potential in which the nuclei move depends on the atomic displacements the potential will be a function of the ξ_i, $V = V(\xi_1 \cdots \xi_{3N})$. The classical definition of the force on the ith atom $F_i = -\partial V/\partial \xi_i$ and Newton's second law leads to $3N$ equations of the form

$$F_i = m_i \ddot{\xi}_i = -\frac{\partial V}{\partial \xi_i} \tag{16.24}$$

Except for very simple systems the complicated form of the potential function results in this set of equations being very difficult to solve.

When one encounters a set of Newtonian equations which present serious difficulties with regard to finding solutions the general procedure is to approach the problem by the use of Lagrangian or Hamiltonian mechanics. As shown in App. S, these two methods employ expressions for the energies of the atoms rather than the forces on the atoms. They also lend themselves to the use of

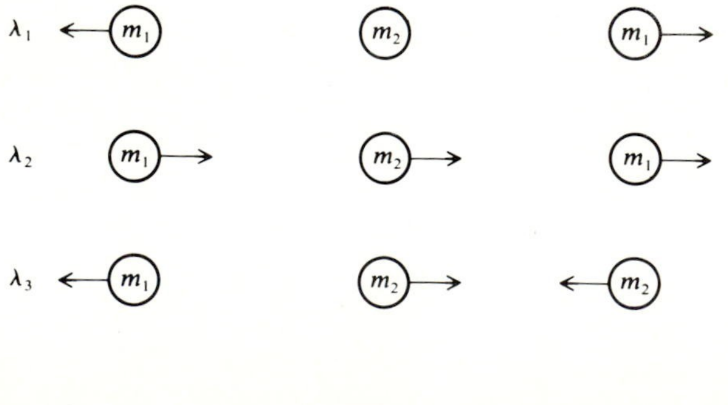

FIGURE 16.2
Normal modes of vibrations of a hypothetical three-atom one-dimensional oscillator.

generalized coordinates which may or may not be Cartesian. Letting q_i be the ith Cartesian displacement coordinate, the use of the Lagrangian $L = T - V$ gives rise to a set of $3N$ equations

$$\frac{d}{dt}\frac{\partial L}{\partial \dot{q}_i} - \frac{\partial L}{\partial q_i} = 0 \ (i = 1, 2, \ldots, 3N) \tag{16.25}$$

It is expedient when evaluating the Lagrangian equations of motion to express the potential as a Taylor series in q_i. This expansion is about the equilibrium positions of the atoms and can be terminated with the third terms if the displacements are small. Using a subscripted zero to denote the evaluations of a term with the nuclei in their equilibrium positions the potential energy is written

$$V = V_0 + \sum_{i=1}^{3N} \left(\frac{\partial V}{\partial q_i}\right)_0 q_i + \frac{1}{2}\sum_{i=1}^{3N}\sum_{j=1}^{3N}\left(\frac{\partial^2 V}{\partial q_i \partial q_j}\right)_0 q_i q_j + \cdots \tag{16.26}$$

The kinetic energy is given by

$$T = \frac{1}{2}\sum_{i=1}^{3N} m_i \dot{q}_i^2 \tag{16.27}$$

The potential term can be simplified by considering the individual terms in light of our knowledge about diatomic oscillators. The term V_0 is the potential of the system with all atoms in their equilibrium positions, and is a constant. Since we always measure potential relative to some point, we can select the equilibrium positions to define that point and set $V_0 = 0$. Since the potential is always a minimum at the equilibrium positions, all of the $(\partial V/\partial \xi_i)_0$ are zero. Hence we have a sum of quadratic terms in the potential function.

The problem can be simplified by the introduction of mass-weighted coordinates $\zeta_i = \sqrt{m_i}\, q_i$, in which case

$$2T = \sum_{i=1}^{3N} \dot{\zeta}_i^2 \tag{16.28}$$

and

$$2V = \sum_{i=1}^{3N}\sum_{j=1}^{3N}\left(\frac{\partial^2 V}{\partial \zeta_i \partial \zeta_j}\right)_0 \zeta_i\zeta_j = \sum_{i=1}^{3N}\sum_{j=1}^{3N} b_{ij}\zeta_i\zeta_j \tag{16.29}$$

It is to be noted that $b_{ij} = b_{ji}$. The substitution of these expressions for T and V into the Lagrangian, Eq. (16.25) leads to a set of $3N$ equations of the form

$$\ddot{\zeta}_i + \sum_{j=1}^{3N} b_{ij}\zeta_j = 0 \qquad (i = 1, 2, \ldots, 3N) \tag{16.30}$$

These constitute a set of second-order differential equations analogous to those of Eqs. (16.6) to (16.8). The general solutions will be

$$\zeta_i = A_{ik}\cos(2\pi\nu_k t - \delta_k) \qquad (i = 1, 2, \ldots, 3N) \tag{16.31}$$

Taking the second derivatives with respect to time,

$$\ddot{\zeta}_i = -4\pi^2\nu_k^2\zeta_i = -\lambda_k\zeta_i \tag{16.32}$$

and substituting these back into the differential equations, Eq. (16.30) gives a set of secular equations

$$\sum_{j=1}^{3N} b_{ij}A_{jk} - \lambda_k A_{ik} = 0 \qquad (i = 1, 2, \ldots, 3N) \tag{16.33}$$

The necessary and sufficient condition for the existence of a nontrivial set of solutions for these equations is that the determinant vanish

$$\begin{vmatrix} b_{11} - \lambda_k & b_{12} & \cdots & b_{1,3N} \\ b_{12} & b_{22} - \lambda_k & \cdots & b_{2,3N} \\ b_{3n,1} & b_{3N,2} & \cdots & b_{3N,3N} - \lambda_k \end{vmatrix} = |b_{ij} - \lambda_k\delta_{ij}| = 0 \tag{16.34}$$

where δ_{ij} is the Kronecker delta. Having considered the full set of $3N$ coordinates in arriving at Eq. (16.34), there will be six zero roots for a nonlinear molecule and five zero roots for a linear molecule. These correspond to the transitional and rotational motions leaving the remaining $3N - 6$ or $3N - 5$ nonzero roots to be associated with the vibrations. Each of these nonzero roots will represent a single normal mode of vibration where all of the atoms move in phase with a frequency $\nu_k = \sqrt{\lambda_k}/2\pi$.

Once the eigenvalues of Eq. (16.34) λ_k are known, they can be substituted back into the secular equations, Eq. (16.33), and the relative magnitudes of the maximum displacements A_{ik} evaluated. This set of equations

$$\begin{aligned} b_{11}A_{1k} + b_{12}A_{2k} + \cdots + b_{1,3N}A_{3N,k} &= \lambda_k A_{1k} \\ b_{21}A_{1k} - b_{22}A_{2k} + \cdots + b_{2,3N}A_{3N,k} &= \lambda_k A_{2k} \\ \vdots \qquad\qquad \vdots \qquad\qquad\qquad \vdots \qquad & \\ b_{3N,1}A_{1k} + b_{3N,2}A_{2k} + \cdots + b_{3N,3N}A_{3N,k} &= \lambda_k A_{3N,k} \end{aligned} \tag{16.35}$$

can only be solved for sets of ratios of the maximum displacements and not their absolute values. Since the primary interest will be in the relative motion of the various atoms, this is of no appreciable deterrence and the same procedure as employed in the first example, that of normalization to give a set of coefficients, can be employed. The normalization is expressed as

$$\sum_{i=1}^{3N} D_{ik}^2 = 1 \tag{16.36}$$

where

$$A_{ik} = c_k D_{ik} \tag{16.37}$$

and c_k is a proportionality constant. If the displacement coefficients D_{ik} are written as column vectors

$$\begin{pmatrix} D_{11} \\ D_{21} \\ \vdots \\ D_{3N\,1} \end{pmatrix} \begin{pmatrix} D_{12} \\ D_{22} \\ \vdots \\ D_{3N\,2} \end{pmatrix} \cdots \begin{pmatrix} D_{1\,3N} \\ D_{2\,3N} \\ \vdots \\ D_{3N\,3N} \end{pmatrix} \tag{16.38}$$

then the components of each vector can be interpreted as being proportional to the corresponding generalized displacement coordinates for a particular normal mode. These column vectors can be used to construct a set of individual displacement vectors which gives the directions and relative displacements for each nuclei for each normal mode.

Looking back at the linear three-atom system, Eqs. (16.19) through (16.23) show the D_{ik} column vectors to be

$$\overset{\lambda_1}{\begin{pmatrix} 1/\sqrt{2} \\ 0 \\ -1/\sqrt{2} \end{pmatrix}} \qquad \overset{\lambda_2}{(\mu_1 + 2\mu_2)^{-1/2}\begin{pmatrix} \mu_2^{1/2} \\ \mu_1^{1/2} \\ \mu_2^{1/2} \end{pmatrix}} \qquad \overset{\lambda_3}{[2(\mu_1 + 2\mu_2)]^{-1/2}\begin{pmatrix} \mu_1^{1/2} \\ -2\mu_2^{1/2} \\ \mu_1^{1/2} \end{pmatrix}} \tag{16.39}$$

It is possible to employ matrix techniques to simplify the vibrational problem for polyatomic molecules. First we will relate the matrix-vector formulation to the systems just discussed, introduce the concepts of normal and internal coordinates, and conclude by discussing applications to the analysis of several systems.

16.3 MATRIX-VECTOR FORMULATION OF THE VIBRATIONAL PROBLEM

The use of a matrix-vector shorthand notation for expression of the vibrational problem is facilitated by the fact that a vector can be represented by a column or row matrix. If the total vibrational motion of a polyatomic molecule containing N atoms is represented in a $3N$-dimensional space by a vector whose components are the $3N$ displacement coordinates the potential energy can be written as

$$V = \tfrac{1}{2}\tilde{\boldsymbol{\zeta}}\underline{\mathbf{B}}\boldsymbol{\zeta} \tag{16.40}$$

where $\tilde{\boldsymbol{\zeta}}$ is the transpose of $\boldsymbol{\zeta}$ (see App. G). Since $\boldsymbol{\zeta}$ is a column vector

$$\boldsymbol{\zeta} = \begin{pmatrix} \zeta_1 \\ \zeta_2 \\ \vdots \\ \zeta_{3N} \end{pmatrix} \tag{16.41}$$

$\tilde{\boldsymbol{\zeta}}$ will be a row vector

$$\tilde{\boldsymbol{\zeta}} = (\zeta_1 \quad \zeta_2 \quad \cdots \quad \zeta_{3N}) \tag{16.42}$$

B is a $3N \times 3N$ matrix whose elements are just the b_{ij} coefficients of Eq. (16.29).

Reference to the kinetic energy expression Eq. (16.28) shows that in the mass-weighted displacement coordinate system it can be expressed as

$$T = \tfrac{1}{2}\tilde{\dot{\zeta}}\dot{\zeta} \tag{16.43}$$

the Lagrangian is

$$L = \tfrac{1}{2}\tilde{\dot{\zeta}}\dot{\zeta} - \tfrac{1}{2}\tilde{\zeta}\underline{\mathbf{B}}\zeta \tag{16.44}$$

and the equations of motion, Eq. (17.30), become

$$\ddot{\zeta} + \underline{\mathbf{B}}\zeta = 0 \tag{16.45}$$

The secular equations are then given by

$$(\underline{\mathbf{B}} - \lambda_k \underline{\mathbf{I}})\mathbf{A}_k = 0 \tag{16.46}$$

where $\underline{\mathbf{I}}$ is the unit matrix and the solutions are

$$\zeta = \mathbf{A}\cos(\lambda_k^{1/2} t - \delta_k) \tag{16.47}$$

For each root λ_k, the displacement amplitudes are found by solving the individual equations represented by Eq. (16.46). The normalization condition corresponding to Eq. (16.36) becomes

$$\mathbf{D}_k \mathbf{D}_k = \underline{\mathbf{1}} \tag{16.48}$$

The normalized column vectors $\mathbf{D}_k$, which are obtained for each root λ_k, are referred to as the eigenvectors of $\underline{\mathbf{B}}$, since the mathematical procedure involved in this discussion is just the classical eigenvalue problem, but using a different formulation than is commonly encountered. Since the matrix $\underline{\mathbf{B}}$ is symmetric with respect to the diagonal, that is, $\underline{\mathbf{B}} = \tilde{\underline{\mathbf{B}}}$, $B_{ij} = B_{ji}$ $(i \neq j)$, the eigenvectors belonging to different eigenvalues are orthogonal, this being expressed as

$$\tilde{\mathbf{D}}_j \mathbf{D}_k = 0 \qquad j \neq k \tag{16.49}$$

These latter two relationships can be summarized as

$$\tilde{\mathbf{D}}_i \mathbf{D}_j = \delta_{ij} \tag{16.50}$$

where δ_{ij} is the Kronecker delta.

16.4 NORMAL COORDINATES

The nature of the expressions just developed for the kinetic and potential energies of a vibrating system of N atoms dictates that in general the solution of the Lagrangian and the resulting equations of motion will be complicated by the fact that the $\underline{\mathbf{B}}$ matrix representing the potential energy contains off-diagonal elements. The kinetic energy term, on the other hand, is a simple squared term and contains no off-diagonal terms. The use of a coordinate system where both the kinetic and potential energy terms contained only diagonal elements would greatly simplify the solution of the vibrational problem. This coordinate system does exist and is referred to as the normal coordinate system.

The normal coordinate Q_k, for which the corresponding root of the secular equation is λ_k, is defined as

$$Q_k = \sum_{i=1}^{3N} D_{ik}\zeta_i = \sum_{i=1}^{3N} \frac{A_{ik}}{(\sum A_{ik}^2)^{1/2}}\zeta_i \tag{16.51}$$

where the D_{ik} are the coefficients of the normalized eigenvectors and the A_{ik} those of the unnormalized ones.

Before considering the general molecule of N atoms let us apply the concepts of normal coordinates to the hypothetical one-dimensional triatomic molecule. Introducing the normalized coefficients given by Eqs. (16.20) to (16.23) into Eq. (16.51) gives

$$Q_1 = \tfrac{1}{2}\zeta_1 + 0\zeta_2 - \tfrac{1}{2}\zeta_3 \tag{16.52}$$

$$Q_2 = [\mu_1 + 2\mu_2]^{-1/2}[\mu_2^{1/2}\zeta_1 + \mu_1^{1/2}\zeta_2 + \mu_2^{1/2}\zeta_3] \tag{16.53}$$

$$Q_3 = [2(\mu_1 + 2\mu_2)]^{-1/2}[\mu_1^{1/2}\zeta_1 - 2\mu_2^{1/2}\zeta_2 + \mu_1^{1/2}\zeta_3] \tag{16.54}$$

The application of some simple algebra shows that this set of coordinates does lead to energy terms which involve only squares, that is, the matrices which represent them are diagonal. Using Eqs. (16.51) to (16.53) and the roots given by Eq. (16.15) shows that

$$\sum_{i=1}^{3} \lambda_i Q_i^2 = \tfrac{1}{2}\mu_1 k(\zeta_1 - \zeta_3)^2 + \tfrac{1}{2}k(\mu_1^{1/2}\zeta_1 - 2\mu_2^{1/2}\zeta_2 + \mu_1^{1/2}\zeta_3)^2$$

$$= k(\xi_1 - \xi_3)^2 + k(\xi_2 - \xi_3)^2 = 2V \tag{16.55}$$

and that

$$\sum_{j=1}^{3} \dot{Q}_i^2 = \frac{1}{\mu_1}\dot{\xi}_1^2 + \frac{1}{\mu_2}\dot{\xi}_2^2 + \frac{1}{\mu_3}\dot{\xi}_3^2 = 2T \tag{16.56}$$

Thus using the normal coordinates as defined yields

$$T = \frac{1}{2}\sum_i \dot{Q}_i^2 \tag{16.57}$$

$$V = \frac{1}{2}\sum_i \lambda_i Q_i^2 \tag{16.58}$$

Some further insight into the nature of normal coordinates can be gained by examining the Hamiltonian written in terms of the Q_i

$$H = T + V = \frac{1}{2}\sum_i \dot{Q}_i^2 + \frac{1}{2}\sum_i \lambda_i Q_i^2 = [\tfrac{1}{2}\dot{Q}_1^2 + \tfrac{1}{2}\lambda_1 Q_1^2] + [\tfrac{1}{2}\dot{Q}_2^2 + \tfrac{1}{2}\lambda_2 Q_2^2] + \cdots \tag{16.59}$$

Note that this is just a sum of terms having the same form as the energy of the one-dimensional harmonic oscillator $\frac{1}{2}m\dot{x}^2 + \frac{1}{2}kx^2$. To a first approximation the total vibrational energy of the molecule can be considered to be composed of N

independent modes of vibration, each of which is described by a normal coordinate. The total motion of the system is the result of superimposing the normal modes.

The application of these concepts to a system of N atoms can be summarized in matrix-vector notation. The normal coordinates are the components of a $3N$-dimensional vector $\mathbf{Q}$, and the energy terms are given by

$$T = \tfrac{1}{2}\tilde{\dot{\mathbf{Q}}}\dot{\mathbf{Q}} \tag{16.60}$$

$$V = \tfrac{1}{2}\tilde{\mathbf{Q}}\underline{\mathbf{\Lambda}}\mathbf{Q} \tag{16.61}$$

where $\underline{\mathbf{\Lambda}}$ has only diagonal elements λ_k. The equations of motion

$$\ddot{Q}_i + \lambda_i Q_i = 0 \tag{16.62}$$

which are derived from the Lagrangian are represented as

$$\ddot{\mathbf{Q}} + \underline{\mathbf{\Lambda}}\mathbf{Q} = 0 \tag{16.63}$$

The determinant of the secular equations is of diagonal form

$$|\underline{\mathbf{\Lambda}} - \lambda\underline{\mathbf{I}}| = 0 \tag{16.64}$$

The solution of this determinant gives $3N$ roots of which $3N - 6$ ($3N - 5$ for a linear molecule) will be nonzero and will correspond to the $3N - 6$ normal modes of vibration with $3N - 6$ associated normal coordinates.

In general it is not possible to write down a set of normal coordinates for a particular system *a priori*. They must be derived from a set of coordinates which can be more directly related to the geometry of the system. We will comment briefly on the problem of a coordinate transformation and then look at the specific transformation from Cartesian to normal coordinates.

Although we are more accustomed to the use of Cartesian coordinates, it is often better to employ a different set to describe the vibrational motion of the atoms within a molecule. The normal coordinates and associated normal modes of vibration are an inherent characteristic of a particular system and are independent of the coordinate system chosen to describe them. For any coordinate system that is selected to represent the motions, it is possible to write a linear transformation to Cartesian coordinates. This transformation is analogous to the procedure discussed during the development of the Euler angles in Chap. 3. Denoting such a set of coordinates by r_i and the mass-weighted Cartesian coordinates by ζ_i, the transformation is written as

$$\begin{pmatrix} \zeta_1 \\ \zeta_2 \\ \vdots \\ \zeta_{3N} \end{pmatrix} = \begin{pmatrix} v_{11} & v_{12} & \cdots & v_{1\,3N} \\ v_{21} & v_{22} & \cdots & v_{2\,3N} \\ \vdots & \vdots & & \vdots \\ v_{3n\,1} & v_{3N\,2} & \cdots & v_{3N\,3N} \end{pmatrix} \begin{pmatrix} r_1 \\ r_2 \\ \vdots \\ r_{3N} \end{pmatrix} \tag{16.65}$$

or

$$\boldsymbol{\zeta} = \underline{\mathbf{V}}\mathbf{R} \tag{16.66}$$

where $\boldsymbol{\zeta}$ and $\mathbf{R}$ are column vectors whose components are the $3N$ coordinates of the two systems and $\underline{\mathbf{V}}$ is a nonsingular transformation matrix. The inverse transformation is given by

$$\mathbf{R} = \underline{\mathbf{V}}^{-1}\zeta \tag{16.67}$$

where $\underline{\mathbf{V}}^{-1}$ is the reciprocal of $\underline{\mathbf{V}}$.

The energies can be expressed in terms of the new set of coordinates by application of the same transformation, hence,

$$V = \tfrac{1}{2}\tilde{\zeta}\underline{\mathbf{B}}\zeta = \tfrac{1}{2}\tilde{\mathbf{R}}\underline{\tilde{\mathbf{V}}\mathbf{B}\mathbf{V}}\mathbf{R} = \tfrac{1}{2}\tilde{\mathbf{R}}\underline{\mathbf{B}}'\mathbf{R} \tag{16.68}$$

In this case the elements of $\underline{\mathbf{B}}'$ are the coefficients for the potential energy expression when it is written in terms of the new coordinates

$$V = \frac{1}{2}\sum_{i=1}^{3N}\sum_{j=1}^{3N} b'_{ij}r_i r_j \tag{16.69}$$

Using an analogous manner, the kinetic energy is given by

$$T = \tfrac{1}{2}\tilde{\dot{\zeta}}\dot{\zeta} = \tilde{\dot{\mathbf{R}}}\underline{\tilde{\mathbf{V}}\mathbf{V}}\dot{\mathbf{R}} = \tilde{\dot{\mathbf{R}}}\underline{\mathbf{X}}\dot{\mathbf{R}} \tag{16.70}$$

where the elements of $\underline{\mathbf{X}}$ are the coefficients in the kinetic energy expression

$$T = \frac{1}{2}\sum_{i=1}^{3N}\sum_{j=1}^{3N} x_{ij}\dot{r}_i\dot{r}_j \tag{16.71}$$

It is to be noted that if $\underline{\mathbf{V}}$ is an orthogonal matrix, that is, $\underline{\tilde{\mathbf{V}}} = \underline{\mathbf{V}}^{-1}$, then $\underline{\tilde{\mathbf{V}}\mathbf{V}} = \underline{\mathbf{V}}^{-1}\underline{\mathbf{V}} = \underline{\mathbf{I}}$ and the transformation of the kinetic energy involves only squared terms. For the general case the secular determinant is of the form

$$|\underline{\mathbf{B}}' - \lambda\underline{\mathbf{X}}| = 0 \tag{16.72}$$

At this point we will look at the transformation which will allow one to transform the potential energy from a set of mass-weighted Cartesian displacement coordinates into the normal coordinates and at the same time only involve the squares of the latter. It will be illustrated by referral to the previously used example of the linear triatomic molecule. The form of the transformation is

$$\zeta = \underline{\mathbf{D}}\mathbf{Q} \tag{16.73}$$

where the elements of $\underline{\mathbf{D}}$ are the components of the normalized eigenvectors of the $\underline{\mathbf{D}}$ matrix, as given by Eq. (16.38), written side by side thus

$$\underline{\mathbf{D}} = \begin{pmatrix} D_{11} & D_{12} & \cdots & D_{1\,3N} \\ D_{21} & D_{22} & \cdots & D_{2\,3N} \\ \vdots & \vdots & & \vdots \\ D_{3N\,1} & D_{3N\,2} & \cdots & D_{3N\,3N} \end{pmatrix} \tag{16.74}$$

The inverse is given by

$$\mathbf{Q} = \underline{\mathbf{D}}^{-1}\zeta = \underline{\tilde{\mathbf{D}}}\zeta \tag{16.75}$$

For the three-atom linear molecule

$$\begin{pmatrix} Q_1 \\ Q_2 \\ Q_3 \end{pmatrix} = \begin{pmatrix} \dfrac{1}{\sqrt{2}} & 0 & -\dfrac{1}{\sqrt{2}} \\ \sqrt{\dfrac{\mu_2}{\mu_1 + 2\mu_2}} & \sqrt{\dfrac{\mu_1}{\mu_1 + 2\mu_2}} & \sqrt{\dfrac{\mu_2}{\mu_1 + 2\mu_2}} \\ \sqrt{\dfrac{\mu_1}{2(\mu_1 + 2\mu_2)}} & \sqrt{\dfrac{\mu_2}{2(\mu_1 + 2\mu_2)}} & \sqrt{\dfrac{\mu_1}{2(\mu_1 + 2\mu_2)}} \end{pmatrix} \begin{pmatrix} \zeta_1 \\ \zeta_2 \\ \zeta_3 \end{pmatrix} \tag{16.76}$$

or

$$Q_1 = \frac{1}{\sqrt{2}}(\zeta_1 - \zeta_3) \tag{16.77}$$

$$Q_2 = \frac{1}{\sqrt{\mu_1 + 2\mu_2}}(\sqrt{\mu_2}\,\zeta_1 + \sqrt{\mu_1}\,\zeta_2 + \sqrt{\mu_2}\,\zeta_3) \tag{16.78}$$

$$Q_3 = \frac{1}{\sqrt{2(\mu_1 + 2\mu_2)}}(\sqrt{\mu_1}\,\zeta_1 - 2\sqrt{\mu_2}\,\zeta_2 + \sqrt{\mu_1}\,\zeta_3) \tag{16.79}$$

These expressions for the Q_i are identical to those derived in Eqs. (16.52) to (16.54). Having already shown that the Q_i do satisfy the condition of occurring only as squares in the energy expressions we have now illustrated an alternate method for obtaining them.

16.5 QUANTUM MECHANICAL IMPLICATIONS

Before proceeding to the application of normal coordinate analysis to some individual systems, let us examine the quantum mechanical implications of the vibrational problem when it is expressed as a simultaneous occurrence of several individual uncoupled normal modes of vibration. This will involve discussions of the symmetry of normal coordinates, the wavefunctions, and the selection rules.

The quantum mechanical behavior of the vibration of a polyatomic molecule is found by solving the Schrödinger equation

$$\mathscr{H}\Psi = E\Psi \tag{16.80}$$

where Ψ is the total vibrational wavefunction of the molecule and $\mathscr{H}$ is the Hamiltonian derived from the classical relationship given by Eq. (16.59). For a normal coordinate Q_i, the operators representing position Q_i and momentum $\dot{Q}_i$ will be Q_i and $-i\hbar(\partial/\partial Q_i)$. The Schrödinger equation is then

$$\left(-\frac{\hbar^2}{2}\sum_{i=1}^{3N}\frac{\partial^2}{\partial Q_i^2} + \frac{1}{2}\sum_{i=1}^{3N}\lambda_i Q_i^2\right)\Psi = E\Psi \tag{16.81}$$

In the previous discussions of the methods of solving multidimensional differential equations it was established that if the equation could be written as a sum of terms, each of which involved only a single coordinate, the wavefunction could be expressed as a product of single coordinate wavefunctions and the energy could be expressed as a sum. Referring to Eq. (16.81) it is seen that the left-hand side is a sum of individual terms of the type, $-(\hbar^2/2)(\partial^2/\partial Q_i^2)+\frac{1}{2}\lambda_i Q_i^2$, hence we will assume that the wavefunction is of the form

$$\Psi = \psi_{n_1}(Q_1)\psi_{n_2}(Q_2)\psi_{n_3}(Q_3)\cdots\psi_{n_{3N}}(Q_{3N}) \tag{16.82}$$

where n_i is the vibrational quantum number of the ith normal mode. Writing the energy as a sum,

$$E = E_{n_1}+E_{n_2}+E_{n_3}+\cdots+E_{n_{3N}} \tag{16.83}$$

the Schrödinger equation separates into $3N$ wave equations of the form

$$\left(-\frac{\hbar^2}{2}\frac{d^2}{dQ_k^2}+\frac{1}{2}\lambda_k Q_k^2\right)\psi_{n_k}(Q_k) = E_{n_k}\psi_{n_k}(Q_k) \tag{16.84}$$

Since there will be six zero roots (five for a linear molecule) which give rise to the translational and rotational wavefunctions; there remain $3N-6(3N-5)$ individual vibrational wave equations, each involving a single normal coordinate.

The form of the equation just derived is the same as that of the diatomic harmonic oscillator except for a change of variable. Allowing for this variable change $\eta^2 \to Q_k^2/\alpha_k$, the solutions can be written by analogy to Eq. (2.16) as

$$\psi_{n_k}(Q_k) = N_{n_k}H_{n_k}(Q_k/\alpha_k^{1/2})e^{-(1/2)(Q_k^2/\alpha_k)} \tag{16.85}$$

where

$$\alpha_k = \hbar/\lambda_k^{1/2} = \hbar/2\pi\nu_k \tag{16.86}$$

$$N_{n_k} = [2^{n_k}n_k!\,\alpha_k^{1/2}\pi^{1/2}]^{-1/2} \tag{16.87}$$

$$n_k = 0, 1, 2, \ldots \tag{16.88}$$

$$H_{n_k}(Q_k/\alpha_k^{1/2}) = \text{Hermite polynomial} \tag{16.89}$$

The Hermite polynomials for the first few values of n_k are given in Table C-1. Using the values from this table the wavefunctions for the first few states of the oscillator in normal mode k can be formulated

$$\psi_0(Q_k) = (\alpha_k\pi)^{-1/4}e^{-(Q_k^2/2\alpha_k)} \tag{16.90}$$

$$\psi_1(Q_k) = \sqrt{2}(\alpha_k^3\pi)^{-1/4}Q_k e^{-(Q_k^2/2\alpha_k)} \tag{16.91}$$

$$\psi_2(Q_k) = (2\sqrt{2})^{-1}(\alpha_k\pi)^{-1/4}\left(\frac{4Q_k^2}{\alpha_k}-2\right)e^{-(Q_k^2/2\alpha_k)} \tag{16.92}$$

$$\psi_3(Q_k) = (4\sqrt{3})^{-1}(\alpha_k^3\pi)^{-1/4}\left(\frac{8Q_k^3}{\alpha_k}-12Q_k\right)e^{-(Q_k^2/2\alpha_k)} \tag{16.93}$$

$$\psi_4(Q_k) = (8\sqrt{6})^{-1}(\alpha_k\pi)^{-1/4}\left(\frac{16Q_k^4}{\alpha_k^2}-\frac{48Q_k^2}{\alpha_k}+12\right)e^{-(Q_k^2/2\alpha_k)} \tag{16.94}$$

It is to be noted that α_k as defined by Eq. (16.66) is not the same as the β used in Chap. 2. The α_k do not involve the reduced mass explicitly since the Q_k are mass-weighted coordinates.

The energy of each normal mode is given by

$$E_{n_k} = (n_k + \tfrac{1}{2})h\nu_k \tag{16.95}$$

For the entire molecule the total vibrational energy is

$$E = \sum_{k=1}^{3N-6} \left(n_k + \frac{1}{2}\right) h\nu_k \tag{16.96}$$

and the total wavefunction is

$$\Psi = \prod_{k=1}^{3N-6} \psi_{n_k}(Q_k) \tag{16.97}$$

16.6 SYMMETRY OF NORMAL COORDINATES

Although it may appear that we are "putting the cart before the horse" it will be instructive at this time to look at the form of the normal coordinates of a simple molecule, without first solving the vibrational problem to obtain them. This will provide a view of the symmetry properties of the vibrations relative to the symmetry of the molecules and as a result provide a better background for the discussion of internal and symmetry coordinates, whose use greatly simplifies normal coordinate analysis. The basic problem will be to investigate the manner in which the normal coordinates change when subjected to the symmetry operations of the point group to which the molecule belongs. The normal coordinates of a bent XY_2 molecule are depicted in Fig. 16.3. The derivation of these coordinates will be considered in detail in Sec. 16.10. The relative motion of the atoms for each normal mode is indicated by the heavy arrows. For this molecule, which belongs to the C_{2v} point group there are four symmetry operations, E, C_2, σ_v, and σ_v'. The effect of any one of these operations on a given normal mode can be ascertained by observation of how the individual atomic displacement vectors change. For example, the operation σ_v, which reflects the atomic displacement vector in a plane bisecting the YXY angle perpendicular to the figure plane,

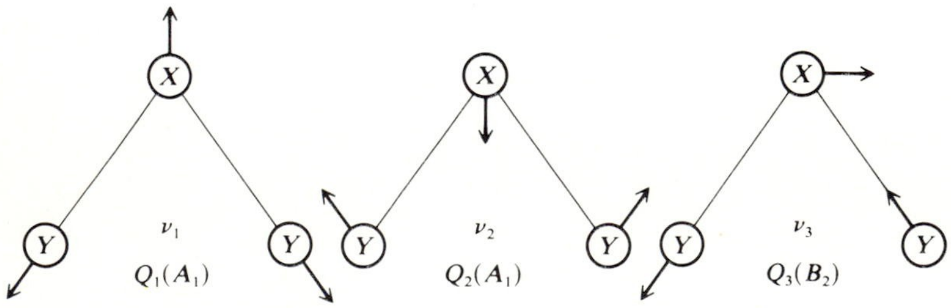

FIGURE 16.3
Normal modes of vibration of a bent XY_2 molecule.

TABLE 16.1
Effect of symmetry operations on the normal modes of a bent XY_2 molecule

	Normal mode		
Operation	Q_1	Q_2	Q_3
E	Q_1	Q_2	Q_3
C_2	Q_1	Q_2	$-Q_3$
σ_v	Q_1	Q_2	$-Q_3$
$\sigma_{v'}$	Q_1	Q_2	Q_3

exchanges the individual displacement vectors for modes 1 and 2 but inverts as well as exchanges the vectors for mode 3. In terms of the normal modes this is expressed as

$$\sigma_v Q_1 = Q_1 \qquad \sigma_v Q_2 = Q_2 \qquad \sigma_v Q_3 = -Q_3 \tag{16.98}$$

The results of all of the symmetry operations acting on the normal coordinates are summarized in Table 16.1. Inspection of the coefficients of the resulting Q_i shows that they are all ± 1. Comparison of the sets of coefficients to the characters of the irreducible representations given in the C_{2v} character table in App. J shows that there is a correspondence of the Q_1 and Q_2 coordinates with the characters of the A_1 representation and of Q_3 with those of the B_2 representation.

In Eq. (16.98), σ_v was used as a symbol denoting a particular transformation of a set of coordinates Q_i. Since σ_v denotes a transformation it in reality represents a matrix which will transform a vector with components Q_i into a new vector. The initial and transformed vectors are given by

$$\mathbf{Q}_I = \begin{pmatrix} Q_1 \\ Q_2 \\ Q_3 \end{pmatrix} \qquad \mathbf{Q}_T = \begin{pmatrix} Q_1 \\ Q_2 \\ -Q_3 \end{pmatrix} \tag{16.99}$$

Reversing the normal thinking regarding matrix multiplication will show the matrix representing the operation σ_v to be

$$\underline{\boldsymbol{\sigma}}_v = \begin{pmatrix} 1 & 0 & 0 \\ 0 & 1 & 0 \\ 0 & 0 & -1 \end{pmatrix} \tag{16.100}$$

Hence,

$$\begin{pmatrix} Q_1 \\ Q_2 \\ -Q_3 \end{pmatrix} = \begin{pmatrix} 1 & 0 & 0 \\ 0 & 1 & 0 \\ 0 & 0 & -1 \end{pmatrix} \begin{pmatrix} Q_1 \\ Q_2 \\ Q_3 \end{pmatrix} \tag{16.101}$$

Since any representation of a group can be expressed as a sum of irreducible representations, we may write the representations Γ_v of the total vibration of the molecule as $\Gamma_v = 2A_1 + B_2$. The techniques of group theory outlined in Chap. 4 provide the method for determining the irreducible representations for any arbitrary representation. If the vibrational representations of a particular group can be determined we should be able to find the irreducible representations to which the normal coordinates belong and thereby deduce their symmetries. Furthermore, if this can be done without first performing a normal mode analysis, some very substantial information which can be used to simplify the analysis will be available. This information can serve as a guide for the selection of an initial set of coordinates, other than Cartesian coordinates, and will provide the basis for the use of a greatly simplified set of potential coefficients.

By assigning a set of three Cartesian displacement vectors to each atom in the molecule the symmetries of the normal modes can be deduced by the application of group theory to this set of vectors. This procedure will be demonstrated by the use of two specific examples. The first of these, an XY_2 molecule with C_{2v} symmetry, will serve to illustrate the general method, and the second, an XY_3 molecule of C_{3v} symmetry, will serve to point out some of the intricacies involved when rotations are by other than 90 or 180°.

For XY_2 the $3N$ Cartesian displacement vectors of the atoms are as shown in Fig. 16.4. Now examine the change in these vectors upon successive application of all of the symmetry elements of the group, write the general transformation matrix for each operation, and determine the trace or character of each matrix. For example, the effect of the σ_v operation is to cause the vectors to undergo the following conversions

$$\begin{array}{lll} \xi_1 \to \xi_7 & \xi_4 \to \xi_4 & \xi_7 \to \xi_1 \\ \xi_2 \to -\xi_8 & \xi_5 \to -\xi_5 & \xi_8 \to -\xi_2 \\ \xi_3 \to \xi_9 & \xi_6 \to \xi_6 & \xi_9 \to \xi_3 \end{array}$$

The matrix representing this conversion can be determined by inspection to be

$$\underline{\boldsymbol{\sigma}}_v \begin{pmatrix} \xi_1 \\ \xi_2 \\ \xi_3 \\ \xi_4 \\ \xi_5 \\ \xi_6 \\ \xi_7 \\ \xi_8 \\ \xi_9 \end{pmatrix} = \begin{pmatrix} 0 & 0 & 0 & 0 & 0 & 0 & 1 & 0 & 0 \\ 0 & 0 & 0 & 0 & 0 & 0 & 0 & -1 & 0 \\ 0 & 0 & 0 & 0 & 0 & 0 & 0 & 0 & 1 \\ 0 & 0 & 0 & 1 & 0 & 0 & 0 & 0 & 0 \\ 0 & 0 & 0 & 0 & -1 & 0 & 0 & 0 & 0 \\ 0 & 0 & 0 & 0 & 0 & 1 & 0 & 0 & 0 \\ 1 & 0 & 0 & 0 & 0 & 0 & 0 & 0 & 0 \\ 0 & -1 & 0 & 0 & 0 & 0 & 0 & 0 & 0 \\ 0 & 0 & 1 & 0 & 0 & 0 & 0 & 0 & 0 \end{pmatrix} \begin{pmatrix} \xi_1 \\ \xi_2 \\ \xi_3 \\ \xi_4 \\ \xi_5 \\ \xi_6 \\ \xi_7 \\ \xi_8 \\ \xi_9 \end{pmatrix} = \begin{pmatrix} \xi_1 \\ -\xi_8 \\ \xi_9 \\ \xi_4 \\ -\xi_5 \\ \xi_6 \\ \xi_1 \\ -\xi_2 \\ \xi_3 \end{pmatrix} \tag{16.102}$$

The character is $\chi(\sigma_v) = 1$. This procedure can be repeated for the other three operations and in an analogous manner the characters are found to be

	E	C_2	σ_v	σ'_v
$\chi(\Gamma_T)$	9	-1	1	3

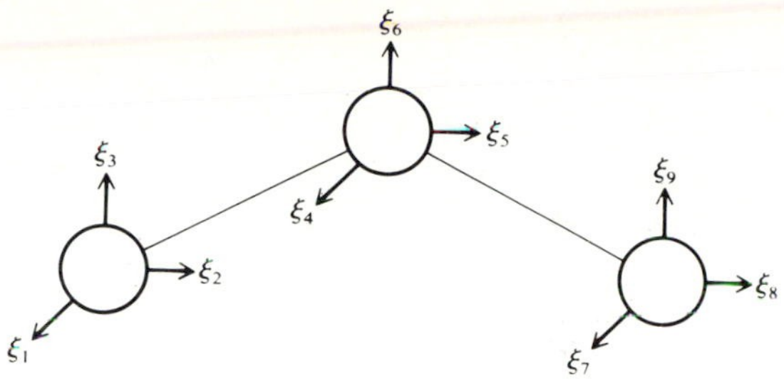

FIGURE 16.4
Cartesian displacement vectors for an XY_2 molecule.

The number of times a_i, that a given irreducible representation Γ_i, is contained in the total representation Γ_T, is found by using the characters of the Γ_i's from the appropriate character table and the characters just derived for Γ_T. For the C_{2v} group, for example,

$$a_1 = a(A_1) = \tfrac{1}{4}(9 \times 1 - 1 \times 1 + 1 \times 1 + 3 \times 1) = 3 \tag{16.103}$$

and

$$\Gamma_T = 3A_1 + A_2 + 2B_1 + 3B_2 \tag{16.104}$$

The representation Γ_T includes all motions of the molecule. Since the interest at this point is only in the vibrational motion we can separate out and discard the contributions due to translation and rotation. The coordinates for translation will be parallel to the Cartesian axis system and will transform in the same manner as the axes themselves. For example, the σ_v operation will leave vectors in the x and z directions unchanged, and reverse the one in the y-direction. This is expressed as

$$\sigma_v \begin{pmatrix} x \\ y \\ z \end{pmatrix} = \begin{pmatrix} 1 & 0 & 0 \\ 0 & -1 & 0 \\ 0 & 0 & 1 \end{pmatrix} \begin{pmatrix} x \\ y \\ z \end{pmatrix} = \begin{pmatrix} x \\ -y \\ z \end{pmatrix} \tag{16.105}$$

and $\chi_{\text{tr}}(\sigma_v) = 1$. Examining the effect on x, y, z by the other symmetry operations gives

	E	C_2	σ_v	σ_v'
$\chi(\Gamma_{\text{tr}})$	3	-1	1	1

which yields

$$\Gamma_{\text{tr}} = A_1 + B_1 + B_2 \tag{16.106}$$

and three of the irreducible representations have been eliminated from further consideration. It is to be noted, looking at the C_{2v} character table, that the coordinates x, y, z are listed in a column opposite these three irreducible representations, hence they could have been eliminated by just referring to the table. In the future this procedure will be followed.

The problem of determining the irreducible representation belonging to the three rotations can be solved by rotating the molecule slightly out of the symmetry

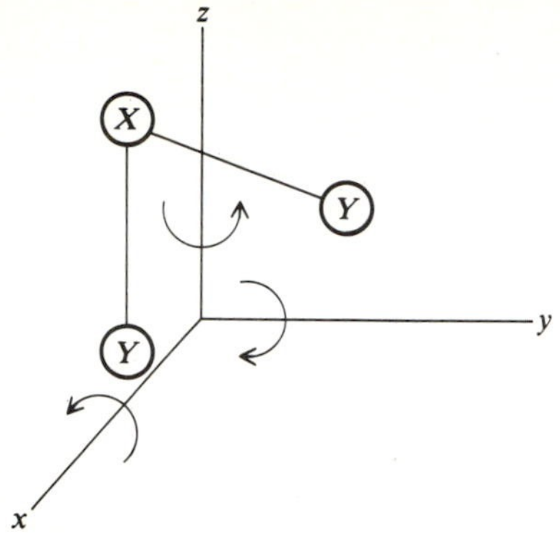

σ_v—xz plane
σ'_v—yz plane
$C2$—z axis

FIGURE 16.5
Reference frame for determination of symmetry of rotational operators.

reference frame, performing the symmetry operations, and comparing the final and initial positions. This is depicted in Fig. 16.5 for a rotation about the x axis, R_x. The E and σ'_v operations leave the configuration unchanged while C_2 and σ_v have the effect of reversing the displacement. The characters are then $\chi(E) = \chi(\sigma'_v) = 1$ and $\chi(C_2) = \chi(\sigma_v) = -1$. Consideration of the displacement rotations about the y and z axes lead to the set of characters

	E	C_2	σ_v	σ'_v
$\chi(\Gamma_{rot})$	3	-1	-1	-1

which yields

$$\Gamma_{rot} = A_2 + B_1 + B_2 \tag{16.107}$$

Another way to arrive at the same conclusions is to consider the rotations to be represented by directed semicircles as also shown in Fig. 16.5, and examine the change in direction of rotation with the application of the symmetry operations. This simple geometric aid comes about from the fact that rotation corresponds to a change in angular momentum. Since angular momentum can be expressed in terms of linear momentum, for example, $L_x = yp_z - zp_y$ and p_q transforms like q, it follows that R_q will transform like L_q. It is also to be noted that the irreducible representations belonging to the rotations are so noted in the character tables by R_q's, so in the future this information can be obtained from that source.

Subtracting Eqs. (16.106) and (16.107) from (16.104) leaves

$$\Gamma_{vib} = \Gamma_T - \Gamma_{Tr} - \Gamma_{rot} = 2A_1 + B_2 \tag{16.108}$$

thus, there will be two normal modes of A_1 symmetry, totally symmetric, and one of B_1 symmetry, symmetric only with respect to E and σ'_v. Reference to Fig. 16.3 shows that this is true and that the normal modes have been indexed accordingly.

The application of the preceding methods of analysis to a molecule involving a three-fold axis is slightly more involved, and since it involves a type transformation that was not encountered in the XY_2 example it is instructive to analyze this type of molecule also. The motion of the atoms in an XY_3 molecule are represented by a set of Cartesian displacement vectors as shown in Fig. 16.6, where the z

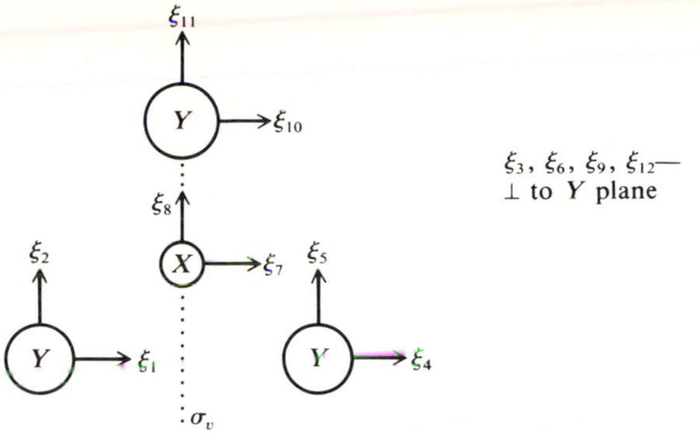

FIGURE 16.6
Cartesian displacement vectors for an XY_3 molecule.

axis is perpendicular to the plane of the three Y atoms and the X atoms is above that plane. The operations of this group are E, $2C_3$ and $3\sigma_v$. When determining the characters the multiple occurrence of the C_3 and σ_v operations must be kept in mind. If we examine the C_3 (counterclockwise) operation we find that, unlike the C_2 operation of the previous example, it does not result in a simple inversion (sign change) of a vector. Reference to Fig. 16.7 and the application of some simple trigonometry shows that rotation through an angle θ gives

$$\xi_7' = \xi_7 \cos\theta + \xi_8 \sin\theta \tag{16.109}$$

$$\xi_8' = -\xi_7 \sin\theta + \xi_8 \cos\theta \tag{16.110}$$

$$\xi_9' = \xi_9 \tag{16.111}$$

Reflecting back for a moment at the previous example where $\theta = 180°$ one observes that this leads to $\xi_5 \to -\xi_5$ and $\xi_4 \to -\xi_4$, which is in agreement with the previous conclusion. For the rotation of the vectors on the X atom this gives

$$\underline{\mathbf{C}}_3 \begin{pmatrix} \xi_7 \\ \xi_8 \\ \xi_9 \end{pmatrix} = \begin{pmatrix} \cos\theta & \sin\theta & 0 \\ -\sin\theta & \cos\theta & 0 \\ 0 & 0 & 1 \end{pmatrix} \begin{pmatrix} \xi_7 \\ \xi_8 \\ \xi_9 \end{pmatrix} = \begin{pmatrix} \xi_7' \\ \xi_8' \\ \xi_9' \end{pmatrix} \tag{16.112}$$

The character for this matrix is $\chi = 1 + 2\cos\theta$. The effect of the C_3 operation on the three off-center atoms will be not only to produce a rotation of the vectors identical to that of the central atom vectors but also to scramble the order of the vectors. The overall effect of the C_3 operation on the 12 displacement vectors

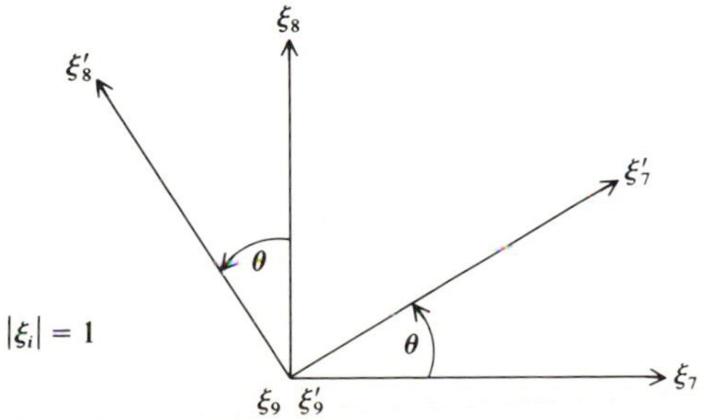

FIGURE 16.7
Effect of rotation of Cartesian vectors by an angle θ.

can be summarized in matrix notation as

$$\underline{\mathbf{C}}_3 \begin{pmatrix} \xi_1 \\ \xi_2 \\ \xi_3 \\ \xi_4 \\ \xi_5 \\ \xi_6 \\ \xi_7 \\ \xi_8 \\ \xi_9 \\ \xi_{10} \\ \xi_{11} \\ \xi_{12} \end{pmatrix} = \left(\begin{array}{c:c:c:c} 0 & A & 0 & 0 \\ \hdashline 0 & 0 & 0 & A \\ \hdashline 0 & 0 & A & 0 \\ \hdashline A & 0 & 0 & 0 \end{array}\right) \begin{pmatrix} \xi_1 \\ \xi_2 \\ \xi_3 \\ \xi_4 \\ \xi_5 \\ \xi_6 \\ \xi_7 \\ \xi_8 \\ \xi_9 \\ \xi_{10} \\ \xi_{11} \\ \xi_{12} \end{pmatrix} \tag{16.113}$$

where each A is the 3×3 transformation matrix of Eq. (16.112) and the 0's are 3×3 blocks of zeros. The character for the C_3 operation is thus $\chi(C_3) = 1 + 2\cos\theta$.

The situation is somewhat simpler for the σ_v operation. The reflection of the molecule in any one of the three symmetry planes is equivalent to its reflections in either of the other two. Looking at the plane denoted by σ_v in Fig. 16.6 the following transformations are noted:

$$\begin{array}{lll} \xi_1 \to -\xi_4 & \xi_4 \to -\xi_1 & \xi_7 \to -\xi_7 \\ \xi_2 \to \xi_5 & \xi_5 \to \xi_2 & \xi_{10} \to -\xi_{10} \\ \xi_3 \to \xi_6 & \xi_6 \to \xi_3 & \end{array}$$

$\xi_9, \xi_8, \xi_{11}, \xi_{12}$ remain unchanged

This transformation is expressed as

$$\underline{\boldsymbol{\sigma}}_v \begin{pmatrix} \xi_1 \\ \xi_2 \\ \xi_3 \\ \xi_4 \\ \xi_5 \\ \xi_6 \\ \xi_7 \\ \xi_8 \\ \xi_9 \\ \xi_{10} \\ \xi_{11} \\ \xi_{12} \end{pmatrix} = \left(\begin{array}{c:c:c:c} 0 & B & 0 & 0 \\ \hdashline B & 0 & 0 & 0 \\ \hdashline 0 & 0 & B & 0 \\ \hdashline 0 & 0 & 0 & B \end{array}\right) \begin{pmatrix} \xi_1 \\ \xi_2 \\ \xi_3 \\ \xi_4 \\ \xi_5 \\ \xi_6 \\ \xi_7 \\ \xi_8 \\ \xi_9 \\ \xi_{10} \\ \xi_{11} \\ \xi_{12} \end{pmatrix} \tag{16.114}$$

where

$$B = \begin{pmatrix} -1 & 0 & 0 \\ 0 & 1 & 0 \\ 0 & 0 & 1 \end{pmatrix} \tag{16.115}$$

and lies on the diagonal only when a reflection leaves a nucleus unchanged. The character for σ_v is $\chi(\sigma_v) = 2$. A reflection of this type is classified as an improper rotation with $\theta = 180°$, and a general expression for the character of such an operation is

$$\chi(R_{\text{imp}}) = -N_R(1 + 2\cos\theta) \tag{16.116}$$

where N_R is the number of nuclei left unchanged by the operation and θ has the following values

Operation	θ
σ	180°
i	0°
S_3	60°
S_4	90°
S_6	120°

A transformation of the type C_n which is a proper rotation has, as a general expression for its character,

$$\chi(R_{\text{prop}}) = N_R(1 + 2\cos\theta) \tag{16.117}$$

where θ has values of

Operation	θ
E	0°
C_2	180°
C_3	120°
C_4	90°
C_5	72°
C_6	60°

The important conclusion of the previous discussion is that there is a way to obtain the characters without going through the details of applying each of the group operations to each of the $3N$ displacement vectors.

The same information, which is conveyed by Eqs. (16.116) and (16.117), may be derived in a somewhat different manner by examining the contributions to the character by the undisplaced atoms as indicated below:

1. A σ operation contributes +1 to $\chi(\sigma)$ for each atom lying on the symmetry plane.

2. A C_n^m operation contributes $1 + 2\cos(2\pi m/n)$ for each atom lying on the axis of rotation.
3. The inversion operation contributes -3 to $\chi(i)$ if the molecule has an atom at the center of symmetry.
4. $\chi(E)$ is always equal to $3N$ since no displacement vectors are changed.

Looking back at the example of the XY_3 pyramidical molecule either method can be used to ascertain that the characters are $\chi(E) = 12$, $\chi(C_3) = 0$, $\chi(\sigma) = 2$. Using the C_{3v} character table in App. J, accounting for the number of elements in the C_3 and σ_v classes, and employing Eq. (4.10) gives $a_1 = a(A_1) = 3$, $a_2 = a(A_2) = 1$ and $a_3 = a(E) = 4$. Therefore $\Gamma_T = 3A_1 + A_2 + 4E$. Using the character table to subtract out the translational and rotational representations, $\Gamma_{Tr} = A_1 + E$ and $\Gamma_{\text{rot}} = A_2 + E$ leaves $\Gamma_{\text{vib}} = 2A_1 + 2E$. Therefore, of the six normal modes of vibration, two will be totally symmetric and the other four will comprise two pairs of degenerate modes. Looking at the pyramidal structure of the molecule it is easy to discern that the two symmetrical modes will consist of a simultaneous symmetric stretching of all XY bonds and a symmetric movement of all YXY angles in an umbrella-like manner. The degenerate modes cannot be as easily surmised and their nature must be found by a complete analysis of the vibration of the molecule.

16.7 DEGENERATE VIBRATIONS

In the discussion of the vibration of the XY_3 molecule it was found that there were four modes of vibrations which belonged to two E-type irreducible representations. These are referred to as degenerate modes. The term, degeneracy, is used in the usual quantum mechanical sense in that it refers to a situation where there are two eigenfunctions (in this case vibrational wavefunctions) belonging to a single eigenvalue (in this case the vibrational energy or frequency). This situation becomes more common in molecules of higher symmetry. If we consider the example of the linear XY_2 molecule and remove the restriction of one-dimensional motion then we have a simple system involving degenerate vibrations.

The linear XY_2 molecule belongs to the $D_{\infty h}$ group, which has a character table of infinite extent. Because of this feature it is more difficult to formally consider molecules of $D_{\infty h}$ and $C_{\infty v}$ symmetry. However, such molecules will in general be limited in their number of atoms and it is possible to incorporate a little intuition along with a formal analysis. Since the motion of the system will consist of either linear bond extensions or deformation from a linear arrangement it is expedient to consider these two motions separately. The linear XY_2 molecule will require four coordinates to specify the vibrations. It will be useful at this point to introduce a simple set of bond-related coordinates, referred to as internal coordinates, to replace the usual Cartesian system. Two of the coordinates will be the XY bond extensions which will fully characterize the linear motion of

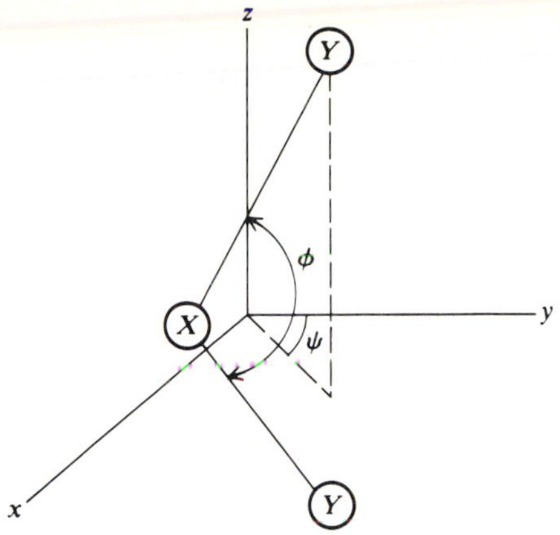

FIGURE 16.8
Vibrational bending motion of a linear molecule.

the molecule. The two remaining coordinates will be associated with the deformation from linearity. There is only one manner in which a three-atom linear molecule can deform from linearity, and that is by a change of the YXY angle. Since the molecule is in reality a three-dimensional system, the specification of this angle alone is insufficient, and the rotation about the symmetry axis must also be given in order to fully characterize the system. This is illustrated in Fig. 16.8, where the two necessary angles are denoted as ϕ and ψ. In this vibrational mode the nuclei will undergo an elliptical-type motion about the z axis, which passes through the center of mass. The eccentricity of the ellipses will depend in the relative phases of the motion. This same motion can equally well be described by means of two mutually perpendicular angles of deformation.

Examining the coordinates specifying the linear motions it is found that any operation of the group will either exchange or leave unchanged the linear deformations, hence the characters χ_r will be either 0 or 2. Referring to the $D_{\infty h}$ character table for the operations to be considered one has,

$$\chi_r(E) = 2 \qquad \chi_r(\sigma_v) = 2 \qquad \chi_r(S_\phi) = 0$$

$$\chi_r(C_\phi) = 2 \qquad \chi_r(i) = 0 \qquad \chi_r(C_2) = 0$$

Without doing any mathematical analysis it can be seen that the only two irreducible representations which can combine to give this set of characters will be Σ_g^+ and Σ_u^+, hence the representations of the linear vibrations have been established. We now examine the effects of the symmetry operations on ϕ. In this case it is somewhat more difficult to visualize the changes but, if the mutually perpendicular bendings are represented by the two curved arrows shown in Fig. 16.9, it is possible to arrive at the right conclusions. The E operation leaves both unchanged. The $\sigma_v(v = yz)$ and C_2 operations reverse the direction of ϕ but leave ϕ' unchanged. The i operation reverses both. The C_ϕ and S_ϕ operations move the curved arrows around the z axis so that they are represented as functions of the undisplaced ones just as was the case for the C_3 operation on the trigonal

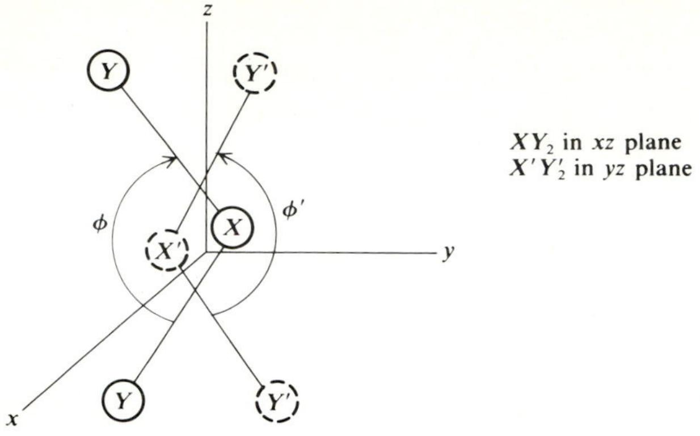

FIGURE 16.9
Angular displacements for degenerate bending vibration.

XY_3 molecule. The characters are

$$\chi_\phi(E) = 2 \qquad \chi_\phi(i) = -2$$
$$\chi_\phi(C_\phi) = 2\cos\phi \qquad \chi_\phi(S_\phi) = 2\cos\phi$$
$$\chi_\phi(\sigma_v) = 0 \qquad \chi_\phi(C_2) = 0$$

Looking at the character table shows that these two modes of vibration belong to the π_u irreducible representation.

The normal modes of the linear XY_2 molecule are summarized in Fig. 16.10. The stretching modes ν_1 and ν_3 are nondegenerate, and the bending mode ν_2 is doubly degenerate. The irreducible representations of each mode and hence of the normal coordinates are given in parenthesis.

This off-axial vibrational motion of linear molecules has the effect of producing a component of angular moment about the molecular symmetry axis, a phenomenon that is not allowed for a rigid linear rotor. It thus constitutes a form of rotational-vibrational interaction. This is important in the analysis of pure rotational spectra of nonrigid linear molecules and will be discussed further in Chap. 17.

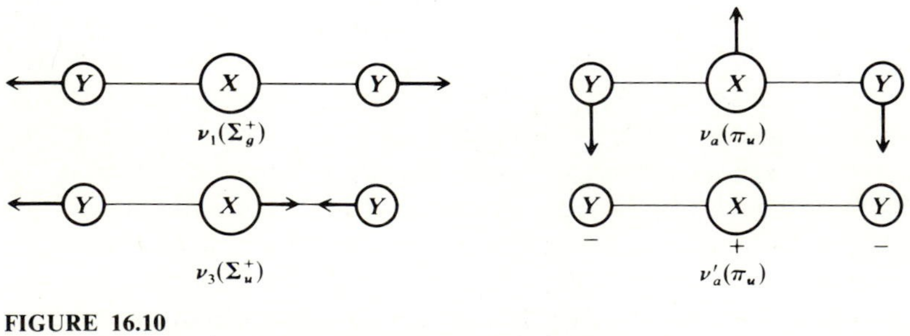

FIGURE 16.10
Fundamental modes of vibration of a linear XY_2 molecule.

16.8 SYMMETRY OF WAVEFUNCTIONS

Having progressed to the point where the symmetry properties of the normal modes can be obtained we will briefly deviate from the main problem of finding the normal modes and relate the information found to the quantum mechanics of the system. Although the methods of vibrational analysis are classical in nature the molecules to which they are applied behave in a quantized manner. Hence, any discussion must always relate to the quantum mechanical properties of the molecules. It is necessary then to examine the selection rules and the nature of the vibrational spectra of polyatomic molecules.

In Sec. 16.5 it was shown that, to a first approximation, the vibrational wavefunction of a molecule could be written as a product of $3N-6$ separate wavefunctions, each being associated with a separate normal coordinate. To a first approximation this separation of normal modes is valid; however, there can be coupling between modes particularly when they are of the same symmetry, are close together in frequency, and involve a mutual atom. Associated with each of the vibrational wavefunctions is an energy which is the sum of the energies of the individual normal modes. Therefore,

$$\psi(v_1, v_2, \ldots, v_{3N-6}) = \prod_{i=1}^{3N-6} \psi_{vi}(Q_i) \tag{16.118}$$

$$E(v_1, v_2, \ldots, v_{3N-6}) = \sum_{i=1}^{3N-6} E_i = \sum_{i=1}^{3N-6} \left(v_i + \frac{1}{2}\right) h\nu_i \tag{16.119}$$

Having found a way to establish the symmetry of the normal coordinates this information can now be used to derive the symmetry of the wavefunctions. This point can be most easily demonstrated by use of an example rather than by use of general expressions like Eqs. (16.118) and (16.119).

For the nonlinear XY_2 molecule the vibrational wavefunction will be

$$\psi(v_1, v_2, v_3) = \psi_{v_1}\psi_{v_2}\psi_{v_3} \tag{16.120}$$

where the ψ_{v_i} are given in Eq. (16.85). The energy of the system is

$$E(v_1, v_2, v_3) = (v_1 + \tfrac{1}{2})h\nu_i + (v_2 + \tfrac{1}{2})h\nu_2 + (v_3 + \tfrac{1}{2})h\nu_3 \tag{16.121}$$

In this case each of the v_i can independently have any integer value including zero. Keeping the same order on the subscripts as was used in Fig. 16.3, assuming that all vibrations are harmonic, and arbitrarily assigning $\nu_2 = 0.6\ \nu_3$ and $\nu_1 = 0.8\ \nu_3$, the energy levels of the molecule will be distributed as shown in Fig. 16.11. The three levels (001), (010), and (100) are called the fundamental levels, while those with two v_i's equal to zero and the other equal to or greater than 2 are referred to as overtone levels. Those with two or three nonzero values of the v_i are called *combination levels.*

In order to determine the symmetry of the wavefunctions, the information necessary for evaluating the zero-nonzero character of the dipole integrals, use is made of the symmetry of the normal coordinates. The three normal coordinates of the bent XY_2 molecule were found to belong to the A_1 and B_2 representations.

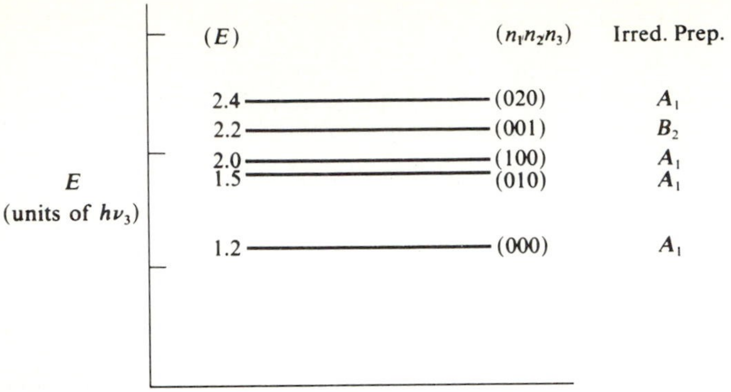

FIGURE 16.11
Vibrational energy levels of a bent XY_2 molecule.

Since these are one-dimensional representations the effect of any operation of the C_{2v} group will be to multiply Q by ± 1. The ground state wavefunction

$$\psi(000) = [\alpha_1 \alpha_2 \alpha_3 \pi^3]^{-1/4} \exp\left[-\frac{1}{2}\left(\frac{Q_1^2}{\alpha_1} + \frac{Q_2^2}{\alpha_2} + \frac{Q_3^2}{\alpha_3} \right) \right] \tag{16.122}$$

contains only terms which are quadratic in the Q_k, and will be invariant to all operations of the group. The characters of the representation to which $\psi(000)$ belongs are all unity, so the representation for $\psi(000)$ is A_1. The fundamental wavefunctions will be

$$\psi(1, 0, 0) = 2[\alpha_1^3 \alpha_2 \alpha_3 \pi^3]^{-1/4} Q_1 \exp\left[-\frac{1}{2}\left(\frac{Q_1^2}{\alpha_1} + \frac{Q_2^2}{\alpha_2} + \frac{Q_3^2}{\alpha_3} \right) \right] \tag{16.123}$$

$$\psi(0, 1, 0) = 2[\alpha_1 \alpha_2^3 \alpha_3 \pi^3]^{-1/4} Q_2 \exp\left[-\frac{1}{2}\left(\frac{Q_1^2}{\alpha_1} + \frac{Q_2^2}{\alpha_2} + \frac{Q_3^2}{\alpha_3} \right) \right] \tag{16.124}$$

$$\psi(0, 0, 1) = 2[\alpha_1 \alpha_2 \alpha_3^3 \pi^3]^{-1/4} Q_3 \exp\left[-\frac{1}{2}\left(\frac{Q_1^2}{\alpha_1} + \frac{Q_2^2}{\alpha_2} + \frac{Q_3^2}{\alpha_3} \right) \right] \tag{16.125}$$

As with $\psi(0, 0, 0)$ the exponential terms contain only squares of the coordinates and will always be invariant to the group operations. Therefore the symmetry will be determined by the preexponential factor, in this case Q_i. The irreducible representation (symmetry species) to which the fundamental wavefunctions belong will be the same as those of the normal coordinates. From this point we can generalize that for any overtone wavefunction the symmetry species will be that of $H_{v_i}(Q_i)$ and furthermore will be odd or even depending on the odd or even nature of v_i.

For combination wavefunctions the procedure for evaluating the symmetry species is analogous to that for fundamentals and overtones except the preexponential factor will be a product of Hermite polynomials. This procedure, while straightforward, can become somewhat involved due to algebraic manipulations. The use of group theory can profitably simplify the problem. The characters of a product function (direct product) are equal to the products of the characters of the initial functions, hence the representation of the product function is the direct product of the representations of the initial functions. This operational procedure plus the fact that the symmetry of an individual wavefunction is that of the corresponding Hermite polynomial provides a method to quickly determine combination wavefunction symmetries. For example

$$\psi(111) = \psi_1(Q_1)\psi_1(Q_2)\psi_1(Q_3) \tag{16.126}$$

and

$$\psi(120) = \psi_1(Q_1)\psi_2(Q_2)\psi_0(Q_3) \tag{16.127}$$

The symmetry species for the initial functions in $\psi(111)$ are A_1, A_1, and B_2 while those for $\psi(120)$ are A_1, A_1, and A_1. Taking the direct products, $A_1 \times A_1 \times B_2 = B_2$ and $A_1 \times A_1 \times A_1 = A_1$. This same procedure is also applicable to overtone wavefunctions. For example, for $\psi(002)$ the direct product is $A_1 \times A_1 \times A_1 = A_1$ and for $\psi(003)$, $A_1 \times A_1 \times B_2 = B_2$.

The concepts presented in the discussion of the example of a bent molecule XY_2 can be summarized as follows and are true in general so long as there are no degenerate modes involved. (1) The ground state wavefunction for any molecule always belongs to the totally symmetric representation. (2) The fundamental wavefunction $\psi_{vk}(Q_k)$ belongs to the same irreducible representation as Q_k itself. (3) The symmetry of overtone wavefunctions alternate, the even ones belonging to the totally symmetric representation and the odd ones belonging to the same representation as Q_k. (4) The irreducible representation of a combination wavefunction is that of the direct product of the individual functions of which it is composed.

The only point of nonconformity encountered regarding the use of these generalizations is with degenerate vibrations. A degenerate vibration will belong to an irreducible representation whose dimension is equal to the order of the degeneracy. In order to better understand this situation the degenerate vibration will be viewed as being composed of two frequencies, ν_i and ν_i', whose magnitudes happen to be identical. For simplicity the discussion will be limited to the doubly degenerate linear XY_2 case. If only one of these frequencies is excited to produce a fundamental say $v_i = 0$, $v_i' = 1$ or $v_i = 1$, $v_i' = 0$, then the fundamental wavefunction is doubly degenerate and will belong to the same representation as the two normal coordinates. The ν_2 and ν_2' vibrations of the XY_2 molecule in Fig. 16.10 are of this type.

Further excitation of the system leads to a somewhat unique situation. If ν_2 and ν_2' were of different magnitudes then the wavefunctions of the overtones $\psi(0, n_2, n_2', 0)$ would be given by $\psi(0200)$ and $\psi(0020)$. However, since they are not, $\psi(0200)$ and $\psi(0020)$ are equivalent and the function $\psi(0110)$ is also an eigenfunction. The overtone state is then triply degenerate. The use of the direct product is not always applicable when degenerate states are present. The representation of the overtone wavefunction may be found by determining the symmetrical product, $\Gamma \otimes \Gamma$, rather than the direct product, $\Gamma \times \Gamma$. The detailed mathematics of this concept is outside the scope of this book, and its introduction and use will be confined to the method for finding such a product when the representations involved are related to the vibrational problem. Continuing with the example of a linear YXY molecule let $\chi(R)$ be the character of the representation of the fundamental of the ν_2 vibrations for the R symmetry operation. For R repeated twice in succession the character is denoted by $\chi(R^2)$. To illustrate these two points consider the π_u representation and the σ_v reflection for the linear XY_2 molecule. In this case $\chi(\sigma_v) = 0$, $\sigma_v^2 = E$, $\chi(E) = 2$, therefore $\chi(\sigma_v^2) = 2$. Consideration of the other operations and the associated characters of this representation shows that the characters of the R^2 operations will be

$$\begin{array}{lll}
\chi(E) = 2 & E^2 = E & \chi(E^2) = 2 \\
\chi(C_\phi) = 2\cos\phi & C_\phi^2 = C_{2\phi} & \chi(C_\phi^2) = 2\cos 2\phi \\
\chi(i) = -2 & i^2 = E & \chi(i^2) = 2 \\
\chi(\sigma_n) = 2 & \sigma_n^2 = E & \chi(\sigma_n^2) = 2 \\
\chi(S_\phi) = 2\cos\phi & S_\phi^2 = C_2 & \chi(S_\phi^2) = 2\cos 2\phi \\
\chi(C_2) = 0 & C_2^2 = E & \chi(C_2^2) = 2
\end{array}$$

The characters of the representation of an overtone of a twofold degenerate vibration are given by the characters of the symmetrical product $\Gamma \otimes \Gamma$ which are

$$\chi_+^{\Gamma\otimes\Gamma}(R) = \tfrac{1}{2}\{[\chi^\Gamma(R)]^2 + \chi^\Gamma(R^2)\} \tag{16.128}$$

An alternate method for finding the nth overtone characters is given by

$$\chi_+^{(n)}(R) = \frac{[1 + (-1)^n]}{2} \tag{16.129}$$

for $\chi(R) = 0$, $\chi(R^2) = 2$, and by

$$\chi_+^{(n)}(R) = \frac{\sin(n+1)\phi'}{\sin\phi'} \tag{16.130}$$

otherwise. When $\phi' = 0$ or π then $\sin(n+1)\phi'/\sin\phi'$ is indeterminate and evaluation of the limit gives $(n+1)$ for $\phi' = 0$ and $[(-1)^n(n+1)]$ for $\phi' = \pi$. Then n in these equations is the number of quanta of excitation possessed by, or the sum of the quantum numbers of, the degenerate mode, and ϕ' is given by

$$2\cos\phi' = \chi(R) \tag{16.131}$$

A detailed examination of a couple of characters for the second overtone, $n = 2$, of the ν_2 vibrations of the YXY molecule will illustrate the use of these relationships. For the C_ϕ operations $\chi(C_\phi) = 2\cos\phi$ and Eq. (16.131) shows that $\phi = \phi'$. Since $\chi(C_\phi^2) = 2\cos 2\phi$, Eq. (16.130) is applicable and gives

$$\chi_+^{(2)}(C_\phi) = \frac{\sin 3\phi}{\sin\phi} = 1 + 2\cos 2\phi \tag{16.132}$$

For the i operation $\chi(i) = -2 = 2\cos\phi'$, so $\phi' = \pi$ and Eq. (16.130) gives

$$\chi_+^{(2)}(i) = [(-1)^2(2+1)] = 3 \qquad (16.133)$$

the use of Eq. (16.128) will give analogous results:

$$\chi_+^{\pi_u\otimes\pi_u}(C_\phi) = \tfrac{1}{2}\{[2\cos\phi]^2 + 2\cos 2\phi\} = 1 + 2\cos 2\phi \qquad (16.134)$$

and

$$\chi_+^{\pi_u\otimes\pi_u}(i) = \tfrac{1}{2}\{[-2]^2 + 2\} = 3 \qquad (16.135)$$

Repetitive use of Eqs. (16.128) to (16.131) provides the balance of the characters: $\chi_+^{(2)}(E) = 3$, $\chi_+^{(2)}(C_2) = 1$, $\chi_+^{(2)}(\sigma_h) = 3$, $\chi_+^{(2)}(\sigma_v) = 1$, and $\chi_+^{(2)}(S_\phi) = 1 + 2\cos 2\phi$. The value of $\chi^{(2)}(E) = 3$ shows this representation to be of dimension 3, hence it is of the correct dimension to accommodate the three possible combinations of quantum numbers. Comparing the characters of the representation

	E	C_ϕ	σ_v	σ_h	i	S_ϕ	C_2
$\Gamma_{\pi_u\otimes\pi_u}$	3	$1+2\cos 2\phi$	1	3	3	$1+2\cos 2\phi$	1

to those of the $D_{\infty h}$ group shows by inspection that

$$\pi_u \otimes \pi_u = \Sigma_g^+ + \Delta_g$$

For higher overtones the procedure is analogous, only in general one must resort to the formal methods of group theory to decompose the representation into its irreducible components.

In this discussion of degenerate vibrations the choice of an example belonging to a more complicated infinite group was deliberate in that by gaining an understanding of this example the student will have the necessary background for dealing with degenerate vibrations of molecules belonging to finite groups as well as with the peculiarities of the infinite ones.

If a degenerate mode contributes to a combination level then the symmetry species of the latter is found by (1) determining the symmetry species of the overtones resulting from each mode, (2) taking the direct product of these representations, and (3) finding the irreducible representations contained in this direct product. For example, consider a linear YXY molecule which exists in an energy state given by, $E = \frac{7}{2}h\nu_1 + \frac{5}{2}h\nu_2$. Proceeding as outlined above,

1. The symmetry species for a $(\frac{7}{2})h\nu_1$ level is that of a (300) wavefunction. Since this is an odd harmonic of a nondegenerate wavefunction it will belong to the same irreducible representation as Q_1, namely A_1.

   The $\frac{5}{2}\,h\nu_2$ is that state having 2 quanta of energy in excess of the ground state, and is the example just discussed with regard to overtones of degenerate vibrations.
2. Since one level belongs to A_1 the direct product $A_1 \times \Gamma_{\pi_u} \times \pi_u = \Gamma_{\pi_u}$, and this was found to be composed of a Σ_g^+ and Δ_g irreducible representation.
3. Since the irreducible representations of the direct product are Σ_g^+ and Δ_g these are the symmetry species for the (320) energy state.

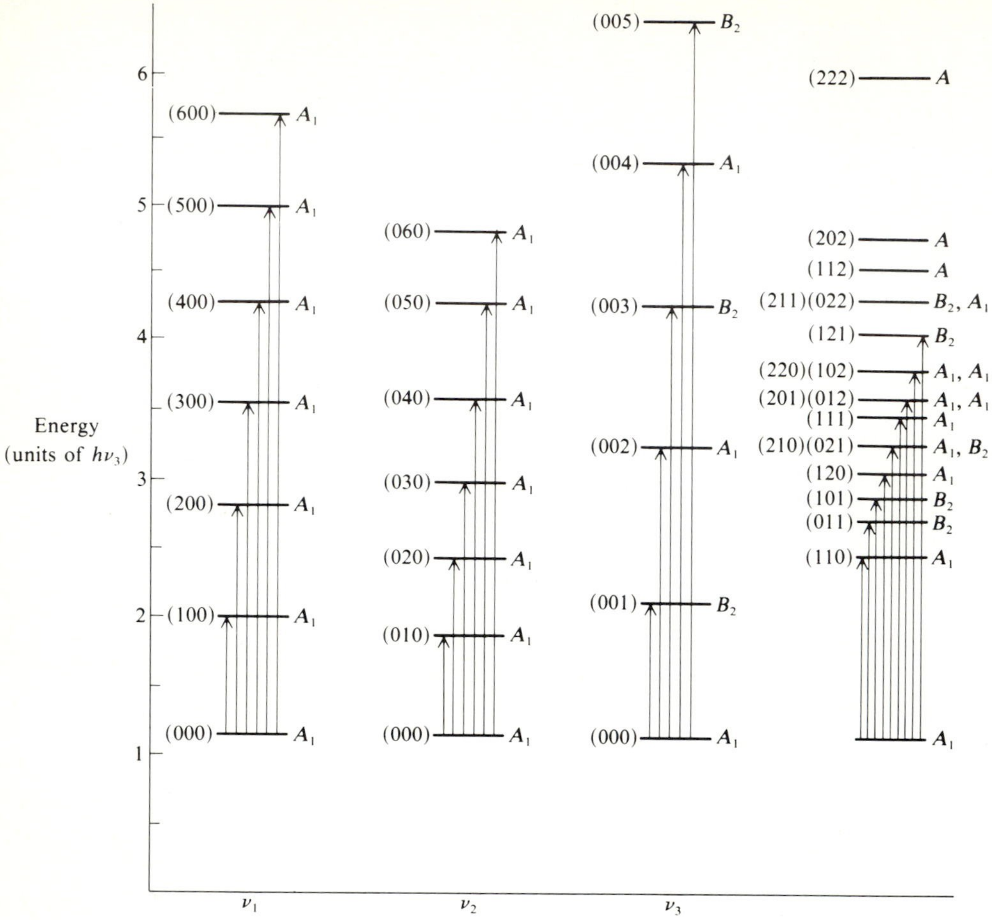

FIGURE 16.12
Energy levels, symmetry species, and allowed transitions for a bent XY_2 molecule ($\nu_1 = 0.8\nu_3$, $\nu_2 = 0.6\nu_3$).

Some of the lower vibrational energy states of the two example molecules discussed are illustrated in Figs. 16.12 and 16.13. The vertical scaling of levels is based on the relative magnitudes of frequencies listed with each figure. The quantum numbers (v_1, v_2, v_3) designating the levels precede and the symmetry species follow each.

16.9 SELECTION RULES

For the diatomic harmonic oscillator, where the potential function is given by a truncated Eq. (16.26), the selection rule obtained by evaluating the dipole moment integral is $\Delta v = \pm 1$. For a homonuclear diatomic molecule having a zero dipole

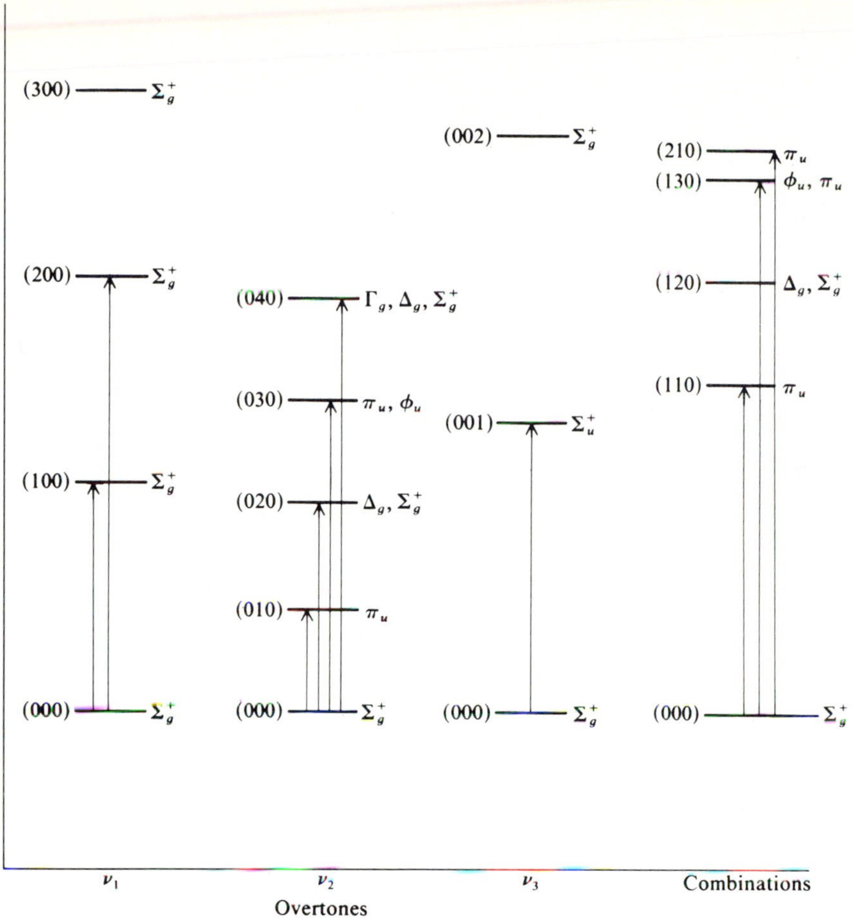

FIGURE 16.13
Energy levels, symmetry species, and allowed transitions for a linear *YXY* molecule ($\nu_1 = 0.75\nu_3$, $\nu_2 = 0.33\nu_3$).

moment as a direct result of the symmetry of the molecule, no vibrational transitions will be observed.

Initially only harmonic behavior will be considered with anharmonicity introduced later. The effect of rotation-vibration interactions will be considered first. For a polyatomic molecule the dipole moment can be expressed in terms of three mutually perpendicular components, lying along a body-fixed Cartesian axis system

$$\mu = \mu_x\hat{\mathbf{i}} + \mu_y\hat{\mathbf{j}} + u_z\hat{\mathbf{k}} \tag{16.136}$$

Expressing the changes in dipole moment with vibration in terms of normal coordinates gives

$$\mu_g = \mu_g^\circ + \sum_{i=1}^{3N-6} \left(\frac{\partial \mu_g}{\partial Q_i}\right)_0 Q_i \qquad g = x, y, z \tag{16.137}$$

since any dipole change $(\partial\mu/\partial Q_i)$ will in general be resolvable into three components. The dipole moment matrix element becomes

$$(N'|\boldsymbol{\mu}|N'')^2 = [(N'|\mu_x|N'')\hat{\mathbf{i}} + (N'|\mu_y|N'')\hat{\mathbf{j}} + (N'|\mu_z|N'')\hat{\mathbf{k}}]^2 \tag{16.138}$$

where N'' and N' denote sets of vibrational quantum numbers for the lower and upper states. Equation (16.138) shows that if any one of the three matrix element components $(N'|\mu_g|N'')$ are nonzero then the total dipole matrix will be nonzero and the transition will be allowed.

Looking at one component we can write

$$(N'|\mu_g|N'') = \int \psi^*_{N'}\mu_g\psi_{N''}\,dv = \mu_g^\circ \int \psi^*_{N'}\psi_{N''}\,dv + \sum_{i=1}^{3N-6}\left(\frac{\partial \mu_g}{\partial Q_i}\right)_0 \int \psi^*_{N'}Q_i\psi_{N''}\,dv \tag{16.139}$$

where $dv = dQ_1 dQ_2 \cdots dQ_{3N-6}$. If ψ_N is taken to be a product of individual normal mode functions [Eq. (16.82)], then the first term in this equation is

$$\mu_g^\circ\left\{\left[\int \psi^*_{v'_1}(Q_1)\psi_{v''_1}(Q_1)dQ_1\right] \times \cdots \times \left[\int \psi^*_{v'_{3N-6}}(Q_{3N-6})\psi_{v''_{3N-6}}(Q_{3N-6})dQ_{3N-6}\right]\right\} \tag{16.140}$$

and the second is

$$\sum_{i=1}^{3N-6}\left(\frac{\partial \mu_g}{\partial Q_i}\right)_0\left\{\left[\int \psi^*_{v'_1}(Q_1)\psi_{v''_1}(Q_1)\,dQ_1\right]\times\cdots\times\left[\int \psi^*_{v'_i}(Q_i)Q_i\psi_{v''_i}(Q_i)\,dQ_i\right]\times\cdots \times\left[\int \psi^*_{v'_{3N-6}}(Q_{3N-6})\psi_{v''_{3N-6}}(Q_{3N-6})\,dQ_{3N-6}\right]\right\} \tag{16.141}$$

The first of these expressions is a product of normalization integrals and will be zero unless $v'_1 = v''_1$, $v'_2 = v''_2, \ldots, v'_{3N-6} = v''_{3N-6}$. This behavior is analogous to that of the diatomic vibrator and shows that the vibrational spectrum is independent of the existence of a permanent dipole moment. Examination of the second expression shows that if $(\partial\mu_g/\partial Q_k)_0$ is nonzero then the conditions for the terms to be nonzero are $v'_1 = v''_1$, $v'_2 = v''_2 \cdots v'_k = v''_k \pm 1, \ldots, v'_{3N-6} = v''_{3N-6}$. The harmonic oscillator selection rule then restricts the vibrational transitions of a polyatomic molecule at $\Delta v_k = \pm 1$. The frequencies of the transitions are the fundamental frequencies of the normal modes. The intensities of the transitions are determined in part by the magnitude of the dipole matrix element. Therefore, as with the diatomic oscillator, strict adherence to the harmonic oscillator selection rules will not permit overtone or combination frequencies.

The effect of the introduction of anharmonicity is comparable to that produced in a diatomic oscillator, except that the wavefunctions $\psi_{v_k}(Q_k)$, for a particular normal mode Q_k, are linear combinations of the harmonic oscillator wavefunctions for that mode, and a mixing of modes can occur. The result is to allow transitions having $\Delta v_k = \pm 1, \pm 2, \pm 3, \ldots$. While this breakdown of the normal harmonic oscillator selection rule leads to the observation of overtone vibrations it will not account entirely for the observation of transitions to combination levels. The selection rules for both overtone and combination transitions may also be found by using a more complete form of Eq. (16.137) which contains the sum of higher-order terms such as $(\partial^2 \mu_g / \partial Q_i \partial Q_j) Q_i Q_j$. A less complicated method involves the use of group theory to establish what are often referred to as the *symmetry selection rules.*

If the integrand of $\int_{-\infty}^{\infty} \psi_{N'}^* \mu_g \psi_{N''} \, dv$ is even, then the value of the integral can be nonzero, otherwise it will be zero. The symmetry species of the integrand will be that of the direct product of the components comprising the integrand. Determination of the symmetry species for the integrand is greatly simplified for vibrational transitions that originate in the ground state ψ_0. The symmetry species of the ground vibrational state of any molecule is always the totally symmetric representation. The direct product of a symmetry species multiplied by itself is totally symmetric; if the species is nondegenerate, or contains the totally symmetric species if it is degenerate. Since ψ_0 belongs to the totally symmetric species, it follows that $\mu_g \psi_{N'}$ must belong to the totally symmetric species. A repetition of this reasoning shows that μ_g and $\psi_{N'}$ must belong to the same symmetry species in order for $\mu_g \psi_{N'}$ to belong to the totally symmetric species. Therefore, one of the dipole components $\mu_g (g = x, y, z)$, must belong to the same symmetry species as $\psi_{N'}$ in order for the transition $\psi_{N'} \leftarrow \psi_0$ to be allowed.

The dipole moment of a collection of partially charged atoms in a molecule is given in general by the vector sum $\boldsymbol{\mu} = \sum_i e_i \mathbf{r}_i$, where e_i is the partial charge on each atom and $\mathbf{r}_i$ is the distance from the center of mass. Writing this equation in terms of its components

$$\mu_g = \sum_i e_i g_i \qquad (g_i = x, y, z) \tag{16.142}$$

Any symmetry operation for a molecule will permute equivalent atoms. Equivalent atoms possess equivalent partial charges so the component μ_g will transform like $\sum g_i$, where the sum is over equivalent atoms. Since any group operation will only permute the indices i, on the g_i's the result will be only to change the order of the terms within the summation. Therefore, the group symmetry operations change $\sum g_i$ in the same manner as they change the g coordinates. The symmetry selection rule can be stated as "transitions from the ground state to any excited state belonging to the same symmetry species as any one of the three Cartesian coordinates are allowed."

The two previous examples will be employed to illustrate the selection rules. For the bent XY_2 molecule, μ_x, μ_y, and μ_z belong to the B_1, B_2, and A_1 species, respectively. The dipole moment or any component induced by vibrations will lie in the plane of the molecule (yz plane) so $\mu_x = 0$, and observed transitions will be limited to those from the ground state to a state belonging to either A_1 or B_2. Reference to Fig. 16.12 indicates that all overtones of all three vibrational modes will occur. Since all of the combination levels will belong to either A_1 or B_2, a transition from the ground state to any combination level is allowed. While all overtones and combinations are allowed by the selection rules, this is no guarantee that the intensities will be adequate for their observation.

Examination of this last point further, by means of an example, can show how the overtone or combination intensities compare to those of the fundamentals. Using the restricted case of an anharmonic oscillator with

$$\psi_0(Q_1) = a\psi_0^\circ(Q_1) + b\psi_1^\circ(Q_1) \tag{16.143}$$

$$\psi_1(Q_1) = a'\psi_0^\circ(Q_1) + b'\psi_1^\circ(Q_1) + c'\psi_2^\circ(Q_1) \tag{16.144}$$

and

$$\psi_2(Q_1) = b''\psi_1^\circ(Q_1) + c''\psi_2^\circ(Q_1) + d''\psi_3^\circ(Q_1) \tag{16.145}$$

where $\psi_v(Q_1)$ is the true wavefunction for the Q_1 normal mode and $\psi_v^\circ\,(Q_1)$ is the harmonic oscillator wavefunction for the same normal mode, dipole elements $(0|\mu|1)$ and $(0|\mu|2)$ become

$$(0|\mu|1) = ab'(0|\mu|1)^\circ + ba'(1|\mu|0)^\circ \tag{16.146a}$$

$$(0|\mu|2) = ab'(0|\mu|1)^\circ + bc''(1|\mu|2)^\circ \tag{16.146b}$$

For a small amount of anharmonicity the general relationships among the coefficients in these expressions will be $a \gg b$, $a' \approx c' \ll b'$, and $b'' \approx d'' \ll c''$. Assuming that the dipole elements involving the harmonic oscillator functions $(v'|\mu|v'')^\circ$ are of comparable magnitude the ratio of the intensities is given by

$$\frac{(0|\mu|1)^2}{(0|\mu|2)^2} = \frac{|ab' + ba'|^2}{|ab'' + bc''|^2} \tag{16.147}$$

But $ab' \gg ba'$ and $ab'' \approx bc''$, therefore the ratio becomes $ab'/2ab'' = b'/2b''$. Since $b'' < b$ (most of ψ_1° is contained in ψ_1) the ratio will be large and the intensity of the first overtone will be much less than that of the fundamental. The intensities of higher overtones are diminished even further.

Looking at the linear *YXY* molecule we find that the symmetry selection rule will provide the same information as can be obtained from an intuitive examination of the system. The symmetry species for the three fundamentals are Σ_g^+, Π_u, and Σ_u^+ for the ν_1, ν_2, and ν_3 normal modes as shown in Fig. 16.10. The symmetry of the ground state is Σ_g^+. The symmetry species of x, y is Π_u and that of z is Σ_u^+. Therefore there can be no transition from Σ_g^+, the ground state, to Σ_g^+, the overtone level of ν_1. This transition is forbidden by symmetry selection rules. This is as expected since ν_1 is a symmetrical stretch and would not produce an oscillating dipole with vibration, that is $(\partial u/\partial Q_1)_0 = 0$.

16.10 INTERNAL AND DISPLACEMENT COORDINATES

The vibrational wavefunctions of a molecule depend explicitly on the normal coordinates so we need to know the exact nature of these in order to be able to

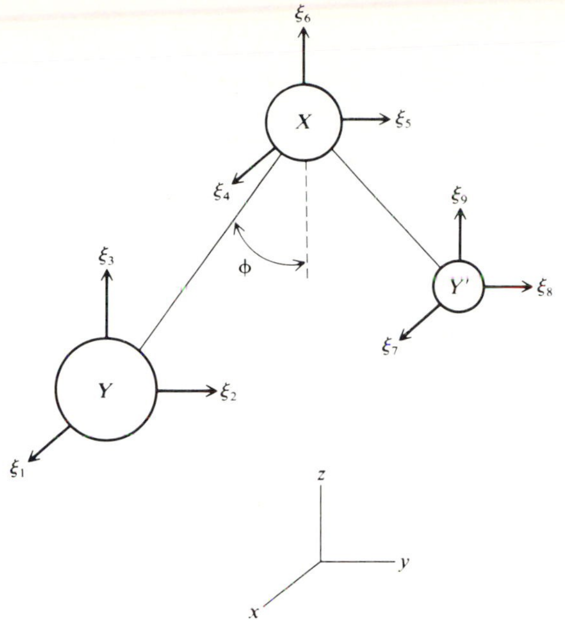

FIGURE 16.14
Vibrational motion in a nonlinear YXY' molecule.

evaluate selection rules and intensities, whose magnitudes depend on the wavefunctions. In Sec. 16.4 it was shown that the normal coordinates could be related to a set of mass-weighted Cartesian coordinates; however, the latter is not always the best set of coordinates with which to perform an analysis. The reason for this is that the b_{ij}'s, which are the coefficients occurring in the potential energy terms, are not easily related to the physical properties of the system. Using as an example the bent YXY' molecule illustrated in Fig. 16.14, we can see the reason. The Cartesian displacements of the atoms are indicated by the $\boldsymbol{\xi}_i$ vectors. The b_{ij}'s would be a measure of the restoring force for displacements associated with a pair of $\boldsymbol{\xi}_i$'s. The motions of the molecule that are most easily related to the structure and properties of the bonds are the stretching and bending of the bonds. The $\boldsymbol{\xi}_i$ displacements will involve a combination of both a change in a bond length r and the bond angle θ. This results in the b_{ij}'s being complicated functions of simpler potential constants that can be associated with the bond stretching and bending and which in turn can be related to the electron distribution and the nature of the bonding within the molecule.

In view of this complication, it will be better to use a set of "internal coordinates" which take advantage of the obvious structural features of the molecule and which utilize a set of readily definable potential constants. For the YXY' molecule the obvious choice of internal coordinates will be the stretching of the two bonds and the bending or deformation of the YXY' angle. If these displacements are represented by r_1, r_2, and θ, and assuming that there are no interactions between the motions, the potential energy will be of the form

$$V = \tfrac{1}{2}F_1 r_1^2 + \tfrac{1}{2}F_2 r_2^2 + \tfrac{1}{2}F_3 \theta^2 \tag{16.148}$$

Considering the molecule as a real mechanical system, one is easily convinced that a motion of the molecule involving any two atoms will affect the motion of any atom attached to either of the first two. This interaction will become more predominant as the masses of the atoms become more nearly equal. When such interactions occur the potential energy expression must be modified to

$$V = \tfrac{1}{2}[F_{11}r_1^2 + F_{22}r_2^2 + F_{33}\theta^2 + 2F_{12}r_1r_2 + 2F_{13}r_1\theta + 2F_{23}r_2\theta] \quad (16.149)$$

For any arbitrary molecular system having $3N - 6$[1] internal coordinates the potential energy will be expressed as

$$V = \tfrac{1}{2}\tilde{\mathbf{R}}\underline{\mathbf{F}}\mathbf{R} \quad (16.150)$$

where $\mathbf{R}$ is a column vector whose components are the $3N - 6$ internal coordinates and $\underline{\mathbf{F}}$ is a $(3N - 6) \times (3N - 6)$ matrix whose terms are the coefficients in the potential energy expression. For the YXY' molecule, if $\underline{\mathbf{F}}$ is of the form

$$\underline{\mathbf{F}} = \begin{pmatrix} F_{11} & 0 & 0 \\ 0 & F_{22} & 0 \\ 0 & 0 & F_{33} \end{pmatrix} \quad (16.151)$$

then

$$V = \frac{1}{2}(r_1 r_2 \theta)\begin{pmatrix} F_{11} & 0 & 0 \\ 0 & F_{22} & 0 \\ 0 & 0 & F_{33} \end{pmatrix}\begin{pmatrix} r_1 \\ r_2 \\ \theta \end{pmatrix} = \frac{1}{2}[F_{11}r_1^2 + F_{22}r_2^2 + F_{33}\theta^2] \quad (16.152)$$

If the $\underline{\mathbf{F}}$ matrix contains all of the off-diagonal elements, the potential energy will be that given by Eq. (16.149). One advantage in the use of the $\underline{\mathbf{F}}$ matrix is that only those potential terms which are necessary for a solution of a particular problem need to be included. In general, the diagonal terms will be the dominant ones and the magnitudes of the off-diagonal terms will be much less, though not always negligible. It is possible to simplify the $\underline{\mathbf{F}}$ matrix for systems where there is no possible interaction between pairs of internal coordinates or where, based on physical grounds, the interaction can be expected to be relatively small.

Since the relationship between the kinetic energy and the internal coordinates is, in general, a more complex problem than that involving the potential energy, we will only discuss the relationships for a simpler molecular type and enumerate those for more complicated systems, leaving the reader to explore more specialized monographs for a definitive treatment. The discussion will proceed by establishing a general matrix relationship and then examining the individual matrix elements for the bent XY_2 molecule.

The kinetic energy in terms of the mass-weighted Cartesian coordinates was given by Eq. (16.70) where the elements of $\dot{\boldsymbol{\zeta}}$ are $\dot{\zeta}_1 = \mu_1^{-1/2}\dot{\xi}_1$, $\dot{\zeta}_2 = \mu_2^{-1/2}\dot{\xi}_2$, etc. Since the coordinates ζ_i and the internal coordinates R_i describe the same motion

[1] Keep in mind that for a linear molecule this factor becomes $(3N - 5)$.

in the same reference frame they will be related by a linear transformation $\underline{U}$ so that

$$\zeta = \underline{U}\mathbf{R} \tag{16.153}$$

where $\mathbf{R}$ is a vector whose components are the internal coordinates, R_i. There will be $3N$ Cartesian coordinates and $3N-6$ internal coordinates so the transformation matrix will not be square, but will contain $3N$ rows and $3N-6$ columns. Figure 16.14 shows that the relationships which can be easily obtained from the geometry of the system are

$$R_1 = r_1 = -\xi_1 \sin\phi - \xi_2 \cos\phi + \xi_3 \sin\phi + \xi_4 \cos\phi \tag{16.154}$$

$$R_2 = r_2 = \xi_5 \sin\phi - \xi_6 \cos\phi - \xi_3 \sin\phi + \xi_4 \cos\phi \tag{16.155}$$

$$R_3 = \Delta\theta = \frac{1}{r_1}[-\xi_1 \cos\phi + \xi_2 \sin\phi + \xi_3 \cos\phi - \xi_4 \sin\phi]$$

where

$$\phi = \frac{\theta}{2} + \frac{1}{r_2}[\xi_5 \cos\phi + \xi_6 \sin\phi - \xi_3 \cos\phi - \xi_4 \sin\phi] \tag{16.156}$$

Introduction of the mass-weighted coordinates $\zeta_i = \sqrt{m_i}\,\xi_i = u_i^{-1/2}\xi_i$ leads to

$$R_1 = -\mu_y\zeta_1 \sin\phi - \mu_y\zeta_2 \cos\phi + \mu_x\zeta_3 \sin\phi + \mu_x\zeta_4 \cos\phi \tag{16.157}$$

$$R_2 = -\mu_y\zeta_5 \sin\phi - \mu_y\zeta_6 \cos\phi - \mu_x\zeta_3 \sin\phi + \mu_x\zeta_4 \cos\phi \tag{16.158}$$

$$R_3 = -\frac{1}{r}[-\mu_x\zeta_1 \cos\phi + \mu_y\zeta_2 \sin\phi - 2\mu_x\zeta_4 \sin\phi$$
$$+\mu_y\zeta_5 \cos\phi + \mu_x\zeta_6 \sin\phi] \tag{16.159}$$

where $\mu_y = \mu_y$, and $r_1 = r_2 = r$. These are the components of the matrix equation

$$\mathbf{R} = \underline{\mathbf{W}}\zeta \tag{16.160}$$

which is the inverse of Eq. (16.153). Since three points define a plane, the vibrational motion of the XY_2 molecule will be restricted to a plane, hence the coefficients of ζ_7, ζ_8, and ζ_9 will be zero. In this case the matrix $\underline{\mathbf{W}}$ has $3N-6$ rows and $3N$ columns. Since $\underline{\mathbf{W}}$ is not a square matrix it cannot be inverted to give the matrix $\underline{U}$. The procedure for obtaining a usable kinetic energy expression in terms of internal coordinates can best be understood by the use of a simple vector diagram.

For the bent XY_2 molecule establish two sets of vectors:

1. For each atom specify a vector $\boldsymbol{\rho}_a$ whose components are either mass-weighted or regular Cartesian displacement vectors for atom a.
2. Equations (16.154) to (16.156) show that

$$R_i = \sum_{j=1}^{3N} B_{ij}\xi_j \tag{16.161}$$

Note that the B_{jj} are trigonometric functions related to the angles between the normal coordinates and the Cartesian coordinates. It is convenient to define a new set of vectors $\mathbf{s}_{ia}$ (i denotes the internal coordinate and a the atom) which satisfy the relation

$$R_i = \sum_{a=1}^{N} \mathbf{s}_{ia} \cdot \boldsymbol{\rho}_a \tag{16.162}$$

The vector $\mathbf{s}_{ia}$ has the physical significance that it denotes the direction in which atom a will move when the molecule undergoes a vibration resulting in a change of the coordinate R_i.

By use of these two vectors to describe the vibrational motion of the molecule we free ourselves from the use of a particular axis system. Figure 16.15 depicts these vectors for the internal stretching coordinates of the bent XY_2 molecule. In this figure $\mathbf{e}_{aa'}$ is a unit vector directed along the bond from atom a to atom a'. The maximum atomic displacements for a bond stretch must be colinear with the bond so it follows that $\mathbf{s}_{1x} = \hat{\mathbf{e}}_{xy} = -\hat{\mathbf{e}}_{yx}$ and $\mathbf{s}_{1y} = \hat{\mathbf{e}}_{yx}$. An analogous situation exists for the stretching of the other bond, giving $\mathbf{s}_{2x} = \hat{\mathbf{e}}_{y'x}$ and $\mathbf{s}_{2y} = \hat{\mathbf{e}}_{xy'}$. Equations (16.157) to (16.159) can be obtained analytically as well as geometrically. The $\boldsymbol{\rho}_a$ vectors, in terms of mass-weighted coordinates, will be

$$\boldsymbol{\rho}_y = \mu_7\zeta_7\hat{\mathbf{i}} + \mu_1\zeta_1\hat{\mathbf{j}} + \mu_2\zeta_2\hat{\mathbf{k}} \tag{16.163}$$

$$\boldsymbol{\rho}_x = \mu_8\zeta_8\hat{\mathbf{i}} + \mu_3\zeta_3\hat{\mathbf{j}} + \mu_4\zeta_4\hat{\mathbf{k}} \tag{16.164}$$

$$\boldsymbol{\rho}_{y'} = \mu_9\zeta_9\hat{\mathbf{i}} + \mu_5\zeta_5\hat{\mathbf{j}} + \mu_6\zeta_6\hat{\mathbf{k}} \tag{16.165}$$

and the unit vectors $\hat{\mathbf{e}}_{aa'}$ are

$$\mathbf{s}_{1y} = \hat{\mathbf{e}}_{xy} = -\sin\phi\mathbf{j} - \cos\phi\mathbf{k} \tag{16.166}$$

$$\mathbf{s}_{2y'} = \hat{\mathbf{e}}_{xy'} = \sin\phi\hat{\mathbf{j}} - \cos\phi\hat{\mathbf{k}} \tag{16.167}$$

$$\mathbf{s}_{1x} = \hat{\mathbf{e}}_{yx} = \sin\phi\hat{\mathbf{j}} + \cos\phi\hat{\mathbf{k}} \tag{16.168}$$

$$\mathbf{s}_{2x} = \hat{\mathbf{e}}_{y'x} = -\sin\phi\hat{\mathbf{j}} + \cos\phi\hat{\mathbf{k}} \tag{16.169}$$

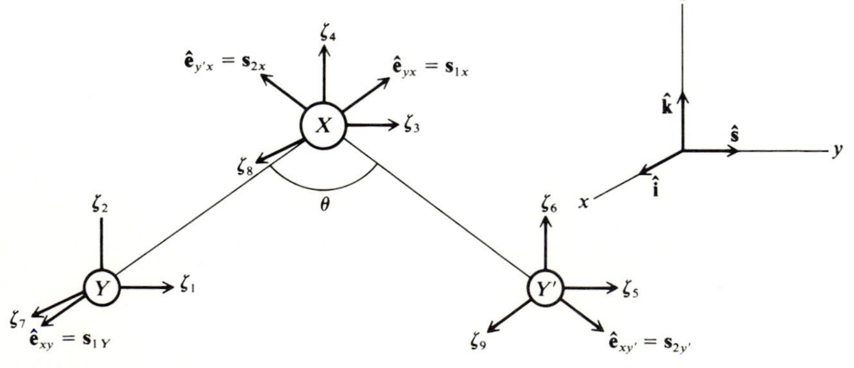

FIGURE 16.15
Vector relationship for bond stretching of a bent YXY' molecule.

R_1 can be obtained from Eq. (16.162)

$$R_1 = \mathbf{s}_{1x} \cdot \boldsymbol{\rho}_x + \mathbf{s}_{1y} \cdot \boldsymbol{\rho}_y \tag{16.170}$$

which, upon substitution of Eqs. (16.163), (16.164), (16.166), and (16.168), becomes

$$R_1 = \sqrt{\mu_3}\,\zeta_3 \sin\phi + \sqrt{\mu_4}\,\zeta_4 \cos\phi - \sqrt{\mu_1}\,\zeta_1 \sin\phi - \sqrt{\mu_2}\,\zeta_2 \cos\phi \tag{16.171}$$

This is identical to Eq. (16.157). In an analogous manner one can show that R_2 is equivalent to Eq. (16.158).

The bending of the HOH angle can be treated in a similar way. The vectors involved in the motion are shown in Fig. 16.16. The internal coordinate in this case is a change in θ, $R_3 = \Delta\theta$, and results in a displacement of all three atoms. Inspection of this motion shows that the vectors $\mathbf{s}_{3i}$ represent the directions of maximum displacement. The total displacement $\Delta\theta$ is only a tiny fraction of θ, so by using small angle relationships one has

$$\Delta\phi = \sin\Delta\phi = \frac{|\mathbf{s}_{3y}|}{r_{yx}} \tag{16.172}$$

A unit displacement in the direction of $\mathbf{s}_{3y}$ will then result in a change in ϕ of $1/r_{xy}$ and the length of $\mathbf{s}_{3y}$ will be $1/r_{xy}$. Symmetry dictates that the same is true for $\mathbf{s}_{3y'}$. The change $\Delta\theta$ will leave the center of gravity unchanged, hence the z displacement of $\mathbf{s}_{3x}$ is opposite and of equal magnitude to the sum of those of $\mathbf{s}_{3y}$ and $\mathbf{s}_{3y'}$. Furthermore, the y displacement of $\mathbf{s}_{3y}$ is equal and opposite to that of $\mathbf{s}_{3y'}$. This leads to the relationships

$$(\mathbf{s}_{3x})_z\hat{\mathbf{k}} = (-\mathbf{s}_{3y} - \mathbf{s}_{3y'})_x\hat{\mathbf{k}} \tag{16.173}$$

$$(\mathbf{s}_{3y})_y\hat{\mathbf{j}} = -(\mathbf{s}_{3y'})_y\hat{\mathbf{j}} \tag{16.174}$$

which combine to give

$$\mathbf{s}_{3x} = -\mathbf{s}_{3y} - \mathbf{s}_{3y'} \tag{16.175}$$

The $\mathbf{s}_{3i}$ vectors can be expressed in terms of the bond-oriented unit vectors $\hat{\mathbf{e}}_{aa'}$. Using Fig. 16.17 one can find the appropriate relationships by use of the Cartesian unit vectors as a reference system. This procedure is necessary since the $\hat{\mathbf{e}}_{aa'}$ do

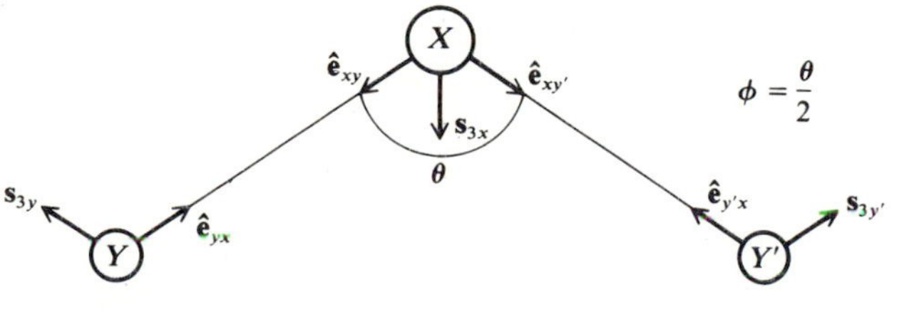

FIGURE 16.16
Vector relations for valence angle bending in a bent YXY_2' molecule.

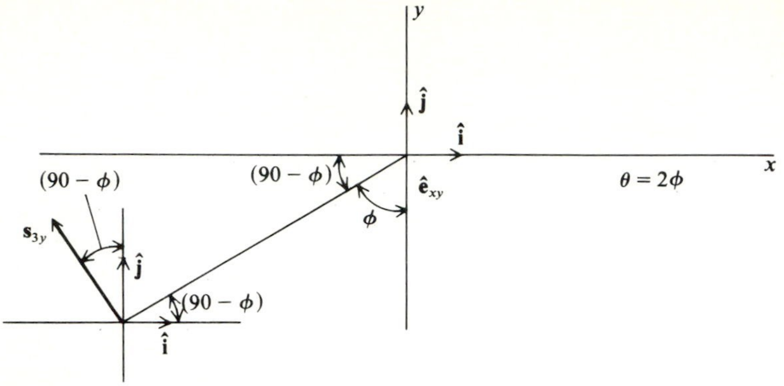

FIGURE 16.17
Detailed vector relations for bent XY_2 molecule.

not form an orthogonal set. Inspection of the figure and knowing $|s_{3y}| = 1/r_{xy}$ gives

$$\mathbf{s}_{3y} = \left[\frac{1}{r_{xy}}\cos(90-\phi)\right]\hat{\mathbf{k}} + \left[-\frac{1}{r_{xy}}\sin(90-\phi)\right]\hat{\mathbf{j}} \tag{16.176}$$

$$\hat{\mathbf{e}}_{xy} = [-\sin(90-\phi)]\hat{\mathbf{k}} + [-\cos(90-\phi)]\hat{\mathbf{j}} \tag{16.177}$$

$$\hat{\mathbf{e}}_{xy'} = [\sin(90-\phi)]\hat{\mathbf{k}} + [\cos(90-\phi)]\hat{\mathbf{j}} \tag{16.178}$$

Using common trigonometric identities to simplify the functions, solving the latter two equations for $\hat{\mathbf{i}}$ and $\hat{\mathbf{j}}$, and substituting the results into Eq. (16.176) leads to

$$\mathbf{s}_{3y} = \frac{1}{2r_{xy}}\left\{\left(\frac{1+\cos\theta}{\sin\theta}\right)(\hat{\mathbf{e}}_{xy} - \hat{\mathbf{e}}_{xy'}) - \left(\frac{\sin\theta}{1+\cos\theta}\right)(\hat{\mathbf{e}}_{xy} + \hat{\mathbf{e}}_{xy'})\right\} \tag{16.179}$$

which then rearranges and reduces to

$$\mathbf{s}_{3y} = \frac{\cos\theta\,\hat{\mathbf{e}}_{xy} - \hat{\mathbf{e}}_{xy'}}{r_{xy}\sin\theta} \tag{16.180}$$

An analogous formulation of $\mathbf{s}_{3y'}$ and the use of Eq. (16.175) gives

$$\mathbf{s}_{3y'} = \frac{\cos\theta\,\hat{\mathbf{e}}_{xy'} - \hat{\mathbf{e}}_{xy}}{r_{xy'}\sin\theta} \tag{16.181}$$

$$\mathbf{s}_{3x} = \frac{(r_{xy} - r_{xy'}\cos\theta)\hat{\mathbf{e}}_{xy} + (r_{xy'} - r_{xy}\cos\theta)\hat{\mathbf{e}}_{xy'}}{r_{xy}r_{xy'}\sin\theta} \tag{16.182}$$

The comparable displacement vectors for motions involving other basic atomic groupings may be evaluated by similar methods, but since the purpose of this book is to present first principles rather than to attempt a consideration of all possible situations, the reader is referred to the more advanced treatises listed at the end of the chapter for such details. The two most important motions

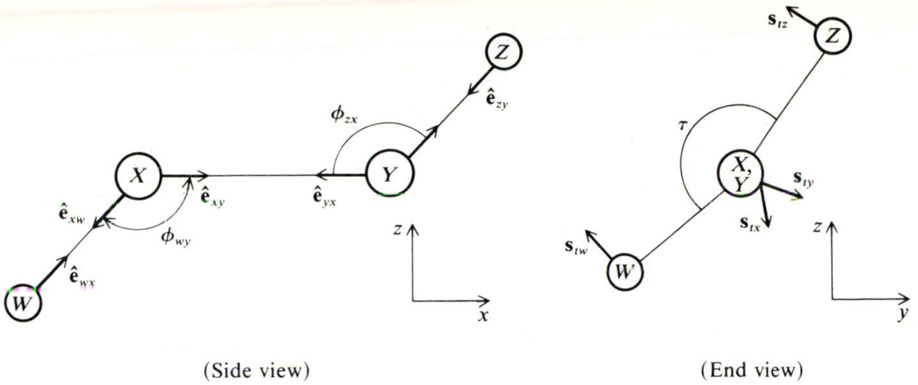

FIGURE 16.18
Vector relations for torsion in a *WXYZ* molecular fragment.

in addition to the stretch and the bend are (1) the torsion, illustrated in Fig. 16.18, with the displacement vectors defined by Eqs. (16.183) to (16.186)

$$\mathbf{s}_{tz} = -\frac{(\hat{\mathbf{e}}_{zy} \times \hat{\mathbf{e}}_{yx})}{r_{zy} \sin^2 \phi_{zx}} \tag{16.183}$$

$$\mathbf{s}_{ty} = \frac{(r_{yx} - r_{zy} \cos \phi_{zx})(\hat{\mathbf{e}}_{zy} \times \hat{\mathbf{e}}_{yx})}{r_{yx} r_{zy} \sin^2 \phi_{zx}} - \frac{(\cos \phi_{wy})(\hat{\mathbf{e}}_{wx} \times \hat{\mathbf{e}}_{xy})}{r_{yx} \sin^2 \phi_{wy}} \tag{16.184}$$

$$\mathbf{s}_{tx} = \frac{(r_{xy} - r_{wx} \cos \phi_{wy})(\hat{\mathbf{e}}_{wx} \times \hat{\mathbf{e}}_{xy})}{r_{xy} r_{wx} \sin^2 \phi_{wy}} - \frac{(\cos \phi_{zx})(\hat{\mathbf{e}}_{zy} \times \hat{\mathbf{e}}_{yx})}{r_{xy} \sin^2 \phi_{zx}} \tag{16.185}$$

$$\mathbf{s}_{tw} = \frac{(\hat{\mathbf{e}}_{wx} \times \hat{\mathbf{e}}_{xy})}{r_{wx} \sin^2 \phi_{wy}} \tag{16.186}$$

and (2) the out-of-plane bend, illustrated in Fig. 16.19, with the displacement vectors given by Eqs. (16.187) to (16.190).

$$\mathbf{s}_{bx} = \frac{1}{r_{wx}} \frac{\hat{\mathbf{e}}_{wy} \times \hat{\mathbf{e}}_{wz}}{\cos \theta \sin \phi_{yz}} - (\tan \theta)\mathbf{e}_{wx} \tag{16.187}$$

$$\mathbf{s}_{by} = \frac{1}{r_{wy}} \frac{\hat{\mathbf{e}}_{wz} \times \hat{\mathbf{e}}_{wx}}{\cos \theta \sin \phi_{yz}} - \frac{\tan \theta}{\sin^2 \phi_{yz}} (\hat{\mathbf{e}}_{wy} - (\cos \phi_{yz})\hat{\mathbf{e}}_{wz}) \tag{16.188}$$

$$\mathbf{s}_{bz} = \frac{1}{r_{wz}} \frac{\hat{\mathbf{e}}_{wx} \times \hat{\mathbf{e}}_{wy}}{\cos \theta \sin \phi_{yz}} - \frac{\tan \theta}{\sin^2 \phi_{yz}} (\hat{\mathbf{e}}_{wz} - (\cos \phi_{yz})\hat{\mathbf{e}}_{wx}) \tag{16.189}$$

$$\mathbf{s}_{bw} = -\mathbf{s}_{bx} - \mathbf{s}_{by} - \mathbf{s}_{bz} \tag{16.190}$$

where

$$\sin \theta = \frac{\hat{\mathbf{e}}_{wy} \times \hat{\mathbf{e}}_{wz} \cdot \hat{\mathbf{e}}_{wx}}{\sin \phi_{yz}} \tag{16.191}$$

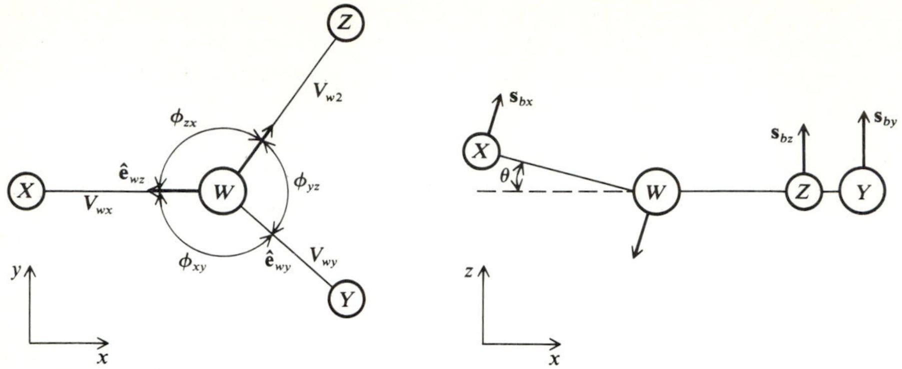

FIGURE 16.19
Vector relations for vibrations in a trigonal $XYZW$ molecular fragment.

The number of displacement coordinates $\mathbf{s}_{ia}$ will be equal to the number of internal coordinates, which is equal to the number of normal modes times the number of atoms, i.e., each atom has an associated displacement vector for each normal vibration. This will be $NX(3N-6)$ for the nonlinear case. For the bent XY_2 molecule this would be $3 \times 3 = 9$ which is confirmed by Figs. 16.15 and 16.16. These displacements contribute in sets of N to form the internal coordinates R_i. The number of each type of internal coordinates can be summarized as follows. Letting

N_b = number of bonds, irrespective of type,

N = number of atoms, and

N_1 = number of atoms having only one bond,

the number of coordinates involving bond-stretching is given by

$$N_s = N_b$$

the number involving planar angle deformation is

$$N_\phi = 4N_b - 3N - N_1$$

and the number of torsional ones is

$$N_\tau = N_b - N_1$$

Each out-of-plane bending reduces N_τ by one. For example:

	Bent XY_2	**Planar XY_3**	**Pyramidal XY_3**	**Bent $XYYX$**
N	3	4	4	4
$3N-6$	3	6	6	6
N_1	2	3	3	2
N_b	2	3	3	3
N_s	2	3	3	3
N_ϕ	1	3	3	2
N_τ	0	0	0	1

We have digressed somewhat from the process of finding a suitable kinetic energy relationship in order to more fully explain the nature of symmetry coordinate, but now we will get back on the main track and continue toward that objective.

16.11 THE $\underline{G}$ MATRIX

The procedure used to relate the kinetic energy of the atoms in a particular vibrational mode to the internal coordinates and the symmetry coordinates employs the $\underset{\sim}{\mathbf{G}}$ matrix whose elements are given by

$$G_{ij} = \sum_{k=1}^{3N} \mu_k B_{ik} B_{jk} \qquad (i, j = 1, 2, 3, \ldots, 3N - 6) \tag{16.192}$$

where $\mu_k = m_k^{-1}$ and the B_{nm} are the elements of the matrix which transforms the Cartesian coordinates to the internal coordinates with k denoting the atom and i, j denoting coordinates. To ensure a clear understanding of the indexing scheme, Eq. (16.162) is written in expanded form for a bent three-atom system

$$R_1 = B_{11}\xi_1 + B_{12}\xi_2 + \cdots + B_{19}\xi_9 \tag{16.193}$$

$$R_2 = B_{21}\xi_1 + B_{22}\xi_2 + \cdots + B_{29}\xi_9 \tag{16.194}$$

$$R_3 = B_{31}\xi_1 + B_{32}\xi_2 + \cdots + B_{39}\xi_9 \tag{16.195}$$

This shows, as was pointed out previously, that the $\underline{\mathbf{B}}$ matrix is a 3×9 or a nonsquare matrix. Note also that by using ξ_i with $i = 1, 2, \ldots, 3N$ one must keep in mind that $i = 1, 2, 3$ refers to atom 1, $i = 4, 5, 6$ to atom 2, etc. The symmetry vectors $\mathbf{s}_{ia}$ are related to the internal coordinates R_i by Eq. (16.162) so, for example, for the bent three-atom system

$$R_1 = \mathbf{s}_{11} \cdot \hat{\boldsymbol{\rho}}_1 + \mathbf{s}_{12} \cdot \hat{\boldsymbol{\rho}}_2 + \mathbf{s}_{13} \cdot \hat{\boldsymbol{\rho}}_3 \tag{16.196}$$

Equations (16.163) to (16.165) and Eqs. (16.166) to (16.169) established the form of $\hat{\boldsymbol{\rho}}_a$ and $\mathbf{s}_{ia}$, so the right-hand side of Eq. (16.196) will be of the form

$$\begin{aligned}
R_1 &= (\beta_{11}\hat{\mathbf{i}} + \beta_{21}\hat{\mathbf{j}} + \beta_{31}\hat{\mathbf{k}}) \cdot (\xi_1\hat{\mathbf{i}} + \xi_2\hat{\mathbf{j}} + \xi_3\hat{\mathbf{k}}) \\
&\quad + (\beta_{12}\hat{\mathbf{i}} + \beta_{22}\hat{\mathbf{j}} + \beta_{32}\hat{\mathbf{k}}) \cdot (\xi_4\hat{\mathbf{i}} + \xi_5\hat{\mathbf{j}} + \xi_6\hat{\mathbf{k}}) \\
&\quad + (\beta_{13}\hat{\mathbf{i}} + \beta_{23}\hat{\mathbf{j}} + \beta_{33}\hat{\mathbf{k}}) \cdot (\xi_7\hat{\mathbf{i}} + \xi_8\hat{\mathbf{j}} + \xi_9\hat{\mathbf{k}}) \\
&= (\beta_{11} + \beta_{12} + \beta_{13})\hat{\mathbf{i}} \cdot (\xi_1 + \xi_4 + \xi_7)\hat{\mathbf{i}} + (\beta_{21} + \beta_{22} + \beta_{23})\hat{\mathbf{j}} \cdot (\xi_2 + \xi_5 + \xi_8)\hat{\mathbf{i}} \\
&\quad + (\beta_{31} + \beta_{32} + \beta_{33})\hat{\mathbf{k}} \cdot (\xi_3 + \xi_6 + \xi_9)\hat{\mathbf{k}}
\end{aligned} \tag{16.197}$$

Comparison of this expression with Eq. (16.193) shows that $\beta_{ij} = B_{ij}$. Therefore the symmetry coordinates may be written

$$\begin{aligned}
\mathbf{s}_{11} &= B_{11}\hat{\mathbf{i}} + B_{12}\hat{\mathbf{j}} + B_{13}\hat{\mathbf{k}} \\
\mathbf{s}_{12} &= B_{12}\hat{\mathbf{i}} + B_{22}\hat{\mathbf{j}} + B_{32}\hat{\mathbf{k}} \\
\mathbf{s}_{33} &= B_{37}\hat{\mathbf{i}} + B_{38}\hat{\mathbf{j}} + B_{39}\hat{\mathbf{k}}
\end{aligned} \tag{16.198}$$

The scalar product of two symmetry coordinate vectors $\mathbf{s}_{11}$ and $\mathbf{s}_{21}$, for example, will be

$$\mathbf{s}_{11} \cdot \mathbf{s}_{21} = (B_{11}\hat{\mathbf{i}} + B_{12}\hat{\mathbf{j}} + B_{13}\hat{\mathbf{k}}) \cdot (B_{21}\hat{\mathbf{i}} + B_{22}\hat{\mathbf{j}} + B_{23}\hat{\mathbf{k}}) \tag{16.199}$$

or

$$\mathbf{s}_{11} \cdot \mathbf{s}_{21} = B_{11}B_{21} + B_{12}B_{22} + B_{13}B_{23} \tag{16.200}$$

Considering all possible scalar products among the $\mathbf{s}_{ia}$ and Eq. (16.192) leads to the redefinition of the $\underline{\mathbf{G}}$ matrix elements as

$$G_{ij} = \sum_{a=1}^{N} \mu_a \mathbf{s}_{ia} \cdot \mathbf{s}_{ja} \tag{16.201}$$

If the G_{ij}, which are related to the displacement vectors (which in turn can be obtained from the motions of simple molecular segments) can be related to the kinetic energy of the molecular vibrations, then they can be employed for the determination of the normal modes of vibration.

The G_{ij} elements can be related to the kinetic energy by considering the kinetic energy expressed in the form

$$T = \tfrac{1}{2}\tilde{\dot{\boldsymbol{\zeta}}}\dot{\boldsymbol{\zeta}} \tag{16.202}$$

where $\dot{\boldsymbol{\zeta}} = (\dot{\zeta}_1, \dot{\zeta}_2, \ldots, \dot{\zeta}_{3N})$ (mass-weighted coordinates). For each coordinate ζ_i, there is an associated conjugate moment p_i, which is related by

$$p_i = \frac{\partial T}{\partial \dot{\zeta}_i} \tag{16.203}$$

Therefore

$$T = \tfrac{1}{2}\tilde{\mathbf{p}}\mathbf{p} \tag{16.204}$$

Equation (16.161) can be expressed in matrix notation as

$$\mathbf{R} = \underline{\mathbf{W}}\boldsymbol{\zeta} \tag{16.205}$$

The kinetic energy is a function of the velocities of the atoms and can be expressed in terms of any set of coordinates, hence using the internal coordinates

$$p_i = \frac{\partial T}{\partial \dot{\zeta}_i} = \sum \frac{\partial T}{\partial R_j} \frac{\partial \dot{R}}{\partial \dot{\zeta}_i} \tag{16.206}$$

Denoting the momenta conjugate to the R_j's by P_j

$$P_j = \frac{\partial T}{\partial \dot{R}_j} \tag{16.207}$$

Using Eq. (16.205) where the elements of $\underline{\mathbf{W}}$ are constants

$$\frac{\partial \mathbf{R}}{\partial \boldsymbol{\zeta}} = \frac{\partial \dot{\mathbf{R}}}{\partial \dot{\boldsymbol{\zeta}}} = \underline{\mathbf{W}} \tag{16.208}$$

or, in terms of the components,

$$\frac{\partial \dot{R}_k}{\partial \dot{\zeta}_i} = \frac{\partial R_k}{\partial \zeta_i} = W_{ki} \tag{16.209}$$

Equation (16.206) then becomes

$$p_i = \sum_k P_k W_{kj} \qquad (j = 1, 2, \ldots, N) \tag{16.210}$$

or in matrix terms

$$\tilde{\mathbf{p}} = \tilde{\mathbf{P}}\underline{\mathbf{W}} \tag{16.211}$$

The momenta components of the system are real so the matrices for **p** and **P** are Hermitian, that is, $\mathbf{p} = \tilde{\mathbf{p}}^*$ and $\mathbf{P} = \tilde{\mathbf{P}}^*$. Equation (16.211) then becomes

$$\mathbf{p}^* = \mathbf{P}^*\underline{\mathbf{W}} \tag{16.212}$$

Taking the transpose of each side, we have

$$\tilde{\mathbf{p}}^* = (\widetilde{\mathbf{P}^*\underline{\mathbf{W}}}) = \underline{\tilde{\mathbf{W}}}\tilde{\mathbf{P}}^* \tag{16.213}$$

which becomes

$$\mathbf{p} = \underline{\tilde{\mathbf{W}}}\mathbf{P} \tag{16.214}$$

Incorporating this expansion along with Eq. (16.211) into Eq. (16.204) gives

$$T = \tfrac{1}{2}\tilde{p}p = \tfrac{1}{2}\tilde{\mathbf{P}}\underline{\mathbf{W}\tilde{\mathbf{W}}}\mathbf{P} \tag{16.215}$$

Expanding Eq. (16.160) to give

$$R_i = \sum_j W_{ij}\zeta_j = \sum_j W_{ij}\mu_j^{-1/2}\xi_j \tag{16.216}$$

and comparing components with these in Eq. (16.161)

$$R_i = \sum_j B_{ij}\xi_j \tag{16.217}$$

shows that

$$W_{ij}\mu_j^{-1/2} = B_{ij} \tag{16.218}$$

The individual elements of $\underline{\mathbf{W}\tilde{\mathbf{W}}}$ will be

$$(\underline{\mathbf{W}\tilde{\mathbf{W}}})_{ij} = \sum_k W_{ik}W_{jk} = \sum_k \mu_k B_{ik}B_{jk} = G_{ij} \tag{16.219}$$

Therefore,

$$\underline{\mathbf{W}\tilde{\mathbf{W}}} = \underline{\mathbf{G}} \tag{16.220}$$

and

$$T = \tfrac{1}{2}\tilde{\mathbf{P}}\underline{\mathbf{G}}\mathbf{P} \tag{16.221}$$

The G_{ij} terms will be relating pairs of momenta which are conjugate to the symmetry coordinates and will exhibit the property $G_{ij} = G_{ji}$, or $\underline{\mathbf{G}}$ will be a symmetric matrix and $\underline{\mathbf{G}} = \tilde{\underline{\mathbf{G}}}$. Beginning with

$$T = \tfrac{1}{2}\tilde{\dot{\zeta}}\dot{\zeta} \tag{16.222}$$

and substituting

$$\dot{\zeta} = \underline{\mathbf{W}}^{-1}\dot{\mathbf{R}} \tag{16.223}$$

gives

$$T = \tfrac{1}{2}(\widetilde{\underline{\mathbf{W}}^{-1}\mathbf{R}})\underline{\mathbf{W}}^{-1}\mathbf{R} = \tfrac{1}{2}\tilde{\mathbf{R}}\tilde{\underline{\mathbf{W}}}^{-1}\underline{\mathbf{W}}^{-1}\mathbf{R} \tag{16.224}$$

Since $\underline{\mathbf{G}} = \underline{\mathbf{W}\tilde{\mathbf{W}}}$, it follows that

$$T = \tfrac{1}{2}\mathbf{R}\underline{\mathbf{G}}^{-1}\mathbf{R} \tag{16.225}$$

the G_{ij} coefficients do relate the kinetic energy to the internal coordinates.

Before we develop the method for combining the $\underline{\mathbf{F}}$ and $\underline{\mathbf{G}}$ matrices to give a set of secular equations from which the normal modes of vibration can be evaluated, let us first look at the evaluation of the $\underline{\mathbf{G}}$ matrix elements. Equation (16.201) relates these coefficients to the displacement vectors. These in turn are evaluated for four common types of molecular segments or motions in Sec. 16.10. Using these relationships the G_{ij} can be evaluated for simple molecules.

This procedure will be illustrated by again using as an example the bent XY_2 molecule whose displacement vectors are given by Eqs. (16.166) to (16.169) and (16.179) to (16.182). Using symmetrical molecules so that $r_{XY} = r_{XY'} = r$, the displacement coordinates become (see Fig. 16.15)

$$\mathbf{s}_{1y} = \hat{\mathbf{e}}_{XY} \tag{16.226}$$

$$\mathbf{s}_{1x} = -\hat{\mathbf{e}}_{XY} \tag{16.227}$$

$$\mathbf{s}_{1y'} = 0 \tag{16.228}$$

$$\mathbf{s}_{2y} = 0 \tag{16.229}$$

$$\mathbf{s}_{2x} = -\hat{\mathbf{e}}_{XY'} \tag{16.230}$$

$$\mathbf{s}_{2y'} = \hat{\mathbf{e}}_{XY'} \tag{16.231}$$

$$\mathbf{s}_{3y} = \frac{\cos\theta\,\hat{\mathbf{e}}_{XY} - \hat{\mathbf{e}}_{XY'}}{r\sin\theta} \tag{16.232}$$

$$\mathbf{s}_{3x} = \frac{r(1\text{-}\cos\theta)(\hat{\mathbf{e}}_{XY} + \hat{\mathbf{e}}_{XY'})}{r^2\sin\theta} \tag{16.233}$$

$$\mathbf{s}_{3y'} = \frac{\cos\theta\,\hat{\mathbf{e}}_{XY'} - \hat{\mathbf{e}}_{XY'}}{r\sin\theta} \tag{16.234}$$

Using Eqs. (16.176) to (16.178)

$$\hat{\mathbf{e}}_{XY}\cdot\hat{\mathbf{e}}_{XY} = 1 \qquad \hat{\mathbf{e}}_{XY'}\cdot\hat{\mathbf{e}}_{XY'} = 1 \qquad \hat{\mathbf{e}}_{XY}\cdot\hat{\mathbf{e}}_{XY'} = \cos\theta \tag{16.235}$$

and

$$G_{11} = \sum_{a=x,y,y'} \mu_a\mathbf{s}_{1a}\cdot\mathbf{s}_{1a}\cdot\mathbf{s}_{1a} = \mu_y\mathbf{s}_{1y}\cdot\mathbf{s}_{1y} + \mu_x\mathbf{s}_{1x}\cdot\mathbf{s}_{1x} + \mu_y\mathbf{s}_{1y'}\cdot\mathbf{s}_{1y'} = \mu_y + \mu_x \tag{16.236}$$

Applying the same technique, the other G_{ij} elements are found to be (note that $\mu_y = \mu_{y'}$)

$$G_{22} = \mu_y + \mu_x \tag{16.237}$$

$$G_{33} = \frac{2\mu_y}{r^2} + \frac{2\mu_x}{r^2}(1 - \cos\theta) \tag{16.238}$$

$$G_{12} = \mu_x \cos\theta \tag{16.239}$$

$$G_{13} = G_{23} = -\frac{\mu_x}{r}\sin\theta \tag{16.240}$$

Appendix T lists the G_{ij} elements for a number of simple cases and a review of important literature sources for obtaining others.

16.12 THE $\underline{\mathbf{GF}}$ MATRIX

The problem of combining the $\underline{\mathbf{F}}$ and $\underline{\mathbf{G}}$ matrices to give a usable set of equations follows the same procedure introduced in Sec. 16.2, using differential equations, or in Sec. 16.3, using the matrix formulation. Using Eqs. (16.150) and (16.225) the Lagrangian is

$$L = T - V = \frac{1}{2}\sum_{i=1}^{3N-6}\sum_{j=1}^{3N-6}\{(G^{-1})_{ij}\dot{R}_i\dot{R}_j - F_{ij}R_iR_j\} \tag{16.241}$$

Using Lagrange's equation of motion (see App. S)

$$\frac{d}{dt}\frac{\partial L}{\partial \dot{R}_k} - \frac{\partial L}{\partial R_k} = 0 \tag{16.242}$$

leads to

$$\sum_{i=1}^{3N-6}(G^{-1})_{ki}\ddot{R}_i - \sum_{i=1}^{N}F_{ik}R_i = 0 \qquad (K = 1, 2, \ldots, 3N-6) \tag{16.243}$$

This is a set of $3N - 6$ second-order differential equations whose solutions are of the form

$$R_i = A_i \cos(2\pi\nu t - \delta) \tag{16.244}$$

Differentiation of this general solution and substitution into Eq. (16.243) leads to a set of secular equations

$$\sum_{i=1}^{3N-6}\{(G^{-1})_{ki}\lambda - F_{ik}\}R_i = 0 \qquad (k = 1, 2, \ldots, 3N-6) \tag{16.245}$$

or

$$\sum_{i=1}^{3N-6}\{F_{ik} - (G^{-1})_{ki}\lambda\}A_i = 0 \qquad (k = 1, 2, \ldots, 3N-6) \tag{16.246}$$

where $\lambda = 4\pi^2\nu^2$. The necessary and sufficient condition that such a set of equations have a nontrivial solution is that the determinant vanish

$$\left| \sum_{i=1}^{N} \{F_{ik} - (G^{-1})_{ki}\lambda\} \right| = 0 \qquad (k = 1, 2, \ldots, N) \tag{16.247}$$

or, in matrix notation,

$$|\underline{\mathbf{F}} - \underline{\mathbf{G}}^{-1}\lambda| = 0 \tag{16.248}$$

The solution of this matrix to give the eigenvalues λ will give the frequencies of the normal modes of vibration. Furthermore, the eigenvalues can be introduced into Eq. (16.246) and the contributions of each symmetry coordinate to a particular normal coordinate can be found. In solving for the normal coordinates only force constants related to bond-stretching and valence angle-bendings are employed. A more useful form of this relationship emerges if Eq. (16.247) is left multiplied by $\underline{\mathbf{G}}$

$$|\underline{\mathbf{GF}} - \underline{\mathbf{GG}}^{-1}\lambda| = |\underline{\mathbf{GF}} - \underline{\mathbf{I}}\lambda| = 0 \tag{16.249}$$

This form has the advantage that only the diagonal terms contain the eigenvalues and the mathematical solution is simplified.

16.13 NORMAL COORDINATES FROM THE GF SECULAR EQUATIONS

Once the eigenvalues of the $\underline{\mathbf{GF}}$ matrix are known, the secular equations can be utilized to obtain the coefficients relating the internal coordinates R_i to the normal coordinates Q_i. The normal coordinates can be expressed in terms of the Cartesian coordinates [see Eq. (16.51)] and the internal coordinates can be expressed in terms of the Cartesian coordinates [see Eq. (16.161)]. Therefore, we would expect a relationship of the form

$$Q_k = \sum_{i=1}^{3N} C_{ik} R_i \tag{16.250}$$

to exist. The secular equations [Eq. (16.246)] involve a set of unknown terms, the A_i's, which are the maximum amplitudes of the internal coordinates R_i. Calculations are usually made using the $\underline{\mathbf{G}}$ matrix rather than the $\underline{\mathbf{G}}^{-1}$ matrix; however, the secular equations involve the latter, so this discussion will begin at that point and the evaluation of the Q_i from $\underline{\mathbf{G}}$ will be considered later.

The secular equations for a particular root λ_k are a set of $3N - 6$ homogeneous linear equation, each of the form

$$[F_{n1} - (G^{-1})_{n1}\lambda_k]A_{1k} + [F_{n2} - (G^{-1})_{n2}\lambda_k]A_{2k} + \cdots + [F_{nn'} - (G^{-1})_{nn'}\lambda_k]A_{nk} = 0 \tag{16.251}$$

These equations may be solved to get a set of independent ratios A_{1k}/A_{2k}, $A_{2k}/A_{3k}, \ldots, A_{(n-1)k}/A_{nk}$. The second subscript on the A_{ik}'s denotes the particular root or mode. Since the solution of the secular equations can only give

$3N-7$ independent ratios one additional condition is needed to uniquely specify the A_{ik}'s. This condition is the normalization

$$\sum_{i=1}^{3N-6} A_{ik}^2 = 1 \qquad k = (1, 2, \ldots, 3N-6) \tag{16.252}$$

The normalization condition can be expressed in a form involving the elements of the $\underline{\mathbf{F}}$ matrix, and hence will contain fewer than the maximum number of terms if there are any zero elements in the $\underline{\mathbf{F}}$ matrix. Writing the relationship between the normal and internal coordinates as

$$R_i = \sum_{k=1}^{3N-6} N_k A_{ik} Q_k \tag{16.253}$$

where N_k is a normalizing factor, and combining this expression with a basic relationship for potential energy in terms of F_{ij}'s and normal coordinates as derived from Eqs. (16.58) and (16.148)

$$2V = \sum_{k=1}^{3N-6} \lambda_k Q_k^2 = \sum_{i=1}^{3N-6} \sum_{j=1}^{3N-6} F_{ij} R_i R_j \tag{16.254}$$

yields

$$N_k^2 \sum_{i=1}^{3N-6} \sum_{j=1}^{3N-6} F_{ij} A_{ik} A_{jk} = \lambda_k \tag{16.255}$$

Further relationships between the vibrational amplitude factors and the force constants are developed in App. U.

Before considering some examples of normal coordinate analysis a summary of the steps in the procedure is presented. There are a variety of paths which may be followed but the primary steps will involve (1) examination of the symmetry of the molecule in order to find the symmetry properties of the normal modes and in order to select a set of internal and/or symmetry coordinates to describe the system, (2) enumeration of the individual $\underline{\mathbf{F}}$ and $\underline{\mathbf{G}}$ matrix elements, (3) Solutions of the determinant $|\underline{\mathbf{GF}} - \underline{\mathbf{I}}\lambda| = 0$ to obtain the normal mode frequencies, and (4) determination of the coefficients in the secular equations in order to specify the nature of the normal modes.

16.14 SELECTED EXAMPLES OF NORMAL MODE ANALYSIS

Although the solution for the normal modes of vibrations of more complicated molecules is of a magnitude to require the use of a digital computer, and the subject of the selection of appropriate force constants is of no small scope, it will nevertheless be beneficial to look at some simple systems that can be treated in a reasonable span. This will enable the reader to get a firmer grasp of the fundamental procedures involved and further develop background to the point where more advanced concepts can be assimilated readily by study of appropriate sources.

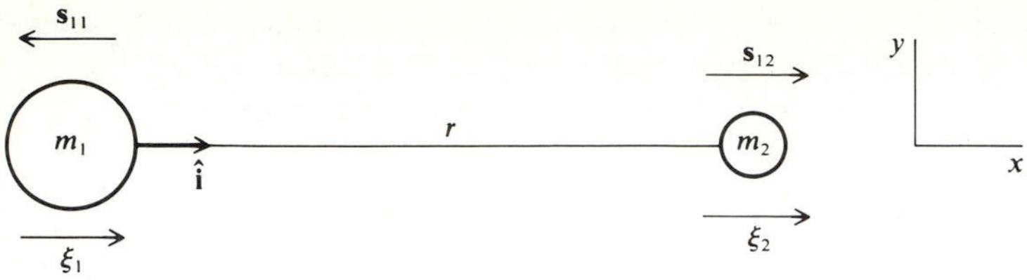

FIGURE 16.20
Vibrational motions of a diatomic molecule.

16.14.1 The Diatomic Molecule

Although the $\underline{FG}$-matrix method is unnecessary to solve the vibrational problem for the diatomic molecule it will be used in order to familiarize the reader with the relationships between actual molecular parameters and the terminology of the theory at the most elementary level. The motion of the system is shown in Fig. 16.20. Since the motion is colinear with the bond, the Cartesian displacements and the displacement vectors will be parallel. The symmetry coordinates, which is the intraatom displacement, can be found by inspection or by use of Eq. (16.162) to be

$$R_1 = \mathbf{s}_{11} \cdot \boldsymbol{\rho}_1 + \mathbf{s}_{12} \cdot \boldsymbol{\rho}_2 \tag{16.256}$$

where $\mathbf{s}_{11} = -\hat{\mathbf{i}}$, $\mathbf{s}_{12} = \hat{\mathbf{i}}$, $\boldsymbol{\rho}_1 = \xi_1\hat{\mathbf{i}}$, and $\boldsymbol{\rho}_2 = \xi_2\hat{\mathbf{i}}$, hence

$$R_1 = \xi_2 - \xi_1 \tag{16.257}$$

The only term in the $\underline{G}$ matrix is found from Eq. (16.201)

$$G_{11} = \mu_1 + \mu_2 \tag{16.258}$$

It is to be noted that this is identical to the G_{11} term derived in the more general treatment ending with Eq. (16.240). There will be a single potential term related to the bond-stretching, hence the only $\underline{F}$ matrix element will be $F_{11} = k$. The secular determinate

$$(\mu_1 + \mu_2)k - \lambda = 0 \tag{16.259}$$

or

$$\lambda = 4\pi^2\nu^2 = (\mu_1 + \mu_2)k = \frac{k}{\mu} \tag{16.260}$$

where $\mu = (m_1m_2)/(m_1 + m_2)$, the reduced mass. Therefore, the normal mode frequency is

$$\nu = \frac{1}{2\pi}\sqrt{\frac{k}{\mu}} \tag{16.261}$$

as was derived in earlier discussions.

Although the problem of finding the normal coordinate is extremely simple in this case we will, nevertheless, follow it through in order to establish the method in the reader's mind before proceeding to more complicated systems. The example of a $^{12}C^{16}O$ molecule with $k = 18.8 \times 10^2$ N/m, $\mu = 6.85$ amu, and $\nu = 6.51 \times 10^{13}\ s^{-1}$, hence $\lambda = 1.67 \times 10^{29}\ s^{-2}$ will be used in order to examine the magnitude of the quantities involved.

Since there is a single root the secular equation, Eq. (16.246) is

$$\{F_{11} - (G^{-1})_{11}\lambda\}A_{11} = 0 \tag{16.262}$$

This equation can only give the relationship $A_{11} \neq 0$, and the normalization condition, Eq. (16.255), is needed to get an explicit value. The use of this latter relationship gives $N_1^2 A_{11}^2 = 1/\mu$. Using Eq. (16.253)

$$R_1 = \frac{Q_1}{\mu^{1/2}} \tag{16.263}$$

or

$$Q_1 = \mu^{1/2} R_1 = (\mu_1 + \mu_2)^{-1/2} R_1 \tag{16.264}$$

Using values for the CO molecule

$$Q_1 = 3.37 \times 10^{-11}(\xi_1 - \xi_2) \tag{16.265}$$

The normal coordinate is equivalent to the actual displacement of the system from equilibrium multiplied by a weighting factor that is a function of the molecular mass.

Having established the general nature of the procedure for the normal mode analysis for the simplest possible molecule, we will next expand our treatment to consider our two previous examples of linear and bent XY_2 molecules. In these discussions we will start at the beginning with each system and work from there, but draw on information previously derived for the system rather than repeat certain specifics. The older literature almost exclusively employs the cgs system of units with the SI system being introduced in more recent work. To provide a familiarity with both systems the example of the linear molecule will employ SI units and that of the bent XY_2 molecule will use cgs units.

16.14.2 The Linear XY_2 Molecule

Unlike the hypothetical one-dimensional system investigated in Sec. 16.2, the real linear XY_2 molecule must be considered as a three-dimensional entity. There will be $3N - 5$ or 4 degrees of vibrational freedom. We will begin by setting up a set of internal coordinates to describe the system. The selection of the internal coordinates can be guided to some extent by the knowledge regarding symmetry-adapted coordinates, as obtained by group theoretical methods. In Sec. 16.7 it was found that this type molecule would have four normal modes belonging to the Σ_g^+, Σ_u^+, and π_u representations. There will be two modes involving only the stretching of the bonds, Σ_g^+ and Σ_u^+, and a degenerate pair due to bending of the molecule. The internal coordinates are selected in a manner so that they can readily be combined to give these modes. The general displacements are shown

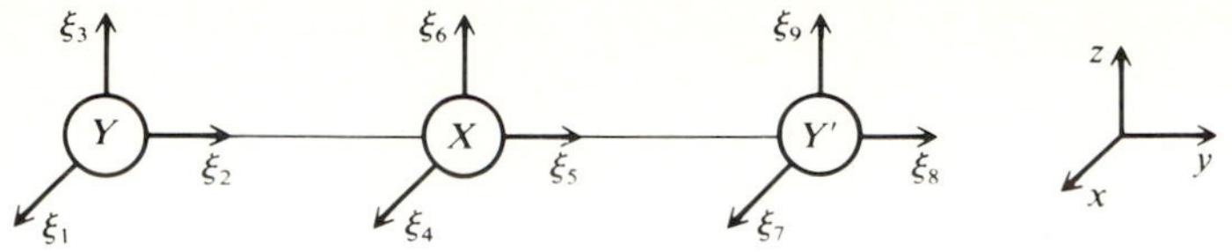

FIGURE 16.21
Coordinates for a linear XY_2 molecule.

in Fig. 16.21 where, although Y is identical to Y', the superscript prime is retained for indexing purposes. Since two modes involve bond-stretching in the y direction the two logical internal coordinates will be

$$R_{YX} = \xi_2 - \xi_5 \tag{16.266}$$

$$R_{XY'} = \xi_5 - \xi_8 \tag{16.267}$$

The introduction of yet another concept can further aid in solving the vibrational problem. This is the concept of internal symmetry coordinates, denoted by S_i. In the previous discussion the internal coordinates were viewed as arising from the motions of molecular segments with no regard to the total structure or symmetry of the molecule. By consideration of both the molecular and normal mode symmetries it is often possible to combine groups of the internal coordinates to produce new internal symmetry coordinates having the symmetry of the molecule or one of its normal modes. In general, these internal symmetry coordinates will be even more simply related to the normal modes than the internal coordinates. Since these new symmetry coordinates serve the same purpose as the internal coordinates, they have the same properties and obey the same relationship as the R_i's of the earlier discussions. An important point to remember is that the number of internal symmetry coordinates and the number of internal coordinates are equal, so if pairs of the latter are combined to give pairs of the former, allowance must be made for proper normalization. Although a molecule may possess a type of symmetry that will allow for the convenient establishment of internal symmetry coordinates, it is not necessarily true that all of its internal coordinates will convert to new symmetry coordinates. Some of the internal symmetry coordinates may be identical to the original internal coordinates. For the linear XY_2 molecule a pair of internal symmetry coordinates that possess the symmetries of two of the normal modes will be

$$\Sigma_g^+ \quad S_1 = \tfrac{1}{2}(R_{yx} + R_{xy'}) = \tfrac{1}{2}(\xi_2 - \xi_8) \tag{16.268}$$

$$\Sigma_u^+ \quad S_2 = \tfrac{1}{2}(R_{yx} + R_{xy'}) = \tfrac{1}{2}(\xi_2 - 2\xi_5 + \xi_8) \tag{16.269}$$

The factor $(\frac{1}{2})$ is the normalization constant obtained when allowance is made for the conservation of the number of coordinates, i.e., when two internal coordinates r_{YX} and r_{XY} were converted to two internal symmetry coordinates.

Figure 16.22 illustrates how the other two internal coordinates are structured. When the molecule bends it will undergo a total angular change of θ_y. Since the

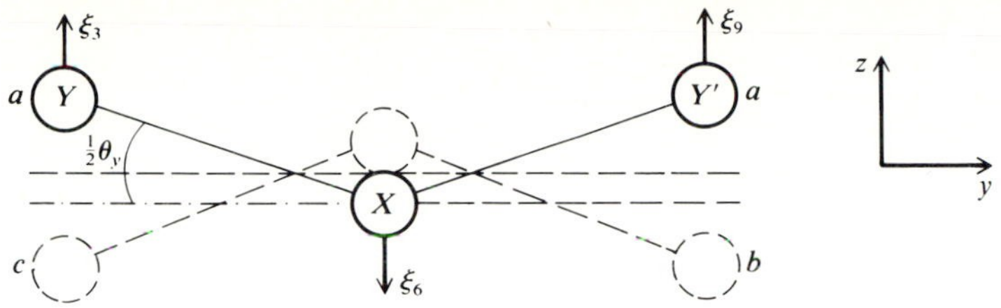

FIGURE 16.22
Bending coordinates for a bent XY_2 molecule.

bending motion is slight and θ_y will be a small angle, it follows that

$$\frac{1}{2}\theta_y = \frac{1}{2}\sin\theta_y = \frac{ac}{r_{yx}} \tag{16.270}$$

But $ac = \xi_3 - \xi_6$ and $ab = \xi_9 - \xi_6$ so

$$\theta_y = \frac{1}{r_{yx}}(\xi_3 - 2\xi_6 + \xi_9) \tag{16.271}$$

There will be an analogous bending in the xz plane denoted by

$$\theta_x = \frac{1}{r_{yx}}(\xi_1 - 2\xi + \xi_7) \tag{16.272}$$

These two coordinates belong to the π_u representation. Due to the degenerate nature of θ_Y and θ_X, they will both contribute in the same manner to the $\underline{\mathbf{G}}$ matrix. We now have a set of internal coordinates expressed in the form of Eq. (16.161), which contains the same B_{ij} elements as Eqs. (16.192), giving the G_{ij} elements.

There is a choice of methods for the evaluation of the $\underline{\mathbf{G}}$ matrix. The B_{ij} elements may be established from the form of the internal coordinates, and from these the G_{ij}'s evaluated. The second method finds the displacement coordinates by use of the formal relationships expressed in Eqs. (16.126) to (16.181), and then the G_{ij}'s are obtained by use of Eq. (16.201). The choice of method depends on the complexity of the problem and the ease of proceeding by one method versus the other. In general, the latter approach is followed because it avoids the dependence on a space-fixed Cartesian coordinate system and instead uses a set of internal coordinates that are more molecular-oriented. For this example the first method provides a simpler approach for a linear system where the Cartesian coordinates are easily visualized. In the discussion of the nonlinear XY_2 molecule we will contrast the methods.

The internal symmetry coordinates are

$$S_1 = \tfrac{1}{2}(\xi_2 - \xi_8) \tag{16.273}$$

$$S_3 = \tfrac{1}{2}(\xi_2 - 2\xi_5 + \xi_8) \tag{16.274}$$

$$S_2 = \frac{1}{r}(\xi_3 - 2\xi_6 + \xi_9) \tag{16.275}$$

$$S_4 = \frac{1}{r}(\xi_1 - 2\xi_4 + \xi_7) \tag{16.276}$$

where $r = r_{yx} = r_{xy'}$. Considering that these equations came from the general matrix relationship

$$\mathbf{S} = \underline{\mathbf{B}}\boldsymbol{\xi} \qquad \text{or} \qquad S_i = \sum_{j=1}^{3N} B_{ij}\xi_i \tag{16.277}$$

the B_{ij} coefficients are evaluated and collected in Table 16.2. Using Eqs. (16.192) and $\mu_y = \mu_{y'}$, the G_{ij} elements are determined and listed in Table 16.3. In this case the $\underline{\mathbf{G}}$ matrix is diagonal, hence $(G^{-1})_{ij} = (G_{ij})^{-1}$ and it is just as convenient to work with either the $\underline{\mathbf{G}}$ or $\underline{\mathbf{G}}^{-1}$ matrix.

Next the elements of the $\underline{\mathbf{F}}$ matrix must be found. Neglecting any possible intersections between nonneighboring atoms the potential energy can be expressed as

$$V = \tfrac{1}{2}k_r(R_{yx}^2 + R_{xy'}^2) + k_\theta(\theta_x^2 + \theta_y^2) \tag{16.278}$$

Using Eqs. (16.268) through (16.276) this becomes

$$V = \tfrac{1}{2}k_r(S_1^2 + S_3^2) + k_\theta(S_2^2 + S_4^2) \tag{16.279}$$

TABLE 16.2
B_{ij} coefficients for a linear XY_2 molecule

$B_{11} = 0$	$B_{21} = 0$	$B_{31} = 0$	$B_{41} = 1/r$
$B_{12} = \frac{1}{2}$	$B_{22} = 0$	$B_{32} = \frac{1}{2}$	$B_{42} = 0$
$B_{13} = 0$	$B_{23} = 1/r$	$B_{33} = 0$	$B_{43} = 0$
$B_{14} = 0$	$B_{24} = 0$	$B_{34} = 0$	$B_{44} = -(2/r)$
$B_{15} = 0$	$B_{25} = 0$	$B_{35} = -\frac{2}{2}$	$B_{45} = 0$
$B_{16} = 0$	$B_{26} = -(2/r)$	$B_{36} = 0$	$B_{46} = 0$
$B_{17} = 0$	$B_{27} = 0$	$B_{37} = 0$	$B_{47} = 1/r$
$B_{18} = -\frac{1}{2}$	$B_{28} = 0$	$B_{38} = \frac{1}{2}$	$B_{48} = 0$
$B_{19} = 0$	$B_{29} = 1/r$	$B_{39} = 0$	$B_{49} = 0$

TABLE 16.3
G Matrix elements for an XY_2 linear molecule

$G_{11} = \mu_y$	$G_{22} = 2(1/r^2)(\mu_y + 2\mu_x)$
$G_{12} = G_{21} = 0$	$G_{23} = G_{32} = 0$
$G_{13} = G_{31} = 0$	$G_{24} = G_{42} = 0$
$G_{14} = G_{41} = 0$	$G_{33} = \mu_y + 2\mu_x$
$G_{34} = G_{45} = 0$	$G_{44} = 2(1/r^2)(\mu_y + 2\mu_x)$

TABLE 16.4
***F*-matrix elements for a linear *XY*$_2$ molecule**

$F_{11} = k_r$	$F_{23} = F_{32} = 0$
$F_{12} = F_{21} = 0$	$F_{24} = F_{42} = 0$
$F_{13} = F_{31} = 0$	$F_{33} = k_r$
$F_{14} = F_{41} = 0$	$F_{34} = F_{43} = 0$
$F_{22} = k_\theta$	$F_{44} = k_\theta$

Since

$$V = \frac{1}{2} \sum_{i=1}^{3N} \sum_{j=1}^{3N} F_{ij} S_i S_j \tag{16.280}$$

the elements of the F matrix are found and listed in Table 16.4. The secular determinant is

$$|\underline{\mathbf{GF}} - \underline{\mathbf{I}}\lambda| = 0 \tag{16.281}$$

or

$$\begin{bmatrix} \mu_y & 0 & 0 & 0 \\ 0 & \dfrac{2(\mu_y + 2\mu_x)}{r^2} & 0 & 0 \\ 0 & 0 & \mu_y + 2\mu_x & 0 \\ 0 & 0 & 0 & \dfrac{2(\mu_y + 2\mu_x)}{r^2} \end{bmatrix} \begin{bmatrix} k_r & 0 & 0 & 0 \\ 0 & k_\theta & 0 & 0 \\ 0 & 0 & k_r & 0 \\ 0 & 0 & 0 & k_\theta \end{bmatrix} - \begin{bmatrix} \lambda & 0 & 0 & 0 \\ 0 & \lambda & 0 & 0 \\ 0 & 0 & \lambda & 0 \\ 0 & 0 & 0 & \lambda \end{bmatrix} = 0 \tag{16.282}$$

and

$$[\mu_y k_r - \lambda][(\mu_y + 2\mu_x)k_r - \lambda][\{2(\mu_y + 2\mu_x)k_\theta / r^2\} - \lambda]^2 = 0 \tag{16.283}$$

The roots are

$$\lambda_1 = \mu_y k_r \tag{16.284}$$

$$\lambda_3 = (\mu_y + 2\mu_x)k_r \tag{16.285}$$

$$\lambda_2 = \lambda_4 = \frac{2k_\theta(\mu_y + 2\mu_x)}{r^2} \tag{16.286}$$

In order to simplify the calculation of the normal coordinates and avoid manipulating equations containing numerous general expressions we will consider a particular XY_2 molecule, $^{12}C^{16}O_2$, for which the molecular parameters are

$$k_r = 17.0 \times 10^2 \text{ N m}^{-1}$$

$$r = 1.16 \times 10^{-10} \text{ m}$$

$$\frac{k_\theta}{r^2} = 0.57 \times 10^2 \text{ N m}^{-1}$$

$$\mu_x = \mu_c = 5.02 \times 10^{25} \text{ kg}^{-1}$$

$$\mu_y = \mu_O = 3.77 \times 10^{25} \text{ kg}^{-1}$$

In this case, since $\nu_i = \lambda_i^{1/2}/2\pi c$, $\nu_1 = 1.344 \times 10^5\ \text{m}^{-1}$ ($1344\ \text{cm}^{-1}$), $\nu_2 = 0.667 \times 10^5\ \text{m}^{-1}$ ($667\ \text{cm}^{-1}$), and $\nu_3 = 2.349 \times 10^5\ \text{m}^{-1}$ ($2349\ \text{cm}^{-1}$).

Since both the $\underline{\mathbf{F}}$ and $\underline{\mathbf{G}}$ matrices are diagonal and $(G^{-1})_{ij} = 1/G_{ij}$, the most convenient form for evaluating the normal coordinates is the secular equation given by Eq. (16.251). Using the nonzero $\underline{\mathbf{F}}$ and $\underline{\mathbf{G}}$ matrix elements the secular equations for the four roots become

$$\begin{array}{llll}
\lambda_1 & (F_{11}G_{11} - \lambda_1)A_{11} = 0 & \lambda_3 & (F_{11}G_{11} - \lambda_3)A_{13} = 0 \\
 & (F_{22}G_{22} - \lambda_1)A_{21} = 0 & & (F_{22}G_{22} - \lambda_3)A_{23} = 0 \\
 & (F_{33}G_{33} - \lambda_1)A_{31} = 0 & & (F_{33}G_{33} - \lambda_3)A_{33} = 0 \\
 & (F_{44}G_{44} - \lambda_1)A_{41} = 0 & & (F_{44}G_{44} - \lambda_3)A_{43} = 0 \\
\lambda_2 & (F_{11}G_{11} - \lambda_2)A_{12} = 0 & \lambda_4 & (F_{11}G_{11} - \lambda_4)A_{14} = 0 \\
 & (F_{22}G_{22} - \lambda_2)A_{22} = 0 & & (F_{22}G_{22} - \lambda_4)A_{24} = 0 \\
 & (F_{33}G_{33} - \lambda_2)A_{32} = 0 & & (F_{33}G_{33} - \lambda_4)A_{34} = 0 \\
 & (F_{44}G_{44} - \lambda_2)A_{42} = 0 & & (F_{44}G_{44} - \lambda_4)A_{44} = 0
\end{array} \tag{16.287}$$

Inserting the values for the λ_i and considering the normalization condition, Eq. (16.252), the only nonzero coefficients are $A_{11} = A_{33} = 1$ and $A_{22} = A_{42} = A_{24} = A_{44} = \frac{1}{2}$. Using Eq. (16.255), the normalizing factors are found to be: $N_1 = \mu = 6.135 \times 10^{12}\ \text{kg}^{-1/2}$, $N_3 = \mu + 2\mu_x = 1.176 \times 10^{13}\ \text{kg}^{-1/2}$, $N_2 = N_4 = 2(\mu_y + 2_{\mu x})/r = 1.1442 \times 10^{23}\ \text{kg}^{-1/2}\ \text{m}^{-1}$. The calculation of the $\underline{\mathbf{W}}^{-1}$ elements necessary for the transformations $\mathbf{Q} = \underline{\mathbf{W}}^{-1}\mathbf{S}$ are found by evaluation of the A'_{ij} parameters using Eqs. (U-27) and (U-28). For each root λ_k there will be a set of eight linear equations. These are simple in form, the first being, for example,

$$\lambda_1^{1/2}A'_{11} + F_{11}A_{11} + F_{12}A_{21} + F_{13}A_{31} + F_{14}A_{41} = 0 \tag{16.288}$$

The solution of these equations gives

$$A'_{21} = A'_{31} = A'_{41} = 0 \tag{16.289}$$

$$A'_{11} = -F_{11}\lambda_i^{-1/2} = -6.74 \times 10^{-12}\ \text{N}^{1/2}\ \text{kg}^{1/2}\ \text{m}^{-1/2} \tag{16.290}$$

$$A'_{13} = A'_{23} = A'_{43} = 0 \tag{16.291}$$

$$A'_{33} = -F_{22}\lambda_3^{-1/2} = -3.51 \times 10^{-12}\ \text{N}^{1/2}\ \text{kg}^{1/2}\ \text{m}^{-1/2} \tag{16.292}$$

$$A'_{12} = A'_{32} = A'_{14} = A'_{34} = 0 \tag{16.293}$$

$$\begin{aligned} A'_{22} = A'_{42} = A'_{24} = A'_{44} &= -F_{33}(2\lambda_2)^{-1/2} \\ &= -4.32 \times 10^{-33}\ \text{N}^{1/2}\ \text{kg}^{1/2}\ \text{m}^{-3/2} \end{aligned} \tag{16.294}$$

The coefficients for the transformation $\mathbf{Q} = \underline{\mathbf{W}}^{-1}\mathbf{S}$ will be

$$(W^{-1})_{jk} = -A'_{jk}N_k\lambda_k^{1/2} \tag{16.295}$$

so

$$\begin{pmatrix} Q_1 \\ Q_2 \\ Q_3 \\ Q_4 \end{pmatrix} = \begin{pmatrix} 1.63 \times 10^{-13} & 0 & 0 & 0 \\ 0 & 1.27 \times 10^{-24} & 0 & 1.27 \times 10^{-24} \\ 0 & 0 & 0.85 \times 10^{-13} & 0 \\ 0 & 1.27 \times 10^{-24} & 0 & 1.27 \times 10^{-24} \end{pmatrix} \begin{pmatrix} \frac{1}{\sqrt{2}}(\xi_2 - \xi_8) \\ \frac{1}{r}(\xi_3 - 2\xi_6 + \xi_9) \\ \frac{1}{\sqrt{2}}(\xi_2 - 2\xi_5 + \xi_8) \\ \frac{1}{r}(\xi_1 - 2\xi_4 + \xi_7) \end{pmatrix}$$

or letting $r = 1.16 \times 10^{-10}$ m

$$Q_1 = 1.15 \times 10^{-13}(\xi_2 - \xi_8) \tag{16.296}$$

$$Q_2 = Q_4 = 1.09 \times 10^{-14}[(\xi_3 - 2\xi_6 + \xi_7) + (\xi_1 - 2\xi_4 + \xi_7)] \tag{16.297}$$

$$Q_3 = 6.01 \times 10^{-14}(\xi_2 - 2\xi_5 + \xi_8) \tag{16.298}$$

There are four important features to note about the results just obtained.

1. The magnitudes of the displacements are small as was initially assumed, being of the order of 1 percent or less of a bond length or angle.
2. The degenerate pair of vibrations end up being mixed in the final transformation. The problem could have been more easily solved by recognizing this degeneracy at the beginning and reducing the problem to use 3×3 rather than 4×4 matrices.
3. Since both the $\underline{\mathbf{F}}$ and $\underline{\mathbf{G}}$ matrices are diagonal, the initial choice of symmetry coordinates is related to the normal coordinates by a constant term. In a case such as this the relation between **S** and **Q** is

$$Q_k = (G^{-1})_{kk} S_k \tag{16.299}$$

   Starting with this relationship could have saved the lengthy procedure for finding **Q**; however, the purpose at this point is to learn the basic relationships for the general case rather than short cuts for a limited number of specific ones.
4. The normal coordinates each belong to a different representation, if the degenerate pair is considered as a single type of motion. This fact dictates the diagonal character of the $\underline{\mathbf{G}}$ matrix. If there is more than one normal mode, exclusive of degenerate pairs, belonging to any irreducible representation, then $\underline{\mathbf{G}}$ will contain off-diagonal elements. It can be reasoned that the selection of a set of $3N - 6$ internal symmetry coordinates, each of which belongs to a different irreducible representation, can provide for an initial simplification of the problem.

16.14.3 The Bent XY_2 Molecule

We will finish the discussion of molecular vibrations by examining the motions of a water molecule. In Sec. 16.6 it was found that for a bent XY_2 molecule there would be three normal modes belonging to the $A_1(2)$ and $B_2(1)$ irreducible representations. With no a priori knowledge of the nature of the normal modes, beyond knowing their symmetries, a set of three internal symmetry coordinates will be constructed. Rather than use the unsymmetrized internal coordinate as discussed in Sec. 16.10 it will be more convenient to combine these to form internal symmetry coordinates that have the symmetries of the irreducible representations of the normal modes. Referring to Fig. 16.14 and the following text we find the necessary coordinate relationship enumerated. Using $\angle$HOH = 104.5°, the three internal symmetry coordinates are given by use of Eqs. (16.154) to (16.156)

$$S_1 = \tfrac{1}{2}(r_{\mathrm{OH}} + r_{\mathrm{OH'}}) = -0.56\xi_1 - 0.43\xi_2 + 0.86\xi_4 + 0.56\xi_5 - 0.43\xi_6 \tag{16.300}$$

$$S_2 = r^{-1}[-0.61\xi_1 + 0.79\xi_2 - 1.58\xi_4 + 0.61\xi_5 + 0.79\xi_6] \tag{16.301}$$

$$S_3 = \tfrac{1}{2}(r_{\mathrm{OH}} - r_{\mathrm{OH'}}) = -0.56\xi_1 - 0.43\xi_2 + 1.12\xi_3 - 0.56\xi_5 + 0.43\xi_6 \tag{16.302}$$

The internal coordinates can be derived analytically by use of a set of displacement vectors $\mathbf{s}_{ia}$. For the angle deformation these vectors are given using Eqs. (16.179) to (16.182).

$$\mathbf{s}_{2\mathrm{H}} = \frac{2(-0.25\hat{\mathbf{e}}_{\mathrm{OH}} - \hat{\mathbf{e}}_{\mathrm{OH'}})}{0.97r} \tag{16.303}$$

$$\mathbf{s}_{2\mathrm{H'}} = \frac{2(-0.25\hat{\mathbf{e}}_{\mathrm{OH'}} - \hat{\mathbf{e}}_{\mathrm{OH}})}{0.97r} \tag{16.304}$$

$$\mathbf{s}_{2\mathrm{O}} = \frac{2 \times 1.25(\hat{\mathbf{e}}_{\mathrm{OH}} + \hat{\mathbf{e}}_{\mathrm{OH'}})}{0.97r} \tag{16.305}$$

For the other two symmetry coordinates Fig. 16.23 is used to visualize the necessary relationships

$$\mathbf{s}_{1\mathrm{H}} = \hat{\mathbf{e}}_{\mathrm{OH}} = \frac{\hat{\mathbf{e}}_{XY}}{2} = \frac{\hat{\mathbf{j}}\sin\phi - \hat{\mathbf{k}}\cos\phi}{2} = -0.56\hat{\mathbf{j}} - 0.43\hat{\mathbf{k}} \tag{16.306}$$

$$\mathbf{s}_{1\mathrm{H'}} = \hat{\mathbf{e}}_{\mathrm{OH'}} = \frac{\hat{\mathbf{e}}_{XY'}}{2} = \frac{\hat{\mathbf{j}}\sin\phi - \hat{\mathbf{k}}\cos\phi}{2} = -0.56\hat{\mathbf{j}} - 0.43\hat{\mathbf{k}} \tag{16.307}$$

$$\mathbf{s}_{1\mathrm{O}} = (-\hat{\mathbf{e}}_{\mathrm{OH}} - \hat{\mathbf{e}}_{\mathrm{OH'}}) = 2\hat{\mathbf{k}}\cos\phi = 0.86\hat{\mathbf{k}} \tag{16.308}$$

$$\mathbf{s}_{3\mathrm{H}} = \hat{\mathbf{e}}'_{\mathrm{OH}} = -0.56\hat{\mathbf{j}} - 0.43\hat{\mathbf{k}} \tag{16.309}$$

$$\mathbf{s}_{3\mathrm{H'}} = -\hat{\mathbf{e}}'_{\mathrm{OH'}} = -0.56\hat{\mathbf{j}} + 0.43\hat{\mathbf{k}} \tag{16.310}$$

$$\mathbf{s}_{3\mathrm{O}} = (-\hat{\mathbf{e}}'_{\mathrm{OH}} + \hat{\mathbf{e}}'_{\mathrm{OH'}})\cos(90 - \phi) = 1.12\hat{\mathbf{j}} \tag{16.311}$$

where $\hat{\mathbf{e}}_{\mathrm{OH}}$, $\hat{\mathbf{e}}_{\mathrm{OH'}}$, $\hat{\mathbf{e}}'_{\mathrm{OH}}$, and $\hat{\mathbf{e}}'_{\mathrm{OH'}}$ are the maximum displacement vectors for S_1 and S_3 while $\hat{\mathbf{e}}_{xy}$ and $\hat{\mathbf{e}}_{xy'}$ are those for R_1 and R_2.

When combining two internal coordinates R_i to give two symmetry coordinates S_i, keep in mind that there is a conservation of coordinates and a need for the inclusion of a normalization factor as a multiplier of the final expressions

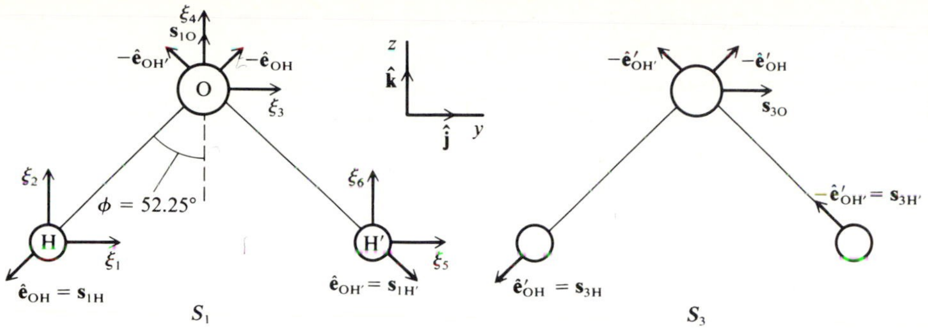

FIGURE 16.23
Displacement coordinates for S_1 and S_2 symmetry coordinates of H_2O (only in-plane coordinates listed).

for the symmetry coordinates. This is most conveniently included in the expressions for the $\mathbf{s}_i$ vectors. The maximum displacements will no longer be the same as they were for the internal coordinates. For example, the maximum displacement of the H atom for the R_1 internal coordinate will be distributed equally between the two symmetry coordinates S_1 and S_2. This can easily be demonstrated by a comparison of coefficients in the two expressions $S_i = \sum_{a=1}^{N} \mathbf{s}_{1a} \cdot \boldsymbol{\rho}_a$ and $s_i = (R_1 + R_2)/2$, where R_i is given by Eq. (16.162) and the $\mathbf{s}_{ia}$ are expressed in terms of the unit vectors $\hat{\mathbf{e}}_{ia}$. Using the previous expressions for the $\mathbf{s}_{ia}$ vectors and the Cartesian vectors

$$\boldsymbol{\rho}_H = \xi_7\mathbf{i} + \xi_1\mathbf{j} + \xi_2\mathbf{k} \tag{16.312}$$

$$\boldsymbol{\rho}_{H'} = \xi_9\mathbf{i} + \xi_5\mathbf{j} + \xi_6\mathbf{k} \tag{16.313}$$

$$\boldsymbol{\rho}_O = \xi_8\mathbf{i} + \xi_3\mathbf{j} + \xi_4\mathbf{k} \tag{16.314}$$

it can be shown that the symmetry coordinates are also derivable from

$$S_i = \sum_{a=1}^{N} \mathbf{s}_{ia} \cdot \boldsymbol{\rho}_a \tag{16.315}$$

For example,

$$S_1 = \mathbf{s}_{1H} \cdot \boldsymbol{\rho}_H + \mathbf{s}_{1H'} \cdot \boldsymbol{\rho}_{H'} + \mathbf{s}_{1O} \cdot \boldsymbol{\rho}_O \tag{16.316}$$

which, upon substitution of Eqs. (16.306) to (16.308) and (16.312) to (16.314) is equal to Eq. (16.300). Since there are two normal modes belonging to one irreducible representation the $\underline{\mathbf{G}}$ matrix will not be diagonal. We can use either of the two methods previously discussed for the evaluation of the G_{ij}. Comparing Eq. (16.160) and Eqs. (16.300) to (16.302) to obtain the B_{ij} coefficients, and using Eq. (16.192) yields, for example,

$$G_{11} = \sum_{k=1}^{3N} \mu_k B_{1k} B_{1k} = \mu_H + 0.75\mu_O \tag{16.317}$$

Using Eq. (16.201) along with the displacement vectors Eqs. (16.306) through

(16.311), the G elements can also be obtained. For example

$$G_{11} = \sum_{a=1}^{3N} \mu_a \mathbf{s}_{1a} \mathbf{s}_{1a} = \mu_H + 0.75\mu_0 \tag{16.318}$$

Using either of these methods the complete set of **G** matrix elements in general notation for an XY_2 molecule with valence angle θ and for water with $\theta = 104.5°$ is found to be

$$G_{11} = \mu_y + \mu_x(1 + \cos\theta) = \mu_H + 0.75\mu_0 \tag{16.319}$$

$$G_{12} = G_{21} = \frac{\sqrt{2}\mu_x}{r} \sin\theta = -\frac{1.37\mu_0}{r} \tag{16.320}$$

$$G_{13} = G_{31} = 0 \tag{16.321}$$

$$G_{22} = \frac{2}{r^2}[\mu_y + \mu_x(1 - \cos\theta)] = \frac{2}{r^2}(\mu_H + 1.25\mu_0) \tag{16.322}$$

$$G_{23} = G_{32} = 0 \tag{16.323}$$

$$G_{33} = \mu_y + \mu_x(1 - \cos\theta) = \mu_H + 1.25\mu_0 \tag{16.324}$$

Considering that the only forces of significance are those involving stretching motions between adjacent atoms and valence angle deformations, the potential energy is written

$$V = \tfrac{1}{2}[k_{\mathrm{OH}} r_{\mathrm{OH}}^2 + k_{\mathrm{OH'}} r_{\mathrm{OH'}}^2 + k\theta^2] \tag{16.325}$$

where θ denotes the deformation of the angle θ. The nonzero elements of the $\underline{\mathbf{F}}$ matrix are

$$F_{11} = F_{33} = k_r \qquad F_{22} = k_\theta \tag{16.326}$$

The secular determinant becomes

$$|\underline{\mathbf{GF}} - \underline{\mathbf{I}}\lambda| = \begin{vmatrix} (\mu_H + 0.75\mu_0)k_r - \lambda & \dfrac{-1.37\mu_0 k_r}{r} & 0 \\ \dfrac{-1.37\mu_0 k_r}{r} & 2k_\theta \dfrac{\mu_H + 1.25\mu_0}{r^2 - \lambda} & 0 \\ 0 & 0 & (\mu_H + 1.25\mu_0)k_r - \lambda \end{vmatrix} = 0 \tag{16.327}$$

This immediately reduces to a linear and a quadratic equation giving the single root

$$\lambda_3 = (\mu_H + 1.25\mu_0)k_r = 0.645 \times 10^{24} k_r \tag{16.328}$$

where the values of the masses have been introduced. The quadratic is solved to give

$$\lambda_1 = \frac{1}{2}\left\{(\mu_H + 0.75\mu_0)k_r + (\mu_H + 1.25\mu_0)\frac{2k_\theta}{r^2} + \left[(\mu_H + 0.75\mu_0)^2 k_r^2 \right.\right.$$
$$\left.\left. + (\mu_H + 1.25\mu_0)^2 \frac{4k_\theta^2}{r^4} - \frac{6k_\theta k_r}{r^2}(\mu_H + 0.75\mu_0)(\mu_H + 1.25\mu_0)\right]^{1/2}\right\}$$
$$= 0.625 \times 10^{24} k_r + (7 \times 10^{-9}) \times 10^{24} \frac{k_\theta}{r^2} \tag{16.329}$$

$$\lambda_2 = \frac{1}{2}\left\{(\mu_H + 0.75\mu_0)k_r + (\mu_H + 1.25\mu_0)\frac{2k_\theta}{r^2} - \left[(\mu_H + 0.75\mu_0)^2 k_r^2 \right.\right.$$
$$\left.\left. + (\mu_H + 1.25\mu_0)^2 \frac{4k_\theta}{r^4} - \frac{6k_\theta k_r}{r^2}(\mu_H + 0.75\mu_0)(\mu_H + 1.25\mu_0)\right]^{1/2}\right\}$$
$$= 1.3 \times 10^{24} \frac{k_\theta}{r^2} - 0.00016 \times 10^{24}\, k_r \tag{16.330}$$

Although the method is not necessary for solutions of this 2×2 case for the solutions of larger eigenvalue problems two useful aids are

$$\lambda_1 + \lambda_2 + \cdots + \lambda_N = \text{trace}\,|\underline{\mathbf{GF}}| \tag{16.331}$$

$$\lambda_1 \times \lambda_2 \times \cdots \times \lambda_N = \det\,|\underline{\mathbf{GF}}| \tag{16.332}$$

If the molecule just discussed is increased in complexity, only to the extent of being comprised of different terminal atoms, and a more complete set of force constants such as given by Eq. (16.149) are employed, then we can encounter a problem which frequently develops when dealing with physical phenomena. It is that of having more unknowns in the analytical relationships than data points for use in their determination. The problem resolves itself into one of a series of successive approximations aided by the use of prior knowledge of simpler systems. Although the procedure to be outlined is somewhat more expansive than necessary for this molecule, it will be pursued in order to acquaint the reader with the general method for attacking such problems.

Before examining the general procedure, it is relevant to examine several related points of interest.

1. For diatomic molecules it was found that $k = 4\pi^2 C^2 \nu^2 \mu$ and that ν was of the order of 50–5000 cm^{-1}. Since ν is inversely proportional to μ and k is primarily a function of bond type, k will be of the order of $(1\text{–}25) \times 10^5$ dyne cm^{-1}.
2. For the secular determinants k_r and k_θ/r^2 must have the same units, that is, dyne cm^{-1}.
3. The force necessary to produce a valence angle deformation will be less than that needed to produce a bond stretch since there is less disturbing of the electron density in the former. It can be assumed that k_θ/r^2 will be small relative to the k_r values in the same molecule.

Using this information it is possible to approximate the roots. For this approximation any off-diagonal terms can also be neglected for the initial trial. Taking $k_r = 10^6$ dyne cm^{-1} and $k_\theta/r^2 = 10^5$ dyne cm^{-1} gives $\lambda_1 = 0.585 \times 10^{30}$ dyne cm^{-1} g^{-1}, $\lambda_2 = 0.175 \times 10^{30}$ dyne cm^{-1} g^{-1}, and $\lambda_3 = 0.645 \times 10^{30}$ dyne cm^{-1} g^{-1} or $\lambda_1 = 4060$ cm^{-1}, $\lambda_2 = 2229$ cm^{-1} and $\lambda_3 = 4273$ cm^{-1}. These approximate values for the frequencies can be used to aid in the assignment of an experimental spectrum and to determine the true values. The observed infrared spectrum of gaseous $H_2^{16}O$ has three strong bands with their centers at 3642 cm^{-1}, 1595 cm^{-1}, and 3756 cm^{-1}. Using these experimental values to calculate $\lambda_1 = 4.736 \times 10^{29}$, $\lambda_2 = 0.901 \times 10^{29}$, and $\lambda_3 = 5.008 \times 10^{29}$, substitution of these values into the secular determinant and solving for the force constants yields $k_r = 7.56 \times 10^5$ dyne cm^{-1} and $k_\theta/r^2 = 0.702 \times 10^5$ dyne cm^{-1} from the quadratic part and $k_r = 7.75 \times 10^5$ dyne cm^{-1} from the linear term. Due to the simplicity of this molecule the force constants thus determined are unambiguous; however, for more complicated systems it is often not possible to obtain a unique set of constants, but rather one can obtain several sets and must be guided by knowledge of known molecular force constants and intuition to make the proper selection.

If we now use the calculated force constants obtained by use of the quadratic part of the determinant, ignore the off-diagonal terms and recalculate these two roots we obtain $\nu_1 = 3649$ cm^{-1} and $\nu_2 = 1597$ cm^{-1}. These are the magnitudes one would have for ν_1 and ν_2 if there was complete independence of the normal modes. The differences between these values and the experimental values indicates the presence of an interaction between two normal modes. The extent of this interaction is reflected in the magnitude of the off-diagonal elements and results in an increase of the displacement of the positions of the absorption bands relative to each other. When the fundamental frequencies are close together the off-diagonal terms are frequently large and the displacements become sizeable.

At this point the frequencies of the vibrations associated with the normal modes and the general form and symmetry of the normal modes have been obtained. The determination of the detailed motion of the individual atoms for each normal mode is a lengthy problem in arithmetic and will not be given in detail, but rather a short outline of the procedure will be given. Since $\underline{\mathbf{G}}^{-1}$ is the inverse of $\underline{\mathbf{G}}$ their elements are related by

$$\sum_{j=1}^{3N-6} G_{ij}(G^{-1})_{ik} = \delta_{ik} \tag{16.333}$$

This expression can be used to find the $\underline{\mathbf{G}}^{-1}$ matrix and the problem of evaluating the $(W^{-1})_{ij}$ element for the transformation $\mathbf{Q} = \underline{\mathbf{W}}^{-1}\mathbf{R}$ proceeds as for the case of the linear XY_2 molecule. The problem is more involved, however, because for H_2O the $\underline{\mathbf{G}}$ matrix is not diagonal and the $(G^{-1})_{ij}$ elements will not simply be the reciprocals of the G_{ij} elements. Performing this inversion for the terms given by Eqs. (16.319) to (16.324) gives

$$(G^{-1})_{11} = \frac{G_{22}}{G_{11}G_{22} - G_{12}^2} = 1.59 \times 10^{-24} \tag{16.334}$$

$$(G^{-1})_{12} = (G^{-1})_{21} = \frac{G_{12}}{G_{12}^2 - G_{11}G_{22}} = -6.0 - 5 \times 10^{-34} \tag{16.335}$$

$$(G^{-1})_{22} = \frac{G_{11}}{G_{11}G_{22} - G_{12}^2} = 7.1 \times 10^{-41} \tag{16.336}$$

$$(G^{-1})_{33} = \frac{1}{G_{33}} = 1.54 \times 10^{-24} \tag{16.337}$$

where the appropriate atomic masses and bond lengths for H_2O have been introduced. Note that if $G_{12} = 0$ these reduce to simple reciprocal relationships and the magnitudes of $(G^{-1})_{11}$ and $(G^{-1})_{22}$ will change by only a small ($<2\%$) amount.

Using the force constants $k_r = 7.56 \times 10^5$ dyne cm^{-1} and $k_\theta = 0.702\ r^2 = 6.44 \times 10^{-12}$ dyne cm^{-1}, the secular equations in the form of Eq. (16.251) are

$$(7.56 \times 10^5 - 1.59 \times 10^{-24}\lambda_k)A_{1k} + 6.05 \times 10^{-34}\lambda_k A_{2k} = 0 \tag{16.338}$$

$$6.05 \times 10^{-34}\lambda_k A_{1k} + (6.44 \times 10^{-12} - 7.1 \times 10^{-41}\lambda_k)A_{2k} = 0 \tag{16.339}$$

$$(7.56 \times 10^5 - 1.54 \times 10^{-24}\lambda_k)A_{3k} = 0 \tag{16.340}$$

Since the off-diagonal $\underline{\mathbf{G}}$ elements are so small compared to the diagonal ones the solution is simplified by neglecting these terms. It must be kept in mind, however, that this is not always possible, particularly when the mass of X and Y are more comparable or $\nu_1 \approx \nu_2$. Successive substituting of the three roots along with the normalization conditions, Eq. (16.252), shows the values of A_{ik} to be $A_{11} = A_{22} = A_{33} = 1$. Equation (16.255) is used to evaluate the normalizing factors as $N_1 = 7.92 \times 10^{11}$, $N_2 = 1.19 \times 10^{20}$, and $N_3 = 8.20 \times 10^{11}$. The A'_{ij} are evaluated by using Eqs. (U-27) and (U-28). The nonzero terms are $A'_{11} = -1.10 \times 10^{-9}$, $A'_{22} = -2.14 \times 10^{-26}$ and $A'_{33} = -1.07 \times 10^{-9}$. Using $(W^{-1})_{ik} = -A'_{ik}\lambda_k^{-1/2}N_k$ provides the transformation $\mathbf{Q} = \underline{\mathbf{W}}^{-1}\mathbf{S}$

$$\begin{pmatrix} Q_1 \\ Q_2 \\ Q_3 \end{pmatrix} = \begin{pmatrix} 1.27 \times 10^{-12} & 0 & 0 \\ 0 & 0.85 \times 10^{-20} & 0 \\ 0 & 0 & 1.24 \times 10^{-12} \end{pmatrix} \begin{pmatrix} S_1 \\ S_2 \\ S_3 \end{pmatrix} \tag{16.341}$$

which, when combined with the symmetry coordinates, Eqs. (16.300) to (16.302), gives for the normal coordinates

$$Q_1 = (-0.71\xi_1 - 0.55\xi_2 + 1.09\xi_4 + 0.71\xi_5 - 0.55\xi_6)10^{-12} \tag{16.342}$$

$$Q_3 = (-0.48\xi_1 - 0.37\xi_2 + 0.73\xi_3 - 0.48\xi_5 + 0.37\xi_6)10^{-12} \tag{16.343}$$

$$Q_2 = (-0.54\xi_1 + 0.70\xi_2 - 1.40\xi_4 + 0.54\xi_5 + 0.70\xi_6)10^{-12} \tag{16.344}$$

It is to be noted that for the case where the $\underline{\mathbf{G}}$ matrix is diagonal $(W^{-1})_{kk} = (G^{-1})_{kk}^{1/2}$. Although the form of the normal modes has been established the actual displacements must be found to complete the analysis. This involves finding the magnitude of the Cartesian displacements ξ_i. Solving for these displacements is a rather formidable and time-consuming project in view of the utility of the end

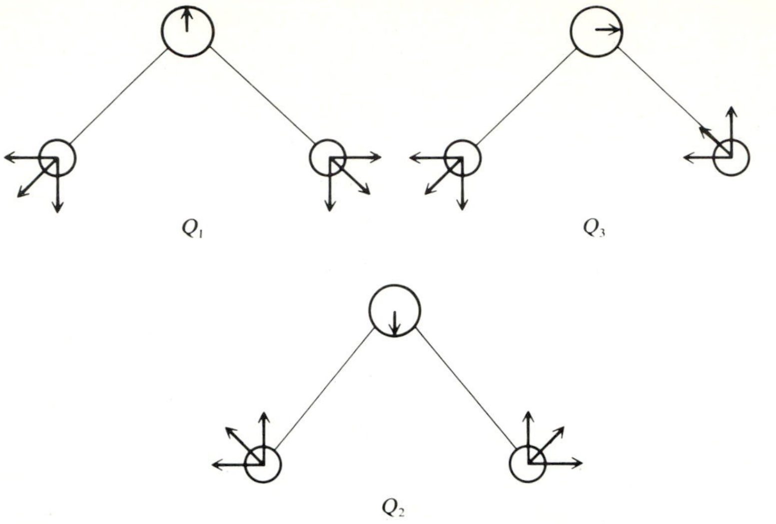

FIGURE 16.24
Normal and mass-weighted coordinates of the H_2O molecule.

results. A somewhat simpler approach is to consider that the motions of the atoms will be mass-weighted, and the relative atomic displacements can be determined by replacing each of the ξ_i's in Eqs. (16.342) to (16.344) by $\mu_i^{1/2}\zeta_i$ to obtain the Q_i in terms of the mass-weighted Cartesian coordinates ζ_i

$$Q_1 = -0.55\zeta_1 - 0.43\zeta_2 + 0.21\zeta_4 + 0.55\zeta_5 - 0.43\zeta_6 \qquad (16.345)$$

$$Q_3 = -0.37\zeta_1 - 0.29\zeta_2 + 0.14\zeta_3 - 0.37\zeta_5 + 0.29\zeta_6 \qquad (16.346)$$

$$Q_2 = -0.42\zeta_1 + 0.54\zeta_2 - 0.27\zeta_4 + 0.42\zeta_5 + 0.54\zeta_6 \qquad (16.347)$$

For any given Q_i the sum of the squares of the coefficients of the ζ_i equal unity, indicating that Q_i is normalized. The mass-weighted atomic displacements for the atoms are as shown in Fig. 16.24.

16.15 RAMAN EFFECT IN POLYATOMIC MOLECULES

The information available from vibrational Raman spectra of polyatomic molecules is analogous to that obtained from normal absorption spectra. The added advantage of using the Raman effect lies in the ability to observe in molecules with inversion centers, transitions not allowed in absorption, and thereby have experimental access to all $3N - 6$ vibrational frequencies of a molecule. In many other cases the selection rules are such that the combination of transitions observed by the two methods complement one another in aiding structure elucidation.

For a polyatomic molecule the polarizability will change with the motion of any atom in any direction. In general any component of the polarizability tensor $\alpha_{gg'}$ can be expressed in terms of the Cartesian displacements for each atom or in terms of the atom displacements associated with the normal modes. Letting x_{ij}, y_{ij}, z_{ij} be the displacements for atom i, in the jth normal mode the polarizability tensor components are given by

$$\alpha_{gg'} = \alpha^\circ_{gg'} \sum_{i=1}^{N} \sum_{j=1}^{N'} x_{ij}\left(\frac{\partial \alpha_{gg'}}{\partial x_{ij}}\right)_0 + y_{ij}\left(\frac{\partial \alpha_{gg'}}{\partial y_{ij}}\right)_0 + z_{ij}\left(\frac{\partial \alpha_{gg'}}{\partial z_{ij}}\right)_0 \qquad (16.348)$$

for a system of N atoms and N' normal modes. It is more convenient to express the changes in the $\alpha_{gg'}$ in terms of the normal coordinates, Q_j, in which case the polarizability tensor components are

$$\alpha_{gg'} = \alpha^\circ_{gg'} + \sum_{j=1}^{N} \left(\frac{\partial \alpha_{gg'}}{\partial Q_j}\right)_0 Q_j \qquad (16.349)$$

where the Q_j are related to normal mode frequencies ν_j by

$$Q_j = Q_j^\circ \cos(2\pi\nu_j t + \delta_j) \qquad (16.350)$$

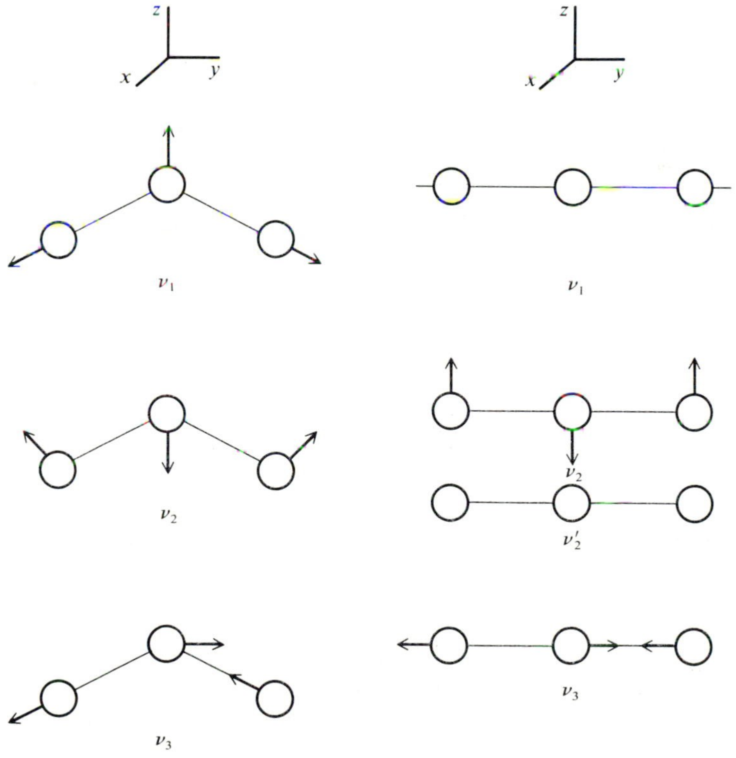

FIGURE 16.25
Normal mode vibrations of (*a*) H_2O and (*b*) CO_2.

The polarizability tensor components are then

$$\alpha_{gg'} = \alpha^\circ_{gg'} + \sum_{j=1}^{N} \left(\frac{\partial \alpha_{gg'}}{\partial Q_j}\right)_0 Q^\circ_j \cos\,(2\pi\nu_j t + \delta_j) \tag{16.351}$$

Using Eq. (13.50) the three induced dipole components are

$$\mu_{Ig} = \alpha_{gg}E_g + \alpha_{gg'}E_{g'} + \alpha_{gg''}E_{g''} \tag{16.352}$$

where g, g', g'' are x, y, z taken in cyclic order. The components of the irradiation field are

$$E_g = E^0_g \cos 2\pi\nu t \tag{16.353}$$

Substituting Eqs. (16.351) and (16.353) into Eq. (16.352) we obtain, taking the x component as an illustration,

$$\begin{aligned}\mu_{Ix} &= (\alpha^\circ_{xx}E^\circ_x + \alpha^\circ_{xy}E^\circ_y\alpha^\circ_{xz}E^\circ_z) \cos 2\pi\nu t \\ &\quad + \sum_{j=1}^{N'} \left\{ \left(\frac{\partial \alpha_{xx}}{\partial Q_j}\right)_0 \xi^\circ_j E^\circ_x + \left(\frac{\partial \alpha_{xy}}{\partial Q_j}\right)_0 \xi^0_j E^0_y + \left(\frac{\partial \alpha_{xz}}{\partial Q_j}\right)_0 \xi^0_j E^0_z \right. \\ &\quad \left. \times \left[\tfrac{1}{2}\cos 2\pi(\nu + \nu_i)t + \tfrac{1}{2}\cos 2\pi(\nu - \nu_i)t\right] \right\}\end{aligned} \tag{16.354}$$

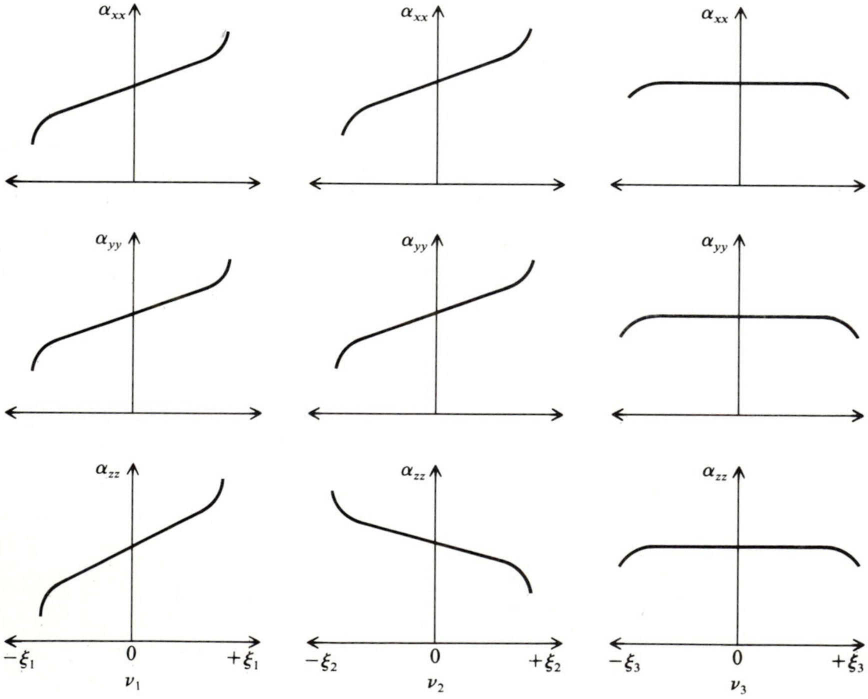

FIGURE 16.26
Variation of principal polarizability tensor components with vibration for H_2O.

Consideration of the other two components gives similar expressions and leads to the conclusion that there will be scattered frequencies of $\nu \pm \nu_j$ if any of the six slopes $(\partial \alpha_{gg'}/\partial Q_j)_0$ are nonzero for the jth normal mode of vibration.

For H_2O and CO_2 the displacements that occur during the normal mode of vibration of each are referenced in Fig. 16.25. The changes in the axial polarizability tensor components of H_2O associated with each normal mode are shown qualitatively in Fig. 16.26. Figure 16.27 illustrates the change in the polarizability ellipsoid of H_2O with each normal vibration. It is clearly seen that the symmetric stretching and bending motion have finite slopes for $(\partial \alpha_{gg}/\partial Q_j)$ at the equilibrium position. The antisymmetric stretch, although appearing to have a zero slope, does in fact produce a change in polarizability by virtue of a

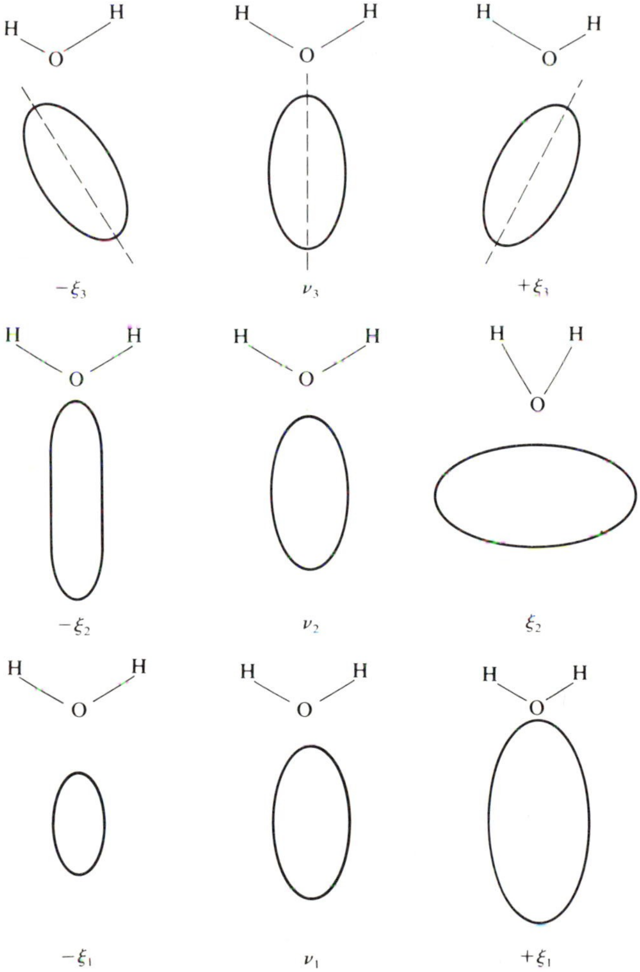

FIGURE 16.27
Variation of polarizability ellipsoid with vibration for H_2O.

change in the axis of the polarizability ellipsoid as qualitatively shown in Fig. 16.27. We thus see that all three of the normal modes of vibration for H_2O should produce Raman scattering. In practice one does not always observe predicted spectral lines, for if all of the slopes $(\partial\alpha_{gg'}/\partial Q_j)$ associated with a particular normal mode are small, the emitted radiation may be of insufficient intensity to detect.

An example of a molecule possessing a center of symmetry where the vibrational Raman effect is not observed for all Q_j is afforded by the CO_2 molecule. The variations of the axial polarizability tensor components and the polarizability ellipsoids are shown in Figs. 16.28 and 16.29, respectively. In this case the symmetric stretching has a finite slope, but the other two modes have zero slopes at the equilibrium position and hence will not give rise to Raman scattering. Note that the two vibrations of CO_2 which are Raman inactive are those which destroy the original symmetry of the molecule $D_{\infty h}$.

As with absorption vibrational spectra the presence of anharmonicity will lead to a breakdown in the normal selection rules, and overtone and combination frequencies are possible if allowed by the presence of nonzero polarizability tensor components. In general, however, the Raman intensities of these transitions are low and they will be very weak compared to the fundamentals.

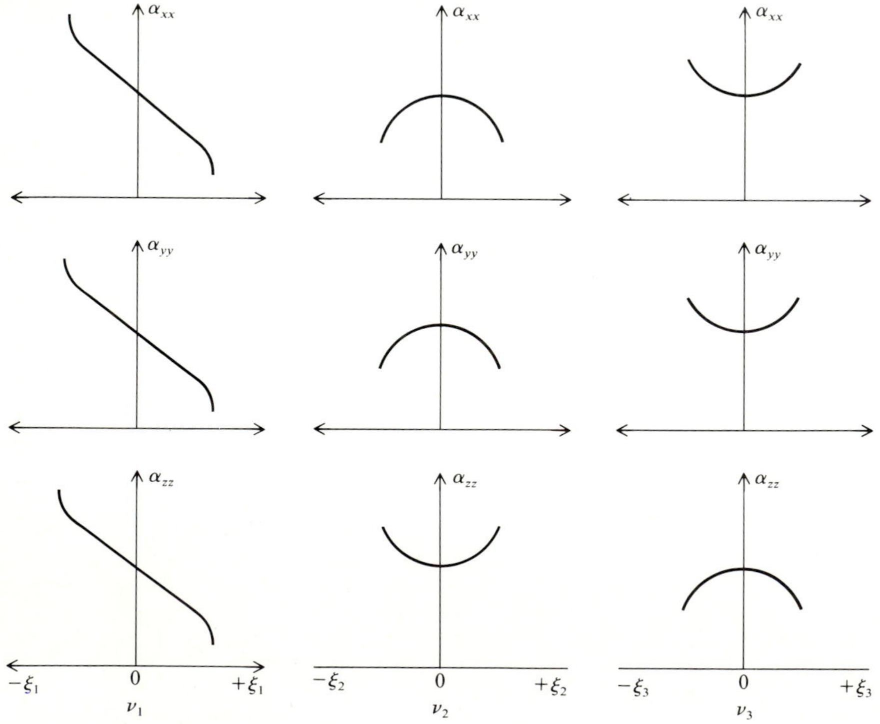

FIGURE 16.28
Variation of principal polarizability tensor components with vibration for CO_2.

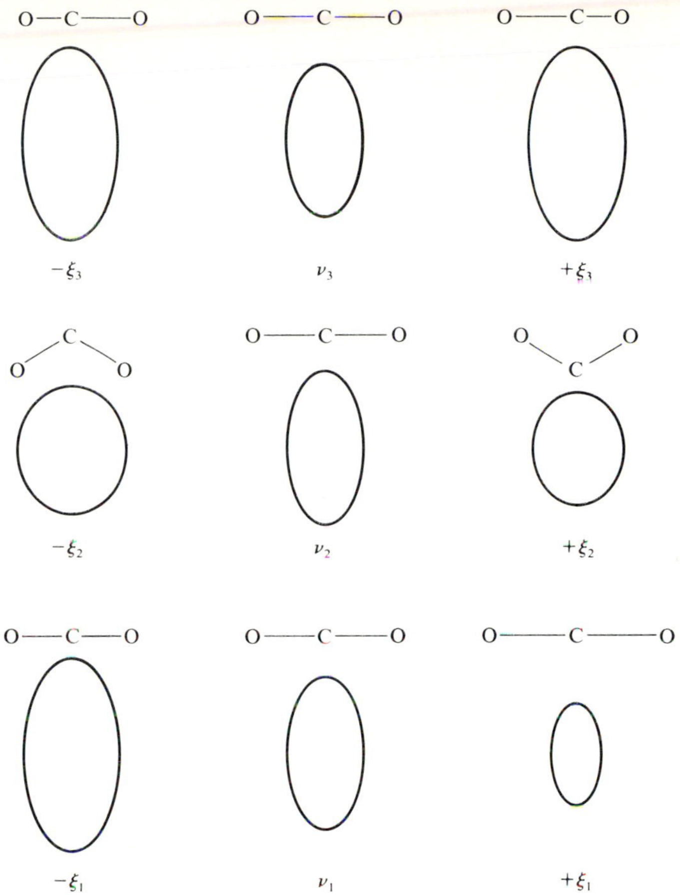

FIGURE 16.29
Variation of polarizability ellipsoid with variation for CO_2.

REFERENCES

General

Allen, H. C., and P. C. Cross, *Molecular Vib-Rotors*, John Wiley, New York, NY, 1963.

Bunker, P. R., *Molecular Symmetry and Spectroscopy*, Academic Press, New York, NY, 1979.

Herzberg, G., *Molecular Spectra and Molecular Structure; II. Infrared and Raman Spectra of Polyatomic Molecules*, Van Nostrand Reinhold, New York, NY, 1945.

King, G. W., *Spectroscopy and Molecular Structure*, Holt, Reinhart & Winston, New York, NY, 1964.

Koningstein, J. A., *Introduction to the Theory of the Raman Effect*, D. Reidel, Dordrecht, NE, 1972.

Long, D. A., *Raman Spectroscopy*, McGraw-Hill, New York, NY, 1977.

Wilson, E. B., J. C. Decius, and P. C. Cross, *Molecular Vibrations*, McGraw-Hill, New York, NY, 1955.

Woodward, L. A., *Introduction to the Theory of Molecular Vibrations and Vibrational Spectroscopy*, Oxford University Press, Oxford, GB, 1972.

PROBLEMS

16.1. Consider the following molecules to be constrained to one dimensional motion (along the symmetry axis) and calculate the classical vibrational frequencies using the approximate internuclear distances and force constants

Molecule (1-2-3)	λ_{12}(A)	λ_{23}(A)	k_{12}(mdyn/A)	k_{23}(mdyn/A)
$^{16}O-^{12}C-^{16}O$	1.2	1.2	15	15
$^{16}O-^{12}C-^{32}S$	1.2	1.5	15	8
$^{32}S-^{12}C-^{32}S$	1.6	1.6	8	8

16.2. For the three systems in Prob. 16.1 calculate the actual Cartesian displacement coordinates for the three examples given.

16.3. For the listed point groups determine the irreducible representations for the indicated symmetrical direct products. (a) C_{3v}, $(e)^2$; (b) D_{3h}, $(e')^2$; (c) D_{6h}, $(e_{1g})^2$; (d) O_h, $(e_g)^2$, $(t_{1g})^2$.

16.4. Using the *FG* matrix method find the numerical values of k and Q_1 for *HD* ($\nu = 3632\ \text{cm}^{-1}$) $^{79}Br^{35}Cl$ ($\nu = 439.5\ \text{cm}^{-1}$), and $H^{79}Br$ ($\nu = 2558.76\ \text{cm}^{-1}$).

16.5. The observed vibrational frequencies for CS_2 are at 397, 658, and 1533 cm^{-1}. Evaluate the roots of the *FG* matrix and determine the force constants and normal modes.

16.6. The observed vibrational frequencies for F_2O are at 461, 836, and 929 cm^{-1}. Determine the roots of the *FG* matrix and find the force constants and normal modes.

16.7. Devise a set of internal symmetry coordinates and write the nonvanishing elements of the G' matrix for H_2CO if it has the planar configuration

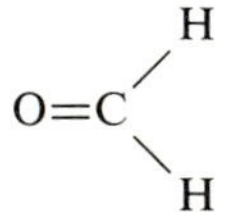

16.8. The N_2F_2 molecule can exist as two isomers, both of which are planar.

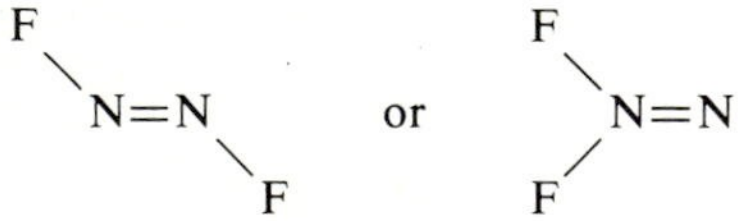

(a) For each isomer determine the symmetry species of the normal modes and see if you can assign atom displacements to each mode using only symmetry considerations. Determine the IR and Raman activity of each mode.

(b) The internal coordinates for the trans isomer are

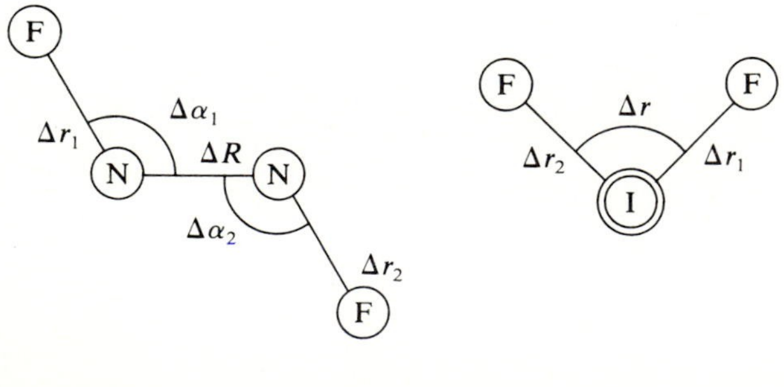

and the molecular parameters are $r_1 = r_2 = 1.44A$, $R = 1.25A$, $\alpha = 115°$ and $\gamma = 0°$. The internal symmetry coordinates are

$$R_1 = \tfrac{1}{2}(\Delta r_1 + \Delta r_2)$$
$$R_2 = \Delta R$$
$$R_3 = \tfrac{1}{2}(\Delta\alpha_1 + \Delta\alpha_2)$$
$$R_4 = \tfrac{1}{2}(\Delta r_1 - \Delta r_2)$$
$$R_5 = \tfrac{1}{2}(\Delta\alpha_1 - \Delta\alpha_2)$$
$$R_6 = \Delta\gamma$$

Determine the symmetry species of each symmetry coordinate.

(*c*) Determine the transformation matrix $\underline{\mathbf{U}}$ for the transformation

$$\underline{\mathbf{G}}' = \underline{\mathbf{UGU}}$$

(*d*) Determine the $\underline{\mathbf{G}}$ matrix for N_2F_2 (trans)

(*e*) Find the $\underline{\mathbf{G}}'$ matrix. (It will be easier at this point if you use the numerical values for bond lengths and angles.)

(*f*) The nonzero $\underline{\mathbf{F}}'$ matrix elements are given by

$F_{11} = f_r + f_{rr}$	$f_r = 4.60$ mdyn/A
$F_{12} = F_{21} = 2f_{rR}$	$f_R = 11.10$ mdyn/A
$F_{13} = F_{31} = rf'_{r\alpha}$	$f_{\alpha'} = 1.02$ mdyn/A
$F_{22} = f_R$	$f_{rr} = 0$ mdyn/A
$F_{23} = F_{32} = 2\ RF'_{R\alpha}$	$f_{rR} = 0.95$ mdyn/A
$F_{33} = rR(f'_\alpha + f'_{\alpha\alpha})$	$f'_{\alpha\alpha} = 0$ mdyn/A
$F_{44} = f_r - f_{rr}$	$f'_{r\alpha} = 0.42$ mdyn/A
$F_{45} = F_{54} = rf'_{r\alpha}$	$f'_{R\alpha} = 0.20$ mdyn/A
$F_{55} = rR(f'_\alpha - f'_{\alpha\alpha})$	$f'_\gamma = 0.29$ mdyn/A
$F_{66} = rRf'_r$	

Calculate the roots of $|G'F' - I\lambda| = 0$ and assign each root to a normal mode.

16.9. For the acetylene molecule constrained to move only along the C_∞ axis

(*a*) Write the Newtonian equations for motion in terms of a mass-weighted Cartesian coordinate set.

(*b*) Using the Newtonian equations derive the secular equations and describe the motion of the system.

(*c*) Solve the secular determinant for the eigenvalues and, given that $k_{CH} = 5$ md/Λ and $k_{cc} = 16$ md/Λ, evaluate the vibrational frequencies.

(*d*) Evaluate the amplitude factors, Λ_{ij}, for each eigenvalue and write the expressions for the normal coordinates of the system.

(*e*) Construct diagrams to show the relative displacements of the atoms in each normal mode.

CHAPTER
17

ROTATION OF POLYATOMIC MOLECULES

17.1 INTRODUCTION

Chapters 11 through 14 discussed the nature of the energy levels and spectra of diatomic molecules undergoing rotation, vibration, and the interaction of the two motions. It was shown that in the analysis of high-resolution rotational and vibrational spectra the interaction of the two motions cannot be neglected.

This chapter will review several aspects of rotational motion in polyatomic molecules. By this time the reader will have gained sufficient background that the treatment will not be as detailed, but will be more in the nature of an outline of the development of certain relationships along with references to more extensive sources. The discussion of the various types of molecules will initially assume the rigid rotor approximation and then extend the discussion by introduction of other interactions as perturbations.

17.2 RIGID LINEAR MOLECULES

The rigid linear molecule is analogous to a diatomic molecule in that it consists of a set of mass points lying on a straight line. It will have a zero moment of inertia about the linear axis and two equal moments about a pair of orthogonal axes lying perpendicular to the linear axis. Denoting the moment of inertia about one of these orthogonal axes as I_B the classical energy of rotation is

$$E = \frac{P^2}{2I_B} \tag{17.1}$$

The expression for the quantized energy levels of a rigid polyatomic linear molecule are identical to those for a rigid diatomic rotor. The matrix elements of the Hamiltonian will be

$$\langle J, K, M|\mathscr{H}|J, K, M\rangle = \frac{1}{2I_B}\langle J, K, M|\mathbf{P}^2|J, K, M\rangle \tag{17.2}$$

The eigenvalues of P^2 will be given by

$$\langle J, K, M|\mathbf{P}^2|J, K, M\rangle = \hbar^2 J(J+1) \tag{17.3}$$

and the energy of the rigid linear rotor becomes

$$E_J = \frac{\hbar^2}{2I_B}J(J+1) = hBJ(J+1) \tag{17.4}$$

The wavefunctions for the linear rotor are the same as those for the diatomic rotor.

A distinct difference exists between the evaluation of molecular parameters for a diatomic and a triatomic linear molecule. Where the internuclear separation of the atoms in a diatomic molecule could be directly determined from the moment of inertia, the same is not true for a triatomic linear molecule since the moment of inertia is dependent on two internuclear separations. The relationship of internuclear distances and masses to the moment of inertia for a triatomic molecule may be formulated by referring to Fig. 17.1. The moment of inertia is

$$I = \sum_i m_i r_i^2 = m_1 z_1^2 + m_2 z_2^2 + m_3 z_3^2 \tag{17.5}$$

The coordinates, with respect to the center of mass z_i, are related to the internuclear separations by

$$r_1 = |z_1| + |z_2| \tag{17.6}$$

$$r_2 = |z_3| - |z_2| \tag{17.7}$$

Even by using the definition of the center of mass, $m_1 z_1 + m_2 z_2 + m_3 z_3 = 0$, the

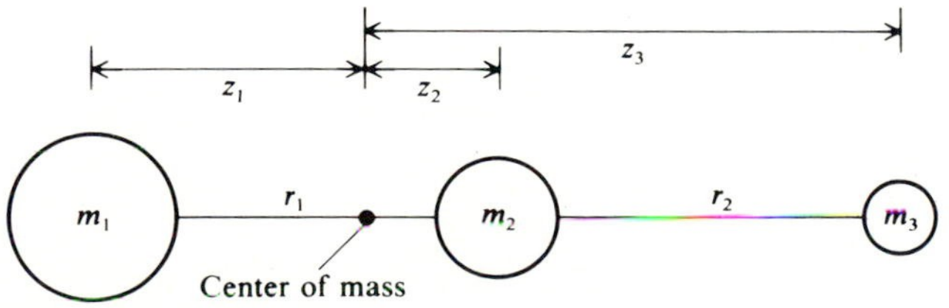

FIGURE 17.1
Linear triatomic molecule.

moment of inertia (which is the experimentally determined property) will be a function of two parameters, either the internuclear separations or a pair of center-of-mass coordinates.

Consideration of the more general properties of the inertial dyadic (see App. L) will provide a method for obtaining complete structural data for this and other molecular systems [1]. The moments of inertia and product of inertia, relative to an arbitrary coordinate system whose origin is not at the center of mass and whose axes are not collinear with the principal axis system, and for a molecule containing i atoms may be written as

$$I_{xx} = \sum_i m_i(y_i^2 + z_i^2) - \frac{\left(\sum_i m_i y_i\right)^2}{M} - \frac{\left(\sum_i m_i z_i\right)^2}{M} \tag{17.8}$$

$$I_{zy} = I_{yz} = -\sum_i m_i y_i z_i + \frac{\left(\sum_i m_i y_i\right)\left(\sum_i m_i z_i\right)}{M} \tag{17.9}$$

where

$$M = \sum_i m_i \tag{17.10}$$

The other terms I_{yy}, I_{xy}, etc., are obtained by cycling x, y, z. Selecting the origin as the center of mass and setting the directions of the axes so that the z axis is collinear with the molecular axis, the product of inertia will vanish and the moments of inertia are the principal ones

$$I = I_{xx} = I_{yy} = I_x = I_y = m_1 z_1^2 + m_2 z_2^2 + m_3 z_3^2 \tag{17.11}$$

$$I_{zz} = I_z = 0 \tag{17.12}$$

It is the principal moment I which is experimentally determined. If an isotopic atom of mass $M_3' = M_3 + \Delta M$ is substituted for M_3 the moment of inertia of the molecule will be changed to I'. If the same coordinate system is used to describe the molecule, I'_{yy} is no longer equal to the principal moment I since isotopic substitution has shifted the center of mass of the molecule. The substitution was on the z axis so the products of inertia are zero and I'_{yy} is given by

$$I'_{yy} = m_1 z_1^2 + m_2 z_2^2 + (m_3 + \Delta m_3) z_3^2 - \frac{[m_1 z_1 + m_2 z_2 + (m_3 + \Delta m_3) z_3]^2}{(M + \Delta m_3)} \tag{17.13}$$

However, since the origin and atom positions have not changed, it follows that $m_1 z_1 + m_2 z_2 + m_3 z_3 = 0$. Therefore

$$I'_{yy} = I_{yy} + \Delta m_3 z_3^2 - \frac{(\Delta m_3 z_3)^2}{(M + \Delta m_3)} \tag{17.14}$$

The products I'_{xy}, etc., are all zero and the moment I'_{yy} is the principal moment of the isotopically substituted molecule. The difference in the two moments,

TABLE 17.1
Rotational transitions and parameters for OCS

Molecule	Transition† frequency, MHz	Mass, amu	B, MHz‡	I, amu-A^2	μ_i, amu
$^{16}O^{12}C^{32}S$	24,325.9	59.96699	6,081.48	83.126	
$^{16}O^{12}C^{33}S$	24,020.2	60.96638	6,005.05	84.184	0.98300
$^{16}O^{12}C^{34}S$	23,731.3	61.96278	5,932.83	85.209	1.9315
$^{16}O^{13}C^{32}S$	24,237.7	60.97034	6,061.92	83.395	0.98683
$^{16}O^{14}C^{32}S$	24,173.0	61.97	6,043.3	83.65	1.9344
$^{18}O^{12}C^{32}S$	22,819.3	61.97123	5,704.83	88.615	1.9394

† All transitions are for the ground vibrational state.

‡ Calculated by neglecting zero point vibrations.

which can be determined experimentally, is given by

$$I' - I = \mu_3 z_3^2 \tag{17.15}$$

where $\mu_3 = M\Delta m_3/(M + \Delta m_3)$. The position of atom no. 3 relative to the center of mass is

$$|z_3| = \frac{1}{\sqrt{\mu_3}}(I' - I)^{1/2} \tag{17.16}$$

Isotopic substitution of other two atoms and the use of the relationship $\sum_i m_i z_i = 0$ provides a method to determine the coordinates z_1 and z_2. Equations (17.6) and (17.7) can then be used to obtain the internuclear distances.

An application of this method is illustrated using the data of Townes, Holden, and Merritt [2] for carbonyl sulfide, OCS. The observed rotational transitions for several isotopically substituted species of OCS are given in Table 17.1. Also included are other parameters necessary for the evaluation of the bond length. The μ_i values given are relative to that of $^{16}O^{12}C^{32}S$ being unity. All frequencies given are for the $J = 2 \leftarrow 1$ transition of each species. By using the

TABLE 17.2
Structural parameters for OCS

Molecular pairs	$z_1(A)$	$z_2(A)$	$z_3(A)$	$r_1 = r_{CS}(A)$	$r_2 = r_{CO}(A)$
$^{16}O^{12}C^{33}S^{16}$-$O^{13}C^{32}S$	1.0374	0.5221	1.6819	1.5595	1.1598
$^{16}O^{12}C^{34}S^{16}$-$O^{13}C^{32}S$	1.0384	0.5221	1.6839	1.5605	1.1618
$^{16}O^{12}C^{33}S^{16}$-$O^{14}C^{32}S$	1.0374	0.518	1.685	1.555	1.167
$^{16}O^{12}C^{34}S^{16}$-$O^{14}C^{32}S$	1.0384	0.518	1.687	1.556	1.169
$^{16}O^{12}C^{33}S^{18}$-$O^{12}C^{32}S$	1.0374	0.5216	1.6823	1.5590	1.1607
$^{16}O^{12}C^{34}S^{18}$-$O^{12}C^{32}S$	1.0384	0.5243	1.6823	1.5627	1.1580
$^{16}O^{13}C^{32}S^{18}$-$O^{12}C^{32}S$	1.0376	0.5221	1.6823	1.5597	1.1602
$^{16}O^{14}C^{32}S^{18}$-$O^{12}C^{32}S$	1.036	0.518	1.6823	1.554	1.164
Average	1.0376	0.5208	1.6834	1.5583	1.1626
Average deviation	±0.0006	±0.0020	±0.0014	±0.0026	±0.0031

moment of inertia for $^{16}O^{12}C^{32}S$ and those of pairs of the other species the intermolecular distances are evaluated as shown in Table 17.2. The bond lengths shown in this table reflect the accuracies of the experimental frequencies, ±0.1 MHz. The table headings are keyed to Figure 17.1 with $M_1 = S$, $M_2 = C$ and $M_3 = O$. Even allowing for the errors in the frequencies of the transitions and for the $^{14}C = 14.0$ mass in the $^{16}O^{14}C^{32}S$ species the average deviations are larger than might be expected. This is due to the fact that no corrections for zero-point vibrations were included in the calculation.

17.3 NONRIGID LINEAR MOLECULES

The results of the calculation of the internuclear distances in OCS have shown that errors can be introduced into calculated values of bond parameters by neglect of zero-point vibrations. This is also evident when one compares the deviations of these calculations to those at the bond parameters given in Table 12.3 for AgCl, where zero-point vibrations were considered. This section will consider the nature of the spectra of triatomic linear molecules due to the nonrigidity.

For any real molecule the rotational spectrum will be affected by the vibrational state, anharmonicity of the vibrations, Coriolis interactions, *l*-type doubling, and Fermi resonance. Detailed discussions regarding these interactions have been presented by H. H. Nielsen [3], A. H. Nielsen [4], and D. M. Dennison [5]. The results of their work and the relationships to spectra will be presented, but the details of the developments will be left for the reader to obtain from the original literature. The discussion of these phenomena will utilize previous results for diatomc vibrators and will be directed only at the triatomic linear molecule.

The effect of centrifugal distortion to a first approximation is analogous to the diatomic case. Since the rotation of the linear molecule is about an axis which is perpendicular to the molecular axis the effect of centrifugal distortion will be to elongate the molecule along its axis. The rotational energy of a linear molecule, allowing for centrifugal distortion and neglecting vibrational interactions will be

$$E_J = B_e J(J + 1) - D_e J^2(J + 1)^2 \qquad (17.17)$$

It is useful to note that the effect of centrifugal distortion can be considered to modify the reciprocal moment. Hence

$$E_J = B_J J(J + 1) \qquad (17.18)$$

where

$$B_J = B_e - D_e J(J + 1) \qquad (17.19)$$

The occurrence of two bonds in the molecule plus the ability of the molecule to bend from its normal linear shape precludes the establishing of a simple relationship between D_e and the vibrational frequency as was developed for the diatomic case. A similar analogy occurs for rotation-vibration interaction if the vibrations are considered to be harmonic. For a linear triatomic molecule there are two linear stretching modes and one doubly degenerate bending mode. Each of these modes can be considered to contribute to changing the moment of inertia of the molecule from that of a rigid rotor. This change can be expressed, following the form used for the diatomic case, as

$$E_{n_1 n_2, \ldots, n_N, J} = B_J(J+1) - \sum_i \alpha_i \left(n_i + \frac{1}{2}\right) J(J+1) \tag{17.20}$$

where the summation is over all vibrational modes. When there are degenerate modes this expression becomes

$$E_{n_1, n_2, \ldots, n_N, J} = B_J(J+1) - \sum_i \alpha_i \left(n_i + \frac{d_i}{2}\right) J(J+1) \tag{17.21}$$

where d_i is the degeneracy of the ith normal mode. Just as in the case of the diatomic molecule, the ground vibrational state, $n_1 = n_2 = \cdots n_i = 0$, will contribute to the effective moment of inertia and

$$E_{0,J} = \left[B_{0,J} - \sum_i \frac{a_i d_i}{2}\right] J(J+1) \tag{17.22}$$

where α_i is the rotation-vibration interaction constant for the ith normal mode.

Equation (17.21) indicates that, insofar as the assumptions made to data are valid, there will be a different effective reciprocal moment for each different vibrational state of the molecule. Looking back at Fig. 16.13, we can see that there would be a manifold of rotational states and a series of rotational transitions for each of the states shown in this diagram. The relative intensities of these series of transitions will directly reflect the population of the vibrational states.

It is generally the situation with molecular spectra that the experimental observations cannot be fully explained by use of only a few simple interactions. Even the triatomic linear molecule is no exception. The two major interactions which must be included are the effects of anharmonicity and the l-type doubling resulting from the Cariolis effect.

If the potential function for the molecule is of the form

$$V = V_{\text{har}} + V_{\text{anhar}} \tag{17.23}$$

where

$$V_{\text{har}} = \tfrac{1}{2}[k_1\xi_1^2 + k_2\xi_2^2 + k + 2k_{12}\xi_1\xi_2 + 2k_{13}\xi_1\xi_3 + 2k_{23}\xi_2\xi_3 + k_4(x^2+y^2)] \tag{17.24}$$

$$V_{\text{anhar}} = k_{111}\xi_1^3 + k_{113}\xi_1^2\xi_3 + k_{112}(\xi_{21}^2 + \xi_{22}^2)\xi_1 + k_{133}\xi_1\xi_3^2 + k_{223}(\xi_{21}^2 + \xi_{22}^2)\xi_3 + k_{333}\xi_3^3 \tag{17.25}$$

and the ξ_i are the displacement coordinates,[1] then the solution of the Schrödinger

[1] The $x^2 + y^2$ term allows for the out-of-line bending.

equations will give the following energy level expression [2]

$$E_{Jl} = E_{Jl}(\text{rot}) + E_{vl}(\text{vibration}) \tag{17.26}$$

where

$$E_{Jl} = h\{[J(J+1) - l_2^2]B_n - [J(J+1) - l_2^2]^2 D\} \tag{17.27}$$

$$\begin{aligned} E_{vl} = h\{&(v_1 + \tfrac{1}{2})\nu_1 + (v_2 + 1)\nu_2 + (v_3 + \tfrac{1}{2})\nu_3 + \chi_{11}(v_1 + \tfrac{1}{2})^2 \\ &+ \chi_{22}(v_2 + 1)^2 + \chi_{33}(v_3 + \tfrac{1}{2})^2(\chi_{21} + \chi_{12})(v_1 + \tfrac{1}{2})(v_2 + 1) \\ &+ (\chi_{13} + \chi_{31})(v_1 + \tfrac{1}{2})(v_3 + \tfrac{1}{2}) + (\chi_{23} + \chi_{32})(v_2 + 1)(v_3 + \tfrac{1}{2}) + \chi_{l_2 l_2} l_2^2\} \end{aligned} \tag{17.28}$$

The constants B_n, D, and ν_i are in units of s^{-1}, and the χ_{ij} are anharmonicity constants. The centrifugal distortion constant can be related to the displacement coordinates. The quantum number l_i which occurs in the solution of the Schrödinger equation is a measure of the angular momentum about the molecular axis. This momentum is due to the molecule being bent out of line during the vibrational motion associated with ν_2. This resulting momentum will modify the overall rotational energy of the molecule. If, for simplicity, the anharmonic terms in Eq. (17.28) are ignored the energy becomes

$$\begin{aligned} E_{Jv} = \{&[J(J+1) - l_2^2]B_v - [J(J+1) - l_2^2]^2 D + (v_1 + \tfrac{1}{2})\nu_1 \\ &+ (v_2 + 1)\nu_2 + (v_3 + \tfrac{1}{2})\nu_3\}h \end{aligned} \tag{17.29}$$

For a transition from $J \to J + 1$, where all molecules remain in the same vibrational state the transition is given by

$$\nu = 2B_v(J+1) - 4D(J+1)[(J+1)^2 - l_2^2] \tag{17.30}$$

where ν is expressed in hertz. As we shall see later in this chapter this expression is similar to that of a rigid symmetric top molecule with l_2 replaced by K, the quantum number denoting the momentum about the symmetry axis. Thus the energy of the nonrigid linear molecule is similar to that encountered for the diatomic molecule except for the terms introduced by the additional bending freedom of the molecule. The solution of the Schrödinger equation has been reviewed in detail by Nielsen [6-8], and the resulting wavefunctions for a linear XYZ molecule are of the form

$$\psi_{Jlv} = \psi_{vl}(\rho, \chi)\psi_{Jl}(\theta, \phi) \tag{17.31}$$

where

$$\psi_{vl}(\rho, \chi) = N_v \rho^{|l|} \exp\left\{\frac{\rho^2}{2} + il\chi\right\} F^{|l|}_{(1/2)(v+|l|)}(\rho^2) \tag{17.32}$$

$$N_{nl} = 2\left[\frac{(v - |l|)!}{2}\right]^{1/2}\left[\frac{(v + |l|)!}{2}\right]^{-3/2}\left(\frac{\pi m \nu_2}{n}\right)^{1/2} \tag{17.33}$$

$$F^{|l|}_{(1/2)(v+|\rho|)}(\rho^2) = \frac{d^{|l|}}{(d\rho^2)^{|l|}}\left\{e^{\rho^2} \frac{d^{(1/2)(v+|l|)}}{(d\rho^2)^{(1/2)(v+|l|)}}[e^{\rho^2(v+|l|)}]\right\} \tag{17.34}$$

(an associate Laguerre polynomial)

$$\rho = (2\pi\sqrt{\nu_2/h})(q_{21}^2 + q_{22}^2)^{1/2} \tag{17.35}$$

$$m = \text{oscillator mass} \tag{17.36}$$

$$\chi = \text{polar angle}$$

$$l = v, v-2, v-4, \ldots, -v \tag{17.37}$$

and the $\psi_{Jl}(\theta, \phi)$ are the same as the symmetric top wavefunctions given by Eq. (17.85) in Sec. 17.4, but with $J = J$, $K = l$, and $M = M$.

Were this the end of the development with regard to linear molecules then their spectra would be like those of diatomic molecules except for the larger number of terms contributing to B_n. However, a linear molecule which is undergoing simultaneous bending and rotation will possess different energies depending on whether the bending tends to be with or opposed to the direction of rotation. This particular interaction of rotation and vibration leads to a removal of the degeneracy in l_2 shown in Eq. (17.29) and gives rise to the phenomenon referred to as l-type doubling. The coupling of rotation and vibration is not limited to the out-of-line bending modes and is the origin of the interaction referred to as *Coriolis forces*. In general, the rotation not only couples with a normal mode but also can serve as an interaction to couple two normal modes. Analytically this results in the terms $\zeta_{21}^2[1/(\nu_1^2 - \nu_2^2)]$ and $\zeta_{23}^2[1/(\nu_2^2 - \nu_3^2)]$, which occur in the relations for the α_i. If ν_1 or ν_3 lies close to ν_2 then these terms can be large and the values of the α_i's, which measure the rotational-vibrational interactions, are large. The proportionality constants for this interaction, ζ_{21} and ζ_{23}, are referred to as the *Coriolis coupling constants.*

Because of the unique nature of the Coriolis effect a more detailed description is warranted. The Coriolis force is an apparent force that occurs since we are studying the molecule in a rotating coordinate system. It can be illustrated by a simple mechanical system. Consider a pair of metal balls held apart by a pinned telescoping rod surrounding a spring. This system is suspended by a string located at the center of mass and the system is given a spin to produce an angular velocity $\boldsymbol{\omega}$. This is illustrated in Fig. 17.2*a*. If, by the use of some magical demon, the pins are removed from the rotating rod, producing the configuration shown in Fig. 17.2*b*, one would observe an increase in the angular velocity to $\boldsymbol{\omega}'$. This constitutes the familiar concept of conservation of angular momentum. If, however, our attention is focused on that part of the system within the solid circle it is observed, since there is no change of mass or position, that the only change is that in the angular velocity with no accompanying momentum change. If this section of the rotating body is moving faster then there must have been some apparent force exerted on it. The centrifugal force in the body is directed radially so it cannot be the apparent force. This is the Coriolis force and has the property of tending to move an object in a rotating system in a direction perpendicular to the direction of motion. The Coriolis force, like the centrifugal force, is only an apparent force, but it shows that to move a particle radially in

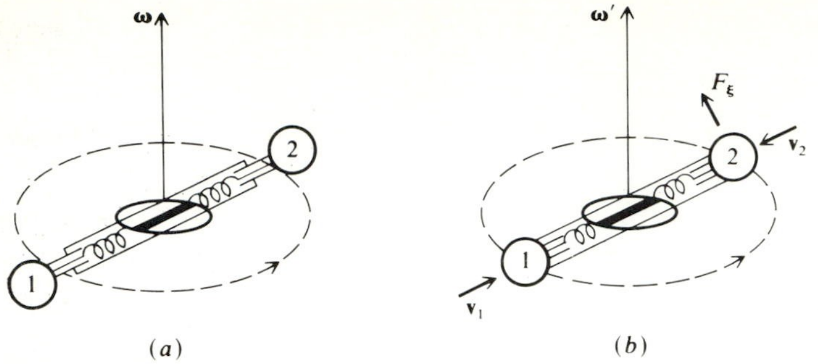

FIGURE 17.2
Conservation of angular momentum and coriolis forces.

a rotating system one must also push it tangentially. The magnitude of this force is given by

$$F_\zeta = \sum 2m_i\mathbf{v}_i \times \boldsymbol{\omega} \tag{17.38}$$

where $\mathbf{v}_i$ is the linear velocity of mass m_i in the moving coordinate system which has an angular velocity $\boldsymbol{\omega}$.

For the diatomic molecule the Coriolis interaction leads to the conservation of angular momentum as the molecule undergoes a periodic change in angular momentum due to the vibration. For triatomic linear molecules it is illustrated in Fig. 17.3. The direction of rotation is shown by the curved arrow with the angular velocity vector $\boldsymbol{\omega}$ directed upward from the plane. The normal mode displacements, the $\mathbf{v}_i$ vectors somewhat exaggerated, are represented by linear solid arrows. The resulting Coriolis force on each atom is indicated by a dotted arrow. Looking at the direction of the Coriolis forces associated with the ν_3

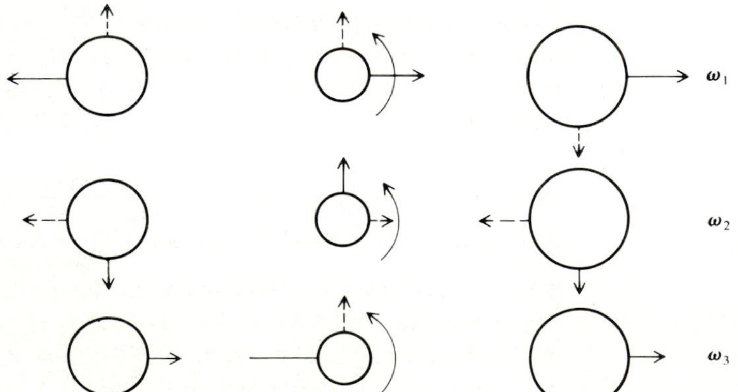

FIGURE 17.3
Coriolis interactions in a linear XYZ molecule.

vibration, we see that they correspond to the directions of the atom motions for one member of the ν_2 degenerate vibrational mode. The net result is that rotation of the molecule tends to couple the ν_2 and ν_3 (also the ν_1 and ν_2) modes. As was mentioned before, this coupling is apparent in the analytical formulation of the rotational-vibrational coupling constants.

There is yet another factor to consider. Although the two degenerate vibrational motions associated with ν_2 are equivalent in a nonrotating system, rotation will interact differently with each. If the degenerate vibration has its displacement vectors perpendicular to those shown in Fig. 17.3, then they will be parallel to $\boldsymbol{\omega}$ and the Coriolis force will vanish. Hence, the coupling of the degenerate vibrational mode with the rotation gives rise to two different energy states. This removal of the system degeneracy is referred to as l-type doubling.

In the solutions which lead to Eqs. (17.27) and (17.28) certain higher-order terms were omitted for simplicity. If the complete solution [8] is examined it is observed that the occurrence of the Coriolis force serves to split the energy levels in the degenerate mode with $|l| = 1$ by the amount $(n_2 + 1)(J + 1)Jq_1$ where

$$q_1 = \frac{B_e^2}{\nu_2}\left[1 + 4\left(\zeta_{21}^2 \frac{\nu_2^2}{\nu_1^2 - \nu_2^2} + \zeta_{23}^2 - \frac{\nu_2^2}{\nu_3^2 - \nu_2^2}\right)\right] \tag{17.39}$$

The rotational frequencies as given by Eq. (17.30) are modified to

$$\nu = \left[2B_n \pm \frac{q_1}{2}(n_2 + 1)\right](J + 1) - 4D(J + 1)[(J + 1)^2 - l^2] \tag{17.40}$$

for the case of $l = 1$. For $l > 1 q_i$ will be quite small and splitting is generally not observed.

The intensities of these transitions are found by using the dipole moment matrix elements as was done for the diatomic molecule. In this case the z axis is taken to be unique so $\mu_x = \mu_y = 0$ and $K \to l$. Evaluating elements of the form $\langle J, l, M | \mu^2 \Phi_{Fz} | J', l, M' \rangle$, where l_1 and l_2 denote the two split states, and summing over the quantum number M, shows the nonzero elements to be

$$\left.\begin{matrix} J + 1, l_1 \leftarrow J, l_1 \\ J + 1, l_2 \leftarrow J, l_2 \end{matrix}\right\} \qquad I(J + 1 \leftarrow J) \propto \mu^2 \frac{(J + 1)^2 - l^2}{(J + 1)(2J + 1)} \tag{17.41}$$

$$J, l_2 \leftarrow J, l_1 \qquad I(l_2 \leftarrow l_1) \propto \mu^2 \frac{l^2}{J(J + 1)} \tag{17.42}$$

There will be two types of transitions possible, $\Delta J = \pm 1$ and $\Delta J = 0$. The notation $(v_1 v_2 v_3)$ is used to denote the vibrational state of the molecule, and appended additional sub- and superscripts i and j denote the lower ($i = 1$) or upper ($i = 2$) component of the doublet belonging to l ($j = l$). A plot of theoretical frequencies for the $J = 2 \leftarrow 1$ transitions and intensities relative to the $(0\ 1_1^1\ 0)$ state being unity is shown in Fig. 17.4. The line representing the (000) state is broken to indicate that at room temperature its intensity would be much larger than any of those belonging to excited states. All of the theoretical frequencies were

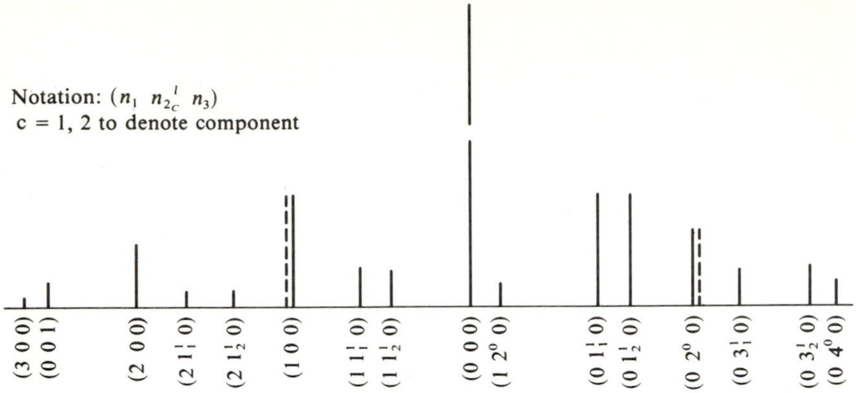

FIGURE 17.4
Theoretical spectrum for a linear *XYZ* molecule.

calculated from Eq. (17.37) with $B_0 = 6000$ MHz, $\nu_1 = 900\ \text{cm}^{-1}$, $\nu_2 = 500\ \text{cm}^{-1}$, $\nu_3 = 2000\ \text{cm}^{-1}$, $q_1 = 6$ MHz, and $D = 0$.

In some instances the vibrational energy levels of a molecule may interact to give another change in the observed spectrum, a Fermi resonance [9]. This will occur if two vibrational states belonging to the same symmetry species have energies lying close together. This will result in a near-degeneracy of the two states and a small displacement of the levels and the rotational transitions associated with the levels. This effect is shown in Fig. 17.4 by the dashed lines which indicate the Fermi resonance-shifted positions for the molecules in the (100) and (02^00) vibrational states. This interaction does not affect all energy levels and is usually quite small when present.

17.4 RIGID SYMMETRIC TOP MOLECULES

The development of an analytical expression for the energy levels of a symmetric top rotor can be approached by use of either wave mechanics or matrix mechanics. This discussion will center about the latter method but a survey of the former and the form and utility of the resulting wavefunctions will be mentioned.

We will begin by developing the classical expression for the energy of rotation of a general molecule containing N atoms. The total motion of a rigid molecule can be described by the relative motion of two coordinate systems, a space-fixed XYZ system and a body-fixed xyz system, as shown in Fig. 17.5. The origin of the xyz system is the center of mass of the molecule and atom i is located in the space-fixed system by a vector $\mathbf{r}_i = X_i\hat{\mathbf{I}} + Y_i\hat{\mathbf{J}} + Z_i\hat{\mathbf{K}}$ and in the body-fixed system by a vector $\boldsymbol{\rho}_i = x_i'\hat{\mathbf{i}} + y_i'\hat{\mathbf{j}} + z_i'\hat{\mathbf{k}}$. The total energy of the system is

$$E = \frac{1}{2}\sum_{i=1}^{N} m_i(\dot{\mathbf{r}}_i \cdot \dot{\mathbf{r}}_i) \tag{17.43}$$

assuming the molecule to be in a field-free space so that the potential is constant.

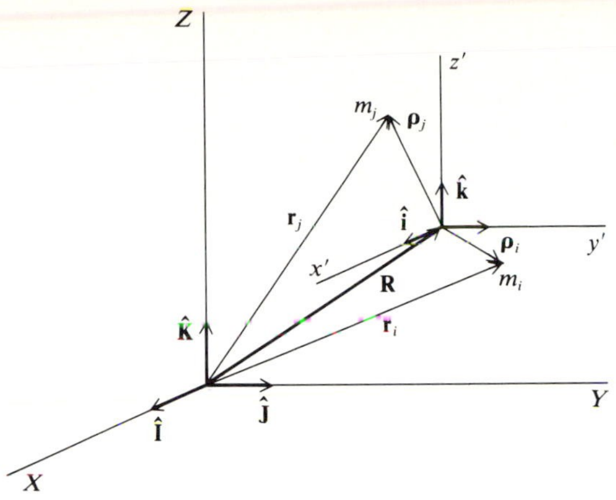

FIGURE 17.5
Coordinates of a rotating molecule.

Since the origin of the body-fixed system is chosen as the center of mass

$$\sum_{i=1}^{N} m_i q_i' = 0 \qquad (q_i' = x_i', y_i', z_i') \tag{17.44}$$

The procedure to be followed will first separate the translational and internal motions as was done for the diatomic molecule in Sec. 2.3 and then obtain the rotational energy by applying restrictions to some of the internal motions. The position and momentum for atom i will be given by

$$\mathbf{r}_i = \mathbf{R} + \boldsymbol{\rho}_i \tag{17.45}$$

$$m_i \dot{\mathbf{r}}_i = m_i \dot{\mathbf{R}} + m_i \dot{\boldsymbol{\rho}}_i \tag{17.46}$$

The total momentum is

$$\sum_{i=1}^{N} m_i \dot{\mathbf{r}}_i = \sum_{i=1}^{N} m_i \dot{\mathbf{R}} + \sum_{i=1}^{N} m_i \dot{\boldsymbol{\rho}}_i \tag{17.47}$$

The energy can be written

$$E = \frac{1}{2} \sum_{i=1}^{N} m_i \mathbf{v}_i \cdot \mathbf{v}_i = \frac{1}{2} \sum_{i=1}^{N} m_i (\dot{\mathbf{R}} + \dot{\boldsymbol{\rho}}_i) \cdot (\dot{\mathbf{R}} + \dot{\boldsymbol{\rho}}_i) \tag{17.48}$$

or

$$E = \frac{1}{2} \sum_{i=1}^{N} m_i \dot{R}^2 + \frac{1}{2} \sum_{i=1}^{N} m_i \dot{\rho}_i^2 + \sum_{i=1}^{N} m_i (\dot{\mathbf{R}} \cdot \dot{\boldsymbol{\rho}}_i) \tag{17.49}$$

$\mathbf{R}$ is independent of the summation so by using $\boldsymbol{\rho}_i = x_i'\mathbf{i} + y_i'\mathbf{j} + z_i'\mathbf{k}$ and Eq. (17.44) the third term becomes

$$\dot{\mathbf{R}} \cdot \left(\hat{\mathbf{i}} \sum_{i=1}^{N} m_i \dot{x}_i' + \hat{\mathbf{j}} \sum_{i=1}^{N} m_i \dot{y}_i' + \hat{\mathbf{k}} \sum_{i=1}^{N} m_i \dot{z}_i' \right) = 0 \tag{17.50}$$

and the energy is given by

$$E = \frac{1}{2} M\dot{R}^2 + \frac{1}{2} \sum_{i=1} m_i\dot{\rho}_i^2 = E_{\text{tr}} + E_{\text{int}} \tag{17.51}$$

where $M = \sum_{i=1}^{N} m_i$. The energy has now been separated into a translation and an internal part.

Next we will relate the motion of an individual mass point to that of the entire assembly. For any vector **G**, which specifies a point in a rotating body, the relationship between the time derivatives of **G** in the two reference frames, space-fixed (F) and body-fixed (g), is

$$(\dot{\mathbf{G}})_F = (\dot{\mathbf{G}})_g + \boldsymbol{\omega} \times \mathbf{G} \tag{17.52}$$

where $\boldsymbol{\omega}$ is the angular velocity of the rotating body. By eliminating any consideration of translational motion a common origin can be used to describe the rotation of the $x'y'z'$ coordinate system relative to the XYZ system. Therefore, for a vector $\mathbf{r}_i$,

$$(\dot{\mathbf{r}}_i)_F = (\dot{\mathbf{r}}_i)_{g'} + \boldsymbol{\omega} \times \mathbf{r}_i \tag{17.53}$$

or

$$(\dot{\mathbf{R}} + \dot{\boldsymbol{\rho}}_i)_F = (\dot{\mathbf{R}} + \dot{\boldsymbol{\rho}}_i)_{g'} + \boldsymbol{\omega} \times (\mathbf{R} + \boldsymbol{\rho}_i) \tag{17.54}$$

Since the origin of the two systems is common, $\mathbf{R} = 0$. A rigid rotor is defined by fixed axes in the rotating body or $(\dot{\boldsymbol{\rho}}_i)_{g'} = 0$, therefore

$$\dot{\boldsymbol{\rho}}_i = \boldsymbol{\omega} \times \boldsymbol{\rho}_i \tag{17.55}$$

and the internal energy is just the rotational energy

$$E_{\text{int}} = E_r = \frac{1}{2} \sum_{i=1}^{N} m_i\dot{\boldsymbol{\rho}}_i \cdot (\boldsymbol{\omega} \times \boldsymbol{\rho}_i) \tag{17.56}$$

Employing vector algebra

$$E_r = \frac{1}{2}\boldsymbol{\omega} \cdot \left[\sum_{i=1}^{N} m_i(\boldsymbol{\rho}_i \times \dot{\boldsymbol{\rho}}_i)\right] = \frac{1}{2}\boldsymbol{\omega} \cdot \left[\sum_{i=1}^{N} m_i(\boldsymbol{\rho}_i \times \boldsymbol{\omega} \times \boldsymbol{\rho}_i)\right] \tag{17.57}$$

$$E_r = \frac{1}{2}\boldsymbol{\omega} \cdot \left[\sum_{i=1}^{N} m_i\boldsymbol{\omega}(\boldsymbol{\rho}_i \cdot \boldsymbol{\rho}_i) - \boldsymbol{\rho}_i(\boldsymbol{\rho}_i \cdot \boldsymbol{\omega})\right] \tag{17.58}$$

$$E_r = \frac{1}{2}\boldsymbol{\omega} \cdot \left[\sum_{i=1}^{N} m_i(\underline{1}\rho_i^2 - \boldsymbol{\rho}_i\boldsymbol{\rho}_i)\right] \cdot \boldsymbol{\omega} = \frac{1}{2}\boldsymbol{\omega} \cdot \underline{\mathbf{II}} \cdot \boldsymbol{\omega} \tag{17.59}$$

where $\underline{1} = \hat{\mathbf{i}}\hat{\mathbf{i}} + \hat{\mathbf{j}}\hat{\mathbf{j}} + \hat{\mathbf{k}}\hat{\mathbf{k}}$. The term $\underline{\mathbf{II}}$ is the inertial tensor discussed in App. L.

Substituting $\boldsymbol{\omega} = \omega_{x'}\hat{\mathbf{i}} + \omega_{y'}\hat{\mathbf{j}} + \omega_{z'}\hat{\mathbf{k}}$ and $\boldsymbol{\rho}_i = x_i'\hat{\mathbf{i}} + y_i'\hat{\mathbf{j}} + z_i'\hat{\mathbf{k}}$ into Eq. (17.58)

$$
\begin{aligned}
E_r = {} & \frac{1}{2}\omega_{x'}^2 \sum_{i=1}^{N} m_i(y_i'^2 + z_i'^2) - \frac{1}{2}\omega_{x'}\omega_{y'} \sum_{i=1}^{N} m_i x_i' y_i' - \frac{1}{2}\omega_{x'}\omega_{z'} \sum_{i=1}^{N} m_i x_i' z_i' \\
& - \frac{1}{2}\omega_{y'}\omega_{x'} \sum_{i=1}^{N} m_i y_i' x_i' + \frac{1}{2}\omega_{y'}^2 \sum_{i=1}^{N} m_i(x_i'^2 + z_i'^2) - \frac{1}{2}\omega_{y'}\omega_{z'} \sum_{i=1}^{N} m_i y_i' z_i' \\
& - \frac{1}{2}\omega_{x'}\omega_{z'} \sum_{i=1}^{N} m_i x_i' z_i' - \frac{1}{2}\omega_{z'}\omega_{y'} \sum_{i=1}^{N} m_i z_i' y_i' + \frac{1}{2}\omega_{z'}^2 \sum_{i=1}^{N} m_i(x_i'^2 + y_i'^2)
\end{aligned}
\tag{17.60}
$$

The coefficients of the $\omega_{g'}$'s are the components of the inertial tensor. There will exist an axis system, the principal axis system, where cross terms such as $\sum_{i=1}^{N} m_i x_i' y_i'$ vanish. Designating this principal axis system by xyz, the classical rotational energy can be expressed as

$$E_r = \tfrac{1}{2}I_{xx}\omega_x^2 + \tfrac{1}{2}I_{yy}\omega_y^2 + \tfrac{1}{2}I_{zz}\omega_z^2 \tag{17.61}$$

where $I_{xx} = \sum_{i=1}^{N} m_i(y_i^2 + z_i^2)$, etc. The angular momentum is related to the angular velocity by $P_g = I_{gg}\omega_g$, so the classical rotational energy expressed in terms of the angular momentum is

$$E_r = \frac{P_x^2}{2I_{xx}} + \frac{P_y^2}{2I_{yy}} + \frac{P_z^2}{2I_{zz}} \tag{17.62}$$

Having developed this expression for the classical energy of the rotating molecule it can serve as the basis for finding the quantum mechanical energies by means of either wave or matrix mechanics. Since the Hamiltonian of the system is a function of the components of angular momenta and the eigenvalues for these components are already known, the nonzero elements of the energy matrix are readily evaluated. The evaluation of these elements is perfectly general and the results are applicable to linear, symmetric top or asymmetric rigid rotors.

The elements of the Hamiltonian (energy) matrix will be

$$\langle J, K, M|\mathscr{H}_r|J'', K'', M''\rangle = \tfrac{1}{2}\langle J, K, M|\frac{P_x^2}{I_{xx}} + \frac{P_y^2}{I_{yy}} + \frac{P_z^2}{I_{zz}}|J'', K'', M''\rangle \tag{17.63}$$

Any single term on the right-hand side of this expression is evaluated using

$$
\begin{aligned}
&\langle J, K, M|\mathbf{P}_g^2|J'', K'', M''\rangle \\
&\quad = \sum_{J'}\sum_{K'}\sum_{M'} \langle J, K, M|\mathbf{P}_g|J', K', M'\rangle\langle J', K', M|\mathbf{P}_g|J'', K'', M''\rangle
\end{aligned}
\tag{17.64}
$$

Using the matrix elements for the angular momentum (Table 3.1) the only nonzero terms in the energy matrix are found to be

$$\langle J, K, M|\mathscr{H}_r|J, K, M\rangle = \frac{\hbar^2}{4}\left\{\frac{2}{I_z} - \frac{1}{I_x} - \frac{1}{I_y}\right\}K^2 + \frac{\hbar^2}{4}\left\{\frac{1}{I_x} + \frac{1}{I_y}\right\}J(J+1) \tag{17.65}$$

$$\langle J, K, M|\mathscr{H}_r|J, K \pm 2, M\rangle$$
$$= \frac{\hbar^2}{8}\{[J(J+1) - K(K \pm 1)][J(J+1) - (K \pm 1)(K \pm 2)]\}^{1/2}\left\{\frac{1}{I_y} - \frac{1}{I_x}\right\} \tag{17.66}$$

where the second subscript on the I_{gg}'s has been dropped for simplicity.

For the diatomic and linear molecules $I_x = I_y = I_B$ and $I_z = 0$. The off-diagonal terms are zero and the energy is given directly by the $(J, K, M|\mathscr{H}_r|J, K, M)$ terms. Note that if $I_z = 0$, $1/I_z = \infty$ and the system would appear to have an infinite rotational energy. This is contrary to experience, hence $K = 0$ for a linear molecule. Since K is a measure of the momentum about the symmetry axis, and for a linear molecule all mass points lie on the symmetry axis, it is necessary that $K = 0$ since the molecule will have no momentum about that axis. The energy levels are then given by

$$E = \frac{\hbar^2}{2I_B}J(J+1) = hBJ(J+1) \tag{17.67}$$

as was derived earlier for the rigid diatomic molecule.

The classical definition of a symmetric top is an object having two equal moments of inertia, that is $I_x = I_y \neq I_z \neq 0$. The off-diagonal elements of the energy matrix are zero since $(1/I_x - 1/I_y)$ will be zero, and the energy will be given by the diagonal $(J, K, M|\mathscr{H}_r|J, K, M)$ elements

$$(J, K, M|\mathscr{H}_r|J, K, M) = \frac{\hbar^2}{2}\left[\frac{1}{I_z} - \frac{1}{I_x}\right]K^2 + \frac{\hbar^2}{2I_x}J(J+1) \tag{17.68}$$

There are two possibilities for the assignment of moments. The first is the prolate top where $I_x > I_z$, normally designated by $I_z \rightarrow I_a$ and $I_x \rightarrow I_b$, and having an energy

$$E = hBJ(J+1) + h(A - B)K^2 \tag{17.69}$$

where $B = h/8\pi^2 I_b$, $A = h/8\pi^2 I_a$. The second or oblate top has $I_x < I_z$, is designated by $I_z \rightarrow I_c$ and $I_x \rightarrow I_b$, and has an energy

$$E = hBJ(J+1) + h(C - B)K^2 \tag{17.70}$$

For the prolate top $A > B$, while for the oblate top $C < B$. This leads to a basic difference in the arrangement of the energy levels as illustrated in Fig. 17.6.

Inspection of this energy-level diagram indicates that the possibilities for transitions to occur are more numerous than in the case of the diatomic molecule. In order to determine the nature of the symmetric top spectrum the selection rules for J and K must be found. The primary aspects of the election rules can be easily established by use of the direction cosine matrix elements, but it is necessary to consider the nature of the symmetric top wavefunctions to completely specify them. We shall derive the basic selection rules for J and K using matrix methods and then examine the symmetry properties of the wavefunctions.

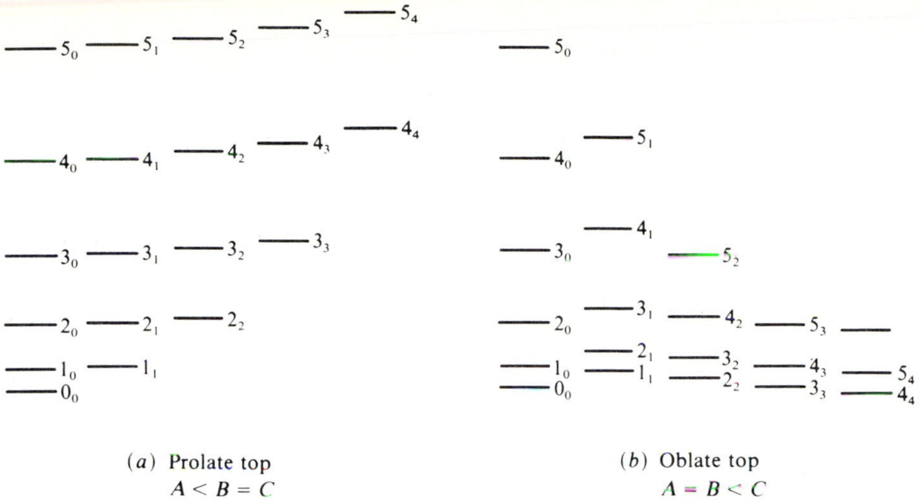

FIGURE 17.6
Qualitative energy-level diagram for symmetric top molecules. (Levels designated by J_K.)

The approach to evaluating selection rules for J and K is similar to that used for the diatomic rotor. The intensities of the symmetric top transitions given as absorption coefficients are

$$\alpha = \frac{8\pi^3 \nu_{JJ'}^2 N_J}{3hc\,\Delta\nu} \langle J, K, M|\boldsymbol{\mu}|J', K', M'\rangle^2 (1 - e^{-h\nu_{JJ}/kT}) \tag{17.71}$$

This equation involves the dipole moment matrix elements $\langle J, K, M|\mu|J'K', M'\rangle^2$, whose zero-nonzero of character will determine the selection rules. In the absence of external fields each state has a $2J + 1$ degeneracy in M so the total dipole moment matrix element is found by summing the three space-fixed orthogonal components over the M values

$$\langle J, K|\boldsymbol{\mu}|J', K'\rangle^2 = \sum_{M=-J}^{J} [\langle J, K, M|\boldsymbol{\mu}_x|J', K', M'\rangle^2 + \langle J, K, M|\boldsymbol{\mu}_y|J', K', M'\rangle^2$$
$$+ \langle J, K, M|\boldsymbol{\mu}_z|J', K', M'\rangle^2] \tag{17.72}$$

The components of the dipole moment relative to the space-fixed coordinates are

$$\mu_F = \mu_x \Phi_{Fx} + \mu_y \Phi_{Fy} + \mu_z \Phi_{Fz} \qquad (F = X, Y, Z) \tag{17.73}$$

The dipole moment of a symmetric top must lie along the symmetry (z) axis so $\mu_x = \mu_y = 0$ and $\mu_z = \mu$. Therefore

$$\mu_F = \mu \Phi_{Fz} \qquad (F = X, Y, Z) \tag{17.74}$$

and Eq. (17.72) reduces to

$$\langle J, K|\boldsymbol{\mu}|J', K'\rangle^2 = \mu^2 \sum_{M=-J}^{J} [\langle J, K, M|\boldsymbol{\Phi}_{Xz}|J', K', M'\rangle^2 + \langle J, K, M|\boldsymbol{\Phi}_{Yz}|J', K', M'\rangle^2 + \langle J, K, M|\boldsymbol{\Phi}_{Zz}|J', K', M'\rangle^2] \tag{17.75}$$

The nonzero elements for Φ_{Fg} are found in Table 3.2, and when substituted into Eq. (17.75) and summed over M gives as the only nonzero elements of $\langle J, K|\mu|J'K'\rangle^2$

$$\langle J, K|\boldsymbol{\mu}|J+1, K\rangle^2 = \mu^2 \frac{(J+1)^2 - K^2}{(J+1)(2J+1)} \tag{17.76}$$

$$\langle J, K|\boldsymbol{\mu}|J, K\rangle^2 = \mu^2 \frac{K^2}{J(J+1)} \tag{17.77}$$

$$\langle J, K|\boldsymbol{\mu}|J-1, K\rangle^2 = \mu^2 \frac{J^2 - K^2}{J(2J+1)} \tag{17.78}$$

The nonzero elements show the selection rules will be $\Delta J = 0, \pm 1$, and $\Delta K = 0$. Using these selection rules and the prolate top as an example the allowed transition frequencies are given by

$$\nu = \frac{E_{J+1,K} - E_{J,K}}{h} = 2B(J+1) \tag{17.79}$$

The most striking point that is immediately evident is that the experimentally measurable transition frequency is a function of only the B rotational constant of the molecule. This is due to the $\Delta K = 0$ selection rule. This rule is a consequence of K being the projection of the angular momentum on the symmetry axis and, since the dipole moment lies along this axis, there is no component of the dipole moment oriented such that it can interact with an external field to produce a change in momentum about the symmetry axis.

A knowledge of the wavefunctions of the symmetric top is necessary to fully understand the various phenomena associated with its rotation and to develop the theory for the asymmetric rotor. The solutions of the Schrödinger equation to give the symmetric top wavefunctions is expedited by using the Euler angles to express the motion. It is shown in App. H that the angular momentum operators relative to the body-fixed axis system, the P_g's, can be expressed in terms of the Euler angles as[2]

$$P_x = -i\hbar\left(-\sin\chi \cot\theta \frac{\partial}{\partial x} + \frac{\sin x}{\sin\theta}\frac{\partial}{\partial\phi} + \cos\chi \frac{\partial}{\partial\theta}\right) \tag{17.80}$$

---

[2] Note that the symbol for the angle about the z-axis has been changed from ψ used in Appendix H to χ in order to avoid confusion with the use of ψ to represent a wavefunction.

$$P_y = -i\hbar\left(-\cos\chi\cot\theta\frac{\partial}{\partial\chi} + \frac{\cos x}{\sin\theta}\frac{\partial}{\partial\phi} + \sin\chi\frac{\partial}{\partial\theta}\right) \tag{17.81}$$

$$P_z = -i\hbar\frac{\partial}{\partial\chi} \tag{17.82}$$

For the symmetric top, $I_{xx} = I_{yy} \neq I_{zz}$ and substitution of these operators into Eq. (17.62) provides the Hamiltonian operator for the symmetric top. The Schrödinger wave equation becomes

$$\frac{1}{\sin\theta}\frac{\partial}{\partial\theta}\left(\sin\theta\frac{\partial\psi}{\partial\theta}\right) + \frac{1}{\sin^2\theta}\frac{\partial^2\psi}{\partial\phi^2} + \left(\frac{\cos^2\theta}{\sin^2\theta} + \frac{C}{B}\right)\frac{\partial^2\psi}{\partial\chi^2} - \frac{2\cos\theta}{\sin^2\theta}\frac{\partial^2\psi}{\partial\chi\partial\phi} + \frac{E}{hB}\psi = 0 \tag{17.83}$$

where $B = h/8\pi^2 I_B$, $I_B = I_{xx}$ and $C = h/8\pi^2 I_C$, $I_C = I_{zz}$.

The solution of this equation, outlined in App. W, gives for the wavefunctions

$$\psi_{JKM} = N_{JKM} z^{(1/2)|K-M|}(1-z)^{(1/2)|K+M|} e^{iM\phi} e^{iK\chi} F \times\left(-J + \frac{\beta}{2} - 1, J + \frac{\beta}{2}, 1 + |K - M|, z\right) \tag{17.84}$$

or

$$\psi_{JKM} = N_{JKM} S_{JKM}(\theta) e^{iM\phi} e^{-iK\chi} \tag{17.85}$$

where $S_{JKM}(\theta)$ and N_{JKM} are given in App. V. A necessary condition for the solutions of the wave equation to have well-behaved solutions is

$$E = hBJ(J+1) + h(C-B)K^2$$

where J is an integer. This, of course, is the same expression obtained by using angular momentum matrix elements. The total angular momentum and the components of angular momentum relative to both the body- and space-fixed axes can be found either by using Eqs. (17.80) to (17.82) to obtain the operators and evaluating the appropriate eigenvalue equation, or by use of the matrix elements from Table 3.1. Using the table we see that

$$P^2 = J(J+1)\hbar^2 \tag{17.86}$$

Solution of the eigenvalue equation $P_z\psi_{JKM} = R\psi_{JKM}$ or evaluation of the matrix element $\langle J, K, M|P_z|J, K, M\rangle$ shows that $R = K\hbar$. $K\hbar$ is the projection of the total angular momentum on the symmetry axis. Similar evaluation of $P_Z\psi_{JKM} = S\psi_{JKM}$ or $\langle J, K, M|P_z|J, K, M\rangle$ shows $S = M\hbar$. $M\hbar$ is the projection of the total angular momentum on the space-fixed Z axis.

Reference to Eq. (V-6) shows that if $K = 0$ it will reduce to the eigenvalue equation for the linear molecule. The wavefunctions for the $K = 0$ states of the symmetric top are then equivalent to those of the linear molecule. The energy, in the absence of any external field, is $2J + 1$-fold degenerate in M and doubly degenerate in K (except for the $K = 0$ level).

The selection rules and the transition intensities for rotating molecules will be affected by the symmetry of the molecule as it relates to the wavefunction. Nuclear spin and statistics of the off-symmetry axis atoms also need to be considered. The general wave equation for a molecule is

$$\mathcal{H}\psi = E_{\text{total}}\psi \tag{17.87}$$

where $\psi = \psi_e\psi_r\psi_v\psi_n$ with ψ_v = vibrational wavefunction, ψ_r = rotational wavefunction, ψ_e = electronic wavefunction, and ψ_n = nuclear wavefunction.

In Chap. 7 it was established that the key factor for determination of selection rules was the vanishing or nonvanishing of the dipole integrals $\int \psi_i\mu_g\psi_j d_v$. The factor μ_g can be expressed as, for example $(q = x, y, z)$ and is an odd function. Therefore if the product $\psi_i\psi_j$ is odd, the integrand will be even and the value of the integral can be nonzero. If the product $\psi_i\psi_j$ is even, then the integral is odd and the value of the integral is zero. Thus, the symmetry of the wavefunction is instrumental in the determination of the selection rules, as was true for the diatomic oscillator. The symmetry selection rules are often expressed as

$$e(+) \leftrightarrow o(-) \qquad e(+) \not\leftrightarrow e(+) \qquad o(-) \not\leftrightarrow o(-)$$

indicating that transitions can occur only between states of opposite symmetry.

The wavefunction referred to the proceeding discussion is the total wavefunction of the molecule. The ground state vibrational wavefunction is symmetric so if we restrict our discussion to molecules in the ground vibrational state, ψ_v can be ignored. The discussion will also be limited to molecules having

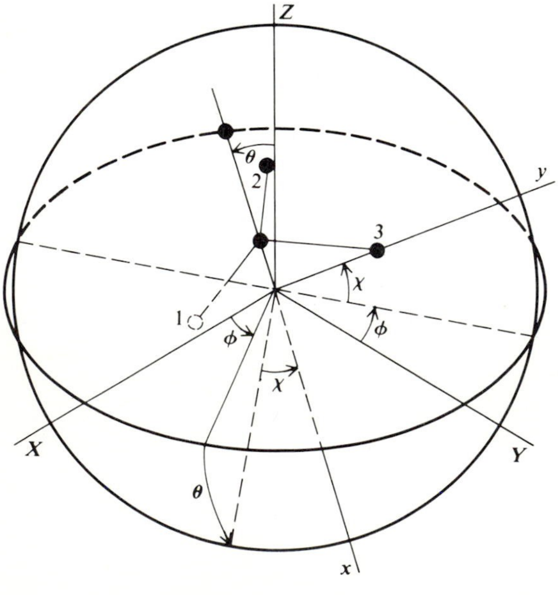

FIGURE 17.7
Symmetric top Eulerian angles.

Σ electronic states, so ψ_e will be even. Therefore, the selection rules for rotational transitions will be determined by the symmetry properties of $\psi_r\psi_n$.

The effects of an inversion of coordinates on the rotational wavefunctions of a symmetric top are visualized by reference to Fig. 17.7 which shows a molecule with C_{3v} symmetry referenced with Euler angles. If the coordinates of each atom in the molecule are inverted through the origin this is equivalent to $\theta \to \theta$, $\phi \to \phi$, and $\chi \to \chi + \pi$. Looking at the wavefunction $\psi_r = N_{JKM}S_{JKM}(\theta)e^{iK\chi}e^{iM\phi}$, it is seen that the only part of the function which is changed by inversion is the $e^{iK\chi}$ which becomes $e^{iK(\chi+\pi)} = e^{iK\chi}e^{iK\pi}$. But, $e^{iK\pi} = (-1)^K$, therefore inversion causes $\psi_r \to (-1)^K\psi_r$. The rotational wavefunctions are odd or even with respect to inversion depending on the nature of K.

While the symmetry properties of ψ_r were determined in a relatively straightforward manner the problem of detailing the effects of nuclear symmetry is substantially more complex. An introduction to the effect of nuclear spins on wavefunctions of homonuclear diatomic molecules was discussed relative to Raman spectra, but to extend that discussion to sets of three or more equivalent nuclei is beyond the intended scope of this presentation. A thorough discussion of this topic has been presented by Townes and Schawlow [10]. The net effect of the nuclear spin statistics will be to cause a K-dependent alteration in the intensities of the transitions. These effects are summarized in Table 17.3 for two common symmetries.

TABLE 17.3
Nuclear spin statistics

Value of K, J-system description†	Weight
C_3 Symmetry‡	
$K = 0$ or $K = 3n$, $J = e$	$\frac{1}{3}(2I+1)(4I^2+4I+3)$
$K \neq 3n$, $J = e$ or o	$\frac{1}{3}(2I+1)(4I^2+4I)$
C_4 Symmetry‡	
$K = 0$, $J = e$, Bose-Einstein statistics	$\frac{1}{2}(2I^2+3I+1)(2I^2+3I+2)$
$K = 0$, $J = o$, Bose-Einstein statistics	$\frac{1}{2}(2I^2+I-1)(2I+1)$
$K = 0$, $J = e$, Fermi-Dirac statistics	$\frac{1}{2}(2I^2+I)(2I^2+I+1)$
$K = 0$, $J = o$, Fermi-Dirac statistics	$\frac{1}{2}(2I^2+3I+1)(2I+3)$
$K \neq 0$, $K = 4n$, $J = e$ or o, Bose-Einstein statistics	$(2I^2+3I+1)(2I^2+I+1)$
$K \neq 0$, $K \neq 4n$, $K = e$, $J = e$ or o, Fermi-Dirac statistics	$(2I^2+3I+1)(2I^2+I+1)$
$K \neq 0$, $K = 4n$, $J = e$ or o, Fermi-Dirac statistics	$(2I^2+I)(2I^2+3I+2)$
$K \neq 0$, $K \neq 4n$, $K = e$, $J = e$ or o, Bose-Einstein statistics	$(2I^2+I)(2I^2+3I+2)$
$K \neq 0$, $K = o$, $J = e$ or o	$(I^2+I)(4I^2+4I+1)$

† n = integer (not zero).

‡ For molecules which do not exhibit inversion.

17.5 INTENSITIES OF SYMMETRIC TOP TRANSITIONS

We have now provided in Eqs. (17.76) to (17.78) and Table 16.1 the necessary information to determine the absorption coefficients for symmetric top molecules. The fraction of molecules in a given state denoted by J, K will be

$$f_{JK} = \frac{N_{JK}}{N_{\text{total}}} = \frac{S(I, K)[(2J+1)e^{-E_{JK}/kT}]}{\sum_{J=0}^{\infty}\sum_{K=-J}^{J} S(I, K)[(2J+1)e^{-E_{JK}/kT}]} \tag{17.88}$$

where $S(I, K)$ is the spin statistical weight factor from Table 16.1 and the $(2J + 1)$ factor is the M degeneracy of the rotational level. At room temperatures $E_{JK} \cong 10^{-25}$ J and $kT \cong 10^{-21}$ J so $E_{JK}/kT \ll 1$. For $x \ll 1$ a summation over e^{-x} can be replaced with an integration and

$$f_{JK} = \frac{S(I, K)[(2J+1)e^{-[BJ(J+1)+(C-B)K^2]h/kT}]}{\int_0^{\infty}\int_{-J}^{J} S(I, K)[(2J+1)e^{-[BJ(J+1)+(C-B)K^2]h/kT}]\, dk\, dJ} \tag{17.89}$$

$S(I, K)$ will depend on the symmetry of the molecule as indicated in Table 17.3. At this point we will limit our consideration to molecules with C_3 symmetry. Evaluating the integral

$$f_{JK} = \frac{S(I, K)[2J+1]}{(4I^2+4I+1)}\left(\frac{B^2Ch^3}{\pi(kT)^3}\right)^{1/2} e^{-E_{JK}/kT} \tag{17.90}$$

For lower values of J where $E_{JK} \ll kT$, the exponential term is close to unity and can be ignored. Using the general expression for the absorption coefficient given in Chap. 7 expanding $\exp\{-h\nu_{\sigma\sigma'}/kT\}$ in a series and neglecting all higher-order terms

$$\alpha = \frac{8\pi^2 N_0 f \langle J, K|\boldsymbol{\mu}|J'K\rangle^2 \nu_0^2}{3ckT}\,\frac{\Delta\nu}{(\nu-\nu_0)^2+(\Delta\nu)^2} \tag{17.91}$$

where $\nu_0 = 2B(J + 1)$. The fraction of molecules in a particular state of interest is given by $f = f_{\text{rot}} f_{\text{vib}} f_{\text{elect}}$. Considering all molecules to be in the lowest electronic state $f_{\text{elec}} = 1$, substitution of Eq. (17.90) into Eq. (17.91) for the transition $J + 1$, $K \leftarrow J$, K gives

$$\alpha = \frac{4\pi\mu^2\nu_0^2 h N_0 f_{\text{vib}} S(I, K)}{(4I^2+4I+1)3ck^2T^2}\left(\frac{C\pi h}{kT}\right)^{1/2}\left(1-\frac{k^2}{(J+1)^2}\right)\left(\frac{\Delta\nu}{(\nu-\nu_0)^2+(\Delta\nu)^2}\right) \tag{17.92}$$

considering that $\nu_0 \sim \nu$. The maximum absorption coefficient, that is, the value at $\nu = \nu_0$ is given by

$$\alpha_{\max} = \frac{12.3\times10^{-20}\nu_0^3 f_{\text{vib}}\mu^2 C^{1/2} S(I, K)}{(4I^2+4I+1)\Delta\nu}\left(1-\frac{K^2}{(J+1)^2}\right) \tag{17.93}$$

when $T = 300$ K. For a typical case where $\nu_0 \cong 10^4$ MHz, $\mu \approx 3$ debye, $\Delta\nu \approx 0.1$ MHz, and $C \cong 10^4$ MHz we have for $J = 5$, $K = 3$ and $I = \frac{1}{2}$, $\alpha_{\max} \approx 10^{-6}$.

This is less than the corresponding coefficient for a linear molecule by a factor of about 10.

17.6 EXAMPLES OF SYMMETRIC TOP ROTATIONAL SPECTRA

A look at the spectrum of $^{14}NF_3$ as reported by Sheridan and Gordy [11] provides us with the opportunity of not only analyzing a simple symmetric top spectrum but also brings in hyperfine interactions which were discussed in Chap. 14. The only additional information which must be introduced is the method for the evaluation of q_J for the symmetric top. The total energy of the molecule, considering that it does not undergo inversion, is given by

$$E = E_{JK} + E_Q \tag{17.94}$$

where all rotational parameters and energies are expressed in hertz and

$$\frac{E_{JK}}{h} = B_0 J(J+1) + (C-B)K^2 - D_J J^2(J+1)^2 - D_{JK} J(J+1)K^2 - D_K K^4 \tag{17.95}$$

$$\frac{E_Q}{h} = \frac{e^2 Q q_J f(I, J, F)}{2I(2I-1)J(2J-1)} \tag{17.96}$$

$$f(I, J, F) = \tfrac{3}{4}C(C+1) - I(I+1)J(J+1) \tag{17.97}$$

$$C = F(F+1) - I(I+1) - J(J+1) \tag{17.98}$$

and D_J, D_{JK}, D_K are centrifugal distortion correction terms. Recalling from Chap. 14 that

$$q_J = q_{zz}\left(\frac{3\cos^2\theta - 1}{2}\right)_{\text{av}} \tag{17.99}$$

and

$$\left(\frac{3\cos^2\theta - 1}{2}\right)_{\text{av}} = \tfrac{1}{2}\langle J, K, M = J\,|\,3\Phi_{Zz}^2 - 1\,|\,J, K, M = J\rangle \tag{17.100}$$

Using Table 3.2 gives

$$E_Q = \frac{e^2 Q q_{zz}\{[3k^2/J(J+1)] - 1\} f(I, J, F)}{2I(2I-1)(2J-1)(2J+3)} \tag{17.101}$$

The observed transitions are shown in Table 17.4. There is only a single $J \to J'$ transition observed, and that for only two values of K, so it will be necessary to arbitrarily eliminate one term from the frequency expression

$$\begin{aligned}\nu &= \frac{(E_{J+1,K,F'} - E_{J,K,F})}{h} \\ &= 2B_0(J+1) - 4D_J(J+1)^3 - 2D_{JK}(J+1)K^2 \\ &\qquad + [E_Q(J+1, K, F') - E_Q(J, K, F)]\end{aligned} \tag{17.102}$$

TABLE 17.4
Observed transitions for $^{14}NF_3$

$J \to J'$	K	$F \to F'$	ν, MHz
$1 \to 2$	0	$0 \to 1$	42,722.16
	0	$1 \to 1$	42,727.39
	0	$1 \to 2$	42,723.28
	0	$2 \to 2$	42,721.73
	0	$2 \to 3$	42,723.94
	1	$0 \to 1$	42,726.60
	1	$1 \to 2$	42,722.16
	1	$2 \to 2$	42,723.28
	1	$2 \to 3$	42,723.94

since the hyperfine splittings can provide an evaluation of only the term in brackets. It has been experimentally observed for other molecules that $D_J < D_{JK}$, so the term with the former will be dropped. Centrifugal distortion is less of a problem in heavy molecules so this will introduce no serious error. Using Eq. (17.98) less the D_J term, to set up appropriate sets of simultaneous equations, one can solve for B_0, D_{JK} and e^2Qq_{zz} obtaining $B_0 = 10{,}680.96$ MHz, $D_{JK} = -0.025$ MHz and $e^2Qq_{zz} = -7.07$ MHz.

We can go one step further and determine the structure from B_0. Note that C cannot be obtained from the observed spectrum. For an XY_3 molecule with C_{3v} symmetry the moments of inertia are found from simple analytical geometric considerations to be

$$I_c = 2m_y r_{xy}^2(1 - \cos\theta) \tag{17.103}$$

$$I_B = m_y r_{xy}^2(1 + \cos\theta) + \frac{m_x m_y r_{xy}^2}{3m_y + m_x}(1 + 2\cos\theta) \tag{17.104}$$

where θ is the YXY angle. Although there are two unknowns r_{NF} and θ in the expression for the experimentally known I_B, the problem is circumvented by the observation of the $J = 2 \leftarrow 1$ transition for $^{15}NF_3$ at 42,517.38 MHz giving for that species $I_B = 10{,}629.35$ MHz. By using the two moments, it is found that $r_{NF} = 1.371$ A and $\theta = 102°9'$.

The use of several isotopic species for the evaluation of molecular constants is further illustrated by the work of Varma and Buckton [12] on germyl cyanide, GeH_3CN. The spectra of the various species were observed under conditions sufficient to resolve the hyperfine structure due to ^{14}N for the more abundant species. Since the molecule is moderately heavy, centrifugal distortion is ignored. The observed transitions and resulting moments are shown in Table 17.5. The analysis of this data is analogous to the procedure followed for NF_3 in the previous example except there are more molecular parameters to be evaluated and their relationships to the moments of inertia differ from Eqs. (17.103) and

TABLE 17.5
Observed rotational transitions and reciprocal moments for germyl cyanide

Isotopic species	$J' \leftarrow J$ Transition frequency, MHz 1 → 2	2 → 3	3 → 4	4 → 5	B_0, MHz
$^{74}GeH_3{}^{12}C^{14}N$	14,478.29†				
	14.475.40‡	21,713.49	28,951.39	36,187.08	3618.97
	14,474.54§				
$^{72}GeH_3{}^{12}C^{14}N$	14,562.18†				
	14,559.82‡	21,839.71	29,119.44	36,349.04	3639.93
	14,558.48§				
$^{70}GeH_3{}^{12}C^{14}N$	14,650.66†				
	14,648.29‡	21,972.37	29,296.18	36,620.33	3662.03
	14,646.91§				
$^{76}GeH_3{}^{12}C^{14}N$	14,395.29	....	28,791.47	35,488.47	3598.95
$^{76}GeH_3{}^{12}C^{15}N$	13,808.97	....	27,618.82	34,522.44	3452.25
$^{74}GeH_3{}^{12}C^{15}N$	13,888.25	....	27,776.53	34,720.48	3472.06
$^{72}GeH_3{}^{12}C^{15}N$	13,971.43	....	27,942.85	34,928.49	3492.86
$^{72}GeH_3{}^{13}C^{14}N$	....	....	28,117.99	....	3597.87
$^{74}GeH_3{}^{13}C^{14}N$	....	....	28,612.07	35,765.05	3576.51

† $F \rightarrow F' = 1 \rightarrow 1 (K = 0)$.
‡ $F \rightarrow F' = 2 \rightarrow 3, 1 \rightarrow 2 (K = 0)$.
§ $F \rightarrow F' = 0 \rightarrow 1, 2 \rightarrow 2 (K = 0)$.

(17.104). For the X_3YZW molecule with C_{3v} symmetry

$$I_C = 2m_y r_{xy}^2 (1 - \cos\theta) \tag{17.105}$$

$$I_B = \frac{3}{2} m_x r_{xy}^2 \sin^2\phi + \frac{m_w(M - m_w)}{M} r_{zw}^2 + \frac{(m_w + m_z)(m_y + 3m_x)}{M} r_{yz}^2 + \frac{3m_x(M - 3m_x)}{M} r_{xy}^2 \cos^2\phi + \frac{2m_w(m_y + 3m_x)}{M} r_{yz} r_{zw} + \frac{6m_w m_x}{M} r_{zw} r_{xy} \cos\phi + \frac{6m_x(m_w + 3m_x)}{M} r_{yz} r_{xy} \cos\phi \tag{17.106}$$

where θ is the XYX angle, ϕ is the XYZ angle, and M is the total mass of the molecule.

Equations (17.105) and (17.106) show that there are five structural parameters to be evaluated. In principle it is possible to set up five simultaneous equations but they would constitute a nonlinear set and would not easily be solved for unambiguous solutions. It would also be possible to use seven equations where the variables are the terms multiplying the mass factors. By using seven moments it would be possible to evaluate these terms and proceed by elimination of various ones to get the desired parameters. For example, if one found values of $r_{xy}^2 \sin^2 \phi = A$ and $r_{xy}^2 \cos^2 \phi = B$ then

$$A + B = r_{xy}^2 (\cos^2 \phi + \sin^2 \phi) = r_{xy}^2 \qquad (17.107)$$

Although this procedure can give reasonable parameters the method of isotopic substitution as discussed in Sec. 17.2 yields better results. Since none of the species studied involve H atom substitution it is not possible to find r_{GeH} by this latter method. Varma and Buckton, by assuming $r_{GeH} = 1.529$ A, evaluated the other structural parameters by the method of isotopic substitution and found them to be $r_{CN} = 1.155$ A, $r_{GeC} = 1.719$ A and $r_{GeN} = 3.074$ A. This example points points out the difficulties of obtaining structural data even for symmetric molecules.

REFERENCES

Specific

1. Kraitchman, J., *Am. J. Phys.*, **21**, 17 (1953).
2. Townes, C. H., A. N. Holden, and R. F. Merritt, *Phys. Rev.*, **74**, 1113 (1948).
3. Nielsen, H. H., *Phys. Rev.*, **60**, 794 (1941).
4. Nielsen, A. H., *J. Chem. Phys.*, **11**, 160 (1943).
5. Dennison, D. M., *Rev. Mod. Phys.*, **3**, 280 (1931); **12**, 175 (1940).
6. Nielsen, H. H., and W. H. Shaffer, *J. Chem. Phys.*, **11**, 140 (1943).
7. ———, *Phys. Rev.*, **75**, 1961 (1949).
8. ———, *Rev. Mod. Ph.*, **23**, 90 (1951).
9. Fermi, E., *Z. Physik*, **71**, 250 (1931).
10. Townes, C. H., and A. L. Schawlow, *Microwave Spectroscopy*, McGraw-Hill, New York, NY, 1955, ch. 3.
11. Sheridan, J., and W. Gordy, *Phys. Rev.*, **79**, 513 (1950).
12. Varma, R., and K. S. Buckton, *J. Chem. Phys.*, **46**, 1565 (1967).

General

T. M. Sugden and C. N. Kenney, *Microwave Spectroscopy of Gases*, Van Nostrand, London, 1965.

C. H. Townes and A. L. Schawlow, *Microwave Spectroscopy*, McGraw-Hill, New York, 1955.

J. E. Wollrab, *Rotational Spectra and Molecular Structure*, Academic Press, New York, 1967.

W. Gordy and R. L. Cook, *Microwave Molecular Spectra*, 2d ed., Interscience Publishers, New York, 1985.

M. W. P. Strandberg, *Microwave Spectroscopy*, Methuen & Co., London, 1954.

PROBLEMS

17.1. For OCSe the following ground state transitions have been observed.

Molecule	Transition	Frequency, MHz
$^{16}O^{12}C^{74}Se$	$3 \leftarrow 2$	24,514.67
$^{16}O^{12}C^{76}Se$	$3 \leftarrow 2$	24,410.58
$^{16}O^{13}C^{78}Se$	$3 \leftarrow 2$	24,030.58
$^{16}O^{13}C^{80}Se$	$3 \leftarrow 2$	23,880.18

Considering OCSe to be a rigid rotor determine the best values for the internuclear distances.

17.2. The following ground state transitions have been observed for SiF_3H (NBS Monograph 70, vol. III)

Molecule	Transition	Frequency, MHz
$^{28}Si^{19}F_3^1H$	$2 \leftarrow 1$	28,831.90
	$3 \leftarrow 2$	43,247.49
	$5 \leftarrow 4$	72,076.8
	$6, 0 \leftarrow 5, 0$	86,490.06
	$6, 1 \leftarrow 5, 1$	86,490.06
	$6, 2 \leftarrow 5, 2$	86,490.67
	$6, 3 \leftarrow 5, 3$	86,491.38
	$6, 4 \leftarrow 5, 4$	86,492.42
	$6, 5 \leftarrow 5, 5$	86,493.80
$^{28}Si^{19}F_3^2H$	$2 \leftarrow 1$	27,560.17
	$3 \leftarrow 2$	41,340.00
	$4 \leftarrow 3$	55,119.4
$^{29}Si^{19}F_3^1H$	$2 \leftarrow 1$	28,782.65
$^{29}Si^{19}F_3^2H$	$3 \leftarrow 2$	41,280.46
$^{30}Si^{19}F_3^1H$	$2 \leftarrow 1$	28,734.80
$^{30}Si^{19}F_3^2H$	$3 \leftarrow 2$	41,222.73

(*a*) Assuming SiF_3H to be a rigid rotor calculate the reciprocal moments and the best structural parameters.

(*b*) Neglecting rotational vibration interactions but allowing for centrifugal distortion calculate the reciprocal moments, distortion constants, and best structural parameters.

17.3. Using the rigid rotor approximation calculate the frequencies for the first six rotational transitions of $^{51}V^{16}O^{14}F_3$, $^{95}No^{16}OF_4$, and $^{119}Sn^{19}F_3^{35}Cl$. In each case assume a reasonable structure for the molecule.

17.4. Calculate the fraction of molecules present in the first six rotational energy states of $^{31}PH_3$ and $^{75}As^{19}F_3$ at 300 K.

17.5. Determine α_{max} for $^{31}PH_3$ and $^{75}AsF_3$ at 300 K.

17.6. If a spectrometer has the capability of observing rotational transitions only for $\alpha_{max} > 10^{-7}$, how many different vibrationally excited states of $^{31}P^{16}O^{19}F_3$ could be observed. For $^{31}P^{16}O^{19}F_3$, $B_0 = 4{,}594$ MHz and the vibrational frequencies are at 345, 473, 485, 873, 900 and 1415 cm^{-1}.

17.7. Consider $AsCl_3$ to be a rigid rotor with the following transitions

$^{75}As^{35}Cl_3$ $\quad \nu_{5\leftarrow4} = 21{,}472$ MHz
$^{75}As^{37}Cl_3$ $\quad \nu_{6\leftarrow5} = 24{,}536$ MHz

From this data obtain the best values for r_{AsCl} and $\angle ClAsCl$.

CHAPTER 18

ELECTRONIC SPECTRA OF POLYATOMIC MOLECULES

18.1 INTRODUCTION

The description of both the electronic structure and the electronic spectra of polyatomic molecules is much more complicated than that reviewed in Chaps. 6 and 15 for diatomic molecules. This is due in part to the increased complexity of the spectra which frequently results in the experiment observation of many closely spaced and nonresolvable spectral transitions. Even with modern high-speed computers, *ab initio* quantum mechanical calculations of electron energies and properties of polyatomic molecules of moderate complexity are very lengthy. In the preceding chapters it has been shown that, for the rotation and vibration of polyatomic molecules, it was possible to formulate a Hamiltonian and solve the resulting eigenvalue equation to give energy expressions which are in excellent agreement with experimental spectra. In the case of polyatomic molecules the formulation of the electronic Hamiltonian is relatively straightforward but it will contain terms resulting from the kinetic energies of the individual nuclei and electrons and terms resulting from the pairwise potential energy interactions of all particles. The difficulty which arises is due to the inseparability of the Schrödinger equation into equations of single variables as is the case for a two-body system. One can thus conclude that, although the solution of Schrödinger equations involving Hamiltonians of this complexity to give energy values in agreement with experiment is a reality, it does not constitute a day-to-day type of calculation, the results of which can be correlated easily with observed spectra. To attempt any detailed discussion of these quantum mechanical techniques would require a separate book of the size of this one.

It has only been within the last two decades that the observation and accurate measurement of highly resolved electronic spectra of polyatomic molecules has become well established. With the exception of a limited number of small (3–5 atom) or very symmetrical molecules, the assignment of even a well-resolved rovibronic spectrum is no small task. It is in light of this background that this chapter is cast in the role of providing a somewhat more general and less mathematical exposition.

In the absence of quick and easy mathematical solutions to polyatomic Schrödinger equations one can resort to semiquantitative molecular orbital theory as was done for diatomic molecules. Thus, the mathematical detail of this chapter will be akin to the level of that in Chaps. 6 and 15. We will, in fact, use the approach of beginning with the results of molecular orbital theory and, except for one initial example, leave the development of such results for the student to pursue in other sources.

In principle, the development of an MO description of polyatomic molecules is analogous to that reviewed in Chap. 6 for diatomic molecules, in that the MOs result from linear combinations of the available atomic orbitals. However, the introduction of three or more centers for orbitals into the system complicated the description. By use of molecular symmetry it is possible not only to order and to classify the electronic energy levels but also to derive appropriate spectral selection rules. Following a moderately detailed example of the development of an MO scheme for an AX_4-type molecule the MO description for several simple polyatomic molecules will be summarized. Then, an explanation of the observed electronic spectra will be developed in view of the MO concepts. This approach, while not one which follows the rigorous mathematical development of molecular quantum mechanics, can provide a reasonable base upon which to develop an understanding of electronic spectra.

18.2 POLYATOMIC MOLECULAR ORBITAL THEORY—AN OVERVIEW

For a polyatomic molecule containing N_n nuclei and N_e electrons the complete wavefunction depends on both the electron coordinates r_i and the nuclear coordinates R_j. The Schrödinger equation will then be of the form

$$\mathscr{H}\Psi(R, r) = E\Psi(R, r) \tag{18.1}$$

where R and r denote the complete sets of nuclear and electronic coordinates. By applying the Born-Oppenheimer approximation the vibrational-rotational energy can be separated and the Schrödinger equation for the electronic motion becomes

$$\mathscr{H}_e\Psi_{e\alpha}(R, r) = E_{e\alpha}(R)\Psi_{e\alpha}(R, r) \tag{18.2}$$

where $E_{e\alpha}(R)$ contains the nuclear coordinate parameters, $\Psi_{e\alpha}(R, r)$ is the electronic wavefunctions for a particular state which is characterized by a set of quantum numbers denoted by α (i.e., a molecular orbital), and the electronic Hamiltonian is

$$\mathscr{H}_e = \frac{-\hbar^2}{2m_e}\sum_i \nabla_i^2 + \sum_{k>l=1}^{N_n} \frac{Z_k Z_l e^2}{R_{kl}} + \sum_{i>j=1}^{N_e} \frac{e^2}{r_{ij}} - \sum_{i=1}^{N_e}\sum_{k=1}^{N_n} \frac{Z_k e^2}{r_{ik}} \tag{18.3}$$

where the Z_k are nuclear charges, the R_{kl} are internuclear distances, the r_{ij} are interelection distances, and the r_{ik} are electron-nuclear distances. The exact solutions of Eq. (18.2) would provide accurate expressions for the energies of the molecular energy states as functions of R and for the molecular electronic wavefunctions. This solution would by nature include the correct contributions of electronic correlation and configuration interactions which are neglected or treated only semiquantitatively in approximation methods. Because of the difficulty in solving such equations, even by approximation methods, we will pursue the discussions of electronic spectra by use of more qualitative molecular orbital schemes as developed by a number of different authors [1-6] and as illustrated in several sources [7, 8]. The selection of the appropriate atomic orbitals for the LCAO MOs is made by the use of group theory [5, 6]. While this will provide reasonable geometric detail it will not necessarily provide the correct ordering of the energy levels. For any particular molecule the ordering will depend on the relative energies of the basis atomic orbitals. We will enumerate the results of MO theory for some selected examples of polyatomic molecules, illustrate by appropriate drawings the symmetry of the combining atomic orbitals and the resulting molecular orbitals, provide a typical energy level scheme which contains the appropriate symmetry notation for the molecule and, where two structures are related by a simple angle change, show the corresponding Walsh diagram [9]. Once the symmetries of the electronic energy levels have been established they may be employed, in a manner analogous to that used in Sec. 15.4, to develop the electronic selection rules.

18.3 MOLECULAR ORBITAL THEORY— AN EXAMPLE

To illustrate the nature of a semiquantitative MO development for a molecule of moderate complexity we will review the case of an AB_4 molecule with D_{4h} symmetry. The A atom will be considered to be a transition metal element having atomic ns, np and $(n-1)d$ orbitals available for MO formation while the B atom orbitals will be restricted to $n's$ and $n'p$ orbitals (n and n' denote principle quantum numbers). The role of symmetry and group theory in the development of the MOs will be reviewed first, then, following a brief summary of the method used to calculate the energy levels, the qualitative MO energy-level scheme will be presented.

The initial operation, following the establishment of the geometry of the system and the coordinate systems for the sets of atomic orbitals, will be to

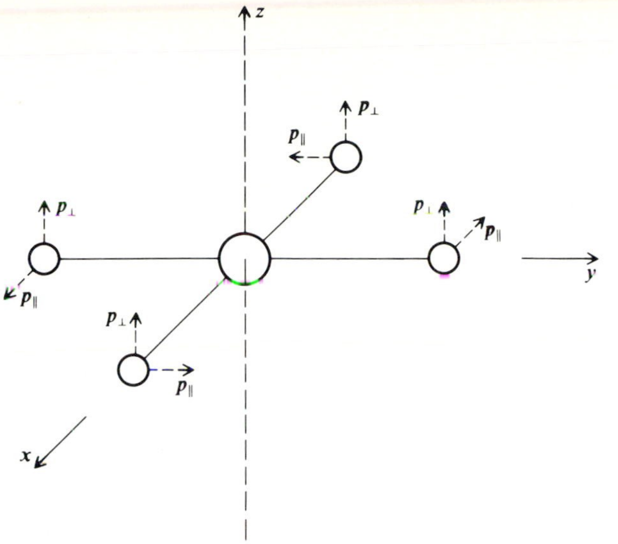

FIGURE 18.1
Geometry and atomic coordinate systems for an AX_4 molecule.

delineate the group representations to which each of the participating AOs belong. Figure 18.1 illustrates one way to establish the coordinate systems.

Reference to the character table for the D_{4h} group (App. J) will directly establish the representations for the AOs on the A atom. The ligand (B atom) orbitals may be synthesized, by use of projection operators, into symmetry-adapted linear combinations [10] (SALCs), which in turn may be combined with central atom AOs of the same symmetry to form MOs. In following this procedure it is simpler if the AOs are initially separated into separate groups of σ or π symmetry. Denoting the basis functions of the A atomic orbitals with s^A, p_i^A and d_j^A, reference to Fig. 18.1 shows that the s^A, p_x^A, and $d_{x^2-y^2}^A$ orbitals can contribute to σ bonds, the p_z^A, $d_{z^2}^A$ d_{yz}^A, and d_{xz}^A can contribute to $\pi_\perp$ bonds (perpendicular to the xy plane) and the p_y^A and d_{xy}^A orbitals can contribute to $\pi_\parallel$ bonds (parallel to the xy plane).

Examination of the ligand orbitals shows that they will fall into four categories:

$$
\begin{array}{ll}
s\sigma & \sigma_1^B, \sigma_2^B, \sigma_3^B, \sigma_4^B \\
p\sigma & P_{x1}^B, P_{x2}^B, P_{x3}^B, P_{x4}^B \\
p\pi_\perp & P_{z1}^B, P_{z2}^B, P_{z3}^B, P_{z4}^B \\
p\pi_\parallel & P_{y1}^B, P_{y2}^B, P_{y3}^B, P_{y4}^B
\end{array}
$$

Consideration of the behavior of these orbitals, when subjected to the operations of the group (+1 = no change, −1 = sign reversal, 0 = change in location), provides the characters of the reducible representations for which these orbitals form

TABLE 18.1
Characters for reducible representations of an AB_4 molecule with D_{4h} symmetry

	Symmetry operation									
Representation	E	$2C$	C_2	$2C_2'$	$2C_2''$	i	$2S_4$	σ_h	$2\sigma_v$	$2\sigma_d$
$\Gamma_A(s)$	1	1	1	1	1	1	1	1	1	1
$\Gamma_A(p)$	3	1	−1	−1	−1	−3	−1	1	1	1
$\Gamma_A(d)$	5	−1	1	1	1	5	−1	1	1	1
$\Gamma_B(s\sigma)$	4	0	0	2	0	0	0	4	2	0
$\Gamma_B(p\sigma)$	4	0	0	2	0	0	0	4	2	0
$\Gamma_B(p\pi_\perp)$	4	0	0	−2	0	0	0	−4	2	0
$\Gamma_B(p\pi_\parallel)$	4	0	0	−2	0	0	0	4	−2	0

a basis. These characters, (Γ), along with those of the representation of A for which the s, $3p$, and $5d$ AOs form a basis are tabulated in Table 18.1. Using Eq. (4.10) these reducible representations can be decomposed into their component irreducible representations can be decomposed into their component irreducible representations

$$\Gamma_A(s) = a_{1g} \tag{18.4}$$

$$\Gamma_A(p) = a_{2u} + e_u \tag{18.5}$$

$$\Gamma_A(d) = a_{1g} + b_{1g} + b_{2g} + e_g \tag{18.6}$$

$$\Gamma_B(s\sigma) = a_{1g} + b_{1g} + e_u \tag{18.7}$$

$$\Gamma_B(p\sigma) = a_{1g} + b_{1g} + e_u \tag{18.8}$$

$$\Gamma_B(p\pi_\perp) = a_{2u} + b_{2u} + e_g \tag{18.9}$$

$$\Gamma_B(p\pi_\parallel) = a_{2g} + b_{2g} + e_u \tag{18.10}$$

The formation of one symmetry-adapted linear combination of ligand orbitals will be reviewed and the remainder of the SALCs listed. Since the d_{xy}^A AO belongs to the b_{2g} representation we will determine the SALC of b_{2g} symmetry formed by projection of one of the p_y^B ligand orbitals.

The projection operator (see Sec. 4.5) will be

$$P_{b_{2g}}(D_{4h}) = E - C_4 - C_4^3 + C_2 - C_2'(1) - C_2'(2) + C_2''(1) + C_2''(2)$$
$$+ i - S_4 - S_4^3 + \sigma_h - \sigma_v - \sigma_v' + \sigma_d + \sigma_d' \tag{18.11}$$

Operation on p_{y1}^B with $P_{b_{2g}}(D_{4h})$ gives

$$P_{b_{2g}}(D_{4h})p_{y1}^B = 4(p_{y1}^B - p_{y2}^B + p_{y3}^B - p_{y4}^B) \tag{18.12}$$

Normalized the SALC becomes

$$\psi_{b_{2g}} = \tfrac{1}{2}(p_{y1}^B - p_{y2}^B + p_{y3}^B - p_{y4}^B) \tag{18.13}$$

Table 18.2 summarizes the symmetries, A atom orbitals, and the ligand SALCs.

TABLE 18.2
Orbital symmetries for an AX_4 molecule with D_{4h} symmetry

Representation	A orbitals	SALCs
a_{1g}	s, d_{z^2}	$\frac{1}{2}(s_1^B + s_2^B + s_3^B + s_4^B)$
		$\frac{1}{2}(p_{x1}^B + p_{x2}^B + p_{x3}^B + p_{x4}^B)$
a_{2g}	.....	$\frac{1}{2}(p_{y1}^B + p_{y2}^B + p_{y3}^B + p_{y4}^B)$
a_{2u}	p_z	$\frac{1}{2}(p_{z1}^B + p_{z2}^B + p_{z3}^B + p_{z4}^B)$
b_{1g}	$d_{x^2-y^2}$	$\frac{1}{2}(s_1^B - s_2^B + s_3^B - s_4^B)$
		$\frac{1}{2}(p_{x1}^B - p_{x2}^B - p_{x3}^B - p_{x4}^B)$
b_{2g}	d_{xy}	$\frac{1}{2}(p_{y1}^B - p_{y2}^B + p_{y3}^B - p_{y4}^B)$
b_{2u}	....	$\frac{1}{2}(p_{z1}^B - p_{z2}^B - p_{z3}^B - p_{z4}^B)$
e_g	d_{xz}, d_{yz}	$\frac{1}{2}(p_{z1}^B - p_{z3}^B), \frac{1}{2}(p_{z2}^B - p_{z4}^B)$
e_u	p_z, p_y	$\frac{1}{\sqrt{2}}(s_1^B - s_3^B), \frac{1}{\sqrt{2}}(s_2^B - s_4^B)$
		$\frac{1}{\sqrt{2}}(p_{x1}^B - p_{x3}^B), \frac{1}{\sqrt{2}}(p_{x2}^B - p_{x4}^B)$
		$\frac{1}{\sqrt{2}}(p_{y1}^B - p_{y3}^B), \frac{1}{\sqrt{2}}(p_{y2}^B - p_{y4}^B)$

The one-electron MOs are next formulated by taking symmetry-equivalent linear combinations of atomic orbitals on atom A and ligand SALCs of the same symmetry. When making these combinations it is necessary to keep in perspective the type of ligand orbitals available. For example, a Cl^- ion or an NH_3 molecule as a ligand will have completely filled π orbitals while a ligand such as the CN^- will have some unfilled π orbitals. One must also consider that, in general, there will be a different number of central atom AOs and ligand SALCs, and not all orbitals of both types will be used in MO formation. In the example we are considering there are 9 central atom AOs and 16 ligand SALCs, so there can be no more than 9 MOs formed by pairing SALCs with the AOs. In cases such as this, the unused orbitals are considered to remain essentially as atomic orbitals which are empty or occupied by lone pairs of electrons.

If it is possible to have 10 molecular orbitals resulting from the combination of 5 ligand SALCs and 5 central atoms AOs then in the most general case the energies of these orbitals will be formed by solving a 10×10 energy determinant whose elements are of the form $H_{ij} - ES_{ij}$, where

$$H_{ij} = \int \psi_i \mathscr{H} \psi_j \, dv \tag{18.14}$$

$$S_{ij} = \int \psi_i \psi_j \, dv \tag{18.15}$$

ψ_i, ψ_j, are either central atom atomic orbitals or ligand SALCs, and $\mathscr{H}$ is the complete Hamiltonian for the molecule in question. From this point one can employ varying degrees of approximation and sophistication to solve for the energies. These can range all the way from the complete neglect of overlap, i.e., all $S_{ij} = 0$ for $i \neq j$, the use of a one-electron Hamiltonian and the incorporation of simple Slater type atomic orbitals, to a very complete Hartree-Fock calculation which employs the calculation of all S_{ij} integrals, the use of a complete Hamiltonian, which will introduce electron correlation, and the use of multiterm Gaussian-type AOs. These varying types of approaches are the subject of extensive quantum mechanical treatments [1-3, 11-16] and will not be pursued in detail other than to complete a qualitative discussion of the AB_4 example.

Irrespective of the level of sophistication of the MO calculations it is possible to make use of molecular symmetry to factor the energy determinant into smaller diagonal blocks, one for each irreducible representation. For example, the energy determinant for the MOs of a_{1g} symmetry will be

$$\begin{vmatrix} H_{ss} - ES_{ss} & H_{sd} - ES_{sd} & H_{sr} - ES_{sr} & H_{st} - ES_{st} \\ H_{ds} - ES_{ds} & H_{dd} - ES_{dd} & H_{dr} - ES_{dr} & H_{dt} - ES_{dt} \\ H_{rs} - ES_{rs} & H_{rd} - ES_{rd} & H_{rr} - ES_{rr} & H_{rt} - ES_{rt} \\ H_{ts} - ES_{ts} & H_{td} - ES_{td} & H_{tr} - ES_{tr} & H_{tt} - ES_{tt} \end{vmatrix} = 0 \qquad (18.16)$$

where $s \equiv s^A$, $d \equiv d_{z^2}^A$, $r \equiv \frac{1}{2}(s_1^B + s_2^B + s_3^B + s_4^B)$ and $t \equiv \frac{1}{2}(p_{x1}^B + p_{x2}^B + p_{x3}^B + p_{x4}^B)$. The form of the four resulting MOs will be

$$\psi_i^{MO}(a_{1g}) = a_{i1}s + a_{i2}d + a_{i3}r + a_{i4}t \qquad (18.17)$$

where the a_{ij} coefficients are found by solving the four secular equations of the form

$$\sum_{j=s,d,r,t} a_{ij}(H_{ij} - ES_{ij}) = 0 \qquad i = s, d, r, t \qquad (18.18)$$

subject to the additional condition

$$\sum_{j=s,d,r,t} a_{ij}^2 = 1 \qquad i = s, d, r, t \qquad (18.19)$$

Looking at Table 18.1 one can determine that there will be seven additional energy determinants of magnitude (symmetry); 1×1 (a_{2g}), 2×2 (a_{2u}), 3×3 (b_{1g})h 2×2 (b_{2g}), 1×1 (b_{2u}), 4×4 (e_g), and 8×8 (e_u). The solution of all eight of the energy determinants will give a total of 25 MOs, some of which are essentially pure SALCs or AOs.

Since it is the orbital energy levels which can be correlated with the observed electronic spectra of molecules the next section will present the qualitative energy-level diagrams for a number of elementary types of molecules. It must be recognized that the ordering of the energy levels of any particular molecule will be a function of the atoms which contribute. Therefore, such diagrams are at best qualitative and do not universally represent the behavior of all molecules of a given type. This point is clearly illustrated by the calculations of Gray and

Ballhausen [17] on AB_4 planar molecules. Their MO energy-level diagrams for molecules where *A* is a metal ion (Ni^{2+}, Pt^{2+}, Pd^{2+}, Au^{3+}) and the *B*s are (1) ligands which have no π orbital systems (Cl^-, Br^-), or (2) ligands which have a π-orbital system (CN^-) are illustrated in Figs. 18.2 and 18.3.

18.4 MOLECULAR ORBITAL SCHEMES, ELECTRON CONFIGURATIONS, AND SYMMETRY OF AH_2-TYPE MOLECULES

For symmetrical triatomic molecules which have either $D_{\infty h}$ or C_{2v} symmetry, the MO scheme is readily developed from group theory considerations and can even be viewed in terms of a "united atom"-"separated atom" correlation diagram. For a linear hydride AH_2, where only the 1*s* AOs of the H atoms are considered,

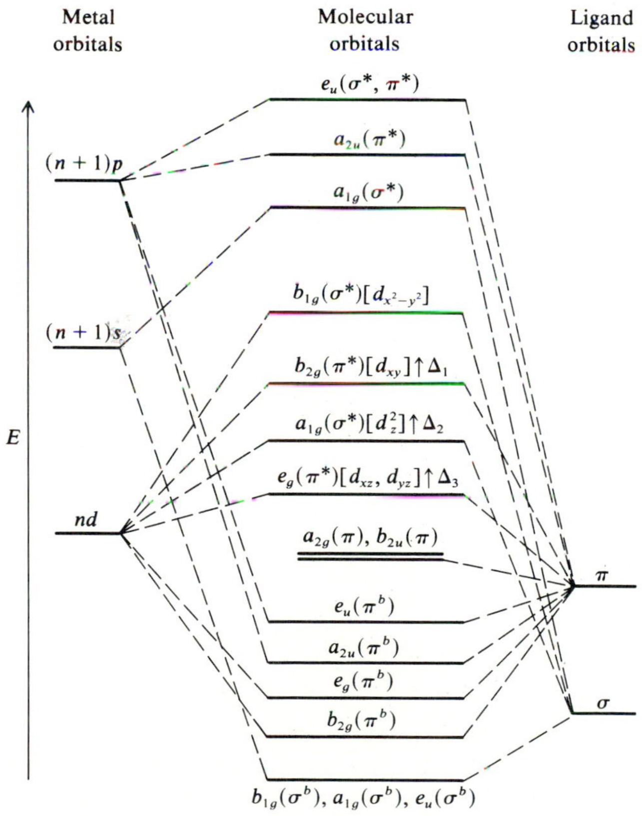

FIGURE 18.2
MO energy-level diagrams for an AB_4 species with *d* orbitals on *A* and no intraligand π-orbital system on the *B*s.

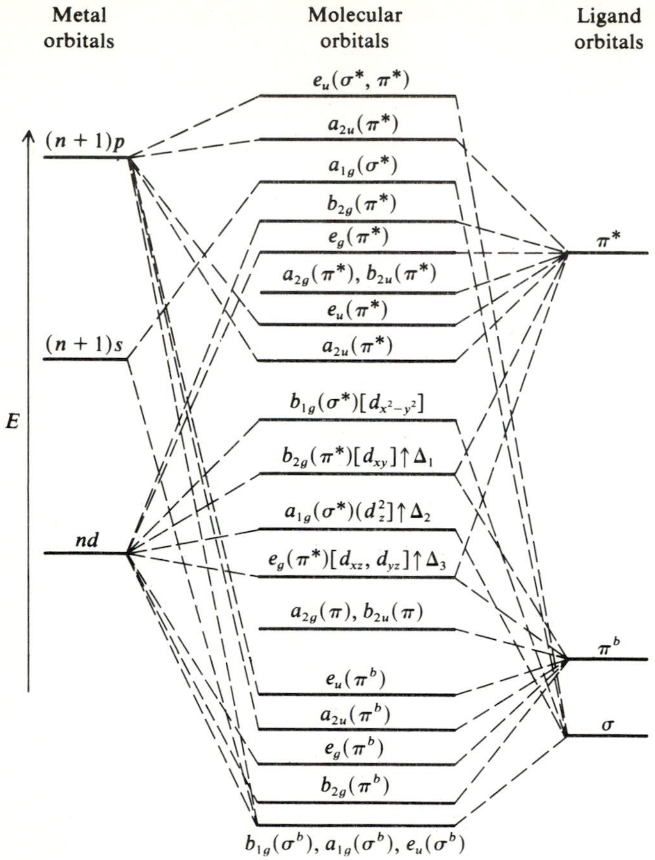

FIGURE 18.3
MO energy-level diagram for an AB_4 species with d-orbitals on A and a π-orbital system on the Bs.

the MOs are formed from the set of s and p AOs on A and the $1s$ AOs on the H atoms. Consulting the $D_{\infty h}$ character table in App. J it is observed that the $2s$ and $2p_z$ AOs in A belong to the σ_g^+ and σ_u^+ representations while the $2p_x$ and $2p_y$ AOs on A belong to the π_u representation. To consider the $1s$ AOs of H we can employ the SALCs, $1s_{\mathrm{H1}} \pm 1s_{\mathrm{H2}}$ which belong to $\sigma_g^+(+)$ and $\sigma_u^+(-)$. The resulting MO scheme, showing the energy levels, symmetry notations, and schematics of the combining AOs is shown in Fig. 18.4.

The normal convention for designating the energy levels is to number MOs of the same symmetry sequentially from that of lowest energy. It should also be noted that AOs which are not involved in bonding are referred to as nonbonding MOs.

The bent AH_2-type molecule has C_{2v} symmetry and the notation changes to conform with the representations of the C_{2v} point group. In this case the s, p_z, p_x, and p_y AOs on A belong to a_1, a_1, b_1, and b_2 representations, respectively.

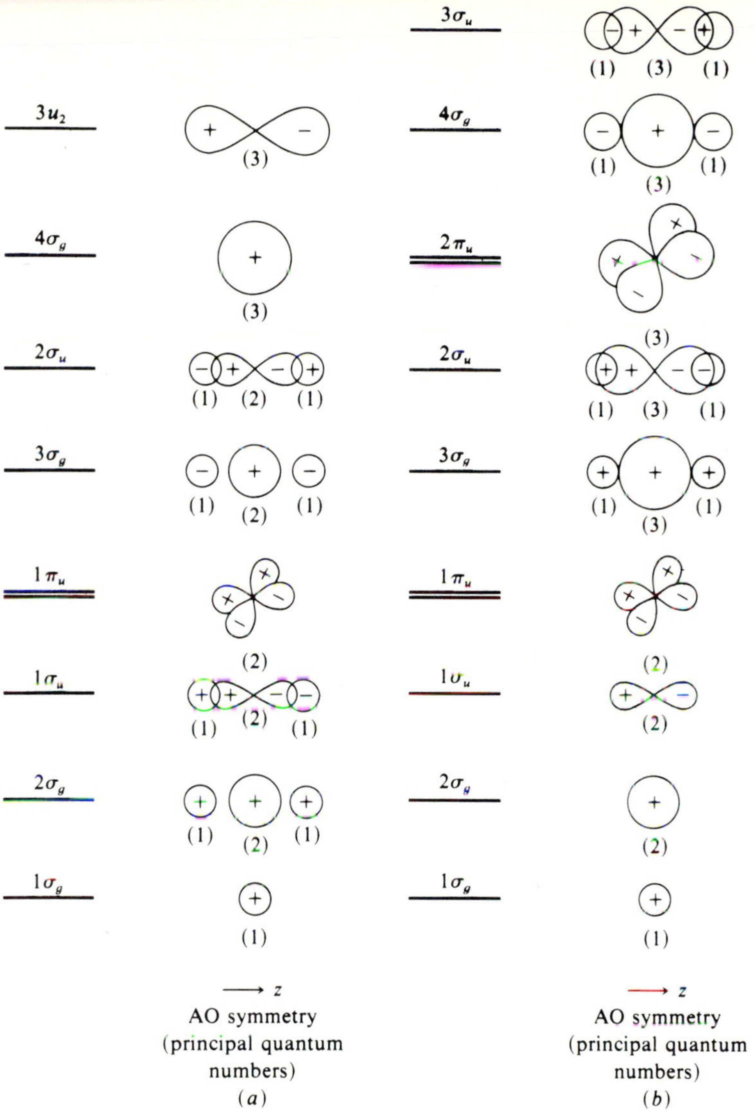

FIGURE 18.4
Molecular orbital scheme for linear AH_2 molecules: (*a*) A = period 2 element; (*b*) A = period 3 element.

The symmetry adapted linear combinations of the H 1*s* AOs will belong to a_1 and b_2 representations. The MO scheme for bent AH_2 molecule is shown in Fig. 18.5.

There are two additional qualitative aids which can be used to describe the ordering of MOs in simple molecules. The first of these is a correlation diagram

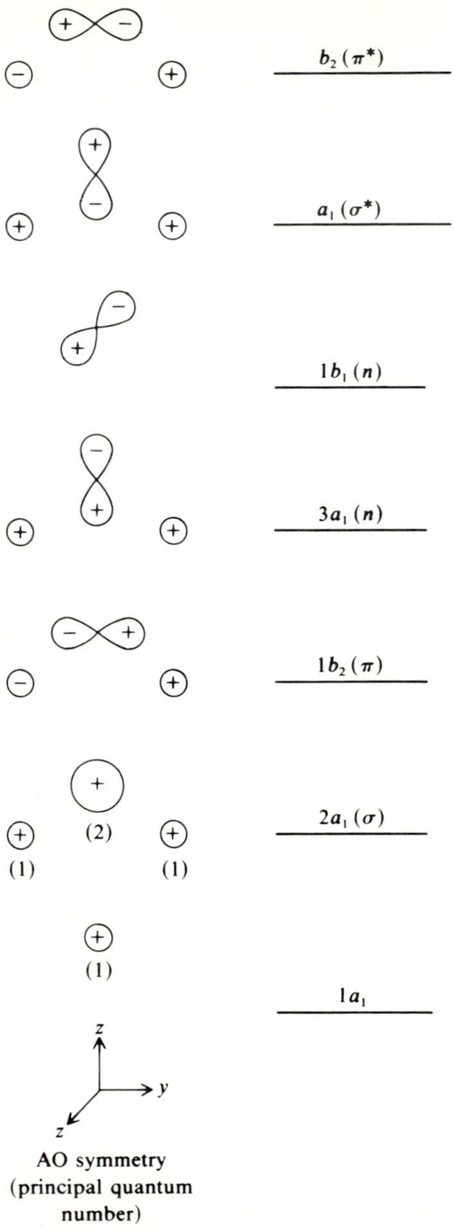

FIGURE 18.5
Molecular orbital scheme for bent AH_2 molecules (A = second period element).

analogous to those developed for diatomic molecules in Chap. 6. Figure 18.6 illustrates a correlation diagram for an AH_2 bent molecule.

A second useful concept, especially for comparing the symmetry and energy levels of similar molecules, is that of Walsh diagrams [9]. These diagrams illustrate the relationships among the MO energy levels as a molecule undergoes a change in a bond angle. Figure 18.7 illustrates this for the AH_2 molecular system.

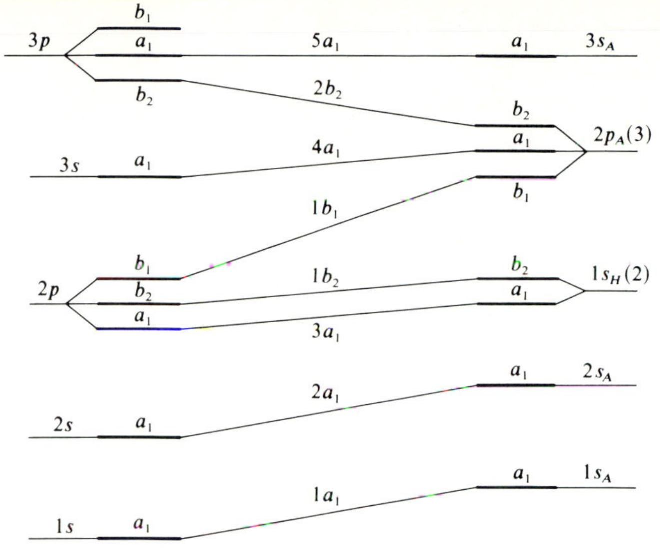

FIGURE 18.6
Correlation diagram for a bent AH_2 molecule where A is a second period element.

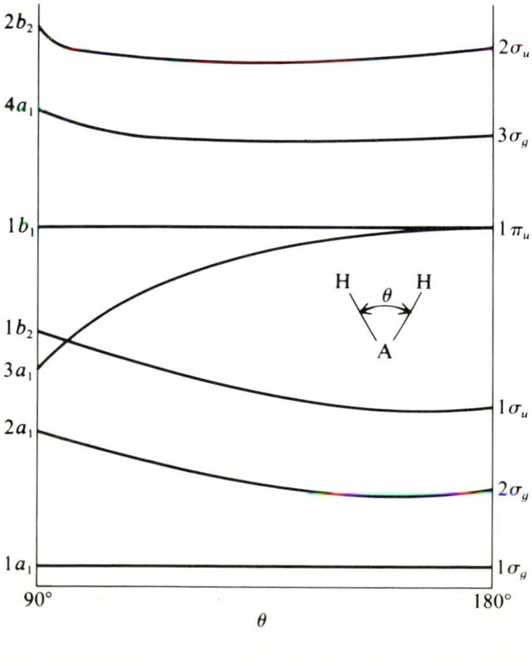

FIGURE 18.7
Walsh diagram for AH_2 molecules.

It was noted earlier that the ordering of MO energy levels will depend on the relative energies of the AOs used as a basis for the MOs. The precise order can be determined, for a given molecule, through a combination of the analysis of the experimental electronic spectrum and quantum mechanical calculations of increasing sophistication.

Up to this point we have illustrated several ways of presenting energy-level diagrams. These have consisted of a simple ordering based on AO symmetries (Figs. 18.4 and 18.5), correlation diagrams (Fig. 18.6) and Walsh diagrams (Fig. 18.7). While each of these presentations has certain advantages the most common method is to use the system illustrated in Figs. 18.2 and 18.3. In these diagrams the available central atom atomic orbitals are presented in order of increasing energy on the left-hand side and all peripheral or ligand SALCs of the same energy and symmetry are referenced to the same vertical position on the right-hand side.

Diagrams of this type will be used to illustrate the MO schemes for several different types of molecules in later sections and are shown in Fig. 18.8 for linear and bent AH_2 molecules.

When referring to MO diagrams of the more general type shown in Fig. 18.8, there are three characteristics to keep in mind.

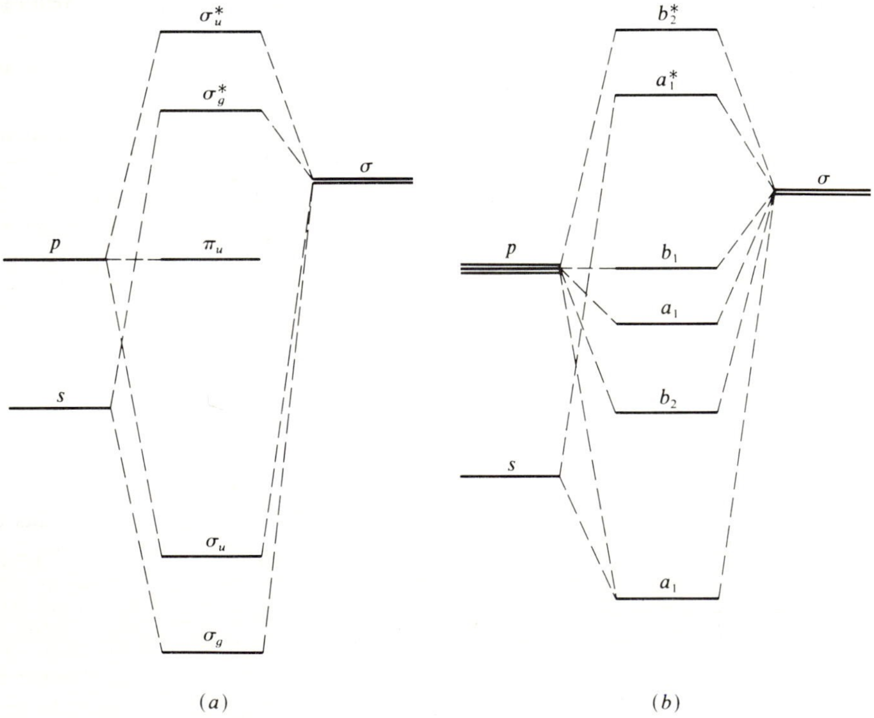

FIGURE 18.8
Molecular orbital energy-level diagram for (*a*) linear AH_2 and (*b*) bent AH_2 molecules.

1. Unless the diagram refers to a particular molecule the sequential numbering preceding the level symbol is often omitted.
2. For specific molecules the ordering of the levels will not necessarily be that shown in a general diagram, this depending on the relative energies of the contributing atomic orbitals.
3. There may be more levels connected (dotted lines) than you would anticipate. This is due to the fact that the MOs do not necessarily contain contributions from only one AO and one SALC. This can be rationalized by viewing the AOs on the central atom to be a set of hybridized AOs which can then combine with the SALCs to form the MOs.

We can revert to diatomic molecules to provide an elementary illustration of this concept. Figure 18.9 illustrates this concept for the Na_2 molecule where *sp*-hydridization is involved.

A further example is afforded by the bent AH_2 molecule for which Fig. 18.10 represents a modification of Fig. 18.9*b* to illustrate the effect of considering the *A* atomic orbitals to be sp^3-hybrids, which then unite with the ligand SALCs to form the MOs. Note that either approach leads to the same set of MO energy levels.

The last point to be examined relative to AH_2 molecules is the shorthand notations which are used to denote the electron arrangement and the determination of the symmetry representation of a configuration. The notation used to show electron occupancy of the MOs in polyatomic molecules is analogous to that used for diatomic molecules. The MOs are denoted using the symbols for the irreducible representations and superscripts to show the number of occupying electrons. Table 18.3 shows some typical configuration for AH_2 molecules.

The symmetry species for a configuration is that of the direct product of the symmetries of the MOs occupied by the electrons. Considering the first excited state of BH_2, for example, one finds that

$$\sigma_g \times \sigma_g \times \sigma_u \times \sigma_u \times \pi_u = \pi_u$$

Since there is a single unpaired electron the molecule will be a doublet state, $^2\pi_u$.

Even for simple polyatomic molecules the ordering of the energy levels is not as straightforward as for diatomics. For example, the CH_2 molecule has a bond angle of about 136° in the ground state. This indicates that there may be a question as to which of the $3a_1$ or $1b_2$ levels is lower. Actually the two levels are sufficiently close that the $^3B_1$ configuration $(2a_1)^2(1b_2)^2(3a_1)^1(1b_1)^1$ with electrons having parallel spins is lower than either the $^1A_1$ configuration $(2a_1)^2(1b_2)^2(3a_1)^2$ or the $^1B_1$ configuration $(2a_1)^2(1b_2)^2(3a_1)^1(1b_1)^1$ with the $3a_1$ and $1b_1$ electrons having paired spins. Another low-lying state which will have the $^1A_1$ configuration $(2a_1)^2(1b_2)^2(1b_1)^2$ is actually a linear conformation $(2\sigma_g)^2(1\sigma_u)^2(1\pi_u)^2$. This ordering of states has been confirmed experimentally.

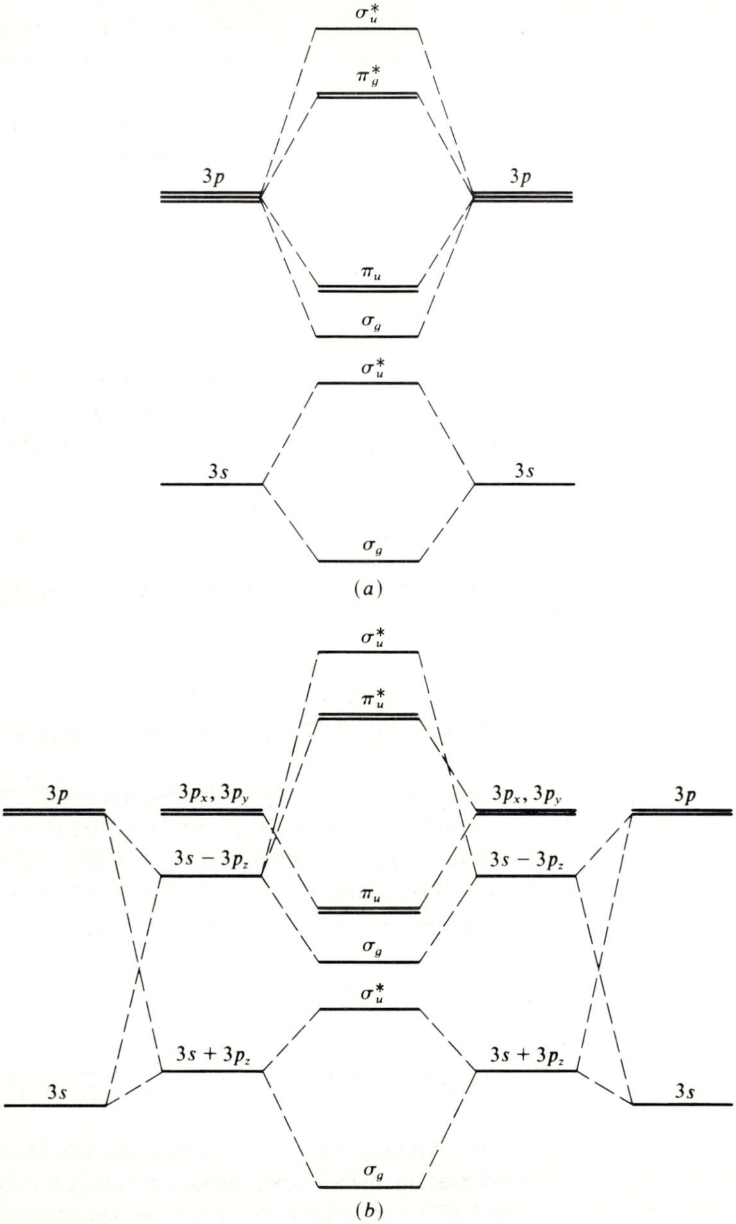

FIGURE 18.9
Molecular orbital energy-level diagram for Na_2 (*a*) without invoking hybridization and (*b*) with the consideration of hybridization.

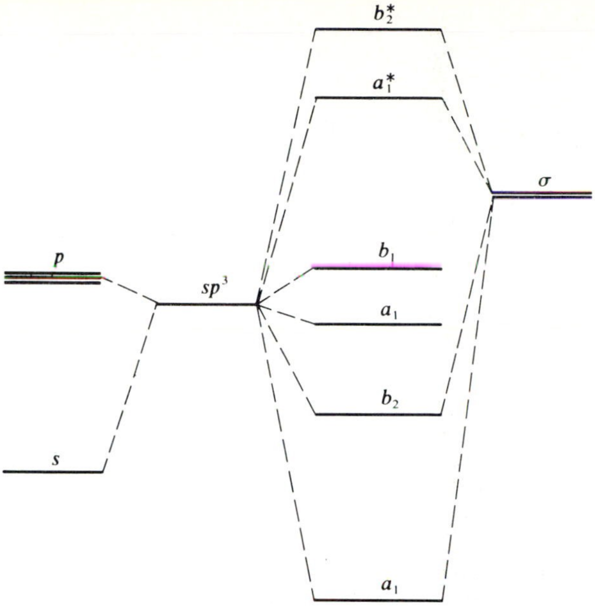

FIGURE 18.10
Molecular orbital energy-level diagrams of a bent AH_2 molecule when hybridization is considered.

At first examination the occurrence of an orbital state $^1A_1$ with paired spins being higher than the $^3B_1$ state with unpaired spins may seem contradictory. However, one must remember that the model on which these simple MO schemes are based does not make allowance for electron reorganization, which is the phenomenon of all MO energies being changed somewhat by the promotion of one or more electrons to higher levels.

Another notation used with polyatomic molecules is a symbol which precedes the symmetry notation, and denotes the ordering of the states. The ground

TABLE 18.3
Electronic configurations and symmetries of some AH_2 molecules

Molecule	No. of electrons	Notes	Configuration†	Symmetry
BeH_2	4	Unstable species—ground state	$(2\sigma_g)^2(1\sigma_u)^2$	$\tilde{X}\,^1\Sigma_g^+$
BH_2	5	Ground state	$(2a_1)^2(1b_2)^2(3a_1)^1$	$\tilde{X}\,^2A_1$
BH_2	5	First excited state	$(2\sigma_g)^2(2\sigma_u)^2(1\pi_u)^1$	$\tilde{A}\,^2\Pi_u$
CH_2	6	Low-energy states	$(2a_1)^2(1b_2)^2(3a_1)^2$	$\tilde{a}\,^1A_1$
	6		$(2a_1)^2(1b_2)^2(3a_1)^1(1b_1)^1$	$\tilde{X}\,^3B_1$
	6		$(2a_1)^2(1b_2)^2(3a_1)^1(1b_1)^1$	$\tilde{b}\,^1B_1$
	6		$(2a_1)^2(1b_2)^2(1b_1)^2$	$\tilde{c}\,^1A_1$
OH_2	8	Ground state	$(2a_1)^2(1b_2)^2(3a_1)^2(1b_1)^2$	$\tilde{X}\,^1A_1$

† Core electrons which do not participate in the bonding are not denoted.

state is denoted by $\tilde{X}$ and successive members of states of comparable multiplicity are denoted by successive alphabetic characters beginning with $\tilde{A}$ or $\tilde{a}$.

18.5 MOLECULAR ORBITAL SCHEMES, ELECTRON CONFIGURATIONS, AND SYMMETRY OF AB_2 MOLECULES

In an AH_2 molecule, substitution of the hydrogen atoms with other atoms beyond the first period rapidly increases the complexity of the MO scheme. This increased complexity arises from the presence of ligand orbitals other than 1*s*. Without detailed development short illustrative summaries, consisting of MO energy level diagrams of the type given in Fig. 18.2, and Walsh diagrams where appropriate, will be presented for selected molecular frameworks.

Figure 18.11 shows a typical MO energy-level scheme for linear and bent AB_2 molecules with no *d* orbitals available on the A atom while Fig. 18.12 shows a similar scheme for molecules with *d* orbitals available on the A atom.

Table 18.4 lists the electronic configurations and bond angles for several common AB_2-type molecules in their ground and low-lying excited states.

Figure 18.13 shows a typical Walsh-type diagram to illustrate the correlation between the MO energy levels of a bent and a linear AB_2-type molecule.

In principle there can be a contribution to any one MO from any central atom AO or SALC of the same symmetry, hence the multitudes of connecting lines in Figs. 18.11 and 18.12. In practice the determination of how many and to what extent individual AOs or SALCs contribute to a particular MO depends on the degree of sophistication of the quantum mechanical calculations. It is also to be recognized that the ordering of the MO energy levels will depend significantly on the relative energies of the AOs and SALCs, both within each type and between the two groups.

18.6 MOLECULAR ORBITAL SCHEMES FOR OTHER AB_n-TYPE MOLECULES

This section will present several typical molecular orbital schemes for some more complex molecules of the AB_n type where $n \geqq 3$. In all diagrams the number of connecting lines between the AOs of atom A or the SALCs of the B atom and the MOs are kept at a minimum to avoid crowding. However, one should keep in mind the discussion at the end of the previous section.

Another point to which attention must be directed involves the notation of MO levels belonging to b_n representations of point groups such as C_{2v}. The designation b_1, b_2 can be reversed, depending on the definition of the axis system employed for defining the molecular plane and that used to establish the symmetry group. The most common notation is for x to belong to b_1.

As with the linear and bent AB_2 types of molecules it is possible to correlate the MO energy levels between other pairs of molecules which are related by a single angular measurement. For example, Fig. 18.18 shows the correlation between the AB_3 planar (D_{3h}) and pyramidal (C_{3h}) molecules [19].

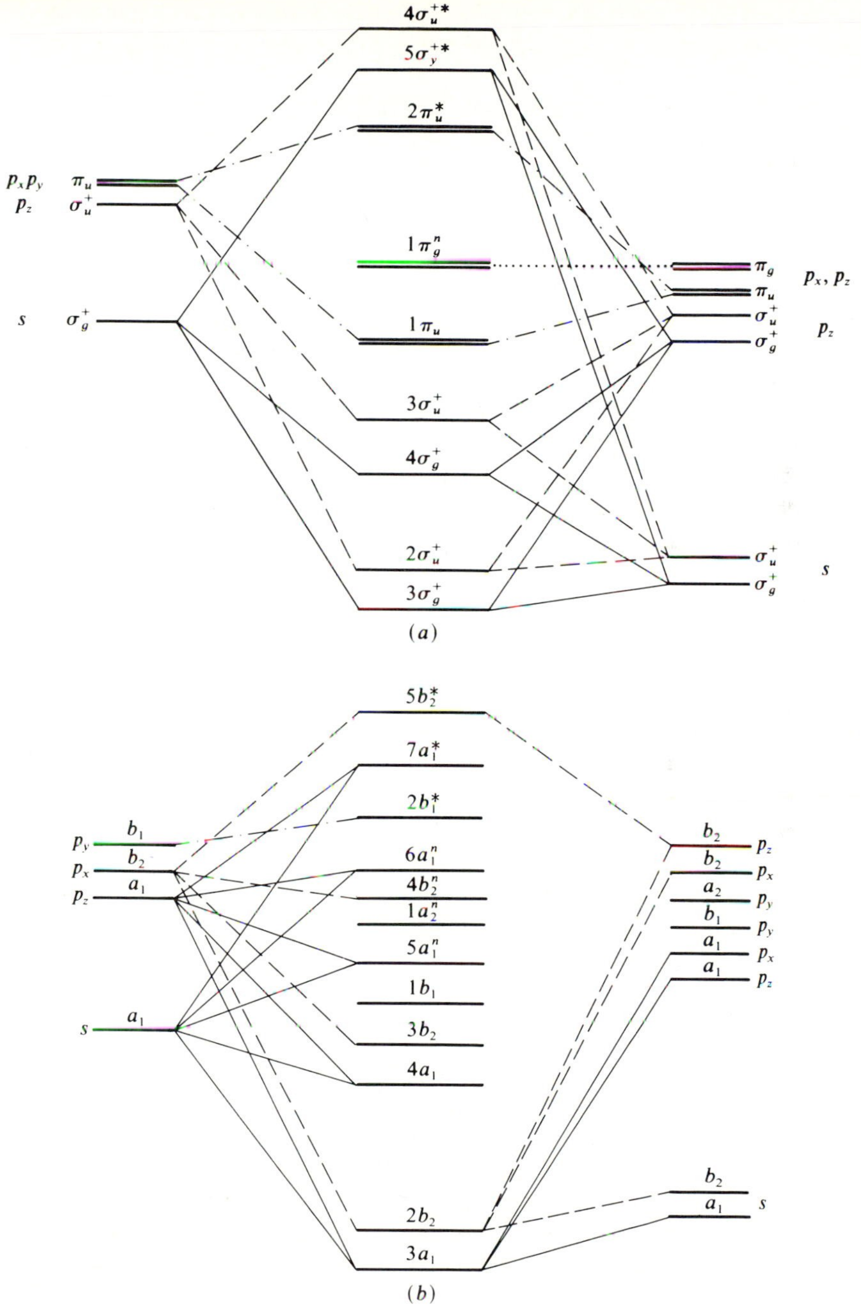

FIGURE 18.11
Typical MO energy-level diagrams for (a) linear ($D_{\infty h}$) and (b) bent (C_{2v}) AB_2-type molecules with no d orbitals on the A atoms ($x \perp AB_2$ plane).

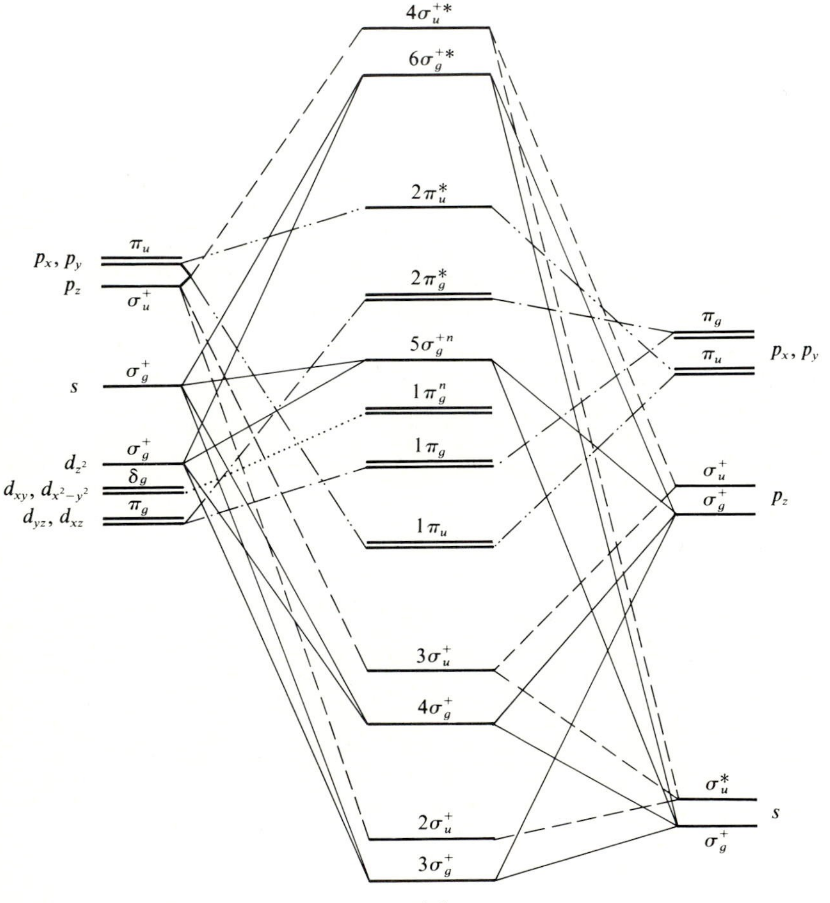

(a)

FIGURE 18.12
Typical MO energy-level diagrams for (*a*) linear ($D_{\infty h}$) and (*b*) bent (C_{2v}) AB_2-type molecules with *d* orbitals on the A atoms. (Some of the AO to MO connecting lines have been omitted for clarity—in general all levels of comparable symmetry are connected as in Fig. 18.10.)

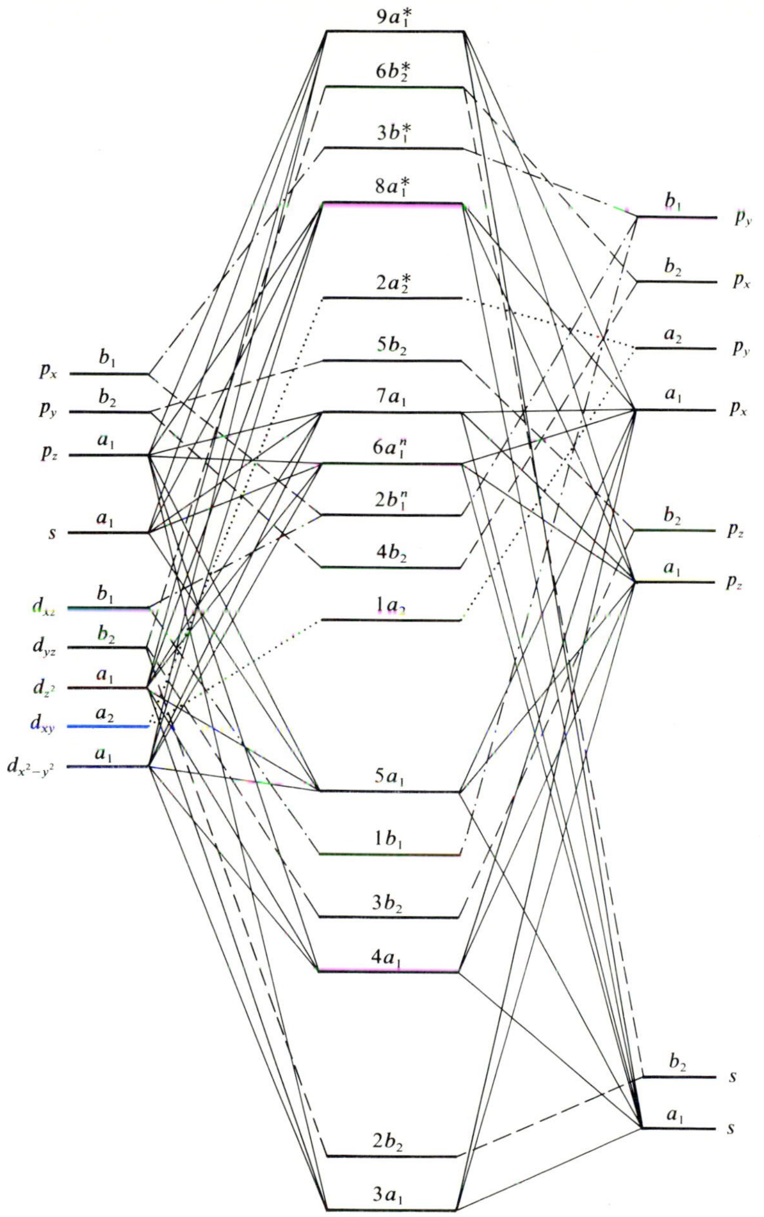

(*b*)

FIGURE 18.12 (continued)

TABLE 18.4

Molecule	No. of electrons	Electron configuration (valence shell only)	$\angle BAB$	State symbol
CO_2	22	$(3\sigma_g)^2(2\sigma_u)^2(4\sigma_g)^2(3\sigma_u)^2(\pi_u)^4(\pi_g)^4$	180°	$\tilde{X}\,^1\Sigma_g^+$
		$(3a_1)^2(2b_2)^2(3b_2)^2(4a_1)^2(4b_1)^2(5a_1)^2(1a_2)^2(4b_2)^1(6a_1)^1$		
C_3	18	$(3\sigma_g)^2(2\sigma_u)^2(4\sigma_g)^2(\pi_u)^4(3\sigma_u)^2$	180°	$\tilde{X}\,^1\Sigma_g^+$
		$(3\sigma_g)^2(2\sigma_u)^2(4\sigma_g)^2(1\pi_u)^4(3\sigma_u)^2(1\pi_g)^1$	180°	$\tilde{A}\,^1\Pi_u$
NO_2	23	$(3a_1)^2(2b_2)^2(3b_2)^2(4a_1)^2(4b_1)^2(5a_1)^2(1a_2)^2(4b_2)^2(6a_1)^1$	134.1°	$\tilde{X}\,^2A_1$
		$(3a_1)^2(2b_2)^2(3b_2)^2(4a_1)^2(4b_1)^2(5a_1)^2(1a_2)^2(4b_2)^1(6a_1)^2$	121°	$\tilde{B}\,^2B_2$

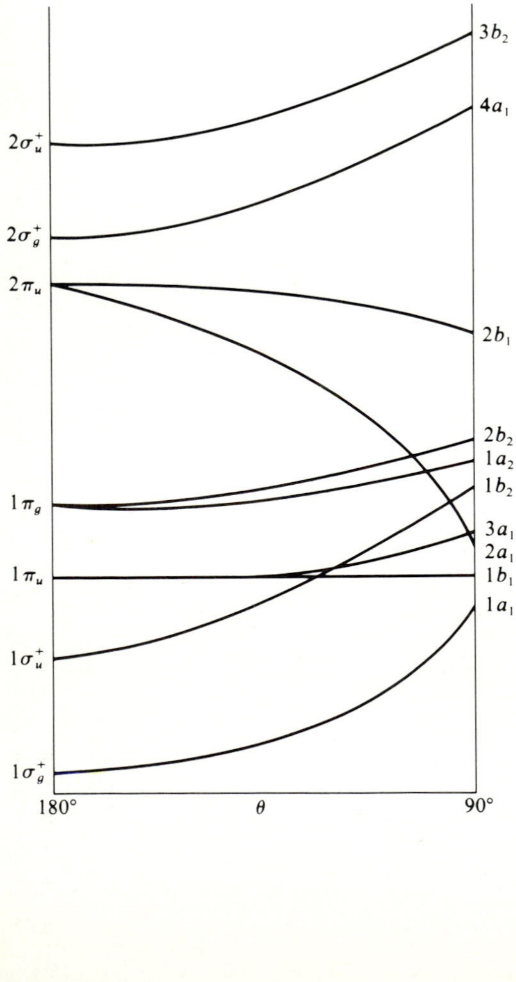

FIGURE 18.13
Walsh-type diagram for AB_2-type molecules.

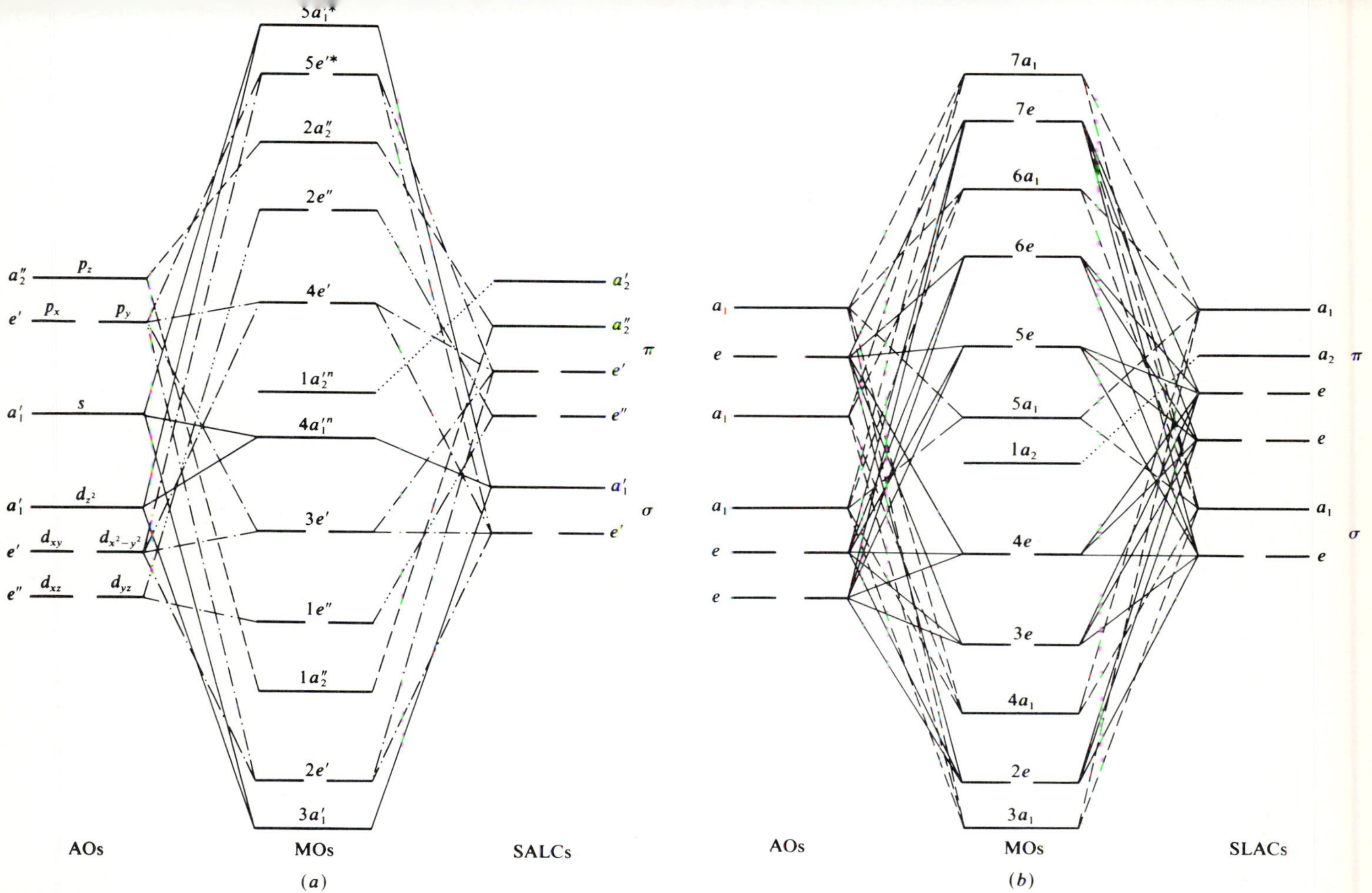

FIGURE 18.14
Typical MO energy-level diagrams for (*a*) planar (D_{3h}) and (*b*) pyramidal (C_{3v}) AB_3-type molecules.

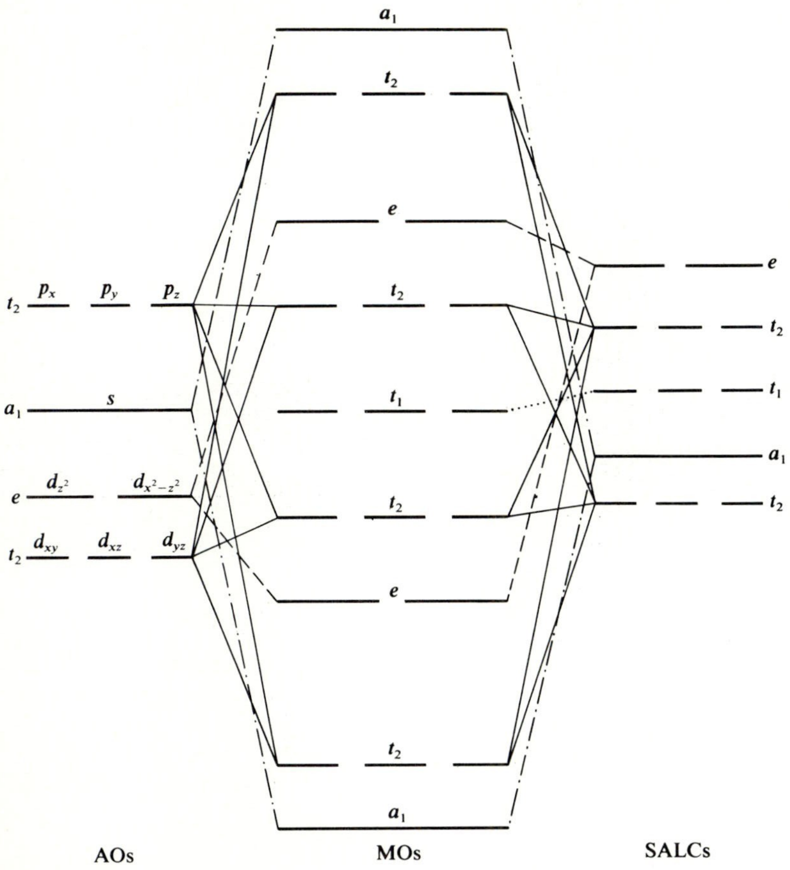

FIGURE 18.15
Typical MO energy-level diagram for tetrahedral (T_d) AB_4 molecules: (a) no π bonding; (b) with π bonding.

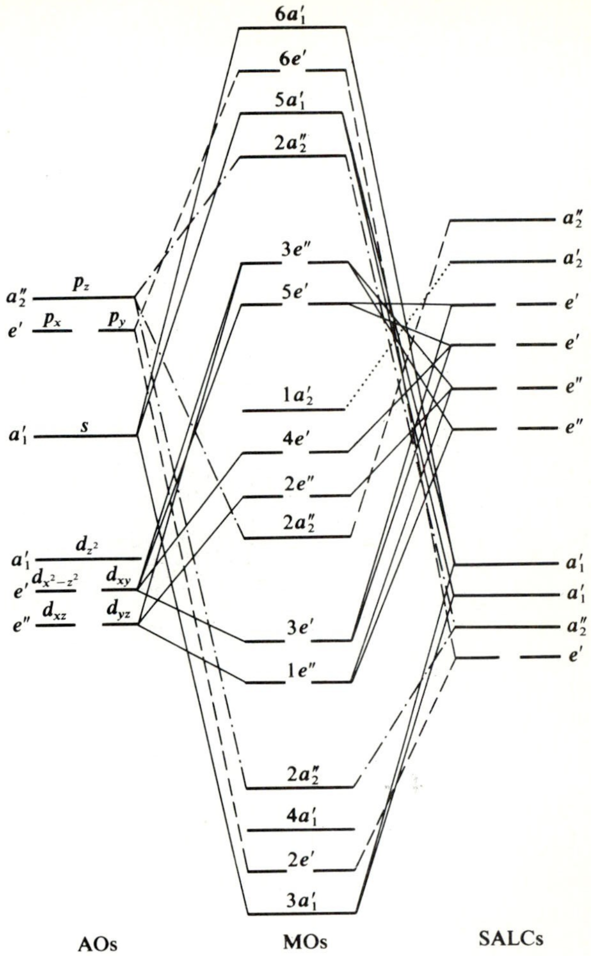

FIGURE 18.16
Typical MO energy-level diagram for trigonal bipyramid (D_{3h}) AB_5 molecules.

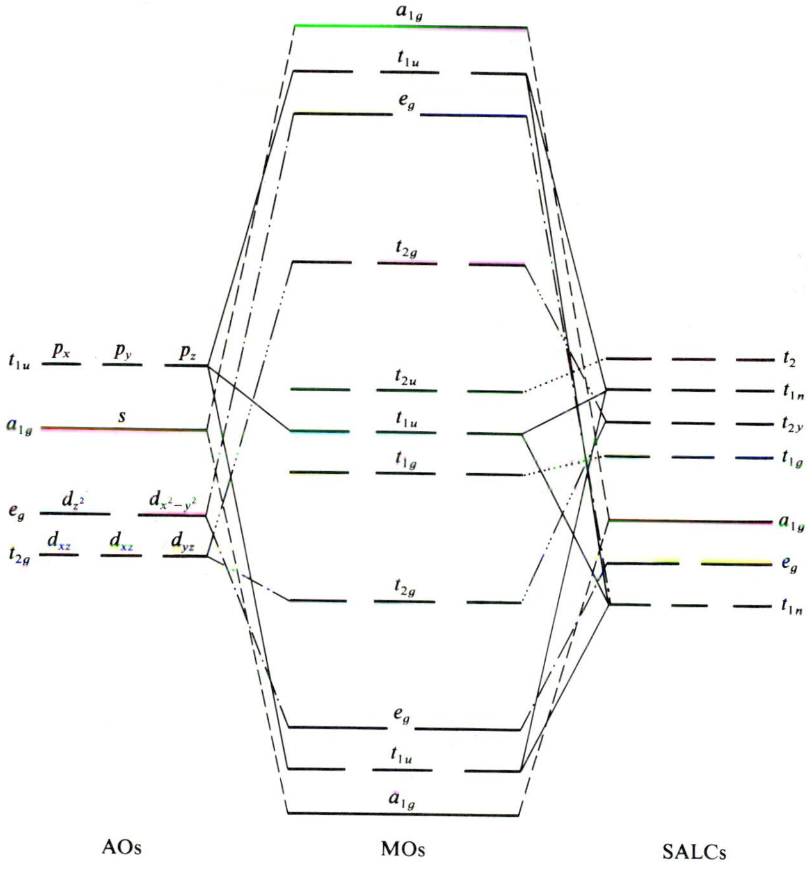

FIGURE 18.17
Typical MO energy-level diagrams for octahedral (O_h) AB_6-type molecules.

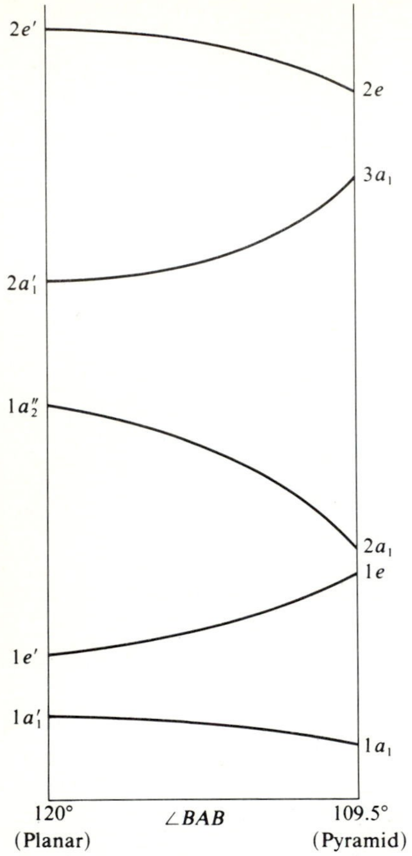

FIGURE 18.18
Walsh-type diagram for AB_3 molecules.

18.7 MOLECULAR ORBITAL SCHEMES FOR SIMPLE NONAROMATIC MOLECULES

In this section we will present, without formal mathematical derivations, typical MO energy level diagrams for several types of small molecules. Although there are relatively few H_2AB type molecules it will be instructive to look at this system because of the wealth of experimental information about the formaldehyde, $H_2C{=}O$, molecule. Figure 18.19 shows the MO energy-level diagram. Since molecules of this type do not have the symmetry characteristics associated with an AB_n-type molecule, which contains a central atom, it is not possible to present MO diagrams of the type shown in Figs. 18.14–18.17. Instead the diagram is a simple vertical scale indicating the ordering and nature of the levels. Next we will review the nature of the H_2AB orbitals and look qualitatively at the ordering of the energy levels.

Consideration of the Lewis structure of H_2AB where A and B are second-period elements leads to the conclusion that there will be three σ bonding orbitals, a π bonding orbital, and two lone pair orbitals. There will be a total of 10

$8a_1 \ (\sigma^*)$

$7a_1 \ (\sigma^*)$

$6a_1 \ (\sigma^*)$

$2b_1 \ (\pi^*)$

$2b_2 \ (n)$

$1b_1 \ (\pi)$

$5a_1 \ (\sigma)$

$1b_2 \ (\sigma)$

$4a_1 \ (\sigma)$

$3a_1 \ (\sigma)$

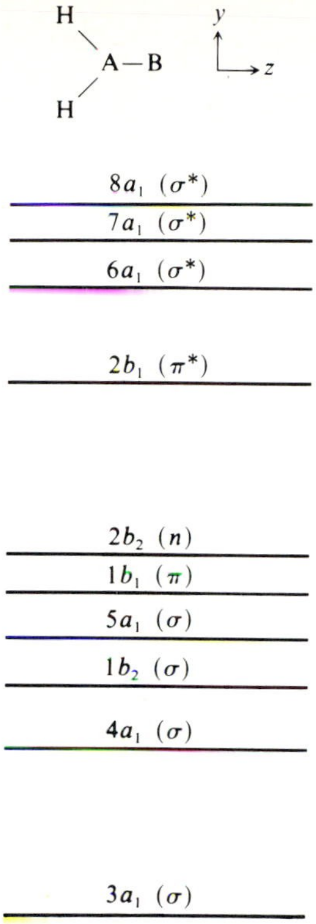

FIGURE 18.19

Typical MO energy-level diagram for an H_2AB-type molecule.

contributing AOs so there will also have to be three σ antibonding orbitals and a π antibonding orbital. The only deviation from this predicted pattern, as shown by the calculations of Walsh [20], is that the lone pair orbital of a_1 symmetry is mixed with the H—A and A—B σ bonding orbitals so that all three orbitals will have mixed σ bonding lone pair character. These three orbitals, along with the σ orbital of b_1 symmetry, will have the lowest energy. The π bonding and antibonding orbitals will be higher and lie below and above the lone pair orbital. The three σ antibonding orbitals of a_1 symmetry will be highest.

    Another small molecular type of interest, because its basic configuration occurs as a part of many larger molecules, is

$$\begin{array}{cc} C & D \\ \diagdown & \diagup \\ \diagup A{=}A \diagdown & \\ B & E \end{array}$$

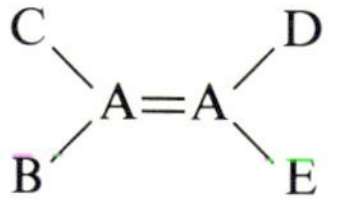

To consider all possible systems of this type is too extensive an undertaking at this point, so the discussion will be limited to $H_2A{=}AH_2$ and $H_2A{=}AHB$ types. The MO schemes for these molecules are shown in Fig. 18.20. The Lewis structure of $H_2A{=}AH_2$ indicates that there should be five σ bonding orbitals and one π bonding orbital. In addition there will be a complement of six antibonding

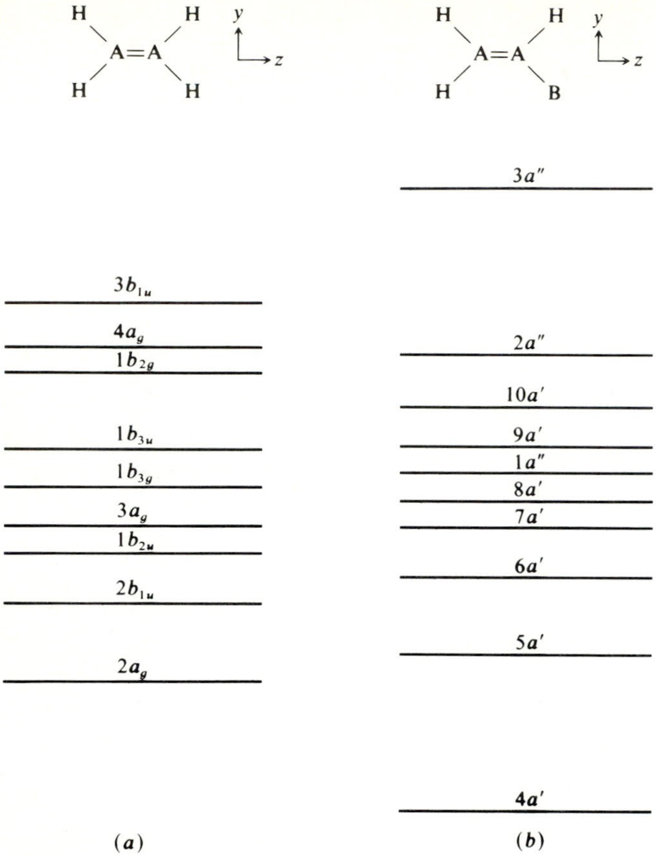

FIGURE 18.20
Typical MO energy-level diagrams for (a) $H_2A{=}AH_2$ and (*b*) $H_2B{=}AHB$ (B = halogen) molecules.

orbitals. The more symmetric configuration of the $H_2C{=}CH_2$ molecule (D_{2h}) allows one to utilize the methods of group theory [7] to establish the symmetry of the MOs. For D_{2h} symmetry the four hydrogen s atomic orbitals form the basis for SALCs belonging to a_g, b_{1g}, b_{2u}, and b_{3u} and the s, p_x, and p_y atomic orbitals on the A atoms transforms as $2a_g + b_{1g} + b_{2u} + 2b_{u3}$. Thus these will be four C—H bonding and four C—H antibonding orbitals with one of each belonging to the a_g, b_{1g}, b_{2u}, and b_{3u} representations. The remaining a_g and b_{3u} orbitals will be the bonding and antibonding A—A orbitals. The composition of the MOs is established by using projection operators to form SALCs and subsequently combining them with the correct symmetry adaptations of the A orbitals. The details of this method and the calculation of orbital energies is reviewed by Pilar [21] and the results are summarized in Fig. 18.20.

If the A atoms are carbon then $H_2C{=}CH_2$ will have an electron configuration of $(2a_g)^2(2b_{1u})^2(1b_{2u})^2(1b_{3u})^2(3a_g)^2(1b_{3u})^2$. The highest orbital, $1b_{3u}$, is a

π orbital. Since the electrons, which will participate in many spectral transitions, will originate from this orbital, Hückel molecular orbital theory is often used to describe this and other systems where the highest filled orbitals are π type.

18.8 MOLECULAR ORBITAL SCHEMES FOR AROMATIC MOLECULES

Hückel MO theory has been found to be particularly useful for discussing conjugated carbon atom systems. Details of its development are to be found elsewhere [22] and we will only review the general nature of the theory and look at the results for a few molecules. The theory is based on four main assumptions.

1. Only π orbitals are considered in setting up the MO scheme. The lower the molecular symmetry the more this becomes an approximation.
2. All overlap integrals $S_{nm} = \int \phi_n \phi_m \, dv$ $(n \neq m)$ where ϕ_i is the atomic p_x orbital on atom i, are considered to vanish.
3. The Coulomb integrals, $H_{nn} = \int \phi_n \mathscr{H} \phi_n \, dv = \alpha$ are assumed to be equal for all carbon atoms.
4. The resonance integrals $H_{nm} = \int \phi_n \mathscr{H} \phi_m \, dv$ are assumed to have a common value β for all adjacent carbon pairs and to be zero for nonadjacent atom pairs.

The MO wavefunctions are then linear combinations of the atomic p_x orbitals ($x \perp$molecular framework). The energies of the resulting MOs are found by solving the secular determinant

$$|H_{ij} - ES_{ij}| = 0 \tag{18.20}$$

after incorporation of the assumptions. For example, the secular determinant for the *s*-trans-buta-1,3-diene molecule with four carbon atoms would be

$$\begin{vmatrix} \alpha\text{-}E & \beta & 0 & 0 \\ \beta & \alpha\text{-}E & \beta & 0 \\ 0 & \beta & \alpha\text{-}E & \beta \\ 0 & 0 & \beta & \alpha\text{-}E \end{vmatrix} = 0 \tag{18.21}$$

For large molecules it is possible to use symmetry to factor the determinant into more manageable blocks. The solution of Eq. (18.21) gives the results shown in Fig. 18.21. The values of α and β, to a first approximation, can be found by evaluation of the Coulomb and resonance integrals. They can also be determined experimentally from the observed spectral transitions. The MO wavefunctions are given by

$$\Psi_i^{\mathrm{MO}} = \sum_j c_{ij} \phi_j \tag{18.22}$$

and the c_{ij} can be found by using the set of secular equations from which the energy determinant was derived.

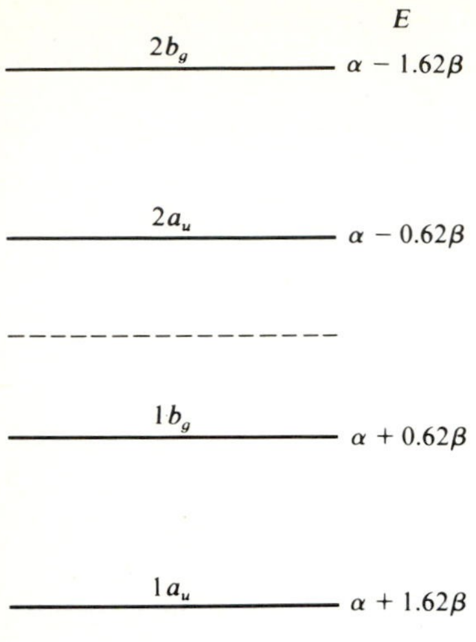

FIGURE 18.21
Hückel MO energy-level scheme for *s*-trans-buta-1,3-diene.

The energy-level scheme shown for buta-1,3-diene in Fig. 18.21 will qualitatively represent the levels for molecules such as *s*-trans-glyoxal, O=CH—CH=O, and *s*-trans-acrolein, H_2C=CH—CH=O. By modifying the value of the Coulomb and resonance integrals to

$$\alpha_x = \alpha_c + h_x\beta_c \tag{18.23}$$

$$\beta_{cx} = k_{cx}\beta_c \tag{18.24}$$

where α_c and β_c are values for a similar all-carbon system, one can apply the theory to molecules containing heteroatoms. The heteroatom constants h_x and β_{cx} are empirical in nature and are generally determined experimentally. Figure 18.22 shows the MO energy-level schemes for some additional common molecules.

In the preceding discussions of this chapter we have provided a summary of important features of molecular electronic structure. This material, although very limited, will serve as a background for the discussion of the spectra of polyatomic molecules.

18.9 SELECTION RULES

The high-resolution electronic spectra of polyatomic molecules will consist of broad bands of transitions whose components are due to simultaneous electronic, vibrational, and rotational transitions, as was the case for diatomic molecules. The primary difference will be added complexity to the vibronic and rovibronic structure due to the occurrence of several vibrational modes and the nonlinearity of most polyatomic molecules. We will first examine the selection rules which govern the basic electronic transitions and delay a discussion of those associated with vibronic and rovibronic transitions until later.

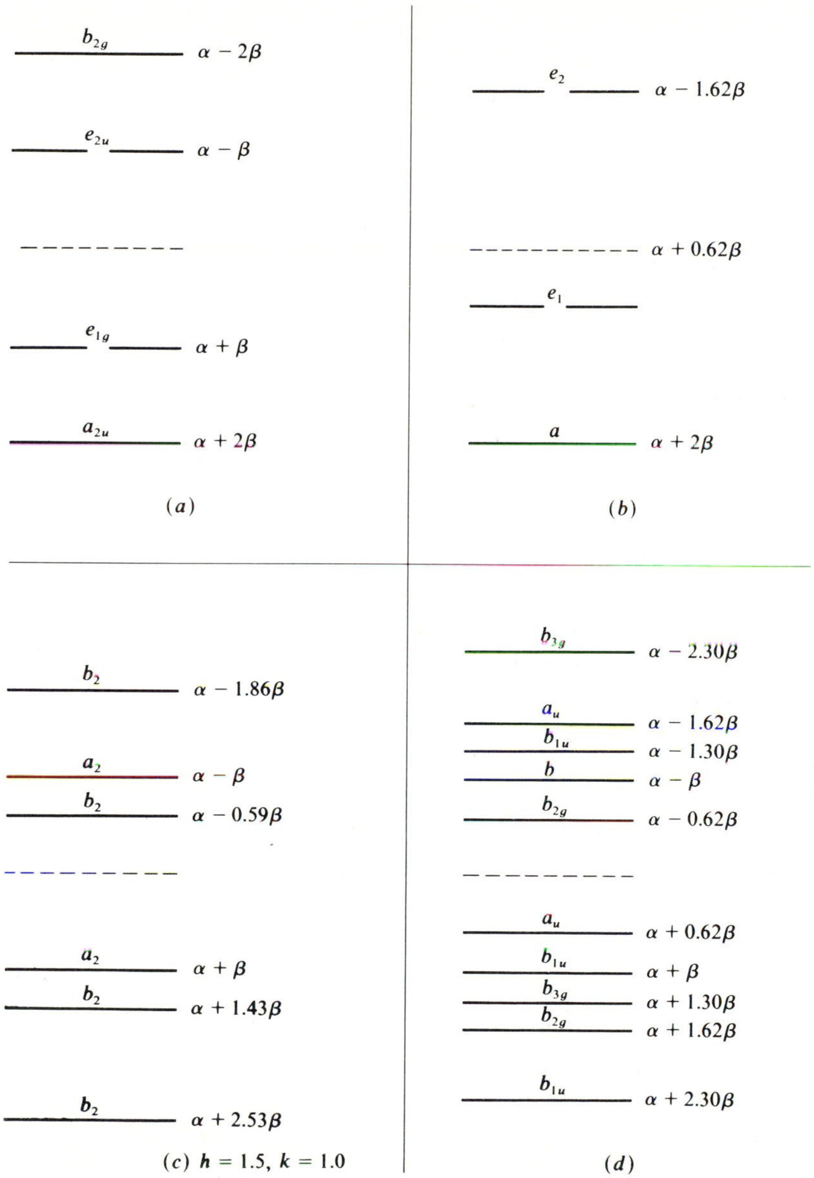

FIGURE 18.22
Hückel MO energy-level schemes for (*a*) benzene (D_{6h}), (*b*) cyclopentadiene (D_{5h}), (*c*) pyridine (C_{2v}), (*d*) naphthalene (D_{2h}).

The general requirement for an electronic transition to be observed is that the dipole integral

$$I = \int \psi'^* \boldsymbol{\mu} \psi'' \, dv \neq 0 \tag{18.25}$$

where ψ' and ψ'' are the total wavefunctions of the upper and lower states, respectively, and $\boldsymbol{\mu}$ is the electric dipole moment operator which contains components due to both electronic displacement μ_e and nuclear displacement μ_n

$$\boldsymbol{\mu} = \boldsymbol{\mu}_e + \boldsymbol{\mu}_n \tag{18.26}$$

The total wavefunction is a product of electronic, vibrational, rotation, and spin functions

$$\psi = \psi_e \psi_v \psi_t \psi_s \tag{18.27}$$

assuming that the spin orbital interaction is small. Substitution of Eqs. (18.26) and (18.27) into (18.25) gives

$$I = \int \psi_e'^* \boldsymbol{\mu}_e \psi_e'' \, dV_e \int \psi_v'^* \psi_v'' \, dV_v \int \psi_r'^* \psi_r'' \, dV_r \int \psi_s'^* \psi_s'' \, dV_s$$

$$+ \int \psi_e'^* \psi_e' \, dV_e \int \psi_v'^* \psi_r'^* \boldsymbol{\mu}_n \psi_v'' \psi_r'' \, dV_v \, dV_r \int \psi_s' \, dV_s \tag{18.28}$$

Since electronic wavefunctions belonging to different orbitals or energy states are orthogonal, the integral $\int \psi_e'^* \psi_e'' \, dV_e$ will always be zero unless $\psi'^* = \psi''$, in which case there is no transition. Therefore the second part of Eq. (18.28) vanishes. The vibrational functions ψ_v' and ψ_v'', and the rotational functions ψ_r' and ψ_r'', are not orthogonal since ψ_v', ψ_r' and ψ_r'', ψ_v'' belong to different electronic states. If the spin orbital interaction in the molecule is small, as is generally the case for light molecules, the spin functions are orthogonal and $\int \psi_s'^* \psi_s''$ will be nonzero only if the spin function is the same for the two states involved in the transition. This translates to the requirement that for an electronic transition the multiplicity does not change or $\Delta S = 0$. Combining the several concepts just discussed leads to the conclusion that the existence of an electronic transition depends on having a nonzero value for

$$I_e = \int \psi_e'^* \boldsymbol{\mu}_e \psi_e'' \, dV_e \tag{18.29}$$

The square of the vibrational overlap integral $\int \psi_v'^* \psi_v'' \, dV_v$, is the Franck-Condon factor discussed in Chap. 15.

The dipole moment $\boldsymbol{\mu}_e$ is a sum of three components

$$\boldsymbol{\mu}_e = \mu_{ex} \hat{\mathbf{i}} + \mu_{ey} \hat{\mathbf{j}} + \mu_{ez} \hat{\mathbf{k}} \tag{18.30}$$

so if any of the integrals

$$I_{eg} = \int \psi_e'^* \mu_{eg} \psi_e'' \, dV_e \qquad g = x, y, z \tag{18.31}$$

are nonzero the transition will be allowed.

Group theory may be employed to determine the zero-nonzero value of I_{eg}. It will be nonzero only if the integrand is symmetric, that is if $\Gamma(\Psi'_e) \times \Gamma(\mu_g) \times \Gamma(\psi''_e)$ belongs to the totally symmetric representation of the molecular point group. The electronic dipole components, being due to the displacement of electronic charge, are given by

$$\mu_{eg} = eg \qquad g = x, y, z \tag{18.32}$$

so the representation for μ_{eg} will be equivalent to that for the coordinate g. This means that the direct product $\Gamma(\psi') \times \Gamma(\psi'')$ must belong to one of the same representations as x, y, or z. If the ground state belongs to the totally symmetric representations, as is frequently the case, then the criterion for the occurrence of a transition is that ψ'' belong to the same representation as either x, y, or z.

If we consider as an example the CH_2 molecule, as given in Table 18.3, and for the moment ignore multiplicity, it is observed that the ground state is of b_1 symmetry while the excited states belong to a_1 and b_1 symmetry. Reference to Fig. 18.7 shows that there is also a higher state of b_2 symmetry. Reference to the C_{2v} characters table shows that x, y, and z belong to b_1, b_2, and a_1 respectively. For $\Gamma(\psi') \times \Gamma(g) \times \Gamma(b_1)$ to be equal to $\Gamma(a_1)$ requires that $\Gamma(\psi') \times \Gamma(g)$ be equal to $\Gamma(b_1)$. Considering the three upper-state symmetries and the three coordinates gives

$$\begin{aligned}
\Gamma(a_1) \times \Gamma(x) &= a_1 \times b_1 = b_1 \\
\Gamma(a_1) \times \Gamma(y) &= a_1 \times b_2 = b_2 \\
\Gamma(a_1) \times \Gamma(z) &= a_1 \times a_1 = a_1 \\
\Gamma(b_1) \times \Gamma(x) &= b_1 \times b_1 = a_1 \\
\Gamma(b_1) \times \Gamma(y) &= b_1 \times b_2 = a_2 \\
\Gamma(b_1) \times \Gamma(z) &= b_1 \times a_1 = b_1 \\
\Gamma(b_2) \times \Gamma(x) &= b_2 \times b_1 = a_2 \\
\Gamma(b_2) \times \Gamma(y) &= b_2 \times b_2 = a_1 \\
\Gamma(b_2) \times \Gamma(z) &= b_2 \times a_1 = b_2
\end{aligned} \tag{18.33}$$

Scanning this collection of products shows that the following transitions will be symmetry-allowed:

$$\begin{aligned}
\tilde{X}B &\leftrightarrow \tilde{a}A_1 \\
\tilde{X}B &\leftrightarrow \tilde{c}A_1 \\
\tilde{X}B_1 &\leftrightarrow \tilde{b}B_1
\end{aligned} \tag{18.34}$$

A more general method can be used to present the symmetry-allowed transitions for a particular point group. This consists of ordering the irreducible representations of possible state functions horizontally and vertically in an array and filling in with g at the crossing points where $\Gamma(\psi') \times \Gamma(\psi'') = \Gamma(g)$ or with 0 to denote any other irreducible representation. Using the C_{2v} group as an example we have

	a_1	a_2	b_1	b_2
a_1	z	0	x	y
a_2	0	z	y	x
b_1	x	y	z	0
b_2	y	x	0	z

We observe that a transition of an electron from the b_1 MO to the b_2^* MO is thus forbidden.

A complicating factor which is frequently encountered in the formulation of symmetry selection rules results from a change in geometry of the molecule as it undergoes a transition from the ground state to an excited state. For instance, a molecule may go from C_{2v} to $D_{\infty h}$ or from C_{3v} to D_{3h}. To determine the allowed transitions in such cases it is necessary to determine which representations in the two point groups correlate. This process is very much like that of correlating vibrational frequencies as the molecular symmetry is changed by atom substitution. The phenomenon will be considered further by referring to the AX_4 molecule.

For the C_{4v} pyramidical configuration the allowed transitions will be given by

C_{4v}	A_1	A_2	B_1	B_2	E
A_1	z	0	0	0	x, y
A_2		z	0	0	x, y
B_1			z	0	x, y
B_2				z	x, y
E					z

while for the C_{4h} planar configuration they are given by

C_{4h}	A_g	B_g	E_g	A_u	B_u	E_u
A_g	0	0	0	z	0	x, y
B_g		0	0	0	z	x, y
E_g			0	x, y	x, y	z
A_u				0	0	0
B_u					0	0
E_u						0

For the C_{4v} and C_{4h} groups the correlation of representations will be [18]

C_{4v}	C_{4h}
A_1	A_g, A_u
A_2	A_g, A_u
B_1	B_g, B_u
B_2	B_g, B_u
E	E_g, E_u

For example, a transition from a_1 and e in a C_{4v} molecule, where the geometry does not change, will become an a_1 and e_u transition when the molecule concurrently changes geometry to C_{4h}.

With all symmetry-allowed transition there must be a simultaneous adherence to the spin selection rule $\Delta S = 0$. Additionally, as shown in the listing for the C_{4h} group, there is another restriction known as the Laporte or parity selection rule which applies to molecules with a center of symmetry. This rule is $g \not\leftrightarrow g$, $u \not\leftrightarrow u$, and $u \leftrightarrow g$.

18.10 GENERAL TYPES OF ELECTRONIC TRANSITIONS

Molecular electronic transitions can be classified by methods which relate to the energies and to the origins of the transitions. They are primarily divided into two general classes—Rydberg and sub-Rydberg transitions. Rydberg transitions are analogous to atomic transitions where the excited electron ends up in an atomic orbital well removed from the nucleus. In molecules they involve the transition of an electron between a low-lying MO, often the ground state, and an MO of very high energy, generally more correctly characterized as a high-energy atomic orbital of one of the constituent atoms. Such transitions are generally observed in the ultraviolet region of the electromagnetic spectrum and involve a change in the principal electronic quantum numbers of the AOs associated with the MOs. In contrast, the sub-Rydberg transitions, which occur in the near ultraviolet and visible regions, result from the excitation of an electron between a bonding or nonbonding MO and an antibonding MO. They generally occur between MOs resulting from AOs belonging to the same principle electronic quantum number.

Sub-Rydberg electronic transitions are also classified by the bonding nature of the MOs for the states involved. This formalism is used to describe orbitals and subsequent electronic transitions when the electrons are considered to be localized into a small region of a molecule. In this method the MOs are denoted σ or π for bonding, n for nonbonding, and σ^* of π^* for antibonding types. For example, the transition of an electron from the highest-filled MO, $\sigma_g 2\text{-}p$, to the lowest-unfilled MO of correct symmetry, $\sigma_u^* 2p$, of N_2 would be designated as a $\sigma \rightarrow \sigma^*$ transition. Transitions from bonding to antibonding MOs will be of the type $\sigma \rightarrow \sigma^*$, which in general lie in the far ultraviolet region, or $\pi \rightarrow \pi^*$, which tend to move toward and into the visible region. Both types of transitions tend to be relatively strong.

Transitions of electrons between nonbonding and antibonding orbitals are designated as $n \rightarrow \sigma^*$ or $n \rightarrow \pi^*$. The $n \rightarrow \sigma^*$ transitions tend to be the more intense of the two and generally lie in the far to midultraviolet.

18.11 LOW RESOLUTION ELECTRONIC SPECTRA OF SELECTED SMALL POLYATOMIC MOLECULES

Before examining the vibronic and rovibronic interactions which can be observed in highly resolved spectra we will look at the general nature of the low-resolution electronic spectra of several types of molecules. The electronic spectra of polyatomic molecules, when observed under low resolution, consist of broad absorption bands which encompass a large number of unresolved vibrational and rotational components. The frequencies of the bands can be related to the MO energy levels of the molecule by using selection rules discussed in Sec. 18.9.

H_2O

Reference to Fig. 18.5 shows that the valence shell electronic structure of H_2O will be $(2a_1)^2(3a_1)^2(1b_2)^2(1b_1)^2$ which is an $\tilde{X}\,^1A_1$, ground state. Typical of the lower-lying excited states is $(2a_1)^2(3a_1)^2(1b_2)^2(1b_1)^1(4a_1)^1$, an $\tilde{A}\,^1B$ state. In this case the highest-occupied MO, the $4a_1$, is an oxygen $3s$ orbital which is essentially Rydberg in character.

Note that this orbital and a number of other Rydberg orbitals have energies lower than the b_2 and a_1 antibonding orbitals in Figure 18.4. It is to be noted

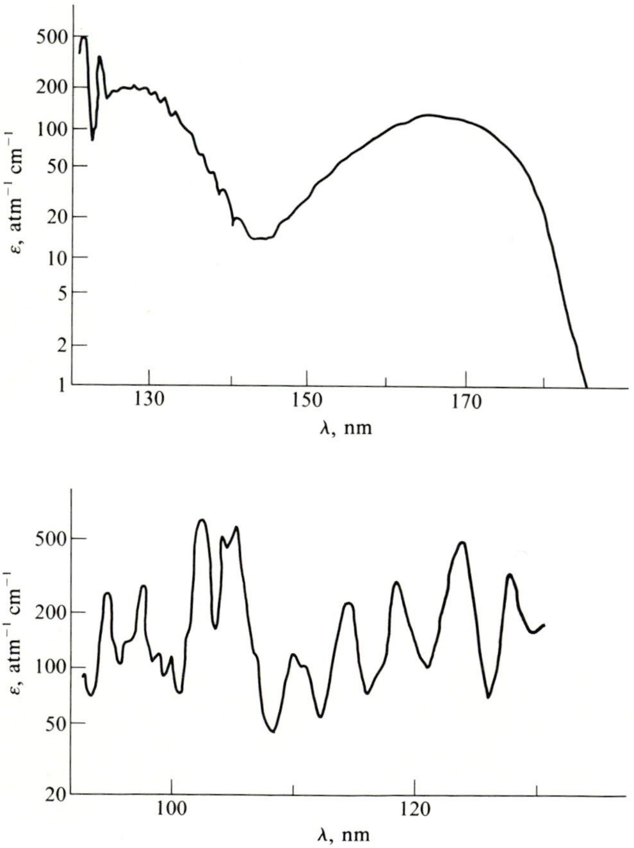

FIGURE 18.23
Electronic absorption spectrum of gaseous H_2O from 90 to 185 nm.

that the $\tilde{X}^1A_1 - \tilde{A}^1B_1$ transition does not involve the loss of an electron from a bonding orbital so one would expect only a slight change in molecular geometry. The promotion of an electron from X^1A_1 to $\tilde{A}^1B_1$ is allowed by symmetry. It has been observed by Watanabe [23], as shown in Fig. 18.23, that the H_2O spectrum contains a broad band extending from 145 to 185 nm. This is the first of a series of transitions where the electron is promoted from the nonbonding $1b_1$ MO to an $ns\ a_1$ orbital ($n \geqq 3$). This series is illustrated in Fig. 18.24. A characteristic of these Rydberg transitions is the ability to express these mean band frequencies by means of a "hydrogenlike" relationship

$$\tilde{\nu}(\text{cm}^{-1}) = 101{,}780 - \frac{R}{(n - 0.05)^2} \qquad (n = 3, 4, \ldots) \tag{18.35}$$

Price [24] has observed another Rydberg series which is much weaker and whose

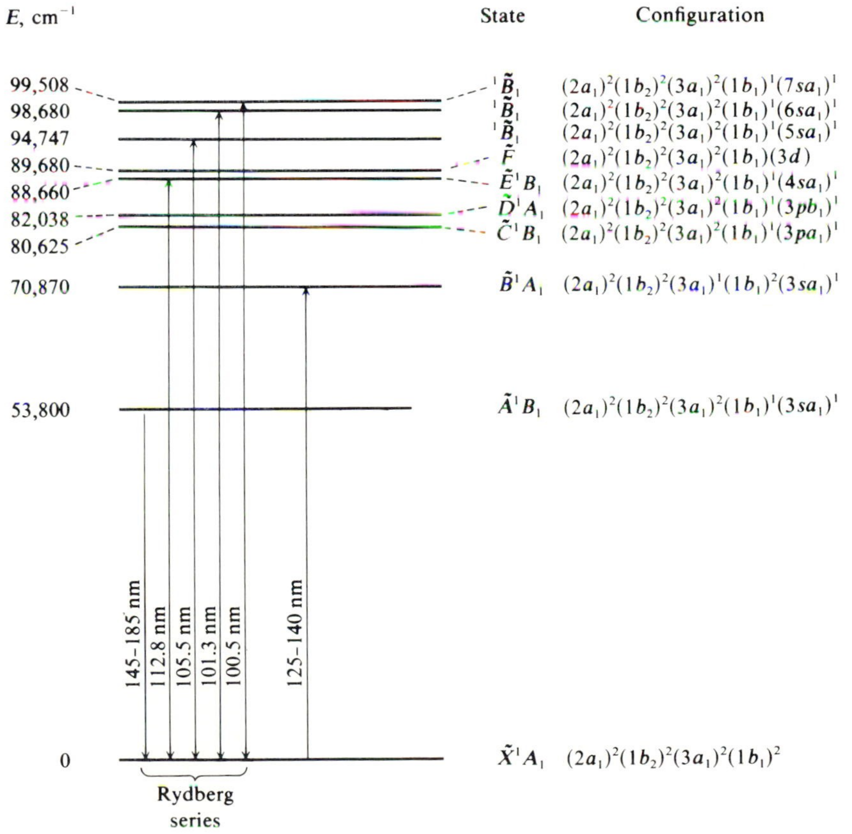

FIGURE 18.24
Electronic energy-level diagram for H_2O. (Vibration and Rotation Not Considered—Levels Based on Band Centers.)

mean band frequencies are given by

$$\tilde{\nu}(\text{cm}^{-1}) = 101{,}780 - \frac{R}{(n-0.7)^2} \qquad (n = 3, 4, \ldots) \tag{18.36}$$

This series contains transitions between the ground state and levels where the configuration of the highest two electrons is $(1b_1)^1(npb_1)^1$.

Examinations of Fig. 18.23 shows that there is a second broad band between 125 and 140 nm [25]. This band exhibits some vibronic structure, which will be ignored for now, since at present the important point to note is that it is not a member of one of the previously mentioned Rydberg series. This band is the lowest member of a series which involves the promotion of a $3a_1$ electron to an $ns\ a_1$ level. There are other minor features of the H_2O spectrum which have not been discussed. It is to be noted that the ionization limit for the two Rydberg series will be 101,780 cm^{-1}, which is considerably below the level of the antibonding $a_1(\sigma^*)$ and $b_1(\pi^*)$ MOs which are approximately at $3.75 \times 10^5\ \text{cm}^{-1}$ and $4.15 \times 10^5\ \text{cm}^{-1}$.

NH_3

The NH_3 molecule provides a good illustration of a change in structure during the course of an electronic transition. As with the H_2O molecule we will consider only the mean band frequencies and ignore the vibronic and rovibronic fine structure. To find the possible transitions it is first necessary to determine the selection rules for both the D_{3h} and C_{3v} configurations. Then, one employs a group correlation chart to ascertain the final transitions. The symmetry selection rule requires that the direct product of the representations of the two states involved in the transition either be equal to or contain a representation of one of the Cartesian coordinates. These are summarized in Table 18.5. The correlation of states between the two configurations is aided by Table W-1. It is also shown by Fig. 18.18. Combining the results of Tables 18.5 and W-1 yields the allowed transitions $a_1 \leftrightarrow a_1'$, $a_1 \leftrightarrow a_2''$, $a_1 \leftrightarrow e'$, $a_1 \leftrightarrow e''$, $a_2 \leftrightarrow a_1''$, $a_2 \leftrightarrow a_2'$, $a_2 \leftrightarrow e'$, $a_2 \leftrightarrow e''$,

TABLE 18.5
Allowed transitions for NH_3 in (*a*) C_{3v} and (*b*) D_{3h} configurations

(*a*)				(*b*)						
C_{3v}	a_1	a_2	e	D_{3h}	a_1'	a_2'	e'	a_1''	a_2''	e''
a_1	z	0	x, y	a_1'	0	0	x, y	0	z	0
a_2		z	x, y	a_2'		0	x, y	z	0	0
e			z	e'			0	0	0	z
				a_1''				0	0	x, y
				a_2''					0	x, y
				e''						0

$e \leftrightarrow e'$, and $e \leftrightarrow e''$. In addition the spin selection rule $\Delta S = 0$, must be considered. Figure 18.25 summarizes the nature of the electronic bands for NH_3. This figure includes only the highest-filled level resulting from use of H 1*s* and the N 2*s* and 2*p* orbitals rather than the complete MO scheme shown in Fig. 18.14. The NH_3 spectrum is similar to that of H_2O in that most of the transitions belong to Rydberg series, the antibonding orbitals lying above the ionization limit of the $3a_1$ electrons. The ground state molecule has C_{3v} symmetry and all higher states, which are illustrated, have D_{3h} symmetry.

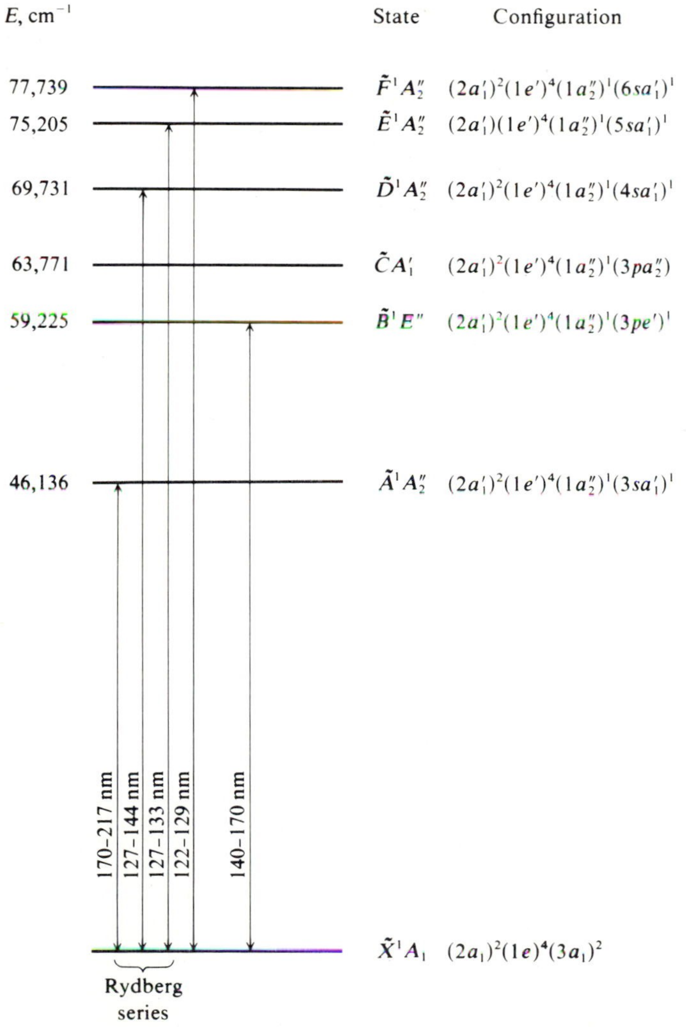

FIGURE 18.25
Energy levels, configurations, and electronic transitions of NH_3.

CO_2

The highest-occupied MO of CO_2 in the ground state is the $1\pi_g$ level. The next level above this is the $2\pi_u^*$ with the energy difference being about 35,000 cm^{-1}. Experiments have shown, however, that the loss of a $1\pi_g$ electron results in a reconfiguration of the molecule into a bent structure with MO occupancy of $(1b_1)^2(5a_1)^2(1a_2)^2(4b_2)^1(6a_1)^1$ at a level 46,000 cm^{-1} above the ground state. Hence the lowest absorption band of CO_2 will lie well into the ultraviolet region. Many other bands have been observed for CO_2 [26]. These all involve excitation to excited linear conformations and several are members of identifiable Rydberg series.

18.12 LOW-RESOLUTION ELECTRONIC SPECTRA OF MOLECULES WITH A LOCALIZED MULTIPLE BOND

To illustrate the nature of spectra of an H_2AB type molecule with A and B multiply bonded we will look at the analysis of the formaldehyde, $H_2C{=}O$, spectrum. This discussion also looks at the situation where some MOs are localized to certain parts of a molecule. The ordering of the MOs in H_2CO is like that given for the general case in Fig. 18.19. With 12 valence electrons the election configuration will be $(3a_1)^2(4a_1)^2(1b_2)^2(5a_1)^2(1b_1)^2(2b_2)^2$. The energy levels for the $2b_2$ and $2b_1$ MOs are separated by approximately 30,000 cm^{-1}, which corresponds to radiation in the near-ultraviolet (333 nm) region. Those for the $1b_1$ and $2b_1$ levels are separated by approximately 49,000 cm^{-1} (204 nm). The symmetry selection rules for C_{2v} molecules, as derived in Sec. 18.9, show that $a_1 \leftrightarrow b_1$, $a_1 \leftrightarrow b_2$, and $a_1 \leftrightarrow a_1$ transitions are allowed, but $a_1 \leftrightarrow a_2$ ones are not. Therefore, one would expect to observe a $\pi \leftrightarrow \pi^*$ transition, where the transition is between the $\ldots(1b_1)^2(2b_2)^2$, $[\tilde{X}\,^1A_1]$, and $\ldots(1b_1)^1(2b_2)^2(2b_1)^1$, $[\tilde{C}\,^1A_1]$ states. This transition is observed as an intense band in the neighborhood of 160 nm. The $n \leftrightarrow \sigma^*$ transition between the $\ldots(1b_1)^2(2b_2)^2$, $[\tilde{X}\,^1A_1]$ and the $\ldots(1b_1)^2(2b_2)(6a_1)$, $[\tilde{B}\,^1B_2(A'')]$ states is allowed and is observed at about 170 nm. It has been determined from experiment that the H_2CO molecule is nonplanar, C_s, in some excited states, for example the $\sigma^*\tilde{B}\,^1B_2$ state, so a complete analysis requires the use of group representation correlations. Although the $a_1 \leftrightarrow a_2$ transitions are electric-dipole forbidden the distortion to nonplanarity in the excited state permits their allowance via the magnetic dipole. Magnetic dipole-allowed transitions occur when the direct product of the representations of the two involved states is either equal to or contains one of the representations for the infinitesimal rotations R_x, R_y, or R_z. Thus a weak $n \leftrightarrow \pi^*$ transition is observed at about 350 nm.

Since the lowest-excited state configuration of H_2CO is $\ldots(1b_1)^2(2b_2)^1(2b_1)^1$ there can be both singlet and triplet states. The triplet state lies at slightly lower energy (25,194 cm^{-1}) than of the singlet state (28,188 cm^{-1}). The transition from the $\tilde{X}\,^1A_1$ ground state to the $^3A_2(A'')$ is normally spin-forbidden, but this selection rule can be broken if there is vibronic mixing of states. For H_2CO a weak $\tilde{X}^1A_1 \leftrightarrow \tilde{a}_3A_1(A'')$ transition has been observed at about 390 nm.

18.13 CHROMOPHORES

Examination of the MOs of H_2CO shows that those from which the electronic transitions originate are predominantly localized about the C=O part of the molecule. It might be suspected that if the H atoms were to be substituted with other atoms or groups then (1) there would not be any drastic changes in these MO energy levels, and (2) the changes observed could be related to the electronegativity of the substituting species. The phenomenon has been born out by experiment as is illustrated by the data in Table 18.6. Also included are the Beer's law extinction coefficients to provide an indication of relative intensities. The utility of this type of information is quite extensive, providing a general method for qualitative identification of compounds. In using it in this manner, however, one must keep in mind that, as with the use of group infrared vibrational frequencies, there are a number of variances such as multiple chromophoric groups in a compound, and solvent effects which must be considered.

18.14 LOW-RESOLUTION SPECTRA OF LINEAR CONJUGATED MOLECULES

In Sec. 18.8 the Hückel MO theory was used to establish the energy-level scheme for the conjugated molecule $CH_2{=}CH{-}CH{=}CH_2$. The lowest electronic transition is due to the promotion of an electron from the filled $1b_g$ level to the unfilled $2a_u$ level. This is a $\pi \leftrightarrow \pi^*$ transition, which is allowed by symmetry. The experimental transition is at 217 nm.

TABLE 18.6
Electronic absorption band frequencies for RRC = O carbonyl compounds

Substituents		Transition					
		$\pi \leftrightarrow \pi^*$		$n \leftrightarrow \sigma^*$		$n \leftrightarrow \pi^*$	
R	*R'*	λ, nm	ε	λ, nm	ε	λ, nm	ε
H	H	160	......	170	......	295	10
CH_3	H	160	20,000	180	10,000	293	12
CH_3	CH_3	165	20,000	190	1,000	280	15
H	OH	138	......	155	......	250	
$H_2C{=}CH$	H	165	......	175	......	235	18
CH_3	OC_2H_5	...	......	...	......	204	60
CH_3	NH_2	...	......	...	......	214	
CH_3	Cl	...	......	...	......	235	53
CH_3	C_2H_5	...	......	...	......	270	16

Hückel MO theory indicates that the energy separation between the $1b_g$ and $2a_u$ levels will be equal to 1.24β. The value of β has been found to be dependent on the average C—C bond length in the molecule. A discussion of such empirical relationships may be found in Streitweiser's monograph on MO theory [3]. The constant β will be relatively constant for a related series of compounds that can differ widely from one type of compound to another. Using the experimental transition frequency gives $\beta = 4.6$ eV.

Another approach to the characterization of spectra of conjugated linear molecules is to use a simple "particle-in-a-box" approach. This technique, referred to as the free electron molecular orbital, FEMO, approach considers each pair of π electrons to occupy a single FEMO. The energy of a particular configuration is given by a "particle-in-a-box" relationship.

$$E = \frac{2h^2}{8ml^2} \sum_{i=1}^{N/2} n_i^2 \tag{18.37}$$

where n_i is the quantum number of the ith level, N is the number of π electrons, and l is the effective length of the molecule. The effective length is found by taking the average length along the bond axis, which is occupied by a $c - c$ segment, to be 140 pm or $1 = 1.4(N - 1)$. Since $\Delta E - hc/\lambda$ Eq. (18.37) will reduce to

$$\lambda(\text{nm}) = 64.5 \frac{(N-1)^2}{N+1} \tag{18.38}$$

For butadiene ($N = 4$) this predicts $\lambda = 116$ nm. Although this differs from the experiment value (217 nm) by nearly a factor of 2 it does indicate that for qualitative work relatively simple theories are frequently useful.

18.15 LOW-RESOLUTION ELECTRONIC SPECTRA OF AROMATIC MOLECULES

Aromatic molecules, both homocyclic and heterocyclic, because of their extensive π electron structure can be described in a reasonable manner by use of Hückel MO theory. We will review the application of the method to benzene, C_6H_6, and compare the results with the observed spectrum. An additional complexity, which is encountered with benzene and did not occur with butadiene, is the occurrence of degenerate energy states. The Hückel MO treatment of benzene is discussed in many monographs on MO theory [3] or group theory [4, 5]. These results are summarized in Fig. 18.26.

Neglecting the σ orbital framework the ground state configuration of benzene is $(a_{2u})^2(e_{1g})^4$, which is an $^1A_{1g}$ state. The excitation of one electron from the ground state to the next lightest one leads to an excited-state configuration of $(a_{2u})^2(e_{1g})^3(e_{2g})^1$. Due to the degeneracies of the e_{1g} and e_{2g} levels it is possible to have several possible state terms. Using the method discussed in Chap. 6, and recognizing that a hole $(e_{1g})^3$ and a single electron $(e_{1g})^1$ can be treated in the same manner, we can determine the possible states. The procedure involves

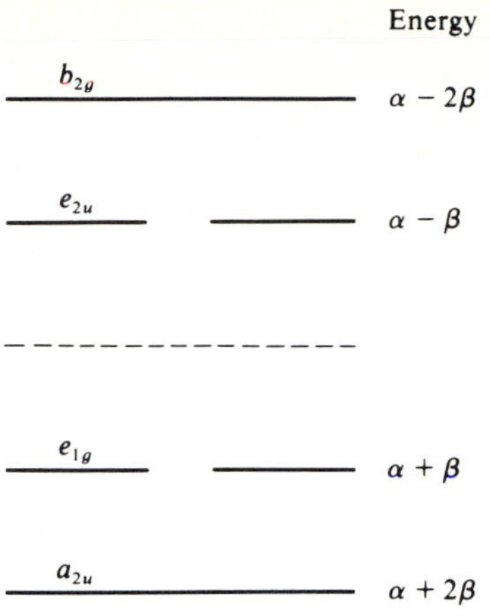

FIGURE 18.26
Hückel MO energy-level diagram for benzene.

finding the components of the direct product of the two occupied degenerate levels

$$e_{1g} \times e_{2u} = b_{1u} + b_{2u} + e_{1u} \tag{18.39}$$

Since the spins of the electrons in the e_{1g} and e_{2u} levels can be either parallel or antiparallel there are a total of six possible terms, ${}^1B_{1u}$, ${}^3B_{1u}$, ${}^1B_{2u}$, ${}^3B_{2u}$, ${}^1E_{1u}$, and ${}^3E_{1u}$. The ground state is a singlet so, for consideration of possible transitions, the three triplet terms will be neglected. If we now examine the direct product of A_{1g} with each of these excited states the following selections rules emerge:

$$A_{1g} \nleftrightarrow B_{1u}$$

$$A_{1g} \nleftrightarrow B_{2u}$$

$$A_{1g} \leftrightarrow E_{1u}$$

Simple Hückel MO theory indicates a degeneracy of the three excited singlet states, but the occurrence of electron interactions results in their possessing different energies in increasing order B_{2u}, B_{1u}, and E_{1u}. We would then expect to observe an $A_{1g} \leftrightarrow E_{1u}$, $\pi \leftrightarrow \pi^*$ transition which, based on MO energy calculations, will be in the neighborhood of 55,000 cm^{-1} (180 nm). There is a strong band observed in the 170–182 nm region ($\varepsilon = 10{,}000$) but there are also two weak bands observed in the 185–205 nm ($\varepsilon = 6300$) and 227–267 nm ($\varepsilon = 230$) regions. These latter two bands are at the positions expected for the forbidden ${}^1A_{1g} \leftrightarrow {}^1B_{1u}$ and ${}^1A_{1g} \leftrightarrow {}^1B_{2u}$ transitions. The molecular framework of the C_6H_6 molecule is not rigid. The molecular vibrations will distort the D_{6h} molecular symmetry to a lower type. Result will be a relaxation of the symmetry selection rules. This, in effect, is due to a breakdown of the Born-Oppenheimer separation of electronic and vibrational motions. The 170–182 m and 227–267 nm bands have been well characterized as belonging to the transitions indicated, but there remains a question as to whether the 185–205 nm band is as assigned or if it is an ${}^1A_{1g} \leftrightarrow {}^1E_{2g}$ transition arising from an upper-state configuration of $(a_{2u})^2(e_{1g})^3(b_{2g})^1$.

The simplified and abbreviated discussion of the few simple polyatomic molecules given in Secs. 18.11 through this one serve to illustrate the general approach to the correlation of electronic spectra and structure. However at the same time they make clear the complexities and problems attendant to the assignment of polyatomic electronic spectra, even without considering the rovibronic fine structure. Some further remarks concerning vibronic and rovibronic interactions will be presented in Sec. 18.17.

18.16 ELECTRONIC SPECTRA OF TRANSITION METAL COMPOUNDS

In the molecules discussed in the preceding five sections the bonding and the MOs considered only involved *s* and *p* electrons. For organic compounds and for compounds of the first- and second-period elements energy considerations dictate that there is little need to consider *d* atomic orbitals unless more sophisticated quantum mechanical calculations are being made. The situation is quite different, however, for compounds of the regular elements of period 3 and higher periods, and in compounds of transition elements. Compounds such as $PF_5(D_{3h})$ and $SF_6(O_h)$ are described in terms of dsp^3 and d^2sp^3 hybridization of the central atom orbitals or by MO schemes of the type shown in Figs. 18.1, 18.2, 18.15, 18.16, and 18.17, where some of the lower MOs have some *d* orbital contributions. For transition metal compounds or ions such as $Co(NH_3)_6^{3+}(O_h)$, $MnO_4^-(T_d)$, or $PtCl_4^{2-}(D_{4h})$ not only do central atom *d* orbitals contribute to the formation of bonding and antibonding MOs but the remaining ones, which often are occupied by electrons, constitute nonbonding orbitals and can contribute to electronic spectral transitions which often fall in the visible region. As was done with the simpler types of molecules we will limit the discussion to a review of a single type of coordination compound, that with a basic octahedral symmetry. In this case the MO scheme is shown in Fig. 18.17.

There are five primary factors which will affect the MO energy levels and hence the observed spectra of octahedral complexes:

1. The number of valence electrons available from the central atom
2. The charge on the central atom
3. The electronegativity of the ligands
4. The availability of π orbitals on the ligands
5. The relative number of different ligand molecules

The precise effect of these different types of factors is discussed in detail in monographs on ligand field theory [27–29] and we will use those results which are necessary to discuss observed spectra.

Initially the discussion of the spectra of transition metal compounds will be limited to systems in which the ligands can only σ bond to the central metal. In this case the MO scheme of Fig. 18.17 simplifies to that shown in Fig. 18.27.

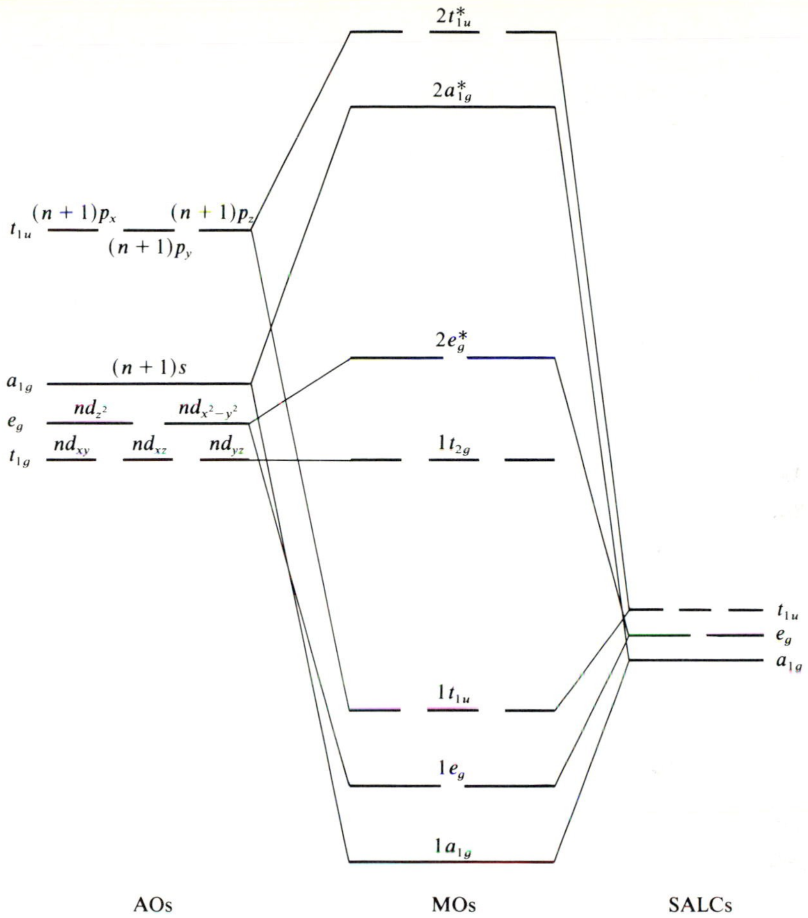

FIGURE 18.27
Molecular orbital energy level diagram for an octahedral complex having only σ bonded ligands.

The lowest three MOs, $1a_{1g}$, $1e_g$, and $1t_{1u}$, are σ-bonding orbitals and are considered to be occupied by six electron pairs donated by the ligands. The electrons which were originally in the metal atom (or ion) d AOs will occupy the $1t_{2g}$ and $2e_g^*$ MOs. The $2e_g^*$ MOs are composed primarily of contributions from the metal d AOs. The separation between the $1t_{1g}$ and the $2e_g^*$ levels is given the symbol Δ_0 or $10Dq$. Its magnitude will depend strongly on the nature of the ligands. The effect of ligand type on the level of the t_{2g} and e_g MOs can be easily rationalized in terms of Fig. 18.28. This figure shows that the distances between the ligands and the lobes of the d_{xy} AO will be greater than those of the $d_{x^2-y^2}$ AO. In terms of simple electrostatic interactions this would indicate a greater repulsion, and hence a higher energy, for the $d_{x^2-y^2}$ AO. This accounts for the $2e_g^*$ level lying above the $1t_{2g}$ as is also predicted from MO theory. If the

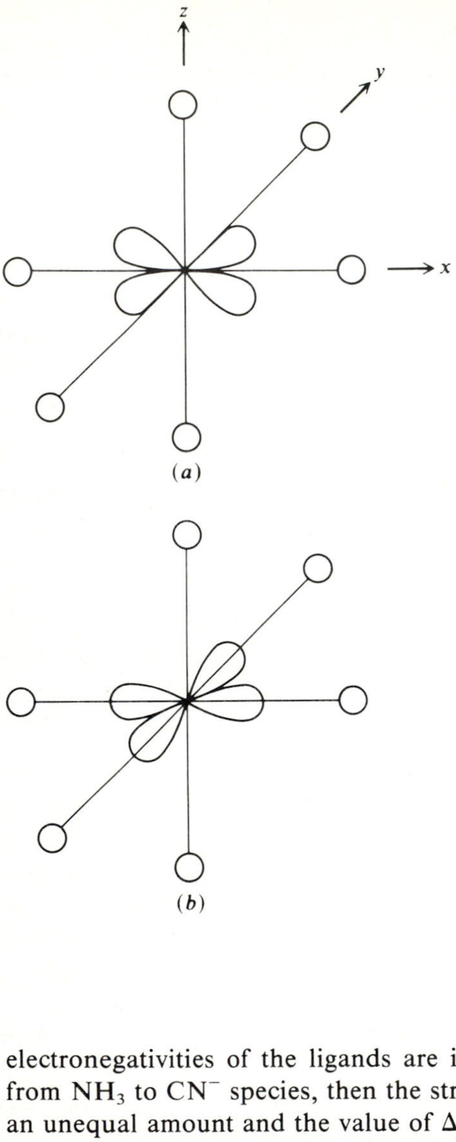

FIGURE 18.28
Position of a set of six octahedrally oriented ligands relative to (*a*) a d_{xy} AO, and (*b*) a $d_{x^2-y^2}$ AO.

electronegativities of the ligands are increased, as when changing the ligands from NH_3 to CN^- species, then the strength of the two interactions increase by an unequal amount and the value of Δ_0 will increase. It has been found experimentally that the value of Δ_0 depends to a much lesser degree on the nature of the central metal atom than on the nature of the ligands. Hence, common ligands can be arranged in a series which reflects their increasing effect on the value of Δ_0. A typical "spectrochemical series" for some common ligands is

$$I^- < Br^- < Cl^- < NO_3^- < F^- < OH^- < H_2O < C_5H_5N <$$
$$< CH_3CN < NH_3 < H_2NCH_2CH_2NH_2 < NO_2^- < CN^-$$

It must be remembered that this is a qualitative ordering and for any given metal atom there can be a reversal of the effect of near members.

Examination of Fig. 18.27 shows that there are basically three types of transitions to be considered. (1) If the highest level occupied by electrons is the $1t_{2g}$ then the lowest energy transition, if allowed, will be the $1t_{2g} \rightarrow 2e_g^*$ or *d-d* transition. (2) At higher energy differences one might expect to find transitions of electrons from the filled σ MOs to the $1t_{2g}$ or $2e_g^*$ MOs. These are referred to as *ligand to metal-charge transfer transitions.* (3) Also, at higher energy difference, there exists the possibility of other charge transfer transitions resulting from an electron being promoted from a $1t_{2g}$ or a $2e_g^*$ to one of the σ antibonding MOs.

The nature of the observed spectra of transition metal complexes will depend not only on the magnitude of Δ_0 as previously indicated but also on electron spin-pairing energies, spin-orbital interactions, and the number of *d* electrons on the metal atom. For the fourth-period transition elements (Ti—Ni) spin-orbital coupling is generally small so we will confine our examples and discussion to compounds of those elements and consider only the other two factors.

We will next consider the effect of spin pairing on MO occupancies. To place a second electron, of antiparallel spin relative to an existing electron, in an MO requires approximately $(18 \pm 6) \times 10^3$ cm^{-1} of energy. Thus, if Δ_0 is less than the spin-pairing energy Δ_s, the occupancy of a $2e_g^*$ MO will be favored over the double occupancy of a $1t_{2g}$ MO. This situation will occur only with electron configurations of d^4 through d^7 since it is always energetically more favorable to add the second and third electrons with parallel spins into the other degenerate levels of $1t_{2g}$. The configuration which result from the spin-pairing energy exceeding Δ_0 is referred to as the *high-spin configuration.* The electron configuration is the dominant factor contributing to the observed magnitude of the magnetic moment of such compounds and ions. This concept is illustrated by Fig. 18.29. The occurrence of unpaired electrons in transition metal compounds leads to their having rich ESR spectra.

In addition to the dominant effect of the type of ligand on the magnitudes of Δ_0 there are other noteworthy experimental observations which are of qualitative value. (1) There is a general increase in Δ_0 as the oxidation state of the metal atom increases. For example [30], for $Fe(H_2O)_6^{2+}$, $\Delta_0 = 10{,}000$ cm^{-1} and for $Fe(H_2O)_6^{3+}$, $\Delta_0 = 14{,}000$ cm^{-1}. (2) There is an increase in Δ_0 as one goes to higher-row transition elements. For example [31], for $Co(NH_3)_6^{3+}$, $\Delta_0 = 23{,}000$ cm^{-1} and for $Rh(NH_3)_6^{3+}$, $\Delta_0 = 34{,}000$ cm^{-1}. (3) The larger Δ_0 values for the higher-row transition elements leads to a predominance of low-spin ($\Delta_0 > \Delta_s$) complexes.

We have examined the general nature of the expected spectra and the qualitative factors which can affect them for transition metal compounds, but there remains the problem of delineating specific energy levels and correlating them with observed spectra. For the majority of transition metal compounds the occurrence of *d* electrons in degenerate MOs results in a ladder of energy states. The interpretation of observed spectra will depend on establishing the dependency of Δ_0 on the energy separations of these states. Since each d^n configuration has a different set of energy states a review of all possible cases, even for octahedral complexes, is beyond the scope of this presentation.

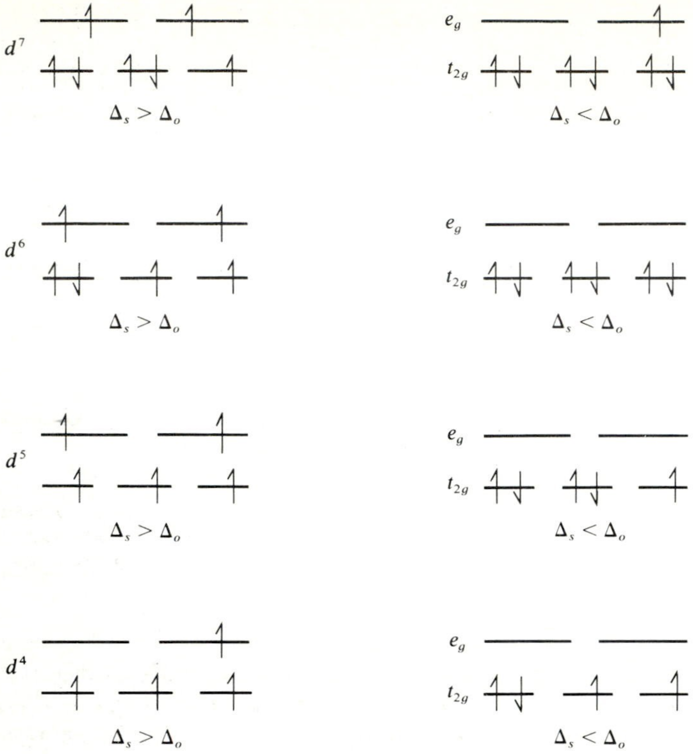

FIGURE 18.29
Spin configurations for d^n electrons in octahedral complexes.

To illustrate the correlations of energy levels and structure the d^2 configurations will be used. The objective will be to develop a correlation diagram that can be used to relate the ordering of the energy levels to the ligand field.

The left-hand side of such a diagram, where $\Delta_0 = 0$, will be the free-metal atom or ion. The right side will be at a sufficiently high ligand field that there will be no electron interactions. We will first examine the very elementary d^1 case to establish methodology. When $\Delta_0 = 0$ one has the free atom or ion and the energy levels are just those due to an atomic d^1 configuration. For a single d electron ($1 = 2$, $s = \frac{1}{2}$) the only possible term is a ^2D. In the presence of an octahedral field the free-ion fivefold degenerate d orbitals will be shifted and split into two degenerate sets of symmetry t_{2g} and e_g. With increasing ligand field the two states will diverge and ultimately reach a maximum separation at a hypothetical infinite separation. Figure 18.30a illustrates the initial change of energy with increasing field and 18.30b shows a typical energy-level diagram.

When more than one d electron is present both the free ion and the complex can have a multitude of terms. The first step in the analysis of the d^2 octahedral species will be to establish the possible terms for the free ion. This procedure is

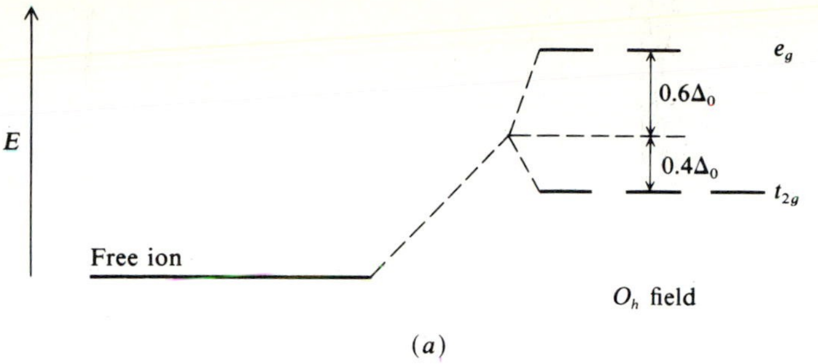

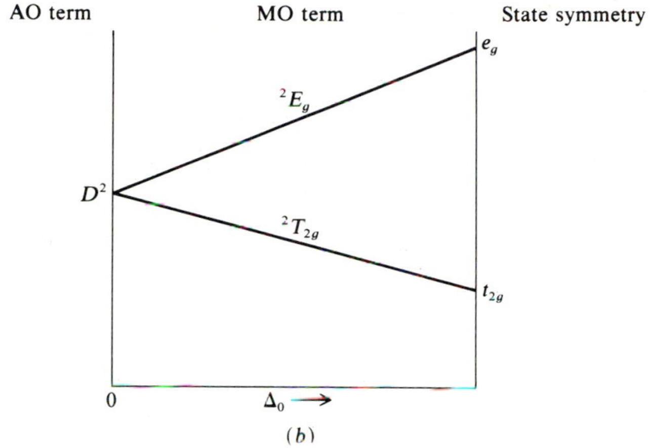

FIGURE 18.30
Energy level for a d^1 configuration: (a) energy change with initial application of an octahedral field; (b) conventional energy-level diagram showing relative changes of energy levels.

outlined in App. M. For a d^2 configuration with identical electrons the possible terms, in increasing order of energy as established by Hund's rules and experimental evidence, will be ^3F, ^1D, ^3P, ^1G, ^1S. The next step is to establish the irreducible representation for each term in O_h symmetry. A complete discussion of this procedure is found in several sources [32, 33]. The procedure employs the octahedral rotational group O whose characters are given in Table 18.7, and

TABLE 18.7
Character table for O group

	E	$6C_4$	$3C_2$	$8C_3$	$6C_2$		
a_1	1	1	1	1	1		$x^2+y^2+z^2$
a_2	1	−1	1	1	−1		
e	2	0	2	−1	0		$(2z^2-x^2-y^2, x^2-y^2)$
t_1	3	1	−1	0	−1	$(R_x, R_y, R_z)(x, y, z)$	(xy, xz, yz)
t_2	3	−1	−1	0	1		

the relationship

$$\chi(\phi) = \frac{\sin(L + \frac{1}{2})\phi}{\sin(\phi/2)} \tag{18.40}$$

where L is the angular momentum quantum number, to give the characters for the representation $\Gamma(\phi)$ of the state. Once the $\chi(\phi)$ are known $\Gamma(\phi)$ can be decomposed into its irreducible representations. Table 18.8 lists the values of $\chi(\phi)$ for different terms and the irreducible representations of each term.

Having established the energy-level configurations for the free ion in a weak field we turn our attention to the infinitely strong field case where there is no electron interaction. In this case there will be three possible configurations: t_{2g}^2, $t_{2g}^1 e_g^1$, or e_g^2. If this infinitely strong field is relaxed then the symmetries of the resulting states are found by finding the irreducible representations contained in the direct product of the infinitely strong field states. For the three possible configurations one has

$$t_{2g} \times t_{2g} = a_{1g} + e_g + t_{1g} + t_{2g}$$

$$t_{2g} \times e_g = 2t_{1g} + 2t_{2g}$$

$$e_g \times e_g = a_{1g} + a_{2g} + e_g$$

It remains to establish the correct multiplicity of each term. For the $t_{2g}^1 e_g^1$ configuration this is simple since the electrons may be placed with spins either paired or unpaired into the t_{2g} and e_g. Therefore ${}^1T_{1g}$, ${}^1T_{2g}$, ${}^3T_{1g}$, and ${}^3T_{2g}$ levels are allowed.

For the e_g^2 and t_{2g}^2 configurations it is necessary to use the method of descending symmetry [32, 33] to find the allowed states. This method involves the lowering of the symmetry of the system so that the degenerate levels are split into either a nondegenerate representation or a sum of nondegenerate representations, such that the spin multiplicities are unaffected. Consultation of the O_h group correlation table, App. W, shows that in the D_{4h} group a_{1g} and b_{1g} correlate with e_g. We now look to see how the two electrons can be placed into the a_{1g} and b_{1g} nondegenerate pair. This is summarized in Fig. 18.31.

TABLE 18.8
Symmetry species of atomic terms in an octahedral field

		$\chi(\phi)$					
Term	L	E	$6C_4$	$3C_2$	$8C_3$	$6C_2$	**Irreducible representations**
S	0	1	1	1	1	1	a_1
P	1	3	1	−1	0	−1	t_1
D	2	5	−1	1	−1	1	$e + t_2$
F	3	7	−1	−1	1	−1	$a_2 + t_1 + t_2$
G	4	9	1	1	0	1	$a_1 + e + t_1 + t_2$

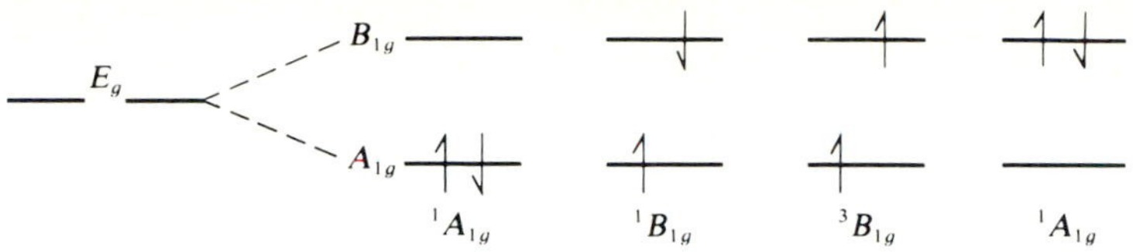

FIGURE 18.31
State occupancies and symmetries for symmetry-lowered states of O_h.

Below the level occupancy diagram in Fig. 18.30 are given the state terms which are determined by taking the direct product of the electron-occupied levels. Reference to Table W-3 shows the following correlations

D_{4h}	O_h
A_{1g}	A_{1g}
B_{1g}	A_{2g}
$A_{1g} + B_{1g}$	E_g

Accounting for the multiplicities we find that the e_g^2 configuration consists of three terms, ^1E_g, $^3A_{2g}$, and $^1A_{1g}$. A similar treatment of the t_{2g}^2 configuration shows that it will be composed of four terms, $^3T_{1g}$, $^1T_{2g}$, ^1E_g, and $^1A_{1g}$. A repetition of the procedure just reviewed allows the remainder of the term for the metal complex to be established and the correlation diagram shown in Fig. 18.32 to be constructed.

Although the details of the variation of energy levels with Δ_0 are well illustrated by Fig. 18.32, a more commonly used method for presentation of this information is the use of the Tanabe-Sugano diagrams [34]. These diagrams have the advantage that the energy of the ground state is represented by the horizontal base line. The axial coordinates are Δ_0/B for the horizontal axis and E/B for the vertical axis, where B is a constant which is characteristic of the metal atom species. The Tanabe-Sugano diagrams for the d^n octahedral configurations are reproduced in App. X. The diagrams for the $d^4 - d^7$ configurations show the behavior of the energy levels when there is a change from, a high-spin to a low-spin configuration. It is interesting to note that the ordering of the energy levels corresponding to particular free atom terms are reversed between d^n and d^{10-n} configurations. Keep in mind that these diagrams only relate to the energies of the d orbitals and d-d transitions.

Having developed the energy level scheme for a d^2 compound we next look at selection rules and allowed transitions. For a d^1 complex in the ground state there will be a single selection in the t_{2g} level or the ground state term will be a $^2T_{2g}$. The direct product for the states involved in a $d - d$ transition will be $T_{2g} \times E_g = 2T_{1g} + 2T_{2g}$. Since x, y, z belong to the T_{1u} representation the $^2T_{2g} \leftrightarrow$ ^2E_g transition is symmetry-forbidden. An example of a d^1 complex is $Ti(H_2O)_6^{3+}$ which has a weak ($\varepsilon < 5$) absorption at 493 nm in the region where $d - d$ transitions are expected. The low intensity of this transition is indicative of its

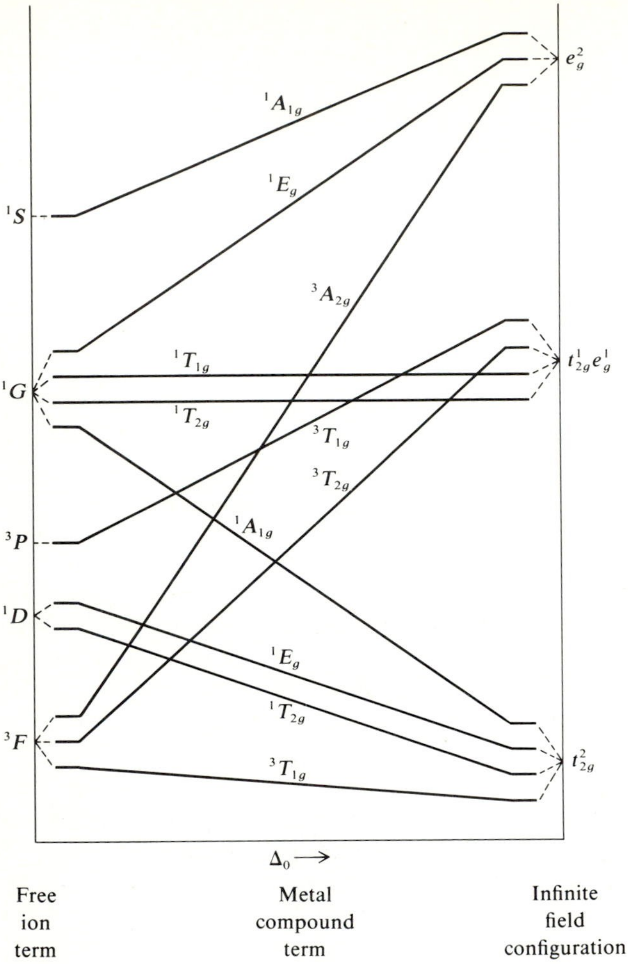

FIGURE 18.32
Correlation diagram for a d^2 configuration in O_h symmetry.

being a "forbidden" transition. In this particular case the breakdown of the normal selection rule is due to Jahn-Teller distortion which removes the degeneracy of the ^2E_g state.

Since it is a common phenomenon in transition metal complexes the Jahn-Teller effect will be briefly reviewed. The Jahn-Teller theorem states that any nonlinear molecule in a state which is orbitally degenerate will spontaneously distort to give a configuration of lower symmetry. It is due to an electronic-vibrational interaction which occurs when the Born-Oppenheimer separation breaks down. It occurs when the molecule has a degenerate electronic state and a nonsymmetric vibrational mode which can distort the molecule to lower symmetry. The result of Jahn-Teller distortion along the Z axis of an octahedral

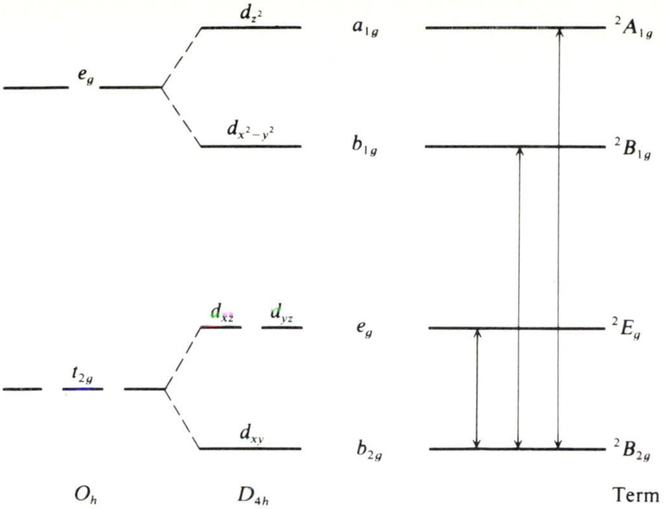

FIGURE 18.33
Effect of Jahn-Teller distortion on an octahedral species.

species is to lower the symmetry to D_{4h}. The effect on the energy levels of the system is illustrated in Fig. 18.33. The left side of the figure illustrates the removal of the orbital degeneracy of the t_{2g} and e_g states while the right side shows the expected transitions. In this figure the splitting of the degenerate levels is exaggerated relative to Δ_0. This is born out by the nonsymmetrical shape of the 493 nm transition of $Ti(H_2O)_6^{3+}$, which is in reality two closely spaced transitions.

The symmetry-allowed transitions are found as usual by comparison of the components of the direct product of the representations of the states involved in the transition to those of the coordinates x, y (e_u) and z (a_{2u}).

$$B_{2g} \times A_{1g} = B_{2g}$$

$$B_{2g} \times B_{1g} = A_{2g}$$

$$B_{2g} \times E_g = E_g$$

As in the case of the $T_{2g} \leftrightarrow E_g$ transition these are also symmetry-forbidden. Since there has been a breakdown of the Born-Oppenheimer separation there is no longer an absence of electronic-vibrational interaction and the selection rule will be governed by the zero-nonzero character of the integral $\int \Psi_e' \Psi_v' \boldsymbol{\mu}_e \Psi_e'' \Psi_v'' dV_e dV_v$.

Since the ground vibrational state for any mode always belongs to the totally symmetric representation, the requirement for this integral to be nonzero, when ψ_v'' is a ground state, is that the direct product

$$\Gamma(\Psi_e'') \times \Gamma(\Psi_e') \times \Gamma(\mu_e) = \Gamma(v)$$

where $\Gamma(v)$ is a representation of one of the vibrational modes. For an octahedral molecule the vibrational modes belong to A_{1g}, E_g, T_{1u}, and T_{2u}. For a D_{4h}

TABLE 18.9
Direct products for the determination of symmetry selection rules for octahedral and D_{4h} distorted octahedral species

	$\Gamma(\psi_e'')$	$\Gamma(\psi_e')$	$\Gamma(\mu_e)$	**Direct product**
O_h	T_{2g}	E_g	T_{1u}	$T_{1u} + T_{2u}$
D_{4h}	B_{2g}	A_{1g}	E_u	E_u
	B_{2g}	B_{1g}	E_u	E_u
	B_{2g}	E_g	E_u	$A_{1u} + A_{2u} + B_{1u} + B_{2u}$
	B_{2g}	A_{1g}	A_{2u}	B_{1u}
	B_{2g}	B_{1g}	A_{2u}	A_{1u}
	B_{2g}	E_g	A_{2u}	E_u

seven-atom molecule the vibrational modes belong to A_{1g}, A_{1g}, B_{1g}, A_{2u}, E_u, B_{2u}, and E_u. Table 18.9 summarizes the possible direct products. Since there are vibrational modes belonging to both T_{1u} and T_{2u}, the transition $T_{2g} \leftrightarrow E_g$ becomes allowed. There are vibrational modes belonging to E_u so all three of the transitions shown in Fig. 18.33 are now symmetry-allowed.

For species containing more than one *d* electron it is generally more difficult to assign the *d-d* spectra. For such systems the Tanabe-Sugano diagrams are very useful as an aid in assignment. As a final example we will examine the spectrum of $V(H_2O)_6^{3+}$ ion which has three transitions ($\varepsilon < 15$) at 17,700, 25,800, and 37,000 cm^{-1}.

This ion is a d^2 species with a ${}^3T_{1g}$ ground state arising from a ${}^3F$ free atom state. Reference to the Tanabe-Sugano diagram for d^2 in App. X indicates that the sequence of triplet levels will be ${}^3T_{2g}$, ${}^3A_{2g}$, ${}^3T_{1g}$ for $Dq/B = 1$ and ${}^3T_{2g}$, ${}^3T_{1g}$, ${}^3A_{2g}$ for $Dq/B = 3$. For this ion it is found that the assignment

$${}^3T_{1g} \leftrightarrow {}^3T_{2g} \quad 17{,}700 \text{ cm}^{-1}$$

$${}^3T_{1g} \leftrightarrow {}^3T_{1g} \quad 25{,}800 \text{ cm}^{-1}$$

$${}^3T_{1g} \leftrightarrow {}^3A_{2g} \quad 37{,}000 \text{ cm}^{-1}$$

can be fitted to the d^2 diagram with a value of $B = 860$ cm^{-1}. The application of the normal symmetry selection rule indicates that all three transitions will be forbidden. As with the $Ti(H_2O)_6^{3+}$, however, the vibronic interaction leads to a breakdown of the symmetry selection rule and the occurrence of these weak transitions.

In addition to the *d-d* transitions just discussed transition metal species often exhibit charge transfer (ligand → metal) spectra. For example, the $[Co(NH_3)_5Cl]^{2+}$ has two strong overlapping bands at 26,500 cm^{-1} ($\varepsilon \approx 2000$) and 34,730 cm^{-1} ($\varepsilon \approx 15{,}800$). These are due to transitions from the upper-filled σ MO, T_{1u}, to a vacancy in a T_{2g} or E_g level. Inspection of the direct products $T_{1u} \times T_{2g} = A_{2u} + E_u + T_{1u} + T_{2u}$ and $T_{1u} \times E_g = T_{1u} + T_{2u}$ shows that, in an octahedral environment, both transitions will be symmetry-allowed and hence will be intense. For the $[Co(NH_3)_5Cl]^{2+}$ species the t_{2g} level is filled, so these transitions must be $T_{1u} \leftrightarrow E_g$. Our discussions, to this point, of the spectra of this species has been based on the assumption of octahedral symmetry, while the species actually has C_{4v} symmetry. The correlation diagram in App. W shows that with this reduction in symmetry, the t_{1u} and e_g levels split into a_1, e and a_1, b_1 levels. Hence, the splitting of the expected single transition for the octahedral case.

For species having ligands with empty π orbitals it is possible to have metal-to-ligand-type of charge-transfer spectra. In this case the transition is due to an electron being promoted from a t_{2g} MO to an empty π^* MO resulting from the ligand orbitals. These are symmetry-allowed transitions and are of high intensity.

18.17 VIBRONIC AND ROVIBRONIC INTERACTIONS

In the liquid state or in solution the spectral fine structure due to the interaction of vibration and rotation with the electronic levels is quenched to the point of giving very broad and nearly structureless transitions. In the gas phase under moderate resolution it is possible to resolve considerable vibronic structure. To resolve the rovibronic structure requires the use of very high resolution optical or laser spectrometers. These three cases are illustrated for benzene in Fig. 18.34.

In principle the features of the rovibronic spectra of polyatomic molecules is analogous to those of diatomic molecules and the analysis should follow the same type of approach. However, there are two complicating factors that increase the difficulty of the analysis by orders of magnitude. (1) For polyatomic molecules there will be $3N - 6$ ($3N - 5$ if linear) normal modes of vibration, many or all of which can interact. Thus instead of having a simple series of vibronic bands there can be as many as $3N - 6$ overlapping and intermixed bands, even if no overtones or combinations are considered. (2) With the exception of a few very symmetric molecules, the majority of polyatomic molecules will be asymmetric rotors. This causes the rotational contribution to be much more complex than that for a diatomic molecule, which exhibits a simple series of almost equally spaced rotational transitions.

It is impossible in the limited space of this chapter to present the detail necessary to discuss the complete analysis of rovibronic spectra. We will provide only enough detail to be able to follow a partial vibrational analysis of one molecule.

For polyatomic molecules each vibrational mode will have its own potential function as is illustrated in Fig. 18.35 for a typical bent AB_2 molecule.

To index the vibrational levels the symbol ν_i^j, where i = number of mode and j = quanta of excitation, is used. Note that ν_0 is the ground state level and is common to all modes. If one were to attempt to diagrammatically illustrate the total vibrational energy scheme on a single figure it would have to be of $3N - 5$ ($3N - 4$ for linear) dimensions. Since this is not possible, even the case

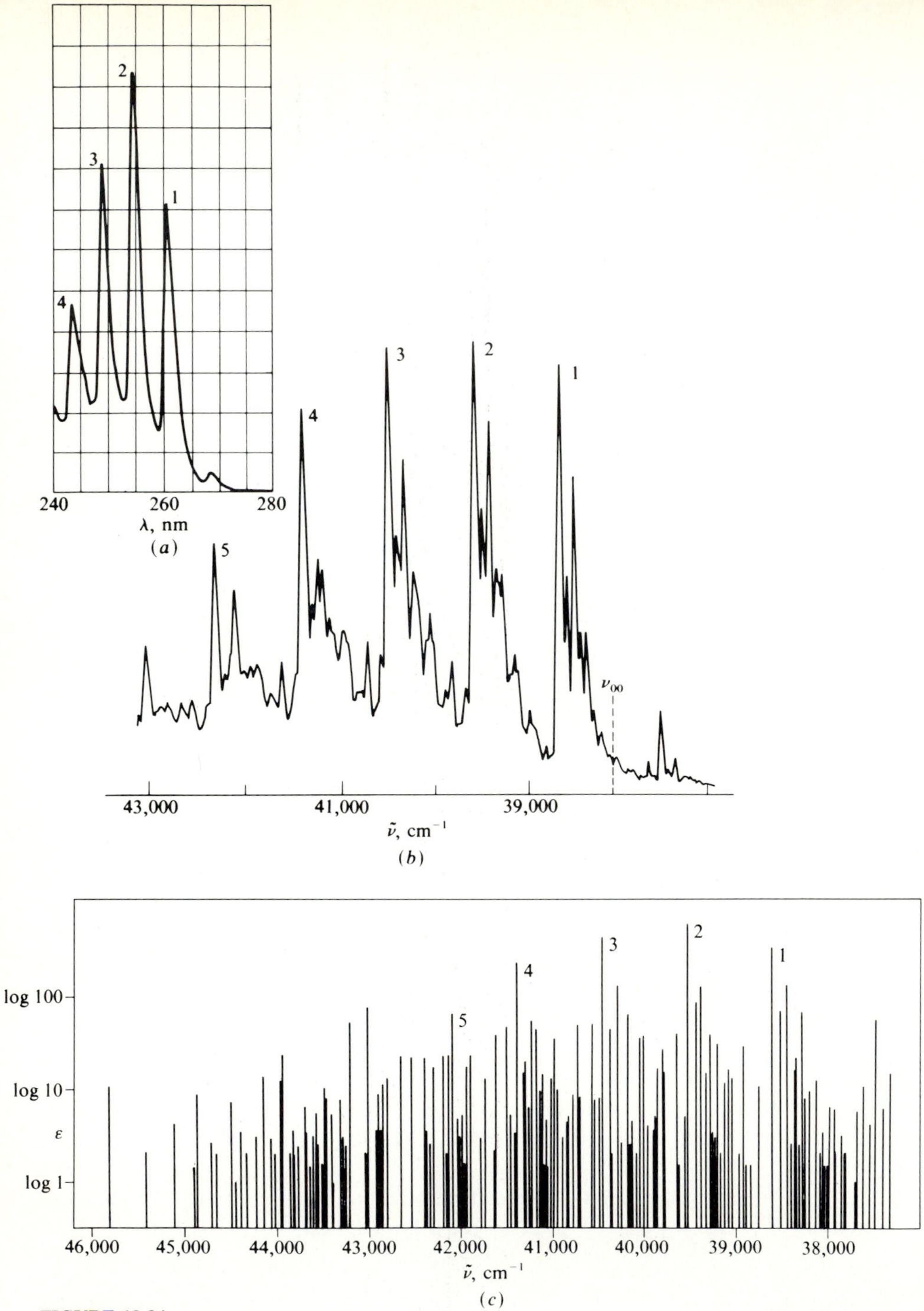

FIGURE 18.34
Electronic absorption spectrum of benzene. (*a*) Liquid. (*b*) Vapor over liquid at 24°C. (*c*) Microphotometer recording of a photographically recorded spectrum from a Zeiss QU18 optical spectrometer-vapor over liquid at 12°C.

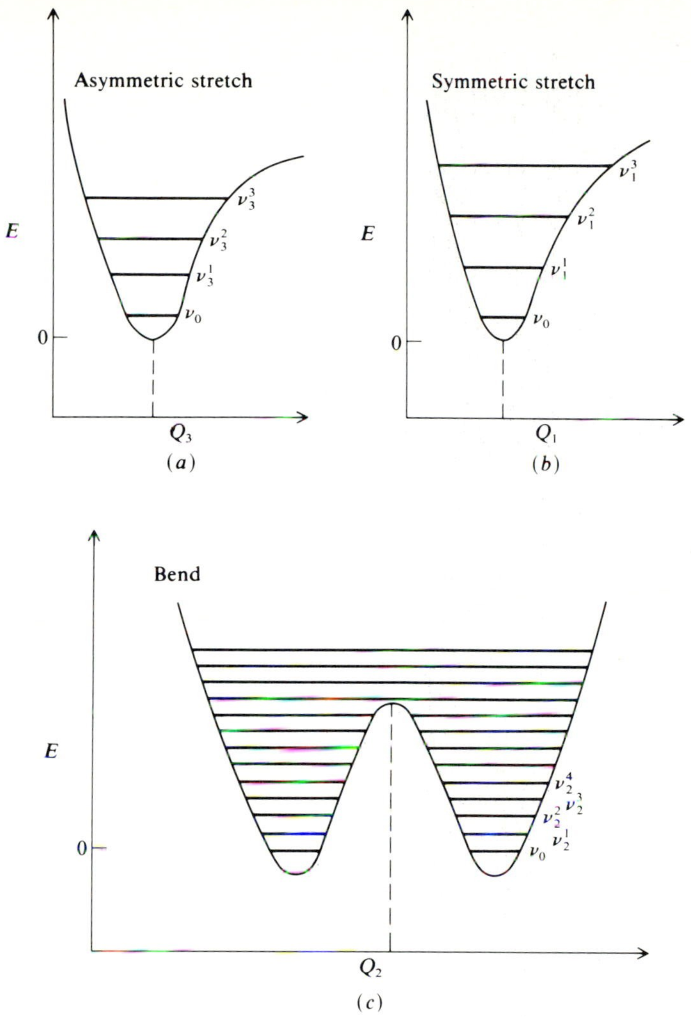

FIGURE 18.35
Potential functions for a typical bent AB_2 molecule: (*a*) symmetric stretch; (*b*) asymmetric stretch; (*c*) bend.

of the AB_2 molecule would require four dimensions, the representation of all vibrational levels on a single arbitrary potential curve is frequently used. Figure 18.36 illustrates this for the bent AB_2 molecule.

For each electronic state there will be a separate potential curve of this type. The electronic spectrum of a molecule will then be a complicated series of transitions which will spread out from the transition between the ground electronic state with all vibrations in their ground states to the excited electronic state with all vibrations remaining in their ground states. This is referred to as the $0-0$ transition. Other transitions are indexed with $\nu_{iv_i''}^{v_i'}$, where ν_i is the indexing number

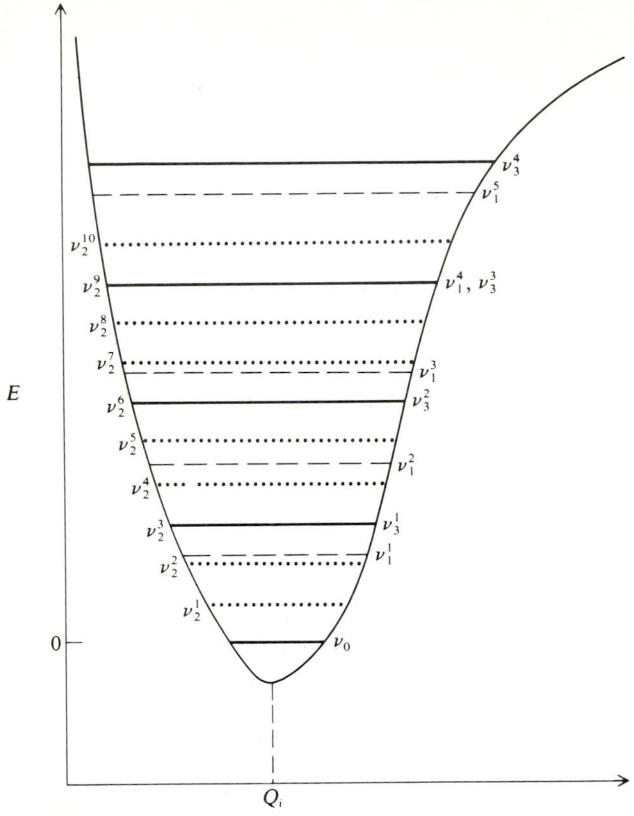

FIGURE 18.36
Arbitrary representation of the three vibrational modes of an AB_2 molecule on a single potential function.

assigned to the ith normal mode, and v_i'' and v_i' are the vibrational quantum numbers for the lower and upper vibrations states respectively. These features are illustrated in Fig. 18.37 where several typical transitions are shown.

We can now visualize the complexity of the energy-level diagram which will emerge for a molecule with a large number of vibrational modes and several closely spaced and accessible electronic levels.

The spectrum of the benzene molecule, for which a reasonable molecular orbital scheme is available (Fig. 18.26), the structure is that of a symmetric top, and the vibrational spectrum is well characterized, will be used to point out some of the main vibronic features. It was mentioned in Sec. 18.15 that there was some ambiguity in the assignment of the benzene bands at 185–205 nm and 227–267 nm. The characterization of either of these bands as belonging to an ${}^1A_{1g} \leftrightarrow {}^1E_{1u}$ is eliminated since $a_{1u} \times e_{1u}$ contains the totally symmetric representation and thus it should be a very intense transition. Both the ${}^1A_{1g} \leftrightarrow {}^1B_{1u}$ and ${}^1A_{1g} \leftrightarrow {}^1B_{2u}$ are symmetry-forbidden transitions whose existence, if observed, would be due to

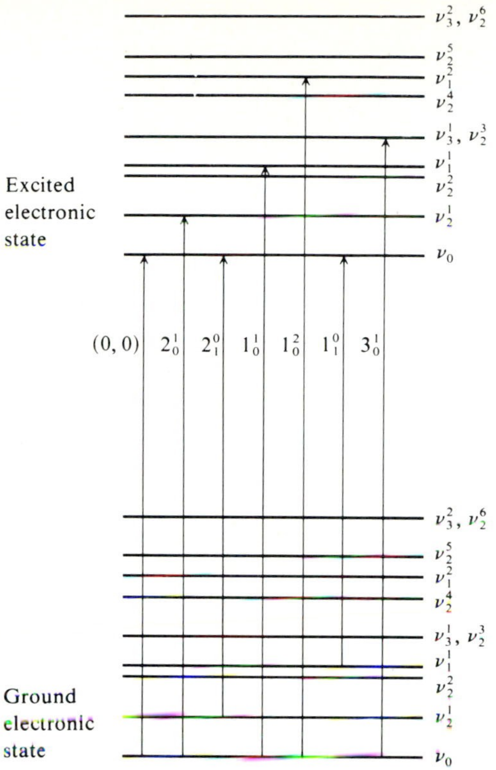

FIGURE 18.37
Typical vibronic transitions for a bent AB_2-type molecule.

vibronic interaction. If we employ the procedure outlined in Sec. 18.16 to determine the allowed vibronic selection rules, Table 18.10 is established.

Benzene has a total of 30 normal modes of which 10 are doubly degenerate thereby giving rise to 20 separate vibrational frequencies. These belong to a_{1g}, a_{2g}, b_{2g}, e_{1g}, e_{2g}, a_{2u}, b_{1u}, b_{2u}, e_{1u}, and e_{2u} representations. Therefore, since there is at least one vibration belonging to the same representation as each of the direct products, both transitions are vibronically allowed.

The assignment of the vibronic spectrum hinges on locating the 0-0 band which for benzene is symmetry-forbidden and will be weak or absent. The

TABLE 18.10
Direct products for the determination of symmetry selection rules for benzene—D_{6h} symmetry

$\Gamma(\psi_e'')$	$\Gamma(\psi_e')$	$\Gamma(u)$	**Direct product**
A_{1g}	B_{1u}	A_{2u}	B_{2g}
A_{1g}	B_{1u}	E_{1u}	E_{2g}
A_{1g}	B_{2u}	A_{2u}	B_{1u}
A_{1g}	B_{2u}	E_{1u}	E_{2g}

assignment has been worked out by Collomon, Dunn, and Mills [35] who have identified the strongest set of peaks as being transitions from the $\tilde{X}\,^1A_{1g}$ ground electronic-vibrational state to a series of vibrationally excited $\tilde{A}\,^1B_{2u}$ excited electronic states. The vibrational states involved are the ν_6'' (e_{2g}) and the $(n\nu_1' + \nu_6')$ (e_{2g}) with the series of transitions being $1_0^0 6_0^1$, $1_0^1 6_0^1$, $1_0^2 6_0^1$, etc. (1, 2, 3, etc. in Fig. 18.34).

The analysis of rovibronic spectra of polyatomic molecules, even when adequate resolution is available, is an exceptionally complex and tedious procedure. It can be aided substantially by prior knowledge of the rotational spectrum of the molecule, just as a knowledge of the vibrational spectrum aids in the assignment of the vibronic spectrum. For many molecules, even with currently available high-resolution techniques, the complexity and density of the observed spectrum prohibits a unique assignment. One method which has been found to be of use is that of band contour analysis [35]. A detailed description of rovibronic spectral analysis is found in the monograph by Herzberg [26].

18.18 FLUORESCENCE, PHOSPHORESCENCE, AND INTERNAL CONVERSION

To conclude this chapter we will briefly consider a set of three related phenomena which are also related to electronic absorption spectra. These were mentioned briefly in Chap. 1 and will now be examined in more detail. The discussion will be qualitative and will use Fig. 18.38 as a point of departure.

The vibrationally nonexcited levels are denoted by S_n for singlet and T_n for triplets. The unlabeled horizontal lines denote vibrational levels of particular electronic states and the slanted line regions denote continua.

Normal vibronic structure for the $S_0 \leftrightarrow S_1$ electronic transition is represented by the vertical connecting lines labeled as A. The phenomenon of fluorescence occurs when the absorption of radiation of sufficient energy ν_0 first results in the promotion of the system to the higher vibrational states of the electronic state S_1. The system then cascades down, B, through the vibrational levels to the S_1 $(\nu = 0)$ level and then decays back, F, to a series of vibrationally excited levels of the ground electronic state S_0 $(\nu = 0, 1, 2, 3, \ldots)$. If one compares the normal absorption spectrum and the fluorescence spectrum their appearances are as shown by the "stick" spectrum of Fig. 18.39. In a real molecular system, as also shown in Fig. 18.39, the mirror image effect is not perfect, principally due to variations in intensities.

Closely related to fluorescence is the phenomenon of phosphorescence, which results from the decay of a system from a triplet state to the ground singlet state, a normally spin-forbidden transition. Because the $T_1 \rightarrow S_0$ transition is spin-forbidden the molecule has a long lifetime in the T_1 state, often up to several minutes, and the phosphorescent emission will continue following the removal of the exciting source. The breakdown of the normal spin selection rule results from a mixing of singlet and triplet wavefunctions due to spin-orbital interactions.

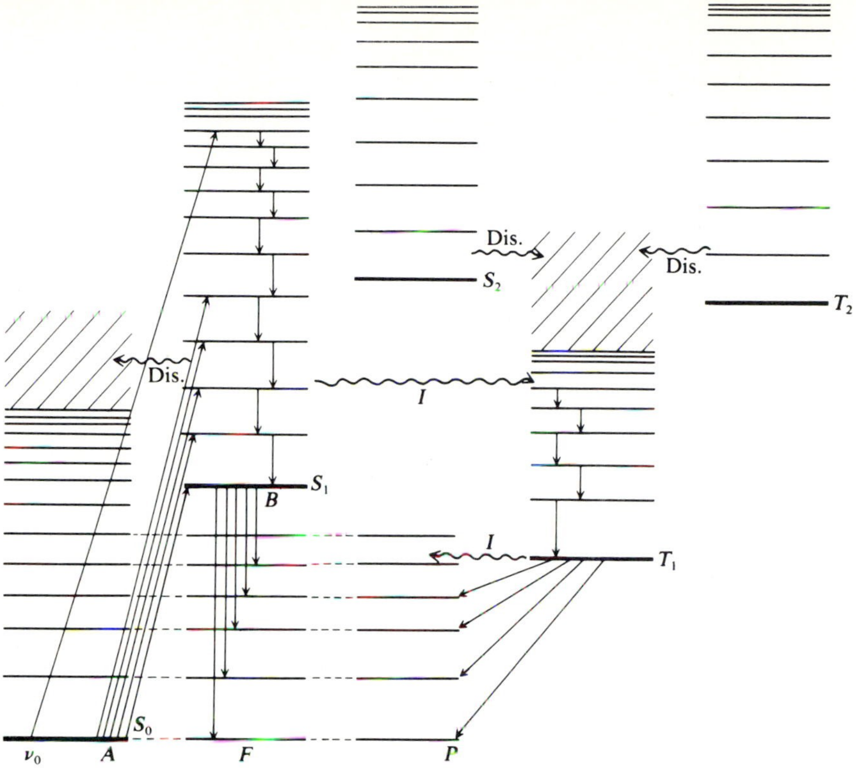

FIGURE 18.38
Energy-level diagram and possible transitions for a polyatomic molecule having a singlet ground state and both singlet and triplet excited states (only electronic and vibrational levels are indicated).

There still remains the problem of providing a mechanism for the system to initially gain access to the T_1 state. The generally accepted mechanisms for this transfer are those of (1) intersystem crossing wherein the potential energy surfaces of the S_1 and T_1 states cross at a level well above S_1 ($\nu = 0$) and there can be an energy transfer, and (2) collision-induced energy transfer in which molecular collisions transfer energy prior to vibrational deactivation. These processes give rise to transfers denoted a *IS* on Fig. 18.38. The phosphorescent transitions are denoted by a *P*.

Another phenomenon illustrated by Fig. 18.38 is that of dissociation (Dis.) and continuous spectra. When the energies of two states (one a discrete quantized set such as the lower part of the S_1 manifold and the other a continuum such as the top part of the S_0 manifold) are comparable, the system may cross over from the excited state S_1 ($\nu = n$) to the continuum, and molecular dissociation, accompanied by a diffuse spectrum, occurs.

It should be apparent to the reader by this time that only "the surface has been scratched" with regard to discussing the many ramifications of the electronic

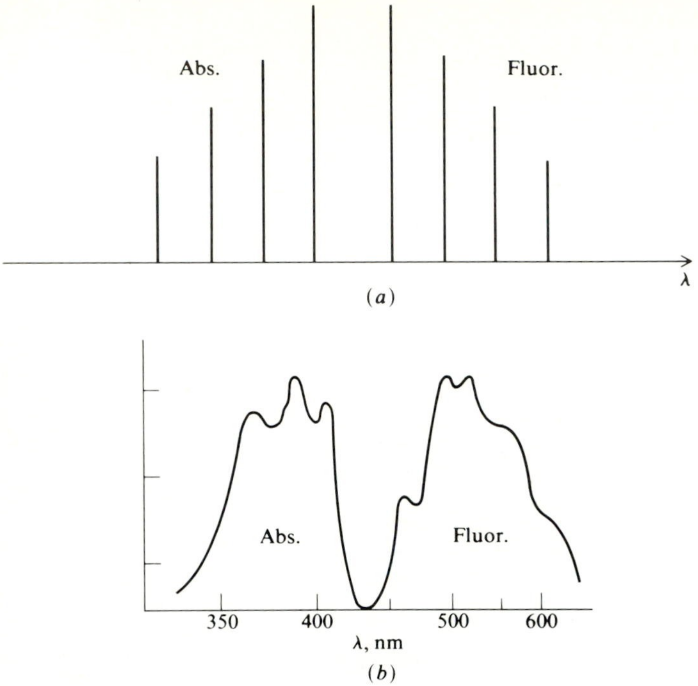

FIGURE 18.39
Comparison of hypothetical absorption and fluorescence spectra.

spectra of polyatomic molecules. The study of this topic beyond the level just presented is best accomplished by consulting a treatise such as that by Herzberg [26].

REFERENCES

Specific

1. Offenhartz, P. O., *Atomic and Molecular Orbital Theory*, McGraw-Hill, New York, N.Y., 1970.
2. Ballhausen, C. J., and H. B. Gray, *Molecular Orbital Theory*, W. A. Benjamin, New York, N.Y., 1965.
3. Streitwiser, A., Jr., *Molecular Orbital Theory for Organic Chemists*, John Wiley, New York, N.Y., 1961.
4. Douglas, B. E., and C. A. Hollingsworth, *Symmetry in Bonding and Spectra*, Academic Press, New York, N.Y., 1985.
5. Cotton, F. A., *Chemical Applications of Group Theory*, 2d ed., John Wiley, New York, N.Y., 1971.
6. Burdett, J. K., *Molecular Shapes*, John Wiley, New York, N.Y., 1980.
7. Jorgensen, W. L., and L. Salem, *The Organic Chemist's Book of Orbitals*, Academic Press, New York, N.Y., 1973.
8. Hout, R. F., Jr., W. J. Pietro, and W. J. Hehre, *A Pictorial Approach to Molecular Structure and Reactivity*, John Wiley, New York, N.Y. 1984.
9. Walsh, A. D., *J. Chem. Soc.*, 2260–2331 (1953).

10. Loc. cit., Cotton, ch. 6.
11. Hamaka, H. F., *Advanced Quantum Chemistry*, Addison-Wesley, Reading, MA, 1965.
12. Kauzmann, W., *Quantum Chemistry*, Academic Press, New York, N.Y., 1957.
13. Levine, I. N., *Quantum Chemistry*, Allyn and Bacon, Rockleigh, N.J., 1970.
14. Pople, J. A., and D. L. Beveridge, *Approximate Molecular Orbital Theory*, McGraw-Hill, New York, N.Y., 1970.
15. Schaefer, Henry F., III, *The Electronic Structure of Atoms and Molecules*, Addison-Wesley, Reading, MA, 1972.
16. Richards, W. G., and D. L. Cooper, *Ab Initio Molecular Orbital Calculations for Chemists*, 2d ed., Oxford University Press, Oxford, G.B., 1982.
17. Gray, H. B., and C. J. Ballhausen, *J. Am. Chem. Soc.*, **85**, 260 (1963).
18. Wilson, E. B., Jr., J. C. Decius, and P. C. Cross, *Molecular Vibrations*, McGraw-Hill, New York, N.Y., 1955, p. 333.
19. Loc. cit., Walsh, p. 2306.
20. Loc. cit., Walsh, p. 2325.
21. Pilar, F. L., *Elementary Quantum Mechanics*, McGraw-Hill, New York, N.Y., 1968, p. 545.
22. McGlynn, S. P., L. G. Vanquickenborne, M. Kinoshita, and D. G. Carroll, *Introduction to Applied Quantum Chemistry*, Holt, Rinehart and Winston, New York, N.Y., 1972, ch. 3.
23. Watanabe, K., and M. Zelikoff, *J. Opt. Soc. Amer.*, **43**, 753 (1953).
24. Price, W. C., *J. Chem. Phys.*, **4**, 147 (1936).
25. Watanabe, K., *J. Chem. Phys.*, **22**, 1564 (1954).
26. Herzberg, G., *Molecular Spectra and Molecular Structure: III. Electronic Spectra and Electronic Structure of Polyatomic Molecules*, Van Nostrand Reinhold, New York, N.Y., 1966.
27. Ballhausen, J. C., *Introduction to Ligand Field Theory*, McGraw-Hill, New York, N.Y., 1962.
28. Orgel, L. E., *An Introduction to Transition Metal Chemistry*, Methuen, London, G.B., 1960.
29. Loc. cit., Douglas, ch. 9.
30. Dunn, T. M., D. S. McClure, and R. G. Pearson, *Some Aspects of Crystal Field Theory*, Harper and Row, N.Y., 1965.
31. Jorgensen, C. K., *Orbitals in Atoms and Molecules*, Academic Press, New York, N.Y., 1962.
32. Loc. cit., Douglas. chs. 5, 9.
33. Loc. cit., Cotton. ch. 9.
34. Tanabe, Y., and S. Sugano, *J. Phys. Soc. Japan*, **9**, 753 (1954).
35. Callomon, J. H., T. M. Dunn, and I. M. Mills, *Phil. Trans. Roy. Soc. (London)*, **259A**, 499 (1966).

General

Hollas, M. J., *High-Resolution Spectroscopy*, Butterworths, London, G.B., 1982.
Jaffe, H. H., and M. Orchin, *Theory and Application of Ultraviolet Spectroscopy*, John Wiley, New York, N.Y., 1962.
See specific references Nos. 13, 26.

PROBLEMS

18.1. Using the appropriate MO diagrams and Walsh diagrams as a guide, determine the electronic configurations of the ground and first three excited electronic states of : (*a*) NH_2, (*b*) CS_2, (*c*) SH_2, (*d*) $H_2C{=}S$, (*e*) CH_3 (planar), (*f*) HCN, and (*g*) $HFC{=}CH_2$.

18.2. Determine the state symbols for each of the electronic configurations in Prob. 18.1.

18.3. For the molecules given in Prob. 18.1, determine which electronic transitions are allowed between the ground and excited states.

18.4. For 1,1′-diethyl-4,4′-carbocyanine iodide

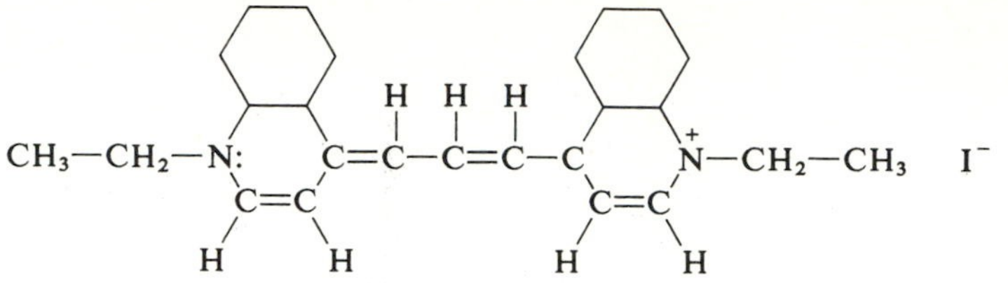

use the FEMO model to calculate the lowest electronic transitions of the π-electrons. How does the calculated value compare to the experimental value of 595 nm? Determine the change in the spectrum if the carbon chain is lengthened by: (*a*) 2 atoms, (*b*) 4 atoms.

18.5. Pyridine has a broad electronic absorption located at 260 nm. (1) Considering pyridine as a two-dimensional "particle-in-a-box" problem, estimate the position of this transition. (2) Using Hückel MO theory, estimate the position of this transition.

18.6. For the case of octahedral symmetry, determine the allowed electronic states (terms) for the following configurations:

Orbital occupancy	Electron configuration
d^1	$(e_g)^1$
d^1	$(t_{2g})^1$
d^3	$(e_g)^3$
d^3	$(t_{2g})^1(e_g)^2$
d^3	$(t_{2g})^2(e_g)^1$
d^3	$(t_{2g})^3$

18.7. The electronic spectrum of HCP consists of the following bands:

$\tilde{a} \leftrightarrow \tilde{X}$	3050–4100A
$\tilde{A} \leftrightarrow \tilde{X}$	2300–3000A
$\tilde{B} \leftrightarrow \tilde{X}$	2580–2780A
$\tilde{C} \leftrightarrow \tilde{X}$	2650–2780A

Determine the state terms and the electron configurations for the ground and the four excited states. ($\tilde{A}$ is $\tilde{C}_s$ symmetry, others are $C_{\infty v}$.)

APPENDIX A

CLASSICAL HARMONIC MOTION

Classical harmonic motion results when the force acting on a body is proportional to the displacement of the body from an equilibrium position. In terms of Newton's first law this is

$$F = ma = -kx \tag{A-1}$$

or

$$m\frac{d^2x}{dt^2} = -kx \tag{A-2}$$

The general solution of this equation is of the form

$$x = A\cos(\omega t + \delta) \tag{A-3}$$

where ω and δ are the angular frequency and phase of the motion. Substitution of this general solution into the initial differential equation shows that $\omega^2 = k/m$, or the linear frequency of the motion is given by

$$\nu = \frac{\omega}{2\pi} = \frac{1}{2\pi}\sqrt{\frac{k}{m}} \tag{A-4}$$

When discussing the motion of diatomic molecules it is convenient to reduce the system to an equivalent but simpler one. For two masses m_1 and m_2, connected

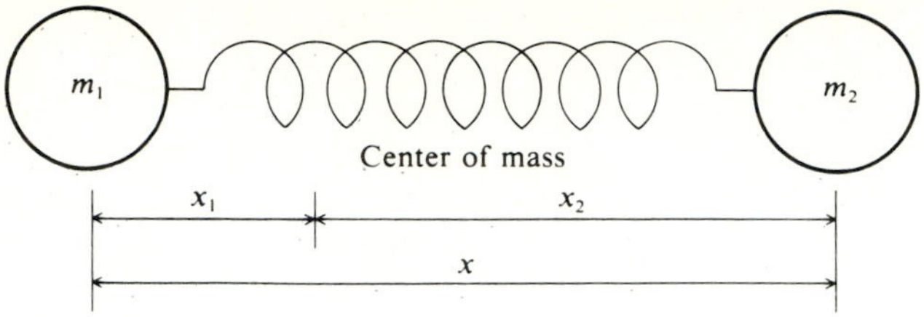

FIGURE A-1
Two-body oscillator.

by a spring having a restoring force k, separated as shown in Fig. A-1 and constrained to move in one dimension, the energy of the system is written as

$$E = \tfrac{1}{2}m_1 v_1^2 + \tfrac{1}{2}m_2 v_2^2 + \tfrac{1}{2}kx^2 \tag{A-5}$$

where v_1 and v_2 are the linear velocities of the two masses and the term $\frac{1}{2}kx^2$ is the potential energy derived from the classical relationship between force and potential

$$F = -\frac{dV}{dx} \tag{A-6}$$

Using the center of mass relationship

$$m_1 x_1 = -m_2 x_2 \tag{A-7}$$

and differentiating with respect to time given

$$m_1 \frac{dx_1}{dt} = -m_2 \frac{dx_2}{dt} \tag{A-8}$$

or

$$m_1 v_1 = -m_2 v_2 \tag{A-9}$$

In terms of the linear momentum $p_i = m_i v_i$ this becomes

$$p_1 = p_2 = p \tag{A-10}$$

Writing the energy in terms of the momenta

$$E = \frac{p_1^2}{2m_1} - \frac{p_2^2}{2m_2} + \frac{1}{2}kx^2 \tag{A-11}$$

or

$$= \frac{1}{2}\left(\frac{m_2 p_1^2 + m_1 p_2^2}{m_1 m_2}\right) + \tfrac{1}{2}kx^2 \tag{A-12}$$

Substitution of Eq. (A-10) gives

$$E = \frac{1}{2}\left(\frac{m_2 + m_1}{m_1 m_2}\right) p^2 + \frac{1}{2}kx^2 \tag{A-13}$$

Letting $\mu = (m_1)(m_2)/(m_1 + m_2)$ (the reduced mass) this becomes

$$E = \frac{p^2}{2\mu} + \frac{1}{2}kx^2 \tag{A-14}$$

This is identical to the equation for the energy of a single mass of magnitude μ subjected to a restoring force of $-kx$, so we have shown that the two systems are identical in behavior.

APPENDIX B

SCHRÖDINGER OPERATORS

Dynamical variable		
Name	**Symbol**	**Quantum mechanical operator**
Coordinates	x, y, z, r, θ, ϕ	x, y, z, r, θ, ϕ
Linear momentum and components	p_x, p_x^2	$-i\hbar\frac{\partial}{\partial x}, -\hbar^2\frac{\partial^2}{\partial x^2}$
	p_y, p_y^2	$-i\hbar\frac{\partial}{\partial y}, -\hbar^2\frac{\partial^2}{\partial y^2}$
	p_z, p_z^2	$-i\hbar\frac{\partial}{\partial z}, -\hbar^2\frac{\partial^2}{\partial z^2}$
	$p^2 = p_x^2 + p_y^2 + p_z^2$	$-\hbar^2\nabla^2 = -\hbar^2\left[\frac{\partial^2}{\partial x^2} + \frac{\partial^2}{\partial y^2} + \frac{\partial^2}{\partial z^2}\right]$

(*Continued*)

Dynamical variable		
Name	**Symbol**	**Quantum mechanical operator**
Angular momentum* and components	$P_x = yp_z - zp_y$	$-i\hbar\left[y\frac{\partial}{\partial z} - z\frac{\partial}{\partial y}\right]$
		$= -i\hbar\left[-\sin\phi\frac{\partial}{\partial\phi} - \cot\theta\cos\phi\frac{\partial}{\partial\phi}\right]$
	$P_y = zp_x - xp_z$	$-i\hbar\left[z\frac{\partial}{\partial x} - x\frac{\partial}{\partial z}\right]$
		$= -i\hbar\left[-\sin\phi\frac{\partial}{\partial\theta} - \cot\theta\cos\phi\frac{\partial}{\partial\phi}\right]$
	$p_z = zp_y - yp_x$	$-i\hbar\left[z\frac{\partial}{\partial y} - y\frac{\partial}{\partial x}\right] = -i\hbar\left[\frac{\partial}{\partial\phi}\right]$
	$P^2 = P_x^2 + P_y^2 + P_z^2$	$-\hbar^2\left[\frac{1}{\sin\theta}\frac{\partial}{\partial\theta}\sin\theta\frac{\partial}{\partial\theta} + \frac{1}{\sin^2\theta}\frac{\partial^2}{\partial\phi^2}\right]$
Kinetic energy	$E_x = \frac{P_x^2}{2m} + \frac{P_v^2}{2m} + \frac{P_z^2}{2m}$	$\frac{-\hbar^2}{2m}\left[\frac{\partial^2}{\partial x^2} + \frac{\partial^2}{\partial y^2} + \frac{\partial^2}{\partial z^2}\right]$
Potential energy	$V(x, y, z)$	$V(x, y, z)$
Total energy (Hamiltonian)	$H = \frac{1}{2}mv^2 + V(x, y, z)$	$\frac{-\hbar^2}{2m}\nabla^2 + V(x, y, z)$
	$= \frac{p^2}{2m} + V(x, y, z)$	

* See App. D for Cartesian–spherical polar relationships.

APPENDIX C

EIGENVALUE EQUATIONS

Since the solutions are necessary for discussion of the harmonic oscillator the general problem of solving differential eigenvalue equations will be examined by using the Hermite differential equation. This differential equation is of the form

$$-\frac{d^2f(x)}{dx^2} + x^2f(x) = af(x) \tag{C-1}$$

The problem is to solve the differential equation for the well-behaved functions $f(x)$ and the constant a.

An instructive manner in which to proceed with the solution of this differential equation is to examine the solutions in the limit of large values of the variable. For the case where x is large it follows that $x^2 \gg a$ and Eq. (C-1) reduces to

$$\frac{d^2f(x)}{dx^2} = x^2f(x) \tag{C-2}$$

The solutions of this equation are

$$f(x) = e^{\pm x^2/2} \tag{C-3}$$

One can verify this by substitution of Eq. (C-3) into Eq. (C-2) and differentiating to give

$$-\frac{d^2e^{\pm x^2/2}}{dx^2} + x^2e^{\pm x^2/2} = -(x^2 \pm 1)e^{\pm x^2/2} + x^2e^{\pm x^2/2} \tag{C-4}$$

If x is large, then $x^2 \gg 1$ and $(x^2 \pm 1) \to x^2$. Therefore

$$\frac{d^2 e^{\pm x^2/2}}{dx^2} = x^2 e^{\pm x^2/2} \qquad \text{(C-5)}$$

and for large x, $e^{\pm(1/2)x^2}$ are solutions. Examination of these solutions shows that for $f(x) = e^{(1/2)x^2}$, $f(x)$ becomes infinite as $x \to \infty$ so it is of no further interest. The function $f(x) = e^{-(1/2)x^2}$ does satisfy the boundary conditions as shown in Fig. C-1. Substitution of $e^{-(1/2)x^2}$ into the eigenvalue equation shows that $e^{-(1/2)x^2}$ is an eigenfunction belonging to the eigenvalue $a = +1$, for any magnitude of x.

We now raise the question: Is this the only acceptable solution or are there others? Substitution of the trial function $f(x) = xe^{-(1/2)x^2}$ shows that it also is an eigenfunction belonging to the eigenvalue $a = +3$. The fact that $e^{-(1/2)x^2}$ and $xe^{-(1/2)x^2}$ are eigenfunctions suggests that the general function $f(x) = H(x)e^{-(1/2)x^2}$, where $H(x)$ is a polynomial in x, may also be an eigenfunction. To determine the form of $H(x)$ when $f(x) = H(x)e^{-(1/2)x^2}$ is an allowed solution one substitutes $H(x)e^{-(1/2)x^2}$ into the eigenvalue equation obtaining

$$\frac{-d^2 H(x)e^{-x^2/2}}{dx^2} + x^2 H(x)e^{-x^2/2} = aH(x)e^{-x^2/2} \qquad \text{(C-6)}$$

Upon differentiation one obtains a second differential equation

$$\frac{d^2 H(x)}{dx^2} - 2x\frac{dH(x)}{dx} + (a-1)H(x) = 0 \qquad \text{(C-7)}$$

which specifies the conditions on $H(x)$ compatible with $H(x)e^{-(1/2)x^2}$ being a solution to the eigenvalue equation. By substituting $(a-1) = 2n$, Eq. (C-7) becomes

$$\frac{d^2 H(x)}{dx^2} - 2x\frac{dH(x)}{dx} + 2nH(x) = 0 \qquad \text{(C-8)}$$

This is the classical differential equation of Hermite. The solutions of this equation are of the form

$$H_n(x) = (-1)^n e^{x^2}\left[\frac{d^n e^{-x^2}}{dx^n}\right] \qquad \text{(C-9)}$$

where n is any positive integer including zero. The solutions are the Hermite

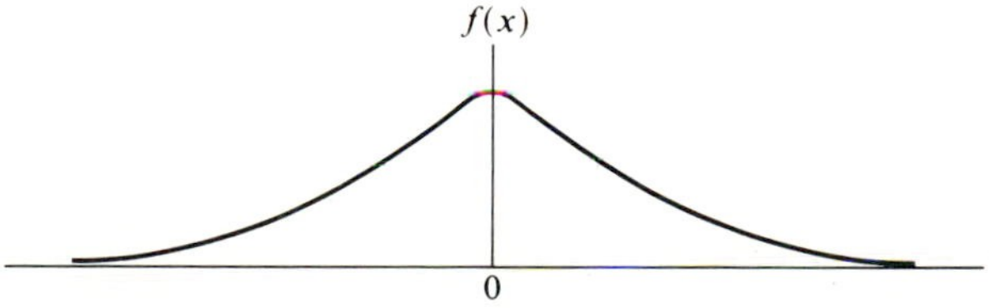

FIGURE C-1
Plot of $f(x) = e^{-(1/2)x^2}$ vs. x.

polynomials which, for the first five values of n, are

$$n = 0 \qquad H_0(x) = 1 \tag{C-10}$$

$$n = 1 \qquad H_1(x) = 2x \tag{C-11}$$

$$n = 2 \qquad H_2(x) = 4x^2 - 2 \tag{C-12}$$

$$n = 3 \qquad H_3(x) = 8x^3 - 12x \tag{C-13}$$

$$n = 4 \qquad H_4(x) = 16x^4 - 48x^2 + 12 \tag{C-14}$$

It is to be noted that the parity of n is related to the powers of x occurring in the polynomial.

In addition to the differential generating function given by Eq. (C-9), the Hermite polynomials can be derived from a generating function

$$e^{[x^2-(z-x)^2]} = \sum_n \frac{Hn(x)}{n!} z^n \tag{C-15}$$

or from the finite series

$$H_n(x) = (2x)^n - \frac{n(n-1)}{1!}(2x)^{n-2} + \frac{n(n-1)(n-2)(n-3)}{2!}(2x^{n-4})$$

$$- \cdots + \frac{(+1)^{k+1} n(n-1)(n-2)\cdots(n-k)}{(k-1)!}(2x)^{n-2k+2} - \cdots \tag{C-16}$$

where k is a running index satisfying the conditions

$$1 \leqq k \leqq \frac{n+2}{2} \qquad n \text{ even} \tag{C-17}$$

$$1 \leqq k \leqq \frac{n+1}{2} \qquad n \text{ odd} \tag{C-18}$$

Derivatives and functions are related by recursion relations

$$H_n'(x) = 2nH_{n-1}(x) \tag{C-19}$$

$$H_n''(x) = 4n(n-1)H_{n-2}(x) \tag{C-20}$$

Two useful integral relations are

$$\int_{-\infty}^{\infty} H_n(x)H_m(x)e^{-x^2}\,dx = \sqrt{\pi}\,2^n n!\,\delta_{nm} \tag{C-21}$$

and

$$\int_{-\infty}^{\infty} H_n(x)xH_m(x)e^{-x^2}\,dx = \sqrt{\pi}[n!\,\delta_{m,n-1} + 2^n(n+1)!\,\delta_{m,n+1}] \tag{C-22}$$

Substitution of a representative function of the polynomial $H(x)e^{-(1/2)x^2}$ into Eq. (C-1)

$$-\frac{d^2[2xe^{-x^2/2}]}{dx^2} + x^2[2xe^{-x^2/2}] = 6xe^{-x^2/2} \tag{C-23}$$

shows that it is an eigenfunction having an eigenvalue of 3. Examination of Eq. (C-22) shows that it differs from the results obtained using $f(x) = xe^{-(1/2)x^2}$ by only a numerical constant. The most general solution to the eigenvalue equation is therefore given by

$$f(x) = N_n H_n(x) e^{-x^2/2} \tag{C-24}$$

where N_n is a numerical constant. A comparison of Eqs. (C-7) and (C-8) shows that

$$a = 2n + 1 \tag{C-25}$$

Since n can only have positive integer values, it follows that the eigenvalues of Eq. (C-1) must be odd positive integers. The magnitude of N_n can be adjusted such that this condition is satisfied. For example, consider the eigenfunction

$$F_1(x) = N_1 H_1(x) e^{-x^2/2} \tag{C-26}$$

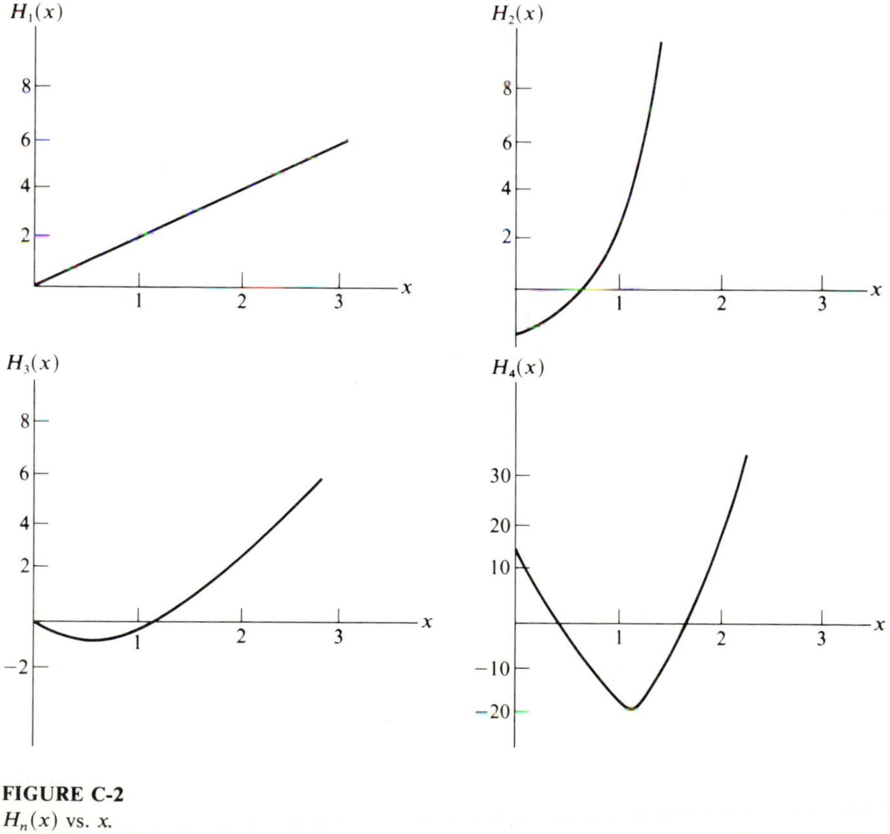

FIGURE C-2
$H_n(x)$ vs. x.

Since $n = 1$ the eigenvalue must be 3, but

$$-\frac{d^2[N_1 2xe^{-x^2/2}]}{dx^2} + x^2[n_1 2xe^{-x^2/2}] = 3[N_1 2xe^{-x^2/2}] \qquad \text{(C-27)}$$

so if $a = 3$, N_1 must be $\frac{1}{2}$. Since any function $N_n f(x)$ is a simultaneous eigenfunction with $f(x)$ so long as N_n is a numerical constant, the magnitude of N_n is somewhat arbitrary insofar as the mathematics of the solution are concerned. It will be found convenient to use particular values for the constants N_n in working with certain systems.

A remaining point is to show that the functions $f(x) = N_n H_n(x) e^{-x^2/2}$ are acceptable solutions insofar as the boundary conditions are concerned. Prior discussion of $e^{-(1/2)x^2}$ (see Fig. C-1) has shown it to be finite, singled-valued, and continuous. Examination of the Hermite polynomials shown in Fig. C-2 shows that except for $H_0(x) = 1$ the higher ones become infinite as $x \to \infty$. We must then compare the rate of convergence to zero of $e^{-(1/2)x^2}$ with the rate of divergence to infinity of the polynomial. Examination of limits shows

$$\lim_{x\to\infty} \frac{x}{e^{x^2/2}} = 0 \qquad \text{(C-28)}$$

or in general

$$\lim_{x\to 0} \frac{H_n(x)}{e^{x^2/2}} = 0 \qquad \text{(C-29)}$$

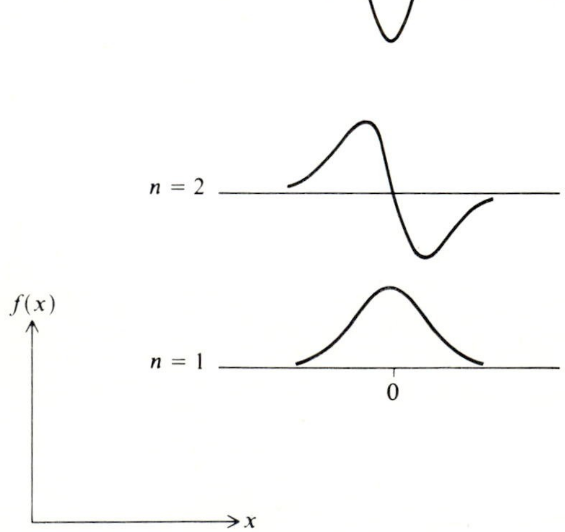

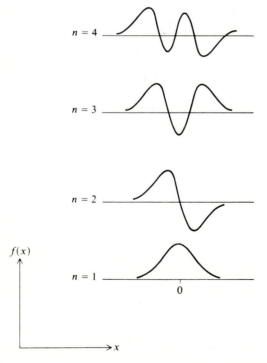

FIGURE C-3
$N_n H_n(x) e^{-(1/2)x^2}$ vs. x.

The form of the first four eigenfunctions, $f(x) = N_n H_n(x) e^{-(1/2)x^2}$, are shown in Fig. C-3. For construction of these curves the value of N_n is chosen so as to normalize the function $f(x)$ to unity, that is

$$\int_{-\infty}^{\infty} |f(x)|^2 \, dx = \int_{-\infty}^{\infty} N_n^2 |H_n(x)|^2 e^{-x^2} \, dx = 1 \qquad \text{(C-30)}$$

or

$$N_n = [2^n n! \sqrt{\pi}]^{-1/2} \qquad \text{(C-31)}$$

APPENDIX D

CLASSICAL ROTATION

Although the motion of any body can be described by its position and linear velocity vectors in Cartesian space it is more convenient when describing angular motion to use spherical polar coordinates, angular velocities, and angular momenta. For a particle of mass m having a linear velocity $\mathbf{v}$ and constrained to move in a plane circular path about the z axis, the angular position is given by the angle ϕ that the position vector $\mathbf{r}$ makes with the x axis. The angular velocity $\boldsymbol{\omega}$ is the time rate of change of ϕ, $d\phi/dt = \boldsymbol{\omega}$. It is represented by a vector $\boldsymbol{\omega}$ perpendicular to the plane of rotation. The linear and angular velocities are related by

$$\mathbf{v} = \boldsymbol{\omega} \times \mathbf{r} \tag{D-1}$$

As seen from Fig. D-1, $\boldsymbol{\omega}$ lies along the z axis, perpendicular to the plane of $\mathbf{v}$ and $\mathbf{r}$.

Angular momentum is defined as

$$\mathbf{P} = \mathbf{r} \times m\mathbf{v} = \mathbf{r} \times \mathbf{p} \tag{D-2}$$

or

$$\mathbf{P} = m\mathbf{r} \times \boldsymbol{\omega} \times \mathbf{r} = m(\mathbf{r} \cdot \mathbf{r})\boldsymbol{\omega} = mr^2\boldsymbol{\omega} = I\boldsymbol{\omega} \tag{D-3}$$

where $I = mr^2$ is the moment of inertia. The kinetic energy of the body is

$$T = \tfrac{1}{2}m\mathbf{v} \cdot \mathbf{v} = \tfrac{1}{2}m\mathbf{v} \cdot (\boldsymbol{\omega} \times \mathbf{r}) = \tfrac{1}{2}m\boldsymbol{\omega} \cdot \mathbf{r} \times \mathbf{v} = \tfrac{1}{2}\boldsymbol{\omega} \cdot \mathbf{P} = \tfrac{1}{2}I\omega^2 \tag{D-4}$$

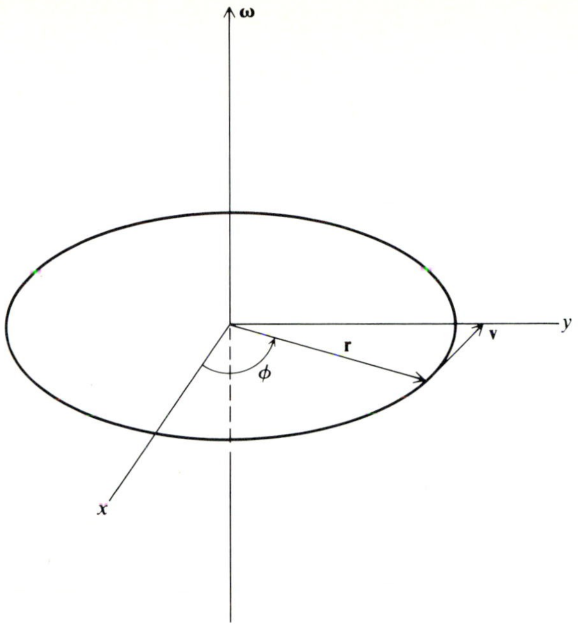

FIGURE D-1
Angular motion.

The same concepts can be extended to a set of particles, each denoted by a mass m_i and a position vector $\mathbf{r}_i$. Hence,

$$\mathbf{v}_i = \boldsymbol{\omega} \times \mathbf{r}_i \tag{D-5}$$

$$\mathbf{P}_i = \mathbf{r}_i \times \mathbf{P}_i \tag{D-6}$$

$$\mathbf{P} = \sum_i \mathbf{P}_i = \sum_i m_i(\mathbf{r}_i \times \mathbf{v}_i) \tag{D-7}$$

This expression can be rewritten as (see App. E)

$$\mathbf{P} = \sum_i m_i[\mathbf{r}_i \times (\boldsymbol{\omega} \times r_i)] = \sum_i m_i r_i^2 \boldsymbol{\omega} = I\boldsymbol{\omega} \tag{D-8}$$

The kinetic energy is similarly expressed as

$$T = \tfrac{1}{2}\sum m_i[(\boldsymbol{\omega} \times \mathbf{r}_i) \cdot \mathbf{v}_i] = \sum_i \frac{1}{2} m_i \boldsymbol{\omega} \cdot (\mathbf{r}_i \times \mathbf{v}_i) = \tfrac{1}{2}\boldsymbol{\omega} \cdot \mathbf{P} = \tfrac{1}{2}I\omega^2 \tag{D-9}$$

Since differential operators are essential to the formulation of wave mechanics and a variety of problems encountered involve rotation, it is necessary to be able to convert differential relationships between Cartesian and spherical polar coordinate systems. The coordinate relationships between the two sets of coordinates are

$$x = r \sin\theta \cos\phi \tag{D-10}$$

$$y = r \sin\theta \sin\phi \tag{D-11}$$

$$z = r \cos\theta \tag{D-12}$$

or

$$r = (x^2 + y^2 + z^2)^{1/2} \tag{D-13}$$

$$\theta = \cos^{-1} \frac{z}{(x^2 + y^2 + z^2)^{1/2}} \tag{D-14}$$

$$\phi = \tan^{-1} \frac{y}{x} \tag{D-15}$$

where x, y, z form a right-handed coordinate system, ϕ is taken counterclockwise from the x axis, and θ is measured downward from the z axis.

Using the calculus of partial derivatives

$$\frac{\partial}{\partial g} = \frac{\partial \theta}{\partial g}\frac{\partial}{\partial \theta} + \frac{\partial r}{\partial g}\frac{\partial}{\partial r} + \frac{\partial \phi}{\partial g}\frac{\partial}{\partial \phi} \qquad (g = x, y, z) \tag{D-16}$$

and applying it to the defining relationships for the coordinates gives

$$\frac{\partial}{\partial x} = \sin\theta\cos\phi\frac{\partial}{\partial r} + \frac{\cos\theta\cos\phi}{r}\frac{\partial}{\partial \theta} - \frac{\sin\phi}{r\sin\theta}\frac{\partial}{\partial \phi} \tag{D-17}$$

$$\frac{\partial}{\partial y} = \sin\theta\sin\phi\frac{\partial}{\partial r} + \frac{\cos\theta\cos\phi}{r}\frac{\partial}{\partial \theta} - \frac{\cos\phi}{r\sin\theta}\frac{\partial}{\partial \phi} \tag{D-18}$$

$$\frac{\partial}{\partial z} = \cos\theta\frac{\partial}{\partial r} - \frac{\sin\theta}{r}\frac{\partial}{\partial \theta} \tag{D-19}$$

Substitution of these $\partial/\partial g$ relations and the squared terms $\partial^2/\partial g^2$ into the Cartesian relationships gives the spherical polar operators summarized in App. B.

APPENDIX E

VECTORS

In simple physical terms a vector is defined as a quantity having both magnitude and direction. Examples of physical properties represented by vectors are velocity, acceleration, and momentum. Mathematically, a vector is a line directed in space, which can be related to its projections onto an orthogonal coordinate system as shown in Fig. E-1 and given by Eq. (E-1).

$$\mathbf{V} = V_x\hat{\mathbf{i}} + V_y\hat{j} + V_z\hat{\mathbf{k}} \tag{E-1}$$

where $\hat{\mathbf{i}}$, $\hat{\mathbf{j}}$, and $\hat{\mathbf{k}}$ have unit lengths. A more frequently used notation for a general vector $\mathbf{r}$ is

$$\mathbf{r} = x\hat{\mathbf{i}} + y\hat{\mathbf{j}} + z\hat{\mathbf{k}} \tag{E-2}$$

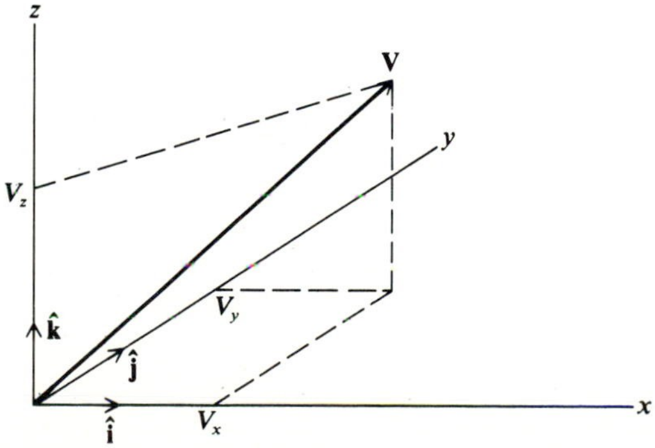

FIGURE E-1
Relation of a vector to its components.

The addition of two vectors $\mathbf{A} = A_x\hat{\mathbf{i}} + A_y\hat{\mathbf{j}} + A_z\hat{\mathbf{k}}$ and $\mathbf{B} = B_x\hat{\mathbf{i}} + B_y\hat{\mathbf{j}} + B_z\hat{\mathbf{k}}$ is given by the relationship

$$\mathbf{C} = \mathbf{A} + \mathbf{B} = (A_x + B_x)\hat{\mathbf{i}} + (A_y + B_y)\hat{\mathbf{j}} + (A_z + B_z)\hat{\mathbf{k}} \tag{E-3}$$

and the relationships between the components are

$$C_g = A_g + B_g \qquad (g = x, y, z) \tag{E-4}$$

Subtraction of two vectors **A** and **B** is most easily considered as the addition of the vectors **A** and −**B**. Therefore

$$\mathbf{D} = \mathbf{A} - \mathbf{B} = (A_x - B_x)\hat{\mathbf{i}} + (A_y - B_y)\hat{\mathbf{j}} + (A_z - B_z)\hat{\mathbf{k}} \tag{E-5}$$

and the relationships between the components are

$$D_g = A_g - B_g \qquad (g = x, y, z) \tag{E-6}$$

There are three types of vector multiplication; multiplication by a scalar, scalar multiplication by another vector, and vector multiplication by another vector. Multiplication of a vector **A** by a scalar s gives a new vector $\mathbf{B} = s\mathbf{A}$ whose direction is the same as **A** but whose magnitude is s times longer.

The scalar or "dot" product of two vectors, $\mathbf{A} \cdot \mathbf{B}$, is defined as

$$\mathbf{A} \cdot \mathbf{B} = |A|\,|B| \cos\theta \tag{E-7}$$

where θ is the angle between **A** and **B**. Since the product $\mathbf{A} \cdot \mathbf{B}$ is a scalar quantity it follows that such multiplication is both commutative

$$\mathbf{A} \cdot \mathbf{B} = \mathbf{B} \cdot \mathbf{A} \tag{E-8}$$

and associative

$$\mathbf{A} \cdot (\mathbf{B} + \mathbf{C}) = \mathbf{A} \cdot \mathbf{B} + \mathbf{A} \cdot \mathbf{C} \tag{E-9}$$

The square of the length of a vector is

$$\mathbf{A} \cdot \mathbf{A} = |A|^2 \cos\theta = A^2 \tag{E-10}$$

Since the unit vectors $\hat{\mathbf{i}}$, $\hat{\mathbf{j}}$, and $\hat{\mathbf{k}}$ are orthogonal and of unit length we have the following relations:

$$\hat{\mathbf{i}} \cdot \hat{\mathbf{i}} = \hat{\mathbf{j}} \cdot \hat{\mathbf{j}} = \hat{\mathbf{k}} \cdot \hat{\mathbf{k}} = 1 \tag{E-11}$$

$$\hat{\mathbf{i}} \cdot \hat{\mathbf{j}} = \hat{\mathbf{j}} \cdot \hat{\mathbf{k}} = \hat{\mathbf{k}} \cdot \hat{\mathbf{i}} = 0 \tag{E-12}$$

The scalar product of two vectors $\mathbf{A} \cdot \mathbf{B}$ is written

$$\mathbf{A} \cdot \mathbf{B} = (A_x\hat{\mathbf{i}} + A_y\hat{\mathbf{j}} + A_z\hat{\mathbf{k}}) \cdot (B_x\hat{\mathbf{i}} + B_y\hat{\mathbf{j}} + B_z\hat{\mathbf{k}}) \tag{E-13}$$

$$\begin{aligned}\mathbf{A} \cdot \mathbf{B} = {} & A_xB_x\hat{\mathbf{i}} \cdot \hat{\mathbf{i}} + A_xB_y\hat{\mathbf{i}} \cdot \hat{\mathbf{j}} + A_xB_z\hat{\mathbf{i}} \cdot \hat{\mathbf{k}} + A_yB_x\hat{\mathbf{j}} \cdot \hat{\mathbf{i}} \\ & + A_yB_y\hat{\mathbf{j}} \cdot \hat{\mathbf{j}} + A_zB_z\hat{\mathbf{j}} \cdot \hat{\mathbf{k}} + A_zB_x\hat{\mathbf{k}} \cdot \hat{\mathbf{i}} + A_zB_y\hat{\mathbf{k}} \cdot \hat{\mathbf{j}} + A_zB_z\hat{\mathbf{k}} \cdot \hat{\mathbf{k}}\end{aligned} \tag{E-14}$$

which, in view of Eqs. (E-11) and (E-12), becomes

$$\mathbf{A} \cdot \mathbf{B} = A_xB_x + A_yB_y + A_zB_z \tag{E-15}$$

The vector product, also known as the cross or skew product of two vectors, is written $\mathbf{C} = \mathbf{A} \times \mathbf{B}$ and is defined as a vector perpendicular to the plane of **A** and **B**, taken in such a direction as to represent the direction of advance of a right-hand screw when rotated from **A** toward **B**. Its magnitude is given by

$$\mathbf{C} = \mathbf{A} \times \mathbf{B} = |A||B| \sin\theta \tag{E-16}$$

Vector multiplication is not commutative since

$$\mathbf{A} \times \mathbf{B} = -\mathbf{B} \times \mathbf{A} \tag{E-17}$$

Vector multiplications is associative, that is

$$\mathbf{A} \times (\mathbf{C} + \mathbf{D}) = \mathbf{A} \times \mathbf{C} + \mathbf{A} \times \mathbf{D} \tag{E-18}$$

Considering the orthogonal nature of the unit vectors $\hat{\mathbf{i}}$, $\hat{\mathbf{j}}$, and $\hat{\mathbf{k}}$, the definition of the vector product, and Eq. (E-17), the following relationships are established:

$$\hat{\mathbf{i}} \times \hat{\mathbf{i}} = \hat{\mathbf{j}} \times \hat{\mathbf{j}} = \hat{\mathbf{k}} \times \hat{\mathbf{k}} = 0 \tag{E-19}$$

$$\hat{\mathbf{i}} \times \hat{\mathbf{j}} = -\hat{\mathbf{j}} \times \hat{\mathbf{i}} = \hat{\mathbf{k}} \tag{E-20}$$

$$\hat{\mathbf{j}} \times \hat{\mathbf{k}} = -\hat{\mathbf{k}} \times \hat{\mathbf{j}} = \hat{\mathbf{i}} \tag{E-21}$$

$$\hat{\mathbf{k}} \times \hat{\mathbf{i}} = -\hat{\mathbf{i}} \times \hat{\mathbf{k}} = \hat{\mathbf{j}} \tag{E-22}$$

The vector product is written as

$$\mathbf{A} \times \mathbf{B} = (A_x\hat{\mathbf{i}} + A_y\hat{\mathbf{j}} + A_z\hat{\mathbf{k}}) \times (B_x\hat{\mathbf{i}} + B_y\hat{\mathbf{j}} + B_z\hat{\mathbf{k}}) \tag{E-23}$$

$$\begin{aligned}\mathbf{A} \times \mathbf{B} = (&A_xB_x\hat{\mathbf{i}} \times \hat{\mathbf{i}} + A_xB_y\hat{\mathbf{i}} \times \hat{\mathbf{j}} + A_xB_z\hat{\mathbf{i}} \times \hat{\mathbf{k}} + A_yB_x\hat{\mathbf{j}} \times \hat{\mathbf{i}} \\ &+ A_yB_y\hat{\mathbf{j}} \times \hat{\mathbf{j}} + A_yB_y\hat{\mathbf{j}} \times \hat{\mathbf{k}} + A_zB_x\hat{\mathbf{k}} \times \hat{\mathbf{i}} + A_zB_y\hat{\mathbf{k}} \times \hat{\mathbf{j}} + A_zB_z\hat{\mathbf{k}} \times \hat{\mathbf{k}})\end{aligned} \tag{E-24}$$

which, in view of Eqs. (E-19) through (E-22), becomes

$$\mathbf{A} \times \mathbf{B} = (A_yB_z - A_zB_y)\hat{\mathbf{i}} + (A_zB_x - A_xB_z)\hat{\mathbf{j}} + (A_xB_y - A_yB_x)\hat{\mathbf{k}} \tag{E-25}$$

This can be written in determinental form as

$$\mathbf{A} \times \mathbf{B} = \begin{vmatrix} \hat{\mathbf{i}} & \hat{\mathbf{j}} & \hat{\mathbf{k}} \\ A_x & A_y & A_z \\ B_x & B_y & B_z \end{vmatrix} \tag{E-26}$$

There are two triple products of importance in vector algebra. The scalar triple product, or mixed triple product, is $(\mathbf{A} \times \mathbf{B}) \cdot \mathbf{C}$ and is given in determinental form as

$$(\mathbf{A} \times \mathbf{B}) \cdot \mathbf{C} = \begin{vmatrix} A_x & A_y & A_z \\ B_x & B_y & B_z \\ C_x & C_y & C_z \end{vmatrix} \tag{E-27}$$

The magnitude of $(\mathbf{A} \times \mathbf{B}) \cdot \mathbf{C}$ is equal to the volume of the parallelepiped defined by the three vectors. If **A**, **B**, and **C** are all nonzero then

$$(\mathbf{A} \times \mathbf{B}) \cdot \mathbf{C} = 0 \tag{E-28}$$

is the necessary and sufficient condition that all three vectors be coplanar.

The scalar triple product is unaffected by interchange of the dot and cross multiplications, or by cyclic permutation of the vectors, that is

$$(\mathbf{A} \times \mathbf{B}) \cdot \mathbf{C} = \mathbf{A} \cdot (\mathbf{B} \times \mathbf{C}) \tag{E-29}$$

and

$$(\mathbf{A} \times \mathbf{B}) \cdot \mathbf{C} = (\mathbf{C} \times \mathbf{A}) \cdot \mathbf{B} = (\mathbf{B} \times \mathbf{C}) \cdot \mathbf{A} \tag{E-30}$$

An interchange of two vectors results in a change of signs

$$(\mathbf{A} \times \mathbf{B}) \cdot \mathbf{C} = -(\mathbf{B} \times \mathbf{A}) \cdot \mathbf{C} \tag{E-31}$$

The vector triple product is written as **A** × (**B** × **C**). This will be a vector normal to the plane of **A** and (**B** × **C**), hence will lie in the plane of **B** and **C**. Since the maximum number of independent vectors one can have in a plane is two, the vector triple product will be of the form

$$\mathbf{A} \times (\mathbf{B} \times \mathbf{C}) = b\mathbf{B} + c\mathbf{C} \tag{E-32}$$

where b and c are scalar coefficients given by

$$\mathbf{A} \times (\mathbf{B} \times \mathbf{C}) = (\mathbf{A} \cdot \mathbf{C})\mathbf{B} + (\mathbf{A} \cdot \mathbf{B})\mathbf{C} \tag{E-33}$$

or for the product (**A** × **B**) × **C** by

$$(\mathbf{A} \times \mathbf{B}) \times \mathbf{C} = (\mathbf{A} \cdot \mathbf{C})\mathbf{B} + (\mathbf{B} \cdot \mathbf{C})\mathbf{A} \tag{E-34}$$

The triple product relationship can be used to simplify more complex vector relationships.

APPENDIX F

POLYNOMIALS

F.1 LEGENDRE AND ASSOCIATED LEGENDRE POLYNOMIALS

The Legendre $P_l(z)$ and associated Legendre polynomials $P_l^{|m|}(z)$ are the respective solutions of the Legendre differential equation

$$\frac{d}{dz}\left\{(1-z^2)\frac{dP_l(z)}{dz}\right\} l(l+1)P_l(z) = 0 \tag{F-1}$$

and the associated Legendre differential equation

$$(1-z^2)\frac{d^2P_l^{|m|}(z)}{dz^2} - 2z\frac{dP_l^{|m|}(z)}{dz} + l(l+1) - \frac{m^2}{(1-z^2)} P_l^{|m|}(z) = 0 \tag{F-2}$$

The Legendre polynomials can be formulated using:

1. A series expansion

$$P_l(z) = \frac{(2l)!}{2^l(l!)^2}\left\{z^l - \frac{l(l-1)}{2(2l-1)}z^{l-2} + \frac{l(l-1)(l-2)(l-3)}{2\cdot 4(2l-1)(2l-3)}z^{l-4} - \cdots\right. \tag{F-3}$$

2. A differential expression

$$P_l(z) = \frac{1}{2^l l!}\frac{d^l(z^2-1)^l}{dz^l} \tag{F-4}$$

3. A generating function

$$(1 - 2xz + x^2)^{-1/2} = \sum_{l=0}^{\infty} P_l(z)x^l \tag{F-5}$$

Useful recursion relations are:

$$P'_{l+1}(z) - zP'_l(z) = (l+1)P_l(z) \tag{F-6}$$

$$(l+1)P_{l+1}(z) - (2l+1)zP_l(z) + lP_{l-1}(z) = 0 \tag{F-7}$$

$$zP'_l(z) - P'_{l-1}(z) = lP_l(z) \tag{F-8}$$

$$P'_{l+1}(z) - P'_{l-1}(z) = (2l+1)P_l(z) \tag{F-9}$$

$$(z^2 - 1)P'_l(z) = l_z P_l(z) - lP_{l-1}(z) \tag{F-10}$$

where the $'$ denotes the derivative with respect to z.

A useful integral relation is

$$\int_{-1}^{1} P_l(z)P_m(z)\,dz = \frac{2}{2l+1}\,\delta_{lm} \tag{F-11}$$

The associate Legendre polynomials can be formulated using:

1. A differential expression

$$P_l^{|m|}(z) = (1 - z^2)^{|m/2|}\frac{d^{|m|}P_l(z)}{dz^{|m|}} \tag{F-12}$$

2. A generating function

$$\frac{(2|m|)!(1 - z^2)^{|m/2|}x^{|m|}}{2^{|m|}(|m|)!(1 - 2xz + x^2)} = \sum_{l=|m|}^{\infty} P_l^{|m|}(z)x^l \tag{F-13}$$

Useful recursion relations are:

$$(1 - z^2)^{1/2}P_l^{|m|+1}(z) = (l - |m|)zP_l^{|m|}(z) - (l + |m|)P_{l-1}^{|m|}(z) = 0 \tag{F-14}$$

$$(2l+1)(1 - z^2)^{1/2}P_l^{|m|}(z) = P_{l-1}^{|m|+1}(z) - P_{l+1}^{|m|+1}(z) \tag{F-15}$$

$$(2l+1)zP_l^{|m|}(z) = (l + |m|)P_{l-1}^{|m|}(z) + (l - |m| + 1)P_{l+1}^{|m|}(z) \tag{F-16}$$

A useful integral relation is

$$\int_{-1}^{1} P_l^{|m|}(z)P_n^{|m|}(z)\,dz = \frac{(l+|m|)!}{(l-|m|)!}\,\frac{2}{(2l+1)}\,\delta_{ln} \tag{F-17}$$

For quantum mechanical problems involving angular coordinates the variable z becomes $\cos\theta$, $dz = \sin\theta d\theta$ and the limits of integration are 0 to π.

F.2 LAGUERRE AND ASSOCIATED LAGUERRE POLYNOMIALS

The Laguerre polynomials $L_n(\rho)$ and the associated Laguerre polynomials $L_{n+l}^{2l+1}(\rho)$ are the respective solutions of the Laguerre differential equation

$$\rho \frac{d^2 L_n(\rho)}{d\rho^2} + (1-\rho)\frac{dL_n(\rho)}{d\rho} + aL_n(\rho) = 0 \tag{F-18}$$

and the associated Laguerre equation

$$\rho \frac{d^2 L_{n+l}^{2l+1}(\rho)}{d\rho^2} + \{2(l+1)-\rho\}\frac{dL_{n+l}^{2l+1}(\rho)}{d\rho} + (n-l-1)L_{n+l}^{2l+1}(\rho) = 0 \tag{F-19}$$

The Laguerre polynomial can be formulated using:

1. A power series

$$L_n(\rho) = (-1)^n \left\{\rho^n \frac{n^2}{1!}\rho^{n-1} + \frac{n^2(n-1)^2}{2!}\rho^{n-2} - \cdots + (-1)^n n!\right\} \tag{F-20}$$

2. A generating function

$$e^{-\rho x/(1-x)}(1-x)^{-1} = \sum_{n=0}^{\infty} \frac{L_n(\rho)x^n}{n!} \tag{F-21}$$

3. A differential expression

$$L_n(\rho) = e^{\rho}\frac{d^n(\rho^n e^{-\rho})}{d\rho^n} \tag{F-22}$$

The recursion relations are:

$$L_{n+1}(\rho) + (\rho - 1 - 2n)L_n(\rho) + n^2 L_{n-1}(\rho) = 0 \tag{F.23}$$

$$L_n'(\rho) - nL_{n-1}'(\rho) + nL_{n-1}(\rho) = 0 \tag{F-24}$$

where the $'$ denotes the derivative with respect to ρ.

An important integral relation is

$$\int_0^{\infty} [L_n(\rho)]^2 e^{-\rho}\, d\rho = (n!)^2 \tag{F-25}$$

The associated Laguerre polynomials can be formulated using:

1. A series expansion

$$L_{n+l}^{2l+1}(\rho) = \sum_{k=0}^{n-l-1} (-1)^{k+1} \frac{\{(n+l)!\}^2 \rho^k}{(n-l-k-1)!(2l+k+1)!k!} \tag{F-26}$$

2. A generating function

$$\frac{(-1)^{2l+1}(1-z)^{-1}z^k}{(1-z)^k} e^{-[z\rho/(1-z)]} = \sum_{n=l+1}^{\infty} L_{n+l}^{2l+1}(\rho)\frac{z^{n+l}}{(n+l)!} \tag{F-27}$$

3. A differential expression

$$L_{n+l}^{2l+1}(\rho) = \frac{d^{2l+1} L_{n+l}(\rho)}{d\rho^{2l+1}} \tag{F-28}$$

An important integral relation is

$$\int_0^\infty e^{-\rho} \rho^q L_k^q(\rho) L_j^q(\rho)\, d\rho = \frac{[(k+q)!]^3}{k!}\, \delta_{kj} \tag{F-29}$$

APPENDIX G

MATRIX MATHEMATICS

G.1 BASIC TERMINOLOGY

Basic definitions and mathematical relationships regarding the form and manipulation of matrices are summarized below. A matrix, represented by a symbol of the form $\underline{\mathbf{A}}$, consists of a rectangular array of numbers. The members of this array are denoted by a_{ij} or $(A)_{ij}$ where j denotes the column and i denotes the row of the array. If the maximum values of i and j are the same then one has a square matrix. If there is only a single index then one has a row or column matrix, both of which denote components of a vector.

Mathematical manipulation of matrices involves combining the elements in particular ways. For the addition or subtraction of two matrices the corresponding elements add or subtract:

$$\underline{\mathbf{C}} = \underline{\mathbf{A}} \pm \underline{\mathbf{B}} \tag{G-1}$$

or

$$C_{ij} = a_{ij} \pm b_{ij} \tag{G-2}$$

The multiplication of two matrices involves summing the products of row and column elements:

$$\underline{\mathbf{D}} = \underline{\mathbf{A}} \times \underline{\mathbf{B}} \qquad \text{or} \qquad d_{ij} = \sum_k a_{ik} b_{kj} \tag{G-3}$$

There are a number of useful manipulations that can be used to generate a particular type of matrix or quantity.

1. The trace of a matrix Tr $\underline{\mathbf{A}}$ is obtained by summing the diagonal elements

$$\mathrm{Tr}\,\underline{\mathbf{A}} = \sum_k a_{kk} \qquad \text{(G-4)}$$

2. The determinant of a matrix, det $\underline{\mathbf{A}}$ or $|\underline{\mathbf{A}}|$, is a conventional determinant formed from the matrix array.
3. The transpose of a matrix, denoted in various discussions by $\tilde{\underline{\mathbf{A}}}$, $\underline{\mathbf{A}}^T$, or $\underline{\mathbf{A}}'$, is formed by interchanging rows and columns

$$(\tilde{\underline{\mathbf{A}}})_{ij} = (\underline{\mathbf{A}})_{ji} \qquad \text{(G-5)}$$

For a diagonal matrix where $(\underline{\mathbf{A}})_{ij} = a_{ij}\delta_{ij}$ it follows that $\tilde{\underline{\mathbf{A}}} = \underline{\mathbf{A}}$.
4. The complex conjugate of a matrix $\underline{\mathbf{A}}^*$ or $\bar{\underline{\mathbf{A}}}$ is formed by taking the complex conjugate of each element

$$(\underline{\mathbf{A}}^*)_{ij} = (\underline{\mathbf{A}})^*_{ij} \qquad \text{(G-6)}$$

5. The transpose conjugate or Hermitian conjugate $\underline{\mathbf{A}}^\dagger$, $\tilde{\underline{\mathbf{A}}}^*$ or $\underline{\mathbf{A}}'$ is formed by taking the complex conjugate of each element of the transpose

$$(\underline{\mathbf{A}}^\dagger)_{ij} = (\tilde{\underline{\mathbf{A}}})^*_{ij} = (\underline{\mathbf{A}})^*_{ji} \qquad \text{(G-7)}$$

6. The adjoint or adjugate of a matrix $\hat{\underline{\mathbf{A}}}$ or adj $\underline{\mathbf{A}}$ is formed by replacing each element with its cofactor and transposing. For example if

$$\underline{\mathbf{A}} = \begin{pmatrix} a_{11} & a_{12} \\ a_{21} & a_{22} \end{pmatrix} \qquad \text{(G-8)}$$

then

$$\hat{\underline{\mathbf{A}}} = \begin{pmatrix} a_{22} & -a_{12} \\ -a_{21} & a_{11} \end{pmatrix} \qquad \text{(G-9)}$$

7. The reciprocal or inverse of a matrix $\underline{\mathbf{A}}^{-1}$ is the adjoint divided by the determinant

$$\underline{\mathbf{A}}^{-1} = \frac{\hat{\underline{\mathbf{A}}}}{\det \underline{\mathbf{A}}} \qquad \text{(G-10)}$$

or

$$(\underline{\mathbf{A}}^{-1})_{ij} = \frac{(\hat{\underline{\mathbf{A}}})_{ij}}{\det \underline{\mathbf{A}}} \qquad \text{(G-11)}$$

8. The unit matrix $\underline{\mathbf{I}}$ is one where $a_{ij} = \delta_{ij}$ for all elements. The product of a matrix and its inverse is the unit matrix

$$\underline{\mathbf{A}}^{-1}\underline{\mathbf{A}} = \underline{\mathbf{I}} \qquad \text{(G-12)}$$

9. A null matrix, $\underline{\mathbf{0}}$ is one with $a_{ij} = 0$ for all ij.

Using the basic relationships just given several types of matrices, which are of use in quantum mechanics, can be defined. These matrices and their qualifying conditions are

Class of matrix	Condition	Element relationship
Symmetric	$\underline{\mathbf{A}} = \tilde{\underline{\mathbf{A}}}$	$A_{ij} = A_{ji}$
Antisymmetric (skew)	$\underline{\mathbf{A}} = -\tilde{\underline{\mathbf{A}}}$	$A_{kk} = 0;\ A_{ij} = -A_{ji}$
Real	$\underline{\mathbf{A}} = \underline{\mathbf{A}}^*$	$A_{ij} = A_{ij}^*$
Pure imaginary	$\underline{\mathbf{A}} = -\underline{\mathbf{A}}^*$	$A_{ij} = \sqrt{-1}\,B_{ij}(B_{ij}$ is real)
Orthogonal	$\underline{\mathbf{A}} = \tilde{\underline{\mathbf{A}}}^{-1}$	$(\tilde{A}A)_{ij} = \delta_{ij}$
Hermitian	$\underline{\mathbf{A}} = \underline{\mathbf{A}}^\dagger$	$A_{ij} = A_{ji}^*$
Unitary	$\underline{\mathbf{A}}^\dagger = \underline{\mathbf{A}}^{-1}$	$A_{ij}^* = A_{ij}^{-1}$
Singular	$\det \underline{\mathbf{A}} = 0$	

G.2 TRANSFORMATIONS

A frequently encountered use of matrices is in the transformation of a vector from one coordinate system to another, under conditions that leave scalar products invariant. This means that if $\mathbf{r} = \underline{\mathbf{A}}\mathbf{s}$ and $\mathbf{p} = \underline{\mathbf{A}}\mathbf{q}$, where $\underline{\mathbf{A}}$ is the transformation matrix, then $\mathbf{r} \cdot \mathbf{p} = \mathbf{s} \cdot \mathbf{q}$. If $\underline{\mathbf{A}}$ has all real elements and satisfies this condition then it is an orthogonal matrix and $\mathbf{r} = \underline{\mathbf{A}}\mathbf{R}$ is an orthogonal transformation. An example of an orthogonal transformation is a clockwise rotation about the z axis of a Cartesian coordinate system. Reference to Fig. G-1 shows that

$$x' = x \cos \theta - y \sin \theta \tag{G-13}$$

$$y' = x \sin \theta + y \cos \theta \tag{G-14}$$

$$z' = z \tag{G-15}$$

Since x, y, z are the components of $\mathbf{R}$ in the X, Y, Z axis system, x', y', z' are the components of $\mathbf{R}$ in the $X'Y'Z'$ axis system, and both sets of components

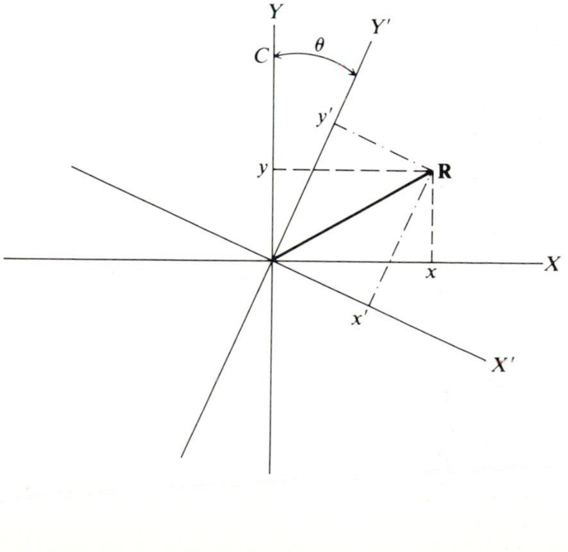

FIGURE G-1
Rotation of axes.

describe the vector **R**, this transformation is written as

$$\begin{pmatrix} x' \\ y' \\ z' \end{pmatrix} = \begin{pmatrix} \cos\theta & -\sin\theta & 0 \\ \sin\theta & \cos\theta & 0 \\ 0 & 0 & 1 \end{pmatrix} \begin{pmatrix} x \\ y \\ z \end{pmatrix} \tag{G-16}$$

It is to be noted that this clockwise rotation of the axis system is equivalent to a counterclockwise vector rotation. The inverse of $\underline{\mathbf{A}}$ is

$$\underline{A}^{-1} = \begin{pmatrix} \cos\theta & \sin\theta & 0 \\ -\sin\theta & \cos\theta & 0 \\ 0 & 0 & 1 \end{pmatrix} \tag{G-17}$$

so that $\underline{\mathbf{A}} \times \underline{\mathbf{A}}^{-1} = \underline{\mathbf{I}}$ and $\underline{\mathbf{A}}$ is orthogonal. Also note that the elements of $\underline{\mathbf{A}}$ are their own cofactors, that is, det $A = 1$.

The necessity to transform operators, as represented by matrices, is also frequently encountered. If there exists an operator $\underline{\mathbf{L}}$, which transforms the vector **p** to **q** in the XYZ-axis system,

$$\underline{\mathbf{L}}\mathbf{p} = \mathbf{q} \tag{G-18}$$

and both **p** and **q** are transformed to the $X'Y'Z'$ system by the transformation matrix $\underline{\mathbf{T}}$ then

$$\mathbf{q}' = \underline{\mathbf{T}}\mathbf{q} \tag{G-19}$$

$$\mathbf{p}'' = \underline{\mathbf{T}}\mathbf{p} \tag{G-20}$$

or

$$\underline{\mathbf{T}}^{-1}\mathbf{q}' = \mathbf{q} \tag{G-21}$$

$$\underline{\mathbf{T}}^{-1}\mathbf{p}' = \mathbf{p} \tag{G-22}$$

Combining these expressions and left multiplying by $\underline{\mathbf{T}}$ gives

$$\mathbf{q}' = \underline{\mathbf{T}}\underline{\mathbf{L}}\underline{\mathbf{T}}^{-1}p' \tag{G-23}$$

or

$$\mathbf{q}' = \underline{\mathbf{L}}'\mathbf{p}' \tag{G-24}$$

where $\underline{\mathbf{L}}'$ is the transformed operator. This is a similarity transformation.

G.3 EIGENVALUES AND DIAGONALIZATION

For a matrix operator there will exist a set of eigenvalues just as there is for a differential operator. There also will be associated with the matrix a set of eigenvectors such that operation on the eigenvectors by the matrix will result in a new set of vectors which may differ in length but not direction. If **e** is an eigenvector of $\underline{\mathbf{P}}$ with components $e_1, e_2, \ldots$ then the eigenvalue equation is

$$\underline{\mathbf{P}}\mathbf{e} = \lambda\mathbf{e} \tag{G-25}$$

where λ is the scalar eigenvalue. This can be written as

$$(\underline{\mathbf{P}} = \lambda \underline{\mathbf{I}})\mathbf{e} = 0 \tag{G-26}$$

which is the abbreviation for a set of linear equations

$$(p_{11} - \lambda)e_1 + p_{12}e_2 + P_{13}e_3 + \cdots + p_{1n}e_n = 0 \tag{G-27}$$

$$p_{21}e_1 + (p_{22} - \lambda)e_2 + p_{23}e_3 + \cdots + p_{2n}e_n = 0 \tag{G-28}$$

$$p_{n1}e_1 + p_{n2}e_2 + p_{n3}e_3 + \cdots + (p_{nn} - \lambda)e_n = 0 \tag{G-29}$$

A nontrivial (i.e., the $e_i \neq 0$) solution of this set of equations exists if det $(\underline{\mathbf{P}} - \lambda \underline{\mathbf{I}}) = 0$. The eigenvalues $\lambda_1, \lambda_2, \ldots, \lambda_n$ are the roots of the nth-order equation obtained by multiplying out the determinant

$$\begin{vmatrix} p_{11} - \lambda & p_{12} & \cdots & p_{1n} \\ p_{21} & p_{22} - \lambda & \cdots & p_{2n} \\ p_{n1} & p_{n2} & \cdots & p_{nn} - \lambda \end{vmatrix} = 0 \tag{G-30}$$

Once the eigenvalues are known one can separately substitute each back into the set of linear equations, use the normalization condition $\sum_i (e_i^k)^2 = 1$, and find the set of eigenvectors for the matrix. The eigenvalues may also be determined by a similarity transformation which diagonalizes the matrix

$$\underline{\mathbf{S}}^{-1}\underline{\mathbf{P}}\underline{\mathbf{S}} = \underline{\mathbf{D}} \tag{G-31}$$

$\underline{\mathbf{D}}$ is then the diagonal matrix

$$\underline{\mathbf{D}} = \begin{pmatrix} \lambda_1 & 0 & 0 & \cdots \\ 0 & \lambda_2 & 0 & \cdots \\ 0 & 0 & \lambda_3 & \cdots \\ \vdots & \vdots & \vdots & \lambda_n \end{pmatrix} \tag{G-32}$$

The columns of the transformation matrix $\underline{\mathbf{S}}$ are the normalized components of the eigenvectors.

A short example will serve to illustrate these points. If

$$\underline{\mathbf{P}} = \begin{pmatrix} 1 & 2 \\ 1 & 1 \end{pmatrix} \tag{G-33}$$

then the eigenvalues are found from

$$\begin{vmatrix} 1 - \lambda & 2 \\ 1 & 1 - \lambda \end{vmatrix} = 0 \tag{G-34}$$

to be $\lambda_k = 1 \pm \sqrt{2}$. Substitution of $\lambda_1 = 1 + \sqrt{2}$ into Eq. (G-27) yields

$$(1 - \lambda_k)e_1^1 + 2e_2^1 = 0 \tag{G-35}$$

which, combined with

$$(e_1^1)^2 + (e_2^1)^2 = 1 \tag{G-36}$$

gives $e_1^1 = \sqrt{\frac{2}{3}}$ and $e_2^1 = \sqrt{\frac{1}{3}}$. Substitution of $\lambda_2 = 1 - \sqrt{2}$ gives $e_1^2 = \sqrt{\frac{2}{3}}$ and $e_2^2 = -\sqrt{\frac{1}{3}}$. The eigenvectors are then

$$\mathbf{e}_1 = \sqrt{\tfrac{2}{3}}\hat{\mathbf{i}} + \sqrt{\tfrac{1}{3}}\hat{\mathbf{j}} \tag{G-37}$$

$$\mathbf{e}_2 = \sqrt{\tfrac{2}{3}}\hat{\mathbf{i}} - \sqrt{\tfrac{1}{3}}\hat{\mathbf{j}} \tag{G-38}$$

The transformation matrix is

$$\underline{\mathbf{S}} = \begin{pmatrix} \sqrt{\frac{2}{3}} & \sqrt{\frac{2}{3}} \\ \sqrt{\frac{1}{3}} & -\sqrt{\frac{1}{3}} \end{pmatrix} \tag{G-39}$$

Taking the inverse of $\underline{\mathbf{S}}$ the similarity transformation becomes

$$\underline{\mathbf{S}}^{-1}\underline{\mathbf{P}}\underline{\mathbf{S}} = \begin{pmatrix} \sqrt{\frac{3}{8}} & \sqrt{\frac{3}{4}} \\ \sqrt{\frac{3}{8}} & -\sqrt{\frac{3}{4}} \end{pmatrix} \begin{pmatrix} 1 & 2 \\ 1 & 1 \end{pmatrix} \begin{pmatrix} \sqrt{\frac{2}{3}} & \sqrt{\frac{2}{3}} \\ \sqrt{\frac{1}{3}} & -\sqrt{\frac{1}{3}} \end{pmatrix} = \begin{pmatrix} 1+\sqrt{2} & 0 \\ 0 & 1-\sqrt{2} \end{pmatrix} \tag{G-40}$$

APPENDIX H

ANGULAR MOMENTUM RELATIONSHIPS WITH EULER ANGLES

The relationships between the components of angular momentum, in either the space-fixed or body-fixed Cartesian axis systems and the Euler angles, can be developed from vector relationships, and the direction cosines expressed in terms of the Euler angles. The latter are given by

$$\Phi_{Xx} = \cos\psi\cos\phi - \cos\theta\sin\phi\sin\psi \tag{H-1}$$

$$\Phi_{Xy} = -\sin\psi\cos\phi - \cos\theta\sin\phi\cos\psi \tag{H-2}$$

$$\Phi_{Xz} = \sin\theta\sin\phi \tag{H-3}$$

$$\Phi_{Yx} = \cos\psi\sin\phi + \cos\theta\cos\phi\sin\psi \tag{H-4}$$

$$\Phi_{Yy} = -\sin\psi\sin\phi + \cos\theta\cos\phi\cos\psi \tag{H-5}$$

$$\Phi_{Yz} = -\sin\theta\cos\phi \tag{H-6}$$

$$\Phi_{Zx} = \sin\theta\sin\psi \tag{H-7}$$

$$\Phi_{Zy} = \sin\theta\cos\psi \tag{H-8}$$

$$\Phi_{Zz} = \cos\theta \tag{H-9}$$

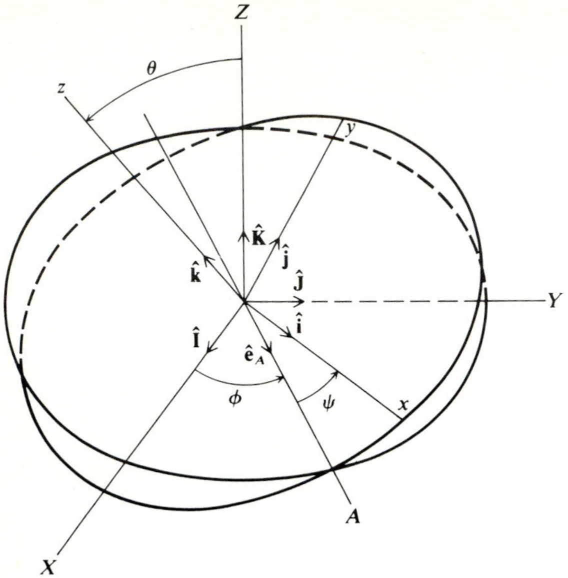

FIGURE H-1
Euler angles.

Looking at Fig. H-1 we can observe that there will be three well-defined angular momentum components, these being

$$P_Z \equiv P_\phi \tag{H-10}$$

$$P_z \equiv P_\psi \tag{H-11}$$

$$P_A \equiv P_\theta \tag{H-12}$$

Since the component G_g of a general vector $\mathbf{G}$ can be expressed as $G_g = \mathbf{G} \cdot \hat{\mathbf{e}}_g$ where $\hat{\mathbf{e}}_g$ is a unit vector lying along the g axis, the component of angular momentum along the nodal axis A becomes

$$P_A = \mathbf{P} \cdot \hat{\mathbf{e}}_A \tag{H-13}$$

Expressing $\mathbf{P}$ in terms of the moving axis components

$$\mathbf{P} = P_x\hat{\mathbf{i}} + P_y\hat{\mathbf{j}} + P_z\hat{\mathbf{k}} \tag{H-14}$$

this becomes

$$P_A = P_x(\hat{\mathbf{i}} \cdot \hat{\mathbf{e}}_A) + P_y(\hat{\mathbf{j}} \cdot \hat{\mathbf{e}}_A) + P_z(\hat{\mathbf{k}} \cdot \hat{\mathbf{e}}_A) \tag{H-15}$$

But

$$\hat{\mathbf{i}} \cdot \hat{\mathbf{e}}_A = \cos\psi \tag{H-16}$$

$$\hat{\mathbf{j}} \cdot \hat{\mathbf{e}}_A = \cos(90 + \psi) = -\sin\psi \tag{H-17}$$

$$\hat{\mathbf{k}} \cdot \hat{\mathbf{e}}_A = \cos 90° = 0 \tag{H-18}$$

Therefore

$$P_A = P_x \cos\psi - P_y \sin\psi \tag{H-19}$$

P_Z can be expressed in terms of the P's in a similar manner.

$$P_Z = \mathbf{P} \cdot \hat{\mathbf{K}} \tag{H-20}$$

or

$$P_Z = P_x(\hat{\mathbf{i}} \cdot \hat{\mathbf{K}}) + P_y(\hat{\mathbf{j}} \cdot \hat{\mathbf{K}}) + P_z(\hat{\mathbf{k}} \cdot \hat{\mathbf{K}}) \tag{H-21}$$

Since

$$\hat{\mathbf{i}} \cdot \hat{\mathbf{K}} = \Phi_{Zx} \qquad \hat{\mathbf{j}} \cdot \hat{\mathbf{K}} = \Phi_{Zy} \qquad \hat{\mathbf{k}} \cdot \hat{\mathbf{K}} = \Phi_{Zz} \tag{H-22}$$

it follows that

$$P_Z = P_x \sin\theta \sin\psi + P_y \sin\theta \cos\psi + P_z \cos\theta \tag{H-23}$$

Equations (H-10)–(H-12), (H-19), and (H-23) can be combined and solved for P_x and P_y giving

$$P_x = \frac{\sin\psi}{\sin\theta}(P_\phi - P_\psi \cos\theta) + P_\theta \cos\psi \tag{H-24}$$

$$P_y = \frac{\cos\psi}{\sin\theta}(P_\phi - P_\psi \cos\theta) - P_\theta \sin\psi \tag{H-25}$$

and

$$P_z = P_\psi \tag{H-26}$$

The components relative to the fixed-axis system can be derived in an analogous manner beginning with

$$\mathbf{P} = P_X\hat{\mathbf{I}} + P_Y\hat{\mathbf{J}} + P_Z\hat{\mathbf{K}} \tag{H-27}$$

For the nodal axis component we have

$$P_A = \hat{\mathbf{P}} \cdot \hat{\mathbf{e}}_A = P_X(\hat{\mathbf{I}} \cdot \hat{\mathbf{e}}_A) + P_Y(\hat{\mathbf{J}} \cdot \hat{\mathbf{e}}_A) + P_Z(\hat{\mathbf{K}} \cdot \hat{\mathbf{e}}_A) \tag{H-28}$$

but

$$\hat{\mathbf{I}} \cdot \hat{\mathbf{e}}_A = \cos\phi \tag{H-29}$$

$$\hat{\mathbf{J}} \cdot \hat{\mathbf{e}}_A = \cos(90 - \phi) = \sin\phi \tag{H-30}$$

$$\hat{\mathbf{K}} \cdot \hat{\mathbf{e}}_A = \cos 90^\circ = 0 \tag{H-31}$$

so

$$P_A = P_X \cos\phi + P_Y \sin\phi \tag{H-32}$$

P_z is now expressed as

$$P_z = \hat{\mathbf{P}} \cdot \hat{\mathbf{k}} = P_z(\hat{\mathbf{I}} \cdot \hat{\mathbf{k}}) + P_y(\hat{\mathbf{J}} \cdot \hat{\mathbf{k}}) + P_Z(\hat{\mathbf{K}} \cdot \hat{\mathbf{k}}) \tag{H-33}$$

and

$$\hat{\mathbf{I}} \cdot \hat{\mathbf{k}} = \Phi_{Xz} \qquad \hat{\mathbf{J}} \cdot \hat{\mathbf{k}} = \Phi_{Yz} \qquad \hat{\mathbf{K}} \cdot \hat{\mathbf{k}} = P_\theta \cos\psi \tag{H-34}$$

so

$$P_z = P_X \sin\theta \sin\phi - P_y \sin\theta \cos\phi + P_Z \cos\theta \tag{H-35}$$

Equations (H-10) to (H-12), (H-32), and (H-35) are now combined to give

$$P_X = \frac{\sin\phi}{\sin\theta}(P_\psi - P_\phi \cos\theta) + P_\theta \cos\phi \tag{H-36}$$

$$P_Y = \frac{\cos\phi}{\sin\theta}(-P_\psi + P_\phi \cos\theta) + P_\theta \sin\phi \tag{H-37}$$

$$P_Z = P_\phi \tag{H-38}$$

Introduction of the appropriate quantum mechanical operators for the momentum components

$$P_\phi \rightarrow -i\hbar\frac{\partial}{\partial\phi} \qquad P_\psi \rightarrow -i\hbar\frac{\partial}{\partial\psi} \qquad P_\theta \rightarrow -i\hbar\frac{\partial}{\partial\theta} \tag{H-39}$$

the operators for the body-fixed angular momentum components become

$$\mathbf{P}_x = -i\hbar\left[\frac{\sin\psi}{\sin\theta}\left(\frac{\partial}{\partial\phi} - \cos\theta\frac{\partial}{\partial\psi}\right) + \cos\psi\frac{\partial}{\partial\theta}\right] \tag{H-40}$$

$$\mathbf{P}_y = -i\hbar\left[\frac{\cos\psi}{\sin\theta}\left(\frac{\partial}{\partial\phi} - \cos\theta\frac{\partial}{\partial\psi}\right) - \sin\psi\frac{\partial}{\partial\theta}\right] \tag{H-41}$$

$$\mathbf{P}_z = -i\hbar\frac{\partial}{\partial\psi} \tag{H-42}$$

and that for the square of the total angular momentum is

$$\mathbf{P}^2 = -\hbar^2\left[\frac{1}{\sin^2\theta}\frac{\partial^2}{\partial\phi^2} - \frac{2\cot\theta}{\sin\theta}\frac{\partial^2}{\partial\phi\partial\psi} + \frac{1}{\sin^2\theta}\frac{\partial^2}{\partial\psi^2} + \frac{1}{\sin\theta}\frac{\partial}{\partial\theta}\sin\theta\frac{\partial}{\partial\theta}\right] \tag{H-43}$$

APPENDIX I

POINT GROUPS

Symmetry operation	(Order)	Schoenflies notation	Hermann-Mauguin symbol	Molecular example	Crystal example
		Cubic systems			
$E, 4C_3, 4C_3^2, 3C_2$	(12)	T	23		$NaClO_3$
$E, 8C_3, 3C_2, 3\sigma_v, i, 8S_6$	(24)	T_h	$m3$		Alums
$E, 6C_4, 8C_3, 3C_2, 6'_{nC2}$	(24)	O	432		None known
$E, 8C_3, 3C_2, 6S_4, 6\sigma_d$	(24)	T_d	$43m$	CH_4	ZnS (blend)
$E, 8C_3, 6C_2, 6C_4, 3C_2', i, 6S_4, 8S_6, 3\sigma_h, 6\sigma_d$	(48)	O_h	$m3m$	SF_6	Cu, Ag, NaCl
		Tetragonal systems			
E, C_4, C_2, C_4^3	(4)	C_4	4		Metaldehyde
E, S_4, C_2, S_4^3	(4)	S_4	4		BPO_4
$E, C_4, C_2, C_4^3, i, S_4^3, \sigma_h, S_4$	(8)	C_{4h}	$4/m$		$CaWO_4$
$E, 2C_4, C_2, 2C_2', 2C_2''$	(8)	D_4	422		$[CH_3NH_3]I$
$E, 2S_4, C_2, 2C_2', 2\sigma_d$	(8)	C_{4v}	$4mm$	SF_5Br	Diaboleite
$E, 2S_4, C_2, 2C_2', 2\sigma_d$	(8)	D_{2d}	$42m$	Allene	Urea, KH_2PO_4
$E, 2C_4, C_2, 2C_2', 2C_2'', i, 2S_4, \sigma_h, 2\sigma_v, 2\sigma_d$	(16)	D_{4h}	$4/mmm$	$Pt(Cl_4^{2-}$ ion	SnO_2, TiO_2

Symmetry operation	(Order)	Schoenflies notation	Hermann-Mauguin symbol	Molecular example	Crystal example
		Orthorhombic systems			
E, C_2, C'_2, C''_2	(4)	D_2	222	$(C_6H_5)_2$-twisted	$MgSO_4 \cdot 7H_2O$
$E, C_2, \sigma_v, \sigma'_v$	(4)	C_{2v}	$mm2$	H_2O	Resorcinol
$E, C_2, C'_2, C''_2, i, \sigma, \sigma', \sigma''$	(8)	D_{2h}	mmm	$CH_2{=}CH_2$	Na_2SO_4
		Monoclinic systems			
E, C_2	(2)	C_2	2	Gauch-CH_2FCH_2F	Tartaric acid
E, σ_h	(2)	C_s	m or 2	HOD	KNO_2
E, C_2, i, σ_h	(4)	C_{2h}	$2/m$	Trans CHF=CHF	$KClO_3$
		Triclinic systems			
E	(1)	C_1	1	Lactic acid	$CaS_2O_3 \cdot 6H_2O$
E, i	(2)	C_i	1	DHFC—CFHD	$CuSO_4 \cdot 5H_2O$
		Trigonal systems			
E, C_3, C_3^2	(3)	C_3	3	Gauch-CH_3CCl_3	$NaIO_5 \cdot 3H_2O$
$E, C_3, C_3^2, i, S_6^5, S_6$	(6)	S_6	3		Na_2SO_3
$E, 2C_3, 3C_2$	(6)	D_3	32		SiO_2 (quartz)
$E, 2C_3, 3\sigma_v$	(6)	C_{3v}	$3m$	NF_3	$KBrO_3$
$E, 2C_3, 3C_2, i, 2S_5, 3\sigma_d$	(12)	D_{3d}	$3m$	Staggered C_2H_6	$CaCO_3$
		Hexagonal systems			
$E, C_6, C_3, C_2, C_3^2, C_6^5$	(6)	C_6	6		CHI_3
$E, C_3, C_3^2, \sigma_h, S_3, S_3^5$	(6)	$C_{3h}(S_3)$	6 or $3/m$	$B(OH)_3$	Very rare
$E, C_6, C_3, C_2, C_3^2, C_6^5,$ $i, S_3^5, S_6^5, \sigma_h, S_6, S_3$	(12)	C_{6h}	$6/m$		Opatite
$E, 2C_6, 2C_3, C_2, 3C'_2, 3C''_2$	(12)	D_6	622		$LiIO_3$
$E, 2C_6, 2C_3, C_2, 3\sigma_v, 3\sigma_d$	(12)	C_{6v}	$6mm$		ZnS (wurtzite)
$E, 2C_3, 3C_2, \sigma_h, 2S_3, 3\sigma_v$	(12)	D_{3h}	$6m2$	BF_3	Very rare
$E, 2C_6, 2C_5, C_2, 3C'_2, 3C''_2,$ $i, 2S_3, 2S_6, \sigma_h, 3\sigma_d, 3\sigma_v$	(24)	D_{6h}	$6/mmm$	C_6H_6	Beryl
		Non-crystallographic systems			
$E, 2C_\infty^\phi, \infty_v^\sigma$	(∞)	$C_{\infty v}$		HCl	
$E, 2C_\infty^\phi, \infty_v^\sigma, i, 2S_\infty^\phi, \infty C_2$	(∞)	$D_{\infty h}$		Cl_2	
$E, 2S_8, 2C_4, 2S_8^3, C_2,$ $4C'_2, 4\sigma_d$	(16)	D_{4d}			
$E, C_5, C_5^2, C_5^3, C_5^4$	(5)	C_5	5		
$E, 2C_5, 2C_5^2, 5C_2$	(10)	D_5			
$E, 2C_5, 2C_5^2, 5\sigma_v$	(10)	C_{5v}			

APPENDIX J

CHARACTER TABLES

$\mathbf{C}_1$	E
A	1

$\mathbf{C}_s$	E	σ_h		
A'	1	1	x, y, R_z	z^2, y^2, z^2, xy
A''	1	−1	z, R_x, R_y	xz, yz

$\mathbf{C}_i$	E	i		
A^g	1	1	R_x, R_y, R_z	$x^2, y^2, z^2, xy, xz, yz$
A^u	1	−1	x, y, z	

$\mathbf{C}_{2v}$	E	C_2	$\sigma_v(xz)$	$\sigma_v(yz)$		
A_1	1	1	1	1	z	x^2, y^2, z^2
A_2	1	1	−1	−1	R_z	xy
B_1	1	−1	1	−1	x, R_y	xz
B_2	1	−1	−1	1	y, R_x	yz

$\mathbf{C}_{3v}$	E	$2C_3$	$3\sigma_v$		
A_1	1	1	1	z	x^2+y^2, z^2
A_2	1	1	−1	R_z	
E	2	−1	0	$(x, y), (R_x, R_y)$	$(x^2-y^2, xy), (xz, yz)$

$\mathbf{C}_{2h}$	E	C_2	i	σ_h		
A_g	1	1	1	1	R_z	x^2, y^2, z^2, xy
A_u	1	−1	−1	−1	z	
B_g	1	−1	1	−1	R_x, R_y	xz, yz
B_u	1	−1	−1	1	x, y	

$\mathbf{C}_{3h}$	E	C_3	C_3^2	σ_h	S_3	S_3^5		
A'	1	1	1	1	1	1	R_z	$z^2, x^2, +y^2$
E'	1	ε	ε^*	1	ε	ε^*	(x, y)	$(xy, x^2, -y^2)$
	1	ε^*	ε	1	ε^*	ε		
A''	1	1	1	−1	−1	−1	z	
E''	1	ε	ε^*	−1	$-\varepsilon$	$-\varepsilon^*$	(R_x, R_y)	(xy, yz)
	1	ε^*	ε	−1	$-\varepsilon^*$	$-\varepsilon$		
	$[\varepsilon = \exp(i2\pi/3)]$							

$\mathbf{C}_{4h}$	E	C_2	C_4	C_3^4	i	S_4	S_4^3	σ_h		
A_g	1	1	1	1	1	1	1	1	R_z	z^2, x^2+y^2
B_g	1	1	1	−1	1	−1	−1	1		xy, x^2-y^2
E_g	1	−1	i	$-i$	1	$-i$	i	−1	(R_x, R_y)	(xz, yz)
	1	−1	$-i$	i	1	i	$-i$	−1		
A_u	1	1	1	1	−1	−1	−1	−1	z	
B_u	1	1	−1	−1	−1	1	1	−1		
E_u	1	−1	i	$-i$	−1	i	$-i$	1	(x, y)	
	1	−1	$-i$	i	−1	$-i$	i	1		

$\mathbf{D}_{2h}$	E	$C_2(x)$	$C_2(y)$	$C_2(z)$	i	$\sigma(xy)$	$\sigma(xz)$	$\sigma(yz)$		
A_g	1	1	1	1	1	1	1	1		x^2, y^2, z^2
B_{1g}	1	−1	−1	1	1	1	−1	−1	R_z	xy
B_{2g}	1	−1	1	−1	1	−1	1	−1	R_y	xz
B_{3g}	1	1	−1	−1	1	−1	−1	1	R_x	yz
A_u	1	1	1	1	−1	−1	−1	−1		
B_{1u}	1	−1	−1	1	−1	−1	1	1	z	
B_{2u}	1	−1	1	−1	−1	1	−1	1	y	
B_{3u}	1	1	−1	−1	−1	1	1	−1	x	

$\mathbf{D}_{3h}$	E	$2C_3$	$3C_2$	σ_h	$2S_3$	$3\sigma_v$		
A_1'	1	1	1	1	1	1		z^2, x^2+y^2
A_2'	1	1	−1	1	1	−1	R_z	
E'	2	−1	0	2	−1	0	(x, y)	(xy, x^2-y^2)
A_1''	1	1	1	−1	−1	−1		
A_2''	1	1	−1	−1	−1	1	z	
E''	2	−1	0	−2	1	0	(R_x, R_y)	(xz, yz)

$\mathbf{D}_{4h}$	E	C_2	$2C_2'$	$2C_2''$	$2C_4$	i	$2S_4$	σ_h	$2\sigma_v$	$2\sigma_d$		
A_{1g}	1	1	1	1	1	1	1	1	1	1		x^2+y^2, z^2
A_{2g}	1	1	−1	−1	1	1	1	1	−1	−1	R_z	
A_{1u}	1	1	1	1	1	−1	−1	−1	−1	−1		
A_{2u}	1	1	−1	−1	1	−1	−1	−1	1	1	z	
B_{1g}	1	1	1	−1	−1	1	−1	1	1	−1		x^2-y^2
B_{2g}	1	1	−1	1	−1	1	−1	1	−1	1		xy
B_{1u}	1	1	1	−1	−1	−1	1	−1	−1	1		
B_{2u}	1	1	−1	1	−1	−1	1	−1	1	−1		
E_g	2	−2	0	0	0	2	0	−2	0	0	(R_x, R_y)	(xz, yz)
E_u	2	−2	0	0	0	−2	0	2	0	0	(x, y)	

$\mathbf{D}_{2d}$	E	C_2	$2C_2'$	$2S_4$	$2\sigma_d$		
A_1	1	1	1	1	1		x^2, x^2+y^2
A_2	1	1	−1	1	−1	R_z	
B_1	1	1	1	−1	−1		x^2-y^2
B_2	1	1	−1	−1	1	z	xy
E	2	−2	0	0	0	$(x, y):(R_xR_y)$	(xz, yz)

$\mathbf{T}_d$	E	$3C_2$	$8C_3$	$6S_4$	$6\sigma_d$		
A_1	1	1	1	1	1		$x^2, +y^2+z^2$
A_2	1	1	1	−1	−1		
E	2	2	−1	0	0		$(2z^2-x^2-y^2, x^2-y^2)$
T_1	3	−1	0	1	−1	(R_x, R_y, R_z)	
T_2	3	−1	0	−1	1	(x, y, z)	(xy, xz, yz)

$\mathbf{O}_h$	E	$8C_3$	$6C_2$	$6C_4$	$3C_2^1$	i	$6S_4$	$8S_6$	$3\sigma_h$	$6\sigma_h$		
A_{1g}	1	1	1	1	1	1	1	1	1	1		$x^2+y^2+z^2$
A_{2g}	1	1	−1	−1	1	1	−1	1	1	−1		
A_{1u}	1	1	1	1	1	−1	−1	−1	−1	−1		
A_{2u}	1	1	−1	−1	1	−1	1	−1	−1	1		
E_g	2	−1	0	0	2	2	0	−1	2	0		$(2z^2-x^2-y^2, x^2-y^2)$
E_u	2	−1	0	0	2	−2	0	1	−2	0		
T_{1g}	3	0	−1	1	−1	3	1	0	−1	−1		
T_{2g}	3	0	1	−1	−1	3	−1	0	−1	1		(xy, yz, xz)
T_{1u}	3	0	−1	1	−1	−3	−1	—	1	1	(x, y, z)	
T_{2u}	3	0	1	−1	−1	−3	1	0	1	−1		

$\mathbf{C}_{\infty v}$	E	$2C_\infty$	$2C_\infty^2$		$\infty\sigma_v$		
Σ^+	1	1	1	···	1	z	z^2, x^2+y^2
Σ^-	1	1	1	···	−1	R_z	
π	2	$2\cos\phi$	$2\cos 2\phi$	···	0	$(x, y), (R_xR_y)$	(yz, xz)
Δ	2	$2\cos 2\phi$	$2\cos 4\phi$	···	0		(xy, x^2-y^2)
Φ	2	$2\cos 3\phi$	$2\cos 6\phi$	···	0		
	···	···	···	···	···		

$\mathbf{D}_{\infty h}$	E	$2C_\infty^\phi$	$2C_\infty^{2\phi}$	···	σ_h	∞C_2	$\infty\sigma_v$	i	$2S_\infty^d$	···		
Σ_g^+	1	1	1	···	1	1	1	1	1	···		z^2, x^2+y^2
Σ_u^+	1	1	1	···	−1	−1	1	−1	−1	···	z	
Σ_g^-	1	1	1	···	1	−1	−1	1	1	···	R_z	
Σ_u^-	1	1	1	···	−1	1	−1	−1	−1	···		
Π_g	2	$2\cos\phi$	$2\cos 2\phi$	···	−2	0	0	−2	$-2\cos\phi$	···		(xy, yz)
Π_u	2	$2\cos\phi$	$2\cos 2\phi$	···	2	0	0	−2	$2\cos\phi$	···	(R_x, R_y)	
Δ_g	2	$2\cos 2\phi$	$2\cos 4\phi$	···	2	0	0	2	$2\cos 2\phi$	···	(x, y)	(xy, x^2-y^2)
Δ_u	2	$2\cos 2\phi$	$2\cos 4\phi$	···	−2	0	0	−2	$-2\cos 2\phi$	···		
Φ_g	2	$2\cos 3\phi$	$2\cos 6\phi$	···	−2	0	0	2	$-2\cos 3\phi$	···		
Φ_u	2	$2\cos 3\phi$	$2\cos 6\phi$	···	2	0	0	−2	$2\cos 3\phi$	···		
···	···	···	···	···	···	···	···	···	···	···	···	

APPENDIX K

FUNDAMENTAL CONSTANTS AND CONVERSION FACTORS

K.1 PHYSICAL CONSTANTS

The fundamental physical constants and the most commonly needed derived constants are given in both the cgs and SI systems of units. The constants are based on the atomic mass scale of $^{12}C = 12.0000$.

Constant	Symbol(s)	Value (uncertainty)	
		SI	cgs
Planck's constant	$\hbar$	$6.626176(36) \times 10^{-34}$ J s	$6.626176(36) \times 10^{-27}$ erg s
Boltzmann constant	k	$1.380662(44) \times 10^{-23}$ J s	$1.380662(44) \times 10^{-16}$ erg K^{-1}
Ideal gas constant	R	$8.20568(26) \times 10^{-5}$ m^3-atm-K^{-1} mol^{-1}	$0.0820568(26)$ 1-atm K^{-1} mol^{-1}
		$8.31441(26)$ J K^{-1} mol^{-1}	$8.31441(26) \times 10^7$ erg K^{-1} mol^{-1}
			$1.985863(12)$ cal K^{-1} mol^{-1}
Avogadro number	N_A, L	$6.022045(31) \times 10^{23}$ molecules mol^{-1}	$6.022045(31) \times 10^{23}$ molecules mol^{-1}
Speed of light	c	$2.99792458(1) \times 10^8$ m s^{-1}	$2.99792458(1) \times 10^{10}$ cm s^{-1}
Atomic mass unit (amu)	M	$1.660565(8) \times 10^{-27}$ kg	$1.660565(8) \times 10^{-24}$ g

Constant	Symbol(s)	Value (uncertainty)	
		SI	cgs
Electron mass	m_e, m	$9.109534(47) \times 10^{-31}$ kg	$9.109534(47) \times 10^{-28}$ g
Electron charge	e		$4.803242(13) \times 10^{-10}$ esu
		$1.6021892(46) \times 10^{-19}$ C	$1.602892(46) \times 10^{-20}$ emu
Bohr radius	a_0	$5.2917706(44) \times 10^{-11}$ m	$5.2917706(44) \times 10^{-9}$ cm
Bohr magneton	β, μ_β	$9.274078(36) \times 10^{-24}$ J T^{-1}	$9.274078(36) \times 10^{-21}$ erg G^{-1}
Free electron g factor	g, g_e	2.0023193134(70)	2.0023193134(70)
Proton mass	m_p	$1.6726485(86) \times 10^{-27}$ kg	$1.6726485(86) \times 10^{-24}$ g
Proton gyromagnetic ratio	γ_p	2.67519×10^8 rad s^{-1} T^{-1}	2.67519×10^4 rad s^{-1} G^{-1}
Nuclear magneton	β_N, β_I	$5.050824(20) \times 10^{-27}$ J T^{-1}	$5.050824(20) \times 10^{-24}$ erg G^{-1}
Fine structure constant	α	$7.2973506(60) \times 10^{-3}$	$7.2973506(60) \times 10^{-3}$
Rydberg constant	R_∞	$1.097373177(83) \times 10^7$ m^{-1}	$1.097373177(83) \times 10^5$ cm^{-1}
Vacuum permeability	μ_0	$4\pi \times 10^{-7}$ H m^{-1}	
Vacuum permittivity	ε_0	$8.854187818(71) \times 10^{-12}$ F m^{-1}	

K.2 CONVERSION FACTORS

Unit	1 Hz	1 cm^{-1}	1 eV	Erg/molecule
1 Hz	1	3.3356×10^{-11}	4.1356×10^{-13}	6.6256×10^{-27}
1 cm^{-1}	2.997924×10^{10}	1	1.2398×10^{-4}	1.9863×10^{-16}
1 eV	2.41804×10^{14}	8.06573×10^3	1	1.60219×10^{-12}
Erg/molecule	1.50929×10^{26}	5.0345×10^{15}	6.2418×10^{11}	1
1 cal/mole	1.40854×10^{10}	0.34976	4.3363×10^{-5}	6.9473×10^{-17}

1 J = 6.24144×10^{18} eV
1 cal = 4.1868 J
1 eV = $1.6021892 \times 10^{-19}$ J
1 J = 10^7 erg
1 C = 0.1c statcoulomb = 0.1 abcoulomb
1 gauss = 10^{-4} Tesla
1 atm = 101,325 Pa = 1.01325×10^6 dyne cm^{-2}
1 V = $c^{-1} \times 10^8$ statvolt

Rotational constant B (MHz) $\times$ I (amu A^2) = 5.053791×10^5
Stark effect constant $\mu(D) \times \mathscr{E}$ (V cm^{-1}) = 0.50344

APPENDIX L

TENSORS

While many physical properties are described by vector quantities, there are some that are anisotropic, such that a force acting on the body can produce a displacement noncollinear with the force. These require more than three components for a complete description. Polarizabilities, susceptibilities, electric field gradients, and inertial moments are some examples of such properties. They can be described by a second rank tensor.

According to the classical definition a tensor of rank (order) n is a quantity $T_{a, b, c, \ldots, n}$ which transforms according to the relationship

$$T'_{a, b, c, \ldots, n} = \sum_{i=1}^{3} \sum_{j=1}^{3} \sum_{k=1}^{3} \cdots \sum_{m=1}^{3} a_{ai} a_{bj} a_{ck} \cdots a_{nm} T_{abc \cdots n} \tag{L-1}$$

where the maximum number of subscripts on $T_{abc \cdots n}$ and the maximum number of coefficients a_{ai}, is n. A tensor will then consist of a set of 3^N terms which can conveniently be displayed as a matrix. For $n = 1$ the general definition becomes

$$T'_a = \sum_{i=1}^{3} a_{ai} T_a \tag{L-2}$$

which is recognized as a vector transformation. Hence a vector is a tensor of rank 1.

In molecular spectroscopy the most commonly encountered tensors, in addition to vectors, are those which describe the properties mentioned in the first paragraph and are of rank 2. A second-rank tensor is often called a dyadic and has the general form

$$\underline{\mathbf{D}} = d_{11}\hat{\mathbf{i}}\hat{\mathbf{i}} + d_{12}\hat{\mathbf{i}}\hat{\mathbf{j}} + d_{13}\hat{\mathbf{i}}\hat{\mathbf{k}} + d_{21}\hat{\mathbf{j}}\hat{\mathbf{i}} + d_{22}\hat{\mathbf{j}}\hat{\mathbf{j}} + d_{23}\hat{\mathbf{j}}\hat{\mathbf{k}} + d_{31}\hat{\mathbf{k}}\hat{\mathbf{i}} + d_{32}\hat{\mathbf{k}}\hat{\mathbf{j}} + d_{33}\hat{\mathbf{k}}\hat{\mathbf{k}} \quad \text{(L-3)}$$

For a symmetric dyadic, one with $d_{ij} = d_{ji}$, there are only six independent components. For a dyadic there will be an axis system in which the cross terms d_{ij} with $j \neq i$, vanish. This principal axis system and the three independent terms d'_{ii} can be found by considering the terms of the dyadic to constitute a 3×3 matrix, diagonalizing the matrix and finding the eigenvectors. The terms of the transformation matrix are the coefficients in the sum

$$D'_{ab} = \sum_{i=1}^{3} \sum_{j=1}^{3} \alpha_{ai}\alpha_{bj}D_{ab} \quad \text{(L-4)}$$

The problem is that of diagonalization and finding eigenvalues as discussed in Chap. 3 and App. G.

An example of a dyadic is the inertial dyadic that represents the inertial moments of a set of point masses. The inertial dyadic is composed of three diagonal terms, called moments of inertia

$$I_{ff} = \sum_{i=1}^{N} m_i(g_i^2 + h_i^2) \qquad (f, g, h = x, y, z \text{ cycled}) \quad \text{(L-5)}$$

and six off-diagonal terms, called products of inertia

$$I_{fg} = I_{gf} = -\sum_{i=1}^{N} m_i f_i g_i \qquad (f, g, h = x, y, z \text{ cycled}) \quad \text{(L-6)}$$

where $f, g, h = x, y, z$ taken in cyclic order to obtain the nine terms. The summation is over all mass points in the system (N) and the coordinates are relative to the center of mass as the origin. For any arbitrary choice for the origin and axis directions of a coordinate system correction terms can be included and the resulting terms of the dyadic become

$$I_{ff} = \sum_{i=1}^{N} m_i(g_i^2 + h_i^2) - \frac{1}{M}\left(\sum_{i=1}^{N} m_i g_i\right)^2 - \frac{1}{M}\left(\sum_{i=1}^{N} m_i h_i\right)^2 \quad \text{(L-7)}$$

$$I_{fg} = -\sum_{i=1}^{N} m_i f_i g_i + \frac{1}{M}\left(\sum_{i=1}^{N} m_i f_i\right)\left(\sum_{i=1}^{N} m_i g_i\right) \quad \text{(L-8)}$$

$$M = \sum_{i=1}^{N} m_i \quad \text{(L-9)}$$

where $f, g, h = x, y, z$ taken in cyclic order as before but x, y, z are now the coordinates in the arbitrary axis system. This formulation is useful for taking advantage of molecular symmetry in the selection of a coordinate system to calculate principal moments of inertia.

The inertial dyadic is then

$$\begin{pmatrix} I_{xx} & I_{xy} & I_{xz} \\ I_{xy} & I_{yy} & I_{yz} \\ I_{xz} & I_{yz} & I_{zz} \end{pmatrix} \tag{L-10}$$

and the principal moments are the eigenvalues of this matrix as found from the determinantal equation

$$\begin{vmatrix} I_{xx} - I & I_{xy} & I_{xz} \\ I_{xy} & I_{yy} - I & I_{yz} \\ I_{xz} & I_{xy} & I_{zz} - I \end{vmatrix} = 0 \tag{L-11}$$

Labeling the principal moments as I_σ ($\sigma = 1, 2, 3$) the eigenvectors which give the directions of the principal axes are found from

$$\mathbf{e}_\sigma = \sum_{i=xyz} A_i^\sigma \hat{\mathbf{e}}_i \qquad (\hat{\mathbf{e}}_i = \hat{\mathbf{i}}, \hat{\mathbf{j}}, \hat{\mathbf{k}}) \tag{L-12}$$

and

$$\sum_{j=x,y,z} I_{ij} A_j^\sigma = I_\sigma A_i^\sigma \qquad (i = x, y, z) \tag{L-13}$$

The three sets of three equations given by this expression, along with the orthonormal conditions

$$\sum_{i=x,y,z} A_i^\sigma A_i^{\sigma'} = \delta_{\sigma\sigma'} \tag{L-14}$$

$$\sum_{\sigma=1}^{3} A_i^\sigma A_j^\sigma = \delta_{ij} \qquad (i = x, y, z; j = x, y, z) \tag{L-15}$$

allow one to determine the three sets of vector components for the $\mathbf{e}_\sigma$ vectors.

A short example of the inertial dyadic concept is provided by an analysis of the hypothetical molecule shown in Fig. L-1. The coordinate system has been

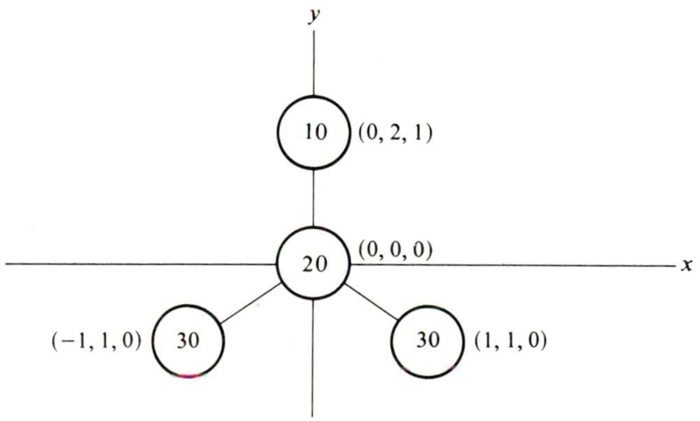

FIGURE L-1
Coordinates of a hypothetical XYZ_2 molecule.

TABLE L-1
Inertial parameters for a hypothetical XYZ_2 molecule

Atom	x_i	y_i	z_i	m_i	x_i^2	y_i^2	z_i^2	$m_i x_i$	$m_i y_i$	$m_i z_i$	$m_i x_i^2$	$m_i y_i^2$	$m_i z_i^2$	$m_i x_i y_i$	$m_i x_i z_i$	$m_i y_i z_i$
X	0	2	1	10	0	4	1	0	20	10	0	40	10	0	0	20
Y	0	0	0	20	0	0	0	0	0	0	0	0	0	0	0	0
Z_1	1	1	0	30	1	1	0	30	30	0	30	30	0	30	0	0
Z_2	−1	1	0	30	1	1	0	−30	30	0	30	30	0	−30	0	0
Σ				90				0	80	10	60	100	10	0	0	20
Σ^2								0	6400	100						

chosen to give the maximum number of equivalent and zero terms. The coordinates are in angstrom units and the masses in amu's. The necessary terms for evaluating the I_{fg}'s are summarized in Table L-1.

Using the parameters from Table L-1 the terms of the dyadic are

$$\begin{aligned}
I_{xx} &= 100 + 10 - \frac{6400}{90} - \frac{100}{90} = 37.8 \\
I_{yy} &= 10 + 60 - \frac{100}{90} - \frac{0}{90} = 68.9 \\
I_{zz} &= 60 + 100 - \frac{0}{90} - \frac{6400}{90} = 89.9 \\
I_{xy} &= -0 + \frac{0 \times 80}{90} = 0 \\
I_{xz} &= -0 + \frac{0 \times 10}{90} = 0 \\
I_{yz} &= -20 + \frac{80 \times 10}{90} = -11.2
\end{aligned} \tag{L-16}$$

The principal moments are found from

$$\begin{vmatrix} 37.8 - I & 0 & 0 \\ 0 & 68.9 - I & -11.2 \\ 0 & -11.2 & 89.9 - I \end{vmatrix} = 0 \tag{L-17}$$

to be

$$I_{XX} = 94.7 \text{ amu } A^2 \tag{L-18}$$

$$I_{YY} = 64.1 \text{ amu } A^2 \tag{L-19}$$

$$I_{ZZ} = 37.8 \text{ amu } A^2 \tag{L-20}$$

From Eq. (L-13)

$$37.8\,A_x^1 + 0\,A_y^1 + 0\,A_z^1 = 94.7\,A_x^1 \tag{L-21}$$

$$0\,A_x^1 + 68.9\,A_y^1 - 11.2\,A_z^1 = 94.7\,A_y^1 \tag{L-22}$$

$$0\,A_x^1 - 11.2\,A_y^1 + 89.9\,A_z^1 = 94.7\,A_z^1 \tag{L-23}$$

Since

$$(A_x^1)^2 + (A_y^1)^2 + (A_z^1)^2 = 1 \tag{L-24}$$

the coefficients of $\mathbf{e}_1$ are $A_x^1 = 0$, $A_y^1 = -0.396$ and $A_z^1 = 0.918$ or

$$\mathbf{e}_1 \equiv \mathbf{e}_X = -0.396\hat{\mathbf{j}} + 0.918\hat{\mathbf{k}} \tag{L-25}$$

From the remaining two principal moments the other eigenvectors are found to be

$$\mathbf{e}_2 \equiv \mathbf{e}_Y = 0.918\hat{\mathbf{j}} - 0.396\hat{\mathbf{k}} \tag{L-26}$$

$$\mathbf{e}_3 \equiv \mathbf{e}_Z = \hat{\mathbf{i}} \tag{L-27}$$

By referring to App. D the concepts regarding rotational motion can be further extended. If we consider the diagonalized inertial dyadic

$$\underline{\mathrm{I}} = I_{XX}\hat{\mathbf{i}}\hat{\mathbf{i}} + I_{YY}\hat{\mathbf{j}}\hat{\mathbf{j}} + I_{ZZ}\hat{\mathbf{k}}\hat{\mathbf{k}} \tag{L-28}$$

then for the rotation of a molecule,

$$\mathbf{P} = \underline{\mathbf{I}} \cdot \boldsymbol{\omega} = I_{XX}\omega_x\hat{\mathbf{i}} + I_{YY}\omega_y\hat{\mathbf{j}} + I_{ZZ}\omega_z\hat{\mathbf{k}} \tag{L-29}$$

and

$$E_k = \tfrac{1}{2}\boldsymbol{\omega} \cdot \underline{\mathbf{I}} \cdot \boldsymbol{\omega} = \tfrac{1}{2}[I_{XX}\omega_x^2 + I_{YY}\omega_y^2 + I_{ZZ}\omega_z^2] \tag{L-30}$$

Since the magnitude of the components of angular momentum are $P_g = I_{GG}\omega_g$ (g; $G = x, y, z$; X, Y, Z) the kinetic energy becomes

$$E_k = \frac{P_x^2}{2I_{XX}} + \frac{P_y^2}{2I_{YY}} + \frac{P_z^2}{2I_{ZZ}} \tag{L-31}$$

APPENDIX M

RUSSELL SAUNDERS COUPLING IN ATOMS

The orbital and spin angular momenta of the electrons in atoms can couple via their respective associated magnetic moments. For light atoms, this coupling first involves the independent coupling of the individual electronic orbital angular moment to form a total resultant orbital angular momentum, and the independent coupling of the individual spin angular momenta to form a total resultant spin momentum. This is followed by a coupling of these two resultant momenta and is referred to as *L-S* or Russell Saunders coupling. This appendix outlines the nature of this coupling by using vector terminology.

The various momenta associated with atoms are:

$\mathbf{l}_i$—orbital angular momentum for electron i
$\mathbf{s}_i$—spin angular momentum for electron i
$\mathbf{j}_i$—total angular momentum for electron i
L—total orbital angular momentum due to all electrons
S—total spin angular momentum due to all electrons
J—total angular momentum due to all electrons

The various electronic states of an atom are denoted by term symbols of the form

$$^{M}L_J$$

where $M = 2S + 1$ and is called the multiplicity. The value of **L** is denoted by

TABLE M-1
Coupling of orbital angular momenta for two electrons

Electrons	l_1	l_2	L	State
ss	0	0	$0\ (l_1 + l_2)$	*S*
sp	0	1	$1\ (l_1 + l_2)$	*P*
pp	1	1	$2\ (l_1 + l_2)$	*D*
			$1\ (l_1 + l_2 - 1)$	*P*
			$0\ (\lvert l_1 - l_2 \rvert)$	*S*
pd	1	2	$3\ (l_1 + l_2)$	*F*
			$2\ (l_1 + l_2 - 1)$	*D*
			$1\ (\lvert l_1 - l_2 \rvert)$	*P*
dd	2	2	$4\ (l_1 + l_2)$	*G*
			$3\ (l_1 + l_2 - 1)$	*F*
			$2\ (l_1 + l_2 - 2)$	*D*
			$1\ (l_1 + l_2 - 3)$	*P*
			$0\ (\lvert l_1 - l_2 \rvert)$	*S*

the symbols $S, P, D, F, G, \ldots$ representing orbital momentum values of $0, 1, 2, 3, 4, \ldots$.

The coupling of the individual electron angular momenta is a vector coupling. This is illustrated in Table M-1 for several examples of coupling of two electrons. This vector coupling for *pp* and *pd* cases are illustrated in Fig. M-1. These concepts can be extended to a larger number of electrons by taking the results of the coupling of two electrons and vectorially adding a third, then repeating the process until all electrons are included. For example, the addition of a *p* electron to the *pp* configuration will lead to the possibility of *F*, *D*, *P*, and *S* states.

The description for the coupling of spin momenta is analogous to that of orbital momenta coupling except the values of the s_i are limited to $\frac{1}{2}$. Therefore two electrons can couple to give $S = \frac{1}{2} + \frac{1}{2} = 1$ or $S = \frac{1}{2} - \frac{1}{2} = 0$ having multiplicities of 3 and 1 respectively.

Russell Saunders coupling is the coupling of the total orbital angular momentum **L** with the total spin momentum **S** to give a total momentum **J**. This is illustrated in Fig. M-2 for a *pd* pair of electrons. In this case $L = 1, 2, 3$ and $S = 0, 1$.

Two points remain to be addressed. One is the selection of allowed terms for a given electron configuration and the second is the ordering of the energy levels represented by each term. The allowed terms are determined by the Pauli exclusion principle which restricts each electron in an atom to having a different set of the four describing quantum numbers n, l, m_l, and m_s. If the electrons being considered all have different values for their principal quantum numbers n_i, then there are no restrictions on the remaining three numbers and all possible terms represent attainable states. If two or more electrons have equivalent n_i

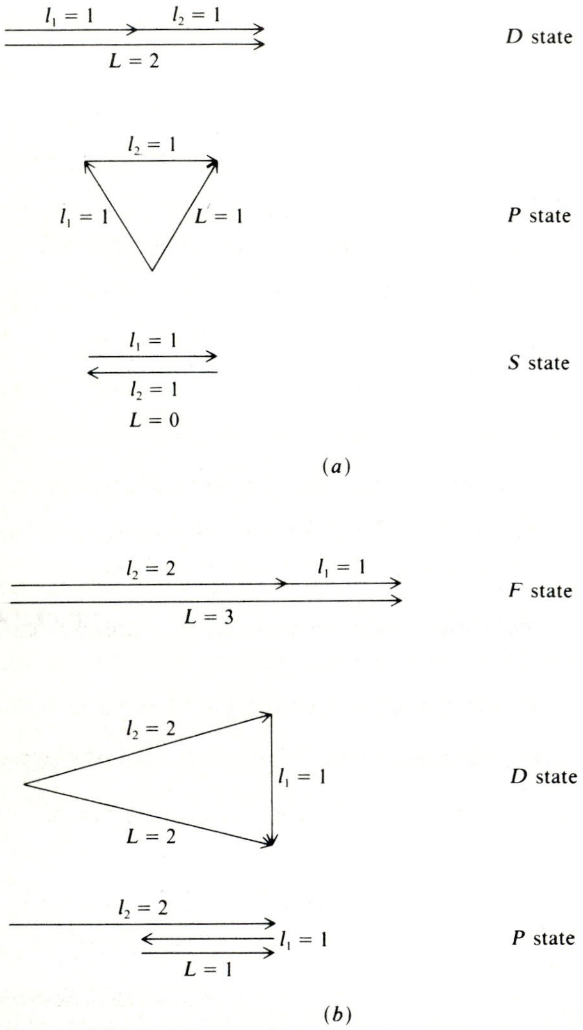

FIGURE M-1
Vector coupling of (*a*) two *p* electrons, and (*b*) a *d* and a *p* electron.

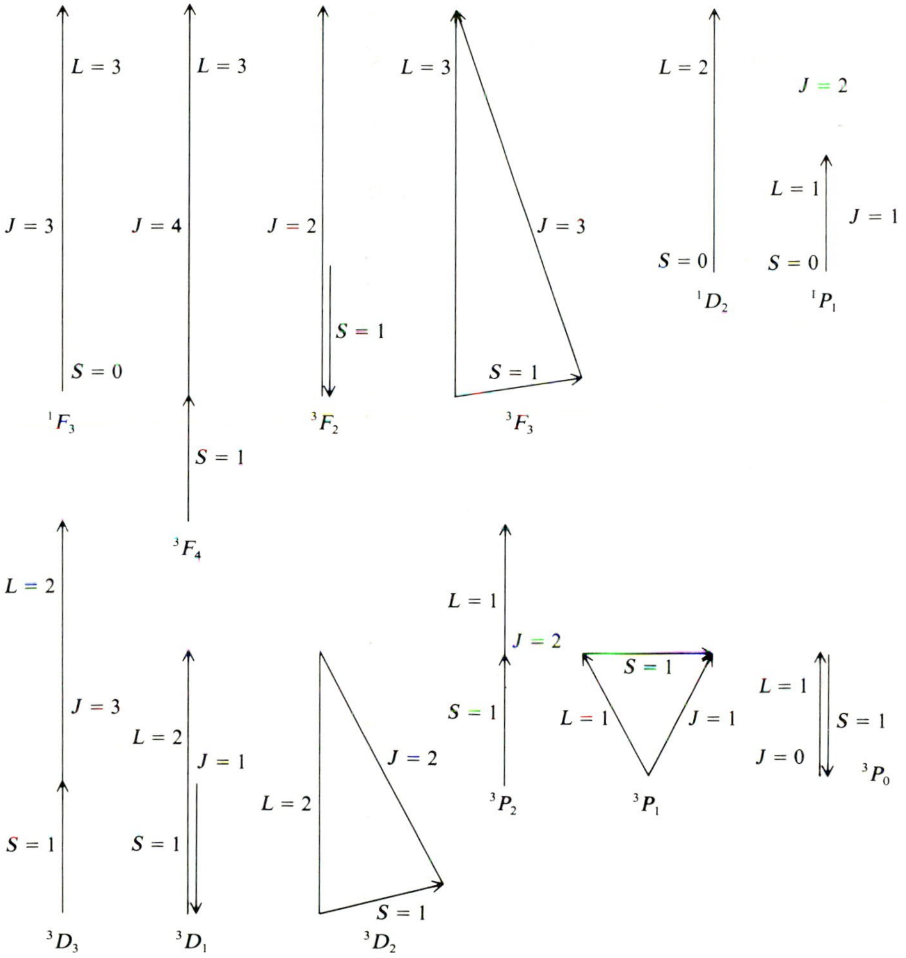

FIGURE M-2
Russell Saunders Coupling for *a p* and *a d* Electron.

TABLE M-2
Allowed states for L-S coupling of two s electrons

Equivalent electrons		Inequivalent electrons	
$1s^2$		$1s2s$	
$n_1 = 1$	$n_2 = 1$	$n_1 = 1$	$n_2 = 2$
$l_1 = 0$	$l_2 = 0$	$l_1 = 0$	$l_2 = 0$
$m_{l_1} = 0$	$m_{l_2} = 0$	$m_{l_1} = 0$	$m_{l_2} = 0$
$s_1 = \frac{1}{2}$	$s_2 = \frac{1}{2}$	$s_1 = \frac{1}{2}$	$s_2 = \frac{1}{2}$
$m_{s_1} = \frac{1}{2}$	$m_{s_2} = -\frac{1}{2}$	$m_{s_1} = \pm\frac{1}{2}$	$m_{s_2} = \pm\frac{1}{2}$
$L = 0$	$S = 0$	$L = 0$	$S = 0, 1$
^1S		$^1S, ^3S$	

values then there will be restrictions on the other numbers and not all states can be realized. This is illustrated for the case of two s electrons in Table M-2.

The analysis of more complex atoms is aided by first finding the possible values of M_L and M_S, the magnetic quantum number associated with L and S, and then constructing a table of allowed microstates which can contribute to the various terms. For example, if a value of $M_L = 2$ were found to exist then it will be possible to associate five of the microstates with one term since M_L can have values of $L, L-1, L-2, \ldots, -L$, and M_S can have values of $S, S-1, S-2, \ldots,$

TABLE M-3
Development of terms for (a) ns^2, $(n+1)s$ and (b) np, $(n+1)p$ systems

Step	m_{l_i} or m_{s_i}	Quantum No.	Relationship	Value
		(a) $ns^2, (n+1)s$		
1	$m_{l_1} = m_{l_2} = m_{l_3} = 0$	M_L	$m_{l_1} + m_{l_2} + m_{l_3} = 0 + 0 + 0$	0
2		L		0
3	$m_{s_1} = \frac{1}{2}, m_{s_2} = -\frac{1}{2}, m_{s_3} = \pm\frac{1}{2}$	M_S	$m_{s_1} + m_{s_2} + m_{s_3} = \frac{1}{2} - \frac{1}{2} \pm \frac{1}{2}$	$\pm\frac{1}{2}$
4		S		$\frac{1}{2}$
Term				^2S
		(b) $np, (n+1)p$		
1	$m_{l_1} = 0, \pm 1; m_{l_2} = 0, \pm 1$	M_L	$m_{l_1} + m_{l_2}$	$2, 1, 0, -1, -2$ $1, 0, -1$ 0
2		L		$2, 1, 0$
3	$m_{s_1} = \pm\frac{1}{2}, m_{s_2} = \pm\frac{1}{2}$	M_S	$m_{s_1} + m_{s_2}$	$1, 0, 1$ 0
4		S		$1, 0$
Terms				$^1D, ^3D, ^1P, ^3P, ^1S, ^3S$

$-S$. Also

$$M_L = \sum_i m_{li} \tag{M-1}$$

and

$$M_S = \sum_i m_{si} \tag{M-2}$$

A four-step procedure will often suffice for the analysis of simple systems but the most expedient way to analyze a more complex one is to establish a table of microstates. These four steps are to find M_L, L, M_S, and S, and are illustrated in Table M-3 for two examples.

Examples of the construction of microstates and resulting terms are shown in Tables M-4 and M-5. In these tables m_s values are denoted by ↑ for $m_s = \frac{1}{2}$ and ↓ for $m_s = -\frac{1}{2}$. States which are not allowed due to the Pauli principle have been lined through.

TABLE M-4
Microstates and terms for an np^2 system

m_l: +1	0	−1	M_L	M_S	Term	Possible Combinations
~~↓↓~~			~~2~~	~~−1~~		
~~↑↑~~			~~2~~	~~1~~		
↑↓			2	0	A ^1D	A $M_L = 2, 1, 0, -1, -2$
↑	↑		1	1	B ^3P	$M_S = 0$
↑	↓		1	0	A ^1D	
↓	↑		1	0	B ^3P	B $M_L = 1, 0, -1$
↓	↓		1	−1	B ^3P	$M_S = 1, 0, -1$
↑		↑	0	1	B ^3P	
↑		↓	0	0	A ^1D	C $M_L = 0$
↓		↑	0	0	B ^3P	$M_S = 0$
↓		↓	0	−1	B ^3P	
	~~↑↑~~		~~0~~	~~1~~		
	↑↓		0	0	C ^1S	
	~~↓↓~~		~~0~~	~~−1~~		
	↑	↓	−1	0	A ^1D	
	↓	↑	−1	0	B ^3P	
	↓	↓	−1	−1	B ^3P	
	↑	↑	−1	1	B ^3P	
		~~↑↑~~	~~−2~~	~~1~~		
		~~↓↓~~	~~−2~~	~~−1~~		
		↑↓	−2	0	A ^1D	
			^1S, ^3P, ^1D allowed			

TABLE M-5
Microstates and terms for an nd^2 system

m_l = 2	1	0	−1	−2	L	S	
↑↓					4	0	A
~~↑↑~~					~~4~~	~~1~~	
~~↓↓~~					~~4~~	~~−1~~	
↑	↓				3	0	A
↑	↑				3	1	B_1
↓	↑				3	0	B_1
↓	↓				3	−1	B_1
↑		↓			2	0	A
↑		↑			2	1	B_2
↓		↑			2	0	B_2
↓		↓			2	−1	B_2
↑			↓		1	0	A
↑			↑		1	1	B_3
↓			↑		1	0	B_3
↓			↓		1	−1	B_3
↑				↓	0	0	A
↑				↑	0	1	B_4
↓				↑	0	0	B_4
↓				↓	0	−1	B_4
	↑↓				2	0	C
	~~↑↑~~				~~2~~	~~1~~	
	~~↓↓~~				~~2~~	~~−1~~	
	↑	↓			1	0	C
	↑	↑			1	1	D_1
	↓	↑			1	0	D_1
	↓	↓			1	−1	D_1
	↑		↓		0	0	C
	↑		↑		0	1	D_2
	↓		↑		0	0	D_2
	↓		↓		0	−1	D_2
	↑			↓	−1	1	A
	↑			↑	−1	1	B_5
	↓			↑	−1	0	B_5
	↓			↓	−1	−1	B_5
		↑↓			0	0	E
		~~↑↑~~			~~0~~	~~1~~	
		~~↓↓~~			~~0~~	~~−1~~	
		↑	↓		−1	0	C
		↑	↑		−1	1	D_3
		↓	↑		−1	0	D_3
		↓	↓		−1	−1	D_3
		↑		↓	−2	0	A
		↑		↑	−2	1	B_7
		↓		↑	−2	0	B_7
		↓		↓	−2	−1	B_7
			↑↓		−2	0	C
			~~↑↑~~		~~−2~~	~~1~~	
			~~↓↓~~		~~−2~~	~~−1~~	
			↑	↓	−3	0	A
			↑	↑	−3	1	B_7
			↓	↑	−3	0	B_7
			↓	↓	−3	−1	B_7
				↑↓	−4	0	A
				~~↑↑~~	~~−4~~	~~1~~	
				~~↓↓~~	~~−4~~	~~−1~~	

(9)	A	$M_L = 4, 3, 2, 1, 0, -1, -2, -3, -4$ $M_S = 0$	$L = 4$ ^1G
(21)	B	$M_L = 3, 2, 1, 0, -1, -2, -3$ $M_S = 1, 0, -1$	$L = 3$ ^3F
(5)	C	$M_L = 2, 1, 0, -1, -2$ $M_S = 0$	$L = 2$ ^1D
(9)	D	$M_L = 1, 0, -1$ $M_S = 1, 0, -1$	$L = 1$ ^3P
(1)	E	$M_L = 0$ $M_S = 0$	$L = 0$ ^1S

TABLE M-6
Values for total angular momentum and resulting terms

L	S	J	M	Term
0	$\frac{1}{2}$	$\frac{1}{2}$	2	$^2S_{1/2}$
0	1	1	3	$^3S_1$
1	1	$\frac{3}{2}$	2	$^2P_{3/2}$
		$\frac{1}{2}$	2	$^2P_{1/2}$
2	1	3	3	$^3D_3$
		2	3	$^3D_2$
		1	3	$^3D_1$

Once the principle term values and multiplicities have been determined one can proceed to finalize the state terms by considering the possible values of J. Due to the vector addition of **L** and **S** to give **J** the allowed values of J will be limited to

$$J = L + S, L + S - 1, L + S - 2, \ldots, |L - S|$$

A few examples are summarized in Table M-6.

The ordering of the energy states is given by Hund's rules.

1. For the ground state configuration terms with maximum multiplicity lie lowest, i.e.,

$$^3P < {}^1D$$

2. For levels of comparable multiplicity the one with the largest L lies lowest, i.e.,

$$^3D < {}^3P$$

3. For levels with equivalent multiplicity and L
   (*a*) The lowest level is that with minimum J if the shell is less than half full.
   (*b*) The lowest level is that with maximum J if the shell is more than half full.

Table M-7 provides some examples.

TABLE M-7
Order of terms for particular electron sets

Electrons	Terms
np^2	$^3P < {}^1D < {}^1S$
$np, (n+1)p$	$^3D < {}^3P < {}^3S < {}^1D < {}^1P < {}^1S$
$3d^2$	$^3F < {}^3P < {}^1G < {}^1D < {}^1S$
	↳ ↳ $^1D_{5/2} < {}^1D_{7/2}$
	$^3F_2 < {}^3F_3 < {}^3F_4$
$3d^8$	$^3F < {}^3P < {}^1G < {}^1D < {}^1S$
	↳ $^3F_4 < {}^3F_3 < {}^3F_2$

APPENDIX N

TABLES OF DIRECT PRODUCTS

Tables are given in a general form. Additions in the form of *g-u* subscripts and prime (′)-double prime (″) must be added following the rules:

1. $g \times g = u \times u = g$
2. $g \times u = u \times g = u$
3. $' \times ' = '' \times '' = '$
4. $' \times '' = '' \times ' = ''$

C_s	A'	A''
A'	A'	A''
A''		A'

C_i	A_g	A_u
A_g	A_g	A_u
A_u		A_g

C_{2v}	A_1	A_2	B_1	B_2
A_1	A_1	A_2	B_1	B_2
A_2		A_1	B_2	B_1
B_1			A_1	A_2
B_2				A_1

$C_{3v}, D_3, D_{3d}, D_{3h}$	A_1	A_2	E
A_1	A_1	A_2	E
A_2		A_1	E
E			$A_1 + (A_2) + E$

C_2, C_{2h}	A	B
A	A	B
B		A

C_3, C_{3h}, S_6	A	E
A	A	E
E		$(A) + A + E$

C_4, C_{4h}, S_4	A	B	E
A	A	B	E
B		A	E
E			$(A)+A+2B$

D_2, D_{2h}	A	B_1	B_2	B_3
A	A	B_1	B_2	B_3
B_1		A	B_3	B_2
B_2			A	B_1
B_3				A

$C_{4v}, D_4, D_{2d}, D_{4h}$	A_1	A_2	B_1	B_2	E
	A_1	A_2	B_1	B_2	E
A_2		A_1	B_2	B_1	E
B_1			A_1	A_2	E
B_2				A_1	E
E					$A_1+(A_2)+B_1+B_2$

O, O_h, T, T_d, T_h	A_1	A_2	E	T_1	T_2
A_1	A_1	A_2	E	T_1	T_2
A_2		A_1	E	T_2	T_1
E			$A_1+(A_2)+E$	T_1+T_2	T_1+T_2
T_1				$A_1+E+(T_1)+T_2$	$A_2+E+T_1+T_2$
T_2					$A_1+E+(T_1)+T_2$

$D_\infty, C_{\infty v}, D_{\infty h}$	Σ^+	Σ^-	Π	Δ	Φ	Γ	...
Σ^+	Σ^+	Σ^-	Π	Δ	Φ	Γ	
Σ^-		Σ^+	Π	Δ	Φ	Γ	
Π			$\Sigma^+ + (\Sigma^-) + \Delta$	$\Pi + \Phi$	$\Delta + \Gamma$	$\Phi + H$	
Δ				$\Sigma^+ + (\Sigma^-) + \Gamma$	$\Pi + H$	$\Delta + I$	
Φ					$\Sigma^+ + (\Sigma^-) + I$	$\Pi + \theta$	
Γ						$\Sigma^+ + (\Sigma^-) + K$	
⋮							

APPENDIX O

PROPORTIONALITY OF SECOND-RANK TENSORS

Proof of the statement: The matrix elements of all traceless, second-rank, symmetric tensors are proportional.

The problem will be to show that Q_{kl} as given by Eq. (10.62) or ∇E_{ij} as given by Eq. (10.46) can be related to the components of I. Let $\mathbf{V}$ be a vector associated with some property of the nucleus or molecular rotation and let it satisfy the same commutation rules with respect to $\mathbf{I}(\mathbf{J})$ as $\mathbf{r}$ or $\mathbf{I}(\mathbf{J})$. These commutation rules are in condensed notation,

$$[\mathbf{V}, \mathbf{I}] = [\mathbf{I}, \mathbf{V}] = -i\hbar \underline{\mathbf{V}} \times \underline{\underline{\mathbf{I}}} \tag{O-1}$$

where

$$\underline{\underline{\mathbf{I}}} = \hat{\mathbf{i}}\hat{\mathbf{i}} + \hat{\mathbf{j}}\hat{\mathbf{j}} + \hat{\mathbf{k}}\hat{\mathbf{k}} \tag{O-2}$$

Let all quantum numbers denoting eigenstate of the Hamiltonian, except those for $\mathbf{I}(\mathbf{J})$, $I(J)$, and $I_z(J_z)$, $m_I(m_J)$ be represented by α. We notice that the concepts being developed are equally applicable to problems involving nuclear or molecular angular momentum. If we are going to relate the vector $\mathbf{r}$ and its components as given by Eqs. (10.45) and (10.46) to $\mathbf{I}$ and its components then we wish to look at matrix elements of the form, letting $m = m_I(m_J)$,

$$\left(\alpha, I, m \left| \frac{3(V_iV_j + V_jV_i)}{2} - \delta_{ij}V^2 \right| \alpha, I, m'\right) \tag{O-3}$$

We thus see that this term will depend on the matrix elements of $\underline{\mathbf{V}}$ and its components. The matrix elements of a general vector are derived from the commutations relationships and the elements of the angular momentum [1, 2, 3], and are

$$(\alpha, I, m|\mathbf{V}|\alpha', I+1, m \pm 1) = \mp(\alpha, I|\mathbf{V}|\alpha', I+1)\tfrac{1}{2}[(I \pm m + 1)(I \pm m + 2)]^{1/2}(\hat{\mathbf{i}} \pm i\hat{\mathbf{j}}) \quad \text{(O-4)}$$

$$(\alpha, I, m|\mathbf{V}|\alpha', I+1, m) = (\alpha, I|\mathbf{V}|\alpha', I+1)[(I+1)^2 - m^2]^{1/2}\hat{\mathbf{k}} \quad \text{(O-5)}$$

$$(\alpha, I, m|\mathbf{V}|\alpha', I, m \pm 1) = (\alpha, I|\mathbf{V}|\alpha', I)\tfrac{1}{2}[(I \mp m)(I \pm m + 1)]^{1/2}(\hat{\mathbf{i}} + i\hat{\mathbf{j}}) \quad \text{(O-6)}$$

$$(\alpha, I, m|\mathbf{V}|\alpha', I, m) = (\alpha, I|\mathbf{V}|\alpha', I)m\hat{\mathbf{k}} \quad \text{(O-7)}$$

$$(\alpha, I, m|\mathbf{V}|\alpha', I-1, m \pm 1) = \pm(\alpha, I|\mathbf{V}|\alpha', I-1)\tfrac{1}{2}[(I \mp m)(I \mp m - 1)]^{1/2}(\hat{\mathbf{i}} \pm i\hat{\mathbf{j}}) \quad \text{(O-8)}$$

$$(\alpha, I, m|\mathbf{V}|\alpha', I-1, m) = (\alpha, I|\mathbf{V}|\alpha', I-1)[I^2 - m^2]^{1/2}\hat{\mathbf{k}} \quad \text{(O-9)}$$

where all of the $(\alpha, I|V|\alpha', I')$ elements are independent of m. Using the conventional matrix summation

$$(\alpha, I, m|V_iV_j|\alpha, I, m') = \sum_{\alpha'}\sum_{I'}\sum_{m''}(\alpha, I, m|\mathbf{V}|\alpha', I', m'')(\alpha', I', m''|\mathbf{V}|\alpha, I, m') \quad \text{(O-10)}$$

Equation O-3 can be written as

$$\begin{aligned}
&\left(\alpha, I, m\left|\frac{3(V_iV_j + V_jV_i)}{2} - \delta_{ij}\mathbf{V}^2\right|\alpha, I, m'\right) \\
&= \Bigg\{\Bigg[\sum_{\alpha'}\{-\tfrac{3}{2}(\alpha, I|\mathbf{V}|\alpha', I+1)(\alpha', I+1|\mathbf{V}|\alpha, I) \\
&\qquad - \tfrac{3}{2}(\alpha, I|\mathbf{V}|\alpha', I-1)(\alpha', I-1|\mathbf{V}|\alpha, I) \\
&\qquad + \tfrac{3}{2}(\alpha, I|\mathbf{V}|\alpha', I)(\alpha', I|\mathbf{V}|\alpha, I)\}\Bigg] \\
&\quad \times \Bigg[\{[(I \mp m)(I \mp m - 1)(I \pm m + 1)(I \pm m + 2)]^{1/2}\} \\
&\qquad \times \{\tfrac{1}{2}[\delta_{i1}\delta_{j1} - \delta_{i2}\delta_{j2} \pm i(\delta_{i1}\delta_{j2} + \delta_{i2}\delta_{j1})]\delta_{m',m\pm 2}\} \\
&\qquad + \{(2m+1)[(I \mp m)(I \pm m + 1)]^{1/2} \\
&\qquad \times \tfrac{1}{2}[\delta_{i3}\delta_{j1} + \delta_{i1}\delta_{j3} \pm i(\delta_{i3}\delta_{j2} + \delta_{i2}\delta_{j3})] \times \delta_{m',m\pm 1}\} \\
&\qquad + \{(\tfrac{2}{3})^{1/2}[3m^2 - I(I+1)] \times (\tfrac{2}{3})^{1/2}[(\delta_{i3}\delta_{j3} - \tfrac{1}{2}\delta_{i1}\delta_{j1} - \tfrac{1}{2}\delta_{i2}\delta_{j2}]\delta_{m'm}\}\Bigg]\Bigg\}
\end{aligned} \quad \text{(O-11)}$$

These matrix elements are of the form

$$\left(\alpha, I, m \left| \frac{3(V_iV_j + V_jV_i)}{2} - \delta_{ij}\mathbf{V}^2 \right| \alpha, I, m'\right) = C(\alpha, I, \alpha', I)F(I, m) \tag{O-12}$$

so we see that the m dependence of the matrix elements is the same regardless of the value of $C(\alpha, I, \alpha', I)$.

The m dependency of any matrix element can readily be found by evaluating Eq. (O-11). For example,

$$\begin{aligned}\left(I, I \left| \frac{3(V_zV_z + V_zV_z)}{2} - \delta_{zz}\mathbf{V}^2 \right| II\right) &= C\{\tfrac{2}{3}[3I^2 - I(I+1)]\} \\ &= C'(2I^2 - I)\end{aligned} \tag{O-14}$$

An alternate method which employs irreducible tensor methods has been given by Tinkham [4] or Silver [5]. Equations O-4 through O-9 may also be obtained from the Wigner-Eckart theorem [4].

1. Condon, E. U., and G. H. Shortley, *The Theory of Atomic Spectra.* Cambridge Press, Cambridge, England, 1953, pp. 59–63.
2. Feenberg, E., and G. E. Pake, *Notes on the Quantum Theory of Angular Momentum,* Addison Wesley, Reading, Mass., 1953, chap. 4.
3. Edmonds, A. R., *Angular Momentum in Quantum Mechanics,* 2d ed., Princeton Univ. Press., Princeton, NJ, 1960.
4. Tinkham, M., *Group Theory and Quantum Mechanics,* McGraw-Hill, New York, N.Y., 1964.
5. Silver, B. L., *Irreducible Tensor Methods,* Academic Press, New York, N.Y., 1976.

APPENDIX P

STEREOGRAPHIC PROJECTIONS

A convenient method for representing a three-dimensional configuration on a two-dimensional surface is by use of a stereographic projection and a Wulff net, the latter being shown in Fig. P-1. The stereographic projection results from the intersection of lines, drawn from one pole to the surface of a sphere, with the equatorial plane of the sphere. This is illustrated in Fig. P-2. An examination of the Wulff net shows that the lines on the net are stereographic projections of the meridians of the sphere.

Stereographic projections are used to represent the external structure of crystals. Such a projection is made by projecting the points on the spherical surface where normals to the crystal faces intersect the sphere. This is illustrated for a section of a cubic crystal in Fig. P-3.

Figure P-4 illustrates a stereographic projection of a typical crystal including not only the projection of the face normals but also projections of the intercepts of symmetry axes.

Another feature of the stereographic projection is that a conic section on the surface of a sphere will project as the same conic surface, i.e., an ellipse will project as an ellipse. This feature makes the concept useful for handling data relating to the location of the zero-splitting loci of Zeeman-split NQR transitions. Figure 10-10 illustrates such a locus for the ^{137}Ba resonance in barium chloride dihydrate.

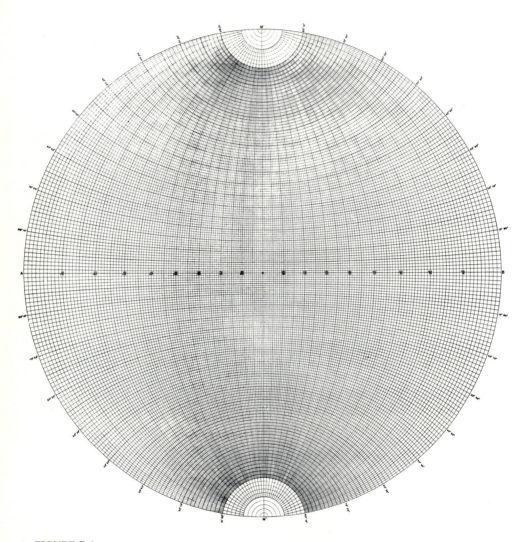

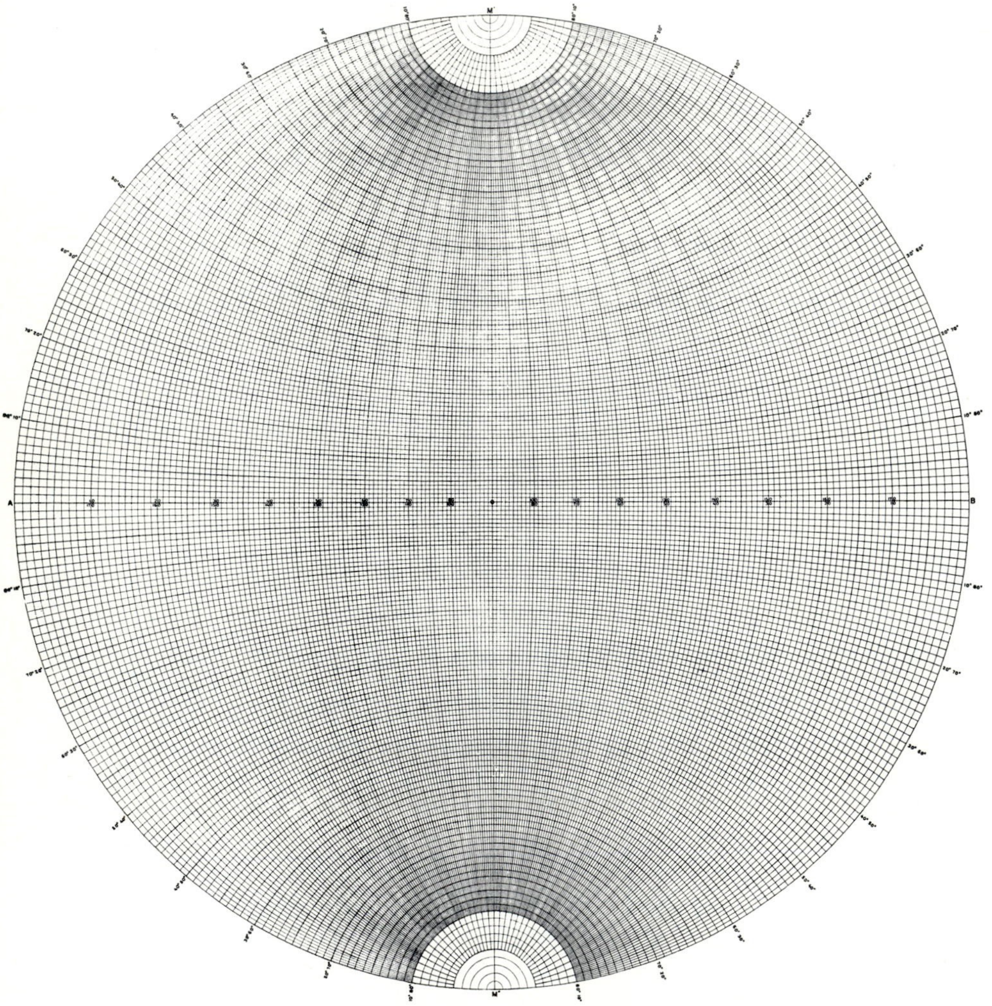

FIGURE P-1
Wulff net.

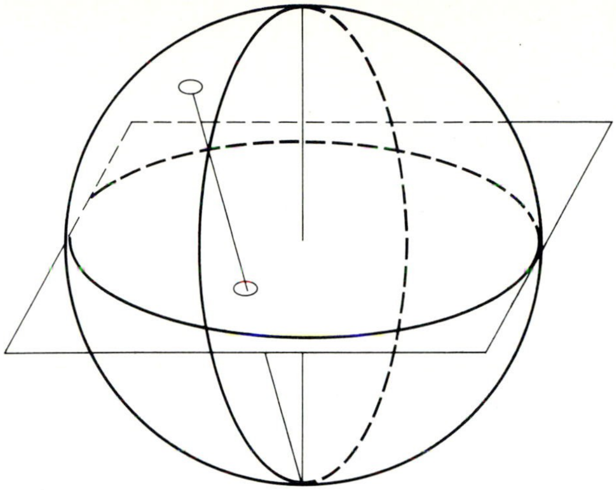

FIGURE P-2
Stereographic projection of a single point.

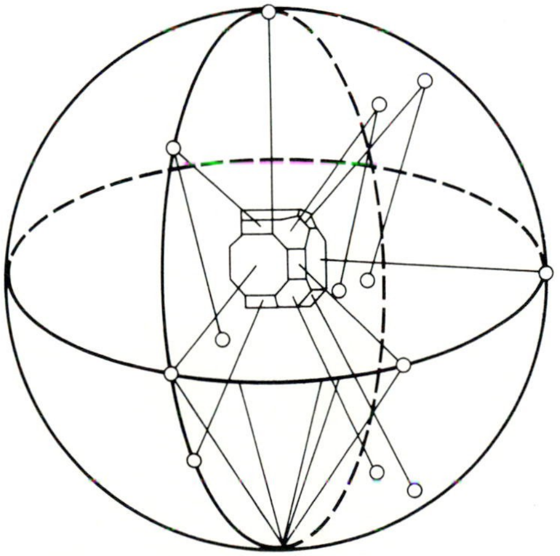

FIGURE P-3
Stereographic projection of a section of a cubic crystal.

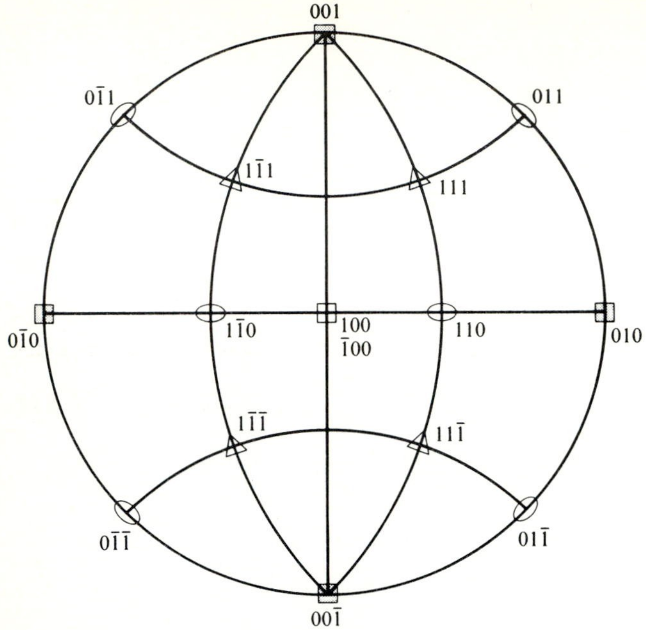

FIGURE P-4
Stereographic projection of a cubic crystal.

GENERAL REFERENCES

Nuffield, E. W., *X-Ray Diffraction Methods*, John Wiley, New York, NY, 1966, ch. 6.
Hartshorne, N. H., and A. Stuart, *Practical Optical Crystallography*, Edward Arnold Ltd., London, GB, 1964, ch. 6.

APPENDIX Q

TRANSFORMATION OF THE LAPLACIAN OPERATOR

The transformation of the Laplacian operator

$$\nabla_i^2 = \frac{\partial^2}{\partial x_i^2} + \frac{\partial^2}{\partial y_i^2} + \frac{\partial^2}{\partial z_i^2} \tag{Q-1}$$

from a system of coordinates based on the individual particle coordinates (x_i, y_i, z_i) to the center of mass coordinates (X, Y, Z) and the relative coordinates (x, y, z) as shown in Fig. 11.1 employs the differential chain rule of calculus. For example the operator $\partial/\partial x_i$ can be written

$$\frac{\partial}{\partial x_1} = \frac{\partial X}{\partial x_1}\frac{\partial}{\partial X} + \frac{\partial x}{\partial x_1}\frac{\partial}{\partial x} + \frac{\partial Y}{\partial x_1}\frac{\partial}{\partial Y} + \frac{\partial y}{\partial x_1}\frac{\partial}{\partial y} + \frac{\partial Z}{\partial x_1}\frac{\partial}{\partial Z} + \frac{\partial z}{\partial x_1}\frac{\partial}{\partial z} \tag{Q-2}$$

Five analogous equations can be written for $\partial/\partial y_1$, $\partial/\partial z_1$, $\partial/\partial x_2$, $\partial/\partial y_2$, and $\partial/\partial z_2$. Employing Eqs. (11.5) through (11.9), one has

$$\frac{\partial}{\partial x_1} = \frac{m_1}{m_1 + m_2}\frac{\partial}{\partial X} - \frac{\partial}{\partial x} \tag{Q-3}$$

and

$$\frac{\partial}{\partial x_2} = \frac{m_2}{m_1 + m_2}\frac{\partial}{\partial X} - \frac{\partial}{\partial x} \tag{Q-4}$$

The second partials become

$$\frac{\partial^2}{\partial x_1^2} = \left(\frac{m_1}{m_1 + m_2}\right)^2 \frac{\partial^2}{\partial X^2} + \frac{\partial^2}{\partial x^2} - 2\left(\frac{m_1}{m_1 + m_2}\right)\frac{\partial^2}{\partial X \partial x} \tag{Q-5}$$

$$\frac{\partial^2}{\partial x_2^2} = \left(\frac{m_2}{m_1 + m_2}\right)^2 \frac{\partial^2}{\partial X^2} + \frac{\partial^2}{\partial x^2} - 2\left(\frac{m_2}{m_1 + m_2}\right)\frac{\partial^2}{\partial X \partial x} \tag{Q-6}$$

Dividing Eq. (Q-5) by m_1, Eq. (Q-6) by m_2, and adding the two together, gives

$$\frac{1}{m_1}\frac{\partial^2}{\partial x_1^2} + \frac{1}{m_2}\frac{\partial^2}{\partial x_2^2} = \frac{1}{m_1 + m_2}\frac{\partial^2}{\partial X^2} + \left(\frac{1}{m_1} + \frac{1}{m_2}\right)\frac{\partial^2}{\partial x^2} \tag{Q-7}$$

In a similar manner it is found that

$$\frac{1}{m_1}\frac{\partial^2}{\partial y_1^2} + \frac{1}{m_2}\frac{\partial^2}{\partial y_2^2} = \frac{1}{m_1 + m_2}\frac{\partial^2}{\partial Y^2} + \left(\frac{1}{m_1} + \frac{1}{m_2}\right)\frac{\partial^2}{\partial y^2} \tag{Q-8}$$

and

$$\frac{1}{m_1}\frac{\partial^2}{\partial z_1^2} + \frac{1}{m_2}\frac{\partial^2}{\partial z_2^2} = \frac{1}{m_1 + m_2}\frac{\partial^2}{\partial Z^2} + \left(\frac{1}{m_1} + \frac{1}{m_2}\right)\frac{\partial^2}{\partial y^2} \tag{Q-9}$$

Figure 11.1 shows that

$$r^2 = (x_2 - x_1)^2 + (y_2 - y_1)^2 + (z_2 - z_1)^2 = x^2 + z^2 + z^2 \tag{Q-10}$$

so

$$V(x_1, y_1, z_1, x_2, y_2, z_2) = V(x, y, z) = V(r) \tag{Q-11}$$

Substituting Eqs. (Q-7), (Q-8), (Q-9), and (Q-11) into the original wave equation Eq. (11.3), one obtains

$$-\frac{\hbar^2}{2}\frac{1}{m_1 + m_2}\nabla_F^2\psi - \frac{\hbar^2}{2\mu}\nabla_g^2\psi + V(g)\psi = E_T\psi \tag{Q-12}$$

where

$$\nabla_F^2 = \frac{\partial^2}{\partial X^2} + \frac{\partial^2}{\partial Y^2} + \frac{\partial^2}{\partial Z^2} \tag{Q-13}$$

$$\nabla_g^2 = \frac{\partial^2}{\partial x^2} + \frac{\partial^2}{\partial y^2} + \frac{\partial^2}{\partial z^2} \tag{Q-14}$$

$$\mu = \frac{m_1 m_2}{m_1 + m_2} \tag{Q-15}$$

$$V(g) = V(x, y, z) \tag{Q-16}$$

(*Note:* The symbols F and g when used as subscripts or variables denote a set of orthogonal space-fixed and orthogonal molecule-oriented axes respectively.)

APPENDIX R

PEKERIS' SOLUTION TO THE DIATOMIC VIBROTOR WAVE EQUATION

Examination of the diatomic vibrotor wave equation

$$\frac{d^2S}{dr^2} + \left\{\frac{-J(J+1)}{r^2} + \frac{8\pi^2\mu}{h^2}\left[E - D - De^{-2a(r-r_e)} + 2De^{-a(r-r_e)}\right]\right\} S = 0 \quad \text{(R-1)}$$

shows the only variable in the equation to be r. If one substitutes $y = e^{-a(r-r_e)}$ and collects a number of constants thus, $A = J(J+1)h^2/8\pi^2\mu r_e^2$, one has

$$\frac{d^2S}{dr^2} + \frac{1}{y}\frac{dS}{dr} + \frac{8\pi^2\mu}{a^2h^2}\left[\frac{E-D}{y} + \frac{2D}{y} - D - \frac{Ar_e^2}{r^2y^2}\right] S = 0 \quad \text{(R-2)}$$

For the case of $A \neq 0$ one can expand $(r/r_e)^2$ in terms of y by use of a Taylor expansion and keep only sufficient terms to give necessary accuracy. Since $y = e^{-a(r-r_e)}$ rearrangement gives $(r_e/r)^2 = [1((\ln y)/ar_e)]^{-2}$. Using the Taylor series expansion about $y = \eta$ of the form

$$f(y) = \sum_{n=0}^{\infty} \frac{1}{n!}\frac{d^nf(\eta)}{dy^n}(y-\eta)^n \quad \text{(R-3)}$$

the expansion about the point $r = r_e$ or $n = 1$ gives the series

$$\left(\frac{r_e}{r}\right)^2 = \frac{1}{[1-(\ln y)/ar_e]^2} = 1 + \frac{2}{ar_e}(y-1) + \left(\frac{-1}{ar_e} + \frac{3}{a^2r_e^2}\right)(y-1)^2 - \cdots \tag{R-4}$$

Using only the first three terms of the series we get

$$\frac{d^2S}{dy^2} + \frac{1}{y}\frac{dS}{dy} + \frac{8\pi^2\mu}{a^2h^2}\left[\frac{E-D-b_0}{y^2} + \frac{2D-b_1}{y} - (D+b_2)\right]S = 0 \tag{R-5}$$

where

$$b_0 = A\left(1 - \frac{3}{ar_e} + \frac{3}{a^2r_e^2}\right) \tag{R-6}$$

$$b_1 = A\left(\frac{4}{ar_e} - \frac{6}{a^2r_e^2}\right) \tag{R-7}$$

$$b_2 = A\left(-\frac{1}{ar_e} + \frac{3}{a^2r_e^2}\right) \tag{R-8}$$

Further simplification results by substituting

$$S(y) = e^{-z/2}z^{d/2}F(z) \tag{R-9}$$

$$g^2 = \frac{8\pi^2\mu}{a^2h^2}(D+b_2) \tag{R-10}$$

$$z = 2gy \tag{R-11}$$

$$d^2 = \frac{-32\pi^2\mu}{a^2h^2}(E-D-b_2) \tag{R-12}$$

so that Eq. (R-5) becomes

$$\frac{d^2F}{dz^2} + \left\{\frac{d+1}{z} - 1\right\}\frac{dF}{dz} + \frac{v}{z}F = 0 \tag{R-13}$$

where

$$v = \frac{4\pi^2\mu}{ga^2h^2}(2D-b_1) - \tfrac{1}{2}(d+1) \tag{R-14}$$

Multiplying Eq. (R-13) by z gives a differential equation of the form

$$z\frac{d^2F}{dz^2} + (d+1+z)\frac{dF}{dz} + vF = 0 \tag{R-15}$$

which is reducible to the Laguerre equation like that encountered in the solution of the hydrogen atom radial equation.

The mathematical solution of the Leguerre differential equation is given in App. F.

The necessary condition for the solutions of Eq. (R-15) to be finite and vanish at the ends of their range is that v be restricted to integer values. Actually this condition is for the boundary conditions $S \to 0$ as $r \to \infty$ and not for the proper condition of $S \to 0$ as $r \to 0$, but prior investigation has shown this approximation to be valid. From Eq. (R-12) the energy of the system can be written as

$$E = -\frac{a^2h^2d^2}{32\pi^2\mu} + D - b_0 \qquad \text{(R-16)}$$

By rearranging Eq. (R-14) to give

$$d = \frac{8\pi^2\mu}{a^2h^2g}(2D - b_1) - (2v + 1) \qquad \text{(R-17)}$$

incorporating Eq. (F-10), and substituting into Eq. (R-16), one gets

$$E = D + b_0 - \frac{(D - \frac{1}{2}b_1)}{(D + b_2)} + \frac{ah(D - \frac{1}{2}b_1)}{\pi[2\mu(D + b_2)]^{1/2}}\left(v + \frac{1}{2}\right) - \frac{a^2h^2}{8\pi^2\mu}\left(v + \frac{1}{2}\right)^2 \qquad \text{(R-18)}$$

By expansions of the third and fourth terms of this equation in a Maclaurin power series of the form

$$f(x) = \sum_{n=0}^{\infty} \frac{1}{n!}\frac{d^nf(0)}{dx^n}x^n \qquad \text{(R-19)}$$

where x is b_1/D or b_2/D, use of Eqs. (R-6), (R-7), (R-8), and (R-10), and the fact that $A = J(J + 1)h^2/8\pi^2\mu r_e^2$ while retaining only those terms to second order in n and J the energy becomes

$$E_{Jv} = \frac{a}{2\pi}\left(\frac{2D}{\mu}\right)^{1/2}\left(v + \frac{1}{2}\right) - \frac{ha^2}{8\pi^2\mu}\left(v + \frac{1}{2}\right)^2 + \frac{hJ(J + 1)}{8\pi^2\mu r_e^2} - \frac{h^3J^2(J + 1)^2}{64\pi^4a^2D\mu^2r_e^6}$$
$$- \left\{\frac{3h^2a}{16\pi^2\mu^{1/2}r_e^2(2D)^{1/2}}\left(\frac{1}{ar_e} - \frac{1}{a^2r_e^2}\right)\left(v + \frac{1}{2}\right)J(J + 1)\right\} \qquad \text{(R-20)}$$

APPENDIX S

CLASSICAL MECHANICS

S.1 LAGRANGIAN MECHANICS

The Lagrangian L is an analytical function, giving the difference between the kinetic energy T and the potential energy V of a system and expressed in terms of coordinates, velocities, and time

$$L = \sum_{i=1}^{N} \sum_{j=1}^{M} T(\dot{q}_{ij}, q_{ij}) - \sum_{i=1}^{N} \sum_{j=1}^{M} V(q_{ij}, t) \tag{S-1}$$

The sum i is over all particles and the sum j is over the coordinates. If we have N particles in a Cartesian system then

$$L = \sum_{k=1}^{N} T(\dot{x}_k, \dot{y}_k, \dot{z}_k) - \sum_{k=1}^{N} V(x_k, y_k, z_k, t) \tag{S-2}$$

For a system of N particles there will be $3N$ equations of motion, Lagrange's equations, derived from

$$\frac{d}{dt}\frac{\partial L}{\partial \dot{q}_i} - \frac{\partial L}{\partial q_i} = 0 \tag{S-3}$$

For example, the nine equations for a set of three particles in a Cartesian system will arise from

$$q_i = x_1, y_1, z_1, x_2, y_2, z_2, x_3, y_3, z_3 \tag{S-4}$$

S.2 HAMILTONIAN MECHANICS

The Hamiltonian, H, is an expression for the total energy of a system in terms of coordinates, momenta, and time. It can be written directly as

$$H = \sum_{i=1}^{N} \sum_{j=1}^{M} \{T(p_{ij}, q_{ij}) + V(q_{ij}, t)\} \tag{S-5}$$

or in terms of the Lagrangian

$$H = \sum_{i=1}^{3N} p_i \dot{q}_i - L \tag{S-6}$$

For N particles in Cartesian coordinates it becomes

$$H = \sum_{k=1}^{N} \{T(p_{x_k}, p_{y_k}, p_{z_k}) + V(x_k, y_k, z_k, t)\} \tag{S-7}$$

For a system of N particles there will be $3N$ pairs of equations, Hamilton's equations, which describe the motion of the system

$$\dot{q}_i = \frac{\partial H}{\partial p_i} \qquad \dot{p}_i = \frac{-\partial H}{\partial q_i} \tag{S-8}$$

The nine pairs for a three-particle system in Cartesian coordinates will arise from $p_i, q_i = p_{x_1}, x_1; p_{y_1}, y_1; p_{z_1}, z_1; p_{x_2}, x_2, \ldots, p_{z_3}, z_3$. The generalized momenta p_i are related to the Lagrangian by

$$p_i = \frac{\partial L}{\partial q_i} \tag{S-9}$$

APPENDIX T

G-MATRIX ELEMENTS

There are two symbolisms that are used to designate G matrix elements. Both types are given in the following listings. The first method uses G^n_{rr}, $G^n_{r\phi}$ and $G^n_{\phi\phi}$ to denote the interactions between two stretching, stretching and bending, and two bending coordinates. The superscript denotes the number of atoms common to the two coordinates. For $n = 1$ the $G^1_{r\phi}$ symbol is not unambiguous since the common atom can be either the end or center atom of the bending coordinate. In this case the symbol is appended by a pair of numbers, $\binom{a}{b}$ which give the number of noncommon atoms to the upper (a) and lower (b) left of the common atom as shown in the accompanying drawing. The second method uses $g(\Delta r_{ij}, \Delta r_{ij})$, $g(\Delta r_{ij}, \Delta\phi_{ijk})$ and $g(\Delta\phi_{ijk}, \Delta\phi_{klm})$. In this method there is no ambiguity since the coordinates are specified by the subscripts on r and ϕ. The order of the subscripts on ϕ are as shown below.

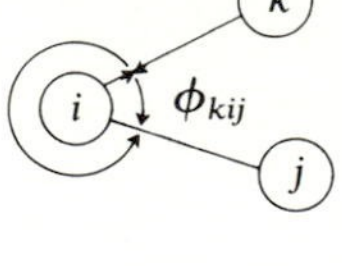

The relationships of the elements to various structural units are shown with heavy circles for the common atoms.

$$G_{rr}^{2} \equiv g(\Delta r_{12}, \Delta r_{12})$$

$$G_{rr}^{1} \equiv g(\Delta r_{12}, \Delta r_{13})$$

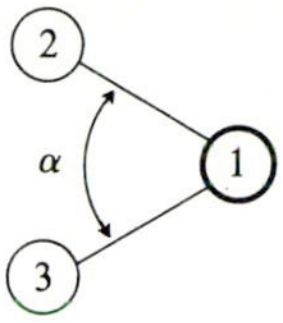

$$G_{r\phi}^{2} \equiv g(\Delta r_{12}, \Delta \phi_{321})$$

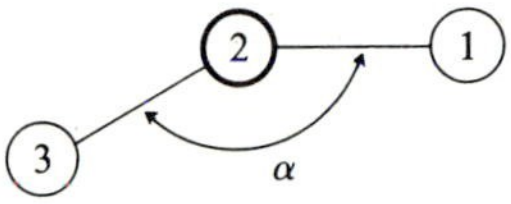

$$G'_{r\phi}\binom{1}{2} \equiv g(\Delta r_{12}, \Delta \phi_{312})$$

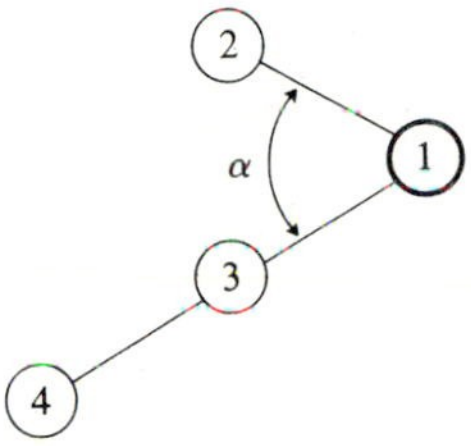

$$G'_{r\phi}\binom{1}{1} \equiv g(\Delta r_{12}, \Delta \phi_{413})$$

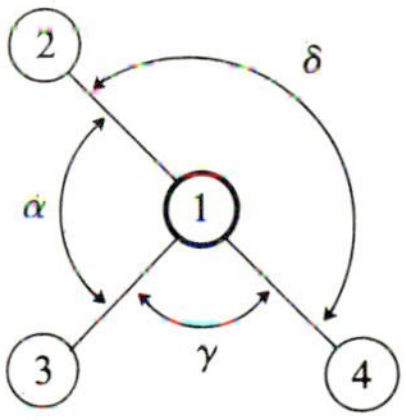

$$G_{\phi\phi}^{1}\binom{2}{2} \equiv g(\Delta \phi_{321}, \Delta \phi_{145})$$

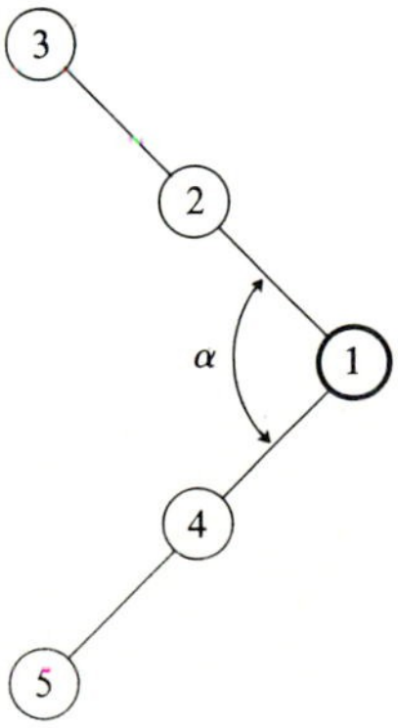

$$G_{\phi\phi}^{3} \equiv g(\Delta \phi_{321}, \Delta \phi_{321})$$

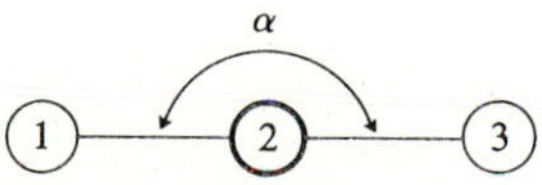

$$G^1_{\phi\phi}\binom{1}{1} \equiv g(\Delta\phi_{312}, \Delta\phi_{514})$$

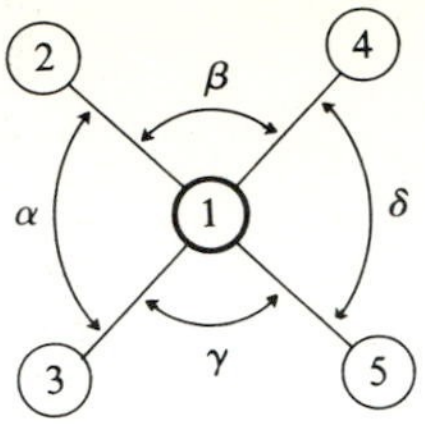

$$G^1_{\phi\phi}\binom{2}{1} \equiv g(\Delta\phi_{214}, \Delta\phi_{415})$$

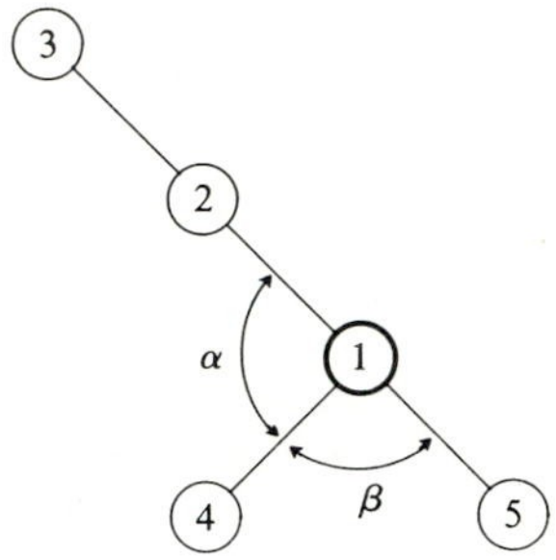

$$G^2_{\phi\phi}\binom{1}{0} \equiv g(\Delta\phi_{123}, \Delta\phi_{314})$$

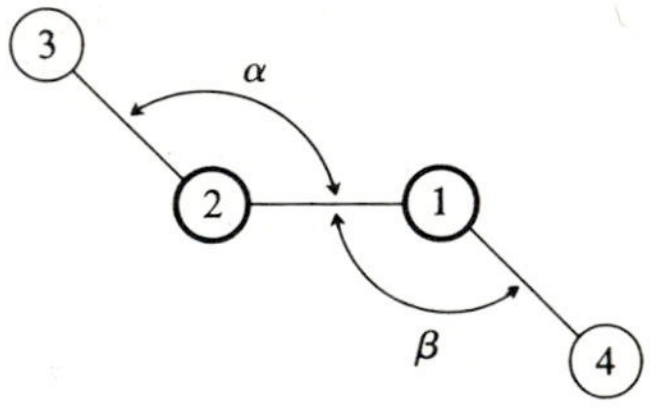

$$G^2_{\phi\phi}\binom{1}{1} \equiv g(\Delta\phi_{123}, \Delta\phi_{421})$$

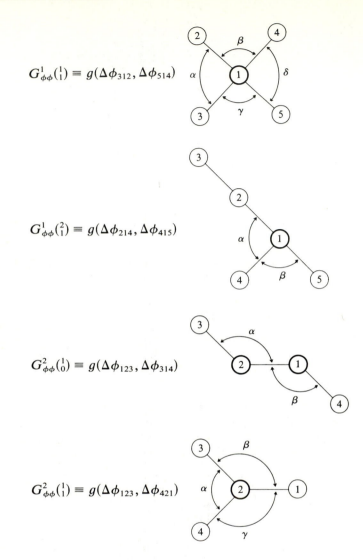

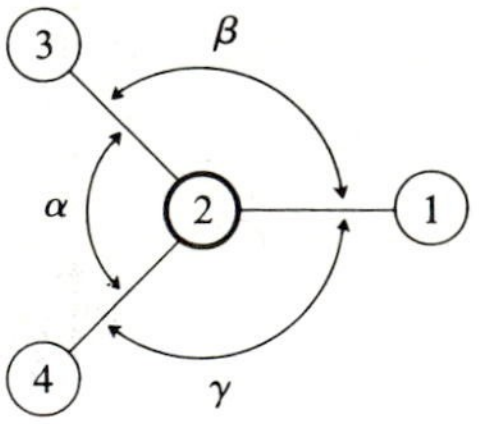

The general formula for the elements are (μ_i = reciprocal of the mass of the ith atom, τ_{ij} = reciprocal of r_{ij}, ϕ_{ihk} = angle between ij and jk bonds, χ_{il} = dihedral angle between planes of ijk and jkl)

$$G^2_{rr} \equiv g(\Delta r_{12}, \Delta r_{12}) = \mu_1 + \mu_2$$

$$G^1_{rr} \equiv g(\Delta r_{12}, \Delta r_{13}) = \mu_1 \cos \alpha$$

$$G^1_{r\phi}\binom{1}{2} \equiv g(\Delta r_{12}, \Delta\alpha) = \tau_{13}\mu_1 \sin \alpha \cos \phi_{134}$$

$$G^2_{r\phi} \equiv g(\Delta r_{23}, \Delta\phi_{321}) = -\tau_{23}\mu_2 \sin \alpha$$

$$G^1_{r\phi}\binom{1}{1} \equiv g(\Delta r_{12}, \Delta\alpha) = -(\tau_{13} \sin \alpha \cos \phi_{234} + \tau_{14} \sin \delta \cos \phi_{243})\mu_1$$

$$G^1_{\phi\phi}\binom{1}{1} \equiv g(\Delta\alpha, \Delta\delta)$$

$$= [(\cos\delta - \cos\phi_{314}\cos\gamma - \cos\beta\cos\phi_{215} + \cos\alpha\cos\beta\cos\gamma)\tau_{12}\tau_{23}$$

$$+ (\cos\phi_{413} - \cos\delta\cos\gamma - \cos\beta\cos\alpha + \cos\phi_{215}\cos\beta\cos\gamma)\tau_{12}\tau_{15}$$

$$+ (\cos\phi_{215} - \cos\alpha\cos\gamma - \cos\beta\cos\delta + \cos\phi_{413}\cos\beta\cos\gamma)\tau_{14}\tau_{13}$$

$$+ (\cos\alpha - \cos\phi_{512}\cos\gamma - \cos\beta\cos\phi_{413} + \cos\delta\cos\beta\cos\gamma)\tau_{14}\tau_{15}]$$

$$\times \left[\frac{\mu_1}{\sin\beta\sin\gamma}\right]$$

$$G^1_{\phi\phi}\binom{2}{1} \equiv g(\Delta\alpha, \Delta\beta) = [(\sin\alpha\sin\beta\cos\chi_{34} - \sin\phi_{215}\cos\chi_{35})\tau_{14}$$

$$+ (\sin\phi_{215}\cos\beta\cos\chi_{35} - \sin\alpha\cos\chi_{34})\tau_{15}]\left[\frac{\tau_{12}\mu_1}{\sin\beta}\right]$$

$$G^1_{\phi\phi}\binom{1}{0} \equiv g(\Delta\alpha, \Delta\beta) = -\tau_{12}\cos\chi_{34}[(\tau_{12} - \tau_{14}\cos\alpha)\mu_1 + (\tau_{12} - \tau_{23}\cos\beta)\mu_2]$$

$$G^1_{\phi\phi}\binom{2}{2} \equiv g(\Delta\phi_{321}, \Delta\phi_{145}) = -(\sin\chi_{25}\sin\chi_{34} + \cos\chi_{25}\cos\chi_{34}\cos\alpha)\tau_{12}\tau_{14}\mu_1$$

$$G^3_{\phi\phi} \equiv g(\Delta\alpha, \Delta\alpha) = \tau_{12}^2\mu_1 + \tau_{23}^2\mu_3 + (\tau_{12}^2 + \tau_{23}^2 - 2\tau_{12}\tau_{23}\cos\alpha)\mu_2$$

$$G^1_{\phi\phi}\binom{1}{1} \equiv g(\Delta\beta, \Delta\gamma) = (\tau_{12}^2\cos\phi_{314})\mu_1$$

$$+ [(\tau_{12} - \tau_{23}\cos\beta - \tau_{24}\cos\gamma)\tau_{12}\cos\phi_{314}$$

$$+ (\sin\beta\sin\gamma\sin\phi_{314} + \cos\alpha\cos\phi_{314})\tau_{23}\tau_{24}]\mu_2$$

These expressions simplify considerably when all angles are either 109°28′ (tetrahedral) or 120° (trigonal). The simplified expressions are tabulated by Wilson, Decius and Cross [1]. Expressions for the nontrigonal, 5-atom planar molecule have been given by Shimanouchi [2]. A complete general treatment has been presented by Decius [3].

---

1. Wilson, E. B., Jr., J. C. Decius, and Paul C. Cross, *Molecular Vibrations*, McGraw-Hill, New York, NY, 1955.
2. T. Shimanouchi, *J. Chem. Phys.*, **25**, 660 (1956).
3. J. C. Decius, *J. Chem. Phys.*, **16**, 1025 (1948); ibid **17**, 1315 (1949).

APPENDIX U

EVALUATION OF VIBRATIONAL AMPLITUDE RELATIONSHIPS

In principle the amplitude factors A_{ij} can be found by numerical solution of Eq. (16.255). However, it will be instructive to show how they can be related to the elements of the $\underline{\mathbf{F}}$ and $\underline{\mathbf{G}}$ matrices in a simple manner. The relationship between the components of internal coordinates R_i and the normal coordinates Q_i as given by Eq. (16.253) may be written as

$$R_i = \sum_{k=1}^{3N-6} W_{ik} Q_k \tag{U-1}$$

where $W_{ik} = N_k A_{ik}$. One observes at this point that if Q_k is considered to be of unit amplitude then W_{ik} is just the amplitude of the displacement R_i for the kth normal mode.

In matrix terminology

$$\underline{\mathbf{R}} = \underline{\mathbf{W}}\,\underline{\mathbf{R}} \tag{U-2}$$

$\underline{\mathbf{W}}$ will not necessarily be orthogonal, $\tilde{\underline{\mathbf{W}}} \neq \underline{\mathbf{W}}^{-1}$, so the inverse transformation $\underline{\mathbf{Q}} = \underline{\mathbf{C}}\mathbf{R}$ can be found only by a complete inversion of $\underline{\mathbf{W}}$, that is by solving the $3N - 6$ equations.

To avoid the need for this inversion the $\underline{\mathbf{G}}$ matrix can be employed. Beginning with the basic definitions for the kinetic and potential energies

$$2T = \tilde{\dot{\mathbf{R}}}\underline{\mathbf{G}}^{-1}\dot{\mathbf{R}} \tag{U-3}$$

$$2V = \tilde{\mathbf{R}}\underline{\mathbf{F}}\mathbf{R} \tag{U-4}$$

and using $\mathbf{R} = \underline{\mathbf{W}}\mathbf{Q}$ and $\tilde{\mathbf{R}} = \tilde{\mathbf{Q}}\underline{\tilde{\mathbf{W}}}$ gives

$$2T = \tilde{\dot{\mathbf{Q}}}\underline{\tilde{\mathbf{W}}}\,\underline{\mathbf{G}}^{-1}\underline{\mathbf{W}}\dot{\mathbf{Q}} \tag{U-5}$$

$$2V = \tilde{\mathbf{Q}}\underline{\tilde{\mathbf{W}}}\,\underline{\mathbf{F}}\,\underline{\mathbf{W}}\mathbf{Q} \tag{U-6}$$

Since

$$2T = \tilde{\mathbf{Q}} \cdot \mathbf{Q} \tag{U-7}$$

and

$$2V = \tilde{\mathbf{Q}}\underline{\mathbf{\Lambda}}\mathbf{Q} \tag{U-8}$$

where $\underline{\mathbf{\Lambda}}$ has only diagonal elements it follows that

$$\underline{\tilde{\mathbf{W}}}\,\underline{\mathbf{F}}\,\underline{\mathbf{W}} = \underline{\mathbf{\Lambda}} \tag{U-9}$$

$$\underline{\tilde{\mathbf{W}}}\,\underline{\mathbf{G}}^{-1}\underline{\mathbf{W}} = \underline{\mathbf{I}} \tag{U-10}$$

Multiplying this last equation by $\underline{\mathbf{W}}^{-1}\underline{\mathbf{G}}$

$$\underline{\tilde{\mathbf{W}}} = \underline{\mathbf{W}}^{-1}\underline{\mathbf{G}} \tag{U-11}$$

followed by substitution into Eq. (U-9) gives

$$\underline{\mathbf{W}}^{-1}\underline{\mathbf{G}}\,\underline{\mathbf{F}}\,\underline{\mathbf{W}} = \underline{\mathbf{\Lambda}} \tag{U-12}$$

Multiplication from the left by $\underline{\mathbf{W}}$ leaves

$$\underline{\mathbf{G}}\,\underline{\mathbf{F}}\,\underline{\mathbf{W}} = \underline{\mathbf{W}}\,\underline{\mathbf{\Lambda}} \tag{U-13}$$

This equation represents a set of $3N - 6$ equations of the form

$$\left(\sum_{i=1}^{3N-6} G_{ni}F_{i1} - \delta_{n1}\lambda_k\right) W_{1k} + \left(\sum_{i=1}^{3N-6} G_{ni}F_{i2} - \delta_{n2}\lambda_k\right) W_{2k} + \cdots + \left(\sum_{i=1}^{3N-6} G_{ni}F_{i,3N-6} - \delta_{n,3N-6}\lambda_k\right) W_{3N-6,k} = 0 \tag{U-14}$$

where $n = 1, 2, \ldots, 3N - 6$ and k denotes a particular root or mode. Since $\underline{\mathbf{\Lambda}}$ is diagonal, an alternate expression becomes

$$(\underline{\mathbf{G}}\,\underline{\mathbf{F}} - \underline{\mathbf{I}}\lambda_k)\mathbf{W}_k = 0 \tag{U-15}$$

where $\mathbf{W}_k$ is a column vector whose elements are those of the kth column of the $\underline{\mathbf{W}}$ matrix.

Left and right multiplication of Eq. (U-13) by $\underline{\mathbf{W}}^{-1}$ gives

$$\underline{\mathbf{W}}^{-1}\underline{\mathbf{G}}\,\underline{\mathbf{F}} = \underline{\mathbf{\Lambda}}\,\underline{\mathbf{W}}^{-1} \tag{U-16}$$

Taking the transpose of both sides, and recalling that $\underline{\mathbf{G}}$, $\underline{\mathbf{F}}$, and $\underline{\boldsymbol{\Lambda}}$ are symmetric, $\underline{\mathbf{G}} = \underline{\tilde{\mathbf{G}}}$, $\underline{\mathbf{F}} = \underline{\tilde{\mathbf{F}}}$, we have

$$\widetilde{\underline{\mathbf{W}}^{-1}\underline{\mathbf{G}}\underline{\mathbf{F}}} = \underline{\tilde{\mathbf{G}}}\underline{\tilde{\mathbf{F}}}\underline{\tilde{\mathbf{W}}}^{-1} = \underline{\tilde{\mathbf{F}}}\underline{\tilde{\mathbf{G}}}\underline{\tilde{\mathbf{W}}}^{-1} = \underline{\mathbf{F}}\underline{\mathbf{G}}\underline{\tilde{\mathbf{W}}}^{-1} = \widetilde{\underline{\boldsymbol{\Lambda}}\underline{\mathbf{W}}^{-1}} = \underline{\tilde{\mathbf{W}}}^{-1}\underline{\boldsymbol{\Lambda}} \qquad \text{(U-17)}$$

This equation gives rise to another set of $3N - 6$ equations of the form

$$\left(\sum_{i=1}^{3N-6} F_{ni}G_{i1} - \delta_{n1}\lambda_k\right)(W^{-1})_{k1} + \left(\sum_{i=1}^{3N-6} F_{ni}G_{i2} - \delta_{n2}\lambda_k\right)(W^{-1})_{k2} + \cdots$$
$$+ \left(\sum_{i=1}^{3N-6} F_{ni}G_{i,3N-6} - \delta_{n,3N-6}\lambda_k\right)(W^{-1})_{k,3N-6} = 0 \qquad \text{(U-18)}$$

Left multiplication of Eq. (U-2) by W^{-1} gives

$$\mathbf{Q} = \underline{\mathbf{W}}^{-1}\mathbf{R} \qquad \text{(U-19)}$$

Hence, the solutions of Eqs. (U-18) for the ratios of the $(W^{-1})_{ij}$ terms coupled with Eq. (U-9) leads to the forms of the normal modes.

A further understanding of the ascertainment of the normalization condition is provided by considering Hamilton's equations (see App. S)

$$\dot{P}_j = -\frac{\partial H}{\partial R_j} \qquad \dot{R}_j = \frac{\partial H}{\partial P_j} \qquad \text{(U-20)}$$

where j denotes a particular internal coordinate and its conjugate momenta. Combining Eqs. (16.150) and 16.221) the Hamiltonian becomes

$$H = \frac{1}{2}\sum_{i=1}^{3N-6}\sum_{j=1}^{3N-6} \{G_{ij}P_iP_j + F_{ij}R_iR_j\} \qquad \text{(U-21)}$$

Applying Hamilton's equations

$$\dot{P}_j = -\sum_{i=1}^{3N-6} F_{ji}R_i \qquad \text{(U-22)}$$

and

$$\dot{R}_j = \sum_{i=1}^{3N-6} G_{ji}P_i \qquad \text{(U-23)}$$

The internal coordinates which are the solutions to the equations of motion are of the form

$$R_j = A_j \cos(\lambda^{1/2}t + \delta) \qquad \text{(U-24)}$$

Since P_j is proportional to R_j

$$P_j = A_j' \sin(\lambda^{1/2}t + \delta) \qquad \text{(U-25)}$$

If these expressions for R_j and P_j are substituted into Hamilton's equations

[Equation (U-21)], then

$$\lambda_k^{1/2} A'_{jk} + \sum_{i=1}^{3N-6} F_{ji} A_{ik} = 0 \qquad (j = 1, 2, \ldots, 3N-6) \tag{U-26}$$

$$\lambda_k^{1/2} A_{jk} + \sum_{i=1}^{3N-6} G_{ji} A'_{ik} = 0 \qquad (j = 1, 2, \ldots, 3N-6) \tag{U-27}$$

The additional subscript k has again been added to denote a particular root. For each mode we then have $2(3N-6)$ simultaneous linear equations. Since we have already found the method for finding the A_{ik}'s we can use these and solve for the A'_{ik}'s. Once these latter coefficients have been related to the $(W^{-1})_{ik}$ terms the problem of obtaining the normal modes will be complete.

Equations (U-27) can be written in matrix terminology as

$$\underline{\mathbf{G}}\,\underline{\mathbf{A}}' = -\underline{\mathbf{A}}\,\underline{\mathbf{K}} \tag{U-28}$$

where $\underline{\mathbf{K}}$ is a diagonal matrix whose elements are the $\lambda_k^{1/2}$'s. The elements of the reciprocal $\underline{\mathbf{K}}^{-1}$ will be the $\lambda_k^{-1/2}$'s. Equations (U-9) and (U-10) may be inverted to give

$$\underline{\mathbf{W}}^{-1} = \underline{\mathbf{A}}^{-1}\underline{\tilde{\mathbf{W}}}\,\underline{\mathbf{F}} \tag{U-29}$$

and

$$\underline{\mathbf{W}} = \underline{\tilde{\mathbf{G}}}\,\underline{\mathbf{W}}^{-1} \tag{U-30}$$

Since

$$\underline{\mathbf{W}} = \underline{\mathbf{A}}\,\underline{\mathbf{N}} \tag{U-31}$$

and

$$\underline{\mathbf{G}}\,\underline{\mathbf{A}}'\underline{\mathbf{K}}^{-1} = -\underline{\mathbf{A}} \tag{U-32}$$

it follows that

$$\underline{\mathbf{W}} = -\underline{\mathbf{G}}\,\underline{\mathbf{A}}'\underline{\mathbf{K}}^{-1}\underline{\mathbf{N}} \tag{U-33}$$

or

$$\underline{\tilde{\mathbf{W}}}^{-1} = \underline{\mathbf{A}}'\underline{\mathbf{K}}^{-1}\underline{\mathbf{N}} \tag{U-34}$$

Therefore the $(\tilde{W}^{-1})_{ik}$ elements are just the A_{ik} elements multiplied by $-\lambda_k^{-1/2} N_k$.

APPENDIX V

SYMMETRIC TOP WAVE EQUATION

If the solution to the Schrödinger equation for the rigid symmetric top rotor

$$\frac{1}{\sin\theta}\frac{\partial}{\partial\theta}\left(\sin\theta\frac{\partial\psi}{\partial\theta}\right)+\frac{1}{\sin^2\theta}\frac{\partial^2\psi}{\partial\phi^2}+\left(\frac{\cos^2\theta}{\sin^2\theta}+\frac{C}{B}\right)\frac{\partial^2\psi}{\partial\chi^2}-\frac{2\cos\theta}{\sin^2\theta}\frac{\partial^2\psi}{\partial\chi\partial\phi}=-\frac{E\psi}{hB} \tag{V-1}$$

is assumed to be separable as was found for the diatomic vibrotor substitution of

$$\psi(\phi,\theta,\chi)=\Phi(\phi)\Theta(\theta)X(\chi) \tag{V-2}$$

will separate the equation into three separate differential equations. Since the variables ϕ and χ appear in the wave equation only in differential terms they are cyclic coordinates and will appear in the wavefunction as exponential terms of the form

$$\Phi(\phi)=N_m e^{M\phi} \tag{V-3}$$

$$X(\chi)=N_K e^{iK\chi} \tag{V-4}$$

where M and K must be integer values so that the functions have identical values at ϕ, $\phi+2\pi$ and χ, $\chi+2\pi$. The wavefunction is then of the form

$$\phi(\phi,\theta,\chi)=N_{JKM}\Theta(\theta)e^{iM\phi}e^{iK\chi} \tag{V-5}$$

which when substituted into Eq. (V-1) gives

$$\frac{1}{\sin\theta}\frac{d}{d\theta}\left(\sin\theta\frac{d\Theta}{d\theta}\right)-\left[\frac{M^2}{\sin^2\theta}+\left(\frac{\cos^2\theta}{\sin^2\theta}+\frac{C}{B}\right)K^2-\frac{2\cos\theta}{\sin^2\theta}KM-\frac{E}{hB}\right]\Theta=0 \tag{V-6}$$

As in the case of the radial equation for the diatomic rotor this equation can be related to a standard form differential equation. Introducing the change of variable

$$z=\tfrac{1}{2}(1-\cos\theta) \tag{V-7}$$

will eliminate the trigonometric functions giving the differential equation

$$\frac{d}{dz}\left[z(1-z)\frac{dS(z)}{dz}\right]+\left[\lambda-\frac{\{M+K(2z-1)\}_2}{4z(1-z)}\right]=0 \tag{V-8}$$

where

$$S(z)=\Theta(\theta) \tag{V-9}$$

and

$$\lambda=\frac{E}{hB}-\frac{C}{B}K^2 \tag{V-10}$$

Substituting

$$S(z)=z^{(1/2)|K-M|}(1-z)^{(1/2)|K+M|}F(z) \tag{V-11}$$

into Eq. (V-8) gives

$$(1-z)\frac{d^2F(z)}{dz^2}+(\alpha-\beta z)\frac{dF(z)}{dz}+\gamma F(z)=0 \tag{V-12}$$

where

$$\alpha=|K-M|-1 \tag{V-13}$$

$$\beta=|K+M|+|K-M|+2 \tag{V-14}$$

$$\gamma=\lambda+K^2-[\tfrac{1}{2}|K+M|+\tfrac{1}{2}|K-M|][\tfrac{1}{2}|K+M|+\tfrac{1}{2}|K-M|+1] \tag{V-15}$$

Equation (V-1) is the hypergeometric differential equation whose solutions are the Jacobi polynomials

$$F(z)=\sum_{n=0}^{\infty}a_nz^n \tag{V-16}$$

where

$$a_{n+1}=\left(\frac{n(n-1)+\beta n-\gamma}{(n+1)(n+\alpha)}\right)a_n \tag{V-17}$$

For $\Theta(\theta)$ to be a satisfactory wavefunction it must be well behaved, which requires that the energy E be given by

$$E=hBJ(J+1)+h(C-B)K^2 \tag{V-18}$$

with $J = n_{\max} + \frac{1}{2}|K + M| + \frac{1}{2}|K - M|$. The value of $n_{\max}$ is the largest value of n for which a_n does not vanish. Since $n_{\max}$, K, and M have been determined to be integers it follows that J must also have integer values. The three quantum members are furthermore restricted, in order to have well behaved solutions, to values of

$$\begin{aligned} J &= 0, 1, 2, \ldots \\ K &= 0, \pm 1, \pm 2, \ldots, \pm J \\ M &= 0, \pm 1, \pm 2, \ldots \pm J \end{aligned}$$

The final explicit form of the total wavefunction is obtained by further consideration of the Jacobi polynomial. Looking at Eq. (V-16) one can see that $F(z)$ is really a function $F(\alpha, \beta, \gamma, z)$. The explicit form of the polynomial is

$$\begin{aligned} F(a, b, c, z) &= 1 + \frac{ab}{c} z + \frac{a(a+1)b(b+1)}{2!c(c+1)} z^2 + \cdots \\ &\quad + \frac{a(a+1)\cdots(a+n-1)b(b+1)\cdots(b+n-1)}{n!c(c+1)\cdots(c+n-1)} z^n + \cdots \end{aligned} \tag{V-19}$$

Including the normalization factor the symmetric top wavefunctions can be written

$$\begin{aligned} &\psi_{JKM} \\ &= N_{JKM} z^{(1/2)|K-M|}(1-z)^{(1/2)|K+M|} e^{iM\phi} e^{iK\chi} F\left(-J + \frac{\beta}{2} - 1, J + \frac{\beta}{2}, 1 + |K - M|, z\right) \end{aligned} \tag{V-20}$$

or in abbreviated form

$$\psi_{JKM} = N_{JKM} e^{iM\phi} e^{iK\chi} S_{JKM}(\theta) \tag{V-21}$$

where

$$N_{JKM} = \left[\frac{(2J+1)(J + \frac{1}{2}|K+M| + \frac{1}{2}|K-M|)! \,/\, (J - \frac{1}{2}|K+M| + \frac{1}{2}|K-M|)!}{8\pi^2 (J - \frac{1}{2}|K+M| - \frac{1}{2}|(K-M|)! \,(|K-M|!)^2 (J + \frac{1}{2}|K+M| - \frac{1}{2}|K-M|)!} \right]^{1/2} \times e^{-i\pi|K-M|/2} \tag{V-22}$$

APPENDIX W

SELECTED GROUP CORRELATION TABLES

TABLE W-1
D_{3d} group

	D_3	C_{3v}	S_6	C_3	C_{2h}	C_2	C_s	C_i
A_{1g}	A_1	A_1	A_g	A	A_g	A	A'	A_g
A_{2g}	A_2	A_2	A_g	A	B_g	B	A''	A_g
E_g	E	E	E_g	E	$A_g + B_g$	$A + B$	$A' + A$	$2A_g$
A_{1u}	A_1	A_2	A_u	A	A_u	A	A''	A_u
A_{2u}	A_2	A_1	A_u	A	B_u	B	A'	A_u
E_u	E	E	E_u	E	$A_u + B_u$	$A + B$	$A' + A''$	$2A_u$

TABLE W-2
D_{3h} group

	C_{3h}	D_3	C_{3v}	C_{2v}	C_3	C_2	$C_2(\sigma_h)$	$C_s(\sigma_v)$
A_1'	A'	A_1	A_1	A_1	A_1	A	A	A'
A_2'	A'	A_2	A_2	B_2	A	B	A'	A''
E'	E'	E	E	$A_1 + B_2$	E	$A + B$	$2A'$	$A' + A''$
A_1''	A''	A_1	A_2	A_2	A	A	A''	A''
A_2''	A''	A_2	A_1	B_1	A	B	A''	A'
E''	E''	E	E	$A_2 + B_1$	E	$A + B$	$2A''$	$A' + A''$

TABLE W-3
O_h group

	O	T_d	D_{4h}	C_{4v}	C_{2v}	D_3	D_{2d}
A_{1g}	A_1	A_1	A_{1g}	A_1	A_1	A_1	A_1
A_{2g}	A_2	A_2	B_{1g}	B_1	A_2	A_2	B_1
E_g	E	E	$A_{1g} + B_{1g}$	$A_1 + B_1$	$A_1 + A_2$	E	$A_1 + B_1$
T_{1g}	T_1	T_1	$A_{2g} + E_g$	$A_2 + E$	$A_2 + B_1 + B_2$	$A_2 + E$	$A_2 + E$
T_{2g}	T_2	T_2	$B_{2g} + E_g$	$B_2 + E$	$A_1 + B_1 + B_2$	$A_1 + E$	$B_2 + E$
A_{1u}	A_1	A_2	A_{1u}	A_2	A_2	A_1	B_1
A_{2u}	A_2	A_1	B_{1u}	B_2	A_1	A_2	A_1
E_u	E	E	$A_{1u} + B_{1u}$	$A_2 + B_2$	$A_1 + A_2$	E	$A_1 + B_1$
T_{1u}	T_1	T_2	$A_{2u} + E_u$	$A_1 + E$	$A_1 + B_1 + B_2$	$A_2 + E$	$B_2 + E$
T_{2u}	T_2	T_1	$B_{2u} + E_u$	$B_1 + E$	$A_2 + B_1 + B_2$	$A_1 + E$	$A_2 + E$

TABLE W-4
T_d group

	T	D_{2d}	C_{2v}	S_4	D_2	C_{2v}	C_3	C_2
A_1	A	A_1	A_1	A	A	A_1	A	A
A_2	A	B_1	A_2	B	A	A_2	A	A
E	E	$A_1 + B_1$	E	$A + B$	$2A$	$A_1 + A_2$	E	$2A$
T_1	T	$A_2 + E$	$A_2 + E$	$A + E$	$B_1 + B_2 + B_3$	$A_2 + B_1 + B_2$	$A + E$	$A + 2B$
T_2	T	$B_2 + E$	$A_1 + E$	$B + E$	$B_1 + B_2 + B_3$	$A_1 + B_1 + B_2$	$A + E$	$A + 2B$

APPENDIX X

TANABE-SUGANO DIAGRAMS

Tanabe-Sugano semiquantitative energy level diagrams for octahedral symmetry

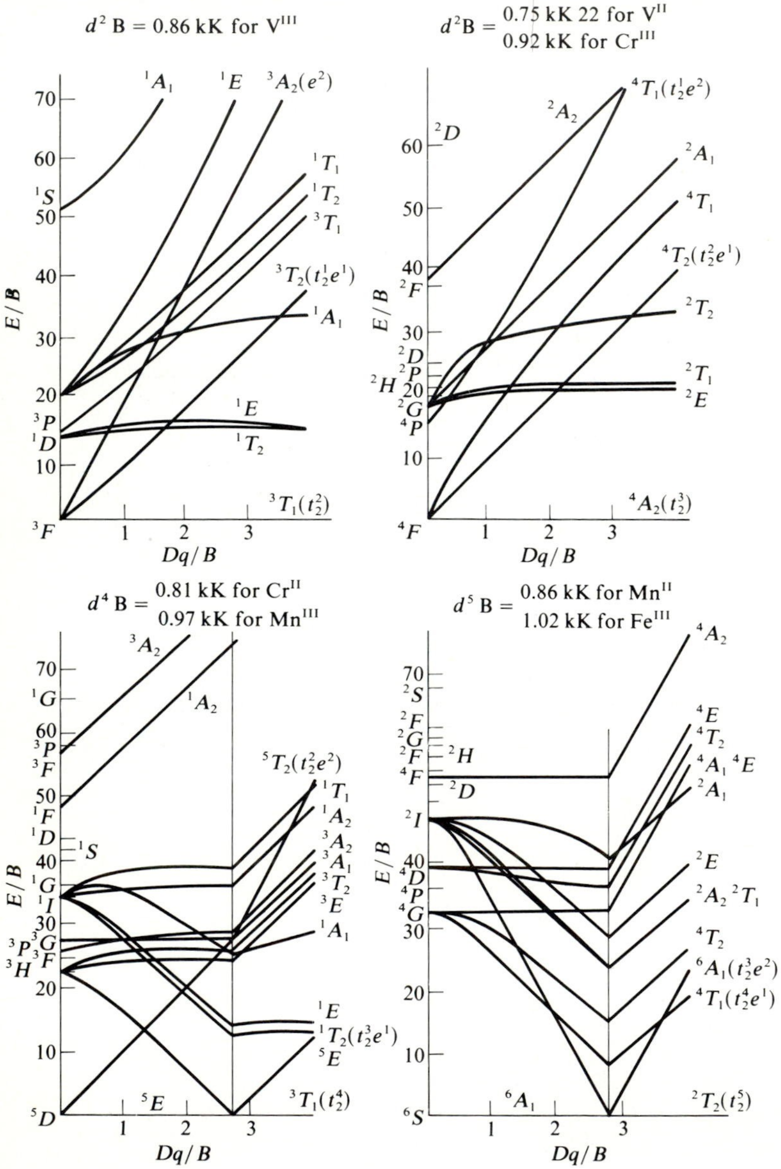
d² B = 0.86 kK for V^III
d²B = 0.75 kK 22 for V^II
0.92 kK for Cr^III
d⁴ B = 0.81 kK for Cr^II
0.97 kK for Mn^III
d⁵ B = 0.86 kK for Mn^II
1.02 kK for Fe^III
E/B
Dq/B

FIGURE X-1

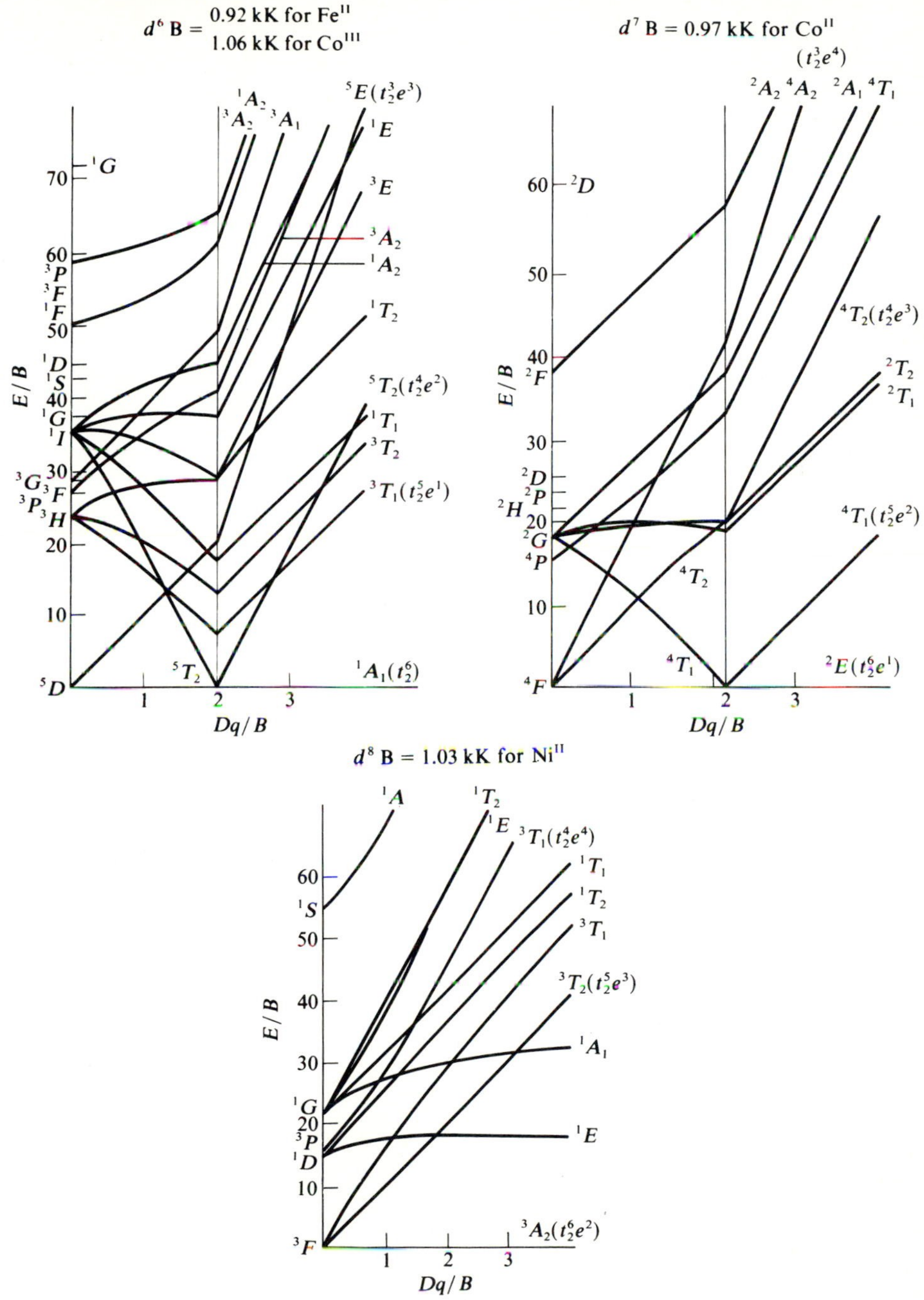

d⁶ B = 0.92 kK for Fe^II
1.06 kK for Co^III
d⁷ B = 0.97 kK for Co^II
d⁸ B = 1.03 kK for Ni^II
E/B
Dq/B

FIGURE X-1

INDEX